Digital Stock

D1499152

Nutrition
Through the Life Cycle
SECOND EDITION

Judith E. Brown
Ph.D., M.P.H., R.D.
University of Minnesota

with

Janet S. Isaacs, PhD., R.D.
Children's National Medical Center

U. Beate Krinke, Ph.D., M.P.H., R.D.
University of Minnesota

Maureen A. Murtaugh, Ph.D., R.D.
University of Utah School of Medicine

Carolyn Sharbaugh, M.S., R.D.
Nutrition Consultant

Jamie Stang, Ph.D., M.P.H., R.D.
University of Minnesota

Nancy H. Wooldridge, M.S., R.D., L.D.
University of Alabama at Birmingham

Photo Disc

THOMSON

WADSWORTH

Australia • Canada • Mexico • Singapore • Spain
United Kingdom • United States

brand X pictures

THOMSON
---*---
WADSWORTH

PUBLISHER: Peter Marshall
DEVELOPMENT EDITOR: Elizabeth Howe
ASSISTANT EDITOR: Elesha Feldman
EDITORIAL ASSISTANT: Lisa Michel
TECHNOLOGY PROJECT MANAGER: Travis Metz
MARKETING MANAGER: Jennifer Somerville
MARKETING ASSISTANT: Melanie Banfield
ADVERTISING PROJECT MANAGER: Shemika Britt
PROJECT MANAGER, EDITORIAL
    PRODUCTION: Cheryll Linthicum
PRINT/MEDIA BUYER: Karen Hunt
PERMISSIONS EDITOR: Joohee Lee
TEXT DESIGNER: Ann Borman
PRODUCTION: Ann Hoffman
COPY EDITOR: Chris Thillen
PROOFREADER: Suzie DeFazio
ILLUSTRATOR: David Farr
INDEXER: Terry Casey
COMPOSITOR: Parkwood Composition
COVER IMAGES: Top right: © Allana Wesley White/Corbis.
    Row 2 left: © Brett Patterson/Corbis; right: ©
    Cameron/Corbis. Row 3 left: © Tom & Dee Ann
    McCarthy/Corbis; center: © Cheque/Corbis; right: ©
    Corbis. Bottom left: © Leland Bobbé/Corbis.
TEXT AND COVER PRINTER:
    Edwards Brothers, Incorporated

COPYRIGHT © 2005 Wadsworth, a division of Thomson
Learning, Inc. Thomson Learning™ is a trademark used
herein under license.

ALL RIGHTS RESERVED. No part of this work covered
by the copyright hereon may be reproduced or used in
any form or by any means—graphic, electronic, or
mechanical, including but not limited to photocopying,
recording, taping, Web distribution, information networks,
or information storage and retrieval systems—without
the written permission of the publisher.

Printed in the United States of America
1 2 3 4 5 6 7 08 07 06 05 04

For more information about our products, contact us at:
**Thomson Learning Academic Resource Center
1-800-423-0563**

For permission to use material from this text, submit a
request online at http://www.thomsonrights.com.
Any additional questions about permissions can be
submitted by email to thomsonrights@thomson.com.

THOMSON WADSWORTH
10 Davis Drive
Belmont, CA 94002-3098
USA

ASIA
Thomson Learning
5 Shenton Way #01-01
UIC Building
Singapore 068808

AUSTRALIA/NEW ZEALAND
Thomson Learning
102 Dodds Street
Southbank, Victoria 3006
Australia

CANADA
Nelson
1120 Birchmount Road
Toronto, Ontario M1K 5G4
Canada

EUROPE/MIDDLE EAST/AFRICA
Thomson Learning
High Holborn House
50/51 Bedford Row
London WC1R 4LR
United Kingdom

LATIN AMERICA
Thomson Learning
Seneca, 53
Colonia Polanco
11560 Mexico D.F.
Mexico

SPAIN/PORTUGAL
Paraninfo
Calle/Magallanes, 25
28015 Madrid, Spain

Library of Congress Control Number: 2004106296
Student Edition: ISBN 0-534-58989-8
Instructor's Edition: ISBN 0-534-58990-1

Contents in Brief

Contents

Chapter 4
Nutrition during Pregnancy 73

Chapter 5
Nutrition during Pregnancy 119
Conditions and Interventions

Chapter 6

Nutrition during Lactation   143

Chapter 7

Nutrition and Lactation   173
Conditions and Interventions

Chapter 10
Toddler and Preschooler Nutrition   241

Chapter 11
Toddler and Preschooler Nutrition   269
Conditions and Interventions

Chapter 12
Child and Preadolescent Nutrition   281

Chapter 13
Child and
Preadolescent Nutrition   307
Conditions and Interventions

Chapter 14
Adolescent Nutrition   325

Preface

It is with a sense of gratitude and pleasure that we offer you the second edition of *Nutrition Through the Life Cycle*. This text was initially developed, and has been revised, with the needs of instructors teaching, and students taking, a two-to-four credit course in life-cycle nutrition in mind. It is written at a level that assumes students have had an introductory nutrition course. Overall, the text is intended to give instructors a tool they can productively use to enhance their teaching efforts, and to give students an engaging and rewarding educational experience they will carry with them throughout their lives.

Authors of *Nutrition Through the Life Cycle* represent a group of experts who are actively engaged in clinical practice, teaching, and research related to nutrition during specific phases of the life cycle. All of us remained totally dedicated to the goals established for the text at its conception: to make the text comprehensive, logically organized, science based, realistic, and relevant to the needs of instructors and students.

Chapter 1 summarizes key elements of introductory nutrition and gives students who need it a chance to update or renew their knowledge. Coverage of the life-cycle phases begins with preconceptional nutrition and continues with each major phase of the life cycle through adulthood and the special needs of the elderly. Each of these 18 chapters was developed based on a common organizational framework that includes key nutrition concepts, prevalence statistics, physiological principles, nutritional needs and recommendations, model programs, case studies, and recommended practices. To meet the knowledge needs of students with the variety of career goals represented in many life-cycle nutrition courses, we include two chapters for each life-cycle phase. The first chapter for each life-cycle phase covers normal nutrition topics, and the second covers nutrition-related conditions and interventions. Every chapter focuses on scientifically based information and employs up-to-date resources and references. Each chapter ends with a list of Web sites and print resources that lead students to reliable information on scientific and applied aspects of life-cycle nutrition.

New to the Second Edition

Writing the first edition of a textbook is like teaching a course for the first time. It is an exciting and intense experience, and you learn a lot about how to do it better the next time. The generous and helpful comments of instructors and reviewers taught us a good deal about how we could make the second edition of the text better than the first.

The second edition of *Nutrition Through the Life Cycle* differs from the first in a number of ways:

- Information on Dietary Reference Intakes released through spring 2004 is incorporated throughout the text.
- Case studies are included in Chapters 2 through 19 and conform to a standard format. Questions for students appear at the end of each case study, and answers to the questions are provided in Appendix D of the text.
- Information in Chapter 1 on basic nutrition is covered in more depth.
- Content on food safety and cross-cultural nutritional considerations by life-cycle phase is increased.
- Web resources listed at the end of each chapter have been expanded and updated.

Readers of the first edition will notice other changes in the second edition. For example, some sections of the text have been reorganized and rewritten; and topics have been added, including the following:

- Additional coverage of vegetarian/vegans throughout each stage of the life cycle
- New information on breast reduction and augmentation and breastfeeding
- New information on functional foods
- More information on use of herbs and breastfeeding
- New material on environmental exposure to toxins and breastfeeding
- More coverage of jaundice and bilirubin metabolism
- New coverage of near-term, immature infants and breastfeeding
- New material on newly developed infant formulas
- New material on introducing solid foods
- More material on prevention of childhood obesity
- New information on use of food and beverage vending machines in schools and more on school lunch programs

- Additional material on pediatric HIV
- More information about dietary phytoestrogens and flaxseed
- New coverage of metabolic syndrome
- Additional coverage of type 2 diabetes
- More coverage of the nutritional concerns of adults with HIV

Instructor Resources

New for the second edition is an *Instructor's Resource CD-ROM* that contains Microsoft PowerPoint™ lecture presentations with artwork; chapter outlines, classroom activities, lecture launchers, Internet exercises, discussion questions, hyperlinks to relevant Web sites, and case studies. The CD-ROM also includes a Test Bank, expanded and improved since the first edition, that includes multiple-choice, true/false, matching, and discussion exercises.

Acknowledgments

Development and revision of *Nutrition Through the Life Cycle* was made possible by Peter Marshall, the Publisher of Wadsworth/Thomson Nutrition and other titles. His excitement for the text is contagious, and his vision for the text delighted us all. A thousand thanks go to Beth Howe, Development Editor for the text, who steadfastly guided us through problem areas while engendering the respect and admiration of all of us. Ann Borman, Senior Production Editor, designed the text and cover. Thanks to her, our unattractive manuscript pages turned miraculously into the text you are holding. Chris Thillen worked diligently with authors to improve the organization and clarity of the text, and her copyediting efforts are greatly appreciated. We were fortunate to have many other capable Wadsworth professionals contribute to this text, including Jennifer Somerville, Marketing Manager; Elesha Feldman, Assistant Editor; and Lisa Michel, Editorial Assistant.

Reviewers

Many thanks to the following reviewers, whose careful reading and thoughtful comments helped enormously in shaping this second edition.

Betty Alford
Texas Woman's University

Leta Aljadir
University of Delaware

Dea Hanson Baxter
Georgia State University

Janet Colson
Middle Tennessee State University

Shelley R. Hancock
The University of Alabama

Dr. Mary Jacob
California State University, Long Beach

Pera Jambazian
California State University, Los Angeles

Tay Seacord Kennedy
Oklahoma State University

Younghee Kim
Bowling Green State University

Barbara Kirks
California State University, Chico

Kaye Stanek Krogstrand
University of Nebraska

Sally Ann Lederman
Columbia University

Richard Lewis
University of Georgia

J. Harriett McCoy
University of Arkansas

Sharon McWhinney
Prairie View A&M University

Robert Reynolds
University of Illinois at Chicago

Sharon Nickols-Richardson
Virginia Polytechnic Institute and State University

Adria Sherman
Rutgers University

Carmen R. Roman-Shriver
Texas Tech University

Joanne Slavin
University of Minnesota

Joanne Spaide
Professor Emeritus, University of Northern Iowa

Diana-Marie Spillman
Miami University, Oxford, Ohio

Wendy Stuhldreher
Slippery Rock University of Pennsylvania

Anne VanBeber
Texas Christian University

Phyllis Moser-Veillon
University of Maryland

Janelle Walter
Baylor University

Doris Wang
University of Minnesota Crookston

Suzy Weems, Ph.D.
Stephen F. Austin State University

Kay Wilder
Point Loma Nazarene College

"To be surprised, to wonder, is to begin to understand."
José Ortega y Gasset

Chapter 1

Nutrition Basics

Chapter Outline

Prepared by **Judith E. Brown**

Key Nutrition Concepts

1. Nutrition is the study of foods, their nutrients and other chemical constituents, and the effects of food constituents on health.

2. Nutrition is an interdisciplinary science.

3. Nutrition recommendations for the public change as new knowledge about nutrition and health relationships is gained.

4. At the core of the science of nutrition are principles that represent basic truths and serve as the foundation of our understanding about nutrition.

5. Healthy individuals require the same nutrients across the life cycle but in differing amounts. Nutritional needs can be met by a wide variety of cultural and religious food practices.

Introduction

Need to freshen up your knowledge of nutrition? Or, do you need to get up to speed on basic nutrition for the course? This chapter presents information about nutrition that paves the way to greater understanding of specific needs and benefits related to nutrition by life-cycle stage.

Nutrition is an interdisciplinary science focused on the study of foods, *nutrients,* and other food constituents and health. The body of knowledge about nutrition is large and is growing rapidly, changing views on what constitutes the best nutrition advice. You are encouraged to refer to nutrition texts and to use the online resources listed at the end of this chapter to fill in any knowledge gaps. You are also encouraged to stay informed in the future and to keep an open mind about the best nutrition advice for many health-related issues. Scientific evidence that drives decisions about nutrition and health changes with time.

This chapter centers on (1) the principles of the science of nutrition, (2) nutrients and other constituents of food, (3) nutritional assessment, (4) public food and nutrition programs, and (5) nationwide priorities for improvements in the public's nutritional health.

Nutrients Chemical substances in foods that are used by the body for growth and health.

Food Security Access at all times to a sufficient supply of safe, nutritious foods.

Food Insecurity Limited or uncertain availability of safe, nutritious foods.

Calorie A unit of measure of the amount of energy supplied by food. Also known as the "kilocalorie," or the "large Calorie."

Principles of the Science of Nutrition

Every field of science is governed by a set of principles that provides the foundation for growth in knowledge. These principles change little with time. Knowledge of the principles of nutrition listed in Table 1.1 will serve as a spring-board to greater understanding of the nutrition and health relationships explored in the chapters to come.

> **PRINCIPLE #1**   Food is a basic need of humans.

Humans need enough food to live and the right assortment of foods for optimal health (Illustration 1.1 on the next page). People who have enough food to meet their needs at all times experience *food security.* They are able to acquire food in socially acceptable ways—without having to scavenge or steal food. *Food insecurity* exists when the availability of safe, nutritious foods, or the ability to acquire them in socially acceptable ways, is limited or uncertain.[1]

> **PRINCIPLE #2**   Foods provide energy (calories), nutrients, and other substances needed for growth and health.

People eat foods for many different reasons. The most compelling reason is the requirement for *calories* (energy), nutrients, and other substances supplied by foods for growth and health.

A calorie is a measure of the amount of energy transferred from food to the body. Because calories are a unit of measure and not a substance actually present in food, they are not considered to be nutrients.

Nutrients are chemical substances in food that the body uses for a variety of functions that support growth, tissue maintenance and repair, and ongoing health. Essentially, every part of our body was once a nutrient consumed in food.

There are six categories of nutrients (Table 1.2). Each category except water consists of a number of different substances.

Table 1.1 Principles of human nutrition

Principle #1   Food is a basic need of humans.

Principle #2   Foods provide energy (calories), nutrients, and other substances needed for growth and health.

Principle #3   Health problems related to nutrition originate within cells.

Principle #4   Poor nutrition can result from both inadequate and excessive levels of nutrient intake.

Principle #5   Humans have adaptive mechanisms for managing fluctuations in food intake.

Principle #6   Malnutrition can result from poor diets and from disease states, genetic factors, or combinations of these causes.

Principle #7   Some groups of people are at higher risk of becoming inadequately nourished than others.

Principle #8   Poor nutrition can influence the development of certain chronic diseases.

Principle #9   Adequacy and balance are key characteristics of a healthy diet.

Principle #10   There are no "good" or "bad" foods.

Illustration 1.1 The need for food is part of Maslow's hierarchy of needs.

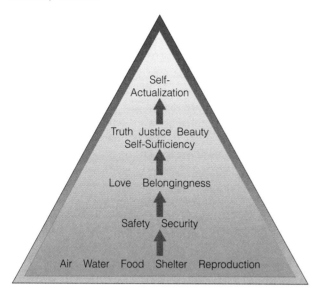

Essential and Nonessential Nutrients

Of the many nutrients required for growth and health, some must be provided by the diet while others can be made by the body.

ESSENTIAL NUTRIENTS Nutrients the body cannot manufacture, or generally produce in sufficient amounts, are referred to as *essential nutrients.* Here *essential* means "required in the diet." All of the following nutrients are considered essential:

- Carbohydrates
- Certain amino acids (the essential amino acids: histidine, isoleucine, leucine, lysine, methionine, phenylalanine, threonine, tryptophan, and valine)
- Linoleic acid and alpha-linolenic acid (essential fatty acids)
- Vitamins
- Minerals
- Water

NONESSENTIAL NUTRIENTS Cholesterol, creatine, and glucose are examples of *nonessential nutrients.* Nonessential nutrients are present in food and used by the body, but they do not have to be part of our diets. Many of the beneficial chemical substances in plants are not considered essential, for example, yet they play important roles in maintaining health.

REQUIREMENTS FOR ESSENTIAL NUTRIENTS
All humans require the same set of essential nutrients, but the amount of nutrients needed varies based on:

- Age
- Body size
- Gender
- Genetic traits
- Growth
- Illness

Table 1.2 The six categories of nutrients

1. **Carbohydrates**   Chemical substances in foods that consist of a single sugar molecule or multiples of sugar molecules in various forms. Sugar and fruit, starchy vegetables, and whole grain products are good dietary sources.
2. **Proteins**   Chemical substances in foods that are made up of chains of amino acids. Animal products and dried beans are examples of protein sources.
3. **Fats (Lipids)**   Components of food that are soluble in fat but not in water. They are more properly referred to as "lipids." Most fats are composed of glycerol attached to three fatty acids. Oil, butter, sausage, and avocado are examples of rich sources of dietary fats.
4. **Vitamins**   Thirteen specific chemical substances that perform specific functions in the body. Vitamins are present in many foods and are essential components of the diet. Vegetables, fruits, and grains are good sources of vitamins.
5. **Minerals**   In the context of nutrition, minerals consist of 15 elements found in foods that perform particular functions in the body. Milk, dark, leafy vegetables, and meat are good sources of minerals.
6. **Water**   An essential component of the diet provided by food and fluid.

- Lifestyle habits (e.g., smoking, alcohol intake)
- Medication use
- Pregnancy and lactation

Amounts of essential nutrients required each day vary a great deal, from cups (for water) to micrograms (for example, for folate and vitamin B_{12}).

Dietary Intake Standards

Dietary intake standards developed for the public cannot take into account all of the factors that influence nutrient needs, but they do account for the major ones of age, gender, growth, and pregnancy and lactation. Intake standards are called Dietary Reference Intakes (DRIs). DRIs have been developed for most of the essential nutrients and will be updated periodically. (These are listed on the inside front covers of this text.) Current DRIs were developed through a joint U.S.–Canadian effort, and the standards apply to both countries. The DRIs are levels of nutrient intake intended for use as reference values for planning and assessing diets for healthy people. They consist of the

Essential Nutrients   Substances required for growth and health that cannot be produced, or produced in sufficient amounts, by the body. They must be obtained from the diet.

Nonessential Nutrients   Nutrients required for growth and health that can be produced by the body from other components of the diet.

Illustration 1.2 Theoretical framework, terms, and abbreviations used in the Dietary Reference Intakes.

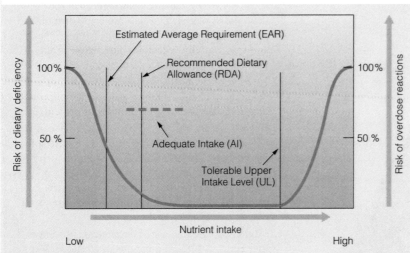

- **Dietary Reference Intakes (DRIs).** This is the general term used for the new nutrient intake standards for healthy people.
- **Recommended Dietary Allowances (RDAs).** These are levels of essential nutrient intake judged to be adequate to meet the known nutrient needs of practically all healthy persons while decreasing the risk of certain chronic diseases.
- **Adequate Intakes (AIs).** These are "tentative" RDAs. AIs are based on less conclusive scientific information than are the RDAs.
- **Estimated Average Requirements (EARs).** These are nutrient intake values that are estimated to meet the requirements of half the healthy individuals in a group. The EARs are used to assess adequacy of intakes of population groups.
- **Tolerable Upper Levels of Intake (ULs).** These are upper limits of nutrient intake compatible with health. The ULs do not reflect desired levels of intake. Rather, they represent total, daily levels of nutrient intake from food, fortified foods, and supplements that should not be exceeded.

Recommended Dietary Allowances (RDAs), which specify intake levels that meet the nutrient needs of over 98% of healthy people, and the other categories of intake standards described in Illustration 1.2. It is recommended that individuals aim for nutrient intakes that approximate the RDAs or Adequate Intake (AI) levels. Estimated Average Requirements (EARs) should be used to examine the possibility of inadequate intakes in individuals and within groups. Additional tests are required to confirm inadequate nutrient intakes and status.[2]

Daily Values (DVs) Scientifically agreed-upon standards for daily intakes of nutrients from the diet developed for use on nutrition labels.

TOLERABLE UPPER LEVELS OF INTAKE (ULs) The DRIs include a table indicating levels of daily nutrient intake from foods, fortified products, and supplements that should not be exceeded. They can be used to assess the safety of high intakes of nutrients, particularly from supplements.

STANDARDS OF NUTRIENT INTAKE FOR NUTRITION LABELS The Nutrition Facts panel on packaged foods uses standard levels of nutrient intakes based on an earlier edition of recommended dietary intake levels.

The levels are known as *Daily Values (DVs)* and are used to identify the amount of a nutrient provided in a serving of food compared to the standard level. The "% DV" listed on nutrition labels represents the percentages of the standards obtained from one serving of the food product. Table 1.3 lists DV standard amounts for nutrients that are mandatory or voluntary components of nutrition labels. Additional information on nutrition labeling is presented later in this chapter.

Carbohydrates

Carbohydrates are used by the body mainly as a source of readily available energy. They consist of the simple sugars (monosaccharides and disaccharides), complex carbohydrates (the polysaccharides), most dietary sources of fiber, and alcohol sugars. Alcohol is a close chemical relative of carbohydrates and is usually considered to be part of this nutrient category. The most basic forms of carbohydrates are single molecules called monosaccharides.

Glucose (also called "blood sugar" and "dextrose"), fructose ("fruit sugar"), and galactose are the most common monosaccharides. Molecules containing two monosaccharides are called disaccharides. The most common disaccharides are

- Sucrose (glucose + fructose, or common table sugar)
- Maltose (glucose + glucose, or malt sugar)
- Lactose (glucose + galactose, or milk sugar)

Complex carbohydrates (also called polysaccharides) are considered "complex" because they have more elaborate chemical structures than the simple sugars. They include

- Starches (the plant form of stored carbohydrate)
- Glycogen (the animal form of stored carbohydrate)
- Most types of fiber

Each type of simple and complex carbohydrate, except fiber, provides 4 calories per gram. Fibers do not count as a source of energy, because they cannot be broken down by human digestive enzymes. The main function of fiber is to provide "bulk" for normal elimination.

Table 1.3 Daily Values (DVs) for nutrition labeling based on intakes of 2000 calories per day in adults and children aged 4 years and above

MANDATORY COMPONENTS OF THE NUTRITION LABEL

Food Component	Daily Value (DV)
Total fat	65 g[a]
Saturated fat	20 g
Cholesterol	300 mg[a]
Sodium	2400 mg
Total carbohydrate	300 g
Dietary fiber	25 g
Vitamin A	5000 IU[a]
Vitamin C	60 mg
Calcium	1000 mg
Iron	18 mg

[a] g = grams; mg = milligrams; IU = international units

It has other beneficial properties, however. High-fiber diets reduce the rate of glucose absorption (a benefit for people with diabetes) and may help prevent cardiovascular disease and some types of cancer.[3]

Nonalcoholic in the beverage sense, alcohol sugars are like simple sugars, except that they include a chemical component of alcohol. Xylitol, mannitol, and sorbitol are common forms of alcohol sugars. Some are very sweet, and only small amounts are needed to sweeten commercial beverages, gums, yogurt, and other products. Unlike the simple sugars, alcohol sugars do not promote tooth decay.

Alcohol (consumed as ethanol) is considered to be part of the carbohydrate family because its chemical structure is similar to that of glucose. It is a product of the fermentation of sugar with yeast. With 7 calories per gram, alcohol has more calories per gram than do other carbohydrates.

RECOMMENDED INTAKE LEVEL Recommended intake of carbohydrates is based on their contribution to total energy intake. It is recommended that 45–65% of calories come from carbohydrates. Added sugar should constitute no more than 25% of total caloric intake. It is recommended that adult females consume between 21 and 25 grams, and males 30–38 grams of total dietary fiber daily.[3]

FOOD SOURCES OF CARBOHYDRATES Carbohydrates are widely distributed in plant foods, while milk is the only important animal source of carbohydrates (lactose). Table 1.4 on the next page lists selected food sources by type of carbohydrate. Additional information about total carbohydrate and fiber content of foods can be found on nutrition information labels on food packages and at this Web address: www.nal.usda.gov/fnic/foodcomp.

Protein

Protein in foods provides the body with *amino acids* used to build and maintain tissues such as muscle, bone, enzymes, and red blood cells. The body can also use protein as a source of energy—it provides 4 calories per gram. However, this is not a primary function of protein. Of the common types of amino acids, nine must be provided by the diet and are classified as "essential amino acids." (These are listed on page 3.) Many different amino acids obtained from food perform important functions, but since the body can manufacture these from other amino acids, they are classified as "nonessential amino acids."

Food sources of protein differ in quality, based on the types of amino acids they contain. Foods of high protein quality include all of the essential amino acids. Protein from milk, cheese, meat, eggs, and other animal products is considered high quality. Plant sources of protein, with the exception of soybeans, do not provide all nine essential amino acids. Combinations of plant foods, such as grains or seeds with dried beans, however, yield high-quality protein. Amino acids found in these individual foods "complement" each other, thus providing a source of high-quality protein.

RECOMMENDED PROTEIN INTAKE DRIs for protein are shown on the inside front cover of this text. In general, proteins should contribute 10–35% of total energy intake.[3] Protein deficiency, although rare in economically developed countries, leads to loss of muscle tissue, weakness, reduced resistance to disease, kidney and heart problems, and the deficiency disease *kwashiorkor* in children. Protein deficiency in adults produces a loss of body tissue protein, heart abnormalities, severe diarrhea, and other health problems.

FOOD SOURCES OF PROTEIN Animal products and dried beans are particularly good sources of protein. These and other food sources of proteins are listed in Table 1.5 on page 7.

Fats (Lipids)

Fats in food share the property of being soluble in fats but not in water. They are actually a subcategory of *lipids*, but this category of macronutrient is referred to as fat in the DRIs.[3] Lipids include fats, oils, and related compounds such as cholesterol. Fats are generally solid at room temperature, whereas oils are usually liquid. Fats and oils are made up of various types of triglycerides (triacylglycerols), which consist of three *fatty acids* attached to

Amino Acids The "building blocks" of protein. Unlike carbohydrates and fats, amino acids contain nitrogen.

Kwashiorkor A disease syndrome in children, primarily caused by protein deficiency. It is generally characterized by edema (or swelling), loss of muscle mass, fatty liver, rough skin, discoloration of the hair, growth retardation, and apathy.

Fatty Acids The fat-soluble components of fats in foods.

Table 1.4 Food sources of carbohydrates

A. SIMPLE SUGARS (MONO- AND DISACCHARIDES)

	Portion Size	Grams of Carbohydrates		Portion Size	Grams of Carbohydrates
Breakfast Cereals			Apple	1 med	16
Raisin Bran	1 c	18	Orange	1 med	14
Corn Pops	1 c	14	Peach	1 med	8
Frosted Cheerios	1 c	13	**Vegetables**		
Bran Flakes	¾ c	5	Corn	½ c	3
Grape-Nuts	½ c	3	Broccoli	½ c	2
Special K	1 c	3	Potato	1 med	1
Wheat Chex	1 c	2	**Beverages**		
Cornflakes	1 c	2	Soft drinks	12 oz	38
Sweeteners			Fruit drinks	1 c	29
Honey	1 tsp	6	Skim milk	1 c	12
Corn syrup	1 tsp	5	Whole milk	1 c	11
Maple syrup	1 tsp	4	**Candy**		
Table sugar	1 tsp	4	Hard candy	1 oz	28
Fruits			Gumdrops	1 oz	25
Watermelon	1 pc (4" × 8")	25	Caramels	1 oz	21
Banana	1 med	21	Milk chocolate	1 oz	16

B. COMPLEX CARBOHYDRATES (STARCHES)

	Portion Size	Grams of Carbohydrates		Portion Size	Grams of Carbohydrates
Grain Products			**Dried Beans**		
Rice, white, cooked	½ c	21	White beans, cooked	½ c	13
Pasta, cooked	½ c	15	Kidney beans, cooked	½ c	12
Oatmeal, cooked	½ c	12	Lima beans, cooked	½ c	11
Cheerios	1 c	11	**Vegetables**		
Cornflakes	1 c	11	Potato	1 med	30
Bread, whole wheat	1 slice	7	Corn	½ c	10
			Broccoli	½ c	2

C. TOTAL FIBER

	Portion Size	Grams of Total Fiber		Portion Size	Grams of Total Fiber
Grain Products			Green peas	½ c	4
Bran Buds	½ c	12	Carrots	½ c	3
All Bran	½ c	10	Potato, with skin	1 med	4
Raisin Bran	1 c	7	Collard greens	½ c	3
Bran Flakes	¾ c	5	Corn	½ c	3
Oatmeal, cooked	1 c	4	Cauliflower	½ c	2
Bread, whole wheat	1 sl	2	**Nuts**		
Fruits			Almonds	½ c	5
Avocado	½ med	7	Peanuts	½ c	3
Raspberries	1 c	5	Peanut butter	2 tbs	2
Mango	1 med	4	**Dried Beans**		
Pear, with skin	1 med	4	Pinto beans, cooked	½ c	10
Orange	1 med	3	Black beans, cooked	½ c	8
Banana	1 med	2	Black-eyed peas, cooked	½ c	8
Vegetables			Navy beans, cooked	½ c	6
Lima beans	½ c	5	Lentils, cooked	½ c	5

glycerol (Illustration 1.3). The number of carbons contained in the fatty acid component of triglycerides varies from 8 to 22.

Fats and oils are a concentrated source of energy, providing 9 calories per gram. Fats perform a number of important functions in the body. They are precursors for cholesterol and sex hormone synthesis, components of cell membranes, vehicles for carrying certain vitamins that are soluble in fats only, and suppliers of the **essential fatty acids** required for growth and health.

Table 1.5 Food sources of protein

	Portion Size	Grams of Protein
Meats		
Beef, lean	3 oz	26
Tuna, in water	3 oz	24
Hamburger, lean	3 oz	24
Chicken, no skin	3 oz	24
Lamb	3 oz	22
Pork chop, lean	3 oz	20
Haddock, broiled	3 oz	19
Egg	1 med	6
Dairy Products		
Cottage cheese, low fat	½ c	14
Yogurt, low fat	1 c	13
Milk, skim	1 c	9
Milk, whole	1 c	8
Swiss cheese	1 oz	8
Cheddar cheese	1 oz	7
Grain Products		
Oatmeal, cooked	½ c	4
Pasta, cooked	½ c	4
Bread	1 slice	2
Rice, white or brown	½ c	2

ESSENTIAL FATTY ACIDS There are two essential fatty acids: linoleic acid and alpha-linolenic acid. Because these fatty acids are essential, they must be supplied in the diet. The central nervous system is particularly rich in derivatives of these two fatty acids. They are found in phospholipids, which—along with cholesterol—are the primary lipids in the brain and other nervous system tissue. Biologically active derivatives of essential fatty acids include *prostaglandins, thromboxanes,* and *prostacyclins.*

Linoleic Acid Linoleic acid is the parent of the omega-6 (or n-6) fatty acid family. One of the major derivatives of linoleic acid is arachidonic acid. Arachidonic acid serves as a primary structural component of the central nervous system. Most vegetable oils, meats, and human milk are good sources of linoleic acid. American diets tend to provide sufficient levels of linoleic acid, and considerable amounts are stored in body fat.

Alpha-Linolenic Acid Alpha-linolenic acid is the parent of the omega-3 (n-3) fatty acid family. It is present in many types of dark green vegetables, vegetable oils, and flaxseed. Derivatives of this essential fatty acid include eicosapentaenoic acid (EPA) and docosahexaenoic acid (DHA). Relatively little EPA and DHA are produced in the body from alpha-linolenic acid because the conversion process is slow.[3] EPA and DHA also enter the body through intake of fatty, cold-water fish and shellfish and human milk. The EPA and DHA content of fish provides health benefits.[4] Regular consumption of fish (two or more meals per week) not only protects against irregular heartbeat, sudden death, and stroke but also reduces high blood pressure and plaque formation in arteries.[5] DHA is found in large amounts in the central nervous system, the retina of the eye, and the testes. The body stores only small amounts of alpha-linolenic acid, EPA, and DHA.[6]

Omega-6 to Omega-3 Fatty Acid Ratio The ratio of omega-6 to omega-3 fatty acid intake is important because the functions of one are adversely modified by the presence of disproportionately high amounts of the other. Although an exact ratio has not been agreed upon, it is thought that people should consume four times (or less) as much omega-6 fatty acids as omega-3 fatty acids. Many Americans regularly consume vegetable oils but eat fish infrequently. Consequently, the ratio between intake of omega-6 and omega-3 fatty acids is over 9 to 1, indicating a need to increase intake of n-3 fatty acids.[7]

SATURATED AND UNSATURATED FATS Fats (lipids) come in two basic types: *saturated* and *unsaturated.* Whether a fat is saturated or not depends on whether it has one or more double bonds between carbon atoms in one or more of the fatty acid components of the fat. If one double bond is present in one or more of the fatty acids, the fat is considered *monounsaturated;* if two or more are present, the fat is *polyunsaturated.*

Glycerol A component of fats that is soluble in water. It is converted to glucose in the body.

Essential Fatty Acids Components of fat that are a required part of the diet (i.e., linoleic and alpha-linolenic acids). Both contain unsaturated fatty acids.

Prostaglandins A group of physiologically active substances derived from the essential fatty acids. They are present in many tissues and perform such functions as the constriction or dilation of blood vessels, and stimulation of smooth muscles and the uterus.

Thromboxanes Biologically active substances produced in platelets that increase platelet aggregation (and therefore promote blood clotting), constrict blood vessels, and increase blood pressure.

Prostacyclins Biologically active substances produced by blood vessel walls that inhibit platelet aggregation (and therefore blood clotting), dilate blood vessels, and reduce blood pressure.

Saturated Fats Fats in which adjacent carbons in the fatty acid component are linked by single bonds only (e.g., –C–C–C–C–).

Unsaturated Fats Fats in which adjacent carbons in one or more fatty acids are linked by one or more double bonds (e.g., –C–C=C–C=C–).

Monounsaturated Fats Fats in which only one pair of adjacent carbons in one or more of its fatty acids is linked by a double bond (e.g., –C–C=C–C–).

Polyunsaturated Fats Fats in which more than one pair of adjacent carbons in one or more of its fatty acids are linked by two or more double bonds (e.g., –C–C=C–C=C–).

Illustration 1.3 Basic structure of a triglyceride.

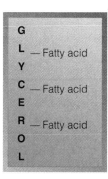

GLYCEROL — Fatty acid — Fatty acid — Fatty acid

Some unsaturated fatty acids are highly unsaturated. Alpha-linolenic acid, for example, contains three double bonds, arachidonic acid four, EPA five, and DHA six. These fatty acids are less stable than fatty acids with fewer double bonds, because double bonds between atoms are weaker than single bonds.

Saturated fats contain no double bonds between carbons and tend to be solid at room temperature. Animal products such as butter, cheese, and meats and two plant oils (coconut and palm) are rich sources of saturated fats. Fat we consume in our diets, whether it contains primarily saturated or unsaturated fatty acids, is generally in the triglyceride (or triacylglycerol) form.

Fats (lipids) also come in these forms:

- Monoglycerides (or monoacylglycerols), consisting of glycerol plus one fatty acid
- Diglycerides (or diacylglycerols), consisting of glycerol and two fatty acids

Although most foods contain both saturated and unsaturated fats, animal foods tend to contain more saturated and less unsaturated fat than plant foods. Saturated fatty acids tend to increase blood levels of LDL cholesterol (the lipoprotein that increases heart-disease risk when present in high levels) whereas unsaturated fatty acids tend to decrease LDL-cholesterol levels.[3]

Hydrogenation and Trans Fats   Oils can be made solid by adding hydrogen to the double bonds of their fatty acids. This process, called hydrogenation, makes some of the fatty acids in oils saturated and enhances storage life and baking qualities. Hydrogenation may alter the molecular structure of the fatty acids, however, changing the naturally occurring *cis* structure to the *trans* form. Trans fatty acids raise blood LDL-cholesterol levels to a greater extent than do saturated fatty acids.[3] Trans fatty acids are naturally present in dairy products and meats, but the primary dietary sources are products made from hydrogenated fats.

| Cholesterol | A fat-soluble, colorless liquid found in animals but not plants. |

CHOLESTEROL   Dietary *cholesterol* is a fatlike, clear liquid substance found in lean and fat components of animal products. Cholesterol is a component of all animal cell membranes, the brain, and the nerves. It is the precursor of estrogen, testosterone, and vitamin D, which is manufactured in the skin upon exposure to sunlight. The body generally produces only one-third of the cholesterol our bodies use, because more than sufficient amounts of cholesterol are provided in most people's diet. The extent to which dietary cholesterol intake modifies blood cholesterol level appears to vary a good deal based on genetic tendencies.[8] Dietary cholesterol intake affects blood cholesterol level substantially less than does saturated fat intake, however.[3] Leading sources of dietary cholesterol are egg yolks, meat, milk and milk products, and fats such as butter.

RECOMMENDED INTAKE OF FATS   Scientific evidence and opinions related to the effects of fat on health have changed substantially in recent years—and so have recommendations for fat intake. In the past, it was recommended that Americans aim for diets providing less than 30% of total calories from fat. Evidence indicating that the *type* of fat consumed is more important to health than is total fat intake has changed this advice. The watchwords for thinking about fat have become "not all fats are created equal: some are better for you than others." Americans are being urged to select food sources of "healthful" or "good" fats while keeping fat intake within the range of 20–35% of total caloric intake. Concerns that high-fat diets encourage the development of obesity have been eased by studies demonstrating that excessive caloric intakes—and not just diets high in fat—are related to weight gain. New recommendations regarding fat intake do not encourage increased fat consumption, but rather emphasize that healthy diets include certain types of fat and that total caloric intake and physical activity are the most important components of weight management.[3]

Fats that elevate LDL-cholesterol levels (which increases the risk of heart disease) are regarded as "unhealthful," while those that lower LDL cholesterol and raise blood levels of HDL cholesterol (the one that helps the body get rid of cholesterol in the blood) are considered healthful.[9,10] The list of unhealthy fats includes trans fats, saturated fats, and cholesterol.[10] Monounsaturated fats, polyunsaturated fats, alpha-linolenic acid, DHA, and EPA are considered healthful fats.

Current recommendations call for consumption of 20–35% of total calories from fat. The AIs for the essential fatty acid linoleic acid are set at 17 grams a day for men and 12 grams for women. AIs for the other essential fatty acid, alpha-linolenic acid, are 1.6 grams per day for men and 1.1 grams for women. It is recommended that people keep their intake of trans fats and saturated fats as low as possible while consuming a nutritionally adequate diet. Americans are being encouraged to increase consumption of EPA and DHA by eating fish more often. They are being urged to reduce saturated fat intake in order to reduce the risk of heart disease.[3]

There is no recommended level of cholesterol intake, because there is no evidence that cholesterol is required in the diet. The body is able to produce enough cholesterol, and people do not develop a cholesterol deficiency disease if it is not consumed. Because blood cholesterol levels tend to increase somewhat as consumption of cholesterol increases, it is recommended that intake should be minimal. Cholesterol intake averages around 250 mg per day in the United States; but a more health-promoting level of intake would be less than 200 mg a day.[11]

FOOD SOURCES OF FAT   The fat content of many foods can be identified by reading the nutrition information labels on food packages. The amount of total fat, saturated fat, trans fat, and cholesterol in a serving of food is listed on the labels. Table 1.6 lists the total fat, saturated fat, unsaturated fat, trans fat, cholesterol, and omega-3–fatty acid contents (EPA and DHA) of selected foods.

Table 1.6 Food sources of fats

A. TOTAL FAT

	Portion Size	Grams of Fat		Portion Size	Grams of Fat
Fats and Oils			Veggie pita	1	17.0
Mayonnaise	1 tbs	11.0	Subway meatball sandwich	1	16.0
Ranch dressing	1 tbs	6.0	Subway turkey sandwich	1	4.0
Vegetable oils	1 tsp	4.7	**Milk and Milk Products**		
Butter	1 tsp	4.0	Cheddar cheese	1 oz	9.5
Margarine	1 tsp	4.0	Milk, whole	1 c	8.5
Meats			American cheese	1 oz	6.0
Sausage	4 links	18.0	Cottage cheese, regular	½ c	5.1
Hot dog	2 oz	17.0	Milk, 2%	1 c	5.0
Hamburger, 21% fat	3 oz	15.0	Milk, 1%	1 c	2.9
Hamburger, 16% fat	3 oz	13.5	Milk, skim	1 c	0.4
Steak, rib eye	3 oz	9.9	Yogurt, frozen	1 c	0.3
Bacon	3 strips	9.0	**Other Foods**		
Steak, round	3 oz	5.2	Avocado	½	15.0
Chicken, baked, no skin	3 oz	4.0	Almonds	1 oz	15.0
Flounder, baked	3 oz	1.0	Cashews	1 oz	13.2
Shrimp, boiled	3 oz	1.0	French fries, small serving	1	10.0
Fast Foods			Taco chips	1 oz (10 chips)	10.0
Whopper	8.9 oz	32.0	Potato chips	1 oz (14 chips)	7.0
Big Mac	6.6 oz	31.4	Peanut butter	1 tbs	6.1
Quarter Pounder with Cheese	6.8 oz	28.6	Egg	1	6.0

B. SATURATED FATS

	Portion Size	Grams of Fat		Portion Size	Grams of Fat
Fats and Oils			Haddock, breaded, fried	3 oz	3.0
Margarine	1 tsp	2.9	Rabbit	3 oz	3.0
Butter	1 tsp	2.4	Pork chop, lean	3 oz	2.7
Salad dressing, Ranch	1 tbs	1.2	Steak, round, lean	3 oz	2.0
Peanut oil	1 tsp	0.9	Turkey, roasted	3 oz	2.0
Olive oil	1 tsp	0.7	Chicken, baked, no skin	3 oz	1.7
Salad dressing, 1000 Island	1 tbs	0.5	Prime rib, lean	3 oz	1.3
Canola oil	1 tsp	0.3	Venison	3 oz	1.1
Milk and Milk Products			Tuna, in water	3 oz	0.4
Cheddar cheese	1 oz	5.9	**Fast Foods**		
American cheese	1 oz	5.5	Croissant w/egg, bacon, & cheese	1	16.0
Milk, whole	1 c	5.1			
Cottage cheese, regular	½ c	3.0	Sausage croissant	1	16.0
Milk, 2%	1 c	2.9	Whopper	1	11.0
Milk, 1%	1 c	1.5	Cheeseburger	1	9.0
Milk, skim	1 c	0.3	Bac'n Cheddar Deluxe	1	8.7
Meats			Taco, regular	1	4.0
Hamburger, 21% fat	3 oz	6.7	Chicken breast sandwich	1	3.0
Sausage, links	4	5.6	**Nuts and Seeds**		
Hot dog	1	4.9	Macadamia nuts	1 oz	3.2
Chicken, fried, with skin	3 oz	3.8	Peanuts, dry roasted	1 oz	1.9
Salami	3 oz	3.6	Sunflower seeds	1 oz	1.6

C. UNSATURATED FATS

	Portion Size	Grams of Fat		Portion Size	Grams of Fat
Fats and Oils			Margarine	1 tsp	2.9
Canola oil	1 tsp	4.1	Butter	1 tsp	1.3
Vegetable oils	1 tsp	3.6			

continued

Table 1.6 Food sources of fats (continued)

C. UNSATURATED FATS (CONTINUED)

	Portion Size	Grams Unsaturated Fats		Portion Size	Grams Unsaturated Fats
Milk and Milk Products			Pork chop, lean	3 oz	5.3
Cottage cheese, regular	½ c	3.0	Turkey, roasted	3 oz	4.5
Cheddar cheese	1 oz	2.9	Tuna, in water	3 oz	0.7
American cheese	1 oz	2.8	Egg	1	5.0
Milk, whole	1 c	2.8	**Nuts and Seeds**		
Meats			Sunflower seeds	1 oz	16.6
Hamburger, 21% fat	3 oz	10.9	Almonds	1 oz	12.6
Haddock, breaded, fried	3 oz	6.5	Peanuts	1 oz	11.3
Chicken, baked, no skin	3 oz	6.0	Cashews	1 oz	10.2

D. TRANS FATS

	Portion Size	Grams Trans Fats		Portion Size	Grams Trans Fats
Fats and Oils			**Milk**		
Margarine, stick	1 tsp	1.3	Whole	1 c	0.2
Margarine, tub (soft)	1 tsp	0.1	**Other Foods**		
Shortening	1 tsp	0.3	Doughnut	1	3.2
Butter	1 tsp	0.1	Danish pastry	1	3.0
Margarine, "no–trans fat"	1 tsp	0	French fries, small serving	1	2.9
Meats			Cookies	2	1.8
Beef	3 oz	0.5	Corn chips	1 oz	1.4
Chicken	3 oz	0.1	Cake	1 slice	1.0
			Crackers	4 squares	0.5

E. CHOLESTEROL

	Portion Size	Milligrams Cholesterol		Portion Size	Milligrams Cholesterol
Fats and Oils			Ostrich, ground	3 oz	63
Butter	1 tsp	10.3	Pork chop, lean	3 oz	60
Vegetable oils, margarine	1 tsp	0	Hamburger, 10% fat	3 oz	60
Meats			Venison	3 oz	48
Brain	3 oz	1476	Wild pig	3 oz	33
Liver	3 oz	470	Goat, roasted	3 oz	32
Egg	1	212	Tuna, in water	3 oz	25
Veal	3 oz	128	**Milk and Milk Products**		
Shrimp	3 oz	107	Ice cream, regular	1 c	56
Prime rib	3 oz	80	Milk, whole	1 c	34
Chicken, baked, no skin	3 oz	75	Milk, 2%	1 c	22
Salmon, broiled	3 oz	74	Yogurt, low fat	1 c	17
Turkey, baked, no skin	3 oz	65	Milk, 1%	1 c	14
Hamburger, 20% fat	3 oz	64	Milk, skim	1 c	7

F. OMEGA-3 (N-3) FATTY ACIDS*

	Portion Size	Grams Omega-3 Fatty Acids		Portion Size	Grams Omega-3 Fatty Acids
Fish and Seafood			Salmon, chinook	3.5 oz	1.4
Sardines in oil	3.5 oz	3.3	Bluefish	3.5 oz	1.2
Mackerel	3.5 oz	2.6	Halibut	3.5 oz	1.2
Salmon, Atlantic, farmed	3.5 oz	2.2	Oysters	3.5 oz	1.1
Lake trout	3.5 oz	2.0	Salmon, pink	3.5 oz	1.0
Herring	3.5 oz	1.8	Trout, rainbow	3.5 oz	1.0
Salmon, sockeye	3.5 oz	1.5	Bass, striped	3.5 oz	0.8
Whitefish, lake	3.5 oz	1.5	Oysters	3.5 oz	0.6
Anchovies	3.5 oz	1.4			

*Grams EPA & DHA.

Table 1.6 Food sources of fats (continued)

F. OMEGA-3 (N-3) FATTY ACIDS* (CONTINUED)					
	Portion Size	Grams Omega-3 Fatty Acids		Portion Size	Grams Omega-3 Fatty Acids
Catfish	3.5 oz	0.5	Crab	3.5 oz	0.3
Pollock	3.5 oz	0.5	Pike, walleye	3.5 oz	0.3
Shrimp	3.5 oz	0.5	Catfish, wild	3.5 oz	0.2
Tuna, white, canned	3.5 oz	1.7	Fish sticks	3.5 oz	0.2
Swordfish	3.5 oz	0.8	Haddock	3.5 oz	0.2
Flounder	3.5 oz	0.5	Lobster	3.5 oz	0.2
Scallops	3.5 oz	0.5	Salmon, red	3.5 oz	0.2
Carp	3.5 oz	0.3	Snapper, red	3.5 oz	0.2

Vitamins

Vitamins are chemical substances in foods that perform specific functions in the body. Thirteen have been discovered so far. They are classified as either fat soluble or water soluble (Table 1.7).

The B-complex vitamins and vitamin C are soluble in water and found dissolved in water in foods. The fat-soluble vitamins consist of vitamins A, D, E, and K and are present in the fat portions of foods. (To remember the fat-soluble vitamins, think of "DEKA" for vitamins D, E, K, and A.) Only these chemical substances are truly vitamins. Substances such as coenzyme Q10, inositol, provitamin B_5 complex, and pangamic acid (vitamin B_{15}) may be called vitamins, but they are not. Except for vitamin B_{12}, water-soluble vitamin stores in the body are limited and run out within a few weeks to a few months after intake becomes inadequate. Fat-soluble vitamins are stored in the body's fat tissues and the liver. These stores can be sizable and last from months to years when intake is low.

Excessive consumption of the fat-soluble vitamins from supplements, especially of vitamins A and D, produces various symptoms of toxicity. High intake of the water-soluble vitamins from supplements can also produce adverse health effects. Toxicity symptoms from water-soluble vitamins,

however, tend to last a shorter time and are more quickly remedied. Vitamin overdoses are very rarely related to food intake.

Vitamins do not provide energy or serve as structural components of the body. Some play critical roles as *coenzymes* in chemical changes that take place in the body known as *metabolism.* Vitamin A is needed to replace the cells that line the mouth and esophagus, thiamin is needed for maintenance of normal appetite, and riboflavin and folate are needed for the synthesis of body proteins. Other vitamins (vitamins C and E, and beta-carotene—a precursor of vitamin A) act as *antioxidants* and perform other functions. By preventing or repairing damage to cells due to oxidation, these vitamins help maintain body tissues and prevent disease.

Primary functions, consequences of deficiency and overdose, primary food sources, and comments about each vitamin are listed in Table 1.8 starting on the next page.

Coenzymes Chemical substances that activate enzymes.

Metabolism The chemical changes that take place in the body. The conversion of glucose to energy or body fat is an example of a metabolic process.

Antioxidants Chemical substances that prevent or repair damage to cells caused by exposure to oxidizing agents such as oxygen, ozone, and smoke and to other oxidizing agents normally produced in the body. Many different antioxidants are found in foods; some are made by the body.

RECOMMENDED INTAKE OF VITAMINS Recommendations for levels of intake of vitamins are presented in the tables on the inside front covers of this text. Note that Tolerable Upper Levels of Intake (ULs) for many vitamins are also given; they represent levels of intake that should not be exceeded. Table 1.9 (pp. 16–19) lists food sources of each vitamin.

Other Substances in Food

"Things don't happen by accident in nature. If you observe it, it has a reason for being there."

Norman Krinsky, Tufts University

There are many substances in foods in addition to nutrients that affect health. Some foods contain naturally occurring

Table 1.7 Vitamin solubility

Water-Soluble Vitamins	Fat-Soluble Vitamins
B-complex vitamins	Vitamin A (retinol,
Thiamin (B_1)	beta-carotene)
Riboflavin (B_2)	Vitamin D (1,25
Niacin (B_3)	dihydroxy-cholecalciferol)
Vitamin B_6	Vitamin E (alpha-tocopherol)
Folate	Vitamin K
Vitamin B_{12}	
Biotin	
Pantothenic acid	
Vitamin C (ascorbic acid)	

Table 1.8 Summary of the vitamins

THE WATER-SOLUBLE VITAMINS

	Primary Functions	Consequences of Deficiency
Thiamin (vitamin B$_1$) AI[a] women: 1.1 mg men: 1.2 mg	• Coenzyme in the metabolism of carbohydrates, alcohol, and some amino acids • Required for the growth and maintenance of nerve and muscle tissues • Required for normal appetite	• Fatigue, weakness • Nerve disorders, mental confusion, apathy • Impaired growth • Swelling • Heart irregularity and failure
Riboflavin (vitamin B$_2$) AI women: 1.1 mg men: 1.3 mg	• Coenzyme involved in energy metabolism of carbohydrates, proteins, and fats • Coenzyme function in cell division • Promotes growth and tissue repair • Promotes normal vision	• Reddened lips, cracks at both corners of the mouth • Fatigue

Niacin (vitamin B$_3$)
RDA   women: 14 mg
      men: 16 mg
   UL: 35 mg (from supplements and fortified foods)
	• Coenzyme involved in energy metabolism
	• Coenzyme required for the synthesis of body fats
	• Helps maintain normal nervous system functions
		• Skin disorders
		• Nervous and mental disorders
		• Diarrhea, indigestion
		• Fatigue

Vitamin B$_6$ (pyridoxine)
AI   women: 1.3 mg
      men: 1.3 mg
      UL: 100 mg
	• Coenzyme involved in amnio acid, glucose, and fatty acid metabolism and neurotransmitter synthesis
	• Coenzyme in the conversion of tryptophan to niacin
	• Required for normal red blood cell formation
	• Required for the synthesis of lipids in the nervous and immune systems
		• Irritability, depression
		• Convulsions, twitching
		• Muscular weakness
		• Dermatitis near the eyes
		• Anemia
		• Kidney stones

Folate (folacin, folic acid)
RDA:
      women: 400 mcg
      men: 400 mcg
   UL: 1000 mcg (from supplements and fortified foods)
	• Required for the conversion of homocysteine to methionine
	• Methyl (CH_3) group donor and coenzyme in DNA synthesis, gene expression and regulation
	• Required for the normal formation of red blood and other cells
		• Megaloblastic cells and anemia
		• Diarrhea, weakness, irritability, paranoid behavior
		• Red, sore tongue
		• Increased blood homocysteine levels
		• Neural tube defects, low birthweight (in pregnancy); increased risk of heart disease and stroke

Vitamin B$_{12}$ (cyanocobalamin) AI women: 2.4 mcg men: 2.4 mcg	• Coenzyme involved in the synthesis of DNA, RNA, and myelin • Required for the conversion of homocysteine to methionine • Needed for normal red blood cell development	• Neurological disorders (nervousness, tingling sensations, brain degeneration) • Pernicious anemia • Increased blood homocysteine levels • Fatigue • Deficiency reported in 39% of adults in one study
Biotin AI women: 30 mcg men: 30 mcg	• Required by enzymes involved in fat, protein, and glycogen metabolism	• Depression, fatigue, nausea • Hair loss, dry and scaly skin • Muscular pain

[a]AI (Adequate Intakes) and RDAs (Recommended Dietary Allowances) are for 19-30-year-olds; UL (Upper Limits) are for 19-70-year-olds, 1997–2002.

Table 1.8 Summary of the vitamins (continued)

THE WATER-SOLUBLE VITAMINS

Consequences of Overdose	Primary Food Sources	Highlights and Comments
• High intakes of thiamin are rapidly excreted by the kidneys. Oral doses of 500 mg/day or less are considered safe.	• Grains and grain products (cereals, rice, pasta, bread) • Ready-to-eat cereals • Pork and ham, liver • Milk, cheese, yogurt • Dried beans and nuts	• Need increases with carbohydrate intake. • There is no "e" on the end of thiamin! • Deficiency rare in the U.S.; may occur in people with alcoholism. • Enriched grains and cereals prevent thiamin deficiency.
• None known. High doses are rapidly excreted by the kidneys.	• Milk, yogurt, cheese • Grains and grain products (cereals, rice, pasta, bread) • Liver, poultry, fish, beef • Eggs	• Destroyed by exposure to light.
• Flushing, headache, cramps, rapid heartbeat, nausea, diarrhea, decreased liver function with doses above 0.5 g per day	• Meats (all types) • Grains and grain products (cereals, rice, pasta, bread) • Dried beans and nuts • Milk, cheese, yogurt • Ready-to-eat cereals • Coffee • Potatoes	• Niacin has a precursor—tryptophan. Tryptophan, an amino acid, is converted to niacin by the body. Much of our niacin intake comes from tryptophan. • High doses raise HDL-cholesterol levels.
• Bone pain, loss of feeling in fingers and toes, muscular weakness, numbness, loss of balance (mimicking multiple sclerosis)	• Oatmeal, bread, breakfast cereals • Bananas, avocados, prunes, tomatoes, potatoes • Chicken, liver • Dried beans • Meats (all types), milk • Green and leafy vegetables	• Vitamins go from B_3 to B_6 because B_4 and B_5 were found to be duplicates of vitamins already identified.
• May cover up signs of vitamin B_{12} deficiency (pernicious anemia)	• Fortified, refined grain products (bread, flour, pasta) • Ready-to-eat cereals • Dark green, leafy vegetables (spinach, collards, romaine) • Broccoli, brussels sprouts • Oranges, bananas, grapefruit • Milk, cheese, yogurt • Dried beans	• Folate means "foliage." It was first discovered in leafy green vegetables. • This vitamin is easily destroyed by heat. • Synthetic form added to fortified grain products is better absorbed than naturally occurring folates.
• None known. Excess vitamin B_{12} is rapidly excreted by the kidneys or is not absorbed into the bloodstream. • Vitamin B_{12} injections may cause a temporary feeling of heightened energy.	• Animal products: beef, lamb, liver, clams, crab, fish, poultry, eggs • Milk and milk products • Ready-to-eat cereals	• Older people and vegans are at risk for vitamin B_{12} deficiency. • Some people become vitamin B_{12} deficient because they are genetically unable to absorb it. • Vitamin B_{12} is found in animal products and microorganisms only.
• None known. Excesses are rapidly excreted.	• Grain and cereal products • Meats, dried beans, cooked eggs • Vegetables	• Deficiency is extremely rare. May be induced by the overconsumption of raw eggs.

continued

Table 1.8 Summary of the vitamins (continued)

THE WATER-SOLUBLE VITAMINS (CONTINUED)

	Primary Functions	Consequences of Deficiency
Pantothenic acid (pantothenate) AI women: 5 mg men: 5 mg	• Coenzyme involved in energy metabolism of carbohydrates and fats • Coenzyme in protein metabolism	• Fatigue, sleep disturbances, numbness, impaired coordination • Vomiting, nausea

Vitamin C (ascorbic acid)
RDA   women: 75 mg
         men: 90 mg
         UL: 2000 mg
	• Required for collagen synthesis
	• Acts as an antioxidant; protects LDL cholesterol, eye tissues, sperm proteins, DNA, and lipids against oxidation
	• Required for the conversion of Fe^{++} to Fe^{+++}
	• Required for neurotransmitters and steroid hormone synthesis
		• Bleeding and bruising easily due to weakened blood vessels, cartilage, and other tissues containing collagen
		• Slow recovery from infections and poor wound healing
		• Fatigue, depression
		• Deficiency reported in 9–24% of adults in one study

THE FAT-SOLUBLE VITAMINS

	Primary Functions	Consequences of Deficiency

Vitamin A
RDA women: 700 mcg
         men: 900 mcg
         UL: 3000 mcg
	• Needed for the formation and maintenance of mucous membranes, skin, bone
	• Needed for vision in dim light
		• Increased susceptibility to infection, increased incidence and severity of infection (including measles)
		• Impaired vision, xerophthalmia, blindness
		• Inability to see in dim light

Vitamin E (alpha-tocopherol)
RDA women: 15 mg
         men: 15 mg
         UL: 1000 mg
	• Acts as an antioxidant, prevents damage to cell membranes in blood cells, lungs, and other tissues by repairing damage caused by free radicals
	• Reduces oxidation of LDL cholesterol
		• Muscle loss, nerve damage
		• Anemia
		• Weakness
		• Many adults may have nonoptimal blood levels

Vitamin D (1,25 dihydroxy-cholecalciferol)
AI   women: 5 mcg
               (200 IU)
         men: 5 mcg
               (200 IU)
         UL: 50 mcg
               (2000 IU)
	• Required for calcium and phosphorus metabolism in the intestines and bone, and for their utilization in bone and teeth formation, nerve and muscle activity	• Weak, deformed bones (children)
		• Loss of calcium from bones (adults), osteoporosis

Table 1.8 Summary of the vitamins (continued)

THE WATER-SOLUBLE VITAMINS (CONTINUED)

Consequences of Overdose	Primary Food Sources	Highlights and Comments
• None known. Excesses are rapidly excreted.	• Many foods, including meats, grains, vegetables, fruits, and milk	• Deficiency is very rare.
• Intakes of 1 g or more per day can cause nausea, cramps, and diarrhea and may increase the risk of kidney stones.	• Fruits: oranges, lemons, limes, strawberries, cantaloupe, honeydew melon, grapefruit, kiwi fruit, mango, papaya • Vegetables: broccoli, green and red peppers, collards, cabbage, tomato, asparagus, potatoes • Ready-to-eat cereals	• Need increases among smokers (to 110–125 mg per day). • Is fragile; easily destroyed by heat and exposure to air. • Supplements may decrease duration and symptoms of colds. • Deficiency may develop within 3 weeks of very low intake.

THE FAT-SOLUBLE VITAMINS

Consequences of Overdose	Primary Food Sources	Highlights and Comments
• Vitamin A toxicity (hypervitamosis A) with acute doses of 500,000 IU, or long-term intake of 50,000 IU per day; limit retinol use in pregnancy to 5000 IU daily • Nausea, irritability, blurred vision, weakness • Increased pressure in the skull, headache • Liver damage • Hair loss, dry skin • Birth defects	• Vitamin A is found in animal products only • Liver, butter, margarine, milk, cheese, eggs • Ready-to-eat cereals	• Beta-carotene is a vitamin A precursor or "provitamin." • Symptoms of vitamin A toxicity may mimic those of brain tumors and liver disease. Vitamin A toxicity is sometimes misdiagnosed because of the similarities in symptoms. • 1 mcg retinol equivalent = 5 IU vitamin A or 6 mcg beta-carotene.
• Intakes of up to 800 IU per day are unrelated to toxic side effects; over 800 IU per day may increase bleeding (blood-clotting time) • Avoid supplement use if aspirin, anticoagulants, or fish oil supplements are taken regularly	• Oils and fats • Salad dressings, mayonnaise, margarine, shortening, butter • Whole grains, wheat germ • Leafy, green vegetables, tomatoes • Nuts and seeds • Eggs	• Vitamin E is destroyed by exposure to oxygen and heat. • Oils naturally contain vitamin E. It's there to protect the fat from breakdown due to free radicals. • Eight forms of vitamin E exist, and each has different antioxidant strengths. • Natural form is better absorbed than synthetic form: 15 IU alpha-tocopherol = 22 IU d-alpha tocopherol (natural form) and 33 IU synthetic vitamin E.
• Mental retardation in young children • Abnormal bone growth and formation • Nausea, diarrhea, irritability, weight loss • Deposition of calcium in organs such as the kidneys, liver, and heart	• Vitamin D–fortified milk and margarine • Butter • Fish • Eggs • Mushrooms • Milk products such as cheese, yogurt, an\d ice cream are generally not fortified with vitamin D	• Vitamin D is manufactured from cholesterol in cells beneath the surface of the skin upon exposure of the skin to sunlight. • Deficiency may be common in ill, homebound, and elderly and hospitalized adults. • Breastfed infants with little sun exposure benefit from vitamin D supplements.

continued

Table 1.8 Summary of the vitamins (continued)

THE FAT-SOLUBLE VITAMINS (CONTINUED)

	Primary Functions	Consequences of Deficiency
Vitamin K (phylloquinone, menaquinone) AI women: 90 mcg men: 120 mcg	• Regulation of synthesis blood-clotting proteins • Aids in the incorporation of calcium into bones	• Bleeding, bruises • Decreased calcium in bones • Deficiency is rare; may be induced by the long-term use (months or more) of antibiotics

(Table 1.8 continues on p. 17→)

aFortified, refined grain products such as bread, rice, pasta, and crackers provide approximately 40 to 60 mcg of folic acid per standard serving.

Table 1.9 Food sources of vitamins

THIAMIN

Food	Amount	Thiamin (milligrams)	Food	Amount	Thiamin (milligrams)
Meats			Rice	½ c	0.1
Pork roast	3 oz	0.8	Bread	1 slice	0.1
Beef	3 oz	0.4	Vegetables		
Ham	3 oz	0.4	Peas	½ c	0.3
Liver	3 oz	0.2	Lima beans	½ c	0.2
Nuts and Seeds			Corn	½ c	0.1
Sunflower seeds	¼ c	0.7	Broccoli	½ c	0.1
Peanuts	¼ c	0.1	Potato	1 med	0.1
Almonds	¼ c	0.1	Fruits		
Grains			Orange juice	1 c	0.2
Bran flakes	1 c (1 oz)	0.6	Orange	1	0.1
Macaroni	½ c	0.1	Avocado	½	0.1

RIBOFLAVIN

Food	Amount	Riboflavin (milligrams)	Food	Amount	Riboflavin (milligrams)
Milk and Milk Products			Beef	3 oz	0.2
Milk	1 c	0.5	Tuna	3 oz	0.1
2% milk	1 c	0.5	Vegetables		
Yogurt, low fat	1 c	0.5	Collard greens	½ c	0.3
Skim milk	1 c	0.4	Broccoli	½ c	0.2
Yogurt	1 c	0.1	Spinach, cooked	½ c	0.1
American cheese	1 oz	0.1	Eggs		
Cheddar cheese	1 oz	0.1	Egg	1	0.2
Meats			Grains		
Liver	3 oz	3.6	Macaroni	½ c	0.1
Pork chop	3 oz	0.3	Bread	1 slice	0.1

NIACIN

Food	Amount	Niacin (milligrams)	Food	Amount	Niacin (milligrams)
Meats			Pork	3 oz	4.5
Liver	3 oz	14.0	Haddock	3 oz	2.7
Tuna	3 oz	10.3	Scallops	3 oz	1.1
Turkey	3 oz	9.5	Nuts and Seeds		
Chicken	3 oz	7.9	Peanuts	1 oz	4.9
Salmon	3 oz	6.9	Vegetables		
Veal	3 oz	5.2	Asparagus	½ c	1.5
Beef (round steak)	3 oz	5.1			

Table 1.8 Summary of the vitamins (continued)

THE FAT-SOLUBLE VITAMINS (CONTINUED)

Consequences of Overdose	Primary Food Sources	Highlights and Comments
• Toxicity is a problem only when synthetic forms of vitamin K are taken in excessive amounts. That may cause liver disease.	• Leafy, green vegetables • Grain products	• Vitamin K is produced by bacteria in the gut. Part of our vitamin K supply comes from these bacteria. • Newborns are given a vitamin K injection because they have "sterile" guts and consequently no vitamin K–producing bacteria.

Table 1.9 Food sources of vitamins (continued)

NIACIN

Food	Amount	Niacin (milligrams)	Food	Amount	Niacin (milligrams)
Grains			Bread, enriched	1 slice	0.7
Wheat germ	1 oz	1.5	Milk and Milk Products		
Brown rice	½ c	1.2	Cottage cheese	½ c	2.6
Noodles, enriched	½ c	1.0	Milk	1 c	1.9
Rice, white, enriched	½ c	1.0			

VITAMIN B₆

Food	Amount	Vitamin B$_6$ (milligrams)	Food	Amount	Vitamin B$_6$ (milligrams)
Meats			Dried beans, cooked	½ c	0.4
Liver	3 oz	0.8	Fruits		
Salmon	3 oz	0.7	Banana	1	0.6
Other fish	3 oz	0.6	Avocado	½	0.4
Chicken	3 oz	0.4	Watermelon	1 c	0.3
Ham	3 oz	0.4	Vegetables		
Hamburger	3 oz	0.4	Turnip greens	½ c	0.7
Veal	3 oz	0.4	Brussels sprouts	½ c	0.4
Pork	3 oz	0.3	Potato	1	0.2
Beef	3 oz	0.2	Sweet potato	½ c	0.2
Eggs			Carrots	½ c	0.2
Egg	1	0.3	Peas	½ c	0.1
Legumes					
Split peas	½ c	0.6			

FOLATE

Food	Amount	Folate (micrograms)	Food	Amount	Folate (Micrograms)
Vegetables			Broccoli	½ c	43
Garbanzo beans	½ c	141	Fruits		
Navy beans	½ c	128	Cantaloupe	¼ whole	100
Asparagus	½ c	120	Orange juice	1 c	87
Brussels sprouts	½ c	116	Orange	1	59
Black-eyed peas	½ c	102	Grains[a]		
Spinach, cooked	½ c	99	Ready-to-eat cereals	1 c/1 oz	100–400
Romaine lettuce	1 c	86	Oatmeal	½ c	97
Lima beans	½ c	71	Noodles	½ c	45
Peas	½ c	70	Wheat germ	2 tbs	40
Collard greens, cooked	½ c	56	Wild rice	½ c	37
Sweet potato	½ c	43			

[a]Fortified, refined grain products such as bread, rice, pasta, and crackers provide approximately 40 to 60 mcg of folic acid per standard serving.

continued

Table 1.9 Food sources of vitamins (continued)

VITAMIN B₁₂

Food	Amount	Vitamin B₁₂ (micrograms)	Food	Amount	Vitamin B₁₂ (micrograms)
Meats			Milk and Milk Products		
Liver	3 oz	6.8	Skim milk	1 c	1.0
Trout	3 oz	3.6	Milk	1 c	0.9
Beef	3 oz	2.2	Yogurt	1 c	0.8
Clams	3 oz	2.0	Cottage cheese	½ c	0.7
Crab	3 oz	1.8	American cheese	1 oz	0.2
Lamb	3 oz	1.8	Cheddar cheese	1 oz	0.2
Tuna	3 oz	1.8	Eggs		
Veal	3 oz	1.7	Egg	1	0.6
Hamburger, regular	3 oz	1.5			

VITAMIN C

Food	Amount	Vitamin C (milligrams)	Food	Amount	Vitamin C (milligrams)
Fruits			Watermelon	1 c	15
Orange juice, vitamin C–fortified	1 c	108	Vegetables		
			Green peppers	½ c	95
Kiwi fruit	1 or ½ c	108	Cauliflower, raw	½ c	75
Grapefruit juice, fresh	1 c	94	Broccoli	½ c	70
Cranberry juice cocktail	1 c	90	Brussels sprouts	½ c	65
Orange	1	85	Collard greens	½ c	48
Strawberries, fresh	1 c	84	Vegetable (V-8) juice	¾ c	45
Orange juice, fresh	1 c	82	Tomato juice	¾ c	33
Cantaloupe	¼ whole	63	Cauliflower, cooked	½ c	30
Grapefruit	1 med	51	Potato	1 med	29
Raspberries, fresh	1 c	31	Tomato	1 med	23

VITAMIN A (RETINOL)

Food Sources of Vitamin A (Retinol)	Amount	Vitamin A Food (micrograms RE)[b]	Sources of Vitamin A (Retinol)	Amount	Vitamin A (micrograms RE)[b]
Meats			2% milk	1 c	139
Liver	3 oz	9124	American cheese	1 oz	82
Salmon	3 oz	53	Whole milk	1 c	76
Tuna	3 oz	14	Swiss cheese	1 oz	65
Eggs			Fats		
Egg	1 med	84	Margarine, fortified	1 tsp	46
Milk and Milk Products			Butter	1 tsp	38
Skim milk, fortified	1 c	149			

VITAMIN A (BETA-CAROTENE)

Food Sources of Beta-Carotene	Amount	Vitamin A Value (micrograms RE)[b]	Food Sources of Beta-Carotene	Amount	Vitamin A Value (micrograms RE)[b]
Vegetables			Fruits		
Pumpkin, canned	½ c	2712	Cantaloupe	¼ whole	430
Sweet potato, canned	½ c	1935	Apricots, canned	½ c	210
Carrots, raw	½ c	1913	Nectarine	1 med	101
Spinach, cooked	½ c	739	Watermelon	1 c	59
Collard greens, cooked	½ c	175	Peaches, canned	½ c	47
Broccoli, cooked	½ c	109	Papaya	½ c	20
Winter squash	½ c	53			
Green peppers	½ c	40			

[b]RE (retinol equivalent) = 3.33 IU.

Table 1.9 Food sources of vitamins (continued)

VITAMIN E

Food	Amount	Vitamin E (IU)[c]	Food	Amount	Vitamin E (IU)[c]
Oils			Collard greens	½ c	3.1
Vegetable oil	1 tbs	6.7	Asparagus	½ c	2.1
Mayonnaise	1 tbs	3.4	Spinach, raw	1 c	1.5
Margarine	1 tbs	2.7	Grains		
Salad dressing	1 tbs	2.2	Wheat germ	2 tbs	4.2
Nuts and Seeds			Bread, whole wheat	1 slice	2.5
Sunflower seeds	¼ c	27.1	Bread, white	1 slice	1.2
Almonds	¼ c	12.7	Seafood		
Peanuts	¼ c	4.9	Crab	3 oz	4.5
Cashews	¼ c	0.7	Shrimp	3 oz	3.7
Vegetables			Fish	3 oz	2.4
Sweet potato	½ c	6.9			

VITAMIN D

Food	Amount	Vitamin D (IU)[d]	Food	Amount	Vitamin D (IU)[d]
Milk			Organ meats		
Milk, whole, low fat, or skim	1 c	100	Beef liver	3 oz	42
Fish and seafood			Chicken liver	3 oz	40
Salmon	3 oz	340	Eggs		
Tuna	3 oz	150	Egg yolk	1	27
Shrimp	3 oz	127			

[c]15 mg alpha-tocopherol = 22 IU d-alpha tocopherol (natural form) and 33 IU synthetic vitamin E.
[d]40 IU = 1 mcg.

toxins, such as poison in puffer fish and solanine in green sections near the skin of some potatoes. Consuming the poison in puffer fish can be lethal; large doses of solanine can interfere with nerve impulses. Some plant pigments, hormones, and other naturally occurring substances that protect plants from insects, oxidization, and other damaging exposures also appear to benefit human health. These substances in plants are referred to as *phytochemicals*, and knowledge about their effects on human health is advancing rapidly. Consumption of foods rich in specific pigments and other phytochemicals, rather than consumption of isolated phytochemicals, may help prevent certain types of cancer, cataracts, type 2 diabetes, hypertension, infections, and heart disease.[12,13,14] High intakes of certain phytochemicals from vegetables, fruits, nuts, seeds, and whole grain products may partially account for lower rates of heart disease and cancer observed in people with high intakes of these foods.[15]

Minerals

Humans require the 15 minerals listed in Table 1.10. Minerals are unlike other nutrients in that they consist of single atoms and carry a charge in solution. The property of being charged (or having an unequal number of electrons and protons) is related to many of the functions of minerals. The charge carried by minerals allows them to combine with other minerals to form stable complexes in bone, teeth, cartilage, and other tissues. In body fluids, charged minerals serve as a source of electrical power that stimulates muscles to contract (e.g., the heart to beat) and nerves to react. Minerals also help the body maintain an adequate amount of water in tissues and control how acidic or basic body fluids remain.

Phytochemicals (*Phyto* = plants) Chemical substances in plants, some of which affect body processes in humans that may benefit health.

The tendency of minerals to form complexes has implications for the absorption of minerals from food. Calcium and zinc, for example, may combine with other minerals in supplements or with dietary fiber and form complexes that cannot be absorbed. Therefore, in general, the proportion of total mineral intake that is absorbed is less than for vitamins.

Functions, consequences of deficiency and overdose, primary food sources, and comments about the 15 minerals needed by humans are summarized in Table 1.11 (pp. 20–23).

Table 1.10 Minerals required by humans

Calcium	Fluoride	Chromium
Phosphorus	Iodine	Molybdenum
Magnesium	Selenium	Sodium
Iron	Copper	Potassium
Zinc	Manganese	Chloride

Table 1.11 Summary of minerals

			Primary Functions	Consequences of Deficiency
Calcium			• Component of bones and teeth	• Poorly mineralized, weak bones (osteoporosis)
AI* women:	1000 mg		• Required for muscle and nerve activity, blood clotting	• Rickets in children
men:	1000 mg			• Osteomalacia (rickets in adults)
UL:	2500 mg			• Stunted growth in children
				• Convulsions, muscle spasms
Phosphorus			• Component of bones and teeth	• Loss of appetite
RDA women:	700 mg		• Component of certain enzymes and other substances involved in energy formation	• Nausea, vomiting
men:	700 mg		• Required for maintenance of acid-base balance of body fluids	• Weakness
UL:	4000 mg			• Confusion
				• Loss of calcium from bones
Magnesium			• Component of bones and teeth	• Stunted growth in children
RDA women:	310 mg		• Needed for nerve activity	• Weakness
men:	400 mg		• Activates enzymes involved in energy and protein formation	• Muscle spasms
UL:	350 mg (from supplements only)			• Personality changes
Iron			• Transports oxygen as a component of hemoglobin in red blood cells	• Iron deficiency
RDA women:	18 mg		• Component of myoglobin (a muscle protein)	• Iron-deficiency anemia
men:	8 mg		• Required for certain reactions involving energy formation	• Weakness, fatigue
UL:	45 mg			• Pale appearance
				• Reduced attention span and resistance to infection
				• Mental retardation, developmental delay in children
Zinc			• Required for the activation of many enzymes involved in the reproduction of proteins	• Growth failure
RDA women:	8 mg		• Component of insulin, many enzymes	• Delayed sexual maturation
men:	11 mg			• Slow wound healing
UL:	40 mg			• Loss of taste and appetite
				• In pregnancy, low-birthweight infants and preterm delivery
Fluoride			• Component of bones and teeth (enamel)	• Tooth decay and other dental diseases
AI women:	3 mg			
men:	4 mg			
UL:	10 mg			
Iodine			• Component of thyroid hormones that help regulate energy production and growth	• Goiter
RDA women:	150 mcg			• Cretinism (mental retardation, hearing loss, growth failure)
men:	150 mcg			
UL:	1100 mcg			

*AIs and RDA are for women and men 19–30 years of age; ULs are for males and females 19–70 years of age, 1997–2004.

Table 1.11 Summary of minerals (continued)

Consequences of Overdose	Primary Food Sources	Highlights and Comments
• Drowsiness • Calcium deposits in kidneys, liver, and other tissues • Suppression of bone remodeling • Decreased zinc absorption	• Milk and milk products (cheese, yogurt) • Broccoli • Dried beans • Calcium-fortified foods (some juices, breakfast cereals, bread, for example)	• The average intake of calcium among U.S. women is approximately 60% of the DRI. • One in four women and one in eight men in the U.S. develop osteoporosis. • Adequate calcium and vitamin D status must be maintained to prevent bone loss.
• Muscle spasms	• Milk and milk products (cheese, yogurt) • Meats • Seeds, nuts • Phosphates added to foods	• Deficiency is generally related to disease processes.
• Diarrhea • Dehydration • Impaired nerve activity due to disrupted utilization of calcium	• Plant foods (dried beans, tofu, peanuts, potatoes, green vegetables) • Milk • Bread • Ready-to-eat cereals • Coffee	• Magnesium is primarily found in plant foods where it is attached to chlorophyll. • Average intake among U.S. adults is below the RDA.
• Hemochromatosis ("iron poisoning") • Vomiting, abdominal pain • Blue coloration of skin • Liver and heart damage, diabetes • Decreased zinc absorption • Atherosclerosis (plaque buildup) in older adults	• Liver, beef, pork • Dried beans • Iron-fortified cereals • Prunes, apricots, raisins • Spinach • Bread • Pasta	• Cooking foods in iron and stainless steel pans increases the iron content of the foods. • Vitamin C, meat, and alcohol increase iron absorption. • Iron deficiency is the most common nutritional deficiency in the world. • Average iron intake of young children and women in the U.S. is low.
• Over 25 mg/day is associated with nausea, vomiting, weakness, fatigue, susceptibility to infection, copper deficiency, and metallic taste in mouth • Increased blood lipids	• Meats (all kinds) • Grains • Nuts • Milk and milk products (cheese, yogurt) • Ready-to-eat cereals • Bread	• Like iron, zinc is better absorbed from meats than from plants. • Marginal zinc deficiency may be common, especially in children. • Zinc supplements may decrease duration and severity of the common cold.
• Fluorosis • Brittle bones • Mottled teeth • Nerve abnormalities	• Fluoridated water and foods and beverages made with it • Tea • Shrimp, crab	• Toothpastes, mouth rinses, and other dental care products may provide fluoride. • Fluoride overdose has been caused by ingestion of fluoridated toothpaste.
• Over 1 mg/day may produce pimples, goiter, and decreased thyroid function	• Iodized salt • Milk and milk products • Seaweed, seafoods • Bread from commercial bakeries	• Iodine deficiency remains a major health problem in some developing countries. • Amount of iodine in plants depends on iodine content of soil. • Most of the iodine in our diet comes from the incidental addition of iodine to foods from cleaning compounds used by food manufacturers.

continued

Table 1.11 Summary of minerals (continued)

	Primary Functions	Consequences of Deficiency
Selenium RDA women: 55 mcg men: 55 mcg UL: 400 mcg	• Acts as an antioxidant in conjunction with vitamin E (protects cells from damage due to exposure to oxygen) • Needed for thyroid hormone production	• Anemia • Muscle pain and tenderness • Keshan disease (heart failure), Kashin-Beck disease (joint disease)
Copper RDA women: 900 mcg men: 900 mcg UL: 10,000 mcg	• Component of enzymes involved in the body's utilization of iron and oxygen • Functions in growth, immunity, cholesterol and glucose utilization, brain development	• Anemia • Seizures • Nerve and bone abnormalities in children • Growth retardation
Manganese AI women: 2.3 mg men: 1.8 mg	• Required for the formation of body fat and bone	• Weight loss • Rash • Nausea and vomiting
Chromium AI women: 35 mcg men: 25 mcg	• Required for the normal utilization of glucose and fat	• Elevated blood glucose and triglyceride levels • Weight loss
Molybdenum RDA women: 45 mcg men: 45 mcg UL: 2000 mcg	• Component of enzymes involved in the transfer of oxygen from one molecule to another	• Rapid heartbeat and breathing • Nausea, vomiting • Coma
Sodium AI adults: 1500 mg UL adults: 2300 mg	• Regulation of acid-base balance in body fluids • Maintenance of water balance in body tissues • Activation of muscles and nerves	• Weakness • Apathy • Poor appetite • Muscle cramps • Headache • Swelling
Potassium AI adults: 4700 mg	• Same as for sodium	• Weakness • Irritability, mental confusion • Irregular heartbeat • Paralysis
Chloride AI adults: 2300 mg	• Component of hydrochloric acid secreted by the stomach (used in digestion) • Maintenance of acid-base balance of body fluids • Maintenance of water balance in the body	• Muscle cramps • Apathy • Poor appetite • Long-term mental retardation in infants

Table 1.11 Summary of minerals (continued)

Consequences of Overdose	Primary Food Sources	Highlights and Comments
• "Selenosis;" symptoms of selenosis are hair and fingernail loss, weakness, liver damage, irritability, and "garlic" or "metallic" breath	• Meats and seafoods • Eggs • Whole grains	• Content of foods depends on amount of selenium in soil, water, and animal feeds. • May play a role in the prevention of some types of cancer.
• Wilson's disease (excessive accumulation of copper in the liver and kidneys) • Vomiting, diarrhea • Tremors • Liver disease	• Bread • Potatoes • Grains • Dried beans • Nuts and seeds • Seafood • Ready-to-eat cereals	• Toxicity can result from copper pipes and cooking pans. • Average intake in the U.S. is below the RDA.
• Infertility in men • Disruptions in the nervous system (psychotic symptoms) • Muscle spasms	• Whole grains • Coffee, tea • Dried beans • Nuts	• Toxicity is related to overexposure to manganese dust in miners.
• Kidney and skin damage	• Whole grains • Wheat germ • Liver, meat • Beer, wine • Oysters	• Toxicity usually results from exposure in chrome-making industries or overuse of supplements. • Supplements do not build muscle mass or increase endurance.
• Loss of copper from the body • Joint pain • Growth failure • Anemia • Gout	• Dried beans • Grains • Dark green vegetables • Liver • Milk and milk products	• Deficiency is extraordinarily rare.
• High blood pressure in susceptible people • Kidney disease • Heart problems	• Foods processed with salt • Cured foods (corned beef, ham, bacon, pickles, sauerkraut) • Table and sea salt • Bread • Milk, cheese • Salad dressing	• Very few foods naturally contain much sodium; processed foods are the leading source. • High-sodium diets are associated with hypertension in "salt-sensitive" people. • Kidney disease, excessive water consumption are related to sodium depletion.
• Irregular heartbeat, heart attack	• Plant foods (potatoes, squash, lima beans, tomatoes, plantains, bananas, oranges, avocados) • Meats • Milk and milk products • Coffee	• Content of vegetables is often reduced in processed foods. • Diuretics (water pills), vomiting, diarrhea may deplete potassium. • Salt substitutes often contain potassium.
• Vomiting	• Same as for sodium (most of the chloride in our diets comes from salt)	• Excessive vomiting and diarrhea may cause chloride deficiency. • Legislation regulating the composition of infant formulas was enacted in response to formula-related chloride deficiency and subsequent mental retardation in infants.

RECOMMENDED INTAKE OF MINERALS Recommendations for intake of minerals are presented in the tables on the inside front covers of this text. Note that Tolerable Upper Levels of Intake for many minerals are also given in a separate table. Table 1.12 lists food sources of each mineral.

Table 1.12 Food sources of minerals

MAGNESIUM

Food	Amount	Magnesium (mg)	Food	Amount	Magnesium (mg)
Legumes			**Vegetables**		
Lentils, cooked	½ c	134	Bean sprouts	½ c	98
Split peas, cooked	½ c	134	Black-eyed peas	½ c	58
Tofu	½ c	130	Spinach, cooked	½ c	48
Nuts			Lima beans	½ c	32
Peanuts	¼ c	247	**Milk and Milk Products**		
Cashews	¼ c	93	Milk	1 c	30
Almonds	¼ c	80	Cheddar cheese	1 oz	8
Grains			American cheese	1 oz	6
Bran buds	1 c	240	**Meats**		
Wild rice, cooked	½ c	119	Chicken	3 oz	25
Breakfast cereal, fortified	1 c	85	Beef	3 oz	20
Wheat germ	2 tbs	45	Pork	3 oz	20

CALCIUM*

Food	Amount	Calcium (mg)	Food	Amount	Calcium (mg)
Milk and Milk Products			Ice milk	1 c	180
Yogurt, low fat	1 c	413	American cheese	1 oz	175
Milk shake			Custard	½ c	150
(low-fat frozen yogurt)	1¼ c	352	Cottage cheese	1½ c	70
Yogurt with fruit, low fat	1 c	315	Cottage cheese, low fat	½ c	69
Skim milk	1 c	301	**Vegetables**		
1% milk	1 c	300	Spinach, cooked	½ c	122
2% milk	1 c	298	Kale	1½ c	47
3.25% milk (whole)	1 c	288	Broccoli	½ c	36
Swiss cheese	1 oz	270	**Legumes**		
Milk shake (whole milk)	1¼ c	250	Tofu	½ c	260
Frozen yogurt, low fat	1 c	248	Dried beans, cooked	½ c	60
Frappuccino	1 c	220	**Foods Fortified with Calcium**		
Cheddar cheese	1 oz	204	Orange juice	1 c	350
Frozen yogurt	1 c	200	Frozen waffles	2	300
Cream soup	1 c	186	Soymilk	1 c	200–400
Pudding	½ c	185	Breakfast cereals	1 c	150–1000
Ice cream	1 c	180			

SELENIUM

Food	Amount	Selenium (mcg)	Food	Amount	Selenium (mcg)
Seafood			Ham	3 oz	29
Lobster	3 oz	66	Beef	3 oz	22
Tuna	3 oz	60	Bacon	3 oz	21
Shrimp	3 oz	54	Chicken	3 oz	18
Oysters	3 oz	48	Lamb	3 oz	14
Fish	3 oz	40	Veal	3 oz	10
Meats			**Eggs**		
Liver	3 oz	56	Egg	1 med	37

*Actually, the richest source of calcium is alligator meat; 3½ ounces contain about 1231 milligrams of calcium. But just try to find it on your grocer's shelf!

Table 1.12 Food sources of minerals (continued)

SODIUM

Food	Amount	Sodium (mg)	Food	Amount	Sodium (mg)
Miscellaneous			Meat loaf	3 oz	555
Salt	1 tsp	2132	Sausage	3 oz	483
Dill pickle	1 (4½ oz)	1930	Hot dog	1	477
Sea salt	1 tsp	1716	Fish, smoked	3 oz	444
Chicken broth	1 c	1571	Bologna	1 oz	370
Ravioli, canned	1 c	1065	**Milk and Milk Products**		
Spaghetti with sauce, canned	1 c	955	Cream soup	1 c	1070
			Cottage cheese	½ c	455
Baking soda	1 tsp	821	American cheese	1 oz	405
Beef broth	1 c	782	Cheese spread	1 oz	274
Gravy	¼ c	720	Parmesan cheese	1 oz	247
Italian dressing	2 tbs	720	Gouda cheese	1 oz	232
Pretzels	5 (1 oz)	500	Cheddar cheese	1 oz	175
Green olives	5	465	Skim milk	1 c	125
Pizza with cheese	1 wedge	455	Whole milk	1 c	120
Soy sauce	1 tsp	444	**Grains**		
Cheese twists	1 c	329	Bran flakes	1 c	363
Bacon	3 slices	303	Cornflakes	1 c	325
French dressing	2 tbs	220	Croissant	1 med	270
Potato chips	1 oz (10 pieces)	200	Bagel	1	260
Catsup	1 tbs	155	English muffin	1	203
Meats			White bread	1 slice	130
Corned beef	3 oz	808	Whole wheat bread	1 slice	130
Ham	3 oz	800	Saltine crackers	4 squares	125
Fish, canned	3 oz	735			

IRON

Food	Amount	Iron (mg)	Food	Amount	Iron (mg)
Meat and Meat Alternates			English muffin	1	1.6
Liver	3 oz	7.5	Rye bread	1 slice	1.0
Round steak	3 oz	3.0	Whole wheat bread	1 slice	0.8
Hamburger, lean	3 oz	3.0	White bread	1 slice	0.6
Baked beans	½ c	3.0	**Fruits**		
Pork	3 oz	2.7	Prune juice	1 c	9.0
White beans	½ c	2.7	Apricots, dried	½ c	2.5
Soybeans	½ c	2.5	Prunes	5 med	2.0
Pork and beans	½ c	2.3	Raisins	¼ c	1.3
Fish	3 oz	1.0	Plums	3 med	1.1
Chicken	3 oz	1.0	**Vegetables**		
Grains			Spinach, cooked	½ c	2.3
Breakfast cereal, iron fortified	1 c	8.0 (4–18)	Lima beans	½ c	2.2
			Black-eyed peas	½ c	1.7
Oatmeal, fortified, cooked	1 c	8.0	Peas	½ c	1.6
			Asparagus	½ c	1.5
Bagel	1	1.7			

ZINC

Food	Amount	Zinc (mg)	Food	Amount	Zinc (mg)
Meats			Lamb	3 oz	3.5
Liver	3 oz	4.6	Turkey ham	3 oz	2.5
Beef	3 oz	4.0	Pork	3 oz	2.4
Crab	½ c	3.5	Chicken	3 oz	2.0

continued

Table 1.12 Food sources of minerals (continued)

ZINC (CONTINUED)

Food	Amount	Zinc (mg)	Food	Amount	Zinc (mg)
Legumes			**Nuts and Seeds**		
Dried beans, cooked	½ c	1.0	Pecans	¼ c	2.0
Split peas, cooked	½ c	0.9	Cashews	¼ c	1.8
Grains			Sunflower seeds	¼ c	1.7
Breakfast cereal, fortified	1 c	1.5–4.0	Peanut butter	2 tbs	0.9
Wheat germ	2 tbs	2.4	**Milk and Milk Products**		
Oatmeal, cooked	1 c	1.2	Cheddar cheese	1 oz	1.1
Bran flakes	1 c	1.0	Whole milk	1 c	0.9
Brown rice, cooked	½ c	0.6	American cheese	1 oz	0.8
White rice	½ c	0.4			

PHOSPHORUS

Food	Amount	Phosphorus (mg)	Food	Amount	Phosphorus (mg)
Milk and Milk Products			**Grains**		
Yogurt	1 c	327	Bran flakes	1 c	180
Skim milk	1 c	250	Shredded wheat	2 large biscuits	81
Whole milk	1 c	250	Whole wheat bread	1 slice	52
Cottage cheese	½ c	150	Noodles, cooked	½ c	47
American cheese	1 oz	130	Rice, cooked	½ c	29
Meats			White bread	1 slice	24
Pork	3 oz	275	**Vegetables**		
Hamburger	3 oz	165	Potato	1 med	101
Tuna	3 oz	162	Corn	½ c	73
Lobster	3 oz	125	Peas	½ c	70
Chicken	3 oz	120	French fries	½ c	61
Nuts and Seeds			Broccoli	½ c	54
Sunflower seeds	¼ c	319	**Other**		
Peanuts	¼ c	141	Milk chocolate	1 oz	66
Pine nuts	¼ c	106	Cola	12 oz	51
Peanut butter	1 tbs	61	Diet cola	12 oz	45

POTASSIUM

Food	Amount	Potassium (mg)	Food	Amount	Potassium (mg)
Vegetables			Hamburger	3 oz	480
Potato	1 med	780	Lamb	3 oz	382
Winter squash	½ c	327	Pork	3 oz	335
Tomato	1 med	300	Chicken	3 oz	208
Celery	1 stalk	270	**Grains**		
Carrots	1 med	245	Bran buds	1 c	1080
Broccoli	½ c	205	Bran flakes	1 c	248
Fruits			Raisin bran	1 c	242
Avocado	½ med	680	Wheat flakes	1 c	96
Orange juice	1 c	469	**Milk and Milk Products**		
Banana	1 med	440	Yogurt	1 c	531
Raisins	¼ c	370	Skim milk	1 c	400
Prunes	4 large	300	Whole milk	1 c	370
Watermelon	1 c	158	**Other**		
Meats			Salt substitutes	1 tsp	1300–2378
Fish	3 oz	500			

Water

Water is the last, but not the least, nutrient category. Adults are about 60–70% water by weight. Water provides the medium in which most chemical reactions take place in the body. It plays a role in energy transformation, the excretion of wastes, and temperature regulation.

People need enough water to replace daily losses from perspiration, urination, and exhalation. In normal weather conditions and with normal physical activity levels, adult males need 15–16 cups of water from fluids and foods each day, and females 11 cups. The need for water is generally met by consuming sufficient fluids to satisfy thirst.[16] The need for water is greater in hot and humid climates, and when physical activity levels are high. People generally consume 75% of the water intake from water and other fluids and 25% from foods.[17] Adequate consumption of water is indicated by the excretion of urine that is pale yellow and normal in volume.[18]

DIETARY SOURCES OF WATER   The best sources of water are tap and bottled water; nonalcoholic beverages such as fruit juice, milk, and vegetable juice; and brothy soups. Alcohol tends to increase water loss though urine, so beverages such as beer and wine are not as "hydrating" as water is. Caffeinated beverages are hydrating in people who are accustomed to consuming them.[19]

> **PRINCIPLE #3**   Health problems related to nutrition originate within cells.

The functions of each cell are maintained by the nutrients it receives. Problems arise when a cell's need for nutrients differs from the amounts that are available. Cells (Illustration 1.4) are the building blocks of tissues (such as bones and muscles), organs (the heart, kidney, and liver, for example), and systems (such as the circulatory and respiratory systems). Normal cell health and functions are maintained when a nutritional and environmental utopia exists within and around cells. This state of optimal, cellular nutrient conditions supports *homeostasis* in the body.

Disruptions in the availability of nutrients, or the presence of harmful substances in the cell's environment, initiate diseases and disorders that eventually affect tissues, organs, and systems. For example, too little folate reduces the conversion of the amino acid methionine to

Illustration 1.4 Schematic representation of the structure and major components of a human cell.

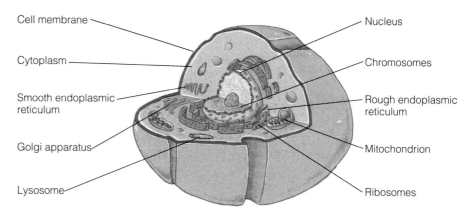

- Cell membrane
- Cytoplasm
- Smooth endoplasmic reticulum
- Golgi apparatus
- Lysosome
- Nucleus
- Chromosomes
- Rough endoplasmic reticulum
- Mitochondrion
- Ribosomes

cysteine. This causes a buildup of an intermediary product called homocysteine. High levels of homocysteine disrupt normal cell processes and enhance the deposition of cholesterol and other materials into artery walls.

> **PRINCIPLE #4**   Poor nutrition can result from both inadequate and excessive levels of nutrient intake.

Each nutrient has a range of intake levels that corresponds to optimum functioning of that nutrient (Illustration 1.5). Intake levels below and above this range are associated with impaired functions.

Inadequate intake of an essential nutrient, if prolonged, results in obvious deficiency diseases. Marginally deficient diets produce subtle changes in behavior or physical condition. If the optimal intake range is exceeded (usually by overdoses of supplements), mild to severe changes in mental and physical functions occur, depending on the

> **Homeostasis**   Constancy of the internal environment. The balance of fluids, nutrients, gases, temperature, and other conditions needed to ensure ongoing, proper functioning of cells and, therefore, all parts of the body.

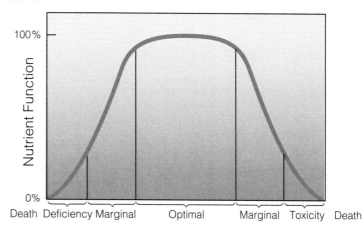

Increasing concentration or intake of nutrient ➡

Y-axis: Nutrient Function (0% to 100%)
X-axis: Death | Deficiency | Marginal | Optimal | Marginal | Toxicity | Death

Illustration 1.5 Nutrient function and consequences by level of intake.

amount of the excess and the nutrient involved. Overt vitamin C deficiency, for example, produces irritability, bleeding gums, pain upon being touched, and failure of bone growth. Marginal deficiency may cause delayed wound healing. The length of time a deficiency or toxicity takes to develop depends on the type and amount of the nutrient consumed and the extent of body nutrient reserves. Intakes of 32 mg/day of vitamin C, or about a third of the RDA for adults (75 mg and 90 mg per day for women and men, respectively), lower blood vitamin C levels to the deficient state within three weeks.[20] On the excessive side, too much supplemental vitamin C causes diarrhea. For nutrients, enough is as good as a feast.

STEPS IN THE DEVELOPMENT OF NUTRIENT DEFICIENCIES AND TOXICITIES   Poor nutrition due to inadequate diets generally develops in the stages outlined in Illustration 1.6. After a period of deficient intake of an essential nutrient, tissue reserves become depleted, and subsequently blood levels of the nutrient decline. When the blood level can no longer supply cells with optimal amounts of nutrients, cell processes change. These changes have a negative effect on the cell's ability to form proteins appropriately, regulate energy formation and use, protect itself from oxidation, or carry out other normal functions. If the deficiency continues, groups of cells malfunction, which leads to problems related to tissue and organ functions. Physical signs of the deficiency may then develop, such as growth failure with protein deficiency or an inability to walk as a result of beriberi (thiamin deficiency). Eventually, some problems produced

by the deficiency can no longer be reversed by increased nutrient intake. Blindness that results from serious vitamin A deficiency, for example, is irreversible.

Excessively high intakes of many essential nutrients produce toxicity diseases. Excessive vitamin A, for example, produces hypervitaminosis A, and selenium overdose leads to selenosis. Signs of toxicity stem from an increased level of the nutrient in the blood and the subsequent oversupply of the nutrient to cells. The high nutrient load upsets the balance needed for optimal cell function. These changes in cell function lead to the signs and symptoms of a toxicity disease.

For both deficiency and toxicity diseases, the best way to correct the problem is at the level of intake. Identifying and fixing intake problems prevents related health problems from developing.

NUTRIENT DEFICIENCIES ARE USUALLY MULTIPLE   Most foods contain many nutrients, so poor diets are generally inadequate in many nutrients. Calcium and vitamin D, for example, are present in milk. Deficiencies of both of these nutrients may develop from a low milk intake and an otherwise poor diet.

The "Ripple Effect"   Dietary changes affect the level of intake of many nutrients. Switching from a high-fat to a low-fat diet, for instance, generally results in a lower intake of calories and higher intake of dietary fiber and vitamins. Consequently, dietary changes introduced for the purpose of improving intake of a particular nutrient produce a "ripple effect" on the intake of other nutrients.

> **PRINCIPLE #5**   Humans have adaptive mechanisms for managing fluctuations in food intake.

Healthy humans have adaptive mechanisms that partially protect the body from poor health due to fluctuations in nutrient intake. These mechanisms act to conserve nutrients when dietary supply is low and to eliminate them when excessively high amounts are present. Dietary surpluses of nutrients such as iron, calcium, vitamin A, and vitamin B_{12} are stored within tissues for later use. In the case of iron and calcium, absorption is also regulated so that the amount absorbed changes in response to the body's need for these nutrients. The body has a low storage capacity for other nutrients, such as vitamin C and water, and excesses are eliminated through urine or stools. Fluctuations in energy intake are primarily regulated by changes in appetite. If too few calories are consumed, however, the body will obtain energy from its glycogen and fat stores. If caloric intakes remain low and a significant amount of body weight is lost, the body down-regulates its need for energy by lowering body temperature and the capacity for physical work. When energy intake exceeds need, the extra is converted to fat—and to a lesser extent, to glycogen—and stored for later use.

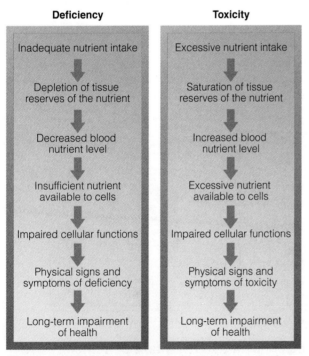

Illustration 1.6 Usual steps in the development of nutrient deficiencies and toxicities.

Although they provide an important buffer, these built-in mechanisms do not protect humans from all the consequences of poor diets. An excessive vitamin A or selenium intake over time, for example, results in toxicity disease; excessive energy intake creates health problems related to obesity; and deficient intakes of other vitamins and minerals compromise health in many ways.

PRINCIPLE #6 Malnutrition can result from poor diets and from disease states, genetic factors, or combinations of these causes.

Malnutrition means "poor nutrition" and results from either inadequate or excessive availability of energy and nutrients. Niacin toxicity, obesity, iron deficiency, and kwashiorkor (protein deficiency in children) are examples of malnutrition.

Malnutrition can result from poor diets as well as from diseases that interfere with the body's ability to use the nutrients consumed. *Primary malnutrition* results when a poor nutritional state is dietary in origin. *Secondary malnutrition,* on the other hand, is precipitated by a disease state, surgical procedure, or medication. Diarrhea, alcoholism, AIDS, and gastrointestinal tract bleeding are examples of conditions that may cause secondary malnutrition.

A portion of the population is susceptible to malnutrition due to genetic factors or, more commonly, to a combination of genetic factors and specific environmental exposures.

NUTRIENT-GENE INTERACTIONS Advances in knowledge about nutrient-gene interactions in health and disease are revolutionizing the science and practice of nutrition. Genes provide the codes for enzyme and other protein synthesis, and consequently affect body functions in a huge number of ways. Although individuals are 99.9% genetically identical, the 0.1% difference in genetic codes makes everyone unique. Variations in gene types (or genotypes) contribute to disease resistance and development, and the way individuals respond to various drugs.[21]

Hundreds of diseases and disorders related to single-gene defects have been identified, and many of these affect nutrient needs. Four examples of such diseases and disorders—including phenylketonuria, celiac disease, lactose intolerance, and hemochromatosis—are described in (Table 1.13).[22] Most diseases and disorders related to genetic makeup, however, are due to interactions among environmental factors, genotype, and gene functions.[23]

Food and nutrient intake is a prominent environmental factor that interacts with genotype and gene function. Certain components of foods can turn genes on or off, and others can compensate for the ill effects of certain genotypes on body processes.[24] Newly identified relationships between dietary components and genes are being announced regularly. Here are a few examples of effects of nutrient-gene interactions on health status:

- Consumption of whole oats lowers blood cholesterol level in some people but not others.
- High alcohol intake during pregnancy in some women sharply increases the risk of fetal alcohol syndrome in her fetus, but the fetuses of other women with different genetic traits are not affected by high alcohol intake.
- Regular consumption of green tea reduces the risk of prostate cancer in certain individuals with a particular genetic trait.

Malnutrition Poor nutrition resulting from an excess or lack of calories or nutrients.

Primary Malnutrition Malnutrition that results directly from inadequate or excessive dietary intake of energy or nutrients.

Secondary Malnutrition Malnutrition that results from a condition (e.g., disease, surgical procedure, medication use) rather than primarily from dietary intake.

Table 1.13 Examples of single-gene disorders that affect nutrient need

PKU (phenylketonuria)	A very rare disorder caused by the lack of the enzyme phenylalanine hydroxylase. Lack of this enzyme causes phenylalanine, an essential amino acid, to build up in the blood. High blood levels of phenylalanine during growth lead to mental retardation, poor growth, and other problems. PKU is treated by low-phenylalanine diets.
Celiac disease	An intestinal malabsorption disorder caused by an inherited intolerance to gluten in wheat, rye, and barley. It causes multiple nutrient deficiencies and is treated with gluten-free diets. Celiac disease is also called "nontropical sprue" and "gluten enteropathy."
Lactose intolerance	A common disorder in adults in many countries resulting from lack of the enzyme lactase. Ingestion of lactose in dairy products causes gas, cramps, and nausea due to the presence of undigested lactose in the gut.
Hemochromatosis	A disorder affecting 1 in 200 people that occurs due to a genetic deficiency of a protein that helps regulate iron absorption. Individuals with hemochromatosis absorb more iron than normal and have excessive levels of body iron. High levels of body iron have toxic effects on tissues such as the liver and heart. The disorder can also be produced by excessive levels of iron intake over time and frequent iron injections or blood transfusions.

- Folate requirements are higher in people with a gene that codes an abnormal enzyme for folate metabolism.
- High-carbohydrate diets or low levels of physical activity increase the likelihood that obesity will develop in genetically susceptible people.[21-25]

Genetic factors alone cannot explain the rapid rise in obesity and type 2 diabetes in the United States, but they do provide clues about needed preventive and therapeutic measures. Personalized modifications of dietary intake based on genotype will become standard practice in health care in the not-too-distant future.[24]

> **PRINCIPLE #7** Some groups of people are at higher risk of becoming inadequately nourished than others.

Women who are pregnant or breastfeeding, infants, children, people who are ill, and frail elderly persons have a greater need for nutrients than healthy adults and elderly people do. As a result, they are at higher risk of becoming inadequately nourished than others. Within these groups, those at highest risk of nutritional insults are the poor. In cases of widespread food shortages, such as those induced by war or natural disaster, the health of these nutritionally vulnerable groups is compromised the soonest and the most.

> **PRINCIPLE #8** Poor nutrition can influence the development of certain chronic diseases.

Poor nutrition not only results in deficiency or toxicity diseases, it also plays an important role in the development of heart disease, hypertension, cancer, stroke, osteoporosis, type 2 diabetes, and obesity. Diets high in saturated and trans fats, for example, are related to the development of heart disease; diets low in vegetables and fruits to heart disease and certain types of cancer; diets low in calcium to osteoporosis; and diets high in sugar to tooth decay.[26] Inadequate and excessive nutrient intakes may contribute to the development of more than one disease—and produce disease by more than one mechanism. The effects of habitually poor diets on chronic disease development often take years to become apparent.[27,28]

> **PRINCIPLE #9** Adequacy and balance are key characteristics of a healthy diet.

Healthy diets contain many different foods that together provide calories and nutrients in amounts that promote the optimal functioning of cells and of the body. A variety of food is required to obtain all the nutrients needed because, although no one food contains them all, many different combinations of food can make up a healthy diet.

Adequate diets are most easily obtained by consuming foods that are good sources of a number of nutrients but not packed with calories. Such foods are considered **nutrient-dense foods**. Those that provide calories and low amounts of nutrients are considered **"empty-calorie" foods**. Vegetables, fruits, lean meats, dried beans, breads, and cereals are nutrient dense. Foods such as beer, chips, candy, pastries, sodas, and fruit drinks lead the list of empty-calorie foods.

> **PRINCIPLE #10** There are no "good" or "bad" foods.

> "All things in nutriment are good or bad relatively."
>
> Hippocrates

Unless you are talking about spoiled food or poisonous mushrooms, there is no such thing as a bad food. There are, however, combinations of foods that add up to an unhealthy diet. Occasional consumption of a hot dog, fried chicken, or a chocolate sundae isn't going to shorten a person's life span. But making them dietary staples, along with other empty-calorie foods, will take its toll.

Nutritional Labeling

In 1990 the U.S. Congress passed legislation establishing requirements for nutrition information, nutrient content claims, and health claims presented on food and dietary supplement labels. This legislation, called the Nutrition Labeling and Education Act, requires that almost all multiple-ingredient foods and **dietary supplements** be labeled with a Nutrition Facts panel (Illustration 1.7). The act also requires that nutrient content and health claims appearing on package labels, such as "trans fat–free" and "helps prevent cancer" qualify based on criteria established by the Food and Drug Administration (FDA).

Nutrition Facts Panel

For foods, the Nutrition Facts panel must list the content of fat, saturated fat, trans fat, cholesterol, sodium, total carbohydrates, fiber, sugars, protein, Vitamins A and C, and calcium and iron in a standard serving. Additional nutrients may be listed on a voluntary basis. If a health claim about a particular nutrient is made for the product, then the product's content of the nutrient addressed in the claim must be shown. Nutrition Facts panels contain a column that lists the % DV (% Daily Value) for each relevant nutrient. This information helps consumers decide, for

Nutrient-Dense Foods Foods that contain relatively high amounts of nutrients compared to their caloric value.

Empty-Calorie Foods Foods that provide an excess of calories relative to their nutrient content.

Dietary Supplements Any product intended to supplement the diet, including vitamin and mineral supplements, proteins, enzymes, amino acids, fish oils, fatty acids, hormones and hormone precursors, and herbs and other plant extracts. In the United States, such products must be labeled "Dietary Supplement."

Illustration 1.7 Example of a Nutrition Facts panel.

Nutrition Facts
Serving Size 1 Entree
Serving Per Container 1

Amount Per Serving

Calories 380 Calories from Fat 170

		%Daily Value
Total Fat 19g		**29%**
Saturated Fat 10g		**50%**
Trans Fat 2g		
Cholesterol 85g		**28%**
Sodium 810mg		**34%**
Total Carbohydrate 33g		**11%**
Dietary Fiber 3g		**12%**
Sugars 5g		
Protein 20g		

Vitamin A 10%	Vitamin C 0%
Calcium 10%	Iron 15%

Percent Daily Values are based on a 2000 calorie diet. Your daily values may be higher or lower depending on your calorie needs:

		Calories	2000	2500
Total Fat	Less Than		65g	80g
Sat Fat	Less Than		20g	25g
Cholesterol	Less Than		300mg	300mg
Sodium	Less Than		2400mg	2400mg
Total Carbohydrate			300g	375g
Dietary Fiber			25g	30g

example, whether the carbohydrate content of a serving of a specific food product is a lot or a little.

Nutrient content claims made on food package labels must meet specific criteria. Products labeled "no trans fat" or "trans fat–free," for example, must contain less than 0.5 grams of trans fat and of saturated fat. Products labeled "low sodium" must contain less than 140 mg sodium per serving.

A variety of health claims for foods have been approved to date; others are pending, and some have not been approved (Table 1.14 on the next page). Companies making health claims for products that have not been approved by the FDA can be fined and the products removed from the market.

Not all foods have to be labeled. Labeling is voluntary for foods with single ingredients such as corn or steak, products sold by local bakeries, and restaurant foods.

Ingredient Label

Food products must list ingredients in an "ingredient label." The list must begin with the ingredient that contributes the greatest amount of weight to the product, and continue with the other ingredients on a weight basis. Ingredients such as milk solids, peanuts, egg whites, sulfites, or nuts that cause allergic or other reactions in some people are particularly important components of the label.

Dietary Supplement Labeling

"You can call anything a dietary supplement, even something you grow in your back yard."

Donna Porter, RD, PhD, Congressional Research Service

Dietary supplements such as herbs, amino acid pills and powders, and vitamin and mineral supplements must show a "Supplement Facts" panel that lists serving size, ingredients, and % DV of essential nutrients contained. Because they do not have to be shown to be safe and effective before they are sold, labels on dietary supplements cannot claim to treat, cure, or prevent disease. They can be labeled with standardized nutrition content claims such as "high in calcium" or "a good source of fiber." They can also be labeled with health claims such as "may reduce the risk of heart disease" if the product qualifies based on nutrition labeling requirements. Dietary supplements can make other claims on product labels not approved by the FDA, such as "supports the immune system" or "helps maintain mental health," as long as the label doesn't state or imply that the product will prevent, cure, or treat disease. If a health claim is made on a dietary supplement label, the label also must present the FDA disclaimer:

> This product has not been evaluated by the FDA. This product is not intended to diagnose, treat, cure, or prevent any disease.

ENRICHMENT AND FORTIFICATION Some foods are labeled as "enriched" or "fortified." These two terms have specific definitions developed prior to the Nutrition Labeling and Education Act. *Enrichment* pertains only to refined grain products and covers some of the vitamins and one of the minerals lost when grains are refined. By law, producers of bread, cornmeal, crackers, flour tortillas, white rice, and other products made with refined grains must use flours enriched with thiamin, riboflavin, niacin, and iron.

Any food can be fortified with added vitamins and minerals, and its manufacturers most often do so on a voluntary basis to enhance product sales. However, some foods must be fortified. Refined grain flours must be fortified with folic acid, milk with vitamin D, and low-fat and skim milk with vitamin D and vitamin A. Although fortification is not required for salt, it is often fortified with iodine. *Fortification* of these foods has contributed substantially to reductions in the incidence of diseases related to inadequate intakes.[29]

Enrichment The replacement of thiamin, riboflavin, niacin, and iron that are lost when grains are refined.

Fortification The addition of one or more vitamins or minerals to a food product.

Herbal Remedies

The FDA considers herbal products to be dietary supplements; they are taken by many people during various stages of the life cycle. Some herbal remedies act like drugs and

Table 1.14 Approved nutrition claims for disease prevention, and claims not approved or pending

TOPICS COVERED BY APPROVED CLAIMS	TOPICS OF CLAIMS NOT APPROVED/PENDING
1. Calcium and osteoporosis	
2. Fats and cancer	
3. Saturated fat and cholesterol and heart disease	
4. Fruits and vegetables and heart disease, cancer	
5. Sodium and hypertension	1. Zinc and infection
6. Folate or folic acid and neural tube defects (newborn malformations such as spina bifida)	2. Dietary fiber and colorectal cancer
7. Whole grains and heart disease	3. 800 mcg folic acid more effective for reducing risk of neural tube defects than lower amounts
8. Soluble fiber in oats and psyllium seed husks and heart disease	4. Calcium and hypertension
9. Sugar alcohols (e.g., xylitol, sorbitol) and tooth decay	5. Alcohol and heart disease
10. Soy protein and heart disease	6. Flaxseed and cancer
11. Whole grain foods and heart disease and certain cancers	7. Omega-3 fatty acids and heart disease
12. Plant stanols and sterols (used in spreads such as Take Control and Benecol) and heart disease	
13. Potassium and high blood pressure and stroke	
14. Fruits, vegetables, high-fiber grain products and heart disease	
15. Soy protein, nuts and heart disease	

have side effects, but they are not considered to be drugs and are loosely regulated. They do not have to be shown to

> **Functional Foods** Generally taken to mean food, fortified foods, and enhanced food products that may have health benefits beyond the effects of essential nutrients they contain.

be safe or effective before they are marketed. Herbs vary substantially in safety and effectiveness—they can have positive, negative, or neutral effects on health. Knowledge of the effects of herbal remedies is far from complete, making it difficult to determine appropriateness of their use in some cases.

Purported effects and side effects of some herbal and similar remedies are listed in Table 1.15. The extent to which herbs included in the table pose a risk to health depends on the amount taken, the duration of use, and the user's age, stage, and health status.

Functional Foods

Also known as "neutraceuticals," *functional foods* include a variety of products that have theoretically been modified to enhance their contribution to a healthy diet.[32] Foods are made "functional" by:

- Taking out potentially harmful components (e.g., cholesterol in egg yolks and lactose in milk)
- Increasing the amount of nutrients and beneficial non-nutrients (e.g., fiber-fortified liquid meals, calcium- and vitamin C–fortified orange juice)
- Adding new beneficial compounds to foods (e.g., "friendly" bacteria to yogurt and other milk products)

Functional foods are not regulated, and no specific standards apply to them.[32] Health claims, however, can be

Table 1.15 Proposed effects and potential side effects of herbal and similar remedies[30,31]

HERB/OTHER REMEDY	PROPOSED EFFECTS	POTENTIAL SIDE EFFECTS
Glucosamine-chondroitin sulphate	Slows progression of osteoarthritis and its pain	Gastrointestinal upset, fatigue, headache
Ginseng	Increases energy, normalizes blood glucose, stimulates immune function, relieves impotence in males, prevents cancer	Insomnia, hyperactivity, hypertension, diarrhea, menstrual dysfunction; interacts with blood thinners*
SAMe	Relieves mild depression, pain relief for arthritis	May trigger manic excitement, nausea
Garlic	Lowers blood cholesterol, relieves colds and other infections	Heartburn, gas, blood thinner
Cholestin	Maintains desirable blood cholesterol levels	Safety of some ingredients unknown
Echinacea	Prevents and treats colds and sore throat	Allergies to plant components
DHEA	Improves memory, mood, physical well-being	Increases risk of breast cancer
Creatine	Sport supplement (increased performance in short, high-intensity events)	Kidney disease
Saw palmetto	Improves urine flow, reduces urgency of urination in men with prostate enlargement	Nausea, abdominal pain
Ginkgo biloba	Increases mental skills, delays progression of Alzheimer's disease, increases blood flow and sexual performance, decreases depression	Nervousness, headache, diarrhea, nausea; interacts with blood thinners
Shark cartilage	Treats lung cancer	Safety unknown
Saint John's wort	Relieves mild to moderate depression	Dry mouth, dizziness, sensitivity to light; interacts with many drugs and chemotherapy
Ephedra (ma huang)	Promotes weight loss, improves respiratory illnesses	Insomnia, headaches, nervousness, seizures, death
Kava (kava-kava)	Relaxation, stress relief, sleep aid, mood enhancer	Liver injury
Black cohosh	Improves menopausal and PMS symptoms	Gastric upset, dizziness, headache, low blood pressure; may increase risk of breast cancer; interacts with antihypertension drugs
Coenzyme Q_{10} (ubiquinone)	Remedy for heart disease, cancer, Parkinson's disease	Nausea, diarrhea, rash, low blood glucose; interacts with some drugs

*Blood thinners include aspirin, warfarin, and coumarin.

made for functional foods given they have been approved by the FDA. Increasingly, the list of functional foods is becoming infiltrated with sports bars, soups, beverages, and cereals spiked with vitamins, minerals, and herbs. Some of the products carry labels with unsubstantiated health claims and may be of little benefit or are potentially unsafe.[33] For these products, the label "functional food" is a marketing term.

Prebiotics and Probiotics The terms *prebiotics* and *probiotics* were derived from *antibiotics* due to their probable effects on increasing resistance to various diseases. They are in a class of functional foods by themselves. *Prebiotics* are fiberlike, indigestible carbohydrates that

are broken down by bacteria in the colon. The breakdown products foster the growth of beneficial bacteria. The digestive tract generally contains over 500 species of microorganisms and 100 trillion bacteria.[34] Some species of bacteria such as *E. coli* may cause disease; others, such as strains of lactobacillus and bifidobacteria, prevent certain diseases.[35] Because they foster the growth of beneficial bacteria, prebiotics are considered "intestinal fertilizer." **Probiotic** is the term for live, beneficial ("friendly") bacteria that enter food products during

Prebiotics Certain fiberlike forms of indigestible carbohydrates that support the growth of beneficial bacteria in the lower intestine. Nicknamed "intestinal fertilizer."

Probiotics Strains of *Lactobacillus* and bifidobacteria that have beneficial effects on the body. Also called "friendly bacteria."

fermentation and aging processes. Those that survive digestive enzymes and acids may start colonies of beneficial bacteria in the digestive tract. Table 1.16 lists foods and other sources of pre- and probiotics.

Prebiotics and probiotics have been credited with important benefits, such as the prevention and treatment of diarrhea and other infections in the gastrointestinal tract; prevention of colon cancer; decreased blood levels of triglycerides, cholesterol, and glucose; and decreased dental caries.[35,36] Prebiotics and probiotics are assumed to be safe because they have been consumed in foods for centuries.[36]

Availability of foods containing prebiotics and probiotics is much more common in Japan and European countries than in Canada or the United States. However, availability of such products is increasing in these countries as research results shed light on their safety and effectiveness.[36]

Meeting Nutritional Needs across the Life Cycle

Healthy individuals require the same nutrients throughout life, but amounts of nutrients needed vary based on age, growth, and development. Nutrient needs during each stage of the life cycle can be met through a variety of foods

---

Table 1.16 Food and other sources of prebiotics and probiotics[30,36]

Probiotics

Fermented or aged milk and milk products
- Yogurt with live culture
- Buttermilk
- Kefir
- Cottage cheese
- Dairy spreads with added inulin

Other fermented products
- Soy sauce
- Tempeh
- Fresh sauerkraut
- Miso

Breast milk

Probiotic tablets, and powders, and nutritional beverages

Prebiotics

Chicory

Jerusalem artichokes

Wheat

Barley

Rye

Onions

Garlic

Leeks

Prebiotic tablets, powders, and nutritional beverages

---

and food practices. There is no one, best diet for everyone. Traditional diets defined by diverse cultures and religions provide the foundation for meeting individuals' nutritional needs and the framework for dietary modification when needed.[37] Although it is inaccurate to say that all or most members of a particular cultural group or religion follow the same dietary practices, groups of individuals may share common beliefs about food and food-intake practices.

Dietary Considerations Based on Ethnicity

People immigrating to the United States and other countries both preserve dietary traditions of their cultural group and integrate cross-cultural adaptations into their dietary practices. The extent to which culturally based food habits change depends to some extent on income, food cost, and ethnic food availability. Immigrant families from El Salvador who live in urban areas of the United States, for example, maintain many cultural food practices from their homeland:

- Breakfast generally consists of fried beans, corn tortillas, occasionally eggs, and sweetened coffee with boiled milk.
- Lunch consists of soup, fried meat, rice or rice with vegetables, corn tortillas, and fruit juice.
- Dinner will offer fried beef or chicken, corn tortillas, rice, dried beans, fruit juice, and black coffee.

Cross-cultural adaptations made by a portion of Salvadorans immigrating to the United States include the addition of french fries, hamburgers, American cheeses, salad dressing, tacos, flour tortillas, and peanut butter to their diets.[38]

Sometimes diets of native populations change when their numbers become overwhelmed by other population groups. A primary example of this phenomenon is represented by changes in traditional dietary practices of Native Americans. In general, traditional diets of Indians in the United States consisted of foods such as buffalo, deer, wild berries and other fruits, corn, turnips, squash, wild potato, and wild rice. Loss of land, buffalo, discrimination, poverty, and food programs that offered refined flour, sugar, salt pork, and other high-fat meats drastically changed what Indians ate, how they lived, and their health status. Activities aimed at bringing back traditional foods and dietary practices are under way among many Indian groups.[39]

Food preferences of African Americans vary widely, but may stem from their cultural food heritage. Historically important foods include corn bread, pork, buttermilk, rice, sweet potatoes, greens, cabbage, salt pork, and fried fish. "Soul foods" make up less of the African American diet now than in the past, but remain foods of choice for special occasions and are the foods most likely to be revered.[40]

Dietary Considerations Based on Religion

Most religions have members in other countries and special dietary laws and practices. For example:

- Hindus may not consume foods such as garlic and onions, which are believed to hinder spiritual development.

- Buddhists in certain countries tend to be vegetarian or to eat fish as their only choice of meat. In countries such as Tibet and Japan, vegetarianism is rare among Buddhists.

- Alcohol is prohibited as part of Sikhism, and meat prepared by kosher or halal methods is avoided.

- The Church of Latter Day Saints, or the Mormon Church, prohibits alcohol and discourages consumption of caffeine. Mormons may eat meat and prize wheat.

- Seventh-Day Adventists tend to follow a strict lacto-ovo vegetarian diet and exclude alcohol and caffeine. Whole grains, vegetables, and fruits are considered to be the base of diets; and dried beans, low-fat dairy products, and eggs may be consumed infrequently.

- Jewish dietary laws require that foods consumed must be kosher, or fit to eat according to Judaic law. Organizations are certified as supplying foods that are kosher. The Jewish calendar includes six fasting days that call for total abstinence from food or drink.

- The Muslim religion has dietary laws that require foods to be halal, or permitted for consumption by Muslims. Pork consumption is not allowed, nor is the consumption of animals slaughtered in the name of any god other than Allah. Slaughterhouses must be under the supervision of a halal certifier in order for meat to be considered fit to eat, although some Muslims will eat other meats. Consuming alcohol is prohibited.[41,42]

Additional information about cultural and religious food practices and beliefs can be obtained directly by getting to know people from a variety of cultures and their dietary preferences. This information can be of great benefit in nutrition education and counseling situations.

Nutritional Assessment

Nutritional assessment of groups and individuals is a prerequisite to planning for the prevention or solution of nutrition-related health problems. It represents a broad area within the field of nutrition and is only highlighted here. Resources related to the selection of appropriate nutritional assessment techniques and their implementation are listed at the end of this chapter.

Nutritional status may be assessed for a population group or for an individual. Community-level assessment identifies a population's status using broad nutrition and health indicators, whereas individual assessment provides the baseline for anticipatory guidance and nutrition intervention.

Community-Level Assessment

A target community's "state of nutritional health" can generally be estimated using existing vital statistics data, seeking the opinions of target group members and local health experts, and making observations. Knowledge of average household incomes; the proportion of families participating in the Food Stamp Program, soup kitchens, school breakfast programs, or food banks; and the age distribution of the group can help identify key nutrition concerns and issues. In large communities, rates of infant mortality, heart diseases, and cancer can reveal whether the incidence of these problems is unusually high.

Information gathered from community-level nutritional assessment can be used to develop community-wide programs addressing specific problem areas, such as childhood obesity or iron-deficiency anemia. Nutrition programs should be integrated into community-based health programs.

Individual-Level Nutritional Assessment

Nutritional assessment of individuals has four major components:

- Clinical/physical assessment
- Dietary assessment
- Anthropometric assessment
- Biochemical assessment

Data from all of these areas are needed to describe a person's nutritional status. Data on height and weight provide information on weight status, for example, and knowledge of blood iron levels tells you something about iron status. It cannot be concluded that people who are normal weight or have good iron status are "well nourished." Single measures do not describe a person's nutritional status.

CLINICAL/PHYSICAL ASSESSMENT A clinical/ physical assessment involves visual inspection of a person by a trained *registered dietitian* or other qualified professional to note features that may be related to malnutrition. Excessive or inadequate body fat, paleness, bruises, and brittle hair are examples of features that may suggest nutrition-related problems. Physical characteristics are nonspecific indicators, but they can support other findings related to nutritional status. They cannot be used as the sole

Registered Dietitian An individual who has acquired knowledge and skills necessary to pass a national registration examination and who participates in continuing professional education.

criterion upon which to base a decision about the presence or absence of a particular nutrition problem.

Dietary Assessment

> "Garbage in, garbage out."

Many methods are used for assessing dietary intake. For clinical purposes, 24-hour dietary recalls and food records analyzed by computer programs are most common. Single, 24-hour recalls and food frequency questionnaires are most useful for estimating dietary intakes for groups, whereas multiple recalls and dietary histories are generally used for assessments of individual diets.

24-HOUR DIETARY RECALLS AND RECORDS Becoming proficient at administering 24-hour recalls takes training and practice. Food records, on the other hand, are completed by clients themselves. These are more accurate if the client has also received some training. Generally, the purpose of assessing an individual's diet is to estimate the person's overall diet quality so that strengths and weaknesses can be identified, or to assess intake of specific nutrients that may be involved in disease states.

Information on at least three days of dietary intake (preferably two usual weekdays and one weekend day) is needed to obtain a reliable estimate of intake by food group, calories, and nutrients. A good approach is to have a trained dietary interviewer administer a 24-hour recall and then have the client record her or his own diet on two other days. The experience of thinking about what was eaten, portion sizes, ingredients, and recipes helps train people to complete their food records accurately. Completed records should be reviewed with the client during a telephone call or clinic session to make sure they are accurate.

DIETARY HISTORY Dietary histories have been used for decades and represent a quantitative method of dietary assessment. They require an interview that is about 1½ hours long and includes a 24-hour dietary recall modified to represent usual intake, careful deliberations over food types and portions, and a cross-check food frequency questionnaire that confirms 24-hour usual dietary intake information. Results must be coded, checked, and processed. Although expensive, diet histories provide more complete and accurate data than most other dietary assessment methods.[46]

FOOD FREQUENCY QUESTIONNAIRES Food frequency questionnaires are often used in epidemiological studies to estimate food and nutrient intake of groups of people. These tools are considered semiquantitative because they force people into describing food intake based on a limited number of food choices and portion sizes (Illustration 1.8).[44] Validated food frequencies are relatively inexpensive to administer and tabulate, and provide good enough estimates of dietary intake to rank people by their food and nutrient intake levels. They tend to underestimate food intake and provide data that is more likely to fail to identify nutrient and health relationships than are quantitative assessment techniques such as the dietary history.[45]

DIETARY ASSESSMENT RESOURCES Several high-quality computer programs and Internet resources are available for dietary assessment. The Interactive Healthy Eating Index (IHEI), which was developed and is maintained by the U.S. Department of Agriculture (USDA), is an example of a high-quality Internet resource. This interactive program provides an analysis of food intake based on the IHEI and evaluates dietary intake based on the Food Guide Pyramid recommendations and guidelines related to intake of fat, cholesterol, sodium, and other nutrients. The Web address for the IHEI is given at the end of this chapter.

Illustration 1.8 Example component of a food frequency questionnaire.

| | Frequency of Consumption | | | | | | | | |
Food	Never or less than once per month	1–3 per month	1 per week	2–4 per week	5–6 per week	1 per day	2–3 per day	4–5 per day	6+ per day
1. a. Broth type soups, 1 cup									
b. Tap water, 1 cup									
c. Sparkling or mineral water, 1 cup serving									
d. Decaffeinated black tea, iced or hot, 1 cup									
e. Herbal tea, (no caffine) iced or hot, 1 cup									
2. Custard or pudding, 1/2 cup									
3. Onions, 1/4 cup, alone or in combination									

SOURCE: J. Brown, University of Minnesota; Diana Project form, adapted from W. Willett's Food Frequency Questionnaire.

All methods of dietary assessment and all computer programs used to analyze food intake have limitations. People often underestimate their food intake and some people have trouble remembering what they ate.[43]

Computer programs may not be updated frequently enough to contain current information on the composition of foods, and they may not include all of the various foods people eat. Specific foods, ingredients, and portion sizes must be carefully recorded and entered if reliable results are to be produced.

Anthropometric Assessment

Individual measures of body size (height, weight, percent body fat, bone density, and head and waist circumferences, for example) are useful in the assessment of nutritional status—if done correctly. Each measure requires use of standard techniques and calibrated instruments by trained personnel. Unfortunately, *anthropometric* measurements are frequently performed and recorded incorrectly in clinical practice. Training on anthropometric measures is often available through public health agencies and programs such as WIC (Special Supplemental Nutrition Program for Women, Infants, and Children), and courses and training sessions are sometimes presented at universities.

Biochemical Assessment

Nutrient and enzyme levels, DNA characteristics, and other biological markers are components of a biochemical assessment of nutritional status. Which biological markers are measured depends on what problems are suspected, based on other evidence. For example, a young child who tires easily, has a short attention span, and does not appear to be consuming sufficient iron based on dietary assessment results may have blood taken for analyses of hemoglobin and serum ferritin (markers of iron status). Suspected inborn errors of metabolism that may underlie nutrient malabsorption may be identified through DNA or other tests. Such results provide specific information on a component of a person's nutritional status and are very helpful in diagnosing a particular condition.

After the fact-finding phase of nutritional assessment, the nutritionist or other professionals must "apply their brains to their clients' problems." There is no "one size fits all" approach to solving nutrition problems—each has to be figured out individually.

Public Food and Nutrition Programs

A variety of federal, state, and local programs are available to provide food and nutrition services to families and individuals. Many communities have nutrition coalitions or partnership groups that collaborate on meeting the food and nutritional needs of community members. Programs representing church-based feeding sites, food shelves, Second Harvest Programs, the Salvation Army, missions, and others are usually a part of local coalitions. These central resources can be identified by contacting the local public health or cooperative extension agency. State-level programs are generally part of large national programs.

Government-sponsored food and nutrition programs are widely available throughout the country. Some programs, such as the School Lunch Program, benefit many children. Other programs are targeted to families and individuals in need. *Need* is generally defined as individual and household incomes below the poverty line. Some programs, such as WIC, have eligibility standards of up to 185% of the poverty line (Table 1.17). Income guidelines change periodically and are several thousand dollars higher for people in Alaska and Hawaii. Federal poverty guidelines can be identified from the Web site **www.ed.gov** by searching "poverty guidelines."

Anthropometry The science of measuring the human body and its parts.

WIC

The Supplemental Food Program for Women, Infants, and Children (WIC) was established in 1972 and is administered by the USDA. The program provides nutrition education and counseling as well as food vouchers for low-income pregnant, postpartum, and breastfeeding women and for low-income children under the age of 5 years. Food vouchers apply to nutritious foods such as fortified breakfast cereals, iron-fortified infant cereals and formula, milk, cheese, eggs, peanut butter, dried beans, and 100% fruit and vegetable juices. Some WIC programs offer vouchers for the purchase of produce at farmer's markets. WIC staff also provide breastfeeding support and referrals to health care and social service providers.

Eligibility for WIC is based on low-income status and the presence of a nutritional risk, such as iron deficiency or underweight. WIC services are provided through approximately 10,000 clinic sites throughout the United

Table 1.17 Income eligibility standards for the WIC program (≤185% of poverty income, 2003–2004)

Household Size	Household Income per Year
1	$16,613
2	22,422
3	28,231
4	34,040
5	39,849
6	45,658
7	51,467
8	57,276
Each additional member	+5,809

States and in American Samoa, Guam, Puerto Rico, and the Virgin Islands. The program serves more than 7 million women and children each year. Nearly half of all infants and a quarter of all young children in the United States participate in the WIC program.

The WIC program has been shown to have a number of positive effects on the health of participants. Infants born to women participating in WIC while pregnant weigh more, and they are less likely to be small at birth and to be born before term than are the infants of non-participating low-income women. Children served by WIC tend to consume more nutritious diets and experience lower rates of iron deficiency than children who are low income but not enrolled in WIC. WIC is cost effective: every dollar invested in WIC prenatal nutrition services saves $3.13 on Medicaid costs for infants during the first 2 years of life.[47]

Table 1.18 presents information on existing federal food and nutrition programs. You can get more information on these programs through the Internet at this address: www.nutrition.gov.

Nationwide Priorities for Improvements in Nutritional Health

Public health initiatives involving population-based improvements in food safety, food availability, and nutritional status have led to major gains in the health status of the country's population. Among the important components of this success story are programs that have expanded the availability of housing; safe food and water; foods fortified with iodine, iron, vitamin D, or folate; fluoridated public water; food assistance; and nutrition education. Since 1900, the average life span of persons in the United States has lengthened by more than 30 years, and

25 years of this gain are estimated to be due to the quiet revolutions that have taken place in public health.[48]

Although the United States spends more on health care than any other country, statistics from the World Health Organization (WHO) indicate that it ranks 26th among developed countries of the world in life expectancy. Today's priorities for improvements in the public's health and longevity center on reducing obesity, infant mortality, smoking, excessive alcohol consumption, accidents, violence, and physical inactivity. Goals for dietary changes are a central part of the nation's overall plan for health improvements. For the two out of three Americans who do not smoke or drink excessively, dietary intakes represent the major environmental influence on long-term health.[27]

Goals for improving the nutritional health of the nation are summarized in the document "Healthy People 2010: Objectives for the Nation" (Table 1.19). Because the seeds of many chronic diseases are planted during pregnancy and childhood, major emphasis is placed on dietary habits early in life.

Food Intake Recommendations

Two major guides offer recommendations for food intake: the Food Guide Pyramid and the "Dietary Guidelines for Americans." The Food Guide Pyramid guides people in the selection of foods from five food groups (Illustration 1.9 on page 40). Each food group has a recommended number of daily servings, and serving numbers vary based on age. Foods within each group are assigned specific portion sizes that correspond to a serving. This food guide is periodically updated.

ETHNIC FOOD GUIDES   The Food Guide Pyramid is ethnically and culturally flexible. It has been adapted to better match the food preferences and practices of Asians,

Table 1.18 Examples of federal food and nutrition programs

Program	Activity
Child and Adult Care Food Program (CACFP)	Reimburses child and adult care organizations in low-income areas for provision of nutritious foods.
Summer Food Service Program	Provides foods to children in low-income areas during the summer.
School Breakfast and Lunch Programs	Provide free breakfasts and reduced-cost or no-cost lunches to children from families who cannot afford to buy them.
Food Stamp Program	Subsidizes food purchases of low-income families and individuals.
WIC (Supplemental Nutrition Program for Women, Infants, and Children)	Serves low-income, high-risk, and pregnant and breastfeeding women and children up to 5 years of age. Provides supplemental nutritious foods and nutrition education as an adjunct to health care.
Head Start Program	Includes nutrition education for children and parents and supplies meals for children in the program.
Team Nutrition	A USDA nutrition education program that brings active learning about nutrition to schoolchildren.

Table 1.19 2010 nutrition objectives for the nation

√ Increase the proportion of adults who are at a healthy weight from 42 to 60%.

√ Reduce the proportion of adults who are obese from 23 to 15%.

√ Reduce the proportion of children and adolescents who are overweight or obese from 11 to 5%.

√ Reduce growth retardation among low-income children under age 5 years from 8 to 5%.

√ Increase the proportion of persons aged 2 years and older who:

- consume at least 2 daily servings of fruit from 28 to 75%.
- consume at least 3 daily servings of vegetables, with at least one-third being dark green or deep yellow vegetables, from 3 to 50%.
- consume at least 6 daily servings of grain products, with at least three being whole grains, from 7 to 50%.
- consume less than 10% of calories from saturated fat from 36 to 75%.
- consume no more than 30% of total calories from fat from 33 to 75%.
- consume 2400 mg or less of sodium daily from 21 to 65%.
- meet dietary recommendations for calcium from 46 to 75%.

√ Reduce iron deficiency among young children and females of childbearing age from 4–11% to 1–7%.

√ Reduce anemia among low-income pregnant females in their third trimester from 29 to 20%.

√ Increase the proportion of worksites that offer nutrition or weight-management classes or counseling from 55 to 85%.

√ Increase the proportion of physician office visits made by patients with the diagnosis of cardiovascular disease, diabetes, osteoporosis, or hyperlipidemia that include counseling or education related to diet and nutrition from 42 to 75%.

√ Increase food security among U.S. households and in so doing reduce hunger rates.

Other Objectives Related to Nutrition

√ Reduce infant mortality from 7.6 to no more than 5 per 1000 live births.

√ Reduce the incidence of spina bifida and other neural tube defects from 7 to 3 per 10,000 live births.

√ Reduce the incidence of birth defects from 1.7 to 1.2 per 1000 live births.

√ Increase the proportion of women who receive preconceptional counseling.

√ Increase the proportion of pregnant women who begin prenatal care in the first trimester from 81 to 90% or more.

√ Reduce low birthweight (<2500 grams) from 7.3 to 5%.

√ Reduce preterm births (<37 weeks) from 9.1 to 7.6%.

√ Increase abstinence from alcohol use by pregnant women from 79 to 95%.

√ Reduce the incidence of fetal alcohol syndrome.

√ Increase the proportion of women who gain weight appropriately during pregnancy.

√ Increase from 60 to 75% the proportion of women who exclusively breastfeed after delivery.

Somalis, Danes, vegetarians, and other groups. The Asian Food Guide Pyramid separates food into monthly, weekly, optional daily, and daily intakes. It focuses on plant foods and de-emphasizes red meat, eggs, poultry, and other animal products.[49] The Guide intended for Somalis highlights rice, water, lean meats, low-fat dairy products, use of vegetable oils rather than lard or animal fat for cooking, and physical activity.[50]

For this topic, as well as others covered in this chapter, feel free to consult the Web sites listed on the next page or nutrition textbooks to expand your understanding of nutrition. It is a broad and deep area of study, and no single chapter can fully address its multiple components. If you understand the material presented here, however, you are well on your way to grasping the concepts and content that follows.

The "Dietary Guidelines for Americans," which is updated every 5 years, recommends nutrition behaviors as well as the types and amounts of food that promote health (Table 1.20). The 2005 guidelines will be available late in 2005 from the Web site www.nal.usda.gov/fnic.

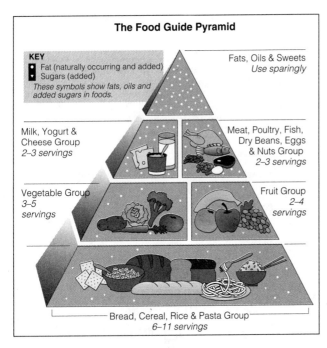

The Food Guide Pyramid

KEY
■ Fat (naturally occurring and added)
▼ Sugars (added)
These symbols show fats, oils and added sugars in foods.

Fats, Oils & Sweets
Use sparingly

Milk, Yogurt & Cheese Group
2–3 servings

Meat, Poultry, Fish, Dry Beans, Eggs & Nuts Group
2–3 servings

Vegetable Group
3–5 servings

Fruit Group
2–4 servings

Bread, Cereal, Rice & Pasta Group
6–11 servings

Illustration 1.9 The Food Guide Pyramid.*
*Revision to be released in 2005 and available from www.nal.usda.gov/fnic.

Table 1.20 Dietary guidelines for Americans, 2000*

- Aim for a healthy weight.
- Be physically active each day.
- Let the Pyramid guide your food choices.
- Choose a variety of grains daily, especially whole grains.
- Choose a variety of fruits and vegetables daily.
- Keep foods safe to eat.
- Choose a diet that is low in saturated fat and cholesterol and moderate in total fat.
- Choose beverages and foods to moderate your intake of sugars.
- Choose and prepare foods with less salt.
- If you drink alcoholic beverages, do so in moderation.

*Revision to be released in 2005 and available from www.nal.usda.gov/fnic.

Resources

Public Nutrition Programs

Start here if you want to identify government resources related to nutrition information, services, and programs.
Web site: www.nutrition.gov

USDA's Food and Nutrition Resource Center

Select Interactive Healthy Eating Index to assess one day's dietary intake. This site also provides the Dietary Guidelines, the Food Guide Pyramid for various ethnic groups, and food composition data.
Web site: www.nal.usda.gov/fnic

American Dietetic Association

Links to a wide variety of resources on nutrition and health.
Web sites: www.eatright.org and www.dietitians.ca

Dietary Guidelines Alliance

Provides suggestions on how to meet the Dietary Guidelines and other information.
Web site: www.ificinfo.health.org/iaay/index.htm

National Academy of Sciences

Want more information on the Dietary Reference Intakes? You can get full reports here.
Web site: www.iom.edu/fnb

General Nutrition Information

Links to high-quality sites in many areas of nutrition.
Web site: www.navigator.tufts.edu

Medline

Get access to scientific journal articles on a variety of nutrition topics.
Web site: www.nlm.nih.gov/medlineplus

Arbor Nutrition Guide

The Arbor Nutrition Guide's home page allows you to link to applied nutrition, clinical nutrition, food science, and food topics, and the Site of the Week. A banner on the home page invites you to register to receive Arbor's Clinical Nutrition Updates, a free service that provides weekly e-mail summaries on a variety of nutrition and health topics.
Web site: www.arborcom.com

Healthy People 2010: Objectives for the Nation and Health Canada

The home page for "Healthy People 2010."
Web site: www.healthypeople.gov
Read about the health status of Canadians and nutrition and other goals for Canada.
Web site: www.hc-sc.gc.ca

References

1. National data on household food security. Nutrition Today 2000;35:4.

2. Summary report on application of the DRIs for dietary assessment. Washington, DC: National Academy of Sciences, 2000.

3. Dietary reference intakes: energy, carbo-hydrate, fiber, fat, fatty acids, cholesterol, protein, and amino acids. Washington, DC: National Academies Press, 2002.

4. Kris-Etherton PM et al. New guidelines focus on fish, fish oil, omega-3 fatty acids. AHA Statement. Circulation, 2002; Oct. 18:1–6.

5. Vanschoonbeck K et al. Fish oil consumption and reduction in arterial disease. J Nutr 2003;133:657–60.

6. Butte NF. Carbohydrate and lipid metabolism in pregnancy: normal compared with gestational diabetes mellitus. Am J Clin Nutr 2000;71:1256S–61S.

7. Kris-Etherton PM et al. Polyunsaturated fatty acids in the food chain in the United States. Am J Clin Nutr 2000;71(suppl): 179S–88S.

8. McNamara DJ. Cholesterol intake and plasma cholesterol: an update. J Am Coll Nutr 1997;16:530–4.

9. Kris-Etherton PM et al. New guidelines focus on fish, fish oil, omega-3 fatty acids. AHA Statement. Circulation, 2002; Oct. 18.

10. Fats: the good and the bad. www.mayoclinic.com/invoke.cfm?id= NU00262, accessed 9/03.

11. Krauss RM et al. Revision 2000: a statement for healthcare providers from the Nutrition Committee of the American Heart Association. J Nutr 2001;131:132–46.

12. Bhathema SJ, Velasquez, MT. Beneficial role of dietary phytoestrogens in obesity and diabetes. Am J Clin Nutr 2002;76: 1191–1201.

13. Mares-Perlman JA et al. The body of evidence to support a protective role for lutein and zeaxanthin in delaying chronic disease. J Nutr 2002;1332;518S–24S.

14. Marcus JB. New age foods for disease prevention. Today's Dietitian 2003; May: 24–30.

15. Craig WJ. Phytochemicals: guardians of our health. J Am Diet Assoc 1997;97: S199–S204.

16. Dietary reference intakes: water and electrolytes. Washington, DC: National Academy Press, 2004.

17. Grandjean AC et al. Hydration: issues for the 21st century. Nutr Rev 2003;61: 261–71.

18. Noakes TD. Too many fluids as bad as too few. BMJ 2003;327:113–4.

19. Kolasa KM. Water: the recommendations we give—are they scientifically accurate? Food and Nutrition Conference and Exhibition, American Dietetic Association, Oct. 2002.

20. Johnston CS, Corte C. People with marginal vitamin C status are at high risk of developing vitamin C deficiency. J Am Diet Assoc 1999;99:854–6.

21. Simopoulos AP. Genetic variation and nutrition. Experimental Biology Annual Meeting, San Diego, CA, April 12, 2003.

22. Guttmacher AE et al. Genomic medicine—a primer. N Engl J Med 2002;347:1512–20.

23. Guttmacher AE et al. Welcome to the genomic era. N Engl J Med 2003; 349:996–8.

24. Shattuck D. Nutritional genomics. J Am Diet Assoc 2003;103:16–18.

25. Murray RF. Nutrigenetics/Nutrigenomics. Experimental Biology Annual Meeting, San Diego, CA, April 12, 2003.

26. Simon HB. Diet and exercise. WebMD Scientific American Medicine 2003;Oct. 14:1–36.

27. Deckelbaum RJ, Fisher EA, Winston M et al. Summary of a scientific conference on preventive nutrition: pediatrics to geriatrics. Circulation 1999;100:450–6.

28. Heaney RP. Long-latency deficiency disease: insights from calcium and vitamin D. Am J Clin Nutr 2003;78:912–19.

29. Park YK et al. History of cereal-grain product fortification in the United States. Nutr Today 2001;36:123–36.

30. Halsted CH. Dietary supplements and functional foods: 2 sides of a coin? Am J Clin Nutr 2003;77(suppl):1001S–7S.

31. DeSmet P. Herbal remedies. N Engl J Med 2002;347:699–704.

32. Hyman P. Claims for functional foods under the current food regulatory scheme. Nutr Today 2002;37:217–19.

33. Hasler CM. Functional foods: benefits, concerns and challenges. J Nutr 2002; 132:3772–81.

34. Beale B. Probiotics: their tiny worlds are under scrutiny. The Scientist 2002; July:20–2.

35. Sanders ME. Probiotics: considerations for human health. Nutr Rev 2003;61:91–7.

36. Brannon CB. Prebiotics: feeding friendly bacteria. Today's Dietitian 2003; Sept.:12–16.

37. Kittler PG, Sucher KP. Diet counseling in a multicultural society. Diabetes Educator 1989;16:127–32.

38. Romero-Gwynn E et al. Dietary patterns and acculturation among immigrants from El Salvador. Nutr Today 2003;35:233–40.

39. Conti K. The four winds nutrition model: a new culturally specific food guide and nutrition approach for the Northern Plains Indians. Seminar presentation, University of Minnesota, Minneapolis, Oct. 28, 2002.

40. Dirks RT, Duran N. African American dietary patterns at the beginning of the 20th century. J Nutr 2001;131:1881–9.

41. McCaffree J. Dietary restrictions of other religions. J Am Diet Assoc 2003;103:912.

42. Eliasi JR, Dwyer JT. Kosher and halal: religious observances affecting dietary intake. J Am Diet Assoc 2003;103:911–12.

43. Scagliusi FB et al. Selective underreporting of energy intake in women: magnitude, determinants, and effect of training. J Am Diet Assoc 2003;103:1306–13.

44. Newby PK et al. Reproducibility and validity of the Diet Quality Index Revised as assessed by use of a food-frequency questionnaire. Am J Clin Nutr 2003;78:941–9.

45. Horner NK et al. Participant characteristics associated with errors in self-reported energy intake from the Women's Health Initiative food-frequency questionnaire. Am J Clin Nutr 2002;76:766–73.

46. Dwyer J. Dietary assessment. In: Modern Nutrition in Health and Disease, Shils ME eds. Philadelphia: Lippincott, Williams & Wilkins, 1998, pp. 937–62.

47. Position of the American Dietetics Association: child and adolescent food and nutrition programs. J Am Diet Assoc 2003; 103:887–92.

48. Public health achievements. Nutrition Today 1999 May/June:2.

49. Asian Food Guide Pyramid; available from www.nal.usda.gov/fnic/etext/ 000023.html, accessed 12/20/03.

50. Somali Language Version of the Food Guide Pyramid, Dept. of Anthropology and Geography at Georgia State University, Atlanta, GA.

"To bring on the menses, recover the flesh by giving her puddings, roast meats, a good wine, fresh air, and sun."

Amusing advice on the treatment of infertility from 1847

Chapter 2

Preconception Nutrition

Chapter Outline

- Introduction
- Preconception Overview
- Reproductive Physiology
- Sources of Disruptions in Fertility
- Nutrition-Related Disruptions in Fertility
- Nutrition and Contraceptives
- Other Preconceptual Nutrition Concerns
- Model Preconceptional Nutrition Programs

Prepared by **Judith E. Brown**

Key Nutrition Concepts

1 Fertility is achieved and maintained by carefully orchestrated, complex processes that can be disrupted by a number of factors related to body composition and dietary intake.

2 Oral contraceptives and contraceptive implants can adversely affect some aspects of nutritional status.

3 Optimal nutritional status prior to pregnancy enhances the likelihood of conception and helps ensure a healthy pregnancy and robust newborn.

Introduction

Human reproduction is the result of a superb orchestration of complex and interrelated genetic, biological, environmental, and behavioral processes. Given favorable states of health, these processes occur smoothly in females and males and set the stage for successful reproduction. However, less than optimal states of health, brought about by conditions such as acute undernutrition or high levels of alcohol intake, can disrupt these finely tuned processes and diminish reproductive capacity. Sometimes conception occurs in the presence of poor nutritional or health status. Such events increase the likelihood that fetal growth and development, and the health of the mother during pregnancy, will be compromised.

This chapter first highlights vital statistics related to the preconception period and presents background information on reproductive physiology. Then the focus is on (1) nutrition and the development and maintenance of the biological capacity to reproduce, (2) nutritional effects of contraceptives, (3) preconceptional nutritional status and the course and outcome of pregnancy, and (4) model programs that promote preconceptional nutritional health. The following chapter addresses the role of nutrition in specific conditions—such as premenstrual syndrome, diabetes eating disorders, celiac disease, and polycystic ovary

Infertility Absence of production of children. Commonly used to mean a biological inability to bear children.

Infecundity Biological inability to bear children after 1 year of unprotected intercourse.

Fertility Actual production of children. The word best applies to specific vital statistic rates, but is commonly taken to mean the ability to bear children.

Fecundity Biological ability to bear children.

Miscarriage Generally defined as the loss of a conceptus in the first 20 weeks of pregnancy. Also called *spontaneous abortion*.

Endocrine A system of ductless glands, such as the thyroid, adrenal glands, ovaries, and testes, that produces secretions that affect body functions.

Immunological Having to do with the immune system and its functions in protecting the body from bacterial, viral, fungal, or other infections and from foreign proteins (i.e., those proteins that differ from proteins normally found in the body).

Subfertility Reduced level of fertility characterized by unusually long time to conception (over 12 months) or repeated, early pregnancy losses.

syndrome—that affect preconceptional health or very early pregnancy outcomes.

Preconception Overview

Approximately 15% of all couples in the Western world are involuntarily childless. They are generally considered to be *infertile*, or more correctly *infecund*.

Fertility refers to the actual production of children, whereas *fecundity* addresses the biological capacity to bear children. *Fertility* best applies to vital statistics on fertility rates, or the number of births per 1000 women of childbearing age (15–44 years in most statistical reports). For example, in 2002 the U.S. fertility rate was 67.8 births per 1000 women aged 15 to 44 years.[1] Even scientists and clinicians rarely use these terms correctly, however, and to do so in this chapter would cause undue confusion. Hence, we will use the familiar meaning of *infertility*.

Infertility is generally defined as the lack of conception after 1 year of unprotected intercourse. This definition leads to a "yes/no" answer about fertility that is misleading. Approximately 40% of couples diagnosed as infertile will conceive a child in 3 years without the help of technology.[2] Given these chances, fertility procedures can be considered successful only if they increase conceptions by over 40%, and many currently do not.[3] Chances of conceiving decrease the longer infertility lasts and as women and men age beyond 35 years.[4]

Healthy couples having regular, unprotected intercourse have a 25 to 30% chance of a diagnosed pregnancy within a given menstrual cycle. However, many more conceptions probably occur. Studies show that 25–30% of conceptions are lost by resorption into the uterine wall within the first 6 weeks after conception.[5] Another 9% are lost by *miscarriage* in the first 20 weeks of pregnancy.[6]

The most common cause of miscarriage is the presence of a severe defect in the fetus. Miscarriages can also be caused by maternal infection, structural abnormalities of the uterus, *endocrine* or *immunological* disturbances, and unknown, random events.[7]

Women who experience multiple miscarriages (sometimes defined as three or more), men who have sperm abnormalities (such as low sperm count or density, malformed sperm or immobile sperm), and women who ovulate infrequently are considered *subfertile*. It is estimated that 18% of married couples in the United States are subfertile due to delayed time to conception (over 12 months) or repeated, early pregnancy losses.[2] A silver lining to subfertility is that the reproductive capacity of one individual can compensate for diminished potential in the other.

2010 Nutrition Objectives for the Nation Related to the Preconceptional Period

National priorities for improvement in health status prior to pregnancy include five related to nutrition (Table 2.1).

Objectives for improving intake of vegetables, calcium, whole grains, and other dietary components apply to men as well as women prior to pregnancy.

Reproductive Physiology

The reproductive systems of females and males (Illustration 2.1 on the next page) begin developing in the first months after conception and continue to grow in size and complexity of function through *puberty*. Females are born with a complement of immature *ova* and males with sperm-producing capabilities. The capacity for reproduction is established during puberty when hormonal changes cause the maturation of the reproductive system over the course of 3 to 5 years.

Approximately 7 million immature ova, or *primordial follicles,* are formed during early fetal development, but only about one-half million per ovary remain by the onset of puberty. During a woman's fertile years, some 400–500 ova will mature and be released for possible fertilization. Very few ova remain by *menopause.*

For men, sperm numbers and viability decrease somewhat after approximately 35 years of age, but are still produced from puberty onward.[4] Because females are born with their lifetime supply of ova, the number with chromosomes damaged by oxidation, radioactive particle exposure, and aging increases with time. Consequently, children born to women older than roughly 35 years of age are more likely to have disorders related to defects in chromosomes than are children born to younger women.[2]

Female Reproductive System

During puberty females develop monthly *menstrual cycles,* the purpose of which is to prepare an ovum for fertilization by sperm and the uterus for implantation of a fertilized egg. Menstrual cycles result from complex interactions among hormones secreted by the hypothalamus, the pituitary gland, and the ovary. Knowledge of hormonal changes during the menstrual cycle is expanding, and the process is more complex than indicated in this presentation, which focuses on nutritional effects on hormonal changes in the menstrual cycle and on fertility.

Menstrual cycles are 28 days long on average, but it is not uncommon for cycles to be several days shorter or longer. The first day of the cycle is when menses, or blood flow, begins. The first half of the cycle is called the *follicular phase;* the last 14 days is the *luteal phase.* Hormonal changes during these phases of the menstrual cycle are shown in Illustration 2.2 on page 47.

HORMONAL EFFECTS DURING THE MENSTRUAL CYCLE   At the beginning of the follicular phase, estrogen stimulates the hypothalamus to secrete *gonadotropin-releasing hormone* (GnRH), which causes the pituitary gland to release the *follicle-stimulating hormone* (FSH) and *luteinizing hormone* (LH). (See Table 2.2 on page 48 for definitions of these hormones.) FSH prompts the growth and maturation of 6–20 follicles, or capsules in the surface of the ovary in which ova mature. The presence of FSH stimulates the production of *estrogen* by cells within the follicles. Estrogen and FSH further stimulate the growth and maturation of follicles while rising LH levels cause cells within the follicles to secrete *progesterone.* Estrogen and progesterone also prompt the uterine wall (or endometrium) to store glycogen and other nutrients and to expand the growth of blood vessels and connective tissue. These changes prepare the uterus for nourishing a conceptus after implantation. Just prior to ovulation, which usually occurs on day 14 of a 28-day menstrual cycle, blood levels of FSH and LH peak. The surge in LH level results in the release of an ovum from a follicle, and voilà! Ovulation occurs.

The luteal phase of the menstrual cycle begins after ovulation. Much of the hormonal activity that regulates biological processes during this half of the cycle is initiated by the cells in the follicle left behind when the egg was released. These cells grow in number and size and form the *corpus luteum* from the original follicle. The corpus luteum secretes large amounts of progesterone and some estrogen. These hormones now inhibit the production of GnRH, and thus the secretion of FSH and LH.

Table 2.1   2010 nutrition objectives for the nation related to preconception

√ Increase the proportion of adults who are at a healthy weight from 42 to 60%.

√ Reduce the proportion of adults who are obese from 23 to 15%.

√ Reduce iron deficiency among young children and females of childbearing age from 4–11% to 1–7%.

√ Reduce the incidence of spina bifida and other neural tube defects from 7 to 3 per 10,000 live births and birth defects from 1.7 to 1.2 per 1000 live births.

√ Increase the proportion of women who receive preconceptional counseling.

Puberty   The period in life during which humans become biologically capable of reproduction.

Ova   Eggs of the female produced and stored within the ovaries (singular = *ovum*).

Menopause   Cessation of the menstrual cycle and reproductive capacity in females.

Menstrual Cycle   An approximately 4-week interval in which hormones direct a buildup of blood and nutrient stores within the wall of the uterus and ovum maturation and release. If the ovum is fertilized by a sperm, the stored blood and nutrients are used to support the growth of the fertilized ovum. If fertilization does not occur, they are released from the uterine wall over a period of 3 to 7 days. The period of blood flow is called the *menses*, or the menstrual period.

Corpus Luteum   (*corpus* = body, *luteum* = yellow) A tissue about 12 inches in diameter formed from the follicle that contained the ovum prior to its release. It produces estrogen and progesterone. The "yellow body" derivation comes from the accumulation of lipid precursors of these hormones in the corpus luteum.

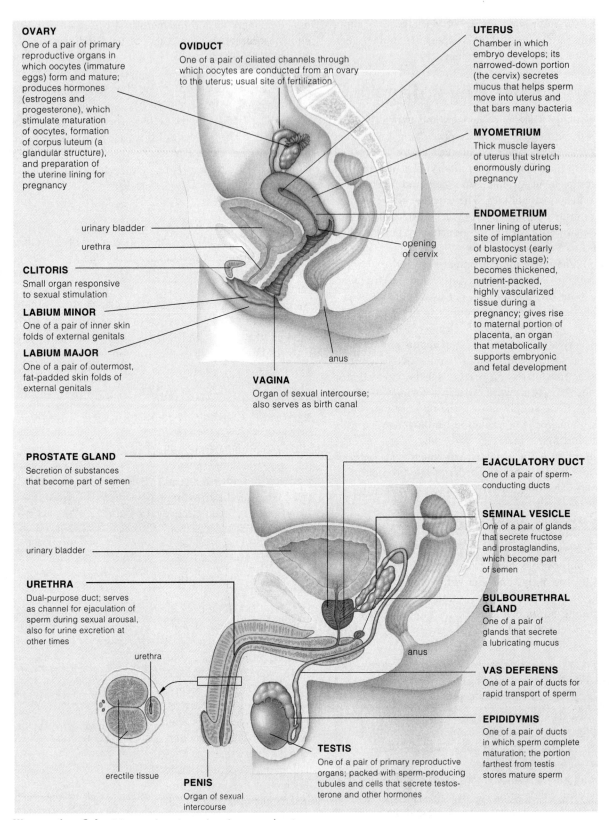

OVARY
One of a pair of primary reproductive organs in which oocytes (immature eggs) form and mature; produces hormones (estrogens and progesterone), which stimulate maturation of oocytes, formation of corpus luteum (a glandular structure), and preparation of the uterine lining for pregnancy

OVIDUCT
One of a pair of ciliated channels through which oocytes are conducted from an ovary to the uterus; usual site of fertilization

urinary bladder

urethra

CLITORIS
Small organ responsive to sexual stimulation

LABIUM MINOR
One of a pair of inner skin folds of external genitals

LABIUM MAJOR
One of a pair of outermost, fat-padded skin folds of external genitals

VAGINA
Organ of sexual intercourse; also serves as birth canal

UTERUS
Chamber in which embryo develops; its narrowed-down portion (the cervix) secretes mucus that helps sperm move into uterus and that bars many bacteria

MYOMETRIUM
Thick muscle layers of uterus that stretch enormously during pregnancy

ENDOMETRIUM
Inner lining of uterus; site of implantation of blastocyst (early embryonic stage); becomes thickened, nutrient-packed, highly vascularized tissue during a pregnancy; gives rise to maternal portion of placenta, an organ that metabolically supports embryonic and fetal development

opening of cervix

anus

PROSTATE GLAND
Secretion of substances that become part of semen

urinary bladder

URETHRA
Dual-purpose duct; serves as channel for ejaculation of sperm during sexual arousal, also for urine excretion at other times

urethra

erectile tissue

PENIS
Organ of sexual intercourse

TESTIS
One of a pair of primary reproductive organs; packed with sperm-producing tubules and cells that secrete testosterone and other hormones

EJACULATORY DUCT
One of a pair of sperm-conducting ducts

SEMINAL VESICLE
One of a pair of glands that secrete fructose and prostaglandins, which become part of semen

BULBOURETHRAL GLAND
One of a pair of glands that secrete a lubricating mucus

VAS DEFERENS
One of a pair of ducts for rapid transport of sperm

EPIDIDYMIS
One of a pair of ducts in which sperm complete maturation; the portion farthest from testis stores mature sperm

anus

Illustration 2.1   Mature female and male reproductive systems.

Without sufficient FSH and LH, ova within follicles do not mature and are not released. (This is also how estrogen and progesterone in some birth control pills inhibit ova maturation and release.) Estrogen and progesterone secreted by the corpus luteum further stimulate the development of the endometrium. If the ovum is not fertilized, the production of hormones by the corpus luteum declines, and blood levels of progesterone and estrogen

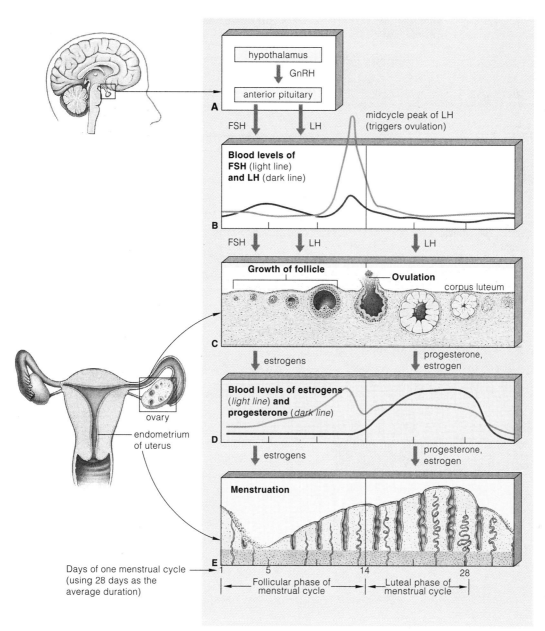

Illustration 2.2
Changes in the ovary and uterus, correlated with changing hormone levels during the follicular and luteal phases of the menstrual cycle.

fall. This decline removes the inhibitory effect of these hormones on GnRH release, and GnRH is again able to stimulate release of FSH for the next cycle of follicle development, and of LH for the stimulation of progesterone and estrogen production. Decreased levels of progesterone and estrogen also cause blood vessels in the uterine wall to constrict, allowing the uterine wall to release its outer layer in the *menstrual flow*. Cramps and other side effects of menstruation can be traced to the production of *prostaglandins* by the uterus. These substances cause the uterus to contract and release the blood and nutrients stored in the uterine wall.

If the ovum is fertilized, it will generally implant in the lining of the uterus within 8 to 10 days. Hormones secreted by the dividing, fertilized egg signal the corpus luteum to increase in size and to continue to produce enough estrogen and progesterone to maintain the nutrient and blood vessel supply in the endometrium. The corpus luteum ceases to function within the first few months of pregnancy, when it is no longer needed for hormone production.

Prostaglandins A group of physiologically active substances derived from the essential fatty acids. They are present in many tissues and perform such functions as the constriction or dilation of blood vessels, and stimulation of smooth muscles and the uterus.

Testes Male reproductive glands located in the scrotum. Also called testicles.

Male Reproductive System

Reproductive capacity in males is established by complex interactions among the hypothalamus, pituitary gland, and **testes**. The process in males is ongoing rather than cyclic. Fluctuating levels of GnRH signal the

Table 2.2 Hormones that affect reproduction

HORMONE	ABBREVIATION	SOURCE	ACTION
Gonadotropin-releasing hormone	GnRH	Hypothalamus	Stimulates release of FSH and LH
Follicle-stimulating hormone	FSH	Pituitary	Stimulates the maturation of ova and sperm
Luteinizing hormone	LH	Pituitary	Stimulates secretion of estrogen, progesterone, and testosterone and growth of the corpus luteum
Estrogen (most abundant form is estradiol)		Ovaries, testes, fat cells, corpus luteum, and placenta (during pregnancy)	Stimulates release of GnRH in follicular phase and inhibits in luteal phase; stimulates thickening of uterine wall during menstrual cycle
Progesterone (progestin, progestogen, and gestagon are similar)		Ovaries and placenta	"Progestational": prepares uterus for fertilized ovum and to maintain a pregnancy; stimulates uterine lining buildup during menstrual cycle; helps stimulate cell division of fertilized ova; inhibits action of testosterone
Testosterone		Mostly by testes	Stimulates maturation of male sex organs and sperm, formation of muscle tissue and other functions

Androgens Types of steroid hormones produced in the testes, ovaries, and adrenal cortex from cholesterol. Some androgens (testosterone, dihydrotestosterone) stimulate development and functioning of male sex organs.

Epididymis Tissues on top of the testes that store sperm.

Semen The penile ejaculate containing a mixture of sperm and secretions from the testes, prostate, and other glands. It is rich in zinc, fructose, and other nutrients. Also called seminal fluid.

Pelvic Inflammatory Disease (PID) A general term applied to infections of the cervix, uterus, fallopian tubes, or ovaries. Occurs predominantly in young women and is generally caused by infection with a sexually transmitted disease, such as gonorrhea or chlamydia, or with intrauterine device (IUD) use.

Endometriosis A disease characterized by the presence of endometrial tissue in abnormal locations, such as deep within the uterine wall, in the ovary, or in other sites within the body. The condition is quite painful and is associated with abnormal menstrual cycles and infertility in 30–40% of affected women.

release of FSH and LH, which trigger the production of testosterone (Table 2.2) by the testes. Testosterone and other *androgens* stimulate the maturation of sperm, which takes 70–80 days. When mature, sperm are transported to the *epididymis* for storage. Upon ejaculation, sperm mix with secretions from the testes, seminal vesicle, prostate, and bulbourethral gland to form *semen.*

Just as some aspects of female reproductive processes remain unclear, scientists have yet to fully elucidate hormonal and other processes involved in male reproduction.

Sources of Disruptions In Fertility

The intricate mechanisms that regulate fertility can be disrupted by many factors, including adverse nutritional exposures, contraceptive use, severe stress, infection, tubal damage and other structural problems, and chromosomal abnormalities (Table 2.3).[8] Conditions that modify fertility appear to affect hormones that regulate ovulation, the presence or length of the luteal phase, sperm production, or the tubular passageways that ova and sperm must travel for conception to occur. Sexually transmitted infections, for example, can result in *pelvic inflammatory disease* (PID), which may lead to scarring and blockage of the fallopian tubes.[9] *Endometriosis* is also a common cause of reduced fertility. It develops when portions of the endometrial wall that build up during menstrual cycles become embedded within other body tissues. Endocrine abnormalities that modify hormonal regulation of fertility are the leading diagnoses related to infertility. "Unknown cause" is the second leading diagnosis, however, and is applied to about one-half of all cases of male and female infertility.[2]

Nutrition-Related Disruptions in Fertility

Disruptions in fertility related to nutritional status include undernutrition, weight loss, obesity, high exercise levels, and intake of specific foods and food components. Nutritional factors generally exert only temporary influence on fertility; normal fertility returns once the problem is corrected.

Undernutrition and Fertility

Does undernutrition decrease fertility in populations? The answer depends on whether the undernutrition is long term (chronic) or short term (acute). Chronic undernutrition appears to reduce fertility by only a small amount.[10]

Table 2.3 Factors related to impaired fertility in women and men

Females and Males	Females	Males
• Weight loss >15% of normal weight	• Recent oral contraceptive use (within 2 months)	• Inadequate zinc status
• Negative energy balance	• Anorexia nervosa, bulimia nervosa	• Inadequate antioxidant status (selenium, vitamins C and E, carotinoids)
• Inadequate body fat	• High caffeine intake (?)[a]	• Heavy metal exposure (lead, mercury, cadmium, manganese)
• Excessive body fat, especially central fat	• High fiber intake	• Halogen (in some pesticides) and glycol exposure (in antifreeze, de-icers)
• Extreme levels of exercise	• Vegetarian diets (?)	• Estrogen exposure (in DDT, PCBs)
• High alcohol intake	• Carotenemia (?)	• Chromosomal abnormalities in sperm
• Endocrine disorders (e.g., hypothyroidism, Cushing's disease)	• Age >35 years	• Sperm defects
• Structural abnormalities of the reproductive tract	• Pelvic inflammatory disease (PID)	• Excessive heat to testes
• Celiac disease	• Endometriosis	• Steroid abuse
• Crowding		
• Severe stress		
• Infection (sexually transmitted diseases)		
• Diabetes		

[a]The "?" indicates that the relationship to impaired fertility is controversial.

Acute undernutrition due to famine or deliberate weight loss in normal-weight women clearly decreases fertility.

CHRONIC UNDERNUTRITION The primary effect of chronic undernutrition on reproduction in women is the birth of small and frail infants who have a high likelihood of death in the first year of life.[11] Infant death rates in developing countries where malnutrition is common often exceed 50 per 1000 live births. By contrast, infant death rates are less than 4 per 1000 live births in countries such as Hong Kong, Japan, and Sweden.[12]

The effect of chronic undernutrition on fertility is difficult to study accurately, and conclusions about relationships will change as more is learned. Investigations of relationships between chronic undernutrition and fertility are complicated by differences in the use of contraception, age of puberty and marriage, breastfeeding duration (longer periods of breastfeeding increase the time to the next pregnancy), access to induced abortions, and social and economic incentives or constraints on family size.[13,14] In less developed countries with poor access to contraception, births per woman average 6 to 8, whereas in developed countries (where contraception is generally available) they average at or below the replacement level of 2.1 births per woman.[15] Without careful study, it might appear fertility is *lower* in better-nourished women in developed countries than in poorly nourished women in less developed countries.

Of the environmental factors that influence fertility, education and child survival appear to be the most important. Fertility rates in poor countries decline substantially as women become educated and as child survival increases.[16,17]

ACUTE UNDERNUTRITION Undernutrition among previously well-nourished women is associated with a dramatic decline in fertility that recovers when food intake does.[11] Periods of feast and famine in the nomadic Kung tribe of Botswana and the Turkana people of Kenya, for example, are associated with major shifts in fertility.[18,19] These groups of hunter-gatherers (although relatively few survive now) experience sharp, seasonal fluctuations in body weight depending on the success of their hunting and foraging for plant foods. Birthrates decline substantially during periods of famine and increase with food availability. When these hunter-gatherers become farmers and food supply is more dependable, body weight increases, activity levels decrease, and pregnancy rates go up.

Other evidence also suggests a connection between undernutrition and infertility. Food shortages in Europe in the seventeenth and eighteenth centuries were accompanied by dramatic declines in birthrates. Famine in Holland during World War II led to calorie intakes of about 1000 per day among women. One out of two women in famine-affected areas stopped menstruating, and the birthrate dropped by 53%. Fertility status improved within 4 months after the end of the famine, but for many women it took as long as a year for their menstrual cycles to return to normal.[20] Similarly, the 1974–1975 famine in Bangladesh resulted in a 40% decline in births.[10]

Famines are associated with more than disruptions in the food supply. They are usually accompanied by low availability of fuel for heating and cooking, poor living conditions, anxiety, fear, and despair. These factors also probably contribute to the declines in fertility observed with famine.

Case Study 2.1

Photo Disc

Cyclic Infertility with Weight Loss and Gain

After four years of experiencing amenorrhea, Tonya seeks medical care to help her become pregnant. She is convinced that her lack of menstrual periods is the cause of her infertility. Tonya's height is 5 feet 5 inches; her weight is 107 pounds, which she has maintained for 4 years (she previously weighed 121 pounds). Her FSH and LH levels are both abnormally low, and she is not ovulating. When the importance and methods of weight gain are explained to her, Tonya agrees to gain some weight. After she regains 7 pounds, her LH level is normal, but her FSH level is still low and the luteal phase of her cycles is abnormally short. When her weight reaches 119 pounds, Tonya's LH and FSH levels, ovulation, and menstrual cycles are normal.

Questions

1. Was Tonya underweight or normal weight based on BMI when she weighed 107 pounds? (Use the BMI chart on the inside back cover of the text to answer this question.)
2. Can you determine Tonya's body fat content based on her BMI?
3. What likely happened to Tonya's average estrogen level when her weight decreased from 121 to 107 pounds?
4. What are two likely reasons Tonya was advised to gain weight to improve her chances of conception rather than being given Clomid or another ovulation-inducing drug?

(Answers are in Appendix D.)

Acute reduction in food intake appears to reduce reproductive capacity by modifying hormonal signals that regulate menstrual cycles in females. It also appears to impair sperm maturation in males.[10]

Body Fat and Fertility

Both low and high amounts of body fat are related to decreased fertility due to alterations in hormone levels. Fat cells produce estrogen and *leptin,* and both of these hormones affect reproductive processes. Estrogen and leptin levels are elevated by high levels of body fat and reduced by low levels of body fat.[21] Weight loss through caloric reduction and increased physical activity is the recommended initial treatment for infertility among women with high levels of body fat.[22] Infertility treatments, such as drugs that induce ovulation, are less effective in obese than in normal-weight women.[23]

> **Leptin** A protein secreted by fat cells that, by binding to specific receptor sites in the hypothalamus, decreases appetite, increases energy expenditure, and stimulates gonadotropin secretion. Leptin levels are elevated by high, and reduced by low, levels of body fat.
>
> **Body Mass Index (BMI)** Weight in kg/height in m². BMIs <18.5 are considered underweight, 18.5–25 normal weight, 25–30 overweight, and BMIs of 30 and higher obesity.
>
> **Amenorrhea** Absence of menstrual cycle.
>
> **Anovulatory Cycles** Menstrual cycles in which ovulation does not occur.

The relationship between low levels of body fat and delayed onset of menstruation and infertility has led to the conclusion that a critical level of body fat is needed to trigger and sustain normal reproductive functions.[21] Specific levels of body fat consistently related to infertility have not been identified. However, fertility is often compromised in women with **body mass indexes (BMIs)** of less than 20 or over 30 kg/m².[24]

Approximately one in three women in the United States is obese, making infertility related to excess body fat a common problem and an important health concern.[25] The topic of body fat and fertility is revisited in more detail in the next chapter.

WEIGHT LOSS AND FERTILITY IN FEMALES

In normal-weight women, weight loss that exceeds approximately 10–15% of usual weight decreases estrogen. Consequences of this hormonal change include **amenorrhea, anovulatory cycles,** and short or absent luteal phases. It is estimated that about 30% of cases of impaired fertility are related to simple weight loss, or "weight-related amenorrhea," as it is called.[11] Hormone levels tend to return to normal when weight is restored to within 95% of previous weight.[26,27] Case Study 2.1 provides an example of the effect of weight loss on fertility.

Weight gain is the recommended first-line treatment for weight-related amenorrhea. In many cases, however, the advice is more easily given than applied. About 10% of underweight women will not consider weight gain and may change health care providers in search of a different solution to infertility.[27]

Treatment of underweight women with Clomid (clomiphene citrate, a drug that induces ovulation) generally does not improve fertility until weight is regained. Fertility may be improved through the use of GnRH, FSH, and other hormones.[23] However, twice as many infants born to underweight women receiving such therapy are small for gestational age compared to infants born to underweight women who gain weight and experience unassisted conception.[11]

The eating disorder anorexia nervosa is associated with similar, but more severe, changes in endocrine and hypothalamic functions than those seen with weight loss in normal-weight women. (This topic is covered in Chapter 3.)

WEIGHT LOSS AND FERTILITY IN MALES
Weight loss decreases fertility in men just as it does in women. In the classic starvation experiments by Keys during World War II,[28] men experiencing a 50% reduction in caloric intake reported substantially reduced sexual drive early in the study. Sperm viability and motility decreased as weight reached 10 to 15% below normal, and sperm production ceased entirely when weight loss exceeded 25% of normal weight. Sperm production and libido returned to normal after weight was regained.

Exercise and Infertility

The adverse effects of intense levels of physical activity on fertility were observed over 40 years ago in female competitive athletes. Since then, a number of studies have shown that young female athletes may experience delayed age at puberty and lack menstrual cycles.[29] Average age of menarche for competitive female athletes and ballet dancers is often delayed by 2 to 4 years. The delay in menarche increases if females begin training for events that require thinness (such as gymnastics) before menarche normally would begin. Very high levels of exercise can also interrupt previously established, normal menstrual cycles.[29] The presence of abnormal cycles reportedly ranges from about 23% in joggers to 86% in female bodybuilders (Table 2.4).[30,31]

Delays and interruptions in normal menstrual cycles appear to result from hor-monal metabolic changes related to high levels of physical activity and caloric deficits.[33] Very high levels of physical activity are related to modified estrogen, LH, FSH, and other hormone levels.[34] Metabolic and hormonal status generally reverts to normal after high levels of training and caloric deficits end.[35]

Some of the hormones involved in fertility impairments perform other important functions in the body, which may also be disrupted. Reduced levels of estrogen that accompany low levels of body fat and amenorrhea, for example, may decrease bone density and increase the risk of shortness, bone fractures, and osteoporosis.[35]

Diet and Fertility

Certain dietary components appear to influence fertility by modifying estrogen, LH, and other hormone levels in women. Vegetarian diets, low fat intake, and high intakes of dietary fiber, soy, caffeine, and alcohol are related to impaired fertility. The effects of several of these factors on fertility status are probably interrelated, but none directly causes infertility.

PLANT FOODS AND FERTILITY   Women who regularly consume plant-based, low-fat, high-fiber diets ($\geq$25 g per day) and no meat appear to have lower circulating levels of estrogen and may be more likely to have irregular menstrual cycles than omnivores. These results apply to vegetarians who are thin, normal weight, or overweight.[36,37] Diets providing less than 20% of calories from fat appear to lengthen menstrual cycles among women in general,[38] and high-fiber diets may reduce estrogen levels.[39]

Consumption of 20–200 mg/day of isoflavones from soy increases menstrual cycle length by about a day. Isoflavones appear to decrease blood levels of gonadotropins, estrogen, and progesterone. Soy intake has not been related to androgen levels nor semen quality in males.[40]

Table 2.4 Incidence of irregular or absent menstrual cycles in female athletes and sedentary women[30,31,32]

	INCIDENCE OF IRREGULAR OR ABSENT MENSTRUAL CYCLES
Joggers (5 to 30 miles per week)	23%
Runners (over 30 miles per week)	34
Long-distance runners (over 70 miles per week)	43
Competitive bodybuilders	86
Noncompetitive bodybuilders	30
Volleyball players	48
Ballet dancers	44
Sedentary women	13

CAROTENEMIA AND FERTILITY Several studies indicate that women with *carotenemia* may experience amenorrhea and menstrual dysfunction.[41,42] Women who consume 12 mg or more of beta-carotene daily for over 6 weeks by constant munching on foods such as dried green pepper flakes or carrots, or by taking "tanning pills," tend to develop carotenemia. Skin color and fertility return to normal within 2 to 6 weeks after high levels of intake are discontinued. Why carotenemia may cause infertility is not known, and the relationship must be considered speculative until additional studies are undertaken. It is not clear that the carotene content of the diet, rather than other substances in plentiful supply in plant foods, accounts for the relationship observed.

CAFFEINE AND FERTILITY Should women concerned about infertility consume coffee and other foods with caffeine? It appears that high intakes of caffeine may prolong the time it takes to become pregnant. In a study of European women, researchers found that the chance of conception within a 10-month interval of unprotected intercourse was half as likely among women who consumed over 4 cups of coffee per day (>500 mg caffeine) versus the conception rate of women who consumed little coffee.[43] Another study reported that intake of over 300 mg of caffeine daily from coffee, sodas, and tea decreased the chance of conceiving by 27% per cycle compared to negligible caffeine intake.[44] In both studies, the effect of caffeine on time to conception was stronger in women who smoked.

It is not known why caffeine appears to prolong time to conception. Nor is it clear if it is caffeine, one or more of the hundreds of other chemical substances in coffee, or an unexamined characteristic of women who consume lots of caffeine that accounts for the apparent effects of caffeine on fertility. The prudent course of action is to advise concerned women who consume more than several cups of regular coffee each day to cut down on coffee and other products high in caffeine. Table 2.5 shows the caffeine content of selected beverages and foods. Women who cut out coffee altogether should be advised to do so gradually, to reduce "caffeine withdrawal headaches" and fatigue.

ALCOHOL AND FERTILITY Alcohol may influence fertility by decreasing estrogen and testosterone levels and disrupting normal menstrual cycles. In a study of 430 Danish couples attempting pregnancy for 6 months, consumption of from 1 to 5 alcohol-containing drinks per week by women was related to a 39% lower chance of conception. Alcoholic beverage consumption of over 10

Carotenemia A condition, caused by ingestion of high amounts of carotenoids (or carotenes) from plant foods, in which the skin turns yellowish orange.

Klinefelter's Syndrome A congenital abnormality in which testes are small and firm, legs abnormally long, and intelligence generally subnormal.

Table 2.5 Caffeine content of foods and beverages	
FOODS AND BEVERAGES	**CAFFEINE (mg)**
Coffee, 1 c	
Drip	137–153
Percolated	97–125
Instant	61–70
Decaffeinated	0.5–4.0
Tea, 1 c	
Brewed 5 minutes	32–176
Instant	40–80
Soft Drinks, 12 oz	
Mountain Dew	54
Coca-Cola	46
Diet Coca-Cola	46
Dr. Pepper	40
Pepsi-Cola	38
Diet Pepsi-Cola	37
Ginger ale	0
7-Up	0
Chocolate Products	
Cocoa, chocolate milk, 1 c	10–17
Milk chocolate, 1 oz	1–15
Chocolate syrup, 2 tb	4

drinks per week was related to a 66% reduction in the probability of conception during the 6-month period.[45] Not all studies show alcohol intake affecting fertility, however; so the effect, if real, may be weak. Nevertheless, because alcohol should not be consumed during pregnancy due to the risk of fetal malformations and mental retardation, it makes sense that women who are attempting conception should not consume alcohol.

Other Factors Contributing to Infertility in Males

The genesis of infertility in males and females is not well understood, but less is known about factors affecting male reproductive capacity than about those affecting females. Examinations of fertility status in males tend to focus on sperm quality, assessed as sperm number (concentration), motility (movement), and morphology (shape).[2] A number of chromosomal abnormalities (*Klinefelter's syndrome,* for example) and environmental toxins have been related to modification of sperm. Certain drugs used for hypertension and cancer, disease processes such as diabetes and atherosclerosis (hardening of the arteries), and endocrine disorders that disrupt testosterone production affect male fertility primarily in ways that involve sperm production or erectile function.

Sperm production appears to be sensitive to a male's exposure to a number of nutritional and other environmental factors (see Table 2.3). Because sperm development occurs over 70 to 80 days, however, there is a delay of up to 3 months between onset of the exposure (such as use of a medication or nutrient toxicity) and the appearance of sperm abnormalities.

LOW ZINC STATUS   Zinc intakes of 5 mg per day or less are associated with decreased semen volume and testosterone levels in experimental studies. Zinc plays a critical role in male reproduction potential, and the concentration of zinc in semen is high. Because zinc is concentrated in semen, it is lost via ejaculated semen. Zinc in seminal fluid is an essential cofactor for enzymes involved in testosterone production, DNA replication, protein synthesis, and cell division. It appears to protect sperm from bacteria and chromosomal damage.

Zinc also plays important roles in sexual organ development in males. Deficiency of zinc prior to and during early adolescence is associated with *hypogonadism* and lack of sexual development, such as reduced growth of the penis, failure of the voice to deepen, and lack of muscle enlargement. Hypogonadism has been observed in some Middle Eastern males due to diets based on whole grains with infrequent consumption of animal products. Zinc in whole grains is poorly absorbed, whereas animal products generally contain good amounts of absorbable zinc.

Supplementation with 10 mg of zinc or more per day reverses zinc-related infertility and sexual maturational problems in males over time.[2,46]

ANTIOXIDANT NUTRIENTS   Sperm are particularly susceptible to oxidative damage because of their high concentration of polyunsaturated fatty acids.

Antioxidants such as selenium, vitamins C and E, beta-carotene, and other carotenoids protect sperm DNA from oxidative damage and promote normal sperm motility and function.[2]

ALCOHOL INTAKE   Impaired fertility is common in alcoholics and is related to direct toxic effects of alcohol on the testes. Light to moderate alcohol consumption does not appear to affect fertility in males.[47]

HEAVY METAL EXPOSURE   Exposure to high levels of lead is related to decreased sperm production and abnormal sperm motility and shape.[48] Inhaled or ingested lead is transported to the pituitary gland, where it appears to disrupt hormonal communications with the testes. The result is lowered testosterone levels and decreased sperm production and motility. The men most likely to be exposed to excess lead tend to be workers in smelting and battery factories.[47]

Mercury can build up in fish living in contaminated waters. Ingestion of fish from waters contaminated with mercury in Hong Kong has been associated with decreased sperm count and abnormal semen.[49] Consumption of fish from the U.S. Great Lakes does not appear to pose similar problems.[50]

Exposure to excess levels of cadmium, manganese, boron, cobalt, copper, nickel, silver, or tin may also affect male fertility, but evidence from human studies is sparse. These metals may build up in male reproductive systems through the inhalation of fumes or dust containing particles or through contaminated water.[47]

HALOGENS   Occupational exposure to pesticides made from halogen compounds (such as dibromochloropropane) has been observed to cause sperm count reduction and male infertility.[47]

GLYCOLS   Glycols are widely used in antifreeze, solvents, and de-icers for airplanes. Occupational exposures to these compounds can decrease fertility in males threefold.[47]

HORMONES   Exposure to synthetic estrogens in pharmaceutical factories and to the estrogen-like substances of DDT (dichlorodiphenyltrichloroethane—an insecticide) PEs (phthalate exters—used as plasticizers) and PCBs (polychlorinated biphenyls—used in transformers) has been found to reduce libido and increase impotence and breast size in males.[47,51] It has been speculated that increased exposure to estrogenic compounds and other pollutants may be responsible for the reported decrease in sperm count in men over recent decades.

> **Hypogonadism**   Atrophy or reduced development of testes or ovaries. Results in immature development of secondary sexual characteristics.
>
> **Scrotum**   A muscular sac containing the testes.

Is sperm count declining in men? The issue is not settled. Approximately 17% of males have low sperm counts, and several studies have noted declines in the number of sperm produced of about 1 to 2% per year over recent decades.[52] Some researchers have attributed declines in sperm counts to estrogenic environmental pollutants.[53] Other studies, however, have not identified declining sperm counts in populations studied.[52,54]

HEAT   Elevating the temperature of the *scrotum* and testes can reduce sperm count. Long-haul truck driving, welding, and foundry work may cause increased temperatures and decreased fertility in some men.[47] Prolonged and frequent exposure to hot-tub water is also thought to reduce sperm count in men.

STEROID ABUSE   Doses of steroids (e.g., testosterone) used by some bodybuilders are up to 40 times higher than therapeutic doses. Side effects of steroid abuse are multiple and include atrophy (shrinking) of the testicles, absence of sperm, and decreased libido. Fertility generally returns to normal after steroid use ends, but other disease risks may linger.[55]

Nutrition And Contraceptives

The contraceptive revolution emerged in full force in the 1960s when use of pills with heavy doses of estradiol (the most biologically active form of estrogen) and progesterone became widespread. The adverse side effects of these oral contraceptives were plentiful and included increased risk of heart attack and stroke, elevated blood lipids, glucose intolerance, weight gain, and folate and vitamin B_6 deficiencies. New generations of oral contraceptives employed increasingly lower hormone doses, and side effects diminished substantially. Nevertheless, some remain. Those related to use of the newer contraceptives are listed in Table 2.6.

> **Venous Thromboembolism**   A blood clot in a vein.

The current generation of oral contraceptives elevates some blood lipids and carries a slight risk of *venous thromboembolism*. Their continuing effect on the formation of blood clots appears to be related to the progesterone content of the pills.[56]

The newest generation of fertility control products for females includes contraceptive implants, patches, and injections. These, too, have a number of nutritional side effects. Development of hormonal contraceptive methods for control of fertility in males lags far behind advances in female contraception.

Oral Contraceptives and Nutritional Status

Many types of birth control pills containing low doses of different forms of estrogen and progesterone are currently available. Use of these pills by healthy, nonsmoking women is not associated with an increased risk of heart attack, glucose intolerance, or nutrient deficiencies, nor is use related to weight gain. Their use is, however, associated with a twofold increased risk of thromboembolism (blood-clot formation) and an increased risk of cervical cancer.[57,58]

The new generation of oral contraceptives increases blood levels of triglycerides by about 30% and total cholesterol levels by approximately 6% on average. HDL cholesterol—the "good" blood cholesterol fraction—is increased slightly by these contraceptives.[59]

Oral contraceptives still decrease blood levels of certain nutrients. Blood levels of vitamin B_{12} were found to be an average of 33% lower in oral contraceptive users versus nonusers in a Canadian study of 14- to 20-year-old females.[61] Serum copper levels were 34 to 55% higher and may be related to the increased risk of blood-clot formation observed among oral contraceptive users.[62]

It is recommended that females who are obese, over the age of 35 years, and smoke, and those who have cardiovascular disease, hypertension, diabetes, or are immobilized, use nonhormonal methods of contraception due to their increased risk of venous thromboembolism.[59] Women who take oral contraceptive pills are cautioned against consuming more than a half-ounce of licorice per day. Genuine black licorice contains glycyrrhizic acid, a substance that gives licorice its distinctive taste. Consumption of several ounces of licorice can increase blood pressure and fluid retention in women using the pill.[63] It is generally also recommended that women stop using oral contraceptive pills about 3 months prior to attempting pregnancy.

Contraceptive Injections

DMPA (depot medroxyprogesterone acetate), or Depo-Provera, is the primary type of injectable contraception used in females. Injections that suppress ovulation are given every 3 months. Although highly effective, Depo-Provera has discontinuation rates that average over 50% within the first year of use. Weight gain is a leading reason for discontinuation (27%); irregular periods (24%), fatigue (23%), headache (25%), and abdominal pain (18%) are also commonly reported reasons for discontinuation.[64,65] Weight gain averages 12 pounds during the first year of Depo-Provera use,[65] but not all studies report weight gain.[66] Long-term use of this contraceptive is related to decreased bone density and blood levels of HDL cholesterol and increased levels of LDL cholesterol and insulin.[67,68]

Contraceptive Implants

Norplant (levonorgestral), the leading contraceptive implant, prevents contraception for up to 7 years.[69] This contraceptive method is highly effective for normal-weight and underweight women, as evidenced by a 1.9% pregnancy rate.[70] Pregnancy rates are 4.2% in obese women, however.[71] High rates of side effects, especially

---

Table 2.6 Nutrition-related side effects of contraceptives

Oral Contraceptives

Increased blood levels of HDL cholesterol (the "good" cholesterol)

Increased blood levels of triglycerides and LDL cholesterol

Increased risk of venous thromboembolism (blood clots)

Decreased blood levels of vitamin B_{12}

Increased blood levels of copper

Contraceptive Injections (Depo-Provera)

Weight gain

Increased blood levels of LDL cholesterol and insulin

Decreased blood levels of HDL cholesterol

Decreased bone density

Contraceptive Implants (Norplant)

Weight gain

erratic bleeding (69%), weight gain (41%), and headaches (30%), lead to early removal of the implant in about half of users.[72] Average weight gain 1 year after the implant has been reported to be 9 pounds.[73]

Contraceptive Patches

The contraception patch releases a type of estrogen and progesterone. Tests indicate it is highly effective and easy to use. The patch is placed on the skin for 3 weeks and then taken off for a week.

Contraceptive patches increase blood levels of cholesterol and triglycerides to a greater extent than do oral contraceptives. On the positive side, use of the patch is related to slight increases in HDL-cholesterol levels. The most commonly reported adverse reactions to contraceptive patches are breast soreness, headaches, application-site reactions, and abdominal pain. Contraceptive patches are less effective in women weighing over 198 pounds (90 kg) than in women who weigh less. Women who should not use a contraceptive patch include those with a history of heart disease, stroke, blood clots, and reproductive cancers.[74] Pregnancy should be separated from use of the patch by at least 6 weeks. Contraceptive patches are relatively expensive. Three patches (good for a month) cost about $35, whereas generic oral contraceptive pills cost approximately $5 per month.

Emerging Forms of Contraceptives

Several types of male contraceptives are under development and testing. One of the products being tested consists of periodic injections of progestin and implanted crystallized testosterone. Initial tests show that the hormones effectively suppress sperm production and are 97% effective in preventing conception. Men receiving the progestin/testosterone contraceptive gained an average of 3% of their initial body weight during the 12-month treatment period. Additional studies on the safety of this new contraceptive are under way.[75] Contraceptive pills for males that suppress sperm production are also being tested.

New types of female contraceptives that may become available include the vaginal ring and contraceptive pills that limit menstrual cycles to four or fewer per year.[76]

Other Preconceptual Nutrition Concerns

Approximately 8 to 10 days after an ovum is fertilized, it implants into the uterine wall. Within the first month after conception, the developing *embryo* will have grown from a single cell to millions of cells, basic structures of organs will have formed, and the blueprint for future growth and development will have been established. All this often happens before women know they are pregnant or attend a prenatal clinic. The time to establish a state of optimal health and nutritional status is before conception.

Very Early Pregnancy Nutrition Exposures

Table 2.7 summarizes major nutritional exposures that adversely affect the growth and development of the embryo and *fetus*.[77] It is important to be aware that any of these conditions, if present preconceptionally, may impair embryonic and fetal growth and development. This chapter touches on the importance of adequate folate intake prior to conception. Chapter 4, on nutrition during pregnancy, expands the folate discussion and returns to the other topics listed in Table 2.7.

> **Embryo** The developing organism from conception through 8 weeks.
>
> **Fetus** The developing organism from 8 weeks after conception to the moment of birth.

FOLATE STATUS PRIOR TO CONCEPTION AND NEURAL TUBE DEFECTS Folate status prior to conception is an important concern because inadequate

---

Table 2.7 Nutritional exposures before and very early in pregnancy that disrupt fetal growth and development

Weight Status
- Underweight increases the risk of maternal complications during pregnancy and the delivery of small and early newborns.
- Obesity increases the risk of clinical complications during pregnancy and delivery of newborns with neural tube defects or excessive body fat.

Nutrient Status
- Insufficient folate intake increases the risk of embryonic development of neural tube defects.
- Excessive vitamin A intake (retinol, retinoic acid) increases the risk the fetus will develop facial and heart abnormalities.
- High maternal blood levels of lead increase the risk of mental retardation in the offspring.
- Iodine deficiency early in pregnancy increases the risk that children will experience impaired mental and physical development.
- Iron deficiency increases the risk of early delivery and development of iron deficiency in the child within the first few years of life.

Alcohol
- Regular intake of alcohol increases the risk of *fetal alcohol syndrome* and *fetal alcohol effects*, both of which include impaired mental and physical development.

Diabetes
- Poorly controlled blood glucose levels early in pregnancy increase the risk of fetal malformations, excessive infant size at birth, and the development of diabetes in the offspring later in life.

folate very early in pregnancy can cause ***neural tube defects*** (NTDs). These defects develop within 21 days after conception—or before many women even know they are pregnant, and well before prenatal care begins.[78] Knowledge of the folate–neural tube defect relationship, and awareness that folate intake is inadequate in many women of childbearing age, prompted efforts to increase folate intake. In particular, efforts are focused on encouraging women to consume folic acid, a highly absorbable, synthetic form of this B vitamin. In 1998, that task was made easier when the Food and Drug Administration mandated that refined grain products such as white bread, grits, crackers, rice, and pasta be fortified with folic acid.

Neural Tube Defects (NTDs) Spina bifida and other malformations of the neural tube. Defects result from incomplete formation of the neural tube during the first month after conception.

Over 30 countries now fortify refined grain products with folic acid, and rates of NTDs have fallen in each of these countries. Rates of NTDs in Nova Scotia, Canada, for example, fell 55% in the 2 years after folic acid fortification became mandatory. Declines in the incidence of NTDs in the United States are reported to be below those achieved in Canada. Countries in the European Union have yet to implement folic acid fortification; rates of NTDs in these countries have not changed.[79]

Although intake of folic acid has increased substantially since fortification, there is still concern that some women are unaware of the importance of consuming folic-acid-fortified refined grain products and other sources of folate prior to pregnancy.[80]

Women can get enough folate by consuming a good basic diet and a fully fortified breakfast cereal (Smart Start, Total, or Product 19, for example) or a regular breakfast cereal (Cheerios, Corn Flakes, Raisin Bran, etc.) and 6 to 8 servings of refined grain products each day. Folic acid supplements (400 mcg per day) can also provide folic acid.

IRON DEFICIENCY Iron deficiency occurs in 12% of U.S. women of childbearing age overall, and among 22% of Mexican American and 19% of African American women. Because of the relatively high prevalence of iron deficiency, many women begin pregnancy in an iron-deficient state. The incidence of iron deficiency in women before pregnancy is related to average dietary intakes of iron of 13 mg per day, a level well below the RDA for iron of 18 mg daily. Improved iron status prior to pregnancy would decrease risks associated with iron deficiency during pregnancy.[81] Iron status of women of childbearing age can be improved by increasing consumption of vitamin C–rich fruits along with plant sources of iron; an iron-fortified cereal; and regular consumption of lean meats, dried beans, and other good sources of iron. Iron supplements generally increase iron status in women with low iron stores.

Recommended Dietary Intakes for Preconceptional Women

Recommendations for food and nutrient intakes for women who may become pregnant differ from those for adult women in general in several ways. It is recommended that women who may become pregnant consume 400 mcg of folic acid from fortified foods or supplements, take no more than 5000 IU of vitamin A (retinol or retinoic acid) from supplements daily, and limit or omit alcohol-containing beverages.[77] Recommendations for nutrient intake given in the Dietary Reference Intakes (DRIs) should be applied, paying careful attention to the Tolerable Upper Intake Levels (see tables on the inside front covers of this text).

Food selection prior to pregnancy should be based on the Food Guide Pyramid recommendations for adults (see Illustration 2.3). The number of servings from each food group within the range of servings given will be determined primarily by a woman's physical activity level and body size. An example menu offering several options for food selection for meals and snacks is given in Table 2.8.

Illustration 2.3 Food Guide Pyramid recommendations for preconceptional women.

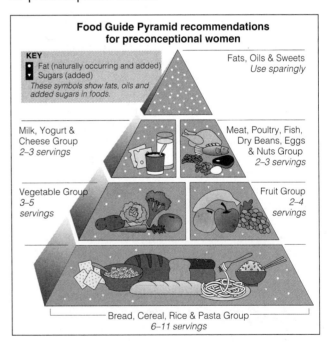

FOOD GROUP	RECOMMENDED NUMBER OF SERVINGS
Bread, cereal, rice, and pasta	6–11
Vegetables	3–5
Fruits	2–4
Milk, yogurt, and cheese	2–3
Meat, poultry, fish, dry beans, eggs, and nuts	2–3

Model Preconceptional Nutrition Programs

This section highlights two model programs, one in the United States and one in Indonesia, related to nutrition during the preconceptional period.

Preconceptional Benefits of WIC

Women are eligible to enter the WIC program (Supplemental Nutrition Program for Women, Infants, and Children) when pregnant. In this model program in California, women were experimentally provided WIC food supplements and nutrition education during *and* between consecutive pregnancies. Women who received WIC benefits during one pregnancy and continuing through the first two months of the next pregnancy had better iron status and delivered newborns with higher birthweights and greater lengths than women who received WIC benefits during pregnancy only.[82] The study demonstrates that low-income women at nutritional risk benefit from WIC services before, as well as during, pregnancy.

Decreasing Iron Deficiency in Preconceptional Women in Indonesia

Approximately one in every two women in Indonesia experiences iron-deficiency anemia during pregnancy. In a unique effort to prevent this problem, the Ministry of Health initiated regulations that require a couple applying for a marriage license to receive advice on iron status from those dispensing the license. All women are now advised to take 30–60 mg of iron along with folic acid in a supplement. Of 344 women studied after the program was initiated, 98% reported that they had purchased and taken iron-folate tablets; 56% had taken at least 30 tablets. The incidence of iron deficiency in this group of women dropped by almost half.[83]

Preconception Care

Increasingly, routine health care visits and educational sessions are being recommended and introduced into health care organizations. Services focus on risk assessment of behaviors such as weight status; dietary intake; folate and iron status; and vitamin, mineral, and herbal supplement use, as well as on the presence of diseases such as diabetes, hypertension, infection, and genetic traits that may be transmitted to offspring. Psychosocial needs should also be addressed as part of preconceptional care, and referrals

made to appropriate services for issues such as eating disorders, abuse, violence, or lack of food or shelter.[84] The desire of couples planning for pregnancy to have a healthy newborn makes the preconceptional period a prime time for positive behavioral changes. It presents opportunities to make lasting improvements in the health and well-being of individuals and families.

Starting pregnancy in the best health status possible can make an important difference to reproductive outcomes. It should be recognized, however, that even in ideal conditions, continued infertility, early pregnancy loss, fetal malformations, and maternal complications will sometimes occur.

Table 2.8 Example menu for preconceptional women that meets the Food Guide Pyramid recommendations; the menu works for men, too, if portion sizes are adjusted upward to meet caloric needs

Breakfast

Fruit juice, 6 oz
Fortified, whole grain breakfast cereal (hot or cold), 1 c
Sliced peaches or banana, ½ c
Skim or low-fat milk, 1 c
Coffee or tea

Lunch

Pork almond ding, 1 c or enchilada with beans, chicken, or beef, 1
Rice, 1 c
Skim or low-fat milk, 1 c

Dinner

Veggie burger, lean beef, poultry, or fish, 3–4 oz
Macaroni salad, pasta, or potato, ½ c
Kale, turnip greens, collards, or spinach, 1 c
Tortilla, pita bread, corn-bread muffin, or whole grain roll, 1
Pumpkin or sweet potato pie, 1 slice, or frozen yogurt, 1 c
Skim or low-fat milk, 1 c

Snacks

Yogurt with fruit
Graham crackers
Dried fruit
Nuts
Seeds
Apple, pear, mango, pineapple, or other fruit
Melted cheese on whole grain toast

Resources

Vital Statistics

The National Center for Health Statistics provides information on fertility and birthrates, and other vital statistics data.
Web site: www.cdc.gov/nchs

Merck Manual of Diagnosis and Therapy

This ever-popular and useful guide is available online and includes information on infertility and related conditions.
Web site: http://www.merck.com/mrkshared/mmanual/home.jsp

Medscape Women's Health Journal

Receive automatic updates on women's health, fertility, and contraception topics by subscribing to this free online journal.
Web site: www.medscape.com/medscape/WomensHealth/journal

The Babycenter Company

A popular source of consumer advice on preconception planning, nutrition, and related topics that is, however, heavily infiltrated with advertisements.
Web site: www.babycenter.com

Public Health Service

A gateway to consumer and services information on fertility clinics and women's health from the U.S. government.
Web site: www.4women.gov

Journal Articles

Look up research articles related to any of the topics covered in this chapter through PubMed.
Web site: www.ncbi.nlm.nih.gov/pubmed

References

1. Hamilton BE et al. Births: preliminary data for 2002. National Vital Statistics Report 2003; 51:1–24.

2. Wong WY et al. Male factor subfertility: possible causes and the impact of nutritional factors. Fertil Steril 2000;73:435–42.

3. Toner JP. Success rates of reproductive techniques improving, Fertil Steril 2002;78:943–50.

4. Ford WC et al. Increasing paternal age is associated with delayed conception in a large population of fertile couples: evidence for declining fecundity in older men. The ALSPAC Study Team (Avon Longitudinal Study of Pregnancy and Childhood). Hum Reprod 2000;15:1703–8.

5. Wilcox AJ et al. Time of implantation of the conceptus and loss of pregnancy. N Engl J Med 1999;340:1796–9.

6. Wilcox AJ et al. Incidence of early loss of pregnancy. N Engl J Med 1988;319:189–94.

7. Kallen B. Epidemiology of human reproduction. Boca Raton, FL: CRC Press; 1988.

8. Warren MP. Effects of undernutrition on reproductive function in the human. Endocrine Reviews 1983;4:363–77.

9. Wiley AS. The ecology of low natural fertility in Ladakh. J Biosoc Sci 1998;30:457.

10. Bongaarts J. Does malnutrition affect fecundity? A summary of evidence. Science 1980;208:564–9.

11. van der Spuy ZM. Nutrition and reproduction. Clin Obstet Gynecol 1985;12:579–604.

12. MacDorman MF et al. Annual summary of vital statistics—2001. Pediatrics 2002;110:1037–50.

13. Rao VG, Sugunan AP, Sehgal SC. Nutritional deficiency disorders and high mortality among children of the Great Andamanese tribe. Natl Med J India 1998;11:65–8.

14. Wood JW. Maternal nutrition and reproduction: why demographers and physiologists disagree about a fundamental relationship. Ann NY Acad Sci 1994;709:101–16.

15. Hirschman C. Why fertility changes. Annu Rev Sociol 1994;20:203–33.

16. Ramakrishnan U et al. Early childhood nutrition, education and fertility milestones in Guatemala. J Nutr 1999;129:2196–202.

17. Pena R et al. Fertility and infant mortality trends in Nicaragua 1964–1993. The role of women's education. J Epidemiol Community Health 1999;53:132–7.

18. Kolata GB. Kung hunter-gatherers: feminism, diet, and birth control. Science 1974; 185:932–4.

19. Leslie PW et al. Evaluation of reproductive function in Turkana women with enzyme immunoassays of urinary hormones in the field. Hum Biol 1996;68:95–117.

20. Stein Z et al. Famine and human development: the Dutch hunger winter of 1944–1945. New York: Oxford University Press, 1975: pp 119–48.

21. Reichlin S. Female fertility and the body fat connection (review). N Engl J Med 2203;348:869–70.

22. Moran LJ, Norman RJ. The obese patient with infertility: a practical approach to diagnosis and treatment. Nutr Clin Care 2002;5:290–7.

23. Steinkampf MP et al. Subcutaneous FSH administration in obese women. Fertil Steril 2003;80:99–102.

24. Zaadstra BM et al. Fat and female fecundity: prospective study of effect of body fat distribution on conception rates. BMJ 1993;306:484–7.

25. Obesity Trends, www.cdc.gov/nccdphp/dnpa/obesity/trend/prev_bmi.htm, accessed 12/20/03.

26. Nylander PPS. The phenomenon of twinning. In: Baron SL, Thomson AM, eds. Obstetrical epidemiology. London: Academic Press; 1983: pp 143–65.

27. Bates GW. Body weight control practice as a cause of infertility. Clin Obstet Gynecol 1985;28:632–44.

28. Keys A et al. The biology of human starvation. Minneapolis: University of Minnesota Press; 1950.

29. Cumming DC et al. Physical activity, nutrition, and reproduction. Ann NY Acad Sci 1994;709:55–76.

30. Lloyd T et al. Interrelationships of diet, athletic activity, menstrual status, and bone density in collegiate women. Am J Clin Nutr 1987;46:681–4.

31. Walberg JL et al. Menstrual function and eating behavior in female recreational weight lifters and competitive body builders. Med Sci Sports Exer 1991;23:30–6.

32. Oian P et al. Menstrual dysfunction in Norwegian top athletes. Acta Obstet Gynecol Scand 1984;63:693–7.

33. Warren MP et al. Hormone therapy and bone loss in dancers with amenorrhea. Fertil Steril 2003;80:398–404.

34. Kraemer RR et al. Follicular and luteal phase hormonal responses to low-volume resistive exercise. Med Sci Sports Exer 1995;27:809–17.

35. Ireland ML, Nattiv A. The female athlete. Philadelphia: W.B. Saunders, 2003.

36. Hill PB et al. Gonadotropin release and meat consumption in vegetarian women. Am J Clin Nutr 1986; 43:37–41.

37. Wyshak G, Snow RC. Fiber consumption and menstrual regularity in young women. J Women's Health 1993;2:295–9.

38. Reichman ME et al. Effect of dietary fat on length of the follicular phase of the men-

strual cycle in a controlled diet setting. J Clin Endocrinol Metab 1992;74:1171–5.

39. Feng W et al. Low follicular estrogen levels in New Zealand women consuming high fibre diets: a risk factor for osteopenia? NZ Med J 1993;106:319–22.

40. Kurzer M. Hormonal effects of soy in premenopausal women. J Nutr 2002; 132:570S–3S.

41. Kemmann E et al. Amenorrhea associated with carotenemia. JAMA 1983;249:926–9.

42. Wentz AC. Body weight and amenorrhea. Obstet Gynecol 1980;56:482–7.

43. Bolumar F et al. Caffeine intake and delayed conception: a European multicenter study on infertility and subfecundity. European study group on infertility subfecundity. Am J Epidemiol 1997;145:324–34.

44. Hatch EE, Bracken MB. Association of delayed conception with caffeine consumption. Am J Epidemiol 1993;138:1082–92.

45. Jensen J et al. The prevalence and etiology of impotence in 101 male hypertensive outpatients. Am J Hypertens 1999;12:271–5.

46. Hunt CD et al. Effects of dietary zinc depletion on seminal volume and zinc loss, serum testosterone concentrations, and sperm morphology in young men. American J Clin Nutr 1992; 56:148–57.

47. Lamb EJ, Bennett S. Epidemiologic studies of male factors in infertility. Ann NY Acad Sci 1994;709:165–78.

48. Benoff S et al. Blood lead levels and infertility in males. Hum Reprod 2003; 18:374–83.

49. Choy CM et al. Infertility, blood mercury concentrations and dietary seafood consumption: a case-control study. Brit J Obstet Gynecol 2002;109:1121–5.

50. Buck GM et al. Consumption of contaminated sport fish from Lake Ontario and time-to-pregnancy. New York State Angler Cohort. Am J Epidemiol 1997;146:949–54.

51. Rozata R et al. Role of environmental estrogens in the deterioration of male factor fertility. Fertil Steril 2002;78:1187–94.

52. Sokol RZ et al. Comparison of two methods for the measurement of sperm concentration. Fertil Steril 2000;73:591–4.

53. Foster PMD et al. Disruption of male reproductive development by antiandrogens. Presented at the 55th Annual Meeting of the American Association for Clinical Chemistry. Philadelphia, PA, July 22, 2003.

54. Joffe M. Time trends in biological fertility in Britain. Lancet 2000;355:1961–5.

55. Lloyd FH et al. Anabolic steroid abuse by body builders and male subfertility. BMJ 1996;313:100–1.

56. Vandenbroucke JP et al. Oral contraceptives and the risk of venous thrombosis. N Engl J Med 2001;344:1527–35.

57. Beral V et al. Oral contraceptives and the risk of cervical cancer. Lancet 2003; 361:1159–67.

58. Kovacs P. The risk of cardiovascular disease with second- and third-generation oral contraceptives. Medscape Women's Health eJournal 2002;7:1–5.

59. Crook D. The metabolic consequences of treating postmenopausal women with non-oral hormone replacement therapy. Br J Obstet Gynaecol 1997;104Suppl16:4–13.

60. Robinson JJ. Nutrition in the reproduction of farm animals. Nutr Res Rev 1990; 3:253–76.

61. Green TJ et al. Oral contraceptives did not affect biochemical folate indexes and homocysteine concentrations in adolescent females. J Am Diet Assoc 1998;98:49–55.

62. Berg G et al. Use of oral contraceptives and serum betacarotene. Eur J Clin Nutr 1997;51:181–7.

63. Sellerberg U et al. Women who take the pill should limit consumption of licorice. www.medscape.com, accessed 2/02.

64. Polaneczky M et al. Early experience with the contraceptive use of depot medroxyprogesterone acetate in an inner-city clinic population. Fam Plann Perspect 1996;28:174–8.

65. Matson SC et al. Physical findings and symptoms of depot medroxyprogesterone acetate use in adolescent females. J Pediatr Adolesc Gynecol 1997;10:18–23.

66. Pelkman C. Hormones and weight change. J Repro Med 2002;47 (suppl):791–4.

67. Wanichsetakul P et al. Bone mineral density at various anatomic sites in women receiving combined oral contraceptives and depot-medroxyprogesterone acetate for contraception. Contraception 2002;65:407–10.

68. Westhoff C. Depot medroxyprogesterone acetate contraception. Metabolic parameters and mood changes. J Reprod Med 1996; 41:401–6.

69. Sivin I et al. Prolonged effectiveness of Norplant® capsule implants: a 7-year study. Contraception 2000;61:187–94.

70. Meirik O, Farley TM, Sivin I. Safety and efficacy of levonorgestrel implant, intrauterine device, and sterilization. Obstet Gynecol 2001;97:539–47.

71. Kaunitz AM. Long-acting hormonal contraception: assessing impact on bone density, weight, and mood. Int J Fertil Womens Med 1999;44:110–7.

72. Haugen MM et al. Patient satisfaction with a levonorgestrel-releasing contraceptive implant. Reasons for and patterns of removal. J Reprod Med 1996;41:849–54.

73. Berenson AB et al. Contraceptive outcomes among adolescents prescribed Norplant implants versus oral contraceptives after one year of use. Am J Obstet Gynecol 1997;176:586–92.

74. Sicat BL. Ortha Evra, a new contraceptive patch. Pharmacotherapy 2003;23:472–80.

75. Turner L et al. Safety and efficacy of a male contraceptive. J Clin Endocrinol Meta 2003;88:4659–67.

76. Parker-Pope T. Doctors push new efforts to eliminate women's periods. Wall Street Journal 2003; June 25:D1.

77. National Academy of Sciences. Nutrition during pregnancy. I. Weight gain.

II. Nutrient supplements. Washington, DC: National Academy Press; 1990.

78. Eskes TK. Open or closed? A world of difference: a history of homocysteine research. Nutr Rev 1998;56:236–44.

79. Oakley G. Inertia on folic acid. BMJ 2003;326:1054.

80. Bailey LB et al. Folic acid supplements and fortification affect the risk for neural tube defects, vascular disease and cancer: Evolving science. J Nutr 2003;133:1961S–8S.

81. Cogswell ME et al. Iron supplement use among women in the United States: science, policy, and practice. J Nutr 2003;133: 1974S–7S.

82. Caan B et al. Benefits associated with WIC supplemental feeding during the interpregnancy interval. Am J Clin Nutr 1987; 45:29–41.

83. Jus'at I et al. Reaching young Indonesian women through marriage registries: an innovative approach for anemia control. J Nutr 2000;130:456S–8S.

84. Dixon D et al. The first prenatal visit. Clin Rv 2000;10:53–74.

"Women's nutritional status before conception influences physiologic events during pregnancy, and nutrition during pregnancy sets the stage for meeting nutritional needs during lactation."

J. C. King[1]

Chapter 3

Preconception Nutrition:
Conditions and Interventions

Chapter Outline

- Introduction
- Premenstrual Syndrome
- Obesity and Fertility
- Eating Disorders and Fertility
- Diabetes Mellitus prior to Pregnancy
- Polycystic Ovary Syndrome
- Inborn Errors of Metabolism
- Herbal Remedies for Fertility-Related Problems

Prepared by **Judith E. Brown**

Key Nutrition Concepts

1   Nutritional and health status directly before and during the first 2 months after conception influences embryonic development and the risk of complications during pregnancy.

2   Nutrition therapy plays an important role in the management of a number of conditions that affect preconceptional health and reproductive outcomes.

Introduction

This chapter addresses specific nutrition-related conditions of women before conception and during the *periconceptional period,* and the interventions that address them. Conditions presented here have important implications for health and well-being, and for reproductive outcomes. We begin with a discussion of premenstrual syndrome and progress to obesity, eating disorders, diabetes, polycystic ovary syndrome, inborn errors of metabolism, and celiac disease. Then we address the role of herbal remedies in the treatment of disorders of menstruation.

Premenstrual Syndrome

It wasn't until 1987 that PMS, or *premenstrual syndrome,* moved from the psychogenic disorder section of medical textbooks to chapters on physiologically based problems. It is diagnosed according to criteria stipulated in the fourth version of the *Diagnostic and Statistical Manual of Mental Disorders (DSM IV).* A standard questionnaire, rather than physical examination or laboratory tests, is used to diagnose PMS. Common physical signs and psychological symptoms of PMS are listed in Table 3.1. For a diagnosis, at least five signs or symptoms of PMS intense enough to disrupt work or social life must be present in three consecutive luteal phases.

PMS is characterized by life-disrupting physiological and psychological changes that begin in the luteal phase of the menstrual cycle and end with menses (menstrual bleeding). Some symptoms of PMS occur in about 40% of women of childbearing age, and they are severe enough to be considered PMS in about 5 to 10% of these women.[2,3] Although up to 70% of women suffer through monthly bouts of lower abdominal cramps, bloating, back pain, headache, food cravings, or

Periconceptional Period Around the time of conception, generally defined as the month before and the month after conception.

Premenstrual Syndrome (*premenstrual* = the period of time preceding menstrual bleeding; *syndrome* = a constellation of symptoms) A condition occurring among women of reproductive age that includes a group of physical, psychological, and behavioral symptoms with onset in the luteal phase and subsiding with menstrual bleeding. Also called premenstrual dysphoric disorder (PMDD).

Dysmenorrhea Painful menstruation due to abdominal cramps, back pain, headache, and/or other symptoms.

irritability, these conditions do not qualify as PMS. They are considered to represent *dysmenorrhea,* a condition related to prostaglandin release near to and during menses.[4]

Premenstrual dysphoric disorder (PDD) is a severe form of PMS. It is characterized by marked mood swings, depressed mood, irritability, anxiety, and physical symptoms (breast tenderness, headache, joint or muscle pain). PDD is diagnosed when three or more of these symptoms occur during at least two consecutive menstrual cycles. The treatment strategy for PDD is the same as that for PMS.[5]

The cause of PMS is unknown, but it is thought to be related to abnormal serotonin activity following ovulation.[2] Almost all remedies tested for PMS show about a 30% decline in symptoms with placebo, so an effective treatment must bring relief to an even higher proportion of women receiving it in experimental studies. Serotonin reuptake inhibitors, which are the active ingredient in some types of antidepressants, effectively reduce PMS symptoms. Decreased caffeine intake; exercise and stress reduction; magnesium, calcium, or vitamin B_6 supplements; and a number of herbal remedies are also used to treat PMS.[5]

Caffeine Intake and PMS

It is commonly recommended that women reduce their intake of coffee and other beverages high in caffeine to decrease PMS symptoms. This recommendation is holding up with time. A study at the University of Oregon demonstrated that PMS symptoms in college women increased in severity as coffee intake increased from 1 cup to 8 to 10 cups a day.[6] Risk of severe symptoms was eight times higher in women consuming the highest average daily amount of coffee (8 to 10 cups) compared to non-coffee drinkers.

Exercise and Stress Reduction

Increasing daily physical activity and reducing daily stressors appear to diminish PMS symptoms in many women.[7] Regular physical activity tends to improve energy level, mood, and feelings of well-being in women with PMS. Relieving stress—by such techniques as sitting comfortably and quietly with eyes closed while relaxing deep muscles, breathing through the nose, and exhaling while silently saying a word such as "one"—appears to decrease

Table 3.1 Common signs and symptoms of PMS[2,3]

PHYSICAL SIGNS	PSYCHOLOGICAL SYMPTOMS
• Fatigue	• Craving for sweet or salty foods
• Abdominal bloating	• Depression
• Swelling of the hands or feet	• Irritability
• Headache	• Mood swings
• Tender breasts	• Anxiety
• Nausea	• Social withdrawal

symptoms. When done for 15 to 20 minutes twice daily over 5 months, this exercise was associated with a 58% improvement in PMS symptoms.[7]

Magnesium, Calcium, and Vitamin B₆ Supplements and PMS Symptoms

Magnesium, calcium, and vitamin B₆ supplements appear to decrease symptoms of PMS in many women. Mechanisms underlying improvements are incompletely understood.

MAGNESIUM Magnesium supplements of 200 mg per day given during two cycles have been shown to decrease swelling, breast tenderness, and abdominal bloating symptoms of PMS. The beneficial response to magnesium was seen during the second month of treatment.[8] The 200-mg daily dose of magnesium is below the Tolerable Upper Intake Level (UL) for magnesium of 350 mg daily and is therefore considered safe.

CALCIUM Calcium supplements of 1200 mg per day for three cycles were found to reduce the PMS symptoms of irritability, depression, anxiety, headaches, and cramps by 48%, versus a reduction of 30% in the placebo group. The effect of calcium increased with duration of supplement use.[9] The UL for calcium is 2500 mg per day.

VITAMIN B₆ The effectiveness of vitamin B₆ in alleviating symptoms of PMS is a matter of controversy. Initial studies were promising; but later studies showed no improvements in symptoms compared to placebo.[10] To help straighten out the controversy, a meta-analysis of studies on the use of B₆ for PMS was undertaken. This study concluded that daily doses of 50–100 mg appear to reduce the severity of premenstrual depressive symptoms.[11] Because of the potential benefit, and the low risk of harm, vitamin B₆ supplements are sometimes used in clinical practice to diminish PMS symptoms. The UL for vitamin B₆ is 100 mg/day; doses recommended should not exceed this level.

Obesity and Fertility

Obesity in men and women impairs fertility. Obesity in men is associated with reduced *sex hormone binding globulin* (SHBG) levels and sperm count.[12] Many obese women experience highly irregular, anovulatory, or no menstrual cycles.[13] Obesity in women is also related to reduced SHBG level, and to increased estrogen, blood glucose, and insulin levels.[14]

Central Body Fat and Fertility

The presence of central body obesity, indicated by a waist circumference of 35 inches or greater, is a risk factor for impaired fertility.[15] In a study of women attending an artificial insemination clinic due to infertility in their partners, conception occurred in just half as many women with high central body fat as in other women.[16] In general, it takes women with high central body fat stores longer to become pregnant than it does women with low levels of central fat.[17]

Fat cells produce estrogen and leptin; the levels of both hormones are elevated by high levels of body fat and reduced by low levels. High and low levels of estrogen are related to abnormal menstrual cycles and infertility. Leptin, which was initially recognized as an appetite-regulating hormone, is now known to affect fertility by stimulating gonadotropin-releasing hormone (GnRH) neurons in the hypothalamus.[18] *Insulin resistance* also plays a role in fertility declines associated with high levels of abdominal body fat. High levels of abdominal fat are closely related to the presence of insulin resistance.[15] (You will be reading more about insulin resistance in this chapter's section on diabetes.)

Weight Loss and Fertility

Weight loss should be the first therapy option for men and women who are obese and infertile.[19] (Read about one woman's experience with weight loss and fertility in Case Study 3.1 on the next page.)

Studies of both women and men have demonstrated that weight loss of 7 to 22 pounds in women with BMIs over 25 kg/m², and of 100 pounds in massively obese men, are related to a return of fertility in most study participants.[13,14,20,21,22] Weight loss in the studies of obese women was accomplished by diet and exercise; the study of massively obese men used gastric bypass surgery. Weight loss produced a drop in testosterone and an increase in SHBG in males; it increased SHBG and decreased estrogen, glucose, and insulin levels in women.

Weight loss is considered the first therapeutic option for infertility in obese people, in part because it is less costly than medications and has many health benefits. This option is also recommended because hormone therapy often does not work in the presence of obesity.[23]

Eating Disorders and Fertility

Both *anorexia nervosa* and *bulimia nervosa* are related to menstrual irregularities and infertility.[24,25] These disorders

---

Sex Hormone Binding Globulin (SHBG) A protein that binds with the sex hormones testosterone and estrogen. Also called steroid hormone binding globulin, because testosterone and estrogen are produced from cholesterol and are thus considered to be steroid hormones. These hormones are inactive when bound to SHBG, but are available for use when needed. Low levels of SHBG are related to increased availability of testosterone and estrogen in the body.

Insulin Resistance A condition in which cells "resist" the action of insulin in facilitating the passage of glucose into cells.

Anorexia Nervosa (*anorexia* = poor appetite; *nervosa* = mental disorder) A disorder characterized by extreme underweight, malnutrition, amenorrhea, low bone density, irrational fear of weight gain, restricted food intake, hyperactivity, and disturbances in body image.

Bulimia Nervosa (*bulimia* = ox hunger) A disorder characterized by repeated bouts of uncontrolled, rapid ingestion of large quantities of food (binge eating) followed by self-induced vomiting, laxatives or diuretic use, fasting, or vigorous exercise in order to prevent weight gain. Binge eating is often followed by feelings of disgust and guilt. Menstrual cycle abnormalities may accompany this disorder.

Case Study 3.1

Photo Disc

Anna Marie's Tale

Exercise can be bad for you—or at least it is for Anna Marie. She and her husband Mark already have two delightful children, full-time jobs, and hectic schedules. Mark wants more children, but Anna Marie is dead set against it. Mark refuses to use contraception and has made Anna Marie promise not to use any, either. Anna Marie makes the promise because she thinks she can avoid pregnancy by staying at her weight of 210 pounds. At this weight, Anna Marie seldom has a menstrual period and figures the odds of conception are slim. For 2 years, Anna Maria's plan for avoiding conception has worked.

Now that the children are older, Anna Maria finds she has a bit of free time, which she uses to indulge her love of swimming. Within months of beginning her program of swimming regularly, however, Anna Marie abandons it. Her menstrual periods have become regular, and her contraception method is lost.

Anna Marie's weight at 210 pounds has remained stable during the months she has been swimming. It appears that her improved level of physical fitness and body fat has improved her fertility status, too.

Questions

1. What was likely the reason for Anna Marie's infertility when she was inactive?
2. Give an example of a hormonal change that may have occurred after Anna Marie began exercising regularly.
3. Name a possible health consequence related to Anna Marie's high weight and lack of menstrual cycles.

(Answers are in Appendix D.)

---

affect about 3 to 5% of young women, and likely twice that many have clinically important symptoms related to these eating disorders.[26] Amenorrhea is a cardinal manifestation of anorexia nervosa, and little bleeding during menses (oligomenorrhea) or amenorrhea may occur in women with bulimia nervosa. Amenorrhea in anorexia nervosa is related to irregular release of GnRH (described in Chapter 2) and very low levels of estrogen. Menses generally returns upon weight gain, but some cases of infertility persist even after normal weight is attained. This effect may be related to continued low levels of body fat, low dietary fat intake, excessive exercise, or other factors.[26]

Interventions for Women with Anorexia Nervosa or Bulimia Nervosa

The primary therapeutic goal for anorexia nervosa is normalization of body weight, and for bulimia nervosa, normalization of eating behaviors.[27] Recommended treatment for anorexia nervosa involves individual, family, or group therapy. Hospitalization may be required in severe cases. Certain psychotherapeutic medications are moderately effec-

tive for treating bulimia nervosa, but cognitive-behavioral therapy is the best, established approach.

Diabetes Mellitus prior to Pregnancy

> **Teratogenic** Exposures that produce malformations in embryos or fetuses.
>
> **Congenital Abnormality** A structural, functional, or metabolic abnormality present at birth. Also called congenital anomalies. These may be caused by environmental or genetic factors, or by a combination of the two. Structural abnormalities are generally referred to as congenital malformations, and metabolic abnormalities as inborn errors of metabolism.

Many women with diabetes mellitus are unaware that the disorder increases the risk of maternal and fetal complications, and they fail to get blood glucose under excellent control prior to conception.[28] High blood glucose levels during the first 2 months of pregnancy are *teratogenic;* they are associated with a two- to threefold increase in *congenital abnormalities* in newborns. Exposure to high blood glucose during the first 2 months in utero is related to malformations of the pelvis, central nervous system, and heart in newborns, as well as to higher rates of miscarriage.[29]

Management approaches to blood glucose control in diabetes depends, in part, on the type of diabetes present. Women may have **type 1 diabetes** (characterized by onset before the adult years) or **type 2 diabetes,** which most often occurs in adults. The most common form of diabetes is type 2—it accounts for 90–95% of all cases of diabetes mellitus.[30] Once thought of as a disease of older adults, type 2 diabetes is becoming increasingly common in young adults. Rates of type 2 diabetes are increasing in all age groups in the United States, but rates of increase are highest in adults in their thirties.[31] Approximately 4 in 5 U.S. adults with type 2 diabetes are overweight or obese.[32] People with type 1 diabetes do not produce any, or enough, insulin and must take insulin. People with type 2, on the other hand, produce insulin but do not utilize it well due to *insulin resistance.*

A Closer Look at Insulin Resistance

Insulin is a hormone produced by the beta cells of the pancreas. It performs a variety of functions, one of which is to increase the passage of glucose from blood into cells by attaching to and stimulating receptors on cell membranes. When insulin is bound to these receptors, enzymes are activated that open cell membrane "doors" to the passage of glucose. With insulin resistance, cells "resist" this process, and that lowers the amount of glucose that is transported into cells. The limited transport of blood glucose into cells triggers the pancreas to produce and release additional insulin. If the capacity of the pancreas to produce increasing quantities of insulin needed for glucose transport into cells is exceeded, blood glucose levels become and remain abnormally high.

The incidence of insulin resistance in adults and children is increasing in the United States and many other countries and is now considered to be a major public health problem.[31] It is estimated that about half of the cases of insulin resistance are related primarily to genetic factors, and half are related to behaviors. Most cases result from an interaction of genetic and environmental factors. Obesity (especially large stores of central body fat), physical inactivity, and small size at birth are related to insulin resistance. The condition is associated with the development of *polycystic ovary syndrome (PCOS), metabolic syndrome,* type 2 diabetes, *gestational diabetes,* and heart disease.[33]

Nutrition Interventions for Type 2 Diabetes

> "There is a substantial gap between recommended diabetes care and the care patients actually receive in the health care setting."[34]

Some people with type 2 diabetes can manage their glucose levels with diet and exercise, whereas others will need an oral medication that increases insulin production or sensitivity, or insulin to further boost glucose absorption

into cells. Some types of oral medications cannot be used during pregnancy, because they also increase insulin in the fetus and cause excessive fetal growth and fat gain.[29]

Individualized diet and exercise recommendations and an educational and follow-up program developed and implemented by registered dietitians, certified diabetes educators (CDE), physicians, and nurses are preferred for diabetes management. Carefully planned and monitored dietary recommendations are a major component of the management of type 2 diabetes.

Individual blood glucose levels vary a good deal in response to diet composition, so dietary prescriptions must be tailored for every person. Diets that provide 45–50% of total calories from carbohydrates, 15–20% of calories from protein, and 30–35% of calories from fat are often employed.[32] Diets developed for people with diabetes emphasize:

- Calorie reduction if weight is high
- Whole grain and high-fiber foods
- Five or more servings of vegetables and fruits each day
- Monounsaturated fats[35]
- Foods that minimize increases in blood glucose and insulin levels (or foods with a low glycemic index)[32]

Type 1 Diabetes A disease characterized by high blood glucose levels resulting from destruction of the insulin-producing cells of the pancreas. This type of diabetes was called juvenile-onset diabetes and insulin-dependent diabetes in the past.

Type 2 Diabetes A disease characterized by high blood glucose levels due to the body's inability to use insulin normally, or to produce enough insulin. In the past this type of diabetes was called adult-onset diabetes and non-insulin-dependent diabetes.

Polycystic Ovary Syndrome (PCOS) (*polycysts* = many cysts; i.e., abnormal sacs with membranous linings) A condition in females characterized by insulin resistance, high blood insulin and testosterone levels, obesity, menstrual dysfunction, amenorrhea, infertility, hirsutism (excess body hair), and acne.

Metabolic Syndrome A constellation of metabolic abnormalities that increase the risk of type 2 diabetes and heart disease. It is characterized by insulin resistance, abdominal obesity, high blood pressure and triglyceride levels, low levels of HDL cholesterol, and impaired glucose tolerance. Also called Syndrome X and insulin-resistance syndrome.

Gestational Diabetes Diabetes first discovered during pregnancy.

Glycemic Index (GI) A measure of the extent to which blood glucose levels are raised by a specific amount of carbohydrate-containing food compared to the same amount of glucose or white bread.

GLYCEMIC INDEX AND GLYCEMIC LOAD

Glycemic index (GI) is a measure of the extent to which 50 grams (about 1¾ ounce) of carbohydrate-containing foods raise 2-hour postprandial blood glucose levels compared to 50 grams of glucose or white bread. Not all expert committees on diabetes recommend the use of low-GI foods as a primary strategy in the dietary management of diabetes.[32,35] However, mounting evidence indicates that low-GI diets are beneficial for the control of blood glucose and insulin levels. Diets that provide low GI carbohydrates and approximately 50 grams of fiber daily are associated with reduced blood levels of glucose, insulin, and triglycerides versus lower-fiber (15 g/day), high-GI diets. (Table 5.7 in Chapter 5 lists the GI of a variety of foods.) Foods with high GI raise blood glucose and insulin levels more than do foods with low GI, and high-GI foods lead to more episodes of hyperglycemia (high blood glucose level) than do diets providing mainly low-GI carbohydrates. Long-term consumption of diets providing mainly high-GI carbohydrate foods appear to increase the risk of type 2 diabetes and coronary heart disease.[36] Although some people have trouble adhering to low-GI diets, they appear to be practical and acceptable in motivated adults with type 2 diabetes.[32]

The blood-glucose-raising effect of carbohydrate-containing foods also depends on the amount of food consumed. So, consuming a high-GI food such as instant mashed potatoes, in small amounts, will raise blood glucose level less than consuming a larger amount of this type of potato. *Glycemic load (GL)* was established to adjust GI values according to the amount of food consumed.

> **Glycemic Load (GL)**   A measure of the extent to which blood glucose levels are raised by a specific amount of carbohydrate-containing food. It is calculated by multiplying the carbohydrate content of an amount of food consumed by the glycemic index of the food, and dividing the result by 100.

The GL of a food is obtained by multiplying the GI of the food by the grams of carbohydrate contained in the amount of food consumed, and then dividing the result by 100. A blueberry muffin, for example, has a GI of 59. If a person ate a 1-ounce muffin containing 11 grams of carbohydrate, the GL would equal (59 × 11)/100, or 6.5. If she ate a 3-ounce muffin with 33 grams of carbohydrate, then the GL would be (59 × 33)/100, or 19.5. The higher the GL of a food, the greater the expected elevation in blood glucose and insulin due to consumption of the food.

Other Components of Nutrition Interventions for Type 2 Diabetes

Individuals with diabetes are at risk for heart disease due to abnormal blood lipid levels. Consequently, diets recommended include foods that improve blood lipid concentrations. Diets that help lower high-LDL cholesterol without lowering concentrations of beneficial HDL cholesterol, and that lower elevated triglyceride levels, follow these guidelines:

- Provide fat mainly in the form of monounsaturated fatty acids (vegetables, olive oil, peanuts and peanut oils, nuts, and seeds).
- Keep intake of saturated fats from animal products below 7% of total calories.
- Limit cholesterol intake to less than 300 mg per day.
- Minimize intake of trans fats from bakery products, fried foods, and snack foods.

Fish intake is often encouraged among people with diabetes. Consuming 2–3 servings of fish per week, or taking fish oil supplements, lowers blood triglycerides in people with elevated levels.[32] A sample menu that incorporates nutritional criteria for type 2 diabetes is given in Chapter 5.

Chromium plays critical roles in glucose and lipid metabolism through its potentiation of insulin action. Trial runs of supplementation with 500–1000 micrograms of chromium daily may be used to test its effects on glucose tolerance. Magnesium supplementation is indicated for people with diabetes who have low serum magnesium levels.[32]

Weight loss and control are often key components of nutrition intervention services for people with type 2 diabetes.

WEIGHT LOSS AND TYPE 2 DIABETES   Weight loss in overweight and obese individuals with diabetes lowers blood glucose levels and blood pressure, improves blood lipid concentrations, and increases insulin sensitivity. The most effective approach to weight loss in people with diabetes tested so far combines reduced caloric intake with exercise and behavioral therapy. Both aerobic and strength-building exercises are recommended. Behavioral therapy generally includes self-monitoring of body weight, diet, and physical activities; stress management; and problem-solving skill building. Selection of acceptable foods and dietary patterns, realistic goals for weight loss, enjoyable physical activities, and helpful behavioral therapies are critical to the long-term success of weight-control efforts. Diet drugs and surgery may be indicated for individuals who are unable to lose weight through behavioral changes.[37]

A concerted effort will be needed to translate new knowledge about diet and nutrition into improved treatment for and prevention of diabetes and its complications.[30]

Reducing the Risk of Type 2 Diabetes

Women who are overweight and have from borderline to high blood glucose levels (or *impaired glucose tolerance*) may be able to reduce their risk of developing type 2 diabetes during pregnancy. Modest levels of weight loss are consistently associated with reduced risk of type 2 diabetes.[38,39] One large study taking place over 3 years found that losses in body weight of about 7% and 150 minutes per week of exercise reduced the risk of developing type 2 diabetes by 50%.[39] Diets rich in whole grain and high-fiber foods are also protective and may facilitate weight loss.[40]

Whether weight loss and dietary changes before pregnancy in women at risk of type 2 diabetes reduce the chances that the disorder will develop during pregnancy is not known. However, improvements in body weight, dietary intake, and physical activity level would confer health benefits and are low-risk interventions. They should be considered for women who are overweight and have impaired glucose tolerance prior to pregnancy.

Polycystic Ovary Syndrome

Case scenario: Lupe is a 28-year-old woman who is 5 feet 3 inches tall and weighs 208 pounds. She and her husband want to start having a family, but her highly irregular periods and a failure to become pregnant as soon as desired have brought her to see her doctor. At the clinic it is determined that Lupe's waist circumference is 38 inches, and her body mass index (BMI) is 37 kg/m². Laboratory tests show that her blood level of insulin and triglycerides are high, and that she is not ovulating. Lupe is diagnosed as having *polycystic ovary syndrome,* or PCOS.

About 10% of women of reproductive age have PCOS, and it is a leading cause of infertility.[41] PCOS is not a disease but rather a syndrome that consists of a variety of clinical signs. Most women with PCOS are infertile due to the absence of ovulation, and they have menstrual irregularities. Characteristically, the outer layer of the ovaries of these women is thick and hard, and it may look yellowish. Due to changes in the ovaries, follicles are unable to develop and ovulation does not occur.[41,42]

Many women with PCOS are obese, and even in women with PCOS who are not obese, levels of intra-abdominal fat are usually high. Some women with PCOS have excess body hair (hirsutism), acne, and high blood levels of insulin, triglycerides, and androgens; and they have low levels of HDL cholesterol (Table 3.2). PCOS is sometimes difficult to diagnose (and may therefore not be treated) because signs and symptoms of the disorder vary considerably among individual women.[41]

The cause of PCOS is still debated, but it looks more and more as though insulin resistance is a pivotal factor in most cases. Less commonly, PCOS is caused by androgen-secreting tumors in the ovaries or adrenal gland, other disorders, and certain medications.[43] It appears that high blood levels of insulin stimulate the ovaries to produce androgens (such as testosterone), and excess androgens disrupt development of follicles.[43] High blood levels of androgens also lead to excess hair growth on the face and other parts of the body, while high insulin levels raise triglyceride and lower HDL-cholesterol levels.[41] PCOS appears to have a strong genetic component. Women whose relatives have type 2 diabetes, central obesity, high blood triglyceride levels, infertility, menstrual problems, and hirsutism are more likely to develop components of the syndrome than are other women.[41] Women with PCOS are at increased risk of gestational and type 2 diabetes and cardiovascular disease.[42]

Nutrition Interventions for Women with PCOS

> "Many women with PCOS have their symptoms ignored for years and have never received the proper diagnosis and treatment. She may have been told 'just lose weight!' and admonished when she found that very hard to do."
>
> Martha McKittrick, RD, CDN, CDE[41]

The primary goal in the treatment of PCOS is to increase insulin sensitivity. A number of insulin-sensitizing drugs, such as metaformin and rosiglitazone, can be used to lower blood insulin levels and reduce the excess production of androgens by the ovaries.[44] Other drugs may be used to stimulate ovulation.[42] For women with PCOS who are obese, weight loss and exercise are primary components of therapy. Weight loss and exercise improve insulin sensitivity, benefit blood lipids and insulin levels, and lower fasting glucose and testosterone levels in women with PCOS. However, women with PCOS have a more difficult time losing weight than the general population. Care must be taken to individualize eating and exercise plans if weight loss is to succeed. PCOS is a long-term health problem that requires a sustainable approach to weight loss and exercise. In addition, women may benefit from knowing more about PCOS, long-term health risks, and why weight loss and exercise are needed.[41]

Moderate and realistic weight-loss goals that consist of small and acceptable changes in behavior work best for the long run. Symptoms of PCOS often improve substantially in women who lose no more than 5–10% of their initial body weight.[41] The weight-loss plan negotiated with the woman should include types of foods that lower insulin need. Consequently, low-fat, high-carbohydrate diets may not be the best option, because they tend to increase insulin and triglyceride levels and lower HDL-cholesterol levels. Suggested diets for PCOS include 45–50% of calories from carbohydrate, 15–20% of calories from protein, and up to 35% of calories from fat. Low-GI sources of carbohydrates are generally employed for women with PCOS.[41] Foods with low GIs include whole grain cereals, fruits, vegetables, yogurt, and dried beans. These foods raise blood glucose and insulin levels to a lesser extent than do high-glycemic foods such as white bread and cereals made from refined grains.[45] In addition, consumption of whole grain foods such as naturally dark breads, high-fiber cold cereals, and cold or cooked whole grain cereals decreases insulin resistance.[46]

Table 3.2 Variation in clinical signs associated with PCOS[41,42]

CLINICAL SIGN	PERCENT OF WOMEN WITH PCOS AFFECTED
Insulin resistance	80%
Excess abdominal fat	80%*
Infertility	70%
Obesity	50–70%
Polycystic ovaries	67–86%
High triglycerides and low HDL-cholesterol levels	60%

*Preliminary estimate

Lean meat, fish, low-fat dairy products, dried beans, peanut butter, soy cheese, and egg whites or egg substitutes are good sources of protein for women with PCOS. Most of the fat in the diet should come from unsaturated fats such as olive oil, peanut oil, flaxseed oil, nuts, seeds, and fish. Intake of saturated fats from foods such as whole milk products, fatty meats, and butter should be limited to 5–10% of total calories. Trans fat intake from stick margarine, fried foods, snack foods and bakery products should be sharply reduced. A multiple-vitamin and mineral supplement providing RDA/AI levels of nutrients should be recommended if the diet is inadequate. Insulin-sensitizing drugs make weight loss a bit easier for some women.[41]

Identifying physical activities that the woman enjoys is a key to development of a successful, long-term exercise plan. The plan should include both aerobic and strength-building exercise to maximize increases in insulin sensitivity.

Counseling for women with PCOS should be supportive and convey a real understanding of the difficulties the woman has probably been through in getting appropriate advice and help in the past.[41]

Inborn Errors of Metabolism

Two inborn errors of metabolism that affect embryonic development or fertility are covered here: *PKU (phenylketonuria)* and *celiac disease.*

PKU (Phenylketonuria)

Phenylketonuria derives its name from the characteristic presence of phenylalanine in the urine of people with this condition. PKU is an inherited problem that causes elevation in blood phenylalanine levels due to low levels or lack of the enzyme phenylalanine hydroxylase. Lack of this enzyme diminishes the conversion of the essential amino acid phenylalanine to tyrosine, a nonessential amino acid, and causes phenylalanine to accumulate in blood. If present during very early pregnancy, high blood levels of phenylalanine impair normal central nervous system development of the embryo. Elevated phenylalanine levels in the first 8 weeks of pregnancy increase the risk of heart defects. The risk increases if high blood levels of phenylalanine are combined with low-protein diets early in pregnancy.[47] Untreated women with PKU have a 92% chance of delivering a newborn with mental retardation, and a 73% chance that the infant will be born with an abnormally small head (microcephaly).[48,49]

PKU (Phenylketonuria) An inherited error in phenylalanine metabolism most commonly caused by a deficiency of phenylalanine hydroxylase, which converts the essential amino acid phenylalanine to the nonessential amino acid tyrosine.

Celiac Disease Malabsorption with fatty stools (steatorrhea) due to an inherited sensitivity to the prolamin portion of gluten in wheat, rye, barley, and perhaps oats. It is often responsible for iron, folate, and zinc deficiencies. Also called celiac sprue and nontropical sprue.

People with undiagnosed PKU may self-select low-protein foods because meat and other rich sources of protein make them light-headed and easily confused. They may find it difficult to comprehend all the information they receive after the inborn error is diagnosed.[50]

Nutrition Intervention for Women with PKU

PKU can be successfully managed by a low-phenylalanine diet, instituted and monitored with the help of an experienced registered dietitian. This diet should be followed throughout life, but it is critical that it be adhered to prior to conception and maintained throughout pregnancy. It usually takes about 4 to 6 months to learn and implement the PKU diet and to lower blood phenylalanine levels.[50] Women who establish normal blood phenylalanine levels before pregnancy and maintain them throughout pregnancy tend to deliver infants with normal intelligence.[51] The later in pregnancy that control is achieved, the more severe mental retardation and other problems become.[48,49]

Celiac Disease

People with celiac disease, an inherited condition, are sensitive to the protein fraction prolamins found in gluten in wheat, rye, barley, and perhaps oats.[52] When gluten is consumed, the sensitivity causes malabsorption of fats and other food components and a flattening of the intestinal lining. Malabsorption can lead to a number of nutrient deficiencies and reduced growth.

Diagnosis of celiac disease is often delayed or missed. It is estimated that 1.3 million Americans have celiac disease, but only a small fraction have been diagnosed.[52] Symptoms of celiac disease tend to be present an average of 11 years before the condition is diagnosed.[53] Quality-of-life improvements are generally seen after the disease is identified and treated.[54]

Signs and symptoms of celiac disease range from mild to severe and vary by age and among individuals. They can include

- Chronic diarrhea
- Intermittent constipation
- Growth failure in children
- Irritability
- Bloating
- Weight loss
- Iron-deficiency anemia
- Nutrient deficiencies
- Infertility[55]

It has become clear that celiac disease is associated with infertility in females and males.[56] In men it can cause delayed sexual maturation and abnormal testosterone utilization. Infertility in females with the condition appears to be related to amenorrhea. Women with untreated celiac disease are at

Case Study 3.2

Celiac Disease

Chloe, age 30, has not had a period for over 2 years. Her gynecological exam turns up no abnormalities, but the hormones she is given to stimulate her menstrual periods do not work. Since the age of 10, Chloe has had painful stomach cramps, frequent diarrhea or constipation, and periodic iron-deficiency anemia. Multiple visits to doctors have failed to find the cause of Chloe's health problems. At around the age of 20, Chloe had begun to wonder if she was a medical anomaly or a hypochondriac.

Still bothered by her health problems and about to be married, Chloe seeks care again. This time she is seen by a nurse practitioner who has just read an article on celiac disease. The nurse sends Chloe to a registered dietitian, who advises Chloe on a gluten-free diet. After faithfully following the diet for a week, Chloe feels better. The cramps, diarrhea, and constipation are much improved, and later on, her menstrual cycles return. She returns to her doctor for a checkup and requests a test for celiac disease. By that time, however, her intestinal biopsy comes back normal because she has been on the diet for months.

Questions

1. What should have been the first clue that Chloe might have celiac disease?
2. What facts provide other clues to the possibility of celiac disease?
3. How long will Chloe have to stay on a gluten-free diet?

(Answers are in Appendix D.)

---

increased risk for miscarriage and fetal growth restriction during pregnancy.[57] Vitamin and mineral deficiencies that accompany untreated celiac disease may contribute to the risk of infertility and compromised pregnancy outcomes.[58] Not all people with celiac disease have overt symptoms, so it may be missed as an underlying cause of infertility. It should be considered in unexplained cases of infertility. If celiac disease is identified, fertility generally returns once gluten is removed from the diet and nutrient deficiencies are corrected.[56] Case Study 3.2 describes the experience of a woman ultimately diagnosed with celiac disease.

Nutrition Management of Women with Celiac Disease

The only treatment for celiac disease is to follow a lifelong, gluten-free diet.[54] Removing gluten from the diet is not as easy as it sounds. Gluten is found in a variety of non-grain foods including hot dogs, deli meats, bouillon, and salad dressing. Counseling by an experienced registered dietitian facilitates the adoption of a gluten-free diet that restores health. An example of a 1-day diet and snack option for an adult with celiac disease is shown in Table 3.3 on the next page. Gluten-free products are available in large grocery and health food stores. Internet resources for people with celiac disease are listed at the end of this chapter.

Herbal Remedies for Fertility-Related Problems

Herbal products and other dietary supplements are increasingly being used to treat fertility-related problems, but much remains to be learned about their safety and effectiveness. Recent evidence indicates that black cohosh, a member of the buttercup family, may reduce pain associated with menstrual periods. Because it may have estrogenic effects, black cohosh may increase the risk of breast cancer.[59] Bee propolis, which consists of plant resins collected by honeybees, appears to have anti-inflammatory activity and has been found to increase pregnancy rates in women with mild endometriosis. Use of the compound was not related to adverse side effects in one study.[60] Herbal extracts from the berries of *Vitex agnus-castus,* or chaste tree (named after the plant monks were said to have chewed to inspire chastity), show promise for safely relieving PMS symptoms. Side effects of *Vitex* include headache, cramps, and skin itching and rash. It is not yet

Table 3.3 Example of one day's diet and snack option for an adult with celiac disease

BREAKFAST	DINNER
Gluten-free bagel with nut butter	Lamb stew (thickened with potato starch)
Sliced bananas in yogurt	with carrots and lentils
Tea	Rice
	Gluten-free cake
	Low-fat milk

LUNCH	SNACK OPTIONS
Gluten-free pasta salad with chicken, broccoli, and tomatoes	Popcorn
Oil and vinegar dressing	Spring rolls with rice paper
Gluten-free roll with margarine	Ice cream
Fresh fruit	Fruit
Low-fat milk	Dark chocolate
	Gluten-free cookies
	String cheese
	Rice cakes

considered safe for women who may become or who are pregnant, or who are taking oral contraceptive pills.[61] Evening primrose oil, which contains high amounts of the essential fatty acids linoleic and alpha-linolenic acid, does not appear to beat a placebo in relation to PMS relief.[62]

Women and men using herbs for fertility problems should inform their health care providers and avoid using herbs if conception is possible.

Resources

Health Topics

High-quality information on eating disorders, PMS, and other conditions as presented in this chapter can be found here.
Web site: www.healthfinder.gov

Nutrition and Women's Health

U.S. Public Health Service site represents the pooled resources of several government agencies and provides updated reports on a wide assortment of nutrition and women's health topics.
Web site: www.4women.gov

PCOS Support

Chat rooms for women with PCOS include the following:
Web site: www.obgyn.net/PCOS/PCOS.asp
Web site: www.PCOSupport.org

Celiac Disease Resources

Support and other resources for people with celiac disease are available from the following Web sites:
The Celiac Disease Foundation: **www.celiac.org**

Celiac Sprue Association, USA: **www.csaceliacs.org**
Gluten Intolerance Group: **www.gluten.net**
Raising Our Celiac Kids: **www.celiackids.com**
Resource guide for gluten-free diets: **www.glutenfreediet.ca**

Diabetes Resources

Support and other resources related to diabetes are available from the following Web sites:
American Diabetes Association: **www.diabetes.org**
The National Institutes of Health: **www.niddk.nih.gov**
Canadian Diabetes Association: **www.diabetes.ca**

Herbal and Other Remedies

Check out the safety and effectiveness of complementary and alternative medical treatments for fertility-related problems at these sites.
Web site: **www.almed.od.nih.gov**
Web site: **http://dietary-supplements.info.nih.gov**

References

1. King, JC. Preface. Am J Clin Nutr 2000;71(suppl):1217S.

2. Freeman EW et al. Differential response to antidepressants in women with premenstrual syndrome/premenstrual dysphoric disorder: a randomized controlled trial. Arch Gen Psychiatry 1999;56:932–9.

3. Ugarriza DN et al. Premenstrual syndrome: diagnosis and intervention. Nurse Pract 1998;23:40, 45, 49–52.

4. Vaitukaitis JL. Premenstrual syndrome (editorial). N Engl J Med 1984;311:1371–3.

5. Grady-Weliky T. Premenstrual dysphoric disorder. N Engl J Med 2003;348:433–8.

6. Daugherty JE. Treatment strategies for premenstrual syndrome. Am Fam Physician 1998;58:183–92, 197–8.

7. Goodale et al. Alleviation of premenstrual syndrome symptoms with the relaxation response. Obstet Gynecol 1990; 75:649–55.

8. Walker AF et al. Magnesium supplementation alleviates premenstrual symptoms of fluid retention. J Women's Health 1998; 7:1157–65.

9. Thys-Jacobs S et al. Calcium carbonate and the premenstrual syndrome: effects on premenstrual and menstrual symptoms. Premenstrual Syndrome Study Group. Am J Obstet Gynecol 1998;179:444–52.

10. Berman MK et al. Vitamin B-6 in premenstrual syndrome. J Am Diet Assoc 1990;90:859–61.

11. Wyatt KM et al. Efficacy of vitamin B_6 in the treatment of premenstrual syndrome: a systematic review. BMJ 1999;318: 1275–81.

12. Kyung NH et al. Effect of carbohydrate supplementation on reproductive hormones during fasting in men. J Clin Endocrinol Metab 1985;60:827–35.

13. Mitchell GW, Rogers J. The influence of weight reduction on amenorrhea in obese women. N Engl J Med 1953;249:835–7.

14. Hollmann M, Runnebaum B, I, G. Effects of weight loss on the hormonal profile in obese, infertile women. Hum Reprod 1996;11:1884–91.

15. Hu FB. Overweight and obesity in women: health risks and consequences. J Women's Health 2003;12:163–72.

16. Zaadstra BM et al. Fat and female fecundity: prospective study of effect of body fat distribution on conception rates. BMJ 1993;306:484–7.

17. Kaye SA et al. Associations of body mass and fat distribution with sex hormone concentrations in postmenopausal women. International J Epidemiol 1990;131:794–803.

18. Reichlin S. Female fertility and the body fat connection (review). N Engl J Med 2003;348:869–70.

19. Moran LJ, Norman RJ. The obese patient with infertility: a practical approach to diagnosis and treatment. Nutr Clin Care 2003;5:290–7.

20. Pasquali R et al. Achievement of near-normal body weight as the prerequisite to normalize sex hormone-binding globulin concentrations in massively obese men. Int J Obes Relat Metab Disord 1997;21:1–5.

21. Clark AM et al. Weight loss in obese infertile women results in improvement in reproductive outcome for all forms of fertility treatment. Hum Reprod 1998;13:1502–5.

22. Clark AM et al. Weight loss results in significant improvement in pregnancy and ovulation rates in anovulatory obese women. Hum Reprod 1995;10:2705–12.

23. Steinkampf MP et al. Subcutaneous FSH administration in obese women. Fertil Steril 2003;80:99–102.

24. Robinson PH. Review article: recognition and treatment of eating disorders in primary and secondary care. Aliment Pharmacol Ther 2000;14:367–77.

25. Warren MP. Effects of undernutrition on reproductive function in the human. Endocrine Rev 1983;4:363–77.

26. Becker AE et al. Eating disorders. N Engl J Med 1999;340:1092–8.

27. Agras WS et al. A multicenter comparison of cognitive behavioral therapy and interpersonal psychotherapy for bulimia nervosa. Arch Gen Psychiatry 2000;57:459–66.

28. Lorber DL. Preconception counseling of women with diabetes. Pract Diabetol 1995:14;12–16.

29. Carr SR. Effect of maternal hyperglycemia on fetal development. American Dietetic Association Annual Meeting and Exhibition, Kansas City, MO, October 10, 1998.

30. Bowman BA, Vinicor F. The knowledge gap in diabetes. Nutr Clin Care 2003; 6:49–50.

31. Health, United States, 2003. www.cdc.gov/nchs.hus.htm, accessed 10/03.

32. Neff LM. Evidence-based dietary recommendations for patients with type 2 diabetes mellitus. Nutr Clin Care 2003;6:51–61.

33. Brunzell JD, Hokanson JE. Dyslipidemia of central obesity and insulin resistance. Diabetes Care 1999;22(suppl)3:C10–3.

34. Saaddine JB et al. A diabetes report card for the United States: quality of care in the 1990s. Ann Intern Med 2002;136:565–74.

35. American Diabetes Association. Evidence-based nutrition principles and recommendations for the treatment and prevention of diabetes and related complications: position statement. Diabetes Care 2003; 26:S51–S61.

36. Foster-Powel K et al. International table of glycemic index and glycemic load values: 2002. Am J Clin Nutr 2002;76:5–56.

37. Bedno SA. Weight loss in diabetes management. Nutr Clin Care 2003;6:62–72.

38. Tuomilehto J et al. Prevention of type 2 diabetes mellitus by changes in lifestyle among subjects with impaired glucose tolerance. N Engl J Med 2001;344:1343–50.

39. Diabetes Prevention Program Research Group. Reduction in the incidence of type 2 diabetes with lifestyle intervention or metformin. N Engl J Med 2002;346: 393–403.

40. McKeown N et al. Whole-grain intake in favorably associated with metabolic risk factors for type 2 diabetes and cardiovascular disease. Am J Clin Nutr 2002;76:390–9.

41. McKittrick M. Diet and polycystic ovary syndrome. Nutr Today 2002;37:63–69.

42. Kovacs P. Metabolic syndrome and PCOS. Medscape Ob/Gyn & Women's Health 2003;8:1–2.

43. Tolstoi LG, Josimovich JB. Weight loss and medication in polycystic ovary syndrome therapy. Nutr Today 2002;37:57–62.

44. Ghazeeri G et al. Effect of rosiglitazone on spontaneous and clomiphene citrate-induced ovulation in women with PCOS. Fertil Sterlit 2003;79:562–6.

45. Brand-Miller J et al. Low glycemic index foods and glycemic control in diabetics. Diabetes Care 2003;26:2261–7.

46. Liese AD et al. Whole grain consumption and insulin sensitivity. Am J Clin Nutr 2003;78:985–71.

47. Michals-Matalon K et al. Maternal phenylketonuria, low protein intake, and congenital heart defects. Am J Obstet Gynecol 2002;187:221–4.

48. Waisbren SE et al. Outcome at age 4 years in offspring of women with maternal phenylketonuria: the Maternal PKU Collaborative Study. Jama 2000;283:756–62.

49. Friedman EG et al. The International Collaborative Study on maternal phenylketonuria: organization, study design and description of the sample. Eur J Pediatr 1996;155(suppl)1:S158–61.

50. Isaacs, J. Metabolic dietitian. Personal communication, 12/10/03.

51. Acosta PB et al. Intake of major nutrients by women in the Maternal Phenylketonuria (MPKU) Study and effects on plasma phenylalanine concentrations. Am J Clin Nutr 2001;73:792–6.

52. Case S, Kaplan CR et al. The new celiac disease. Today's Dietitian 2003;Feb:39–43.

53. Green PHR et al. Characteristics of adult celiac disease in the USA: results of a national survey. Am J Gastroenterol 2001; 96:126–31.

54. Green PHR et al. Characteristics of adult celiac disease in the USA: results of a national survey. Am J Gastroent 2001; 96:126–31.

55. Scully RE et al. Presentation of a case of celiac disease. N Engl J Med 2001; 345:276–81.

56. Sher AV, Mantovani A. A risk factor for female fertility: celiac disease. Gynecol Endocrinol 2000;14:454–63.

57. Eliakim R, Sherer DM. Celiac disease: fertility and pregnancy. Gynecol Obstet Invest 2001;51:3–7.

58. Sher KS et al. Infertility, obstetric and gynecological problems in celiac sprue. Dig Dis 1994;12:186–90.

59. Thomas PR. Managing menopause naturally? Nutr Today 2003;38:191–7.

60. Hitt E. Bee propolis may improve infertility associated with mild endometriosis. American Society of Reproductive Medicine Annual Meeting, abstract 0-84, presented Oct. 13, 2003.

61. Vitex or chaste tree; it helps balance women's cycles. Environ Nutr 1999;22:8.

62. Collins A et al. Essential fatty acids in the treatment of premenstrual syndrome. Obstet Gynecol 1993;81:93–8.

"Everyone is kneaded out
of the same dough but not
baked in the same oven."
Yiddish proverb

Chapter 4

Nutrition during Pregnancy

Chapter Outline

- Introduction
- The Status of Pregnancy Outcomes
- Reducing Infant Mortality and Morbidity
- Physiology of Pregnancy
- Embryonic and Fetal Growth and Development
- Pregnancy Weight Gain
- Nutrition and the Course and Outcome of
  Pregnancy
- Healthy Diets for Pregnancy
- Exercise and Pregnancy Outcome
- Food Safety Issues during Pregnancy
- Common Health Problems during Pregnancy
- Model Nutrition Programs for Risk
  Reduction in Pregnancy

Prepared by **Judith E. Brown**

Key Nutrition Concepts

1   Many aspects of nutritional status, such as dietary intake, supplement use, and weight change, influence the course and outcome of pregnancy.

2   The fetus is not a parasite; it depends on the mother's nutrient intake to meet its nutritional needs.

3   Periods of rapid growth and development of fetal organs and tissues occur during specific times throughout pregnancy. Essential nutrients must be available in required amounts during these times for fetal growth and development to proceed optimally.

4   The risk of heart disease, diabetes, hypertension, and other health problems during adulthood may be influenced by maternal nutrition during pregnancy.

Introduction

The 9 months of pregnancy represent the most intense period of growth and development humans ever experience. How well these processes go depends on many factors, most of which are modifiable. Of the factors affecting fetal growth and development that are within our control to change, nutritional status stands out. At no other time in life are the benefits of optimal nutritional status more obvious than during pregnancy.

This chapter addresses the status of pregnancy outcomes in the United States and other countries. It covers physiological changes that take place to accommodate pregnancy, and the impact of these changes on maternal nutritional needs. The chapter presents the roles of nutrition in fostering fetal growth, development, and long-term health; it also covers dietary supplement use and weight-gain recommendations. The discussion goes on to consider common problems during pregnancy that can be addressed with nutritional remedies. We begin by highlighting vital statistic reports that clearly show a need for improving pregnancy outcomes in the United States.

The Status of Pregnancy Outcomes

The status of reproductive outcomes in the United States and other economically developed countries is routinely assessed through examination of a particular set of vital statistics data called *natality statistics* (*natality* means "related to birth"). Natality statistics summarize important information about the occurrence of pregnancy complications and harmful behaviors, in addition to infant mortality (death) and morbidity (illness) rates within a specific population. These data are used to identify problems in need of resolution and to identify progress in meeting national goals for improvement in the course and outcome of pregnancy.

Illustration 4.1 presents a time line of key intervals and events before, during, and after pregnancy. Specific time points and periods are labeled. Table 4.1 shows terms characterizing and changing rates of different natality statistics. They are frequently referred to in this chapter. Also referred to often in this chapter are weights in grams (g) and kilograms (kg), as well as in pounds and ounces. There are 448 grams in a pound and 2.2 pounds in a kilogram. (FYI: a table of measurement abbreviations and equivalents is located inside the back cover of this textbook.)

Infant Mortality

"Infant mortality is a mirror of a population's physical health and socioeconomic status."

Infant mortality reflects the general health status of a population to a considerable degree because so many of the envi-

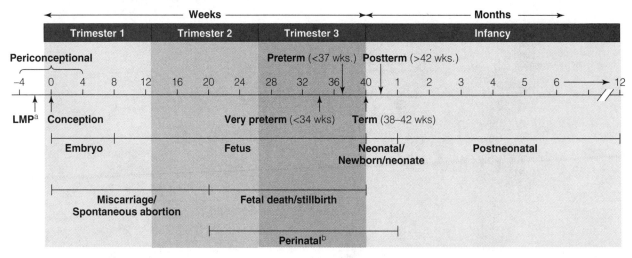

aLMP = last menstrual period

bPerinatal definition varies from 20 to 24 weeks gestation to 7 to 28 days after birth.

Illustration 4.1 Time-related terms before, during, and after pregnancy.

Table 4.1 Natality statistics: Rates, definitions, and trends in the rates in the United States[1]

| | RATES | | |
	1995	2001	Definition
Maternal mortality	7.1	7.1	Deaths/100,000 live births
Fetal deaths (stillbirths)	7.0	6.6	Deaths/1000 pregnancies over 20 weeks gestation
Perinatal mortality	7.6	7.0	Deaths/1000 deliveries over 20 weeks gestation to 7 days after birth
Neonatal mortality	4.9	4.6	Deaths from delivery to 28 days/1000 live births
Postneonatal mortality	2.7	2.3	Deaths from 28 days after birth to 1 year/1000 live births
Infant mortality	7.6	6.8	Deaths from birth to age 1 year/1000 live births
Preterm	11.0	11.8	Births <37 weeks gestation/100 live births
Very preterm	1.9	1.9	Births <34 weeks gestation/100 live births
Low birthweight	7.3	7.8	Newborn weights <2500 g (5 lb 8 oz)/100 live births
Very low birthweight	1.4	1.5	Newborn weights <1500 g (3 lb 4 oz)/100 live births
Multifetal pregnancies			
Twins	1 in 40	1 in 34	Number of twin births/total live births
Triplets+	1 in 784	1 in 555	Number of triplets plus higher-order multiple births/total live births
Adolescent pregnancies	56.8	45.9	Births /1000 females aged 15 to 19 years

ronmental factors that affect the health of pregnant women and newborns also affect the health of the rest of the population.[2] The graph presented in Illustration 4.2 demonstrates this point. This historical view of infant mortality rates indicates that population-wide improvements in social circumstances, infectious disease control, and availability of safe and nutritious foods have corresponded to greater reductions in infant mortality than have technological advances in medical care.[2] Small improvements in infant mortality in the past few decades in the United States are largely due to technological advances in medical care that save ill newborns. High-level medical care has not favorably affected the need for such care, however. The United States spends more money on health care than any other nation, yet it ranks 27th in the international comparison of infant mortality rates (Table 4.2 on the next page).

Infant mortality reflects deaths during the first year of life. However, two-thirds of deaths to *liveborn infants* occur within the first month after birth, or during the neonatal period. Deaths in the first month

Liveborn Infant The World Health Organization developed a standard definition of *liveborn* to be used by all countries when assessing an infant's status at birth. By this definition, a liveborn infant is the outcome of delivery when a completely expelled or extracted fetus breathes, or shows any sign of life such as beating of the heart, pulsation of the umbilical cord, or definite movement of voluntary muscles, whether or not the cord has been cut or the placenta is still attached.

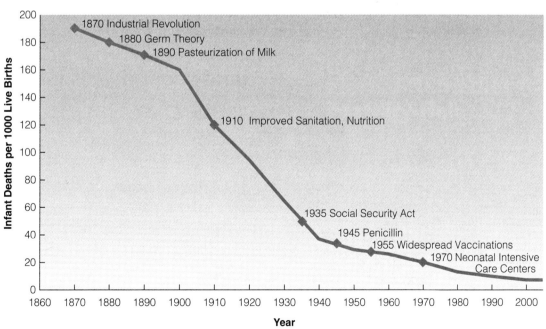

Illustration 4.2 Chronology of events related to declines in infant mortality in the United States.

SOURCE: Judith E. Brown, 2001.

Table 4.2 A partial listing of the International rankings of infant mortality rates in 1999[3]

COUNTRY	INFANT MORTALITY RATE
Hong Kong	3.1
Japan	3.2
Sweden	3.4
Singapore	3.5
Norway	3.9
Denmark	4.2
Finland	4.2
Austria	4.4
France	4.4
Germany	4.5
Czech Republic	4.6
Switzerland	4.6
Spain	4.7
Netherlands	5.2
Canada	5.3
Belgium	5.5
Greece	5.5
Ireland	5.5
Australia	5.6
Portugal	5.6
England and Wales	5.8
Israel	5.8
Italy	6.0
New Zealand	6.1
Scotland	6.2
Cuba	6.4
United States	7.0

Low Birthweight, Preterm Delivery, and Infant Mortality

Infants born low birthweight or preterm are at substantially higher risk of dying in the first year of life than are larger and older newborns. Low-birthweight infants, for example, make up 7.8% of all births, yet comprise 66% of all infant deaths. The 11.8% of newborns delivered prior to 37 weeks of pregnancy similarly account for a disproportionately large number of infant deaths.[3] Low-birthweight and preterm infant outcomes are intertwined in that the shorter the pregnancy, the less newborns tend to weigh. Table 4.3 shows increases in birthweight with the duration of pregnancy and birthweight-specific infant mortality rates.

Rates of low birthweight in the United States have trended slowly upward since 1983 and have remained approximately two times higher in African American infants than in other infants. The high rate of low birthweight in African American infants clearly represents a problem in need of resolution.

Reducing Infant Mortality and Morbidity

Deaths and illnesses associated with low-birthweight and preterm infants can be reduced through improvements in the birthweight of newborns. Infants weighing 3500 to 4500 grams at birth (or 7 lb 12 oz to 10 lb) are least likely to die within the first year of life, as well as in the perinatal, neonatal, and postneonatal periods.[4] Newborns weighing 3500 to 4500 grams are also at an advantage as a group in relation to overall health status and subsequent mental development. They are less likely to develop heart disease, diabetes, lung disease, and hypertension later in life.[5] Reducing the proportion of infants born small or early would clearly decrease infant mortality.

after birth and fetal deaths (or stillbirths) result largely from health problems that develop in the mother or fetus during pregnancy. Together they constitute the 11th leading cause of death in the United States.[1]

Table 4.3 Range of birthweights by gestational age, U.S.[1]

BIRTHWEIGHT		WEEKS GESTATION	INFANT MORTALITY RATE
Pounds (lb) and Ounces (oz)	Grams		
<1 lb 2 oz	<500	<22	846
1 lb 2 oz–2 lb 3 oz	500–999	22–27	316
2 lb 3 oz–3 lb 5 oz	1000–1499	27–29	62
3 lb 5 oz–4 lb 6 oz	1500–1999	29–31	28
4 lb 6 oz–5 lb 8 oz	2000–2499	31–33	12
5 lb 8 oz–6 lb 10 oz	2500–2999	33–36	4.6
6 lb 10 oz–7 lb 11 oz	3000–3499	36–40	2.4
7 lb 11 oz–8 lb 13 oz	3500–3999	40+	1.7
8 lb 13 oz–9 lb 14 oz	4000–4499	40+	1.5
9 lb 14 oz–11 lb	4500–4999	40+	2.5
>11 lb	5000+	40+	—

Table 4.4 Health objectives for the nation related to pregnant women and infants

- Reduce anemia among low-income pregnant females in their third trimester from 29 to 20%.
- Reduce infant mortality from 7.6 to no more than 5 per 1000 live births.
- Reduce the incidence of spina bifida and other neural tube defects from 7 to 3 per 10,000 live births.
- Reduce low birthweight (<2500 grams) from 7.3 to 5%.
- Reduce preterm births (<37 weeks) from 9.1 to 7.6%.
- Increase abstinence from alcohol use by pregnant women from 79 to 95%.
- Reduce the incidence of fetal alcohol syndrome.
- Increase the proportion of women who gain weight appropriately during pregnancy.

Health Objectives for the Year 2010

National health objectives for pregnant women and newborns focus on the reduction of low birthweight, preterm delivery, and infant mortality. A number of the objectives are related to improvements in nutritional status (Table 4.4).

Physiology of Pregnancy

Conception triggers thousands of complex and sequenced biological changes that transform two united cells into a member of the next generation of human beings. The rapidity with which structures and functions develop in mother and fetus and the time-critical nature of energy and nutrient needs make maternal nutritional status a key element of successful reproduction.

Pregnancy begins at conception; that occurs approximately 14 days before a woman's next menstrual period is scheduled to begin. Assessed from conception, pregnancy averages 38 weeks, or 266 days, in length. Most commonly, however, pregnancy duration is given as 40 weeks (280 days) because it is measured from the date of the first day of the last menstrual period (LMP). Consequently, the common way of measuring pregnancy duration includes two nonpregnant weeks at the beginning. The anticipated date of delivery is denoted by the ancient terminology of "estimated date of confinement," or EDC. Assessment of duration of pregnancy as weeks from conception is correctly termed *gestational age,* whereas time in pregnancy estimated from LMP reflects *menstrual age.* It is particularly important to get these terms straight during early fetal development, when a 2-week error in duration of pregnancy

may mean miscalculating the timing of nutrient-related events in pregnancy.

Maternal Physiology

Changes in maternal physiology during pregnancy are so profound that they were previously considered abnormal and in need of correction. Doctors routinely advised pregnant women to follow low-sodium diets to reduce fluid retention; restricted their patients' weight gain and dietary intake to prevent complications at delivery; and prescribed excessive levels of iron and other supplements to bring blood nutrient levels back up to "normal." We now know that what is considered normal physiological status of nonpregnant women cannot be considered normal for women who are pregnant. Fortunately, it is now understood that attempts to bring maternal physiological changes back to nonpregnant levels may cause more harm than good to the pregnancy.

Changes in maternal body composition and functions occur in a specific sequence during pregnancy. The order of the sequence is absolute because the successful completion of each change depends on the one before it. Because maternal physiological changes set the stage for fetal growth and development, they begin in earnest within a week after conception.[6]

The sequence of physiological changes taking place during pregnancy is listed in Table 4.5. The table indicates the timing of maximal rates of change in maternal tissues, the *placenta,* and fetal weight across pregnancy. To provide the fetus with sufficient energy, nutrients, and oxygen for growth, the mother must first expand the volume of plasma that can be circulated. Maternal nutrient stores are accumulated next. These stores are established in advance of the time they will be needed to support large gains in fetal weight. Similarly, the maximal rate of placental growth is timed to precede that of fetal weight gain. This sequence of events ensures that the placenta is fully prepared for the high level of functioning that will be needed as fetal weight increases most rapidly. Fetuses depend on the functioning

Placenta A disk-shaped organ of nutrient and gas interchange between mother and fetus. At term, the placenta weighs about 15% of the weight of the fetus.

Table 4.5 Sequence of tissue development and approximate gestational week of maximal rates of change in maternal systems, the placenta, and fetus during pregnancy[6]

Tissue	Sequence of Development	Gestational Week of Maximal Rate of Growth
Maternal plasma volume	1	20
Maternal nutrient stores	2	20
Placental weight	3	31
Uterine blood flow	4	37
Fetal weight	5	37

Table 4.6 Summary of maternal anabolic and catabolic phases of pregnancy[7,8,11]

Maternal Anabolic Phase 0–20 Weeks	Maternal Catabolic Phase 20+ Weeks
Blood volume expansion, increased cardiac output	Mobilization of fat and nutrient stores
Buildup of fat, nutrient, and liver glycogen stores	Increased production and blood levels of glucose, triglycerides, and fatty acids; decreased liver glycogen stores
Growth of some maternal organs	Accelerated fasting metabolism
Increased appetite, food intake (positive caloric balance)	Increased appetite and food intake decline somewhat near term
Decreased exercise tolerance	Increased exercise tolerance
Increased levels of anabolic hormones	Increased levels of catabolic hormones

of multiple systems, established well in advance of their maximal rates of growth and development. Abnormalities in the development of any of these physiological systems can modify fetal growth and development.

Normal Physiological Changes during Pregnancy

Physiological changes in pregnancy can be divided into two basic groups: those occurring in the first half of pregnancy and those in the second half. In general, physiological changes in the first half are considered "maternal anabolic" changes because they build the capacity of the mother's body to deliver relatively large quantities of blood, oxygen, and nutrients to the fetus in the second half of pregnancy. The second half is a time of "maternal catabolic" changes in which energy and nutrient stores, and the heightened capacity to deliver stored energy and nutrients to the fetus, predominate (Table 4.6). Approximately 10% of fetal growth is accomplished in the first half of pregnancy, and the remaining 90% occurs in the second half.[8]

The list of physiological changes that normally occur during pregnancy is extensive (Table 4.7), and such changes affect every maternal organ and system. Changes that are most directly related to maternal energy and nutrient needs are discussed further.

Edema Swelling (usually of the legs and feet, but can also extend throughout the body) due to an accumulation of extracellular fluid.

Steroid Hormones Hormones such as progesterone, estrogen, and testosterone produced primarily from cholesterol.

BODY WATER CHANGES A woman's body gains a good deal of water during pregnancy, primarily due to increased volumes of plasma and extracellular fluid, as well as amniotic fluid.[18] Total body water increases in pregnancy range from 7 to 10 liters (approximately 7 to 10 quarts, or about 2 to 2½ gallons). About two-thirds of the expansion is intracellular (blood and body tissues) and one-third is extracellular (fluid in spaces between cells).[6] Plasma volume begins to increase within a few weeks after conception and reaches a maximum at approximately 34 weeks. Early pregnancy surges in plasma volume appear to be the primary reason that pregnant women feel tired and become exhausted easily when undertaking exercise performed routinely prior to pregnancy. Fatigue associated with plasma volume increases in the second and third months of pregnancy declines as other compensatory physiological adjustments are made.

Gains in body water vary a good deal among women during normal pregnancy. High gains are associated with increasing degrees of *edema* and weight gain. If not accompanied by hypertension, edema generally reflects a healthy expansion of plasma volume. Birthweight is strongly related to plasma volume: generally, the greater the expansion, the greater the newborn size.[6] The increased volume of water in the blood is responsible for the "dilution effect" of pregnancy on blood concentrations of some vitamins and minerals. Blood levels of fat-soluble vitamins tend to increase in pregnancy, whereas levels of the water-soluble vitamins tend to decrease. Vitamin supplement use can modify these relationships.[9]

HORMONAL CHANGES Many physiological changes in pregnancy are modulated by hormones produced by the placenta (Table 4.8 and Illustration 4.3 on page 80). The placenta serves many roles, but a key one is the production of *steroid hormones,* such as progesterone and estrogen. The placenta is also the main supplier of many other hormones needed to support the physiological changes of pregnancy.

MATERNAL NUTRIENT METABOLISM Adjustments in maternal nutrient metabolism are apparent within the first few weeks after conception and progress throughout pregnancy.[8] The changes are sequenced and interrelated in complex ways that are appreciated but not yet fully understood.

Carbohydrate Metabolism Many adjustments in carbohydrate metabolism are made during pregnancy that promote the availability of glucose to the fetus. Glucose is the fetus's preferred fuel, even though fats can be utilized for energy. Continued availability of a fetal supply of glucose is accomplished primarily through metabolic changes that promote maternal insulin resistance. These changes, sometimes referred to as the *diabetogenic effect of preg-*

Table 4.7 Normal changes in maternal physiology during pregnancy[6,7]

Blood Volume Expansion
- Blood volume increases 20%
- Plasma volume increases 50%
- Edema (occurs in 60–75% of women)

Hemodilution
- Concentrations of most vitamins and minerals in blood decrease

Blood Lipid Levels
- Increased concentrations of cholesterol, LDL cholesterol, triglycerides, HDL cholesterol

Blood Glucose Levels
- Increased insulin resistance (increased plasma levels of glucose and insulin)

Maternal Organ and Tissue Enlargement
- Heart, thyroid, liver, kidneys, uterus, breasts, adipose tissue

Circulatory System
- Increased cardiac output through increased heart rate and stroke volume (30–50%)
- Increased heart rate (16% or 6 beats/min)
- Decreased blood pressure in the first half of pregnancy (−9%), followed by a return to nonpregnancy levels in the second half

Respiratory System
- Increased tidal volume, or the amount of air inhaled and exhaled (30–40%)
- Increased oxygen consumption (10%)

Food Intake
- Increased appetite and food intake; weight gain
- Taste and odor changes, modification in preference for some foods
- Increased thirst

Gastrointestinal Changes
- Relaxed gastrointestinal tract muscle tone
- Increased gastric and intestinal transit time
- Nausea (70%), vomiting (40%)
- Heartburn
- Constipation

Kidney Changes
- Increased glomerular filtration rate (50–60%)
- Increased sodium conservation
- Increased nutrient spillage into urine; protein is conserved
- Increased risk of urinary tract infection

Immune System
- Suppressed immunity
- Increased risk of urinary and reproductive tract infection

Basal metabolism
- Increased basal metabolic rate in second half of pregnancy
- Increased body temperature

Hormones
- Placental secretions of large amounts of hormones needed to support physiological changes of pregnancy

nancy, make normal pregnant women slightly carbohydrate intolerant in the third trimester of pregnancy.[12] Illustration 4.4 on the next page provides an example of the normal levels of plasma glucose and insulin during late pregnancy compared to prepregnancy levels.

Carbohydrate metabolism in the first half of pregnancy is characterized by estrogen- and progesterone-stimulated increases in insulin production and conversion of glucose to glycogen and fat. In the second half, rising levels of hCS and prolactin from the mother's pituitary gland inhibit the conversion of glucose to glycogen and fat.[11] At the same time, insulin resistance builds in the mother, increasing her reliance on fats for energy. Decreased conversion of glucose to glycogen and fat, lowered maternal utilization of glucose, and increased liver production of glucose help to ensure that a constant supply of glucose for fetal growth and development is available in the second half of pregnancy.

Fasting maternal blood glucose levels decline in the third trimester due to increased utilization of glucose by the rapidly growing fetus. However, post-meal blood glucose concentrations are elevated and remain higher longer than before pregnancy.[12]

Accelerated Fasting Metabolism Maternal metabolism is rapidly converted toward *glucogenic amino acid* utilization, fat oxidation, and increased production of *ketones* with fasts that last longer than 12 hours. Decreased levels of plasma glucose and insulin, and increased levels of triglycerides, free fatty acids, and ketones are seen hours before they occur in nonpregnant fasting women. The rapid conversion to fasting metabolism allows pregnant women to use primarily stored fat for energy while sparing glucose and amino acids for fetal use.[12]

Although these metabolic adaptations help ensure a constant fetal supply of glucose, fasting eventually increases the dependence of the fetus on ketone bodies for

Glucogenic Amino Acids Amino acids such as alanine and glutamate that can be converted to glucose.

Ketones Metabolic by-products of the breakdown of fatty acids in energy formation. β-hydroxybutyric acid, acetoacetic acid, and acetone are the major ketones, or "ketone bodies."

Table 4.8 Key placental hormones and examples of their roles in pregnancy[10,11]

Human chorionic gonadotropin (hCG): maintains early pregnancy by stimulating the corpus luteum to produce estrogen and progesterone. It stimulates growth of the endometrium. The placenta produces estrogen and progesterone after the first two months of pregnancy.

Progesterone: maintains the implant; stimulates growth of the endometrium and its secretion of nutrients; relaxes smooth muscles of the uterine blood vessels and gastrointestinal tract; stimulates breast development, promotes lipid deposition.

Estrogen: increases lipid formation and storage, protein synthesis, and uterine blood flow; prompts uterine and breast duct development; promotes ligament flexibility.

Human chorionic somatotropin (hCS): increases maternal insulin resistance to maintain glucose availability for fetal use; promotes protein synthesis and the breakdown of fat for energy for maternal use.

Leptin: may participate in the regulation of appetite and lipid metabolism, weight gain, and utilization of fat stores.

new maternal and fetal tissues. It is estimated that 925 grams (2 pounds) of protein is accumulated during pregnancy.[14] To some extent the increased need for protein is met through reduced levels of nitrogen excretion and the conservation of amino acids for protein tissue synthesis. There is no evidence, however, that the mother's body stores protein early in pregnancy in order to meet fetal needs for protein later in pregnancy. Maternal and fetal needs for protein are primarily fulfilled by the mother's intake of protein during pregnancy.[8]

Fat Metabolism Multiple changes occur in the body's utilization of fats during pregnancy. Overall, changes in lipid metabolism promote the accumulation of maternal fat stores in the first half of pregnancy and enhance fat mobilization in the second half.[12] In addition to seeing increasing maternal reliance on fat stores for energy as pregnancy progresses, we see blood levels of many lipoproteins increase dramatically (Table 4.9). Plasma triglyceride levels increase first and most dramatically, reaching three times nonpregnant levels by term.[8,15] Cholesterol-containing lipoproteins, phospholipids, and fatty acids also increase, but to a lesser extent than do triglycerides. The increased cholesterol supply is used by the placenta for steroid hormone synthesis, and by the fetus for nerve and cell membrane formation.[12] Small increases in HDL cholesterol in pregnancy appear to decline within a year postpartum and

energy. Prolonged fetal utilization of ketones, such as occurs in women with poorly controlled diabetes or in those who lose weight during part or all of pregnancy, is associated with reduced growth and impaired intellectual development of the offspring.[13]

Protein Metabolism Nitrogen and protein are needed in increased amounts during pregnancy for synthesis of

Illustration 4.3 Changes in maternal plasma concentration of hormones during pregnancy.

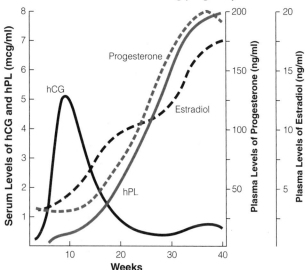

SOURCE: From Pedro Rosso. "Nutrition and Metabolism in Pregnancy: Mother and Fetus." Copyright © 1990 by Oxford University Press, Inc. Used by permission of Oxford University Press, Inc.

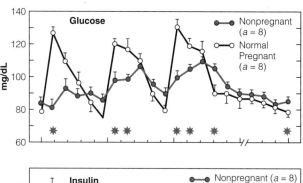

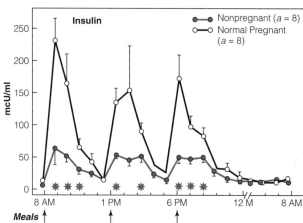

Illustration 4.4 Plasma glucose and insulin levels in nonpregnant women and in women near term.

SOURCE: Reprinted from American Journal of Obstetrics and Gynecology 140(6): 730–736. R. L. Phelps, et al., © 1981, with permission from Elsevier.

Table 4.9 Changes in cholesterol and triglyceride levels during pregnancy[15,16]

Trimester	Cholesterol mmol/L	Triglycerides mmol/L
1	5.78	1.19
2	6.88	1.32
3	8.14	2.58
Nonpregnant	5.11	0.80

*To convert mmol/L to mg/dL, see the Measurement Abbreviations and Equivalents table located inside the back cover of the textbook.

remain lower than prepregnancy levels. It is speculated that declines in HDL cholesterol after pregnancy may contribute to an increased risk of heart disease in women. Other changes in serum lipids appear to revert to prepregnancy levels postpartum.[16]

By the third trimester of pregnancy most women have a lipid profile that would be considered atherogenic, if not for pregnancy. These blood lipid changes are normal, however, which is why blood lipid screening is not recommended during pregnancy.[16] Normal changes in blood lipid levels during pregnancy appear to be unrelated to maternal dietary intake.[17]

Mineral Metabolism   Impressive changes in mineral metabolism occur during pregnancy. Calcium metabolism is characterized by an increased rate of bone turnover and reformation.[18] Elevated levels of body water and tissue synthesis during pregnancy are accompanied by increased requirements for sodium and other minerals. Sodium metabolism is delicately balanced during pregnancy to promote an accumulation of sodium by the mother, placenta, and fetus. This is accomplished by changes in the kidneys that increase aldosterone secretion and the retention of sodium. This normal change in pregnancy renders attempts to prevent and treat high blood pressure in pregnancy by reducing sodium intake ineffective and potentially harmful. Sodium restriction may overstress mechanisms that act to conserve sodium and lead to functional and growth impairments due to sodium depletion.[19]

The Placenta

The word *placenta* is derived from the Latin word for *cake*. The placenta, with its round, disklike shape, looks somewhat like a cake—the more so the more active the imagination. The placenta develops from embryonic tissue and is larger than the fetus for most of pregnancy. Development of the placenta precedes fetal development.

Functions of the placenta include hormone and enzyme production, nutrient and gas exchange between the mother and fetus, and removal of waste products from the fetus. Its structure, including a double lining of cells

separating maternal and fetal blood, acts as a barrier to some harmful compounds, and it governs the rate of passage of nutrients and other substances into and out of the fetal circulation (Illustration 4.5). The barrier role of the placenta is better described as a fence than as a filter that guards the fetus against all things harmful. Many potentially harmful substances (alcohol, excessive levels of some vitamins, drugs, and certain viruses, for example) do pass through the placenta to the fetus. The placenta is a barrier to the passage of maternal red blood cells, bacteria, and many large proteins. The placenta also prevents the mixing of fetal and maternal blood until delivery, when ruptures in blood vessels may occur.

NUTRIENT TRANSFER   The placenta uses 30–40% of the glucose delivered by the maternal circulation. If nutrient supply is low, the placenta fulfills its needs before nutrients are made available to the fetus. If nutrient supplies fall short of meeting placental needs, functioning of the placenta is compromised to sustain the nutrient supply and health of the mother.[20]

Nutrient transfer across the placenta depends on a number of factors, including

- The size and the charge of molecules available for transport
- Lipid solubility of the particles being transported
- The concentration of nutrients in maternal and fetal blood

Small molecules with little or no charge (water, for example) and lipids (cholesterol and ketones, for instance) pass through the placenta most easily, while large molecules (e.g., insulin and enzymes) aren't transferred at all. Nutrient exchange between the mother and fetus is unregulated

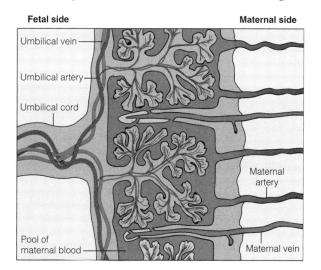

Illustration 4.5 Structure of the placenta. Maternal arteries and veins are part of the maternal circulation, whereas umbilical arteries and veins are part of the fetal circulation. Blood enters the fetus through umbilical veins and exits through umbilical arteries.

Table 4.10 Mechanisms of nutrient transport across the placenta[11,21]

MECHANISM	EXAMPLES OF NUTRIENTS
Passive diffusion (also called *simple diffusion*) Nutrients transferred from blood with higher concentration levels to blood with lower concentration levels	Water, some amino acids and glucose, free fatty acids, ketones, vitamins E and K,[a] some minerals (sodium, chloride), gases
Facilitated diffusion Receptors ("carriers") on cell membranes increase the rate of nutrient transfer	Some glucose, iron, vitamins A and D
Active transport Energy (from ATP) and cell membrane receptors	Water-soluble vitamins, some minerals (calcium, zinc, iron, potassium) and amino acids required for transfer
Endocytosis (also called *pinocytosis*) Nutrients and other molecules are engulfed by placenta membrane and released into fetal blood supply	Immunoglobulins, albumin

[a]Vitamin K crosses the placenta slowly and to a limited degree.

for some nutrients, oxygen, and carbon dioxide; it is highly regulated for other nutrients. Nutrient transfer based on concentration gradients determined by the levels of the nutrient in the maternal and the fetal blood is unregulated. In these cases, nutrients cross placenta membranes by simple diffusion from blood with high concentration of the nutrient to blood with lower concentration.

Three primary mechanisms regulate nutrient transfer: facilitated diffusion, active transport, and endocytosis (or pinocytosis). Table 4.10 summarizes mechanisms of nutrient transfer across the placenta and provides examples of nutrients transported by each specific mechanism as they are known.

The fetus receives small amounts of water and other nutrients from ingestion of *amniotic fluid.* By the second half of pregnancy the fetus is able to swallow and absorb water, minerals, nitrogenous waste products, and other substances in amniotic fluid.

Amniotic Fluid The fluid contained in the amniotic sac that surrounds the fetus in the uterus.

Growth Increase in an organism's size through cell multiplication (hyperplasia) and enlargement of cell size (hypertrophy).

Development Progression of the physical and mental capabilities of an organism through growth and differentiation of organs and tissues, and integration of functions.

THE FETUS IS NOT A PARASITE   The fetus is not a "parasite"—it cannot take whatever nutrients it needs from the mother's body at the mother's expense. When maternal nutrient intakes fall below optimum levels or adjustment thresholds, fetal growth and development are compromised more than maternal health.[22] In general, nutrients will first be used to support maternal nutrient needs for her health and physiological changes, and next

for placental development, before they become available at optimal levels to the fetus. For example:

- Underweight women gaining the same amount of weight as normal-weight women tend to deliver smaller infants and to retain more of the weight gained during pregnancy at the expense of fetal growth.[11]
- Fetal growth tends to be reduced in pregnant teenagers who gain height during pregnancy compared to fetal growth in teens who do not grow during pregnancy.[23]
- Vitamin and mineral deficiencies and toxicities in newborns have been observed in women who showed no signs of deficiency or toxicity diseases during pregnancy.[11]

If the fetus did act as a parasite, it would harm the mother for its own benefit. Rather, the fetus is generally harmed more by poor maternal nutritional status than is the mother.[21]

Embryonic and Fetal Growth and Development

8-40 weeks

The rate of human *growth* and *development* is higher during gestation than at any time thereafter. If the rate of weight gain achieved in the 9 months of gestation continued after delivery, infants would weigh about 160 pounds at their first birthdays and be 20 feet tall by age 20! Table 4.11 provides an overview of embryonic and fetal growth and development during pregnancy.

Table 4.11 Notes on normal embryonic and fetal growth and development[11,24]

Day 1	Conception; one cell called the zygote exists.	Week 9	Embryo now considered a fetus.
Day 2–3	Eight cells have formed (called the morula) and enter the uterine cavity.	Month 3	Weighs an ounce; primitive egg and sperm cells developed, hard palate fuses, breathes in amniotic fluid.
Day 6–8	The morula becomes fluid filled and is renamed the blastocyst. The blastocyst is comprised of 250 cells, and cell differentaion begins.	Month 4	Weighs about 6 ounces; placenta diameter is 3 inches.
Day 10	Embryo implants into the uterine wall, where glycogen is accumulating.	Month 5	Weighs about a pound, 11 inches long; skeleton begins to calcify, hair grows.
Day 12	Embryo is composed of thousands of cells; differentiation well under way. Utero placental ciculation being formed.	Month 6	14 inches long; fat accumulation begins, permanent teeth buds form; lungs, gastrointestinal tract, and kidneys formed but are not fully functional.
Week 4 (21–28 days)	¼ inch long; rudimentary head, trunk, arms; heart "practices" beating; spinal cord and two major brain lobes present.		
Week 5 (28–35 days)	Rudimentary kidney, liver, circulatory system, eyes, ears, mouth, hands, arms, and gastrointestinal tract; heart beats 65 times per minute circulating its own, newly formed blood.	Month 7	Gains ½ to 1 ounce per day.
Week 7 (49–56 days)	½ inch long, weighs 2–3 grams; brain sends impulses, gastrointestinal tract produces enzymes, kidney eliminates some waste products, liver produces red blood cells, muscles work. (Approximately 25% of blastocysts and embryos will be lost before 7 weeks.)	Months 8 and 9	Gains about an ounce per day; stores fat, glycogen, iron, folate, B_6 and B_{12}, riboflavin, calcium, magnesium, vitamins A, E, D; functions of organs continue to develop. Growth rate declines near term. Placenta weighs 500–650 g (1–1½ lbs) at term.

Critical Periods of Growth and Development

Fetal growth and development proceed along genetically determined pathways in which cells are programmed to multiply, *differentiate,* and establish long-term functional levels during set time intervals. Such time intervals are known as *critical periods* and are most intense during the first 2 months after conception, when a majority of organs and tissues form. On the whole, critical periods represent a "one-way street," because it is not possible to reverse directions and correct errors in growth or development that occurred during a previous critical period. Consequently, adverse effects of nutritional and other insults occurring during critical periods of growth and development persist throughout life.[25]

HYPERPLASIA Critical periods of growth and development are characterized by hyperplasia, or an increase in cell multiplication. Because every human cell has a specific amount of DNA, periods of hyperplasia can be determined by noting times during gestation when the DNA content of specific organs and tissues increases sharply. The critical period of rapid cell multiplication of the forebrain, for example, is between 10 and 20 weeks of gestation (Illustration 4.6 on the following page).

The brain is the first organ that develops in humans, and along with the rest of the central nervous system, it is

Differentiation Cellular acquisition of one or more characteristics or functions different from that of the original cells.

Critical Periods Preprogrammed time periods during embryonic and fetal development when specific cells, organs, and tissues are formed and integrated, or functional levels established. Also called *sensitive periods.*

Illustration 4.6 The critical period of cell multiplication of the forebrain. Growth in cell numbers is indicated by increases in DNA content of a given amount of tissue.

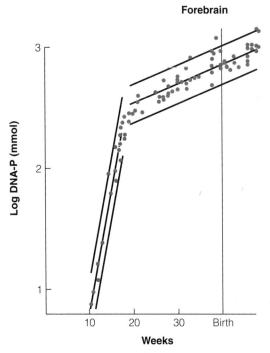

SOURCE: From "Quantitative Growth and Development of Human Brain" by J. Dobbing and J. Sands, in Archives of Disease of Children, 48(10):757–767. © 1973 BMJ Publishing Group. Reprinted with permission.

given priority access to energy, nutrient, and oxygen supplies. Thus, in conditions of low energy, nutrient, and oxygen availability, the needs of the central nervous system will be met before those of other fetal tissues such as the liver or muscles. The heart and adrenal glands come next after the central nervous system in the hierarchy of targets for preferential nutrient delivery.[20]

Deficits or excesses in nutrients supplied to the embryo and fetus during critical periods of cell multiplication can produce lifelong defects in organ and tissue structure and function. The organ or tissue undergoing critical periods of growth at the time of the adverse exposure will be affected most.[26] For example, the neural tube develops into the brain and spinal cord during weeks 3 and 4 after conception. If folate supplies are inadequate during this critical period of growth, permanent defects in brain or spinal cord formation occur, regardless of folate availability at other times. Other tissues—such as the pancreas, which does not undergo rapid cell multiplication until the third trimester of pregnancy—do not appear to be affected by the early shortage of folate.

Some degree of hyperplasia takes place in a number of organs and tissues in the first year or two after birth and during the adolescent growth spurt. Cells of the central nervous system, for instance, continue to multiply for about two years after birth, but at a much slower pace than early in pregnancy. Skeletal and muscle cells increase in number during the adolescent growth spurt.[27]

In utero and early life changes in DNA content of the brain have been investigated in fetuses, infants, and young children dying from non-nutritional causes and from undernutrition. Illustration 4.7 presents results from one such study that show deficits in DNA content (or cell number) in the brains of children dying of protein-energy malnutrition versus those dying from accidents. Deficits in DNA were apparent a few months after birth, indicating that severe malnutrition early in pregnancy reduced brain cell number in utero.[28]

HYPERPLASIA AND HYPERTROPHY   Cell multiplication continues at a lower rate after critical periods of cell multiplication and is accompanied by increases in the size of cells. This phase of growth can be seen in Illustration 4.6, where it begins around 20 weeks in the forebrain when the rate of increase in DNA content slows. Cell size increases mainly due to an accumulation of protein and lipids inside of cells. Consequently, increases in cell size can be determined by measuring the protein or lipid content of cells. Specialized functions of cells, such as production of digestive enzymes by cells within the small intestine or neurotransmitters by nerve cells, occur along with increases in cell number and size.[11]

HYPERTROPHY   Periods of hyperplasia-hypertrophy are followed by hypertrophy only. During this phase cells continue to accumulate protein and lipids, and functional levels continue to grow in sophistication, but cells no longer multiply. Reductions in cell size caused by unfavorable nutrient environments or other conditions are associated with deficits in organ and tissue functions, such as reduced mental capabilities or declines in muscular coordination. Such functional changes can often be reduced or reversed later if deficits are corrected.[25]

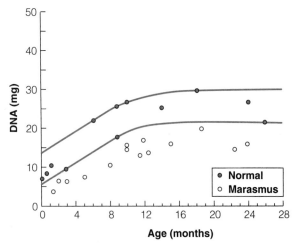

Illustration 4.7 DNA content of the cerebellum of the human brain in young children dying from non-nutritional causes and from undernutrition.

SOURCE: Reprinted from M. Winick, "Malnutrition and Brain Development," Journal of Pediatrics 74(6):667–679, © 1969, with permission from Elsevier.

MATURATION The last phase of growth and development is maturation—the stabilization of cell number and size. This phase occurs after tissues and organs are fully developed later in life.

Fetal Body Composition

The fetus undergoes marked changes in body composition during pregnancy (Table 4.12). The general trend is toward progressive increases in fat, protein, and mineral content. Some of the most drastic changes take place in the last 5 weeks of pregnancy, when fat and mineral content increase substantially.

Variation in Fetal Growth

Given a healthy mother and fetal access to needed amounts of energy, nutrients, and oxygen and freedom from toxins, fetal genetic growth potential is achieved.[30] However, as evidenced by the relatively high rate of low birthweight in the United States, optimal conditions required for achievement of genetic growth potential often do not exist during pregnancy. Variations in fetal growth and development are not generally due to genetic causes but rather to environmental factors such as energy, nutrient, and oxygen availability, and to conditions that interfere with genetically programmed growth and development.[20] Insulin-like growth factor-1 (IGF-1) is the primary growth stimulator of the fetus. It promotes uptake of nutrients by the fetus and inhibits fetal tissue breakdown. Levels of IGF-1 are sensitive to maternal nutrition; it is decreased by undernutrition. Low levels of IGF-1 decrease muscle and skeletal mass and produce asymmetrical growth.[20] Factors such as prepregnancy underweight and shortness, low weight gain during pregnancy, poor dietary intakes, smoking, drug abuse, and certain clinical complications of pregnancy are associated with reduced fetal growth.[31]

Risk of illness and death varies substantially with size at birth and is particularly high for newborns experiencing intrauterine growth retardation (IUGR).[26] For a portion of newborns, smallness at birth is normal and may reflect familial genetic traits. Because IUGR is complicated to determine, it is usually approximated by assessment of size for gestational age using a reference standard (Table 4.13 on the next page). Infants are generally considered likely to have experienced intrauterine growth retardation if their weight for gestational age or length is low. Newborns whose weight is less than the 10th percentile for gestational age are considered *small for gestational age,* or SGA. This determination is further categorized into

disproportionately small for gestational age (dSGA) and *proportionately small for gestational age* (pSGA). Newborns who weigh less than the 10th percentile of weight for gestational age but have normal length and head circumference for age are considered dSGA. If weight, length, and head circumference are less than the 10th percentile for gestational age, then the newborn is considered pSGA.[30] Approximately two-thirds of SGA newborns in the United States are disproportionately small, and one-third are proportionately small.[30]

dSGA Infants who are disproportionately small for gestational age look skinny, wasted, and wrinkly. They tend to have small abdominal circumferences, reflecting a lack of glycogen stores in the liver, and little body fat. It appears that these infants have experienced in utero malnutrition in the third trimester of pregnancy and that it compromised liver glycogen and fat storage. Short-term episodes of malnutrition, such as maternal weight loss or low weight gain late in pregnancy that compromise energy, nutrient, or oxygen availability appear to be related to dSGA.[30] These infants generally have smaller organ sizes but the normal number of cells in organs and tissues. Infants who are dSGA are at risk of developing the "hypos" after birth (hypoglycemia, hypocalcemia, hypomagnesiumenia, and hypothermia). If the period of maternal undernutrition was short, dSGA infants can experience good catch-up growth with nutritional rehabilitation.[20] Unfortunately, disproportionately small infants appear to be at greater risk for heart disease, hypertension, and type 2 diabetes in the adult years.[26] (We return to this issue later in this chapter.)

> **Small for Gestational Age (SGA)** Newborn weight is ≤10th percentile for gestational age. Also called *small for date* (SFD).
>
> **Disproportionately Small for Gestational Age (dSGA)** Newborn weight is ≤10th percentile of weight for gestational age; length and head circumference are normal. Also called *asymmetrical SGA.*
>
> **Proportionately Small for Gestational Age (pSGA)** Newborn weight, length, and head circumference are ≤10th percentile for gestational age. Also called *symmetrical SGA.*

Table 4.12 Estimated changes in body composition of the fetus by time in pregnancy[11,29]

Component	10 Weeks	20 Weeks	30 Weeks	40 Weeks
Body weight, g	10	300	1667	3450
Water, g	<9	263	1364	700
Protein, g	<1	22	134	446
Fat, g	<1	26	66	525
Sodium, meq	<1	32	136	243
Potassium, meq	<1	12	75	170
Calcium, g	<1	1	10	28
Magnesium, mg	<1	5	31	76
Iron, mg	<1	17	104	278
Zinc, mg	<1	6	26	53

Table 4.13 Percentiles of weight for newborn gestational age

Gestational Age (wk)	5th Pctl	10th Pctl	50th Pctl	90th Pctl	95th Pctl
20	249	275	412	772	912
21	280	314	433	790	957
22	330	376	496	826	1023
23	385	440	582	882	1107
24	435	498	674	977	1223
25	480	558	779	1138	1397
26	529	625	899	1362	1640
27	591	702	1035	1635	1927
28	670	798	1196	1977	2237
29	772	925	1394	2361	2553
30	910	1085	1637	2710	2847
31	1088	1278	1918	2986	3108
32	1294	1495	2203	3200	3338
33	1513	1725	2458	3370	3536
34	1735	1950	2667	3502	3697
35	1950	2159	2831	3596	3812
36	2156	2354	2974	3668	3888
37	2357	2541	3117	3755	3956
38	2543	2714	3263	3867	4027
39	2685	2852	3400	3980	4107
40	2761	2929	3495	4060	4185
41	2777	2948	3527	4094	4217
42	2764	2935	3522	4098	4213
43	2741	2907	3505	4096	4178
44	2724	2885	3491	4096	4122

Pctl = percentile

SOURCE: From Obstetrics and Gynecology, Vol. 87, No. 2, 1996, pp. 163–168, table 2. Copyright © 1996. Reprinted by permission of Lippincott, Williams & Wilkins.

out life than do infants born *appropriate for gestational age* (AGA) or *large for gestational age* (LGA).[32]

The goal of nutritional rehabilitation for pSGA infants should be catch-up in weight and length, and not just weight. This goal appears to be easier to reach if pSGA infants are breastfed. Excessive weight gain by pSGA infants appears to increase the risk of obesity and insulin-resistance-related disorders, such as hypertension and type 2 diabetes, later in life.[33]

LGA Newborns with weights greater than the 90th percentile for gestational age are considered to be large for gestational age. About 1–2% of U.S. newborns are LGA. Although it is difficult to predict LGA, it appears to be related to prepregnancy obesity, poorly controlled diabetes in pregnancy, excessive weight gain in pregnancy (over 44 pounds), and possibly other factors.

Except for infants born to women with poorly controlled diabetes during pregnancy or other health problems, LGA newborns experience far lower illness and death rates than do SGA infants, and tend to be taller later in life.[34] Delivery and postpartum complications in mothers, however, tend to be higher with LGA newborns, and include increased rates of operative delivery, *shoulder dystocia,* and postpartum hemorrhage.[35]

pSGA Proportionately SGA newborns look small but well proportioned. It is believed that these infants have experienced long-term malnutrition in utero, due to factors such as prepregnancy underweight, consistently low rates of maternal weight gain in pregnancy and the corresponding inadequate dietary intake, or chronic exposure to alcohol.[11]

Because nutritional insults existed during critical periods of growth early in pregnancy, pSGA infants generally have a reduced number of cells in organs and tissues. These babies tend to exhibit fewer health problems at birth than do dSGA infants, but catch-up growth is poorer, even with nutritional rehabilitation.[42] On average, pSGA infants remain shorter and lighter, and have smaller head circumferences through-

Appropriate for Gestational Age (AGA) Weight, length, and head circumference are between the 10th and 90th percentiles for gestational age.

Large for Gestational Age (LGA) Weight for gestational age exceeds the 90th percentile for gestational age. Also defined as birthweight greater than 4500 g (≥10 lb) and referred to as *excessively sized for gestational age,* or *macrosomic.*

Shoulder Dystocia Blockage or difficulty of delivery due to obstruction of the birth canal by the infant's shoulders.

PONDERAL INDEX Increasingly, ponderal index (PI) is being used to assess the appropriateness of newborn size. Like body mass index, it is a measure of weight-for-length, but it is calculated by dividing weight by the cube of length (PI = weight in grams/cm^3 × 1000). Values between approximately 23 and 25 reflect normal weight-for-length; lower values represent thinness and higher values heaviness at birth. Wider adoption of PI to assess newborn size in practice awaits development of standard values indicative of increased health risk related to thinness and heaviness at birth.

Nutrition, Miscarriages, and Preterm Delivery

Several other pregnancy outcomes are related in part to maternal nutrition. Highlighted here are the roles played by nutrition in miscarriage and preterm delivery.

MISCARRIAGES Approximately 30% of implanted embryos are lost by reabsorption into the uterus or expulsion before 20 weeks of conception. Roughly a third of these losses are recognized as a miscarriage.[36] Such early losses of embryos and fetuses are thought to be primarily caused by genetic, uterine, or hormonal abnormalities, reproductive tract infections, or to tissue rejection due to immune system disorders.[37] Intakes of caffeine over 500 mg per day (or the equivalent of about 4 cups of brewed coffee), especially in women with nausea and vomiting of pregnancy, sharply increase the risk of miscarriage.[38] The presence of nausea and vomiting is otherwise related to a very low risk of miscarriage.[39]

PRETERM DELIVERY Infants born preterm are at greater risk than other infants of death, neurological problems reflected later in low IQ scores, congenital malformations, and chronic health problems such as *cerebral palsy*. The risk for these outcomes increases rapidly as gestational age at birth decreases. Infants born very preterm (<34 weeks) commonly have problems related to growth, digestion, respiration, and other conditions due to immaturity.[40] Low stores of fat, essential fatty acids, glycogen, and other nutrients in very preterm infants may also interfere with growth and health after delivery. Additionally, breast-milk content of riboflavin and vitamins A, C, and B_{12} may be low in women who have inadequate intake of these vitamins during the third trimester of pregnancy.[41,42,43]

Although preterm delivery is a major health problem in the United States, its etiology remains unclear, and the search for effective prevention programs continues. A portion of preterm deliveries appears to be related to genital tract infections, insufficient uterine-placental blood flow, placental abruption (bleeding into the uterus), prepregnancy underweight, low weight gain in pregnancy, short interpregnancy interval (<6 months), and high levels of psychological or social stress. It is also fairly common in women who have previously delivered preterm.[44,45] Improvements in prenatal care for women at risk of preterm delivery—such as close supervision of the pregnancy, inclusion of nutritional counseling, encouragement of adequate weight gain in underweight and normal-weight women, and home visits—appear to decrease the risk of preterm delivery somewhat. Calcium, essential fatty acids, and fish oil supplements have been used to prevent preterm labor, but results of studies on their effectiveness are mixed.[44]

The Fetal-Origins Hypothesis of Chronic Disease Risk

"The implications of the associations between fetal nutrition and adult disease are immense, and if substantiated, demand intense scrutiny of current prenatal nutrition policies."

J. King, 2000[46]

In the last decade, thinking about chronic disease risk has changed substantially. In contrast to the earlier idea that disease risk begins during childhood or in the adult years,

Table 4.14 Diseases and other conditions in adults related to smallness or thinness at birth[26,47]

Allergies	Mood disorders
Autoimmune diseases	Obesity
Bronchitis	Ovarian cancer
Cardiovascular disease	Polycystic ovary syndrome
Decreased bone mineral content	Schizophrenia
	Short stature
Gestational diabetes	Stroke
Hypertension	Subfertility in males
Kidney disease	Suicide
Metabolic syndrome	Type 2 diabetes

studies testing the *fetal-origins hypothesis* indicate that risks may begin in utero. The concept that chronic disease risk may be established in utero is strongly supported by animal studies and by epidemiological investigations in humans.[5] Much of the evidence that relates in utero exposures to later disease risk in humans comes from studies showing increased risk for diseases such as heart disease, hypertension, type 2 diabetes, gestational diabetes, and chronic bronchitis in small, short, and thin newborns (Table 4.14). Because newborn size is a primary indicator, it is generally thought that maternal nutrition may underlie the development of later disease risk.[26]

Cerebral Palsy A group of disorders characterized by impaired muscle activity and coordination present at birth or developed during early childhood.

Fetal-Origins Hypothesis The theory that exposures to adverse nutritional and other conditions during critical or sensitive periods of growth and development can permanently affect body structures and functions. Such changes may predispose individuals to cardiovascular diseases, type 2 diabetes, hypertension, and other disorders later in life. Also called *metabolic programming* and the *Barker Hypothesis*.

Relatively small reductions in newborn weight or disproportions in size have been related to increased later disease risk. Risk of cardiovascular disease (heart disease and stroke), for instance, is associated with birthweights below 7.5 pounds (3360 g)—weights that are often considered "normal." Results of the U.S. Nurses Study, which compared newborn birthweight to risk of cardiovascular disease in adults aged 46 to 71 years, are provided in Table 4.15 on the next page to illustrate this point. Infants at risk for later chronic disease include those born at weights below that genetically programmed, even if birthweights are considered "normal."[47] Additionally, disease risk in people born small, short, or thin may be exacerbated by later, excessive weight gain.[48]

How Is Disease Risk Increased in Utero?

Given less than optimal growing conditions during gestation, fetal tissues make adaptations to cope with energy and nutrient shortages and excesses, and with harmful substances. Although such adaptations may support in utero survival, they may increase the risk of disease later

Table 4.15 Association of birthweight with the risk of cardiovascular disease in the U.S. Nurses Study[49]

BIRTHWEIGHT	RELATIVE RISK OF:	
	Heart Disease	Stroke
<5 lb (2240 g)	1.5	2.3
5–5½ lb (2240–2500 g)	1.3	1.4
5½–7 lb (2500–3136 g)	1.1	1.3
7–8½ lb (3136–3808 g)	1.0	1.0
8½–10 lb (3808–4480 g)	1.0	1.0
>10 lb (>4480 g)	0.7	0.7

in life. Adaptations in structures and functions that occur during critical periods of fetal growth and development may result in *programming,* or permanent changes in the structure and function of organs and tissues. Altered newborn size appears to be an outward sign that such programming has occurred.

THE WHY AND HOW OF PROGRAMMING The ability to modify structures and functions in response to available energy or nutrient supply is advantageous to the fetus in several ways. Adaptations could help the fetus survive by changing its requirements for energy or nutrients, and they may also biologically prepare the fetus for similar nutritional circumstances after birth. Fetal adaptations that lower energy or nutrient needs in response to a low supply, for example, may promote survival in an environment of limited food after birth. IGF-1, for instance, is sensitive to maternal nutrition. Levels of IGF-1 may fall in response to maternal undernutrition to match fetal nutrient utilization with nutrient availability.[47]

> **Programming** The process by which exposure to adverse nutritional or other conditions during sensitive periods of growth and development produces long-term effects on body structures, functions, and disease risk.

EXAMPLE OF A POTENTIAL MECHANISM Take the example of a fetus exposed to a low supply of glucose early in pregnancy. This circumstance would hinder central nervous system development and threaten fetal survival. Mechanisms are set in place, however, that triage available glucose to the central nervous system (CNS). This change represents an adaptation by the body to how glucose utilization is programmed to operate.

What adaptations are made to ensure the CNS gets priority access to glucose? Mechanisms behind adaptations have been mostly investigated in animals, so their application to humans is unclear. However, animal studies indicate that the expression of genes that produce insulin receptors on muscle cell membranes may be suppressed in response to a low availability of glucose. This increases insulin resistance and decreases uptake of glucose by muscle cells, and reduces their growth. It would also increase the availability of glucose for CNS development.

Adaptations that decrease muscle utilization of glucose and reduce muscle size may serve the offspring well later in life if food availability and intake are limited. If food is abundant and food intake is high, however, such adaptations may lead to elevated blood levels of glucose and insulin. These changes may increase the risk of obesity, type 2 diabetes, gestational diabetes, and other disorders associated with insulin resistance.[47]

Increased susceptibility to insulin resistance and weight gain in infants experiencing nutritional insults in utero has been attributed to a "thrifty phenotype," or genetic functional types programmed in utero that act to conserve energy. Alternatively, it is proposed that an inherited thrifty genotype may produce the same effects as the thrifty phenotype, and account for the relationship between abnormal or reduced fetal growth and later energy conservation.[47]

Other In Utero Exposures That May Be Related to Later Disease Risk

Some studies have shown a link between maternal nutritional exposures during pregnancy and later disease risk in infants with a wide range of birthweights. Low weight gain around mid-pregnancy, for example, has been associated with higher blood pressure in children, and low levels of maternal body fat during pregnancy with increased risk of heart disease in offspring.[50] Children born to women receiving a calcium supplement during pregnancy have been found to have lower blood pressure than children born to women given a placebo.[51] Obviously, not all programmed changes that affect later disease risk impair fetal growth. In addition, chronic disease risk appears to be lower among infants born small due to preterm delivery alone than for infants born small, thin, or short for gestational age.[5]

Non-nutritional factors, such as hypertension during pregnancy and genetic abnormalities, may also interfere with delivery of an optimal supply of energy and nutrients to the fetus during critical periods of growth and development. Conditions that alter the hormonal milieu of the placenta and fetus may also program later disease risk.[5]

Limitations of the Fetal-Origins Hypothesis

The hypothesis that maternal and fetal nutrition exposures influence later disease risk is gaining support and recognition. Many questions are unanswered, however. Which specific nutritional exposures are responsible for increased disease risk? When do the vulnerable periods of fetal sensitivity to poor nutrition occur? To what extent are genetic factors responsible for the relationships observed? The implications of the associations between maternal and fetal nutrition and adult disease risk are immense, and if substantiated, demand intense scrutiny of current prenatal nutrition policies and recommendations.[46]

Pregnancy Weight Gain

"Any obstetrician who allows a woman to lose her attractiveness (i.e., gain too much weight) is depriving her of many things that make for her mental well-being, her husband's contentment, and her own personal satisfaction."

Loughran, American Journal of Obstetrics and Gynecology, 1946

Weight gain during pregnancy is an important consideration because newborn weight and health status tend to increase as weight gain increases. Birthweights of infants born to women with weight gains of 15 pounds (7 kg) for example, average 3100 grams (6 lb 14 oz). This weight is about 500 grams less than the average birthweight of 3600 grams (8 lb) in women gaining 30 pounds (13.6 kg). Rates of low birthweight are higher in women gaining too little weight during pregnancy.[21] Weight gain during pregnancy is an indicator of plasma volume expansion and positive calorie balance, and provides a rough index of dietary adequacy.[52]

Multiple studies show broad agreement on amounts of weight gain that are related to the birth of infants with weights that place them within the lowest category of risk for death or health problems.[31] Yet how much weight should be gained during pregnancy remains a hotly debated topic. Earlier in the last century, when gains were routinely restricted to 15 or 20 pounds, weight gain in pregnancy was seen as the cause of pregnancy hypertension, difficult deliveries, and obesity in women. Pregnant women would be placed on low-calorie diets and given diuretics and amphetamines and urged to use saccharin to limit weight gain.[53]

Although none of these notions have been shown to be true, weight gain during pregnancy still represents a prickly issue. Weight gain and body weight are not only a matter of health, but are also closely linked to some people's view of what is socially acceptable.

Psychological and sociological biases related to body weight and shape in women are an important reason to apply recommendations for weight gain in pregnancy based on scientific studies and consensus.

Pregnancy Weight Gain Recommendations

Current recommendations for weight gain in pregnancy are based primarily on gains associated with the birth of healthy-sized newborns (approximately 3500–4500 g or 7 lb 13 oz to 10 lb).[31] As shown in Illustration 4.8, however, prepregnancy weight status influences the relationship between weight gain and birthweight. The higher the weight before pregnancy, the lower the weight gain needed to produce healthy-sized infants. Recommended weight gains for women entering pregnancy underweight, normal weight, overweight, and obese are displayed in Table 4.16.[31]

Illustration 4.8 Pregnancy weight gain by prepregnancy weight status and birthweight.

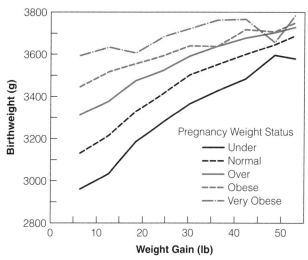

SOURCE: From Clinical Nutrition, Vol. 7, Fig. 1, p. 186, J. E. Brown, © 1988. Reprinted by permission from Elsevier.

Because underweight women tend to retain some of the weight gained in pregnancy for their own needs, they need to gain more weight in pregnancy than do other women. Overweight and obese women, on the other hand, are able to use a portion of their energy stores to support fetal growth, so they need to gain less. A separate and higher pregnancy weight gain recommendation is given for women expecting twins.

Young adolescents and women who are at the low end of the prepregnancy weight ranges are encouraged to gain near the upper end of their weight-gain range, and short women and those at the higher end of the prepregnancy weight ranges ought to gain near the lower end of the range. Because factors such as duration of gestation, smoking, maternal health status, *gravida,* and *parity* also influence birthweight, gaining a certain amount of weight during

> **Gravida**   Number of pregnancies a woman has experienced.
>
> **Parity**   The number of previous deliveries experienced by a woman; *nulliparous* = no previous deliveries, *primiparous* = one previous delivery, *multiparous* = two or more previous deliveries. Women who have delivered infants are considered to be "parous."

Table 4.16 Pregnancy weight gain recommendations[31]

Prepregnancy Weight Status	Recommended Gain*
Underweight, BMI <19.8	28–40 lb (12.7–18.2 kg)
Normal weight, BMI 19.8–26.0	25–35 lb (11.4–15.9 kg)
Overweight, BMI >26.0–29.0	15–25 lb (6.8–11.4 kg)
Obese, BMI >29.0	15 lb (6.8 kg) at least
Twin pregnancy	35–45 lb (15.9–20.5 kg)

*Young adolescents should achieve gains at the upper end of ranges, short women at the lower end.

pregnancy does not guarantee that newborns will be a healthy size. It does improve the chances that this will happen, however.

Approximately one-third of U.S. women gain within the recommended weight ranges during pregnancy.[54] For all except the obese, women who gain within the recommended ranges are approximately half as likely to deliver low-birthweight or SGA newborns as are women who gain less. Rates of LGA newborns, Caesarean-section deliveries, and postpartum weight retention tend to be higher when pregnancy weight gain exceeds that recommended.[54]

Restriction of pregnancy weight gain to levels below the recommended ranges is not recommended. It does not decrease the risk of pregnancy-related hypertension and is associated with increased infant death and low birthweight, and poorer offspring growth and development.[19] In addition, low weight gain in pregnancy may increase the risk that infants will develop heart disease, type 2 diabetes, hypertension, and other types of chronic disease later in life.[26]

RATE OF PREGNANCY WEIGHT GAIN   Rates at which weight is gained during pregnancy appear to be as important to newborn outcomes as is total weight gain. Low rates of gain in the first trimester of pregnancy may down-regulate fetal growth and result in reduced birthweight and thinness.[55] For underweight and normal-weight women, rates of gain of less than 0.5 pound (0.25 kg) per week in the second half of pregnancy, and of less than 0.75 pound (0.37 kg) per week in the third trimester of pregnancy, double the risk of preterm delivery and SGA newborns. For overweight and obese women, rates of gain of less than 0.5 pound (0.25 kg) per week in the third trimester also double the risk of preterm birth.[56] Third-trimester weight gains exceeding approximately 1.5 pounds a week (0.7 kg), however, add little to birthweight in normal-weight and heavier women, and may increase postpartum weight retention.[57]

Rate of weight gain is generally highest around mid-pregnancy—which is prior to the time the fetus gains most of its weight (Illustration 4.9). In general, the pattern of gain should be within a few pounds of that represented by the weight-gain curves shown in Illustration 4.10.[31] Some weight (3 to 5 pounds) should be gained in the first trimester, followed by gradual and consistent gains thereafter. The rate of weight gain often slows a bit a few weeks prior to delivery, but as is the case for the rest of pregnancy, weight should not be lost until after delivery.[55]

Composition of Weight Gain in Pregnancy

A question often asked by pregnant women is, "Where does the weight gain go?" Where the weight gain generally goes by time in pregnancy is shown in Table 4.17 on page 92. The fetus actually comprises only about a third of the total weight gained during pregnancy in women who enter pregnancy at normal weight or underweight. Most of the rest of the weight is accounted for by the increased weight of maternal tissues.

BODY FAT CHANGES   Pregnant women store a significant amount of body fat in normal pregnancy in order to meet their own and the fetus's energy needs, and quite likely to prepare for the energy demands of lactation. Body fat stores increase the most between 10 and 20 weeks of pregnancy, or before fetal energy requirements are highest. Levels of stored fat tend to decrease before the end of pregnancy. Only 0.5 kg of the approximately 3.5 kg of fat stored during pregnancy is deposited in the fetus.[8,25]

Postpartum Weight Retention

Concern about the role of pregnancy weight gain in fostering long-term maternal obesity has increased in the United States, along with the rising incidence of obesity in adults. Increased weight after pregnancy appears to be related to a variety of factors, including excessively high weight gain in pregnancy (over 45 lb, or 20.5 kg), weight gain after delivery, and low activity levels.[58] High blood levels of insulin early in pregnancy, and levels of leptin, have been related to increased weight gain during pregnancy. Levels of both hormones are related to diet.[59,60]

Women tend to lose about 15 pounds the day of delivery, but subsequent weight loss is highly variable.[58] On average, however, women who gain within the recommended ranges of weight gain are 2.0 pounds (0.9 kg) heavier 1 year after delivery than they were before pregnancy.[61] This gain is slightly above the amount of weight women tend to gain with age.[66] Postpartum weight retention tends to be slightly less in women who breastfeed for

Illustration 4.9 Rates of maternal and fetal weight gain during pregnancy.

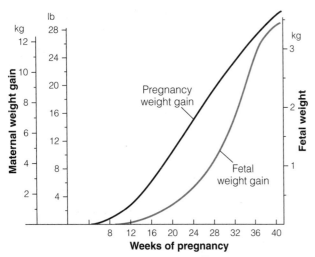

SOURCE: Curves drawn by Judith E. Brown, 2002.[55]

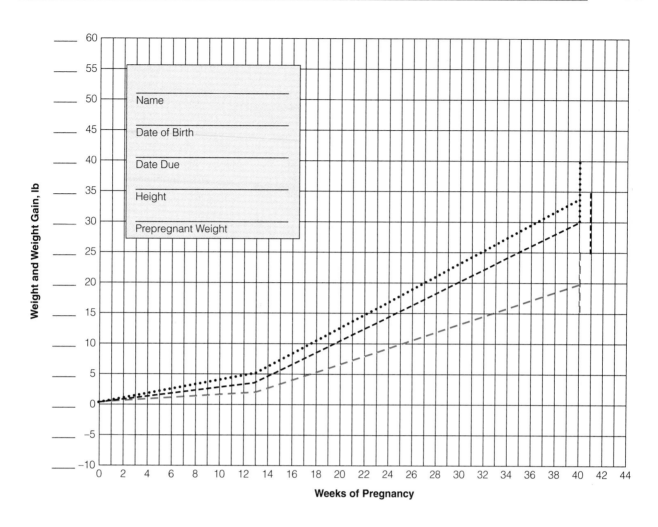

Name

Date of Birth

Date Due

Height

Prepregnant Weight

Weight of Pregnancy

Weight Record			
Date	Weeks of Gestation	Weight	Notes

Please bring this chart with you to each prenatal visit.

Illustration 4.10 The Institute of Medicine's prenatal weight-gain graph.

SOURCE: From *Nutrition during Pregnancy and Lactation*. National Academy of Sciences, 1992. Reprinted with permission of National Academy Press.

Table 4.17 Components of weight gain during pregnancy for healthy, normal-weight women delivering a 3500-gram (about 8 lb) infant at term[6,8,11,29]

	WEIGHT GAIN, GRAMS			
Component	10 Weeks	20 Weeks	30 Weeks	40 Weeks
Fetus	5	300	1500	3550
Placenta	20	170	430	670
Uterus	140	320	600	1120
Amniotic fluid	30	350	750	896
Breasts	45	180	360	448
Blood supply	100	600	1300	1344
Extracellular fluid	0	265	803	3200
Maternal fat stores	315	2135	3640	3500
Total weight gain at term = 14.7 kg or 32 lb				

at least 6 months after pregnancy.[63] Women who gain less than the recommended amount of weight gain in pregnancy do not retain less weight on average after pregnancy than do women who gain within the ranges.[61] Postpartum weight can be reduced by identifying high weight gainers during pregnancy and getting the women identified involved in an exercise and healthy eating program.[64]

Nutrition and the Course and Outcome of Pregnancy

> "A mother who wishes her child to have black eyes should frequently eat mice."
>
> Prenatal diet folklore from Ancient Rome[65]

The history of beliefs about the effects of maternal diet on the course and outcome of pregnancy is rife with superstition, ill-founded and hazardous conclusions, and unhelpful suggestions. Societies have shared a belief in the importance of "eating right" during pregnancy for the child's sake, but actual knowledge about maternal nutrition and the course and outcome of pregnancy has been acquired only relatively recently.

Famine and Pregnancy Outcome

Much of the scientific interest in the effects of maternal nutrition on the course and outcome of pregnancy comes from studies done in the first half of the twentieth century. Ecological studies on effects of famines in Europe and Japan during World War II on the course and outcome of pregnancy demonstrated potential negative, as well as positive, effects of food intake on fertility and newborn outcomes.

THE DUTCH HUNGER WINTER, 1943–1944 As mentioned briefly in Chapter 2, people in many parts of Holland experienced severe food shortages for an 8-month period during World War II due to enemy occupation of major cities. Although people in Holland were generally well nourished and had a reasonable standard of living before the disaster, conditions rapidly deteriorated during the famine. In addition to intakes that averaged only about 1100 cal and 34 g of protein per day, fuel was in low supply and the winter harsh.

Carefully kept records by health officials showed a sharp decline in pregnancy rates of over 50% during the famine, an effect attributed to absent and irregular menstrual periods. Average birthweight declined by 372 g (13 oz), delivery of low-birthweight infants increased by 50%, and rates of infant deaths increased. Birthweight did not fully "catch up" in infants born to women exposed to famine early in pregnancy, even if they received enough food later in pregnancy. This result supports the notion that the fetal growth trajectory may be established early in pregnancy and that early nutritional deprivations limit fetal growth regardless of food intake later in pregnancy.[66]

Although the Dutch famine was associated with major declines in fertility and newborn health and survival, the rather good nutritional status of women prior to the famine likely protected pregnant women and their infants from more severe disruptions in health. Normal fertility status and newborn outcomes returned within a year after the famine ended.[67]

Studies undertaken in the last 30 years on adults who were born to women during the hunger winter (the Dutch famine cohort) show relationships between the timing of famine during pregnancy and adult offspring health outcomes. Examples of relationships identified are shown in Table 4.18.

THE SEIGE OF LENINGRAD, 1942 Unlike people in Holland, the population in Leningrad (now called St. Petersburg) had experienced moderate deprivations in nutritional status and quality of life prior to the famine. As was the case for pregnant women in Holland, the famine in Leningrad resulted in average intakes of approximately 1100 cal per day. Infertility and low-birthweight rates increased over 50%, infant death rates rose, and birthweights dropped by an average of 535 g (1.2 lb) during the famine.[70] Rates of pSGA newborns also increased, suggesting that the poor nutritional status of women coming into pregnancy and persistent undernutrition during pregnancy interfered with critical periods of fetal growth.

FOOD SHORTAGES IN JAPAN Effects of World War II–associated food shortages on reproductive outcomes in Japan were similar to those observed in Holland. Japanese women tended to be well nourished prior to the shortages. Lack of food before and during pregnancy was

Table 4.18 Exposure to the Dutch World War II famine by time in pregnancy and adult offspring health risks[68,69]

PERIOD OF FAMINE		
First Trimester	First and/or Second Trimester	Second Half of Pregnancy
Schizophrenia	Antisocial personality disorder	Decreased glucose tolerance
High LDL and low HDL cholesterol		
High body weight and central body fat		
Infertility		
Neural tube defects		

reflected in decreased fertility status among women and in reductions in birthweight that averaged 200 g.

Social and economic improvements occurring in Japan after the war led to increased availability of many foods, including animal products. This higher plane of nutrition achieved during the postwar years in Japan was accompanied by major increases in newborn size and the "growing up" of Japanese children. In a trend that continues today, subsequent generations of Japanese adults averaged 2 inches taller than the previous generation.[71] Infant mortality in Japan, which ranked among the highest for industrialized nations prior to World War II, declined incredibly after the war and remains well below rates in the United States and in a number of other developed countries.[72]

Food shortages continue to occur in various parts of the world and to adversely affect fertility and the course and outcome of pregnancy. Effects have become predictable, such that declines in fertility and newborn size and vitality are viewed as part of the consequences of such disasters. For example, the siege of Sarajevo, which decreased food availability during 1993–1994, led to reduced caloric and nutrient intakes during pregnancy, reduced maternal weight gain and newborn weights, and increased rates of perinatal mortality and congenital anomalies.[73] Birthweight did not fully catch up in infants born to women exposed to famine early in pregnancy, even if they received enough food later in pregnancy. This result supports the notion that the fetal growth trajectory may be established early in pregnancy and that early nutritional deprivations limit fetal growth regardless of food intake later in pregnancy.[66]

Contemporary Prenatal Nutrition Research Results

"Faulty nutrition, not just malnutrition, can influence fetal health."

Bertha Burke, 1948[74]

Carefully conducted studies of diet and pregnancy outcome in the first half of the 20th century began the era of scientifically based recommendations on nutrition and pregnancy. The now classic studies conducted by Bertha Burke at Harvard in the 1940s were particularly influential.[74,75] These studies showed that diet quality during pregnancy, assessed using diet histories, was strongly related to newborn health status. Newborns assessed as having optimal physical condition by pediatricians were found to be much more common among women consuming high-quality diets, whereas those with the poorest physical condition were born to women with the poorest-quality diets. Average birthweight of newborns assessed as being in optimal physical condition was 7 pounds, 15 ounces in females, and 8 pounds, 8 ounces in males.[75] Although Burke's studies did not show that high-quality pregnancy diets by themselves were responsible for robust newborn health, they provided some of the first evidence that prenatal diet quality may strongly influence pregnancy outcome.

Thousands of other studies on the effects of nutrition on the course and outcome of pregnancy are now available. The following sections highlight research results and recommendations related to calories, key nutrients, and other substances in food that influence the course and outcome of pregnancy.

Energy Requirement in Pregnancy

Energy requirements increase during pregnancy, mainly due to increased maternal body mass and fetal growth. The additional requirements can be allocated to different maternal and fetal tissues by estimating the amount of oxygen used (or "consumed") by the various tissues. Illustration 4.11 on the next page shows the results of work on oxygen consumption during pregnancy undertaken by Hytten and Chamberlain.[76] Approximately one-third of the increased calorie need in pregnancy is related to increased work of the heart, and another third to increased energy needs for respiration and accretion of breast tissue, uterine muscles, and the placenta. The fetus accounts for about a third of the increased energy needs of pregnancy.

The increased need for energy in pregnancy averages 300 cal a day, or a total of 80,000 cal.[6] The DRIs for energy intake for pregnancy are +340 kcal per day for the second trimester and +452 kcal per day for the third trimester of pregnancy. Caloric intake recommendations represent a rough estimate that by no means applies to all women.

Additional energy requirements of women have been found by different studies to range from 210–570 cal a day.[8] The need for additional calories during pregnancy may be a good deal lower in women who perform little

Illustration 4.11 Components of increased oxygen consumption in normal pregnancy.

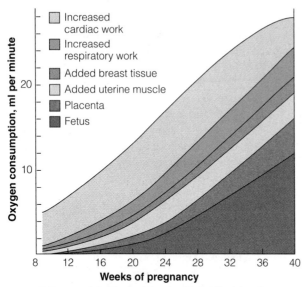

SOURCE: F. Hytten and G. Chamberlain, eds., Clinical Physiology in Obstetrics. Reprinted by permission of Blackwell Science Ltd.

Illustration 4.12 Estimated caloric balance in pregnancy through 6–8 weeks postpartum.

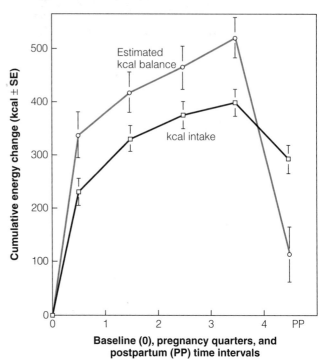

SOURCE: From Clinical Perinatology, 24(2):433–449, by J. E. Brown and E. S. B. Khan, © 1997. Reprinted by permission of W. B. Saunders Co.

exercise, and higher in women who are very active. Low levels of energy expenditure from physical activity are common in the first trimester of pregnancy, and the energy savings may produce a positive caloric balance even though a woman's caloric intake hasn't changed much. Contrary to a previous belief, energy needs of pregnant women do not appear to be affected by "metabolic efficiencies" of pregnancy that decrease caloric need.[77]

Illustration 4.12 shows the difference between caloric (kcal) intake and estimated caloric balance throughout pregnancy in a group of women served by a health maintenance organization.[77] The graph indicates that estimated caloric balance is higher than caloric intake throughout pregnancy and becomes negative postpartum. The positive caloric balance observed during pregnancy is due to the fact that women consumed more calories than they expended in physical activity and basal metabolism.

ASSESSMENT OF CALORIC INTAKE Adequacy of calorie intake is most easily assessed in practice by pregnancy weight gain. Rates of gain in women who do not have noticeable edema are a good indicator of caloric balance. Women who are losing weight are in negative balance, and those gaining weight are in positive caloric balance. Adjustments in rate of weight gain can be approached by modifying physical activity level, calorie intake, or both. Women should maintain a positive caloric balance and rate of weight gain throughout pregnancy. Meal skipping or fasting during pregnancy are not advised. Whether women should restrict their food and fluid intake during labor, however, is hotly debated.

FASTING IN LABOR: IS IT NECESSARY? As in every area of medicine, some obstetrical practices are based

more on ritual and intuition than scientific evidence. The practice of restricting fluid and food intake during labor appears to be one such practice, because there is no evidence that it improves outcomes of mothers or infants. Labor is a high-energy, Olympic-caliber event. Yet many women are made to fast their way through it, dependent on intravenous fluid and glucose along with occasional sips of water or ice chips. This is the practice in most U.S. hospitals, but the routine in hospitals in other countries allows unrestricted fluids, or fluids and certain solid foods.[78]

The motivation behind fluid and food restriction during labor is based on the concern that anesthesia (if needed) will delay stomach emptying and increase the risk of aspiration—the regurgitation and inhalation of stomach contents. This situation is very rare but can lead to the development of pneumonia. The odds of dying from aspiration in labor, however, are less than those for being struck by lightning—less than 1 in 600,000. In addition, an empty stomach does not guarantee that aspiration will not occur. Hydrochloric acid and other digestive juices normally present in an empty stomach can enter the lungs if aspiration occurs. There is also evidence that intravenous glucose solutions given to laboring women may cause the fetus to have high blood glucose levels, which then drop rapidly after delivery.[78] It is possible that allowing women to eat high-carbohydrate, pre-event-type foods—like athletes do—and to drink hydrating fluids during labor would help them maintain their strength throughout this physically demanding event.

Practices related to the duration and nature of fasting before and after surgeries, including C-sections, are changing. It is now concluded that women undergoing C-sections should be allowed to drink clear fluids (water, tea, coffee, soft drinks) up to 2 hours before surgery and to eat a meal 8 hours before. Introducing food and fluids within 6–8 hours after C-sections has been found to improve bowel mobility, decrease bowel distention and pain, and shorten hospital stays.[79]

Carbohydrate Intake during Pregnancy

Approximately 50–65% of total caloric intake during pregnancy should come from carbohydrate. Women should consume a minimum of 175 grams carbohydrate to meet the fetal brain's need for glucose. On average, women in the United States consume 53% of calories (269 grams) from carbohydrates during pregnancy.[80] High-fiber foods generally provide a variety of beneficial phytochemicals and a hefty measure of protection against constipation.[81]

ARTIFICIAL SWEETENERS   There is no evidence that consumption of aspartame (Nutrasweet) or acesulfame K (Sunette) is harmful in pregnancy.[82] Diet soft drinks and other artificially sweetened beverages and foods are often poor sources of nutrients, however, and may displace other, more nutrient-dense foods in the diet.

Alcohol and Pregnancy Outcome

Prenatal exposure to alcohol is a leading, preventable cause of birth defects, mental retardation, and developmental disorders.[83] Approximately 1 in 12 women consume alcohol during pregnancy, and 1 in 30 consume five or more drinks on one occasion at least monthly.[84] It is estimated that approximately 50,000 infants are born each year in the United States with some degree of alcohol-related damage.[85]

Alcohol consumed by a woman easily crosses the placenta to the fetus. Because the fetus has yet to fully develop enzymes that break it down, alcohol lingers in the fetal circulation. This situation, combined with the fact that the fetus is smaller and has far less blood than the mother does, increases the harmful effects of alcohol on the fetus as compared to the mother. Alcohol exposure during critical periods of growth and development can permanently impair organ and tissue formation, growth, health, and mental development.

Poor dietary intakes of some women who consume alcohol regularly in pregnancy, as well as the negative effect of alcohol on the availability of certain nutrients, may also contribute to the harmful effects of alcohol exposure during pregnancy.[86]

Consumption of four or more drinks a day, or occasional episodes of consumption of five or more drinks in a row, is considered to represent heavy alcohol intake during pregnancy. Heavy drinking during pregnancy increases

Table 4.19 Approximate incidence of structural abnormalities of the brain in 5- to 14-week-old fetuses exposed to alcohol

Maternal alcohol intake, drinks	Abnormal brain structure
13 to 31/day	100%
6.3 to 13/day	83%
2 to 6.3 occasionally	29%
≤2/day	0%

SOURCE: Table is based on information presented by Konovalov et al., 1997.[87]

the risk of miscarriage, stillbirth, and infant death within the first month after delivery.[85] Approximately 40% of the fetuses born to women who drink heavily early in pregnancy will develop fetal alcohol syndrome (FAS). The likelihood that the fetus will be affected by FAS increases as the number of drinks consumed early in pregnancy increases (Table 4.19).

FAS   FAS was first described in 1973 and consists of pSGA, mental retardation, and a set of common malformations (Illustration 4.13). Diagnosis of FAS is difficult, however, because many of the facial features are not unique to the syndrome. Short noses, flat nasal bridges, and thin upper lips, for example, are also sometimes normal facial features.

Children with FAS tend to have poor coordination, short attention span, and behavioral problems, and they

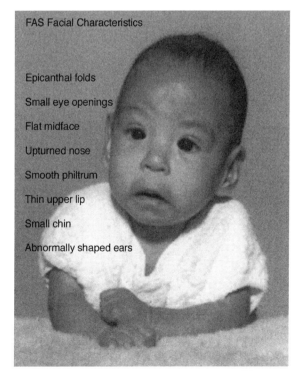

FAS Facial Characteristics

Epicanthal folds
Small eye openings
Flat midface
Upturned nose
Smooth philtrum
Thin upper lip
Small chin
Abnormally shaped ears

Illustration 4.13 Features of FAS in children.[86]

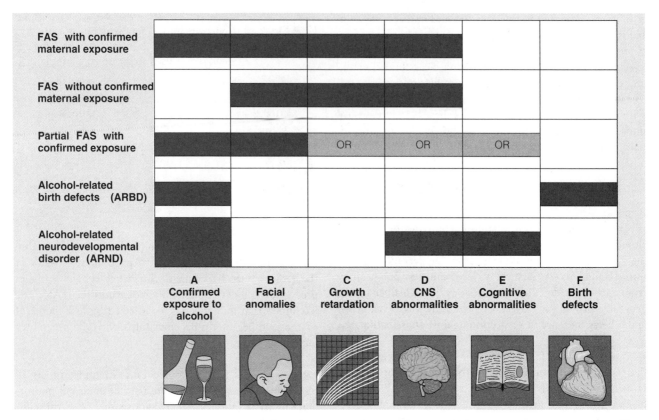

Illustration 4.14 Diagnostic classification of FAS, alcohol-related birth defects, and neurodevelopmental disorders.[83]

SOURCE: Adapted from Fetal Alcohol Syndrome: Diagnosis, Epidemiology, Prevention, and Treatment, 1996;4–5. Reprinted by permission of the American Academy of Pediatrics.

remain small for their age. Adults with FAS often find it difficult to hold a job or live independently.[85]

FETAL ALCOHOL EFFECTS (FAE) Approximately 10 times the number of infants born with FAS have lesser degrees of alcohol-related damage known as FAE. This is characterized by mental and behavioral abnormalities but not malformations.[83]

Alcohol-Related Birth Defects and Neurodevelopmental Disorders Fetal exposure to alcohol has been linked to many other neurodevelopmental problems since it was first described in 1973. To reflect advances in knowledge, it has been recommended that the terms *alcohol-related birth defects* (ARBD) and *alcohol-related neurodevelopmental disorders* (ARND) replace the term *FAE*. The changes (summarized in Illustration 4.14) emphasize the wide range of neurodevelopmental problems such as mental retardation, aggressiveness, destructiveness, nervousness, and short attention span associated with intrauterine exposure to excess levels of alcohol.[83]

Protein Requirement

The recommended protein intake for pregnancy is +25 g per day, or 71 g daily for females aged 14 and older. On average, pregnant women in the United States consume 78 g of protein daily.[80]

Protein content of nonvegetarian diets can be simply estimated by evaluating women's usual daily intake of major sources of protein. A tool for estimating protein intake is shown in Table 4.20.

Vegetarian Diets in Pregnancy

"The topic of vegetarian dietary practices often brings with it a variety of images and attitudes regarding those who follow such practices. Those attitudes may have limited, if any, basis in actual fact."

Patricia Johnston, 1988[89]

Nutrient needs in pregnancy may be met by many different types of diets, including those that omit animal products.[90] It is the type and amount of food consumed, not the label placed on it, that determines the appropriateness of dietary intake during pregnancy.

A food guide for pregnant women who are vegetarian can be found in Table 4.21. Diets of pregnant vegetarians are sometimes low in vitamins B$_{12}$ and D, calcium, zinc, omega-3 fatty acids (or n-3 fatty acids), and riboflavin due to the lack of consumption of rich food sources of these nutrients. Table 4.22 on page 98 provides a list of plant-based food sources of four of these key nutrients. Vitamin

Table 4.20 Tool for estimating protein intake

Food	Protein, grams	How much protein is there in this usual day's diet?	
Milk, 1 c	8		
Cheese, 1 oz	7	2 sl toast	6
Egg, 1	7	1 c milk	8
Meat, 1 oz	7	3 oz tuna	21
Dried beans, 1 c	13	2 sl bread	6
Bread, 1 sl or oz	3	2 oz chicken	14
		1 oz cheese	7
		2 tortillas	6
		½ c refried beans	7
		Total g protein =	75

Table 4.21 Vegetarian food guide adapted for pregnant women[90,91,92]

FOOD GROUP	SERVINGS PER DAY
A. Grains	
Whole grain bread, 1 slice	6–11
Cooked grains, ½ c	
Fortified cold cereals, 1 oz	
Fortified cooked cereals, ½ c	
Corn, ½ c	
Pasta, ½ c	
Tortillas, 1 small	
Crackers, 4 small	
B. Legumes, Nuts, Seeds, Dairy	5–7
Dried beans, cooked, ½ c	
Peas, ½ c	
Soy products, ½ c or 2–3 oz	
Soynuts, ¼ c	
nut and seed butter, 2 tb	
Nuts and seeds, ¼ c	
Eggs, 1	
Cow's milk, 1 c	
Cheese, 1 oz	
Yogurt, ½ c	
Fortified soymilk, 1 c	
C. Vegetables	4
Cooked vegetables, ½ c	
Raw vegetables, 1 c	
Vegetable juice, ½ c	
D. Fruits	2
Medium-sized fruit, 1	
Cut-up raw or cooked, ½ c	
Fruit juice, ½ c	
Dried fruit, ¼ c	
E. Fats, Oils, and Sweets	2+ depending on caloric need
Mayonaise, oil margarine, 1 tb	
Honey, syrup, jams, jellies, sugar, 1 tb	

B_{12} deficiency during pregnancy may not become apparent until after delivery. Two cases of neurological impairment and growth failure due to maternal B_{12} deficiency were identified in 4- to 8-month-old infants in Georgia in 2001. Both infants were born to women who followed a vegetarian diet during pregnancy.[93] Overall, the incidence of iron deficiency is the same in vegetarian and nonvegetarian pregnant women.[90]

Protein intake is generally adequate in the vegetarian diet, but may be low in vegans. Protein needs are met by vegetarians who regularly consume a variety of plant sources of protein and meet energy needs. In pregnant women who consume no animal products, the variety of plant protein sources needs to include complementary sources of protein daily. Protein sources that complement each other, or provide a complete source of protein, include legumes (such as lentils, chickpeas, black-eyed peas, black beans, and lima beans) and grains (corn, rice, bulgur, and barley, for example). Protein need is about 30% higher in vegetarians than nonvegetarians due to the lower essential amino acid content and digestibility of plant protein (except for soy).[90]

Availability of vegetarian food products in large groceries and organic food-stores has expanded substantially in the past few years. Vegetarians can now select veggie burgers, meat analog entrees, meals-in-a-cup, and frozen desserts from food-store shelves. Fortified juice, soymilks, breakfast cereal, and meat substitutes are available and contribute substantially to vegetarians' intake of vitamins B_{12} and D and calcium. DHA derived from algae can be used to provide a source of this omega-3 fatty acid in diets of vegetarian pregnant women who do not consume fish or seafood.[90] Walnuts, dark green vegetables, and flaxseed provide alpha-linolenic acid, but little of it is converted to DHA by the body.[94]

Computerized nutrient analysis of several days of usual food intake may be especially helpful in vegetarian diets due to the variability of dietary practices.[90] Evaluation of rate of weight gain in pregnancy is generally a good way to assess the adequacy of energy intake. Case Study 4.1 on page 99 is related to the dietary assessment results of a pregnant, vegan woman.

Maternal Intake of Essential Fatty Acids and Pregnancy Outcome

The essential fatty acids, linoleic acid and alpha-linolenic acid, were presented in Chapter 1. They are referred to again here because inadequate maternal intake of these fatty acids and their derivatives may impair fetal growth and development. Recent research has focused on the importance of maternal intake of alpha-linolenic acid and its derivative, docosahexaenoic acid (DHA), in fetal vision development and learning ability. Adequacy of maternal intake of DHA from foods is being emphasized because it does not appear that sufficient amounts of DHA can be

Table 4.22 Examples of sources of iron, calcium, vitamin D, and vitamin B$_{12}$ in vegetarian diets[90,91,92]

IRON	
Food	Iron, mg
A. Legumes, Cooked	
Tofu, firm, ½ c	6.6
Soybeans, ½ c	4.4
Lentils, ½ c	3.3
Kidney beans, ½ c	2.6
Chickpeas, ½ c	2.4
Tempeh, ½ c	2.2
Black beans, ½ c	1.7
B. Nuts and Seeds	
Dried pumpkin and squash seeds, ¼ c	5.2
Sesame tahini, 2 Tb	2.7
Toasted sunflower seeds, ¼ c	2.3
Cashews, ¼ c	2.1
C. Fortified Breakfast Cereals	
Ready-to-eat cereals, 1 oz	2.1–18
Cream of wheat, cooked, ½ c	5.1
Instant oatmeal, cooked, ½ c	4.2
Quinoa, cooked, ½ c	2.1

CALCIUM	
Food	Calcium, mg
A. Legumes	
Fortified soy yogurt, ½ c	367
Tofu, firm, calcium set, ½ c	120–430
Fortified soymilk, 1 c	200–400
Soybean greens, ½ c	130
Tempeh, ½ c	92
Soybeans, cooked, ½ c	88
Navy beans, cooked, ½ c	64
Black beans, cooked, ½ c	46
B. Nuts and Seeds	
Sesame tahini, 2 tb	128
Almonds, ¼ c	88
Almond butter, 2 tb	86
Filberts (hazelnuts), chopped, ¼ c	64
Brazil nuts, ¼ c (2)	62

CALCIUM (CONTINUED)	
Food	Calcium, mg
C. Vegetables, Cooked	
Spinach, ½ c	122
Kale, ½ c	47
Broccoli, ½ c	36
Bok choy (Chinese cabbage), ½ c	36
D. Dairy Products	
Yogurt, ½ c	207
Cow's milk, 1 c	288
Swiss cheese, 1 oz	270
Frappuccino, 1 c	220
Cheddar cheese, 1 oz	204
Ice Cream, 1 c	180

VITAMIN D	
Food	Vitamin D, mcg
Cow's milk, 1 c	2.5
Fortified soymilk, 1 c	1–3
Fortified breakfast cereals, 1 oz	0.5–1.0
Egg yolk, 1 large	0.6

VITAMIN B$_{12}$	
Food	Vitamin B$_{12}$, mcg
Fortified breakfast cereals, 1 oz	0.6–6.0
Fortified soymilk, 1 c	0.8–3.2
Nutritional yeast, 1 tb	1.5
Fortified veggie soy meats, 3 oz	1.5–3.6
Cow's milk, 1 c	0.8–1.0
Egg, 1 large	0.5

produced by the body from alpha-linolenic acid to meet fetal needs. Only 1–4% of alpha-linolenic acid is converted to DHA by the body.[94] Eicosapentaenoic acid (EPA) is also a derivative of alpha-linolenic acid and is present in certain foods. EPA is readily converted to DHA by the body.[95]

DHA AND VISION DEVELOPMENT Knowledge of the effects of DHA on vision development comes primarily from studies involving preterm infants, but may apply to term infants as well. Preterm infants are likely to be born with inadequate DHA in the retina, the part of the eye that receives light rays. DHA is incorporated into the retina at a high rate during retinal development in the final months of pregnancy and the first 6 months after birth.[12] Lack of DHA during retinal development appears to reduce the sharpness of vision and slows visual recognition of objects. Neurological development measured by IQ score and problem-solving skills may also benefit from the presence of adequate levels of DHA during fetal and infant development.[95]

Human milk tends to be a good source of DHA; infant formulas are not, unless fortified. It is suspected that the DHA content of human milk is related to the superior intellectual performance of preterm infants given human milk versus those fed formula. Women consuming vegetarian diets that exclude fish and other seafood are likely to have low levels of DHA available for fetal growth and development.[95]

Case Study 4.1

Vegan Diet during Pregnancy

Photo Disc

Ms. Lederman, a healthy 32-year-old woman entering her 13th week of pregnancy, asks her doctor for a referral to a dietitian to discuss her vegan diet. She receives the referral, and while making an appointment with the nutrition consulting service, is asked to record her food intake for 3 days prior to the appointment. Ms. L follows the instructions she was given and carefully completes a 3-day food record. Prior to the appointment, she sends her food record to the dietitian she will be seeing.

During the appointment, the dietitian learns that Ms. L started pregnancy at normal weight, has gained 3 pounds so far in pregnancy, has no history of iron or another nutrient deficiency, and is experiencing a normal course of pregnancy. Ms. L has been a vegan since the age of 16, and although she believes it is good for her health, she worries that her baby may not be getting the nutrients she or he needs. Ms. L wears sunscreen whenever she goes outside, so she makes little or no vitamin D in her skin. She makes sure to combine plant sources of protein (usually dried beans and grains) so she'll consume complete sources of protein every day.

Results of the dietary analysis performed by the dietitian showed the following average calorie and nutrient intake levels:

Kcal: 2237
Protein, g: 71
Linoleic acid (n-6 fatty acids), g: 15.2
Alpha-linolenic acid (n-3 fatty acids), g: 0.54
Vitamin B_{12}, mcg: 2.1
Vitamin D, mcg: 3 (120 IU)
Zinc, mg: 15

Questions

1. Is Ms. L consuming enough protein?
2. Based on the information presented, which nutrients are consumed in amounts that are below the DRI standard for pregnancy?
3. Suggest three types of food Ms. L could consume to bring up her intake of the nutrients identified in question 2.

(Answers are in Appendix D.)

DIETARY INTAKE RECOMMENDATIONS RELATED TO ESSENTIAL FATTY ACIDS The DRI for linoleic acid is 13 g, and for alpha-linolenic acid 1.4 g per day during pregnancy. It is recommended that adults consume 0.65 g DHA and EPA daily.[96] Intake of DHA and EPA in U.S. pregnant women and adults in general is substantially less than 0.65 g per day.[94]

Physiological effects of the essential fatty acids are interrelated, and high intakes of one combined with low intakes of the other can be detrimental to health. Consequently, a balanced intake of the essential fatty acids is required for their optimal functioning. For most North Americans, intake of linoleic acid exceeds that of alpha-linolenic acid by too wide a margin. Americans are being urged to increase their intake of food sources of alpha-linolenic acid, EPA, and DHA (Table 1.6 in Chapter 1). Fish and omega-3 fatty acid–fortified eggs are good sources of EPA and DHA. Plant oils high in the omega-3 fatty acids (EPA and DHA) are being developed.[97]

The Need for Water during Pregnancy

The large increase in water need during pregnancy is generally met by increased levels of thirst. On average, women consume about 9 cups of fluid daily during pregnancy.[98] Women who engage in physical activity in hot and humid climates should drink enough to keep urine light colored and normal in volume. Water, diluted fruit juice, iced tea, and other unsweetened beverages are good choices for staying hydrated.

Folate and Pregnancy Outcome

Inadequate folate during pregnancy has long been associated with anemia in pregnancy and reduced fetal growth.[99] Only during the last two decades, however, has the broad spectrum of effects of folate been recognized. Discoveries of the multiple effects of inadequate folate intake on the development of congenital abnormalities and clinical complications of pregnancy represent some of the most important advances in our knowledge about nutrition and pregnancy.

FOLATE BACKGROUNDER The term *folate* encompasses all compounds that have the properties of folic acid and includes monoglutamate and polyglutamate forms of the vitamin. The monoglutamate form of folate is represented primarily by folic acid, a synthetic form of folate used in fortified foods and supplements. A similar monoglutamate form of folate naturally occurs in a few foods. Food sources of folate contain primarily the polyglutamate form of folate. The two major types of folates are often distinguished by referring to the monoglutamates as folic acid and the polyglutamates as dietary folate.

Bioavailability of folic acid and dietary folate differs substantially. Folic acid is nearly 100% bioavailable if taken in a supplement on an empty stomach, and 85% bioavailable if consumed with food or in fortified foods. Naturally occurring folates are 50% bioavailable on average.[100]

Folate requirements increase dramatically during pregnancy due to the extensive organ and tissue growth that takes place.

FUNCTIONS OF FOLATE Folate is a methyl group (CH_3) donor and enzyme cofactor in metabolic reactions involved in the synthesis of DNA, gene expression, and gene regulation. Deficiency of folate impairs these processes, leading to abnormal cell division and tissue formation.[101] Folate serves as a methyl donor in the conversion of homocysteine to the amino acid methionine. The conversion of homocysteine to methionine depends primarily on three enzymes and folate, vitamin B_{12}, and vitamin B_6 cofactors. Lack of folate in particular, and less commonly a lack of vitamin B_{12}, as well as genetic abnormalities in the enzymes can lead to an accumulation of homocysteine. This may result in methionine shortage at a crucial stage of fetal development. High cellular and plasma levels of homocysteine may also contribute to the development of some of the adverse effects of folate inadequacy.[102] A common genetic defect has been identified in the enzyme 5,10-methylene tetrahydrofolate reductase (MTHFR). This variant of the normal enzyme reduces the level of activity of MTHFR by about half. Variant forms of MTHFR and defects in methionine synthase are thought to be present in approximately 30% of the population.[102]

Folate and Congenital Abnormalities

> "There is a widespread belief that congenital malformations are always the result of defective genes. Perhaps a nutritional deficiency resulting in a defective gene leads to the same congenital abnormality."
>
> R. D. Mussey, 1949[65]

Researchers have known since the 1950s that low and high intakes of certain vitamins and minerals cause congenital abnormalities in laboratory animals. They have also known that neural tube defects, brain and heart defects, and cleft palate can be caused by feeding pregnant rats folate-deficient diets.[103] Firmly held beliefs that only severe malnutrition affects fetal growth and that genetic errors are the sole cause of congenital abnormalities delayed recognition of the importance of folate to human pregnancy.[104]

Neural tube defects (NTDs) are malformations of the spinal cord and brain. There are three major types of NTDs:

- Spina bifida is marked by the spinal cord failing to close, leaving a gap where spinal fluid collects during pregnancy (see Illustration 4.15). Paralysis below the gap in the spinal cord occurs in severe cases.

- Anencephaly is the absence of the brain or spinal cord.

- Encephalocele is characterized by the protrusion of the brain through the skull.

It is now well accepted that inadequate availability of folate between 21 and 27 days after conception (when the embryo is only 2–3 mm in length) can interrupt normal cell differentiation and cause NTDs.[105] Neural tube defects are among the most common types of congenital abnormalities identified in infants, with approximately 4000 pregnancies affected each year in the United States.[106] NTDs

Illustration 4.15 A newborn child with spina bifida.

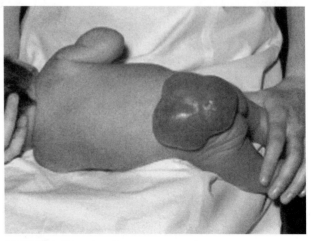

SOURCE: Photo Researchers, Inc.

Illustration 4.16 Mean red cell folate level in preconceptional women by level of intake of various sources of folate.

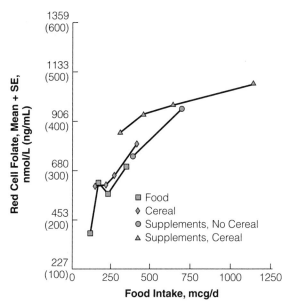

SOURCE: JAMA, Vol. 277, No. 6, p. 551 (Feb 19, 1997).

are among the most preventable types of congenital abnormalities that exist.[107] Approximately 70% of cases of NTDs can be prevented by consumption of adequate folate before and during very early pregnancy.[99]

FOLATE STATUS OF WOMEN IN THE UNITED STATES Folate status is assessed by serum and red cell folate levels. Of the two measures, red cell folate levels are the preferred indicator because they represent long-term folate intake, whereas serum folate levels reflect only recent intake. Levels of red cell folate of over 300 ng/mL (or 680 nmol/L) are associated with very low risk of NTDs.[108] These levels of red cell folate can generally be achieved by folic acid intakes that average 400 mcg daily.[109] As shown in Illustration 4.16, red cell folate levels are higher among women who consume folic-acid-fortified cereals or supplements compared to women consuming folate from food only.[109]

Folate status in women of childbearing age has improved since the advent of folic-acid fortification of refined grain products in 1998. Average levels of red cell folate in U.S. women have increased from 181 to 315 ng/mL since fortification began.[107] Low levels of intake of folic-acid-fortified grain products and breakfast cereals still leave some women with too little folate, however.[110]

DIETARY SOURCES OF FOLATE Many vegetables and fruits are good sources of folate (see Table 1.9 in Chapter 1), but only a few foods contain the highly bioavailable form of folate. Table 4.23 lists some foods that naturally contain the highly bioavailable, monoglu-

mate form of folate and foods that provide folic acid through fortification.

Adequacy of folic acid intake before and during pregnancy can be estimated by adding up the amount of folic acid in foods typically consumed in the daily diet using the data in the table. Whole grain products including breads and pastas, brown rice, oatmeal, shredded wheat, and organic grain products may or may not be fortified with folic acid. You have to check food labels to find that out.

RECOMMENDED INTAKE OF FOLATE Due to variation in folate bioavailability, the DRI for folate takes into consideration a measure called *dietary folate equivalents,* or DFE. One DFE equals any of the following:

- 1 mcg food folate
- 0.6 mcg folic acid consumed in fortified foods or a supplement taken with food
- 0.5 mcg of folic acid taken as a supplement on an empty stomach

Folic acid taken in a supplement without food provides twice the dietary folate equivalents as does an equivalent amount of folate from food.

It is recommended that women consume 600 mcg DFE of folate per day during pregnancy and include 400 mcg folic acid from fortified foods or supplements.[100] The remaining 200 mcg DFE should be obtained from vegetables and fruits. These nutrient-dense foods provide an average of 40 mcg of folate per serving.[109] Because NTDs develop before women may realize they are pregnant, adequate folate should be consumed several months prior to, as well as throughout, pregnancy.

Women who have previously delivered an infant affected by an NTD are being urged to take 4000 mcg (4.0 mg) of folic acid in a supplement to reduce the risk of recurrence.[106] This dose, however, may be much higher

Table 4.23 Food sources of folic acid

AMOUNT		FOLIC ACID, mcg
A. Foods		
Orange	1	40
Orange juice	6 oz	82
Pineapple juice	6 oz	44
Papaya juice	6 oz	40
Dried beans	½ c	50
B. Fortified Foods		
Highly fortified breakfast cereals[a]	1 c or 1 oz	400
Breakfast cereals	1 c or 1 oz	100
Bread, roll	1 sl or 1 oz	40
Pasta	½ c	30
Rice	½ c	30

[a]Includes All Bran, Complete, Crispix, Healthy Choice, Just Right, Product 19, Smart Start, Special K, Total, and Life cereals.

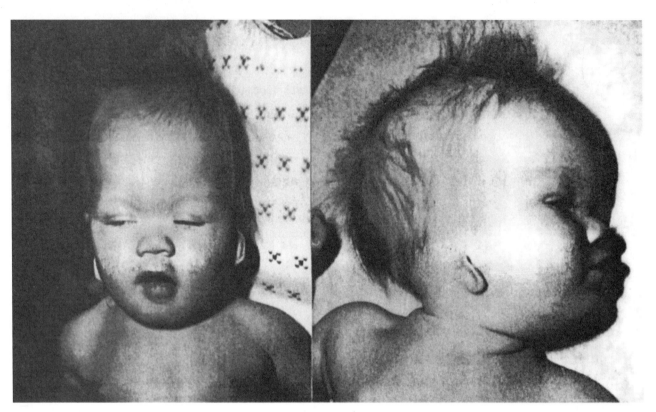

Illustration 4.17 An 8-month-old infant exposed to high levels of retinoic acid in utero. Note the high forehead, flat nasal bridge, and malformed ear.

SOURCE: Used by permission of Harcourt Health Sciences, Inc. Lott IT et al., Fetal hydrocephalus and ear abnormalities associated with maternal use of isotretinoin, J Pediatr 1984;105:597–600.

than needed based on results of clinical trials.[111] The upper limit for intake of folic acid from fortified foods and supplements is set at 1000 mcg per day. There is no upper limit for folate consumed in its naturally occurring form in foods. The 1000 mcg level represents an amount of folic acid that may mask the neurological signs of vitamin B_{12} deficiency. If left untreated, vitamin B_{12} deficiency leads to irreversible neurological damage.[106]

Vitamin A and Pregnancy Outcome

Vitamin A is a key nutrient in pregnancy because it plays important roles in reactions involved in cell differentiation. Deficiency of this vitamin is rare in pregnant women in industrialized countries, but it is a major problem in many developing nations. Vitamin A deficiency that occurs early in pregnancy can produce malformations of fetal lungs, urinary tract, and heart.[112]

Of more concern than vitamin A deficiency in the United States are problems associated with excessive intakes of vitamin A in the form of retinol or retinoic acid (but not beta-carotene). Intakes of these forms of vitamin A of over 10,000 IU per day, and the use of medications such as Accutane and Retin-A for acne and wrinkle treatment, increase the risk of fetal abnormalities. Effects are particularly striking in infants born to women using Accutane or Retin-A early in pregnancy (see Illustration 4.17). Fetal exposure to the high doses of retinoic acid in these drugs tends to develop "retinoic acid syndrome." Features of this syndrome include small ears or no ears, abnormal or missing ear canals, brain malformation, and heart defects.[11]

Due to the potential toxicity of retinol, it is recommended that women take no more than 5000 IU of vitamin A as retinol from supplements during pregnancy.[113] Most supplements made today contain beta-carotene rather than retinol. Although women are issued strong warnings not to take Retin-A or Accutane if pregnancy is possible, ill-timed use continues to occur to some extent, as does the retinoic acid syndrome.

Vitamin D Requirement

Several groups of women are at risk for vitamin D deficiency in pregnancy. Included are women who:

- Consume small amounts of vitamin D–fortified milk, or consume raw milk
- Rarely expose their skin directly to the sun
- Consistently use sunblock

- Have dark skin
- Are vegans

The primary effect of insufficient vitamin D during pregnancy is poor fetal bone formation due to poor utilization of calcium.[11] Infants born to women with inadequate vitamin D status in pregnancy tend to be small, have poorly calcified bones and abnormal enamel, and show low blood levels of calcium after birth.[114]

The need for vitamin D (5 mcg/day) can be met by including 3 cups of vitamin D–fortified milk in the daily diet, or generally by exposing the face, hands, and arms to sunlight for ½ to 2 hours per week. Winter sunlight is too weak in northern climates to permit vitamin D formation in the skin, however.[114]

Calcium Requirements in Pregnancy

Calcium metabolism changes meaningfully during pregnancy. Absorption of calcium from food increases, excretion of calcium in urine likewise increases, and bone mineral turnover takes place at a higher rate.[115] The additional requirement for calcium in the last quarter of pregnancy is approximately 300 mg per day and may be obtained by increased absorption and by release of calcium from bone.[8] (Calcium is not taken from the teeth, however.) Calcium lost from bones appears to be replaced after pregnancy in women with adequate intakes of calcium and vitamin D.[46] Inadequate calcium intake has been related to increased blood pressure during pregnancy, decreased subsequent bone remineralization, increased blood pressure of infants, and decreased breast milk concentration of calcium.[115]

CALCIUM AND THE RELEASE OF LEAD FROM BONES Lead is transferred from the mother to the fetus and can disrupt normal development of the central nervous system.[116] Pregnant women who do not consume enough calcium show greater increases in blood lead levels than women who consume 1000 mg (the DRI for calcium) or more per day. Bone tissues contain about 95% of the body's lead content, and the lead is released into the bloodstream when bones demineralize. Bone tissues demineralize to a greater extent in pregnant women who fail to consume adequate calcium.[116]

Calcium needs during pregnancy can be met by drinking 3 cups of milk or calcium-fortified soymilk, or 2 cups of calcium-fortified orange juice plus a cup of milk, or by choosing a sufficient number of other good sources of calcium daily. (Table 1.12 in Chapter 1 lists food sources of calcium.)

Fluoride

Teeth develop in utero, so why isn't it recommended that pregnant women consume sufficient fluoride so that the fetus builds cavity-resistant teeth? This is a logical question, but it has the answer "because only trace amounts of fluoride pass through the placenta to the fetus." Children of pregnant women given fluoride supplements during pregnancy have the same rates of dental caries as do children of women who did not receive supplements.[117]

Iron Status and the Course and Outcome of Pregnancy

Iron status is a leading topic of discussion in prenatal nutrition because the need for iron increases substantially; women require about 1000 mg (1 g) of additional iron during pregnancy:

- 300 mg is used by the fetus and placenta.
- 250 mg is lost at delivery.
- 450 mg is used to increase red blood cell mass.

Maternal iron stores get a boost after delivery when iron liberated during the breakdown of surplus red blood cells is recycled.[118]

Approximately 12% of women enter pregnancy with *iron deficiency* and little stored iron, and consequently are at risk of developing *iron-deficiency anemia* in pregnancy.[119]

IRON-DEFICIENCY ANEMIA IN PREGNANCY
Over recent decades, rates of iron-deficiency anemia in pregnancy have remained high in women in developing as well as developed countries (Table 4.24). Iron-deficiency anemia at the beginning of pregnancy increases the risk of preterm delivery and low-birth-weight infants by two to three times.[120] Iron deficiency during pregnancy is related to lower scores on intelligence, language, gross motor, and attention tests in affected children at the age of 5 years.[121] The mechanisms underlying these effects are unknown, but they may be related to decreased oxygen delivery to the placenta and fetus, increased rates of infection, and altered neurotransmitter function or nerve formation in the brain in iron-deficient women.[121] Iron

Iron Deficiency A condition marked by depleted iron stores. It is characterized by weakness, fatigue, short attention span, poor appetite, increased susceptibility to infection, and irritability.

Iron-Deficiency Anemia A condition often marked by low hemoglobin level. It is characterized by the signs of iron deficiency plus paleness, exhaustion, and a rapid heart rate.

Table 4.24 Estimates of the incidence of iron-deficiency anemia in women in developing and developed countries[118,119]

| | % WITH IRON-DEFICIENCY ANEMIA | |
	Developing Countries	Developed Countries
Nonpregnant	43	12
Pregnant	56	18

deficiency often occurs toward the end of pregnancy even among women who enter pregnancy with some iron stores. It is far more common than iron-deficiency anemia.

ASSESSMENT OF IRON STATUS Red cell mass increases substantially (30%) in pregnancy. However, plasma volume expands more (by about 50%). The higher increase in plasma volume compared to red cell mass makes it appear that amounts of hemoglobin, ferritin, and packed red blood cells have decreased.[7] They have not decreased but rather have become diluted by the large increase in plasma volume. Hemoglobin concentration normally decreases until the middle of the second trimester and then rises somewhat in the third. It is not necessary to prevent normal declines in hemoglobin level during pregnancy.[123]

Due to the dilution effects of increased plasma volume, changes in hemoglobin levels tend to be more indicative of plasma volume expansion than of iron status.[120] Low levels of hemoglobin or serum ferritin may be associated with high plasma volume expansion (hypervolemia), and high hemoglobin levels are related to low plasma volume expansion (hypovolemia). Low levels of plasma volume expansion are associated with reduced fetal growth, whereas newborns tend to be larger in women with higher levels of plasma volume expansion.[122]

The Centers for Disease Control have developed standard hemoglobin levels to be used in the identification of iron-deficiency anemia in pregnant women. These standards (shown in Table 4.25) represent levels below the fifth percentile of hemoglobin values in pregnancy.[124]

By trimester, hemoglobin levels indicative of iron-deficiency anemia are

- <11.0 g/dL in the first and third trimesters
- <10.5 g/dL in the second trimester

Serum ferritin cut-points indicative of iron-deficiency anemia in pregnancy have also been developed:[125]

	SERUM FERRITIN, ng/mL
Normal	>35
Depleted Stores	<20
Iron Deficiency	≤15

Hemoglobin and serum ferritin are the most commonly employed measures of iron status in pregnant women.[31]

The diagnosis of iron-deficiency anemia is more complicated than often thought. No single test of iron status is totally accurate, because (1) many factors, including infection and inflammatory disease, affect iron status; and (2) each test measures a different aspect of iron status. It is best to base the diagnosis of iron-deficiency anemia on results of several tests.[124]

Women entering pregnancy with adequate iron stores tend to absorb about 10% of total iron ingested; those with low stores absorb more—about 20% of the iron consumed.

Table 4.25 CDC's gestational age-specific cutoffs for anemia in pregnancy[124]

Gestational Weeks	Hemoglobin (g/dL) Indicating Anemia[a]
12	<11.0
16	<10.6
20	<10.5
24	<10.5
28	<10.7
32	<11.0
36	<11.4
40	<11.9

[a]For women living in high altitudes, hemoglobin values should be increased by 0.2 g/dL for every 1000 feet above 3000 and by 0.3 g/dL for every 1000 feet above 7000. For cigarette smokers, hemoglobin values should be adjusted upward by 0.3 g/dL.

The largest percentage of iron absorption, 40%, occurs in women who enter pregnancy with iron-deficiency anemia.

Iron absorption from foods and supplements is enhanced in women with low iron stores during pregnancy, and absorption increases as pregnancy progresses.[31] Absorption is highest after the 30th week of pregnancy, when the greatest amount of iron transfer to the fetus occurs. The amount of iron the placenta can transfer to the fetus, however, is limited in women who are iron deficient. Maternal iron depletion in pregnancy decreases fetal iron stores, increases the risk that infants will develop iron-deficiency anemia, and is associated with development of maternal postpartum depression.[43,126]

PROS AND CONS OF IRON SUPPLEMENTATION Absorption of iron from multimineral supplements is substantially lower than is iron absorption from supplements containing iron only. For example, women given a multimineral supplement containing iron, calcium, and magnesium absorb less than 5% of the iron, whereas women given a similar dose of iron in a supplement containing iron only absorb over twice that much.[127]

The amount of iron absorbed from supplements depends primarily on women's need for iron and the amount of iron in the supplement. As can be seen in Illustration 4.18, the amount of iron absorbed from supplements decreases substantially as the dose of iron increases. Although controversial, one theory states that the small level of improvement in iron absorption that occurs with doses of iron over 30 or 60 mg may be too little to justify their use.[123] In addition, the acceptance of high levels of iron supplementation by women is often poor, due to side effects related to the ingestion of high amounts of iron. Nausea, cramps, gas, and constipation are associated with the presence of free iron in the intestines, and these side effects increase as doses of supplemental iron increase (Table 4.26). Side effects experienced in using iron supplements are a major reason that women fail to take them.[120]

Illustration 4.18 Effect of dose of supplemental iron on iron absorption in women during pregnancy.

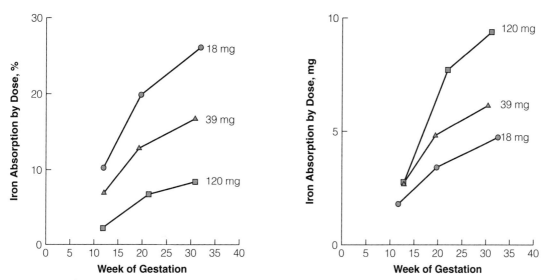

SOURCE: From Pedro Rosso, "Nutrition and Metabolism in Pregnancy: Mother and Fetus." Copyright © 1990 by Oxford University Press, Inc. Used by permission of Oxford University Press, Inc.

A relatively new concern about high-dose iron supplements is emerging. Iron supplements providing 60 mg or more iron per day regularly expose the intestinal mucosa to free iron radicals. The oxidizing effects of iron radicals cause inflammation and mitochondrial damage in cells.[130] In addition, iron doses over 30 mg per day decrease zinc absorption and lower zinc status.[31]

Amounts of elemental iron in supplements vary depending on the form of the iron compound in the supplement (Table 4.27). The proportion of iron absorbed from a constant amount of iron from each type of supplement listed in Table 4.27 is approximately equal.[131]

Since 1997 the FDA has required that the iron content of multivitamin and mineral supplements not exceed 30 mg. This action limits the excess iron that will remain in the gut after ingestion of multivitamin and mineral supplements with iron. Higher doses of iron have to be provided as single-ingredient iron supplements.

RECOMMENDATIONS RELATED TO IRON SUPPLEMENTATION IN PREGNANCY   It is generally recommended that pregnant women in the United States take a 30-mg iron supplement daily after the 12th week of pregnancy.[132] Women with iron-deficiency anemia are often given 60–180 mg of iron per day.[13,132] Iron

supplementation during pregnancy, and usual clinical practices related to it, are subjects of heated debate.

Although iron supplements increase iron stores and help prevent anemia in many women who take them, some pregnant women do not benefit from iron supplements because they do not need them.[123] Unused iron supplements, when stored and later found by young children, pose a risk of iron poisoning. Because iron status early in pregnancy is associated with adverse newborn outcomes, iron supplements may be given too late in pregnancy to achieve optimal improvements in outcomes.[123]

Proposed Alternatives to Routine Iron Supplementation   It has been suggested that women's iron status be assessed at the first prenatal visit to determine if there is a need for iron supplements. A 30-mg iron supplement would be indicated when hemoglobin levels are <11 g/dL, or if serum ferritin levels are <30 mcg/L. Women with higher values would be monitored for iron status but not given a supplement.[131]

RECOMMENDED INTAKE OF IRON DURING PREGNANCY   The increased need for iron can be met by intakes that lead to an additional 3.7 mg absorbed iron per day on average throughout pregnancy. This is a large

Table 4.26 Increased occurrence of side effects in women by supplemental iron dose[128,129]

Dose of Iron, mg/day	Side Effects
60	32%
120	40%
240	72%

Table 4.27 Percent of elemental iron by weight in various types of iron supplements

Supplement Type	Iron Content
Ferrous sulfate	20%
Ferrous gluconate	12%
Ferrous fumarate	32%

increase, especially considering that nonpregnant women consuming the DRI for iron (18 mg) absorb only around 1.8 mg of iron daily. Given an ongoing need for 1.8 mg of absorbed iron a day, and the additional need of 3.7 mg of iron daily for pregnancy, the total need for absorbed iron during pregnancy is 5.5 mg daily. Assuming 20% of iron consumed is absorbed, average iron consumption of 27 mg per day (the DRI for iron for pregnancy) will meet the iron needs of pregnancy. The Upper Limit for iron intake during pregnancy is set at 45 mg per day.

Zinc Requirement in Pregnancy

Zinc functions as a cofactor for many enzymes, including those involved in protein synthesis. Like iron, the bioavailability of zinc is higher in meats and low in plants, especially whole grains. Low levels of zinc intake, reliance on whole grain cereals for dietary zinc, and doses of iron supplements over 30 mg daily all decrease zinc status.[133]

A firm consensus on effects of zinc deficiency in pregnancy has not been reached, although animal studies clearly show that it is related to growth retardation and malformations.[134] Levels of serum zinc representative of marginal deficiency have been associated with preterm delivery, intrapartum hemorrhage, infections, and prolonged labor in human studies. It is difficult to know if these results reflect true relationships, however, because serum zinc does not appear to be a very good marker of zinc status.[133]

Iodine and Pregnancy Outcome

Iodine is needed for the synthesis of thyroid hormones, and deficiency of it early in pregnancy can lead to *hypothyroidism* in the offspring. Hypothyroidism in infants is endemic in parts of southern and eastern Europe, Asia, Africa, and Latin America.[135] The incidence of infant hypothyroidism has been found to decrease by over 70% when at-risk women in developing countries are given iodine supplements before or in the first half of pregnancy.

> **Hypothyroidism** A condition characterized by growth impairment and mental retardation and deafness when caused by inadequate maternal intake of iodine during pregnancy. Used to be called *cretinism*.

Rates of infant deaths are also substantially improved, as is the psychomotor development of the offspring.[136] Iodine supplementation in the second half of pregnancy does not improve infant outcomes.[135]

The Need for Sodium during Pregnancy

Sodium plays a critical role in maintaining the body's water balance. Requirements for it increase markedly during pregnancy due to plasma volume expansion. But the need for increased amounts of sodium in pregnancy hasn't always been appreciated. Thirty years ago in the United States, it was accepted practice to put all pregnant women on low-sodium diets. (Routine sodium restriction is still practiced in some European countries.) It was then thought that sodium increased water retention and blood pressure, and that sodium restriction would prevent edema and high blood pressure. We now know this isn't accurate and that inadequate sodium intake can complicate the course and outcome of pregnancy.[137] Sodium restriction during pregnancy may exhaust sodium conservation mechanisms and lead to excessive sodium loss.[138]

Sodium restriction is not indicated in normal pregnancy or for the control of edema or high blood pressure that develops in pregnancy. Women should be given no advice on modifying sodium intake; if they ask about it, they should be advised to consume salt "to taste."[139]

Caffeine Use in Pregnancy

Caffeine has long been suspected of causing adverse effects in pregnant women because it increases heart rate, acts as a diuretic, and stimulates the central nervous system. It easily passes from maternal to fetal blood and lingers in the fetus longer than in maternal blood because the fetus excretes it more slowly.[140]

Due to its high caffeine level, coffee has generally been the target of investigations on the effects of caffeine intake on pregnancy outcome. Coffee, however, contains hundreds of substances, and some of these have effects similar to those of caffeine. So, although conclusions about caffeine's effects on pregnancy are largely based on coffee intake, it is possible that other components of coffee are responsible for effects observed. Coffee is by far the largest contributor to caffeine intake in most people,[141] and pregnant women consume on average 170 mg caffeine from coffee per day in pregnancy.[142] (Table 2.4 in Chapter 2 provides a list of the caffeine content of beverages and foods.)

Despite the possibilities, caffeine and coffee intake during pregnancy have not been related to an increased risk of fetal malformations, reduced fetal growth, labor or delivery complications, preterm delivery, or developmental problems in offspring.[140,141]

No long-term consequences of coffee intake during pregnancy have been observed in children 7 years later. Children of women who drank coffee during pregnancy have been found to have similar levels of intellectual and neuromotor development when compared to non-coffee drinkers.[142] High levels of coffee and caffeine intake (over approximately 500 mg per day) have, however, been related to miscarriage.[40] It is generally concluded that intake of up to 4 cups of coffee per day during pregnancy is safe.

Healthy Diets for Pregnancy

Healthy diets for women during pregnancy are described in terms of calories and nutrient intake, and by food choices. Such diets have a number of characteristics in common (Table 4.28).

Table 4.28 Basics of a good diet for normal pregnancy

Good pregnancy diets:
1. Provide sufficient calories to support appropriate rates of weight gain.
2. Follow the Food Guide Pyramid food group recommendations.
3. Provide all essential nutrients at recommended levels of intake from the diet (with the possible exception of iron).
4. Include 400 mcg of folic acid daily.
5. Provide sufficient dietary fiber (28 g/day)
6. Include 9 cups of fluid daily.
7. Include salt "to taste."
8. Exclude alcohol and limit coffee intake to ≤4 cups per day.
9. Are satisfying and enjoyable.

Adequacy of caloric intake during pregnancy is generally based on rate of weight gain, but for nutrients it is based on the DRIs (Table 4.29 on the next page). Nutrient intakes during pregnancy should approximate those given in the DRI table, and food intake should correspond to the recommended types and quantity of food recommended in the Food Guide Pyramid. (The Food Guide Pyramid recommendations are presented in an upcoming section on dietary assessment in pregnancy.)

Effect of Taste and Smell Changes on Dietary Intake during Pregnancy

No inner voice directs women to consume foods that provide needed nutrients during pregnancy. Pregnant women may, however, develop food preferences and aversions due to changes in the sense of taste and smell; and they may experience *pica.*

Changes in the way certain foods taste, and the odor of foods and other substances, affect two out of three women during pregnancy. If asked to recall, many previously pregnant women could tell you which foods tasted really good to them, and which odors made them feel queasy to even think about. Increased preference for foods such as sweets, fruits, salty foods, and dairy products are common.[15] The odors of meat being cooked, coffee, perfume, cigarette smoke, and gasoline are common nasal offenders and may stimulate episodes of nausea.[130] The biological bases for such changes are not known, but they are suspected of being related to hormonal changes of pregnancy.

Pica

"Pica permits the mind no rest until it is satisfied."

F. W. Craig, 1935

Classified as an eating disorder, pica affects over half of pregnant women in some locations of the southern part of the United States. It is more common in African Americans than in other ethnic groups, and it is common enough to be considered a normal behavior in some countries.[144] Historically, one type of pica—*geophagia*—was thought to provide women with additional minerals and to ease gastrointestinal upsets. The cause of pica remains a mystery.[144]

Nonfood items most commonly craved and consumed by pregnant women with pica include ice or freezer frost (*pagophagia*), laundry starch or cornstarch (*amylophagia*), baking soda and powder, and clay or dirt (geophagia). Women experiencing pica are more likely to be iron deficient than those who don't, and iron-deficiency anemia is especially common among pregnant women who compulsively consume ice or freezer frost.[145] It is not clear, however, whether iron deficiency leads to pica or if pica leads to iron deficiency.

Pica does not appear to be related to newborn weight or preterm delivery. It can, however, complicate control of gestational diabetes if starch is eaten, and it has caused lead poisoning, intestinal obstruction, and parasitic infestation of the gastrointestinal tract.[145] Women with amylophagia sometimes accept powdered milk as an alternative to laundry starch or cornstarch, and treating anemia often stops the craving for ice or freezer frost.

PICA An eating disorder characterized by the compulsion to eat substances that are not food.

Geophagia Compulsive consumption of clay or dirt.

Pagophagia Compulsive consumption of ice or freezer frost.

Amylophagia Compulsive consumption of laundry starch or cornstarch.

Assessment of Dietary Intake during Pregnancy

Routine assessment of dietary practices is recommended for all pregnant women to determine the need for an improved diet or vitamin and mineral supplements.[31] Dietary assessment in pregnancy should cover usual dietary intake, dietary supplement use, and weight-gain progress. For best results, several days of accurately recorded, usual intake should be used.

Several levels of dietary assessment can be undertaken. (Internet resources for dietary assessment are listed at the end of the chapter.) Which assessment level is best primarily depends on the skill level of the health professional responsible for interpreting the results. Results of food-based assessments are rather straightforward to interpret, whereas computerized assessments of levels of nutrient intake are more complex.

The Food Guide Pyramid provides a good way to assess overall quality of usual dietary intake. Table 4.30 on page 109 presents an example of a recording form that can be used to assess dietary quality based on the Food Guide

Table 4.29 Dietary Reference Intakes (DRIs) for pregnant and nonpregnant women aged 19–30 years*

	Pregnant	Nonpregnant	Upper Limit (UL)
Energy, kcal			
• 2nd trimester	+350	2403	—
• 3rd trimester	+452		
Protein, gm	71	46	—
Linoleic acid, g	13	12	—
Alpha-linolenic acid, g	1.4	1.1	—
Vitamin A, mcg	770	700	3000
Vitamin C, mg	85	75	2000
Vitamin D, mcg[a]	5	5	50
Vitamin E, mg	15	15	1000[c]
Vitamin K, mcg	90	90	—
Thiamin, mg	1.4	1.1	—
Riboflavin, mg	1.4	1.1	—
Niacin, mg	18	14	35[c]
Vitamin B$_6$	1.9	1.3	100
Folate, mcg[b]	600	400	1000[c,d]
Vitamin B$_{12}$, mcg	2.6	2.4	—
Pantothenic acid, mcg	6	5	—
Biotin, mcg	30	30	—
Choline, g	450	425	3.5
Calcium, mg	1000	1000	2500
Chromium, mcg	30	25	—
Copper, mcg	1000	900	10,000
Fluoride, mg	3	3	10
Iodine, mcg	220	150	1100
Iron, mg	27	18	45
Magnesium, mg	350	310	350[c]
Manganese, mg	2	1.8	11
Molybdenum, mcg	50	45	2000
Phosphorus, mg	700	700	3500
Selenium, mcg	60	55	400
Zinc, mg	11	8	40

*DRIs for females <19 and >30 years are listed inside the front covers of this book.

a 1 mcg = 40 IU vitamin D; DRI applies in the absence of adequate sunlight.

b As Dietary Folate Equivalent (DFE). 1 DFE = 1 mcg food folate = 0.6 mcg folic acid from fortified food or supplement consumed with food = 0.5 mcg of a supplement taken on an empty stomach.

c UL applies to intake from supplements or synthetic form only.

d Applies to intake of folic acid.

Pyramid. Computerized analysis, given accurate records and entry of dietary intake and a high-quality nutrient database, provides results useful for estimating the quantity of calories and nutrients consumed. Detailed knowledge of dietary intake is particularly useful for women at risk of nutrient inadequacies or excesses, and for women with conditions such as gestational diabetes, food intolerances, multiple fetuses, and other special dietary needs.

Evaluation of dietary supplement use includes an examination of types and amounts of supplements consumed. Levels of intake exceeding the Upper Limits (ULs) for pregnancy, as well as the use of herbs and other dietary supplements not known to be safe for pregnancy, should be identified.

CULTURAL CONSIDERATIONS

People tend to be attached to existing food preferences, many of which may have deep cultural roots. Dietary recommendations will differ for Native Alaskans accustomed to a diet based on wild game; for Cambodians, Vietnamese, and Somalis who may think no meal is complete without rice; and for lactose-intolerant individuals.

The belief that consumption of certain foods "marks" the baby is common in many cultures. People may think, for example, that a woman who loves mangos and eats lots of them during pregnancy may have a baby born with a "mango-shaped" birthmark. Some cultures would hold that the baby will also have learned to love mangos because mom ate them often while pregnant.

Dietary recommendations that are not consistent with a person's usual dietary practices and beliefs, or that are not viewed as acceptable or even preferred by the woman, are least likely to be effective. For best results, dietary adjustments recommended for each individual pregnant woman should take into account her usual practices and preferences.

Vitamin and Mineral Supplementation during Pregnancy

The only supplement routinely recommended for pregnant women is iron.[31] Vitamin and mineral supplements may be indicated for specific individuals, however. An appropriate vitamin or mineral supplement is recommended when dietary and clinical assessments determine that a woman is deficient in a particular nutrient. It is also recommended that certain groups of women at high risk for poor dietary intake or increased nutrient need be given a multivitamin and mineral supplement in place of the iron supplement.

Individuals in these groups include women who:

- Have a poor-quality, unchangeable diet
- Are pregnant with multiple fetuses
- Are vegan
- Smoke cigarettes
- Have iron-deficiency anemia
- Use illicit drugs or abuse alcohol

Table 4.30 Recording and evaluation form for dietary intake in pregnancy using the Food Guide Pyramid recommendations[146]

Food Group	Serving Sizes	Recommended Servings	Actual Servings Consumed	Difference
Breads, Cereals, Rice, and Pasta		6–11		
• Bread	1 slice or 1 oz			
• Roll, biscuit	1 oz			
• Muffin, bagel	1 oz			
• Tortilla	1 oz			
• Ready-to-eat cereal	1 cup or 1 oz			
• Pasta, rice, cooked cereal	½ cup			
Vegetables		4–5		
• Raw	1 cup			
• Cooked	½ cup			
• Juice	¾ cup or 6 oz			
Fruits		3–4		
• Fresh	1 piece			
• Canned or cooked	½ cup			
• Juice	¾ cup or 6 oz			
Milk, Yogurt, Cheese		3+		
• Milk, fortified soymilk	1 cup			
• Yogurt	1 cup			
• Cheese	1½ oz			
• Cottage cheese	½ cup			
Meat, Poultry, Fish, Dried Beans, Eggs, Nuts		2–3 (6–7 oz)		
• Meat, fish, poultry	2–3 oz			
• Dried beans, cooked	½ cup			
• Tofu	½ cup			
• Eggs	2			
• Peanut butter	4 tb			
• Nuts and seeds	½ cup or 3 oz			
Fats, Oils, and Sweets				
Based on caloric need				

The multivitamin and mineral supplement recommended for use during pregnancy has the following composition:

VITAMINS	MINERALS
Vitamin B_6, 2 mg	Iron, 30 mg
Folate, 200 mcg	Zinc, 15 mg
Vitamin C, 50 mg	Copper, 2 mg
Vitamin D, 5 mcg	Calcium, 250 mg

Herbal Remedies and Pregnancy

Herbal remedies are becoming commonly used during pregnancy, although very little is known about their safety and effectiveness. Little is known because herbs are rarely tested in pregnant women due to concerns about potential damage to the fetus.

About one-third of commonly used herbal remedies have been deemed unsafe for use by pregnant women.[147] Table 4.31 provides a list of some of these herbs.

Advice to use herbal remedies during pregnancy appears to be based primarily on their traditional use in different societies. This strategy for assessing the safety of herbs doesn't always work. Some herbs considered safe based on traditional use have been found to produce malformations in animal studies.[147] Others, such as blue cohosh, which was previously thought to safely induce uterine contractions, may increase the risk of heart failure in the baby.[149] Ginseng, the most commonly used herb in the world, has been found to cause malformations in rat embryos.[150] Peppermint tea and ginger root, taken for nausea, appear to be safe.[147]

Manufacturers of herbal remedies do not have to prove they are safe for use by pregnant women. However,

Table 4.31 Herbs to avoid in pregnancy[148,149,150]

Aloe vera	Ergot
Anise	Feverfew
Black cohosh	Ginkgo
Black haw	Ginseng
Blue cohosh	Juniper
Borage	Kava
Buckthorn	Licorice
Comfrey	Pennyroyal
Cotton root	Raspberry leaf
Dandelion leaf	Saw palmetto
Ephedra, ma huang	Senna

the FDA does advise that claims related to pregnancy not be made for herbal supplements.

Exercise and Pregnancy Outcome

"Exercise is no longer simply being allowed during pregnancy, it is actively being encouraged."

K. Johnson[151]

There is no evidence that moderate or vigorous exercise undertaken by healthy women consuming high-quality diets and gaining appropriate amounts of weight is harmful to mother or fetus.[80] The bulk of evidence indicates that exercise during pregnancy benefits both the mother and her fetus. Women who exercise regularly during pregnancy feel healthier and have an enhanced sense of well-being, and their labors appear to be somewhat shorter than is the case for women who do not exercise.[152]

Researchers and practitioners are beginning to focus more on the advantages of exercise than on possible dis-advantages. Women who exercise regularly during pregnancy reduce their risk of developing gestational diabetes, pregnancy-induced hypertension, low back pain, excessive weight gain, and blood clots.[80,151]

Exercise during pregnancy can reduce fetal growth in women who are poorly nourished and gain little weight in pregnancy. It is also important for women to avoid dehydration by drinking plenty of fluids while exercising and not to become overheated during physical activity.[80]

Is it safe to begin an exercise program during pregnancy? Not only is it generally safe, it is being encouraged.[151] Beginning an exercise program during pregnancy

L. Monocytogenes, or Listeria   A foodborne bacterial infection that can lead to preterm delivery and stillbirth in pregnant women. Listeria infection is commonly associated with the ingestion of soft cheeses, unpasteurized milk, ready-to-eat deli meats, and hot dogs.

T. Gondii, or Loxoplasmosis   A parasitic infection that can impair fetal brain development. The source of the infection is often hands contaminated with soil or the contents of a cat litter box; or raw or partially cooked pork, lamb, or venison.

may improve fetal growth. This effect was shown in a study involving non-exercising pregnant women who began to exercise at 8 weeks of pregnancy. Women participated in three to five weight-bearing exercise sessions a week until delivery. Placenta function was better, and newborn weight and length greater, in exercising women compared to women who did not exercise.[153]

Exercise Recommendations for Pregnant Women

Exercise recommendations for pregnant women are similar to those for other healthy women. Pregnant women should exercise 3 to 5 times a week for 30 minutes at a heart rate that achieves 60–70% VO_2 max (Table 4.32). Exercise should begin with 5 to 15 minutes of warm-up stretches and movements and end with the same length of cooldown activities. Recommended types of exercise include walking, cycling, swimming, jogging, and dancing. Better left until after pregnancy are activities such as water and snow skiing, surfing, mountain climbing, scuba diving, and horseback riding. Switching to non-weight-bearing exercises is advised toward the end of pregnancy.[80]

Food Safety Issues during Pregnancy

Foodborne illness can be devastating during pregnancy. Increased progesterone levels that normally occur decrease pregnant women's ability to resist infectious diseases, so they are more susceptible to the effects of foodborne infections.[154] One particularly important foodborne illness is caused by *Listeria monocytogenes.* The placenta does not protect the fetus from listeria infection in the mother. Listerosis during pregnancy is associated with spontaneous abortion and stillbirth.[154] To prevent this foodborne infection, pregnant women should not eat raw fish, oysters, unpasteurized cheese, raw or undercooked meat, or unpasteurized milk. Luncheon meats, hot dogs, and other processed meats should be stored correctly.

The protozoan *Toxoplasma gondii* also causes serious effects in pregnant women and their fetuses. This protozoan can be transferred from mother to fetus and cause mental retardation, blindness, seizures, and death.[155] Sources of *T. gondii* include raw and undercooked meats,

Table 4.32 Target heart rates for healthy pregnant women[80]

Age, years	Heart Rate
<20	140–155
20–29	135–150
30–39	130–145
40+	125–140

the surface of fruits and vegetables, and cat litter. Cats that eat wild animals and undercooked meats can become infected and transfer the infection via stools left in their litter boxes.[155]

Mercury Contamination

Fish have come under fire as a potential source of mercury overload due to contamination of waters and fish by fungicides, fossil fuel exhaust, and products used in smelting plants, pulp and paper mills, leather-tanning facilities, and chemical manufacturing plants. Mercury, which passes from the mother's blood to the fetus, is a fetal neurotoxin that can produce mild to severe effects on fetal brain development. Fetuses exposed to high amounts of mercury can develop mental retardation, hearing loss, numbness, and seizures. Pregnant women are generally only slightly affected by the mercury overload. However, it accumulates in the mother's tissues and may increase fetal exposure to mercury during pregnancy and lactation.[156]

High levels of mercury are most likely to be present in the muscles of large, long-lived predatory fish such as shark, swordfish, tilefish, albacore tuna, walleye, pickerel, and bass. Mercury content of bottom feeders, such as carp, channel catfish, and white sucker is generally less than half the amount found in predatory fish. Other fish that tend to have low mercury content include "light" (not white) tuna, haddock, tilapia, salmon, cod, pollack, and sole. Shrimp, lobster, and crab generally have low mercury content, too.[157]

The FDA advises that pregnant women and women who may become pregnant not eat shark, swordfish, king mackerel, or tilefish. During pregnancy, weekly consumption of 12 ounces or less of a variety of cooked, low-mercury-content fish appears to be safe.[154] The World Wide Web address for local fish advisories is listed at the end of this chapter.

Common Health Problems during Pregnancy

Some of the physiological changes that occur in pregnancy are accompanied by side effects that can dull the bliss of expecting a child by making women feel physically miserable. Common ailments of pregnancy, such as nausea and vomiting, heartburn, and constipation, are generally more amenable to prevention than to treatment, but often can be relieved through dietary measures.

Nausea and Vomiting

Nausea occurs in about 7 in 10 pregnancies, and vomiting in 4 of 10. The conditions are so common that they are considered a normal part of pregnancy. Unless severe or prolonged, nausea and vomiting during pregnancy are associated with a reduction in risk of miscarriage of greater than 60% and with healthy newborn outcomes.[39] The cause of nausea and vomiting is not yet clear, but they are thought to be related to increased levels of human chorionic gonadotropin, progesterone, estrogen, or other hormones early in pregnancy.[158]

In the past, the nausea and vomiting of pregnancy was called "morning sickness," because it was thought to occur mostly after waking up. It actually occurs at all times of day and tends to begin during the fourth week after conception. In many women, nausea and vomiting suddenly disappear, and for most (but not all) women the symptoms end at around the 10th week. Iron supplements may aggravate nausea and vomiting when taken in the first trimester of pregnancy.[26]

HYPEREMESIS GRAVIDARUM   Between 1 and 2% of pregnant women with nausea and vomiting develop *hyperemesis gravidarum* (more commonly called hyperemesis).[159] Hyperemesis is characterized by severe nausea and vomiting that last throughout much of pregnancy. It can be debilitating. In addition to the mother feeling very sick, frequent vomiting can lead to weight loss, electrolyte imbalances, and dehydration. Women with hyperemesis who gain weight normally during pregnancy (about 30 pounds total) are not at increased risk of delivering small infants, but women who gain less (21–22 pounds) are.[160]

MANAGEMENT OF NAUSEA AND VOMITING
Many approaches to the treatment of nausea and vomiting are used in clinical practice, but only a few are considered safe and effective. Dietary interventions represent the safest method, primarily because the short- and long-term safety of many drugs and herbal remedies early in pregnancy is unclear.[161] Here are some general recommendations for women experiencing nausea and vomiting:

- Continue to gain weight.
- Separate liquid and solid food intake.
- Avoid odors and foods that trigger nausea.
- Select foods that are well tolerated.

Many women find that hard-boiled eggs, potato chips, popcorn, yogurt, crackers, and other high-carbohydrate foods are well tolerated. Personal support and understanding are important components of counseling women with nausea and vomiting. Care should be taken to individualize dietary advice based on each woman's food preferences and tolerances. Women with hyperemesis may require rehydration therapy to restore fluids and electrolyte balance.

Periodically, articles will appear in the popular press claiming that nausea and vomiting are caused by certain foods, and that women should avoid them to protect their fetus from harmful substances in the foods. Not too long ago it was claimed that bitter-tasting vegetables, for example, should be avoided. When put to the test, this notion was found to be groundless.[162] Theoretical claims that certain foods elicit nausea and vomiting in order to protect the fetus from harmful effects of the food should be considered unreliable until proven in scientific studies.

Vitamin B_6 supplements (10 mg three times a day) reduce the severity of nausea in many women.[163] (The Tolerable Upper Intake Level for vitamin B_6 in pregnancy is 100 mg per day.)

Heartburn

Pregnancy is accompanied by relaxation of gastrointestinal tract muscles. This effect is attributed primarily to progesterone. Relaxation of the muscular valve known as the cardiac or lower esophageal sphincter at the top of the stomach is thought to be the principal reason for the 30–50% incidence of heartburn in women during pregnancy. The loose upper valve may allow stomach contents to be pushed back into the esophagus.[164]

MANAGEMENT OF HEARTBURN   Dietary advice for the prevention and management of heartburn includes:

- Ingest small meals frequently.
- Do not go to bed with a full stomach.
- Avoid foods that seem to make heartburn worse.

Elevating the upper body during sleep, and not bending down so your head is below the waist, also reduce gastric reflux. Antacid tablets, which act locally in the stomach, are often recommended; but heartburn pills are not.[164]

Constipation

Relaxed gastrointestinal muscle tone is thought to be primarily responsible for the increased incidence of constipation and hemorrhoids in pregnancy. The best way to prevent these maladies is to consume approximately 30 grams of dietary fiber daily.[81] (Food sources of fiber are listed in Table 1.4C in Chapter 1, page 6.) Laxative pills are not recommended for use by pregnant women, but bulk-forming fiber in products such as Metamucil, Citrucel, and Perdiem are considered safe and effective for the prevention and treatment of constipation.[164] Women should drink a cup or more of water along with the fiber supplement.

Model Nutrition Programs for Risk Reduction in Pregnancy

"Pregnancy may be the most sensitive period of the life-cycle in which intervention may reap the greatest benefits."

A. Prentice[115]

Two programs that have been shown to substantially improve pregnancy outcomes are highlighted in this section. First is the intervention program offered by the Montreal Diet Dispensary (MDD); second is the Supplemental Nutrition Program for Women, Infants, and Children (WIC).

The Montreal Diet Dispensary

The Montreal Diet Dispensary (MDD) has served low-income, high-risk pregnant females with nutritional assessment and intervention services since the early 1900s. Part of the rationale for the WIC program in the United States was based on the successes of the MDD program. The program is located in a large, comfortable house (see Illustration 4.19) in urban Montreal. Clients are warmly welcomed into a nonthreatening, relaxed setting.

Developed as an adjunct to routine prenatal care, the MDD intervention strategy has four major components:

1. Assess the usual dietary intake and risk profile of each pregnant woman, including calories, protein, and selected vitamin and mineral adequacy; also assess stress level.
2. Determine individual nutritional rehabilitation needs based on results of the assessment.
3. Teach clients the importance of optimal nutrition and about changes that should be made through practical examples.
4. Provide regular follow-up and supervision.

The MDD dietitians are carefully trained and hold the interests of their clients first in their hearts. They treat clients with respect, openness, and affection; they also address client needs, such as transportation or emergency food or housing. Staff interactions with clients are nonjudgmental in nature and include positive feedback and praise for dietary changes and other successes of clients.

The initial client visit to the MDD takes about 75 minutes, and follow-up visits are scheduled at 2-week intervals for 40 minutes each. Women are identified as undernourished if their protein intake falls below that rec-

Illustration 4.19 The Montreal Diet Dispensary.

Judith Brown

ommended for pregnancy, and an additional protein allowance is added to the diet. Women who are underweight are given an additional daily allowance of 20 grams protein and 200 calories for each additional pound of weight gain needed to achieve a maximum of 2 pounds per week. Women identified as being under excessive stress (such as having a partner in jail, being homeless, or being abused) receive an additional allowance of protein and calories and lots of positive attention. Food supplements, including milk and eggs, and vitamin supplements are provided to women who need them.

IMPACT OF MDD SERVICES   Multiple studies have shown that women receiving MDD services have higher-birthweight infants (+107 grams), fewer low-birthweight infants (−50%), and infants with lower rates of perinatal mortality than is the case for similar women not receiving MDD services.[165,166]

The program is cost effective in relation to savings on newborn critical care, and programs based on MDD services have spread across Canada. Expenditures per client average $450. The program is primarily supported by Centraide of Greater Montreal, provincial and federal programs, and other contributions.[167]

The WIC Program

The WIC program represents an outstanding example of a successful public program intended to serve the nutritional needs of low-income women and families. It is cited as a model program in several other chapters and is described in Chapter 1.

In operation since 1974, WIC provides nutritional assessment, education and counseling, food supplements, and access to health services to over 6 million participants. WIC serves low-income pregnant, postpartum, and breastfeeding women; and children up to 5 years of age who are at nutritional risk. Supplemental food provided to women includes milk, ready-to-eat cereals, dried beans, fruit juice, and cheese; some programs offer vouchers for farmer's markets.

Participation in WIC is related to reduced rates of iron-deficiency anemia in pregnancy, higher-birthweight infants, decreased low-birthweight infants, and lower rates of iron-deficiency anemia in women after delivery. For each dollar invested in WIC, approximately $3 in health care costs are saved. Internet addresses leading to additional information about WIC are listed in the "Resources" section at the end of this chapter.

Resources

Pregnancy Resources and Information
Visit the Women's Health Resource Center for access to journal articles and summaries; information on pregnancy, growth, and development; and health care and diversity information.
Web site: **www.medscape.com**
The Bureau of Maternal and Child Health Web site provides information on programs for pregnant women, hot topics, and announcements of new publications.
Web site: **http://mchlibrary.info**
The National Library of Medicine Web site offers extensive coverage of scientific journal articles, summaries, and educational resources from a variety of reputable organizations on pregnancy, nutrition, diet, and disorders of pregnancy.
Web site: **www.nlm.nih.gov/medlineplus**

Fish Advisories
This site links to local freshwater fish advisories.
Web site: **www.epa.gov/ost/fish**

Health Canada
Nutrition Web site that provides access to Health Canada's policy statement on nutrient needs of pregnant women.
Web site: **www.hc-sc.gc.ca/nutrition**

U.S. Government
This USDA site provides links to vast resources related to education materials for WIC, vegetarian diets, health fraud, and other topics.
Web site: **www.nal.usda.gov/fnic**
The U.S. Government's site, which provides information on food and nutrition programs and eligibility; links to scientific references; and information about the nutrition needs of infants, children, adults, and seniors.
Web site: **www.nutrition.gov**

Vegetarian Diets
Useful information on vegetarian diets, resources, and organizations can be obtained through these Web sites:
Web site: **www.nal.usda.gov/fnic**
Web site: **www.llu.edu/llu/vegetarian/vegnews.htm**
Web site: **www.vegetariannutrition.net**
Web site: **www.vrg.org**

Cultural Diversity
Provides information on ethnic diets as well as cultural and health beliefs of a wide range of population groups immigrating to the United States and Canada.
Web site: **healthlinks.washington.edu/clinical/ethnomed**

WIC
USDA provides access to information about the WIC program, the WIC Works Food Safety Resource List, and other resources.
Web site: **www.nal.usda.gov/fnic**

Dietary Analysis
Select "Health Eating Index" and run dietary intake records one day at a time. Analyzes diet by food groups and selected nutrients.
Web site: **www.usda.gov/cnpp**
From the University of Illinois Food Science and Nutrition Department, this "Nutrition Analysis Tool" is free for noncommercial use. Provides in-depth food constituent results.
Web site: **www.nat.uiuc.edu**
This USDA site is the best one for food composition data.
Web site: **www.nal.usda.gov/fnic/foodcomp**

References

1. Health United States, 2000. Vol. 2001; and Deaths: Preliminary Data for 2002; www.cdc.gov/nchs, accessed 2/14/04.

2. Wegman ME. Infant mortality in the 20th century, dramatic but uneven progress. J Nutr 2001;131:401S–8S.

3. MacDorman MF et al. Annual summary of vital statistics. Pediatrics 2002;110:1037–56.

4. Williams RL, Creasy RK, Cunningham GC et al. Fetal growth and perinatal viability in California. Obstet Gynecol 1982; 59:624–32.

5. Godfrey KM. Maternal regulation of fetal development and health in adult life. Eur J Obstet Gynecol Repro Biol 1998; 78:141–50.

6. Hytten FE, Leitch I. The physiology of human pregnancy. Oxford: Blackwell Scientific Publications; 1971.

7. Cruikshank DP et al. Maternal physiology in pregnancy. In: Gabbe SG et al., eds. Obstetrics: normal and problem pregnancies. New York: Churchill Livingstone; 1996: pp 91–109.

8. King JC. Physiology of pregnancy and nutrient metabolism. Am J Clin Nutr 2000; 71:1218S–25S.

9. Baker H et al. Vitamin profile of 563 gravidas during trimesters of pregnancy. J. Am Coll Nutr 2002;21:33–7.

10. Fazleabas AT. Update and advances in reproductive biology. Experimental Biology Annual Meeting, San Diego, April 11, 2003.

11. Rosso P. Nutrition and metabolism in pregnancy. New York: Oxford University Press; 1990: pp 117–118, 125, 150–151. Winick M. Nutrition, pregnancy, and early infancy. Baltimore: Williams & Wilkins; 1989: p 182.

12. Butte NF. Carbohydrate and lipid metabolism in pregnancy: normal compared with gestational diabetes mellitus. Am J Clin Nutr 2000;71:1256S–61S.

13. Naeye RL, Chez RA. Effects of maternal acetonuria and low pregnancy weight gain on children's psychomotor development. Am J Obstet Gynecol 1981;139:189–93.

14. Duggleby SL, Jackson AA. Higher weight at birth is related to decreased amino acid oxidation during pregnancy. Am J Clin Nutr 2002;76:852–7.

15. van Stiphout WAHJ, Hofman A, de Bruijn AM. Serum lipids in young women before, during, and after pregnancy. Am J Epidemiol 1987;126:922–8.

16. Martin U, Davies C, Hayavi S et al. Is normal pregnancy atherogenic? Clin Sci (Colch) 1999;96:421–5.

17. Ortega RM, Gaspar MJ, Cantero M. Influence of maternal serum lipids and maternal diet during the third trimester of pregnancy on umbilical cord blood lipids in two populations of Spanish newborns. Int J Vitam Nutr Res 1996;66:250–7.

18. King JC. Effect of reproduction on the bioavailability of calcium, zinc and selenium. J Nutr 2001;131:1355S–8S.

19. American College of Obstetricians and Gynecologists. Nutrition during pregnancy. ACOG Technical Bulletin No. 179 1993:1–7.

20. Gluckman PD. Fetal origins of adult disease: insulin resistance. Medscape Ob/Gyn & Women's Health 2003;8:4–5.

21. Rosso P, Cramoy C. Nutrition and pregnancy. In: Winick M, ed. Human nutrition: pre- and post-natal development. New York: Plenum Press; 1979: pp 133–228.

22. King JC. The risk of maternal nutritional depletion and poor outcomes increases in early or closely spaced pregnancies. J Nutr 2003;133:1732S–6S.

23. Scholl TO, Stein TP, Smith WK. Leptin and maternal growth during adolescent pregnancy. Am J Clin Nutr 2000;72:1542–7.

24. Norwitz ER et al. Implantation and the survival of early pregnancy. N Engl J Med 2001;345:1400–8.

25. Rozovski SJ, Winick M. Nutrition and cellular growth. In: Winick M, ed. Human nutrition: pre- and postnatal development. New York: Plenum Press; 1979: pp 61–102.

26. Godfrey KM, Barker DJ. Fetal nutrition and adult disease. Am J Clin Nutr 2000; 71:1344S–52S.

27. Smith DW. Growth and its disorders: basics and standards, approach and classifications, growth deficiency disorders, growth excess disorders, obesity. In: Schaeffer AJ, Markowtiz M, eds. Major problems in clinical pediatrics. Vol. 15. Philadelphia: W. B. Saunders Company; 1979: pp 134–169.

28. Winick M. Malnutrition and brain development. J Pediatr 1969;74:667–79.

29. Ziegler EE, O'Donnell AM, Nelson SE, Fomon SJ. Body composition of the reference fetus. Growth 1976;40:329–41.

30. Rosso P, Winick M. Intrauterine growth retardation. A new systematic approach based on the clinical and biochemical characteristics of this condition. J Perinat Med 1974;2:147–60.

31. National Academy of Sciences. Nutrition during pregnancy. I. Weight gain. II. Nutrient supplements. Washington, DC: National Academy Press; 1990.

32. Hediger ML, Overpeck MD, Maurer KR et al. Growth of infants and young children born small or large for gestational age: findings from the Third National Health and Nutrition Examination Survey. Arch Pediatr Adolesc Med 1998;152:1225–31.

33. Lucas A, Fewtrell MS, Davies PSW et al. Breastfeeding and catch-up growth in infants born small for gestational age. Acta Paediatr 1997;86:564–9.

34. Luo ZC, Albertsson-Wikland K, Karlberg J. Length and body mass index at birth and target height influences on patterns of postnatal growth in children born small for gestational age. Pediatrics 1998;102:E72.

35. Lazer S, Biale Y, Mazor M et al. Complications associated with the macrosomic fetus. J Repro Med 1986;31:501–5.

36. Kallen B. Endpoints studied in the epidemiology of reproduction. Epidemiology of human reproduction. Boca Raton, FL: CRC Press; 1988: pp 142–9.

37. Tempfer C et al. The etiology of habitual abortion: a review. Geburtshilfe und Frauenheilkunde 2000;60:604–8.

38. Wen W, Shu XO, Jacobs DR Jr. et al. The associations of maternal caffeine consumption and nausea with spontaneous abortion. Epidemiology 2001;12:38–42.

39. Weigel RM, Weigel MM. Nausea and vomiting of early pregnancy and pregnancy outcome. A meta-analytical review. Br J Obstet Gynecol 1989;96:1312–18.

40. Kramer MS, Demissie K, Yang H et al. The contribution of mild and moderate preterm birth to infant mortality. Fetal and Infant Health Study Group of the Canadian Perinatal Surveillance System. JAMA 2000; 284:843–9.

41. Chery C et al. Hyperhomocysteinemia is related to a decreased level of vitamin B_{12} in the second and third trimesters of pregnancy. Clin Chem Lab Med 2002;40:1105–8.

42. Ortega RM et al. Riboflavin levels in maternal milk: the influence of vitamin B_2 status during the third trimester of pregnancy. J Am Coll Nutr 1999;18:324–9.

43. Sommer A et al. Assessment and control of vitamin A deficiency: the Annecy Accords. J Nutr 2002;132:2845S–50S.

44. Goldenberg RL, Rouse DJ. Prevention of premature birth. N Eng J Med 1998; 339:313–20.

45. Hobel C, Culhane J. Role of psychosocial and nutritional stress on poor pregnancy outcome. J Nutr 2003; 133:1709S–17S.

46. King JC. Preface. Am J Clin Nutr 2000;71:1217S.

47. Gluckman PD, Pinal CS. Regulation of fetal growth by the somatotrophic axis. J Nutr 2003;133:1741S–6S.

48. Lucas A, Fewtrell MS, Cole TJ. Fetal origins of adult disease—the hypothesis revisited. BMJ 1999;319:245–9.

49. Rich-Edwards JW, Stampfer MJ, Manson JE et al. Birth weight and risk of cardiovascular disease in a cohort of women followed up since 1976. BMJ 1997;315:396–400.

50. Kuzawa CW, Adair LS. Lipid profiles in adolescent Filipinos: relation to birth weight and maternal energy status during pregnancy. Am J Clin Nutr 2003;77:960–6.

51. Belizan JM, Villar J, Bergel E et al. Long-term effect of calcium supplementation during pregnancy on the blood pressure of offspring; follow up of a randomized controlled trial. BMJ 1997;315:281–5.

52. Villar J, Cossio TG. Nutritional factors associated with low birth weight and short gestational age. Clin Nutr 1986;5:78–85.

53. Eastman NJ. Expectant motherhood. Boston: Little, Brown and Company; 1947: p 198.

54. Parker JD, Abrams B. Prenatal weight gain advice: an examination of the recent prenatal weight gain recommendations of the Institute of Medicine. Obstet Gynecol 1992;79:664–9.

55. Brown JE, Murtaugh MA, Jacobs DR et al. Variation in newborn size by trimester weight change in pregnancy. Am J Clin Nutr 2002;76:205–9.

56. Carmichael SL, Abrams B. A critical review of the relationship between gestational weight gain and preterm delivery. Obstet Gynecol 1997;89:865–73.

57. To WW, Cheung W. The relationship between weight gain in pregnancy, birth-weight and postpartum weight retention. Aust N Z J Obstet Gynaecol 1998;38:176–9.

58. Abrams B. Prenatal weight gain and postpartum weight retention: a delicate balance (editorial). Am J Public Health 1993; 83:1082–4.

59. King JC, Mukherjea R. Leptin targeted in research on obesity, pregnancy. ARS News Service, USDA, www.ars.usda.gov, accessed 10/02.

60. Scholl TO et al. Insulin and the "thrifty" woman: the influence of insulin during pregnancy on gestational weight gain and postpartum weight retention. Maternal Child Health Journal 2002;6:255–61.

61. Keppel KG, Taffel SM. Pregnancy related weight gain and retention: implications of the 1990 Institute of Medicine guidelines. Am J Public Health 1993;83:1100–3.

62. Brown JE, Kaye SA, Folsom AR. Parity-related weight change in women. Int J Obes 1992;16:627–31.

63. Gigante DP, Victora CG, Barros FC. Breast-feeding has a limited long-term effect on anthropometry and body composition of Brazilian mothers. J Nutr 2001;131:78–84.

64. Polley A et al. Randomized controlled trial to prevent weight gain in pregnant women. Int J Obes 2002;26:1494–1502.

65. Mussey RD. Nutrition and human reproduction: an historical review. Am J Obstet Gynecol 1949;58:1037–48.

66. Stein AD, Ravelli AC, Lumey LH. Famine, third-trimester pregnancy weight gain, and intrauterine growth: the Dutch Famine Birth Cohort Study. Hum Biol 1995;67:135–50.

67. Smith CA. Effects of maternal undernutrition upon the newborn infant in Holland (1944–1945). J Pediatr 1947;30:229–43.

68. Susser M, Stein Z. Timing in prenatal nutrition: a reprise of the Dutch Famine Study. Nutr Rev 1994;52:84–94.

69. Roseboom TJ, van der Meulen JH, Osmond C et al. Plasma lipid profiles in adults after prenatal exposure to the Dutch famine. Am J Clin Nutr 2000;72:1101–6.

70. Antonov AN. Children born during the siege of Leningrad in 1942. J Pediatr 1947; 30:250–9.

71. Gruenwald P, Funakawa H, Mitani S et al. Influence of environmental factors on fetal growth in man. Lancet 1967;I:1026–8.

72. Guyer B, MacDorman MF, Martin JA et al. Annual summary of vital statistics—1997. Pediatrics 1998;102:1333–49.

73. Simic S, Idrizbegovic S, Jaganjac N et al. Nutritional effects of the siege on newborn babies in Sarajevo. Eur J Clin Nutr 1995; 49(suppl)2:S33–6.

74. Burke BS. Nutritional needs in pregnancy in relation to nutritional intakes as shown by dietary histories. Obstetr Gynecol Survey 1948;3:716–30.

75. Burke BS, Harding VV, Stuart HC. Nutrition studies during pregnancy. IV. Relation of protein content of mother's diet during pregnancy to birth length, birth weight, and condition of infant at birth. J Pediatr 1943;23:506–15.

76. Hytten F, Chamberlain G. Clinical physiology in obstetrics. Oxford: Blackwell Scientific Publications; 1980: pp 92–130.

77. Brown JE, Kahn ESB. Maternal nutrition and the outcome of pregnancy: a renaissance in research. Clin Perinatol 1997;24:433–49.

78. Sleutel M, Golden SS. Fasting in labor: relic or requirement. J Obstet Gynecol Neonatal Nurs 1999;28:507–12.

79. Mangesi L, Hofmeyr GJ. Early compared with delayed oral fluids and food after caesarean section. Cochrane Database Syst Rev 2002;3:CD003515.

80. Dietary reference intakes: energy, carbohydrate, fiber, fat, fatty acids, cholesterol, protein, and amino acids. Washington, DC: National Academies Press, 2002.

81. Anderson JW. Health implications of wheat fiber. Am J Clin Nutr 1985;41:1103–12.

82. Duffy VB, Anderson GH. Position of the American Dietetic Association: use of nutritive and nonnutritive sweeteners. J Amer Diet Assoc 1998;98:580–7.

83. American Academy of Pediatrics. Committee on Substance Abuse and Committee on Children with Disabilities. Fetal alcohol syndrome and alcohol-related neurodevelopmental disorders. Pediatrics 2000;106:358–61.

84. Ventura SJ, Martin JA, Curtin SC et al. Births: final data for 1998. Natl Vital Stat Rep 2000;48:1–100.

85. March of Dimes Birth Defects Foundation. Public health education information sheet: drinking alcohol during pregnancy, 1997.

86. Polygenis D, Wharton S, Malmberg C et al. Moderate alcohol consumption during pregnancy and the incidence of fetal malformations: a meta-analysis. Neurotoxicol Teratol 1998;20:61–7.

87. Konovalov HV, Kovetsky NS, Bobryshev YV et al. Disorders of brain development in the progeny of mothers who used alcohol during pregnancy. Early Hum Dev 1997; 48:153–66.

88. Food and Nutrition Board. National Academy of Sciences (Institute of Medicine) NRC, Subcommittee on the 10th Edition of the RDAs, Commission on Life Sciences. Recommended dietary allowances. Washington, DC: National Academy Press, 1989.

89. Johnston PK. Counseling the pregnant vegetarian. Am J Clin Nutr 1988;48:901–5.

90. Position of the American Dietetic Association and Dietitians of Canada: vegetarian diets. J Am Diet Assoc 2003;103:748–65.

91. Messina V et al. A new food guide for North American vegetarians. J Am Diet Assoc 2003;103:771–5.

92. Haddad EH et al. Vegetarian food guide pyramid: a conceptual framework. Am J Clin Nutr 1999;70(suppl):615S–19S.

93. Neurologic impairment in children associated with maternal dietary deficiency of cobalamin—Georgia, 2001. MMWR 22003;52(04):61–4.

94. Innis SM, Elias SL. Intakes of essential n-6 and n-3 polyunsaturated fatty acids among pregnant Canadian women. Am J Clin Nutr 2003;77:473–8.

95. Conner WE. Importance of n-3 fatty acids in health and disease. Am J Clin Nutr 2000;71:171S–5S.

96. Kraus RM et al. Revision 2000: a statement for healthcare providers from the Nutrition Committee of the American Heart Association. J Nutr 2001;131:132–46.

97. Ursin VM. Modification of plant lipids for human health: development of functional land-based omega-3 fatty acids. J Nutr 2003;1003:4271–4.

98. Ershow AG, Brown LM, Cantor KP. Intake of tapwater and total water by pregnant and lactating women. Am J Public Health 1991;81:328–34.

99. Lumley J, Watson L, Watson M et al. Periconceptional supplementation with folate and/or multivitamins for preventing neural tube defects. Cochrane Database Syst Rev 2000;2.

100. Suitor CW, Bailey LB. Dietary folate equivalents: interpretation and application. J Amer Diet Assoc 2000;100:88–94.

101. Bailey LB. Evaluation of a new Recommended Dietary Allowance for folate. J Amer Diet Assoc 1992;92:463–71.

102. Fodinger M, Wagner OF, Horl WH et al. Recent insights into the molecular genetics of the homocysteine metabolism. Kidney Int 2001;59(suppl)78:S238–42.

103. Warkany J. Production of congenital malformations by dietary measures. JAMA 1958;168:2020–3.

104. Smithells RW. Availability of folic acid. Lancet 1984;1:508.

105. Eskes TKAB. From birth to conception. Open or closed. Eur J Obstet Gynecol Reprod Biol 1998;78:169–77.

106. Folic acid. MMWR Morb Mortal Wkly Rep 2001;50:185–9.

107. Folate status in women of childbearing age—United States, 1999. MMWR Morb Mortal Wkly Rep 2000;49:962–5.

108. Daly LE, Kirke PN, Molloy A et al. Folate levels and neural tube defects. JAMA 1995;274:1698–702.

109. Brown JE, Jacobs DR, Jr., Hartman TJ et al. Predictors of red cell folate level in women attempting pregnancy. JAMA 1997;277:548–52.

110. Cuskelly GJ, McNulty H, Scott JM. Fortification with low amounts of folic acid makes a significant difference in folate status in young women: implications for the prevention of neural tube defects. Am J Clin Nutr 1999;70:234–9.

111. Shaw GM, Schaffer D, Velie EM et al. Periconceptional vitamin use, dietary folate, and the occurrence of neural tube defects. Epidemiology 1995;6:219–26.

112. Zile MH. Function of vitamin A in vertebrate embryonic development. J Nutr 2001;131:705–8.

113. Smithells D. Vitamins in early pregnancy. BMJ 1996;313:128–9.

114. Specker BL. Do North American women need supplemental vitamin D during pregnancy or lactation? Am J Clin Nutr 1994;59:484S–91S.

115. Prentice A. Maternal calcium metabolism and bone mineral status. Am J Clin Nutr 2000;71:1312S–6S.

116. Hertz-Picciotto I, Schramm M, Watt-Morse M et al. Patterns and determinants of blood lead during pregnancy. Am J Epidemiol 2000;152:829–37.

117. Leverett DH, Adair SM, Vaughan BW et al. Randomized clinical trial of the effect of prenatal fluoride supplements in preventing dental caries. Caries Res 1997;31:174–9.

118. Fairbanks VF. Iron in medicine and nutrition. In: Modern Nutrition in Health and Disease, Shils ME et al., eds. 9th ed. Philadelphia: Lippincott, Williams & Wilkins, 1999; pp 193–221.

119. Cogswell ME et al. Iron supplement use among women in the United States: science, policy, and practice. J Nutr 2003;133: 1974S–7S.

120. Allen LH. Biological mechanisms that might underlie iron's effects on fetal growth and preterm birth. J Nutr 2001;131:581S–9S.

121. Tamura T et al. Cord serum ferritin concentrations and mental and psychomotor development of children at five years of age. J Pediatrics 2002;140:165–70.

122. Allen LH. Anemia and iron deficiency: effects on pregnancy outcome. Am J Clin Nutr 2000;71:1280S–4S.

123. Beard JL. Iron requirements in adolescent females. J Nutr 2000;130:440S–2S.

124. Recommendations to prevent and control iron deficiency in the United States. Centers for Disease Control and Prevention. MMWR Morb Mortal Wkly Rep 1998;47:1–29.

125. Godel JC, Pabst HF, Hodges PE et al. Iron status and pregnancy in a northern Canadian population: relationship to diet and iron supplementation. Can J Public Health 1992;83:339–43.

126. Sweet DG, Savage G, Tubman TR et al. Study of maternal influences on fetal iron status at term using cord blood transferring receptors. Arch Dis Child Fetal Neonatal Ed 2001;84:F40–3.

127. Seligman PA, Caskey JH, Frazier JL et al. Measurements of iron absorption from prenatal multivitamin–mineral supplements. Obstet Gynecol 1983;61:356–62.

128. Dawson EB, Dawson R, Behrens J et al. Iron in prenatal multivitamin/multimineral supplements. Bioavailability. J Reprod Med 1998;43:133–40.

129. Singh K, Fong YF, Kuperan P. A comparison between intravenous iron polymaltose complex (ferrum hausmann) and oral ferrous fumarate in the treatment of iron deficiency anaemia in pregnancy. Eur J Haematol 1998;60:119–24.

130. Casanueva E, Viteri FE. Iron and oxidative stress in pregnancy. J Nutr 2003; 133:1700S–8S.

131. Iron deficiency anemia: recommended guidelines for prevention, detection, and management among US children and women of childbearing age. Washington, DC: National Academy Press, 1993.

132. Cogswell ME et al. Iron supplementation during pregnancy, anemia, and birth weight: a randomized controlled trial. Am J Clin Nutr 2003;78:773–81.

133. King JC. Determinants of maternal zinc status during pregnancy. Am J Clin Nutr 2000;71:1334S–43S.

134. Tamura T, Goldenberg RL, Johnston KE et al. Maternal plasma zinc concentrations and pregnancy outcome. Am J Clin Nutr 2000;71:109–13.

135. Xue-Yi C, Xin-Min J, Zhi-Hong D et al. Timing of vulnerability of the brain to iodine deficiency in endemic cretinism. N Eng J Med 1994;331:1739–44.

136. Mahomed K, Gulmezoglu AM. Maternal iodine supplements in areas of deficiency (Cochrane Review). The Cochrane Library. Issue 3. Oxford: Update Software, 1999.

137. Delemarre FM, Steegers EA, Berendes JN et al. Eclampsia despite strict dietary sodium restriction. Gynecol Obstet Invest 2001;51:64–5.

138. Pike RL, Gursky DS. Further evidence of deleterious effects produced by sodium restriction during pregnancy. Am J Clin Nutr 1970;23:883–9.

139. Duley L, Henderson-Smart D. Reduced salt intake compared to normal dietary salt, or high intake, in pregnancy. Cochrane Database Syst Rev 2000;2.

140. Nehlig A, Debry G. Consequences on the newborn of chronic maternal consumption of coffee during gestation and lactation: a review. J Am Coll Nutr 1994;13:6–21.

141. Hinds TS, West WL, Knight EM et al. The effect of caffeine on pregnancy outcome variables. Nutr Rev 1996;54:203–7.

142. Barr HM, Streissguth AP. Caffeine use during pregnancy and child outcome: a 7-year prospective study. Neurotoxicol Teratol 1991;13:441–8.

143. Patten-Hitt E. Nausea during pregnancy linked to keen sense of smell. Reuters Health 2000.

144. Johns T, Duquette M. Detoxification and mineral supplementation as functions of geophagy. Am J Clin Nutr 1991;53:448–56.

145. Rainville AJ. Pica practices of pregnant women are associated with lower maternal hemoglobin level at delivery. J Amer Diet Assoc 1998;98:293–6.

146. Healthful eating during pregnancy: follow the Food Guide Pyramid. www.usda.gov/fnic, accessed 10/03.

147. FDA statement concerning structure/function rule and pregnancy claims. HHS Statement, U.S. Department of Health and Human Services 2000; 65 FR 1000; January 6, 2000; Docket No. 98N-0044.

148. Belew C. Herbs and the childbearing woman. Guidelines for midwives. J Nurse Midwifery 1999;44:231–52.

149. Jones FA. Herbs: useful plants. J R Soc Med 1996;89:717–9.

150. Meisler JG. Toward optimal health: the experts discuss the use of botanicals by women. J Women's Health 2003;12:847–52.

151. Johnson K. Pregnancy exercise recommendations growing more liberal. Medscape Ob/Gyn & Women's Health 2003;8:1–2.

152. Artal R, Sherman C. Exercise during pregnancy—safe and beneficial for most. Physician and Sports Medicine 1999;27: 51–56.

153. Clapp JF, III, Kim H, Burciu B et al. Beginning regular exercise in early pregnancy: effect on fetoplacental growth. Am J Obstet Gynecol 2000;183:1484–8.

154. Position of the American Dietetics Association: Food Safety. J Am Diet Assoc 2003;103:1203–18.

155. Soto C. Toxoplasmosis in pregnancy. Clin Rev 2002;12:51–6.

156. EPA fish advisory update. Vol. 2001, 2001.

157. Schober SE et al. Fish consumption of blood mercury levels in women of childbearing age in the United States. JAMA 2003; 289:1667–74.

158. Erick M. Hyperolfaction and hyperemesis gravidarum: what is the relationship? Nutr Rev 1995;53:289–95.

159. Tsang IS, Katz VL, Wells SD. Maternal and fetal outcomes in hyperemesis gravidarum. Int J Gynaecol Obstet 1996; 55:231–5.

160. Gross S, Librach C, Cecutti A. Maternal weight loss associated with hyperemesis gravidarum: a predictor of fetal outcome. Am J Obstet Gynecol 1989;160:906–9.

161. Scialli A. Revived focus on treatment for morning sickness. Medscape Wire 2000.

162. Brown JE, Kahn ES, Hartman TJ. Profet, profits and proof: do nausea and vomiting of early pregnancy protect women from "harmful" vegetables? Am J Obstet Gynecol 1997;176:179–81.

163. Jewel D, Young G. Interventions for nausea and vomiting in early pregnancy. Cochrane Database Syst Rev 2002; (1):CD000145.

164. Baron TH, Ramirez B, Richter JE. Gastrointestinal motility disorders during pregnancy. Ann Intern Med 1993;118:366–75.

165. Higgins AC, Moxley JE, Pencharz PB et al. Impact of the Higgins Nutrition Intervention Program on birth weight: a within-mother analysis. J Amer Diet Assoc 1989; 89:1097–103.

166. Dubois S, Coulombe C, Pencharz P et al. Ability of the Higgins Nutrition Intervention Program to improve adolescent pregnancy outcome. J Amer Diet Assoc 1997; 97:871–8.

167. Duquette MP, Director of the Montreal Diet Dispensary. Personal communication, 2000.

"Women at greater risk
of adverse birth
outcomes benefit the
most from educational
health care messages."
M. D. Kogan et al. 1994

Chapter 5

Nutrition during Pregnancy:

Conditions and Interventions

Chapter Outline

- Introduction
- Hypertensive Disorders of Pregnancy
- Diabetes in Pregnancy
- Multifetal Pregnancies
- HIV/AIDS during Pregnancy
- Eating Disorders in Pregnancy
- Nutrition and Adolescent Pregnancy
- Evidence-Based Practice

Prepared by **Judith E. Brown**

Key Nutrition Concepts

1   Some complications of pregnancy are related to women's nutritional status.

2   Nutritional interventions for a number of complications of pregnancy can benefit maternal and infant health outcomes.

3   Nutritional interventions during pregnancy should be based on scientific evidence that supports their safety, effectiveness, and affordability.

Introduction

"Practice is science touched with emotion."

Stephen Paget, 1909, *Confessio Medici*

Almost all healthy women expect that their pregnancies will proceed normally and that they will be rewarded at delivery with a healthy newborn. For the vast majority of pregnancies, this expectation is fulfilled. For other women, however, the path to a healthy newborn is strewn with obstacles in the form of health problems that women bring into or develop during pregnancy. This chapter addresses a number of these health conditions and the role of nutrition in their etiology and management. The specific health conditions presented are hypertensive disorders of pregnancy, preexisting and gestational diabetes, obesity, multifetal pregnancy, HIV/AIDS, eating disorders, and adolescent pregnancy.

Hypertensive Disorders of Pregnancy

Hypertensive disorders of pregnancy are the second leading cause of maternal mortality in the United States. They affect 6 to 8% of pregnancies and contribute significantly to stillbirths, fetal and newborn deaths, and other adverse outcomes of pregnancy. The causes of most cases of hypertension during pregnancy remain unknown, and cures for these disorders remain elusive.[1]

Several types of hypertensive disorders in pregnancy have been identified (Table 5.1). In the past, the major types of hypertensive disorders in pregnancy were grouped under the heading "pregnancy-induced hypertension," or PIH. This terminology is being phased out in favor of the classification scheme for hypertensive disorders of pregnancy presented in Table 5.1. Use of edema as an indicator of hypertensive disorders is fading because edema is not predictive of adverse pregnancy outcomes.[1] In addition, new recommendations for the diagnosis of hypertensive

---

Table 5.1 Definitions and features of hypertensive disorders of pregnancy*[1]

Chronic Hypertension

Hypertension that is present before pregnancy or diagnosed before 20 weeks of pregnancy. Hypertension is defined as blood pressure ≥140 mm Hg systolic or ≥90 mm Hg diastolic blood pressure.

Hypertension first diagnosed during pregnancy that does not resolve after pregnancy is also classified as chronic hypertension.

Gestational Hypertension

This condition exists when elevated blood pressure levels are detected for the first time after mid-pregnancy. It is not accompanied by proteinuria. If blood pressure returns to normal by 12 weeks postpartum, the condition is considered to be transient hypertension of pregnancy. If it remains elevated, then the woman is considered to have chronic hypertension.

Women with gestational hypertension are at lower risk for poor pregnancy outcomes than are women with preeclampsia.

Preeclampsia-Eclampsia

A pregnancy-specific syndrome that usually occurs after 20 weeks gestation (but that may occur earlier) in previously normotensive women. It is determined by increased blood pressure during pregnancy to ≥140 mm Hg systolic or ≥90 mm Hg diastolic and is accompanied by proteinuria. In the absence of proteinuria, the disease is highly suspected when increased blood pressure is accompanied by headache, blurred vision, abdominal pain, low platelet count, and abnormal liver enzyme values.

- Proteinuria is defined as the urinary excretion of ≥0.3 grams of protein in a 24-hour urine specimen. This usually correlates well with readings of ≥30 mg/dL protein, or ≥2 on dipstick readings taken in samples from women free of urinary tract infection. In the absence of urinary tract infection, proteinuria is a manifestation of kidney damage.
- Eclampsia is defined as the occurrence of seizures that cannot be attributed to other causes in women with preeclampsia.

Preeclampsia Superimposed on Chronic Hypertension

This disorder is characterized by the development of proteinuria during pregnancy in women with chronic hypertension. In women with hypertension and proteinuria before 20 weeks of pregnancy, it is indicated by a sudden increase in proteinuria, blood pressure, or abnormal platelet or liver enzyme levels.

*Blood pressure values used to determine status should be based on two or more measurements of blood pressure in relaxed settings.

disorders of pregnancy no longer include specific levels of increase in blood pressure, such as 30 mm Hg systolic and 15 mm Hg diastolic, as have been used in the past.[1]

Chronic Hypertension

The incidence of chronic hypertension—or that diagnosed before 20 weeks after conception, as opposed to hypertension that develops later in pregnancy—ranges from 1 to 5% depending on the population studied. The condition is more likely to occur in African Americans, obese women, women over 35 years of age, and women who experienced high blood pressure in a previous pregnancy.[2]

Women with mild hypertension may be taken off anti-hypertension medications preconceptionally or early in pregnancy, because the drugs do not appear to improve the course or outcome of pregnancy.[3] Mild hypertension in healthy women that does not become worse during pregnancy appears to pose few risks to maternal and newborn health. Pregnancies among women with blood pressures ≥160/110 mm Hg—either or both values—are associated with an increased risk of fetal death, preterm delivery, and fetal growth retardation. Selection of the proper anti-hypertension medicines for women during pregnancy reduces these risks somewhat. Some anti-hypertension medicines, though, reduce maternal blood sodium levels and limit plasma volume expansion. This, in turn, decreases fetal growth.

NUTRITIONAL INTERVENTIONS FOR WOMEN WITH CHRONIC HYPERTENSION IN PREGNANCY Preconceptional and pregnancy diets of women with hypertension should be carefully monitored with the aim of achieving adequate and balanced diets for pregnancy. Weight-gain recommendations are the same as for other pregnant women.

Women with salt-sensitive hypertension, or hypertension that responds to dietary sodium intake, must be managed along a fine line between consuming too much sodium for good blood pressure control, and consuming too little at the potential cost of impaired fetal growth.[3] For women with hypertension that was managed successfully in part by a low-sodium diet prior to pregnancy, continuing that dietary approach is generally recommended.[1]

Although exercise is beneficial to pregnant and nonpregnant women alike, it is not clear whether it is safe for women with chronic hypertension. Therefore, until more is known about its safety, aerobic exercise is discouraged for women with chronic hypertension.[1]

Gestational Hypertension

Gestational hypertension is generally diagnosed after 20 weeks of pregnancy and represents a less serious condition than preeclampsia. Unlike those suffering from preeclampsia, women with gestational diabetes have nei-

ther elevated serum insulin levels nor proteinuria.[4] Women with this disorder are at greater risk for hypertension and stroke later in life.[5]

Preeclampsia-Eclampsia

Preeclampsia-eclampsia represents a syndrome characterized by:

- Deficits in *prostacyclin* relative to *thromboxane*
- Blood vessel spasms and constriction
- Increased blood pressure
- Adverse maternal immune system responses to the placenta
- Platelet aggregation and blood coagulation
- Alterations of hormonal and other systems related to blood volume and pressure control
- Oxidative tissue damage and inflammation
- Alterations in calcium regulatory hormones
- Reduced calcium excretion[1,6]

Virtually all maternal organs can be affected in preeclampsia. Organs most affected by small blood clots, vasoconstriction, and reduced blood flow are the placenta and the mother's kidney, liver, and brain.[6]

Eclampsia can be a life-threatening condition and one that is difficult to predict. Eclamptic seizures appear to be related to hypertension, the tendency of blood to clot, and spasms of and damage to blood vessels in the brain. It complicates about 1 in 2000 pregnancies.[7]

Signs and symptoms of preeclampsia range from mild to severe (Table 5.2), as do the health consequences (Table 5.3 on the next page). The cause

> **Prostacyclin** A potent inhibitor of platelet aggregation and a powerful vasodilator and blood pressure reducer derived from n-3 fatty acids.
>
> **Thromboxane** The parent of a group of thromboxanes derived from the n-6 fatty acid arachidonic acid. Thromboxane increases platelet aggregation and constricts blood vessels, causing blood pressure to increase.

Table 5.2 Signs and symptoms of preeclampsia[9,10,11]

- Hypertension
- Increased urinary protein (albumin)
- Decreased plasma volume expansion (hemoglobin levels >13 g/dL)
- Low urine output
- Persistent and severe headaches
- Sensitivity of the eyes to bright light
- Blurred vision
- Abdominal pain
- Nausea
- Increased platelet aggregation, vasoconstriction related to increased thromboxane levels and decreased levels of prostacyclin

Table 5.3 Outcomes related to the existence of preeclampsia during pregnancy[2,12,13]

Mother
- Early delivery by Cesarean section
- Acute renal dysfunction
- Increased risk of gestational diabetes, hypertension, and type 2 diabetes later in life
- Abruptio placenta (rupture of the placenta)

Newborn
- Growth restriction
- Respiratory distress syndrome

Table 5.4 A sampling of factors that may influence the development of preeclampsia in women with insulin resistance[4,12,14]

Prepregnancy
- Elevated serum insulin, free fatty acids, triglycerides
- Disrupted platelet functions

Pregnancy
- Genetic predisposition to hypertension
- Hormonal and metabolic changes
- Renal disease
- Nutritional status

of preeclampsia is unknown but appears to originate from abnormal implantation and vascularization of the placenta.[6] The only cure is delivery.[1] Signs and symptoms of preeclampsia generally disappear rapidly after delivery, but eclampsia may occur within 12 days following delivery.[8]

PREECLAMPSIA AND INSULIN RESISTANCE
Growing evidence indicates that the onset and effects of preeclampsia are at least partially mediated by insulin resistance.[14] (For background information on insulin resistance, see Chapter 3.) The evidence is based on the presence of features of the insulin resistance syndrome in women with preeclampsia and the fact that women with insulin resistance are at increased risk of developing preeclampsia.[6]

Shared features of preeclampsia and insulin resistance include elevated serum insulin, hypertension, dyslipidemia (reduced HDL cholesterol and elevated free fatty acids and triglycerides), relative glucose intolerance, disruption of platelet functions and coagulation abnormalities, atherosclerotic changes, and obesity.[15] A number of these conditions may be "clinically silent" or unnoticed in women with preeclampsia but represent persistent alterations in body functions. Abnormalities associated with insulin resistance may worsen during pregnancy due to the profound hormonal and other metabolic changes that occur. The likelihood that this situation will arise appears to be increased in women with insulin resistance who are genetically predisposed to developing hypertension (Table 5.4).[12]

Elevated fasting insulin of twice normal values and postprandial levels that are four times higher than normal for pregnancy are found in preeclampsia and may be accompanied by normal glucose levels. The simultaneous presence of normal glucose and high insulin levels indicates that women can still produce enough insulin to compensate for low cell membrane sensitivity to the action of insulin. The ability of pancreatic beta cells to continue to

Endothelium The layer of flat cells lining blood and lymph vessels. These cells produce a variety of proteins that play a role in blood pressure regulation and body fluid distribution.

produce high enough levels of insulin to overcome the effects of insulin resistance may eventually end, however.

Women with preeclampsia are at increased risk for developing diabetes during pregnancy and type 2 diabetes later in life. In addition, about 15% of women with gestational diabetes and 30% of women with type 2 diabetes before pregnancy will develop preeclampsia. High insulin levels found in women with insulin resistance favor the development of hypertension, atherosclerosis, heart disease, stroke, and increased levels of body fat in the years following pregnancy.[12]

High blood pressure associated with insulin resistance may participate in physiological mechanisms that impair functions of the *endothelium* and modify the balance among various types of prostaglandins, such as prostacyclin and thromboxane.[15] As presented in Chapter 2, prostaglandins are physiologically active substances made from the fatty acid arachidonic acid. Found in many tissues, they constrict or dilate blood vessels, among other functions.

RISK FACTORS FOR THE DEVELOPMENT OF PREECLAMPSIA
The roots of preeclampsia lie very early in pregnancy, but as yet there is no reliable means of identifying women who will develop it before the condition is established.[8] However, women with insulin resistance, obesity, or other characteristics listed in Table 5.5 are at increased risk for developing the disease. Characteristics listed confer at least a twofold increase in risk.

Increased rates of preterm delivery and low birthweight in infants born to women with preeclampsia are partly related to clinical decisions to deliver fetuses early in order to treat the disease. Most infants born to women with this disorder are normal weight, however, and some newborns are large for gestational age. Variations in birthweight associated with preeclampsia appear be related to the severity of the disease in individual women.[18]

The risk of developing hypertension associated with preeclampsia is higher in women who were born small for gestational age (SGA). It appears that growth restriction

Table 5.5 Risk factors for preeclampsia[1,3,16,17]

- First pregnancy (nulliparous)
- Obesity, especially high levels of central body fat
- Underweight
- Mother's smallness at birth
- African Americans, American Indians
- History of preeclampsia
- Preexisting diabetes mellitus
- Age over 35 years
- Multifetal pregnancy
- Insulin resistance
- Chronic hypertension
- Renal disease
- History of preeclampsia in the mother's or father's mother
- High blood levels of homocysteine
- Inadequate diet (possibly related to inadequate vitamins C and E, calcium, zinc, and the n-3 fatty acids)

Table 5.6 Status of effectiveness of preventive measures for preeclampsia[1,3,21]

- Antihypertensive medications: Controversial, may not improve pregnancy outcomes
- Low-dose aspirin: May reduce risk of preeclampsia and improve outcomes to some extent
- Calcium supplementation: Reduces the incidence of preeclampsia in high-risk women
- Magnesium supplementation: Does not appear to prevent preeclampsia
- Fish oils (n-3 fatty acids): No reduction in preeclampsia in high-risk women
- Vitamins C and E: May prevent preeclampsia
- High- or low-protein diets: Ineffective for prevention of preeclampsia
- Low-sodium diets: Ineffective for the prevention of preeclampsia
- Weight loss: Ineffective for the prevention of preeclampsia

in utero may impair mechanisms involved in the regulation of blood pressure and increase the probability that high blood pressure will develop with the physiological stresses of pregnancy.[19] Interestingly, women or men whose mothers had preeclampsia while pregnant with them are more likely to have children born from pregnancies complicated by preeclampsia.[20]

Several drugs (metformin and troglitazone, for example) enhance insulin sensitivity, but their safety for treating insulin resistance in pregnancy is not clear. Insulin sensitivity can be improved by weight loss after pregnancy, exercise, and diet modification.[12]

NUTRIENT INTAKE AND PREECLAMPSIA
Inadequate nutrient status has been investigated as a potential cause of preeclampsia, and various nutrients have been given to women with the aim of reducing its occurrence. Some of the results are promising. Table 5.6 provides an overview of nutritional and other remedies that have been tested for their usefulness in preventing preeclampsia.

Calcium and Preeclampsia   Calcium plays a role in the maintenance of normal blood vessel tone and has been implicated in the etiology of hypertension in adults. Whether it plays a role in the development of preeclampsia in pregnant women continues to be investigated.

Randomized, placebo-controlled clinical trials have found that intakes around the DRI for calcium in pregnancy (1000 mg per day), or 1 to 2 grams of supplemental calcium daily (1000–2000 mg) decrease blood pressure and preeclampsia in pregnant women. (The UL for cal-

cium in pregnant women is 2.5 grams.) Results of various trials indicate that calcium supplements decrease blood pressure in pregnant women in general by about 20%. Women at high risk of preeclampsia, however, experience a 78% reduction in occurrence, and women with low, baseline intakes of calcium a 68% reduction in preeclampsia. Calcium does not appear to affect the risk of preterm delivery in women in general, but a 56% reduction is seen in women at high risk of developing preeclampsia.[22] In a randomized, placebo-controlled trial, women at risk of preeclampsia given 450 mg of linoleic acid and 600 mg of calcium per day in the third trimester experienced a 75% reduction of preeclampsia. Infants born to women taking the supplements weighed an average of 124 grams more than those of women who received the placebo.[23]

Studies have also found that increased calcium intakes from food or supplements ranging from 375 to 2000 mg per day during pregnancy may decrease blood pressure levels in children.[24]

Vitamin C and E and the Prevention of Preeclampsia
Supplementation with vitamin C (1000 mg) and E (400 IU) from 16 to 22 weeks of gestation to delivery shows promise in reducing oxidative damage caused by preeclampsia,[25] and has been found to reduce its occurrence by over half.[26] Decisions on the use of antioxidant supplementation to prevent preeclampsia await the results of additional clinical trials.

Effects of calcium and antioxidant nutrients on the risk of preeclampsia may depend on when the increased intakes begin. It is possible that protective effects are diminished by initiating these changes too late in pregnancy, or after about 20 weeks.[2] Such options are being seriously considered because the effectiveness of current therapies for treating preeclampsia is controversial.[16]

Folic Acid, Hyperhomocysteinemia, and Preeclampsia
Women with preeclampsia tend to experience many of the same clinical features as do women at risk of heart disease. Such conditions include insulin resistance and atherosclerotic changes in blood vessels (of the placenta in pregnant women). High blood levels of homocysteine are related to these same conditions, giving rise to the theory that folic acid and hyperhomocysteinemia may be related to preeclampsia.[27] Women with elevated blood levels of homocysteine are over four times more likely to have preeclampsia or eclampsia than are women with low homocysteine levels.[24] Although it is not known whether adequate intake of folic acid reduces preeclampsia or its symptoms, it does appear to normalize plasma homocysteine levels in pregnant women with preeclampsia.[28]

High concentrations of homocysteine in pregnancy are also associated with repeated spontaneous abortion and placenta abruption.[29]

n-3 Fatty Acids and the Risk of Preeclampsia A number of trials of the effectiveness of n-3 fatty acid supplements or fish consumption on the occurrence of preeclampsia have been reported. It is speculated that prostacyclin levels would be increased and thromboxane levels decreased by these fatty acids, and that vasoconstriction would be lowered as a result. Results of the trials are mixed, however. It appears that n-3 fatty acids may increase birthweight and gestational age at delivery somewhat but may not affect the occurrence of preeclampsia.[16]

OTHER NUTRITIONAL THEORIES RELATED TO PREECLAMPSIA Through the years, various nutritional causes of preeclampsia have been proposed, and corresponding nutritional remedies implemented in practice— even in the absence of supportive studies. These theories have included the belief that low-protein diets, high and rapid weight gain in pregnancy, and high salt intakes prompt the development of preeclampsia.[6] In one situation, belief in the safety and potential benefits of a rice and fruit diet motivated a physician to recommend it for the prevention of preeclampsia. At least one woman who followed the advice ended up winning a malpractice lawsuit. The grossly inadequate diet was found to be related to her daughter's mental retardation.[30] The bottom line is that it is unwise to test unsubstantiated theories about nutrition and diseases such as preeclampsia on women and their fetuses.

Preeclampsia Case Presentation

Signs, symptoms, severity, and causes of preeclampsia vary from woman to woman. Therefore, appropriate interventions for women presenting differing aspects of the syndrome are best designed on a case-by-case basis.[18]

By way of example, Case Study 5.1 describes the course of preeclampsia in one woman experiencing the condition.

Nutritional Recommendations and Interventions for Preeclampsia

In the best of circumstances, dietary interventions for preeclampsia would begin prior to pregnancy. This approach might give women the opportunities to decrease body weight and stores of central body fat, become physically fit, and consume a diet that reduces the need for insulin. Short of those circumstances, dietary recommendations and interventions should begin as early in pregnancy as possible and target at-risk women.

Nutritional and physical activity recommendations that may benefit women at risk of preeclampsia are within the normal scope of practice:

- 1000 mg per day of dietary calcium
- 600 DFE of folate, which includes 400 mcg of folic acid daily
- Five or more servings of vegetables and fruits daily
- No restriction of sodium intake (with the possible exception of some women with chronic hypertension)
- Consumption of the assortment of other basic foods recommended in the Food Guide Pyramid
- Moderate exercise (for example, walking, swimming, noncompetitive tennis, or dancing for 30 minutes three times a week) unless medically contraindicated
- Weight gain that follows recommendations based on prepregnancy weight status
- Three regular meals and snacks daily
- Consumption of low-glycemic-index rather than high-glycemic-index carbohydrate foods.[1,31,32,33,34]

The glycemic index (GI) of various food sources of carbohydrates is shown in Table 5.7 on page 126. Additional information about GI and issues related to its use in practice can be found in Chapter 3. Iron supplements, especially if taken in high doses, may aggravate inflammation by increasing the body's free radical load.[36] Women with preeclampsia should not be given high-dose iron supplements.

Whether supplemental calcium or vitamins C and E should be recommended to women at risk of preeclampsia isn't clear, but doses below the Tolerable Upper Level of Intake (UL) for pregnancy are likely safe. Before these supplements can be confidently recommended or ignored, additional information is needed on their benefits to outcomes such as intrauterine growth retardation and perinatal death associated with preeclampsia. The multivitamin and mineral supplement recommended by the Institute of Medicine (IOM)[37] and discussed previously should be given to women considered to be at risk of nutrient shortages.

Case Study 5.1

A Case of Preeclampsia

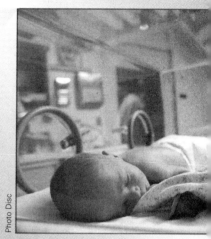

Photo Disc

Susan became pregnant when she was 31 years old and weighed 209 pounds (95 kg). She experienced hypertension and renal disease in her early 20s, but has been free of these disorders for 5 years before pregnancy. Six months before the current pregnancy, Susan spontaneously aborted an 11-week-old fetus affected by congenital malformations. She has a strong family history of heart disease; both her mother and 44-year-old brother died of heart attacks.

At 17 weeks gestation, Susan was identified as having proteinuria and was excreting 2300 mg protein in her urine in a 24-hour period. It was reasoned that the protein loss was related to scarring in tubules of the kidney due to her earlier bout with renal disease. By 21 weeks of pregnancy, her blood pressure had increased to 152/100 mm Hg. Ultrasound revealed an active fetus who was at the 25th percentile for weight by gestational age. Susan was given antihypertension medications, a multivitamin and mineral supplement, and aspirin, but at 23 weeks her blood pressure was still high.

Susan was admitted to a hospital at 27 weeks because the proteinuria and hypertension had worsened. Protein excretion had increased to 7 grams a day, and blood pressure remained high but did not increase further. The level of protein excretion was now thought to be an effect of the preeclampsia, in addition to that related to previous kidney damage. Due to her family history of heart disease, Susan's homocysteine levels were tested and found to be elevated. She was not experiencing the symptoms of severe headache, blurred vision, abdominal pain, or nausea that are commonly associated with severe preeclampsia. She also was found to have normal platelet and liver enzyme values, and normal kidney function. A diagnosis of preeclampsia was made and confirmed by renal biopsy.

Susan developed a low platelet count and elevated liver enzymes during week 28, and a Cesarean-section delivery was performed. A severely growth retarded 655-gram fetus and 133-gram placenta were delivered. The infant had a long hospital stay but was eventually able to go home.

Questions

1. What clinical finding might be involved in Susan's previous miscarriage of a malformed fetus, her high homocysteine level, and the severity of her preeclampsia?
2. Name three disorders Susan is at risk of developing due to her history of preeclampsia.
3. If you were to develop a nutrition intervention plan for Susan and could focus on only one health problem, what would it be?

(Answers are in Appendix D.)

Diabetes in Pregnancy

Diabetes is the second leading complication in pregnancy.[38] It has several forms:

- Gestational diabetes
- Type 2 diabetes
- Type 1 diabetes
- Other specific types[39]

Gestational diabetes and type 2 diabetes are part of the same disease. This chapter focuses on gestational diabetes, which develops during pregnancy. Information related to type 1 diabetes, or insulin-dependent diabetes, is presented as well.

Gestational Diabetes

"The difference between gestational and type 2 diabetes may be the moment of detection."

Branchtein[41]

Table 5.7 Glycemic index (GI) of selected foods[35]

HIGH GI		MEDIUM GI		LOW GI	
glucose	100	Cheerios	74	muesli	48
French bread	95	popcorn	72	green peas	48
scone	92	watermelon	72	pasta	48
potato, baked	85	Grape Nuts	71	carrots, raw	47
potato, instant mashed	85	wheat bread	70	Cassava	46
Corn Chex	83	white bread	70	lactose	46
pretzel	83	orange soda	68	milk chocolate	43
Rice Krispies	82	sucrose	68	All Bran	42
cornflakes	81	croissant	67	orange	42
Corn Pops	80	Cream of Wheat	66	peach	42
Gatorade	78	couscous	65	apple juice	40
jelly beans	78	chapati	62	plum	39
doughnut, cake	76	sweet potato	61	apple	38
waffle, frozen	76	muffin, blueberry	59	pear	38
French fries	75	Coca-Cola	58	tomato juice	38
shredded wheat	75	rice, white or brown	60	yam	37
		honey	55	dried beans	25
		oatmeal	54	grapefruit	25
		corn	53	cherries	22
		cracked wheat bread	53	fructose	19
		orange juice	52	xylitol	8
		banana	52	hummus	6
		mango	51		
		potato, boiled	50		

Over 3% of pregnant women develop *gestational diabetes* during pregnancy, and the incidence is increasing along with obesity.[40] The cause of gestational diabetes is controversial, but it is considered to be a form of non-insulin-dependent diabetes, or type 2.[42] It is not clear whether the disorder occurs primarily due to increasing insulin resistance and exaggerated liver production of glucose late in pregnancy, or to limits on the ability of beta cells to produce enough insulin to overcome the effects of increased insulin resistance in pregnancy.[43] Gestational diabetes in underweight and normal-weight women appears to be related to insulin resistance combined with a deficit in insulin production, whereas insulin resistance—not inadequate insulin production—may underlie it in obese women.[44] Women who develop gestational diabetes appear to enter pregnancy with a predisposition to insulin resistance and type 2 diabetes that is expressed due to physiological changes that occur during pregnancy.

The insulin resistance brought into pregnancy, or the tendency to develop it, may be clinically silent in that glucose levels may not be elevated and blood pressure may be normal. However, high blood levels of glucose and other signs related to increased insulin resistance develop as pregnancy progresses. Women with gestational diabetes develop elevated levels not only of blood glucose but also

Gestational Diabetes Carbohydrate intolerance with onset or first recognition in pregnancy.

of triglycerides, fatty acids, and sometimes blood pressure. In some cases, gestational diabetes appears to be related to exaggerated metabolic changes favoring elevated blood glucose levels[43] or reduced insulin output.[45] High maternal blood glucose levels reach the fetus (Illustration 5.1) and cause the fetus to increase insulin production to lower

Illustration 5.1 Concentrations of fetal blood glucose following an intravenous dose of glucose to the mother.

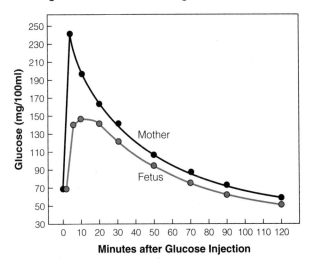

SOURCE: F. Hytten and G. Chamberlain, eds., Clinical Physiology in Obstetrics, Blackwell, 1980. Reprinted by permission; and based on data from Coultart, et al. Reprinted by permission.

Table 5.8 Adverse outcomes associated with gestational diabetes[12,43,46]

Mother
- Cesarean delivery to prevent shoulder dystocia
- Increased risk for preeclampsia during pregnancy
- Increased risk of type 2 diabetes, hypertension, and obesity later in life
- Increased risk for gestational diabetes in a subsequent pregnancy

Offspring
- Stillbirth
- Spontaneous abortion
- Macrosomia (>10 pounds or >4500 grams)
- Neonatal hypoglycemia
- Increased risk of insulin resistance, type 2 diabetes, high blood pressure, and obesity later in life

it. The higher the level of blood glucose received, the larger the fetal output of insulin.

Consequences of Poorly Controlled Gestational Diabetes

Potential consequences associated with gestational diabetes are summarized in Table 5.8. Exposure to high insulin levels in utero leads to increased glucose uptake into cells and the conversion of glucose to triglycerides. These changes increase fetal formation of fat and muscle tissue and may program metabolic adaptations, increasing the likelihood that insulin resistance, type 2 diabetes, high blood pressure, and obesity will develop later in life.[43] The chances that these disorders will occur increase with higher maternal levels of glucose and triglycerides during pregnancy.[46]

Effects of high maternal levels of glucose and triglycerides are particularly striking in the Pima Indians of Arizona. Fetal exposure to poorly controlled maternal diabetes incurs a tenfold increase in the risk that children will develop type 2 diabetes. Offspring of diabetic Pima mothers are heavier at birth, have higher body mass index (BMI) throughout childhood, and have 7–20 times greater incidence of type 2 diabetes in early adulthood. Although risks of these conditions increase in offspring of women with poorly controlled diabetes in general, the pronounced effect in Pima Indians is likely due to a strong genetic tendency toward insulin resistance and obesity.[47]

The end of pregnancy initially restores insulin sensitivity in most women with gestational diabetes. However, a degree of insulin resistance often remains.[12] Close to half of women with gestational diabetes in a previous pregnancy will develop it in a subsequent pregnancy.[48] The cumulative incidence of type 2 diabetes in women who experienced gestational diabetes is approximately 50% after 5 years.

Women with weight gain after pregnancy and repeated pregnancies continue to experience insulin insufficiency and resistance; this group is even at higher risk of developing type 2 diabetes later in life. Among women who have experienced gestational diabetes, those requiring insulin therapy have higher blood pressure than women whose gestational diabetes was controlled with diet and exercise.[12]

Risk Factors for Gestational Diabetes

Both type 2 and gestational diabetes are linked to multiple inherited predispositions and their environmental triggers, such as excess body fat and low physical activity levels.[12] About half of women who develop gestational diabetes have no identified risk for the disease, however.[49] Risk factors for gestational diabetes are outlined in Table 5.9.

Diagnosis of Gestational Diabetes

The diagnosis of gestational diabetes is principally based on blood glucose levels, and disagreement continues about what levels of blood glucose are most likely to predict which women are at significant risk. The disagreement exists because the relationship between maternal blood glucose levels and adverse perinatal outcomes is continuous. Consequently, cut-points selected for blood glucose levels diagnostic of gestational diabetes are arbitrary and subject to change.[52] Current standards for blood glucose levels indicative of risk for gestational diabetes, as well as characteristics of women which place them at potential risk, are outlined next.

CRITERIA FOR THE DIAGNOSIS OF GESTATIONAL DIABETES Glucose screening is recommended for women at high risk of gestational diabetes at

Table 5.9 Risk factors for gestational diabetes[48–52]
- Obesity, especially high levels of central body fat
- Weight gain between pregnancies
- Underweight
- Age over 35 years
- Native American, Hispanic, African American, South or East Asian, Pacific Islander, Indigenous Australian ancestry
- Family history of gestational diabetes
- History of delivery of a macrosomic newborn (>4500 g or 10 lbs)
- Chronic hypertension
- Mother was SGA at birth
- History of gestational diabetes in a previous pregnancy
- Diabetes in pregnant women's mothers during the pregnancy with them and LGA at birth

the initial visit or as soon as possible thereafter. High risk is identified in women who have one or more of the following:

- Marked obesity
- Diabetes in a mother, father, sister, or brother
- History of glucose intolerance
- Previous macrosomic infant
- Current glucosuria

A 50-gram oral glucose challenge test is generally used for blood glucose screening. This test can be done without fasting. Blood is collected 1 hour after the glucose load is consumed and tested for glucose content. This test should be followed by an oral glucose tolerance test if glucose level is high, or ≥130 mg/dL (7.2 mmol/L). (You can convert mg/dL to millimoles per liter, or mmol/L, by multiplying mg/dL by 0.05551.)

The oral glucose tolerance test (OGTT) is the basis for the diagnosis of most cases of gestational diabetes. It can be bypassed among women with very high glucose screening results and treatment started. A 100-gram glucose, 3-hour test is used for the OGTT. The practice of "loading women up" with a high-carbohydrate diet 3 days in advance of the test is no longer recommended. It does not alter results of the OGTT nor reduce the proportion of women incorrectly diagnosed with gestational diabetes.[53] The beverage provided for this test is very sweet and is supposed to be consumed in 5 minutes. Many women find it hard to swallow the liquid in that time because it makes them feel queasy. New glucose-load products, such as flavored soft drinks, jellybeans, and gelatin, are also being used.[54] A diagnosis of gestational diabetes is made when two or more values for venous serum or plasma glucose concentrations exceed these levels:

Overnight fast	95 mg/dL
One hour after glucose load	180 mg/dL
Two hours after glucose load	155 mg/dL
Three hours after glucose load	140 mg/dL

A 75-gram oral glucose load can also be used, but the 3-hour glucose level will likely differ from results using the 100-gram glucose load test. Results indicating two or more abnormal glucose values using a 75-gram load are diagnostic of gestational diabetes. Because of their increased risk for preeclampsia, women with gestational diabetes should be closely monitored for preeclampsia.[51]

A plasma glucose screening between 24 and 28 weeks of pregnancy is recommended for women at "average risk" and for high-risk women not determined by glucose screen to have elevated glucose levels earlier. Average risk is defined as women who fit neither the low- nor the high-risk profile.

Glucose screens are not recommended for women at low risk, defined as:

- Age <25 years
- Member of a low-risk ethnic group (those other than Hispanic, African American, South or East

Asian, Pacific Islander, Native American, or Indigenous Australian)
- No diabetes in first-degree relatives
- Normal prepregnancy weight and weight gain during pregnancy
- No history of glucose intolerance
- No prior poor obstetrical outcomes.[51]

Women with gestational diabetes may notice an increased level of thirst (especially in the morning), an increased volume of urine, and other signs related to high blood glucose levels.[55] Urinary glucose cannot be used to diagnose nor monitor gestational diabetes, because the results do not correspond to blood glucose levels.[56]

Treatment of Gestational Diabetes

A team approach to caring for women with diabetes in pregnancy is advised. Such teams often consist of an obstetrician, a registered dietitian who is also a certified diabetes educator, a nurse educator, and an endocrinologist.[57] The primary approach to treatment is medical nutrition therapy that begins with attempts to normalize blood glucose levels with diet and exercise.

If postprandial glucose levels remain high 2 weeks after institution of the diet and exercise plan, then insulin injections will be added. Postprandial blood glucose rather than fasting glucose levels are related to fetal overgrowth and are the main indicators of adequacy of blood glucose control.[43]

Medical nutrition therapy has been shown to effectively normalize blood glucose levels and to decrease the risk of adverse perinatal outcomes. Results shown in Table 5.10 demonstrate the effect and the usefulness of identifying and intervening upon women with gestational diabetes. It can also be noted from the results that a higher proportion of large newborns occurs even with medical nutrition therapy, but that the incidence is substantially less than in women with untreated gestational diabetes.

Blood glucose levels can be brought down by low caloric intakes. However, accelerated rates of starvation

Table 5.10 Comparison of outcomes of unrecognized and diet-treated gestational diabetes

	GESTATIONAL DIABETES		
Outcome	Unrecognized	Diet Treated	Controls
LGA (>90th percentile)	44%	9%	5%
Macrosomia (>4500 g)	44%	15%	8%
Shoulder dystocia	25%	3%	3%
Birth trauma	25%	0%	0%

SOURCE: Data from Adams, 1998.[56]

metabolism during pregnancy, as well as potentially deleterious effects of resulting ketonemia on fetal development, exclude this approach to blood glucose control.[58] Correspondingly, restriction of pregnancy weight gain to below recommended amounts is not advised.[59] Aggressive treatment of gestational diabetes that excessively limits caloric intake and weight gain increases the risk of SGA newborns.[60] On the other hand, excessively high caloric balances and weight gains are of concern because they increase the risk of macrosomia.[61]

Type 2 diabetes in nonpregnant individuals is often treated with sulfonylurea oral medications. These drugs cannot be used in pregnancy, because they cross the placenta and stimulate fetal insulin production. Other types of oral medications are being tested for use among women with gestational diabetes.[62]

Presentation of a Case Study

No two women with gestational diabetes share the same history, risks, needs, and response to treatment. Case Study 5.2, presented on page 130, represents an individual's experience with the disorder.

Exercise Benefits and Recommendations

Insulin resistance is decreased and blood glucose control enhanced by regular aerobic exercise such as walking, jogging, biking, golfing, hiking, swimming, and moderate weight lifting. This appears to be the case as well in women with gestational diabetes. Weight lifting with the arms 3 days a week for 20 minutes per session for 6 weeks, and exercising on a recumbent bicycle at 50% VO_2 max for 45 minutes three times a week, have been found to normalize blood glucose levels in some women.[63]

Levels of exercise that approximate 50–60% of VO_2 max, or maximal oxygen uptake, are most often recommended for women with gestational diabetes. These levels are estimated in practice using a formula for heart rates associated with various levels of VO_2 max. The formula is 220 − age × 0.50 (for 50% of VO_2 max) = heartbeats per minute. In the case of a 29-year-old, the estimated heart rate at 50% of VO_2 max would be 220 − 29 × 0.50, or 96 beats per minute. This would be the maximum heart rate she should experience while exercising. Levels of exercise should make women become slightly sweaty but not overheated, dehydrated, or exhausted.[59]

Nutritional Management of Women with Gestational Diabetes

Primary outcome goals for women with gestational diabetes are well-controlled blood glucose levels and a healthy newborn. Other goals include the normalization of carbohydrate metabolism and a reduction in the mother's and offspring's subsequent risk of diabetes,

hypertension, heart disease, and obesity.[59] For most women, diet and exercise changes will be the primary way to achieve these goals. In other women, supplemental insulin will also help achieve these goals.

Except in cases of very high glucose levels, dietary changes are given a 2-week trial before insulin is used. A 2-week trial of diet and exercise changes is generally undertaken due to the general impression that blood glucose control not achieved during a 2-week trial is unlikely to be achieved given a longer period of time.[64] Blood glucose control is more likely to require insulin in obese women than in normal-weight or underweight women.[9]

The following are components of the nutritional management of women with gestational diabetes:

- Assessing dietary and exercise habits
- Developing an individualized diet and exercise plan
- Monitoring weight gain
- Interpreting blood glucose and urinary ketone results
- Ensuring follow-up during pregnancy and postpartum[34,63]

Women with type 2 diabetes coming into pregnancy are managed in much the same fashion as are women with gestational diabetes, only nutritional care begins earlier. Ideally, normal blood glucose levels should be established prior to conception and then maintained in good control through pregnancy. Diet and exercise plans for women with type 2 diabetes can often be based on what has worked in the past, thus simplifying planning to needs associated with pregnancy.

THE DIET PLAN   In general, diets developed for women with gestational diabetes emphasize

- Whole grain breads and cereals, vegetables, fruits, and high-fiber foods
- Limited intake of simple sugars and foods and beverages that contain them
- Low-GI foods, or carbohydrate foods that do not greatly raise glucose levels
- Monounsaturated fats
- Three regular meals and snacks daily

Dietary planning is based around a calculated level of caloric need. These initial estimates of caloric need are intended to meet both maternal and fetal demand for energy while limiting increases in blood glucose levels. They are based on the pregnant woman's current weight and her need to gain weight during pregnancy. Estimated levels of caloric need according to women's current weight status are shown in Table 5.11 on page 131.

Calories are generally distributed among the meals and three snacks. Lunch serves as the largest meal, and breakfast and snacks are limited to 10 to 15% of total calories.[59]

Case Study 5.2

Photo Disc

Elizabeth's Story: Gestational Diabetes

Elizabeth is a 36-year-old Hispanic woman in her second pregnancy. The first pregnancy resulted in the birth of a 3450-gram (7.7 lb) healthy infant. Before her current pregnancy, Elizabeth was of normal-weight status with a BMI of 24.5 kg/m². Pregnancy has progressed normally through the first 26 weeks, and her weight gain of 20 pounds is within the normal limits.

Because Elizabeth is considered to be at "average risk," she receives a 50-g oral glucose screening test at her 26-week appointment. She is referred for a 100-g oral glucose tolerance test (OGTT) because the screening result was 147 mg/dL glucose. The next day's OGTT results identify the following glucose levels:

Fasting	90 mg/dL
1-hour	195 mg/dL
2-hour	163 mg/dL
3-hour	135 mg/dL

Based on elevated 1- and 2-hour results, a diagnosis of gestational diabetes is made. Elizabeth's certified nurse-midwife sees her the next day to answer her questions and provide education on gestational diabetes and its implications for her care. The midwife also advises Elizabeth on her newborn's risk for hypoglycemia and the need to have her glucose levels tested again after pregnancy. She stresses the importance of keeping good control of blood glucose levels, of monitoring her blood glucose levels regularly, and of carefully following the diet and exercise plan that will be worked out.

Elizabeth is advised that her glucose will be tested again after she has followed the diet and exercise plan for a week, and that if her blood glucose levels remain high after 2 weeks, she will be put on insulin injections. Elizabeth is given an appointment with the clinic's dietitian and certified diabetes educator. After performing a dietary assessment and learning about her food and exercise preferences, the dietitian estimates Elizabeth's caloric need. Elizabeth's current body weight of 69 kg (152 lb) is multiplied by 30 calories to yield an estimate of a need for 2070 calories per day. Together, Elizabeth and the dietitian negotiate a food plan including 48% of calories from carbohydrate in three meals and two snacks daily. Elizabeth also agrees to maintain her current, moderate level of physical activity.

Elizabeth is then instructed on blood glucose assessment using a glucometer that stores the results in its built-in memory, and on the procedures for fasting urinary ketone testing. She is asked to call the dietitian or midwife if questions or problems arise.

The thought of having insulin injections if the diet plan doesn't work is a major motivation for Elizabeth. She carefully follows her diet plan, recording her food intake and testing her blood levels of glucose four times a day. She also performs a dipstick test for urinary ketones first thing in the morning. When she returns to the clinic the next week, she is rewarded by normal glucose values—but she has lost a half-pound, and she had elevated urine levels of ketones twice during the past week. The dietitian reevaluates Elizabeth's diet plan, adding a bedtime snack, and increases the caloric level of the plan by 400 calories a day to allow for weight gain.

Elizabeth shows no signs of developing hypertension or other complications, and fetal growth appears to be normal based on ultrasound results. Consequently, the consulting endocrinologist and obstetrician decide there is no need for them to see Elizabeth. A week after starting the revised diet plan, Elizabeth has gained a pound; she continues to gain about 1 pound per week for the next 12 weeks while maintaining acceptable blood glucose and urinary ketone levels most of the time.

Elizabeth goes into labor at the beginning of the 39th week of pregnancy. After 6 hours of labor, Elizabeth delivers a healthy, 3550-gram (7.9 lb) girl whose blood glucose levels remain within the normal range through 2 hours after birth. Two months after delivery, Elizabeth returns to the clinic for a follow-up glucose test, and the 2-hour OGTT test results are normal. She is concerned, however, that she is not losing weight quickly enough; so she asks for an appointment with the dietitian.

After following the diet and exercise plan negotiated with the dietitian for 5 months, Elizabeth is 2 pounds below her prepregnancy weight and physically fit. She feels great and is determined to keep her weight under control and to stay physically active. Type 2 diabetes or hypertension is not going to be in her future if she has anything to do with it!

Questions

1. Does Elizabeth have insulin resistance?
2. What's the primary reason that Elizabeth lost weight during her first week on her new diet plan?
3. If Elizabeth continues to eat a healthful diet, exercise, and not gain weight, will she be able to prevent type 2 diabetes from developing?

(Answers are in Appendix D.)

Caloric levels and meal and snack plans are considered to be starting points and often require modifications after results of blood glucose home monitoring tests are known. Caloric levels for overweight and obese women are particularly likely to initially fall short of need,[65] and morning blood glucose levels are most likely to be high.[66] Reduction of calories from carbohydrate is indicated at breakfast for women with high morning glucose levels.

Specific recommendations for the distribution of calories from carbohydrate and fat primarily depend on individual eating habits and the effect of carbohydrate and fat on blood glucose levels. The distribution of calories utilized in practice, however, commonly falls into these ranges:

- 40–50% of calories from carbohydrate
- 30–40% from fat
- 20% from protein

The relatively low carbohydrate, high-fat diet decreases the need for insulin by lowering the amount of glucose absorbed from food, and blunts postprandial increases in blood glucose and insulin levels (Table 5.12 on the next page). The addition of high-fiber foods to diet plans may also enhance blood glucose control. These changes in turn reduce fetal overgrowth and other adverse effects of insulin resistance and high levels of glucose and insulin.[63]

Consumption of Foods with a Low Glycemic Index

Whether low-GI foods benefit women with diabetes in pregnancy has been much debated and is somewhat controversial. Low-GI foods help women sustain modest improvements in blood glucose levels and decrease insulin requirements.[32,68] Illustration 5.2 on the following page demonstrates this point by showing blood glucose levels after a meal containing white bread (GI = 70) or spaghetti (GI = 48) is consumed.

EXAMPLE MEAL PLANS Individualized diet plans for women with gestational diabetes include a variety of foods that correspond to the preferences and needs of women. Two examples of such diet plans are shown in Table 5.13 on page 133. One menu provides approximately 2200 calories, the other 2400. Both menus include low-GI food sources of carbohydrate and the nutrients needed by women during pregnancy.

Table 5.11 Estimating levels of caloric need in women with gestational diabetes[63]

Current Weight Status	Definition	Body Weight, kcal/kg
Underweight	<80% of average weight	35–40
Normal weight	80–120% of average weight	30
Overweight	120–150% of average weight	24
Obese	>150% of average weight	12–15

Table 5.12 Effects of 6 weeks of low- and high-carbohydrate diets on maternal and newborn outcomes in women with gestational diabetes[67]

	DIET	
OUTCOME	Low Carbohydrate (<42% of total calories)	High Carbohydrate (>45% of total calories)
Postprandial glucose values	110 mg/dL	132 mg/dL
Fasting glucose	92 mg/dl	94 mg/dL
Insulin requirement	5%	33%
Urinary ketones	10%	0%
Birthweight, g	3694	3890
LGA	9%	42%
Cesarean delivery due to LGA fetus	3%	48%

URINARY KETONE TESTING Women with gestational diabetes are generally instructed to monitor urinary ketone levels using dipsticks. In the past all dipsticks used were insensitive to β-hydroxybutyrate, the primary ketone spilled into urine. Sticks that detect β-hydroxybutyrate are becoming increasingly available and should be used. When interpreting results of urinary ketone tests, keep in mind that 10–20% of pregnant women spill ketones after an overnight fast.[59] This means the severity and consistency of positive findings for urinary ketones should be considered.

Postpartum Follow-Up

The relatively low carbohydrate, high-fat diet instituted during pregnancy is generally changed to a higher-carbohydrate, lower-fat diet after delivery. Emphasis on low-GI and high-fiber foods, along with gradual weight loss, may also be components of postpartum nutrition counseling.

Prevention of Gestational Diabetes

Reducing overweight and obesity, increasing physical activity, and decreasing insulin resistance prior to pregnancy are important components of reducing the risk of gestational diabetes.[65] Screening programs that identify women with insulin resistance or glucose intolerance prior to pregnancy have also been advocated for risk reduction.[42] The risk of type 2 diabetes after pregnancy can be reduced substantially by healthful eating, aerobic and resistance exercise, and maintenance of normal weight.[69]

Type 1 Diabetes during Pregnancy

Women with type 1 diabetes have deficient insulin output and must rely on insulin injections or an insulin pump to meet their need for insulin. Type 1 diabetes represents a potentially more hazardous condition to mother and fetus than do most cases of gestational diabetes.

Type 1 diabetes places women at risk of kidney disease, hypertension, and other complications of pregnancy.

Newborns of women with this type of diabetes are at increased risk of mortality, of being SGA or LGA, and of experiencing hypoglycemia and other problems within 12 hours after birth. Hypoglycemia occurs in about half of macrosomic infants.[70] Coming into pregnancy with this type of diabetes also increases by threefold (from 2–3% to 6–9%) the risk of congenital malformations of the pelvis, central nervous system, and heart in offspring.[71] Good control of blood glucose and diets rich in antioxidants reduce the risk of malformations. Maintenance of normal glucose levels from the start of pregnancy decreases the risk of fetal malformations and macrosomia.[71]

Blood glucose control from the beginning of pregnancy is also important because the fetal growth trajectory may be largely determined in the first half of pregnancy. Exposure to high amounts of glucose and insulin when the fetal growth trajectory is being established may set the "metabolic stage" for fetal accumulation of fat and lean tissue later in pregnancy.[72] Even relatively low elevations in blood glucose levels can meaningfully increase birth-

Illustration 5.2 Blood glucose response in people with type 2 diabetes to meals containing white bread or spaghetti.

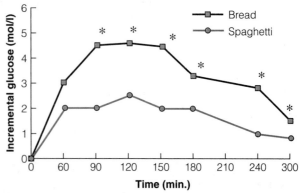

SOURCE: G. Riccardi and A. A. Rivellese, Diabetes: nutrition in prevention and management. Nutr Metab Cardiovasc Dis 1999;9:33–6. Reproduced with permission of Medikal Press S.r.l.

Table 5.13 Examples of three-meal, three-snack 1-day menus at two caloric and carbohydrate levels for women with gestational diabetes

2220-CALORIE DIET			2400-CALORIE DIET		
	Carbohydrates, g	Calories		Carbohydrates, g	Calories
Breakfast			**Breakfast**		
All Bran, ½ c	22	80	Complete Wheat		
2% milk, ½ c	6	61	Bran Flakes, ¾ c	23	90
Mozzarella cheese			2% milk, ½ c	6	61
stick, 1 oz	1	78	Egg, 1	1	74
Black coffee, tea			Black coffee, tea		
Morning Snack			**Morning Snack**		
Oat bran bagel, ½	19	98	Peanuts, 2 oz	10	326
Sugar-free, low-fat			Carrot, 1	7	31
yogurt, 1 c	17	155	Graham crackers,		
Lunch			4 small or 1 sheet	11	59
Tuna salad, ½ c	19	192	**Lunch**		
Whole grain			Beef or chicken		
bread, 2 sl	24	130	burrito, 1	33	255
Carrot and celery			Salsa, ½ c	7	33
sticks, 1 c	7	31	Black beans, 1 c	40	228
Potato salad, ½ c	14	179	Apple, 1	21	81
Orange, 1	15	62	Black coffee, tea,		
Black coffee, tea,			water, or diet soda		
water, or diet soda			**Midday Snack**		
Midday Snack			Banana, ½	28	55
Peaches canned in			2% milk, 1 c	12	121
juice, ½ c	29	109	**Dinner**		
2% milk, 1 c	12	121	Lean pork chop, 4 oz	0	263
Dinner			Pinto beans, 1 c	22	116
Lean roast beef, 3 oz	0	188	Corn bread, 1 oz	12	92
Broccoli, 1 c with	10	50	Margarine, 1 tsp	1	33
Cheese (melted), 1 oz	1	105	Garden salad, 2 cup	0	10
Roll, 2 oz	30	167	Feta cheese, 1 oz	1	74
Margarine, 2 tsp	0	67	Salad dressing, 2 tb	3	104
Grapes, 15	14	53	Black coffee, tea,		
Black coffee, tea, water,			water, or diet soda		
or diet soda			**Bedtime Snack**		
Bedtime Snack			Peanut butter, 2 tb	12	190
Hard-boiled egg, 1	1	74	Rice cake, 1	8	35
Saltine crackers, 4	8	50	2% milk, 1 c	12	121
2% milk, 1 c	12	121	**Total:**	**270 g**	**2442**
Total:	**261 g**	**2,171**		or 44% of total calories	
	or 48% of calories				

SOURCE: Developed by Judith Brown, 2001.

weight.[57] Unfortunately, only 10% of women with type 1 diabetes receive preconceptional care.[71]

Availability of a variety of new insulins, the insulin pump, and self-monitoring technology has revolutionized the care of type 1 diabetes during pregnancy.

NUTRITIONAL MANAGEMENT OF TYPE 1 DIABETES IN PREGNANCY Primary goals for the nutritional management of type 1 diabetes in pregnancy are continual control of blood glucose levels, nutritional adequacy of dietary intake, achievement of recommended amounts of weight gain, and a healthy mother and newborn. Careful home monitoring of glucose levels and adjustments in dietary intake, exercise, and insulin dose based on the results are key events that increase the likelihood of reaching these goals. Monitoring urinary ketones is particularly important in women with type 1 diabetes because they are more prone to developing ketosis than are women with gestational diabetes.[39] Inclusion of ample amounts of dietary fiber (25–35 g per day) reduces insulin requirements in many women with type 1 diabetes in pregnancy.[66]

Multifetal Pregnancies

Rates of multifetal pregnancy in the United States have increased markedly since 1980. Twin births, which accounted for 1 in 56 births in 1980, constituted 1 in 34 births in 2001. Rates of triplet and higher-order multiple births (referred to as *triplet+* births) have increased from 1 in 2941 to 1 in 551 births.[73] Only one in five triplet+ pregnancies are spontaneously conceived. Rates of twin and higher-order births are highest by far in women 45–54 years old (one in five births), the age group most likely to receive medical and technological interventions to achieve pregnancy.[73]

The progressively older ages at which U.S. women are bearing children also contribute to rising rates of multifetal pregnancies. The chances of a spontaneous multifetal pregnancy increase with age after about 35 years. Rates of spontaneous multifetal pregnancy also increase with increasing weight status. For example, the rate of twin pregnancy is about two times higher in obese than in underweight women.[74] Rates of triplet+ pregnancies are headed downward due to improved assisted reproductive technologies that reduce higher-order, multifetal pregnancies.[73]

Upward trends in low birthweight and preterm delivery in the United States over recent years have been strongly influenced by the upsurge in multiple births. Only 3% of newborns are from multifetal pregnancies, yet they account for 21% of all low-birthweight newborns, 14% of preterm births, and 13% of infant deaths.[76]

Background Information about Multiple Fetuses

The most common type of multifetal pregnancies, those with twin fetuses, come in several types and levels of risk. Twins are dizygotic (DZ) if two eggs were fertilized, and monozygotic (MZ) if one egg was. Monozygotic twins result when the fertilized and rapidly dividing egg splits in two within days after conception. The term *identical* is often used to describe MZ twins, and *fraternal* denotes DZ twins. These terms are misleading, so the preferred terms are *monozygotic* and *dizygotic*.[77,78] About 70% of twins are DZ, and 30% are MZ.

Monozygotic twins are always the same sex, whereas DZ twins are the same sex half the time and different sexes half the time. Monozygotic twins are genetically identical in almost all ways, but they are seldom absolutely identical. Genetic differences in pairs of MZ twins can result from chromosome abnormalities in one twin, unequal genetic expression of maternally and paternally derived genes, and environmental effects on gene expression. Rates of MZ twins are remarkably stable across population groups and do not appear to be influenced by heredity.[79]

Dizygotic twins represent individuals with differing genetic "fingerprints." The incidence of DZ twin pregnancies is influenced both by inherited and environmental factors. Rates of DZ twins vary among racial groups and by country. Rates tend to decrease in populations during famine and to increase when food availability and nutritional status improve.[80] Periconceptional vitamin and mineral supplement use has also been related to an increased incidence of DZ twin pregnancy.[81]

Twins also vary in the number of placentas; some are born having used the same placenta, but more commonly each fetus has its own. Twins may share a common amniotic sac and one of the membranes around the sac (the chorion), or have separate amniotic sacs and membranes (Illustration 5.3). Twins at highest risk of death, malformations, growth retardation, short gestation, and other serious problems are those that share the same amniotic sac and chorion, and to a lesser degree, MZ twins in general.[77,82]

Determining twin type is not always an easy task during or after pregnancy. Definitive diagnoses of tough cases can be made through DNA fingerprinting.[82]

IN UTERO GROWTH OF TWINS AND TRIPLETS
Fetal growth patterns of twins and triplets compared to singleton fetuses are shown in Illustration 5.4. Rates of weight gain for each group of fetuses are the same until about 28 weeks of gestation. Rates of weight gain begin to decline in twin and triplet fetuses after that point, however, and remain lower until delivery. Variations in birthweight of twin and triplet newborns appear to be related to factors that affect fetal growth after 28 weeks of pregnancy.[82]

THE VANISHING TWIN PHENOMENON   The disappearance of embryos within 13 weeks of conception is not unusual. It has been estimated that 6 to 12% of pregnancies begin as twins, but that only about 3% result in the birth of twins. Most fetal losses silently occur by absorption into the uterus within the first 8 weeks after conception. The prognosis for continued viability of a pregnancy associated with a vanishing twin tends to be good.[83]

Risks Associated with Multifetal Pregnancy

Singleton pregnancy is the biological norm for humans, so it may be expected that multifetal pregnancy would be accompanied by increasing health risks (Table 5.14).[84] Multifetal pregnancies present substantial risks to both mother and fetuses, and the risks increase as the number of fetuses increase (Table 5.15 on page 136). Newborns from twin pregnancies at lowest risk of death in the perinatal period weigh between 3000 and 3500 grams (6.7 to 7.8 lb) at birth and are born at 37–39 weeks gestation. Triplets tend to do best when they weigh over 2000 grams (4.5 lb) and are born at 34–35 weeks gestation.[84]

Illustration 5.3 Twins (a) with two amniotic sacs, two chorions, and two placentas; (b) with one amniotic sac, chorion, and placenta; and (c) with two amniotic sacs, one chorion, and fused placentas.

SOURCE: From G. Martins in Hebammen-Lelubuch, 4th Ed. Copyright © 1983 by Thieme. Reprinted with permission.

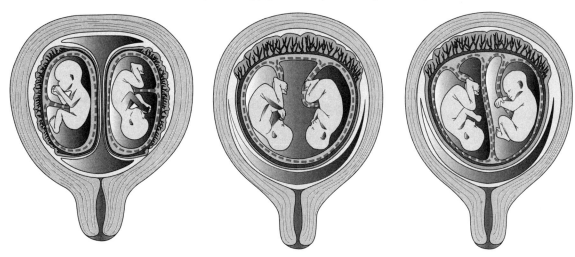

Unfortunately, these outcomes do not represent the usual. Data presented in Table 5.16 on the next page show that median weights of twins born at 37, 38, and 39 weeks gestation fall below the 3000- to 3500-gram range. However, the 3000- to 3500-gram birthweight range for twins, and the >2000-gram mark for triplet newborns, can serve as goals for the provision of nutrition services.

Interventions and Services for Risk Reduction

Special multidisciplinary programs that offer women with multifetal pregnancy a consistent, main provider of care, preterm prevention education, increased attention to nutritional needs, and intensive follow-up achieve better pregnancy outcomes than does routine prenatal care.[85] Rates of very low birthweight (≤1500 g or ≤3.3 lb) have been reported to be substantially lower (6% versus 26%), neonatal intensive care admissions three times lower (13% versus 38%), and perinatal mortality strikingly lower (1% versus 8%) among women who receive such services.[87] Interventions offered by the Montreal Diet Dispensary, which focuses on improving the nutritional status and well-being of the pregnant women served, have been shown to substantially reduce poor outcomes compared to those for similar women not receiving the services. Improvements include a 27% reduction in the rate of low birthweight, 47% decline in very low birthweight, 32%

Illustration 5.4 Rates of fetal weight gain in singleton, twin, and triplet fetuses.

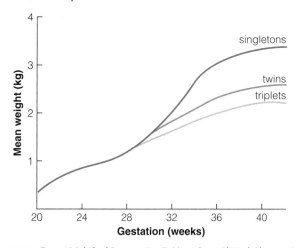

SOURCE: From "Multifetal Pregnancies: Epidemiology, Clinical Characteristics, and Management," by M. Smith-Levitin, et al., in Maternal-Fetal Medicine: Principles and Practice, 3rd Ed., R. K. Creasy and R. Resnick (Eds.), p. 589–601. Copyright © 1994. Reprinted by permission of W. B. Saunders Co.

Table 5.14 Risks to mother and fetuses associated with multifetal pregnancy[84]

Pregnant Women
- Preeclampsia
- Iron-deficiency anemia
- Gestational diabetes
- Hyperemesis gravidarum
- Placenta previa
- Kidney disease
- Fetal loss
- Preterm delivery
- Cesarean delivery

Newborns
- Neonatal death
- Congenital abnormalities
- Respiratory distress syndrome
- Intraventricular hemorrhage
- Cerebral palsy

Table 5.15 Average birthweight and gestational age at delivery, and low-birthweight rates, of singleton, twin, and triplet newborns[75,84]

	Mean Birthweight	Mean Gestational Age	Low-Birthweight Rate
Singletons	3440 g (7.7 lb)	39–40 weeks	6%
Twins	2400 g (5.4 lb)	37 weeks	54%
Triplets	1800 g (4.0 lb)	33–34 weeks	90%

lower rate of preterm delivery, and a 79% drop in mortality during the first seven days after birth.[88]

Nutrition and the Outcome of Multifetal Pregnancy

Nutritional factors are suspected of playing a major role in the course and outcome of multifetal pregnancy, but much remains to be learned. Of the nutritional factors that influence multifetal pregnancy, weight gain during twin pregnancy has been studied most.

WEIGHT GAIN IN MULTIFETAL PREGNANCY

As with singleton pregnancy, weight gain in multifetal pregnancy is linearly related to birthweight, and weight gains associated with newborn weight vary based on prepregnancy weight status (Table 5.17).[84] The Institute of Medicine recommends that women with twins gain 35 to 45 pounds (15.9 to 20.5 kg). It is advised that underweight women gain toward the upper end of this range, and overweight and obese women near the lower end.[37]

Table 5.16 Median birthweight for gestational age at delivery of twins

Gestational Age, weeks	Birthweight
28	995 g (2.2 lb)
29	1145 g (2.6 lb)
30	1300 g (2.9 lb)
31	1445 g (3.2 lb)
32	1580 g (3.5 lb)
33	1750 g (3.9 lb)
34	1905 g (4.3 lb)
35	2165 g (4.8 lb)
36	2275 g (5.1 lb)
37	2430 g (5.4 lb)
38	2565 g (5.7 lb)
39	2680 g (6.0 lb)
40	2810 g (6.3 lb)
41	2685 g (6.0 lb)

SOURCE: Data from Cohen SB, 1997.[86]

RATE OF WEIGHT GAIN IN TWIN PREGNANCY

A positive rate of weight gain in the first half of twin pregnancy is strongly associated with increased birthweight.[89,91] On the other hand, weight loss after 28 weeks of pregnancy increases the risk of preterm delivery by threefold.[90]

Recommended rates of weight gain for women with twin pregnancy are

- 0.5 pounds (0.2 kg) per week in the first trimester
- 1.5 pounds (0.7 kg) per week in the second and third trimesters[84]

WEIGHT GAIN IN TRIPLET PREGNANCY

Several studies have examined the relationship between weight gain and birthweight in women with triplets. The general result is that weight gains of about 50 pounds (22.7 kg) correspond to healthy-sized triplets. Rates of gain related to a total weight gain of 50 pounds in women who will average 33 to 34 weeks of gestation are 1.5 pounds (0.7 kg) per week or more, starting as early in pregnancy as possible.[84]

Dietary Intake in Twin Pregnancy

Ensuring "adequate nutrition" is widely acknowledged to be a key component of prenatal care for women with multifetal pregnancy. However, it is not clear what constitutes adequate nutrition. Energy and nutrient needs clearly increase during multifetal pregnancy due to increased levels of maternal blood volume, extracellular fluid, and uterine, placental, and fetal growth. These increases place

Table 5.17 Prepregnancy weight status and weight-gain relationships in twin pregnancy[74]

Prepregnancy Weight Status	Weight Gain Related to Birthweights of >2500 g (5.5 lb)
Underweight	44.2 lb (20.1 kg)
Normal weight	40.9 lb (18.6 kg)
Overweight	37.8 lb (17.2 kg)
Obese	37.2 lb (16.9 kg)
Very obese	29.2 lb (13.3 kg)

higher energy and nutrient demands on the mother in terms of the nutritional costs of building and maintaining these tissues. Although their newborns are smaller, women with twins still produce around 5000 g (11.2 lb) of fetal weight, and women with triplets 5400 g (13.4 lb) or more.

Evidence of higher caloric need for tissue maintenance and growth in multifetal than singleton pregnancy comes from studies that show increased weight gain and a quicker onset of starvation metabolism in women expecting more than one newborn. Reduced rates of twin deliveries, as well as the higher incidence of twins in overweight and obese women, imply that energy status is an important factor in multifetal pregnancy.[84] Whereas it is obvious that energy and nutrient needs are higher in multifetal than singleton pregnancy, levels of energy balance and nutrient intake associated with optimal outcomes of multifetal pregnancy have not been quantitated.

One prospective study of dietary intakes of women pregnant with twins has been undertaken.[93] Results from this study indicate that women with twins enter pregnancy with higher average caloric intakes (2030 versus 1789 cal per day) and consume an average of 265 cal more per day during pregnancy than women with singleton pregnancy. Nutrient intakes during pregnancy are also higher in women bearing twins than with singleton pregnancy.

Several studies have concluded that the need for specific nutrients is increased during multifetal pregnancy. The need for essential fatty acids (linoleic and alpha-linolenic acid) appears to be increased in multifetal pregnancy. Poor essential fatty acid status is related to neurologic abnormalities and vision impairments in twin offspring.[92] Requirements for iron and calcium have also been found to be increased based on the magnitude of physiological changes that take place in multifetal pregnancy. Levels of essential fatty acids, iron, or calcium required by women to meet these increased needs are unknown, however.[84]

VITAMIN AND MINERAL SUPPLEMENTS AND MULTIFETAL PREGNANCY Benefits and hazards of multivitamin and mineral supplement use in multifetal pregnancy have not been reported. Consequently, whether they should be provided—and if so, in what dose levels—is unknown. Levels of nutrient intake exceeding the DRI Tolerable Upper Intake Levels should be avoided.

Nutritional Recommendations for Women with Multifetal Pregnancy

Due to the lack of study results, nutritional recommendations for women with multifetal pregnancy are largely based on logical assumptions and theories (Table 5.18). It is reasoned, for example, that caloric needs for twin pregnancy can be extrapolated from weight gain. Theoretically, to achieve a 40-pound (18.2 kg) weight gain, or 10 pounds (4.5 kg) more than in singleton pregnancy, women with twins would need to consume approximately 35,000 cal

---

Table 5.18 "Best practice" recommendations for nutrition during multifetal pregnancy[84]

Weight Gain

Twin pregnancy: Overall gain of 35–45 pounds (15.9–20.5 kg) underweight women should gain at the upper end of this range, and overweight and obese women at the lower end

- First trimester: 4–6 pounds (1.8–2.7 kg)
- Second and third trimesters: 1.5 pounds (0.7 kg) per week

Triplet pregnancy: Overall gain of approximately 50 pounds (22.7 kg)

- Gain of 1.5 pounds (0.7 kg) per week through pregnancy

Daily Food Intake

Twin pregnancy

- Breads, cereal, rice, pasta: >6 servings
- Vegetables: 3+ servings
- Fruits: 2+ servings
- Meat, poultry, fish, dry beans, eggs, nuts: 3+ servings
- Milk, yogurt, cheese: 3+ servings
- Fats and sugars: Use oils rich in essential fatty acids (safflower oil, sunflower oil, walnut oil, and canola oil, for example).

Triplet pregnancy

- Foods intake from the Food Guide Pyramid groups should be consumed at a level that promotes targeted weight gain

Caloric Intake

Twin pregnancy

- 450 calories above prepregnancy intake, or the amount consistent with targeted weight gain progress

Triplet pregnancy

- Caloric intake levels should promote targeted weight gain progress

Nutrient Intake

Twin and triplet pregnancy

- DRI levels or somewhat more than these levels
- Intakes should be lower than ULs

Vitamin and Mineral Supplements

Twin pregnancy

- Use a supplement providing:

iron: 30 mg	vitamin B_6: 2 mg
zinc: 15 mg	folate: 300 mcg
copper: 2 mg	vitamin C: 50 mg
calcium: 250 mg	vitamin D: 200 IU

Triplet pregnancy

- Provide a supplement containing at least the above levels for twin pregnancy while avoiding excessive amounts

more during pregnancy than do women with singleton pregnancies. This increase would amount to about 150 cal per day above the level for singleton pregnancy, or an average of 450 more cal per day than prepregnancy. To achieve higher rates of gain, underweight women may need a higher level of intake, and overweight and obese women lower levels. Energy needs will also vary by energy expenditure levels. As for singleton pregnancy, adequacy of caloric intake can be estimated by weight-gain progress.[84]

Food-intake recommendations for women with multifetal pregnancy are primarily estimated based on assumptions related to caloric and nutrient needs. Women with multifetal pregnancy likely benefit from diets selected from the Food Guide Pyramid groups and nutrient intakes that meet or slightly exceed the DRIs.

RECOMMENDATIONS FROM THE POPULAR PRESS   Web sites, books, and pamphlets are available that provide ample amounts of scientifically unsupported "guesses" about food and nutrient requirements of women with multifetal pregnancy. Even if presented with steely resolution, any advice that strays from current scientifically based wisdom about nutritional needs of women during pregnancy should be sidestepped.

HIV/AIDS during Pregnancy

The world first became aware of acquired immunodeficiency syndrome (AIDS) in the summer of 1981. It was caused by a newly recognized microbe, the human immunodeficiency virus (HIV). Since then, over 100,000 women of childbearing age in the United States have been diagnosed with AIDS,[95] and 15 million worldwide. Transmission of the virus during pregnancy and delivery is a major route to the spread of the infection. Approximately 20% of children with HIV/AIDS are infected during pregnancy or delivery, and 14–21% during breastfeeding.[93]

Treatment of HIV/AIDS

The primary focus of care for pregnant women with HIV/AIDS is the prevention of transmission of the virus to the fetus and infant. The transmission of the AIDS virus from mother to child is negligible if treatment is provided before, during, and after pregnancy.[93] In developing countries where there is not enough money to purchase the drugs, transmission rates can be substantially reduced by giving the mother a short course of a specific anti-HIV drug before delivery.[94]

Consequences of HIV/AIDS during Pregnancy

Disease processes such as compromised immune system functions related to HIV/AIDS progress during pregnancy, but it does not appear that the infection itself is related to adverse pregnancy outcomes. Although adverse pregnancy outcomes such as preterm delivery, fetal growth retardation, and low birthweight tend to be higher in women with HIV/AIDS, differences are most closely related to poverty, poor food availability, compromised health status, and the coexistence of other infections.[95]

Nutritional Factors and HIV/AIDS during Pregnancy

HIV/AIDS is related to poor nutritional status that further compromises the body's ability to fight infections. The disease can lead to nutrient losses and fat malabsorption due to diarrhea, and inflammatory responses to the infection cause the loss of lean muscle mass. Risks of inadequate nutrient status of a wide variety of vitamins and minerals increase as the disease progresses. Nutritional needs increase the most during the later stages of HIV/AIDS as diarrhea, wasting, and reductions in CD4 counts (a measure of white blood cells that help the body fight infection) increase. New drugs used to treat HIV/AIDS are associated with increased insulin resistance and the accumulation of central body fat.[96]

Results of a number of studies have indicated that vitamin A and multiple-vitamin supplementation during pregnancy may reduce the transmission of HIV/AIDS to offspring and improve pregnancy outcomes of AIDS-infected women in developing countries.[93,97] Early and appropriate nutrition intervention may retard progression of the disease during pregnancy. Although nutrition intervention to correct nutritional problems is recommended for women with HIV/AIDS, routine supplementation may not decrease transmission or improve the course and outcome of pregnancy in well-nourished women.

The compromised immune status of women with HIV/AIDS, and further decreases in immune response during pregnancy, mean that women with the disease are at high risk of developing foodborne infections during pregnancy. Risk of infection originating from foods can be decreased if raw or uncooked meats and seafood and unpasteurized milk products and honey are not consumed. Safe food-handling practices at home can also reduce the risk of foodborne infection.

Nutritional Management of Women with HIV/AIDS during Pregnancy

Goals for the nutritional management of women with HIV/AIDS include

- Maintenance of a positive nitrogen balance and preservation of lean muscle mass
- Adequate intake of energy and nutrients to support maternal physiological changes and fetal growth and development
- Correction of elements of poor nutritional status identified by nutritional assessment

- Avoidance of foodborne infections
- Delivery of a healthy newborn

Nutrient requirements of HIV-infected pregnant women are not known, but it is suspected that energy and nutrient needs will be somewhat higher due to the effects of the virus on the body.[96,98]

Insufficient information exists to provide specific standards for nutritional care for women with HIV/AIDS during pregnancy. Consequently, nutritional recommendations for women with HIV/AIDS are consistent with recommendations for pregnant women in general. As is the case for other pregnant women, foods should be the primary source of nutrients in women with HIV/AIDS.[96,98]

Eating Disorders in Pregnancy

Eating disorders represent relatively rare conditions in pregnancy because many women with such disorders are subfertile or infertile. Such disorders can have far-reaching effects on both mother and fetus, however, when they do occur.[99] The eating disorder most commonly observed among pregnant women in the United States is bulimia nervosa, or a condition marked by both severe food restriction and bingeing and purging.[99] It is estimated that 1–3% of adolescents and young women in the United States have this condition.[100] Women with bulimia nervosa exhibit poorly controlled eating patterns marked by recurrent episodes of binge eating. To prevent weight gain, women will induce vomiting, use laxatives, exercise intensely, or fast after binges. Self-worth in women with bulimia nervosa is usually closely tied to their weight and shape. A history of sexual abuse is common among women with this eating disorder, as well as in women with anorexia nervosa.[101]

Pregnancy is rarely suspected in women with anorexia nervosa, because amenorrhea is a diagnostic criteria for the disorder. Nonetheless, women with anorexia nervosa may occasionally ovulate and become at risk for conception. To women with anorexia nervosa, body weight is of utmost importance; and they are generally fully dedicated to achieving extreme thinness. Adolescents and women with this condition will refuse to eat, even when ravenously hungry; limit their food choices to low-calorie foods only; and exercise excessively.[101]

Eating disorder symptoms often subside during the second and third trimesters of pregnancy, but they rarely vanish altogether. Symptoms tend to return after delivery, sometimes to a more severe extent than was the case prior to pregnancy.[102]

Consequences of Eating Disorders in Pregnancy

Women with eating disorders during pregnancy are at higher risk for spontaneous abortion, hypertension, and difficult deliveries than are women without an eating disorder. Pregnancy weight gain is generally below the recommended amounts, and newborns tend to be smaller and to experience higher rates of neonatal complications.[103]

Treatment of Women with Eating Disorders during Pregnancy

It is recommended that pregnant women with eating disorders be referred to an eating disorders clinic or specialist. Most large communities have special clinics and programs for women with eating disorders, and they commonly use a team approach to problem solving around the eating disorder. Nutritionists or dietitians often participate in these services because they are knowledgeable about the woman's individual nutritional needs and those of pregnancy.

Health professionals serving women with eating disorders in pregnancy can facilitate open communication and behavioral change by gently encouraging women to talk about their eating disorder, fears, and concerns.[102]

Nutritional Interventions for Women with Eating Disorders

Behavioral changes required for improvements in nutritional status and weight gain in women with eating disorders are most likely to work when the changes are considered acceptable to the women with the disorder. Frequently, the health professional presents the types of changes that need to be made and explains why, and then works with the woman to develop specific plans accomplishing these changes.

Nutrition and Adolescent Pregnancy

Between 1991 and 2001, rates of pregnancy among teens 15–19 years old fell 26%. Adolescent pregnancies hit a record low of 43 births per 1000 females aged 15–19 years in 2002.[73] Although the rates are declining, the United States continues to have one of the highest rates of adolescent pregnancy of all developed countries.[75] Adolescents are at higher risk for a number of clinical complications and other unfavorable outcomes compared to adult women (Table 5.19 on the next page). The downward shift in birthweight observed in newborns of adolescent mothers is related to increased rates of perinatal mortality, and suggests the existence of unfavorable nutrition and health conditions during pregnancy.[104]

The extent to which increased rates of poor outcomes in pregnant teens are associated with biological immaturity or with lifestyle factors such as drug use, smoking, and poor dietary intakes (that influence health status) is unclear. Age-related differences in outcome diminish substantially when potentially harmful lifestyle factors are

Table 5.19 Risks associated with adolescent pregnancy[104,108]
• Low birthweight
• Perinatal death
• Cesarean delivery
• Cephalopelvic disproportion (head too large for birth canal)
• Preeclampsia
• Iron-deficiency anemia
• Delayed, reduced educational achievement
• Low income

taken into account, diminishing the theory that biological immaturity accounts for the differences. Very young adolescents becoming pregnant within a few years after the onset of menstruation may be at risk due to biological immaturity. They tend to have shorter gestations and a higher likelihood of cephalopelvic disproportion.[104] Poorly nourished, growing adolescent mothers may compete with the fetus for calories and nutrients—and win.[103]

Growth during Adolescent Pregnancy

Young teens who are growing when pregnancy occurs continue to gain height and weight during pregnancy—but at the expense of fetal growth. Teens who continue to grow during pregnancy give birth, on average, to infants that weigh 155 g less than infants of adult women, even if they gain more weight than adults do.[106] Young adolescents gain more maternal fat tissue during the last trimester of pregnancy and retain more weight postpartum than do nongrowing teens. Growing teens experience a surge in blood leptin levels during the last trimester, which may decrease maternal use of fat stores and increase utilization of glucose by the mother. Increased use of glucose by the mother appears to decrease energy ability to the fetus.[107]

Dietary Recommendations for Pregnant Adolescents

Recommendations for pregnant adolescents are basically the same as for older pregnant women, with a few exceptions. Recommendations for weight gain and protein intake are the same, but young adolescents may need more calories to support their own growth as well as that of the fetus. Caloric need should be met by a nutrient-dense diet and lead to rates of weight gain that follow those recommended. Pregnant adolescents have a higher requirement for calcium.[105] The DRI for pregnant teens for calcium is 1300 mg per day, or 300 mg higher than for adult pregnant women. This increased need can be met by the consumption of 4 daily servings of milk and milk products, combined with a varied, basic diet.

The importance of lifestyle and other environmental factors to pregnancy outcome in teens emphasizes the need for special, comprehensive teen pregnancy health care programs. Because most pregnant adolescents have low income, referral to appropriate food and nutrition programs and other assistance related to health care, housing, and education should be core components of services.

Evidence-Based Practice

"Enormous amounts of new knowledge are barreling down the information highway, but they are not arriving at the doorsteps of our patients."

Claude Lenfant, National Institutes of Health[109]

The clinical, nutritional management of the conditions covered in this chapter, as well as other complications during pregnancy, is not entirely evidence based. Such practices are a problem when they burden women and families with costs, call for dietary changes not known to work, or potentially cause harm. Practices not based on evidence, which likely pose little risk or burden and may potentially be of help, should nonetheless be carefully evaluated. Outdated practices often linger far too long, at the expense of missed opportunities for real improvements.

Use of practices not supported by scientific evidence should always be questioned and confirmed to represent "best practice" insofar as that can be determined. To know what best practice is requires vigilant attention to scientific developments related to the nutritional management of clinical conditions during pregnancy.

Resources

Clinical Conditions of Pregnancy

The U.S. Preventive Health Services Task Force periodically reviews and updates recommendations for gestational and other forms of diabetes. Updates can be accessed at this site:
Web site: **www.ahrq.gov/clinic/serfiles.htm**
Visit the Women's Health Resource Center for access to journal articles and summaries and health care information.
Web site: **www.medscape.com**

The National Library of Medicine's Web site offers extensive coverage of scientific journal articles, summaries, and educational resources from a variety of reputable organizations on disorders of pregnancy.
Web site: **www.nlm.nih.gov/medlineplus**
The Canadian Diabetes Association's Web site section "About Diabetes" clearly presents facts on gestational diabetes.
Web site: **www.diabetes.ca**

References

1. Report of the National High Blood Pressure Education Program Working Group on High Blood Pressure in Pregnancy. Am J Obstet Gynecol 2000;183:S1–S22.

2. Sibai BM et al. Risk factors for preeclampsia, abruptio placentae, and adverse neonatal outcomes among women with chronic hypertension. National Institute of Child Health and Human Development Network of Maternal-Fetal Medicine Units. N Engl J Med 1998;339:667–71.

3. Haddad B, Sibai BM. Chronic hypertension in pregnancy. Ann Med 1999;31:246–52.

4. Solomon CG et al. Higher cholesterol and insulin levels in pregnancy are associated with increased risk for pregnancy-induced hypertension. Am J Hypertens 1999;12:276–82.

5. Smith CS et al. High blood pressure during pregnancy and heart disease risks later in life. BMJ 2003;326:845–9.

6. Roberts JM et al. Nutrient involvement in preeclampsia. J Nutr 2003;133:1684S–92S.

7. Prevent recurrent eclamptic seizures with magnesium sulfate, an unconventional anticonvulsant. Drug Ther Perspect 2000;16:6–8.

8. Pipkin FB. Risk factors for preeclampsia (editorial). N Engl J Med 2001;344:925–6.

9. Galtier-Dereure F et al. Obesity and pregnancy: complications and cost. Am J Clin Nutr 2000;71:1242S–8S.

10. Ritchie LD, King JC. Dietary calcium and pregnancy-induced hypertension: is there a relation? Am J Clin Nutr 2000;71:1371S–4S.

11. Velzing-Aarts FV et al. Umbilical vessels of preeclamptic women have low contents of both n-3 and n-6 long-chain polyunsaturated fatty acids. Am J Clin Nutr 1999;69:293–8.

12. Berkowitz KM. Insulin resistance and preeclampsia. Clin Perinatol 1998;25:873–85.

13. Hauth JC, Ewell MG et al. Pregnancy outcomes in healthy nulliparas who developed hypertension. Calcium for Preeclampsia Prevention Study Group. Obstet Gynecol 2000;95:24–8.

14. Inners KE, Wimsatt JH. Pregnancy-induced hypertension and insulin resistance: evidence for a connection. Acta Obstet Gynecol Scand 1999;78:263–84.

15. Kaaja R et al. Evidence of a state of increased insulin resistance in preeclampsia. Metabolism 1999;48:892–6.

16. Villar J et al. Nutritional interventions during pregnancy for the prevention or treatment of maternal morbidity and preterm delivery: an overview or randomized controlled trials. J Nutr 2003;133:1606S–25S.

17. Coomarasamy A et al. Effect of low-dose aspirin on the incidence of preeclampsia. Obstet Gynecol 2003;101:1319–32.

18. Xiong X et al. Association of preeclampsia with high birth weight for age. Am J Obstet Gynecol 2000;183:148–55.

19. Klebanoff MA et al. Maternal size at birth and the development of hypertension during pregnancy: a test of the Barker hypothesis. Arch Intern Med 1999;159:1607–12.

20. Esplin MS et al. Paternal and maternal components of the predisposition to preeclampsia. N Engl J Med 2001;344:867–72.

21. Duley L et al. Antiplatelet drugs for prevention of pre-eclampsia and its consequences: systematic review. BMJ 2001;322:329–33.

22. Atallah AN et al. Calcium supplementation during pregnancy for preventing hypertensive disorders and related problems. Cochrane Database Syst Rev 2000;2.

23. Herrera JA et al. Prevention of preeclampsia by linoleic acid and calcium supplementation: a randomized controlled trial. Obstet Gynecol 1998;91:585–90.

24. Belizan JM et al. Long-term effect of calcium supplementation during pregnancy on the blood pressure of offspring; follow-up of a randomized controlled trial. BMJ 1997;315:281–5.

25. Chambers JC et al. Association of maternal endothelial dysfunction with preeclampsia. JAMA 2001;285:1607–12.

26. Chappell LC et al. Effect of antioxidants on the occurrence of pre-eclampsia in women at increased risk: a randomized trial. Lancet 1999;354:810–6.

27. Rajkovic A et al. Plasma homocyst(e)ine concentrations in eclamptic and preeclamptic African women postpartum. Obstet Gynecol 1999;94:355–60.

28. Leeda M et al. Effects of folic acid and vitamin B$_6$ supplementation on women with hyperhomocysteinemia and a history of preeclampsia or fetal growth restriction. Am J Obstet Gynecol 1998;179:135–9.

29. Scholl TO, Johnson WG. Folic acid: influence on the outcome of pregnancy. Am J Clin Nutr 2000;71:1295S–303S.

30. Woman wins lawsuit on nutritional malpractice: Community Nutrition Institute Weekly Report, 1977:5.

31. Yeo S et al. Effect of exercise on blood pressure in pregnant women with a high risk of gestational hypertensive disorders. J Reprod Med 2000;45:293–8.

32. Brand-Miller J et al. Low-glycemic-index diets in the management of diabetes: a meta-analysis of randomized controlled trials. Diabetes Care 2003;26:2261–67.

33. Roberts JM et al. Nutrient involvement in preeclampsia. J Nutr 2003;133:1684S–92S.

34. Hernandez-Diaz S et al. Risk of gestational hypertension in relation to folic acid supplementation during pregnancy. Am J Epidemiol 2002;156:806–12.

35. Foster-Powell K et al. International table of glycemic index and glycemic load values: 2002. Am J Clin Nutr 2002;76:5–56.

36. Rayman MP et al. Iron supplementation in preeclampsia. Am J Obstet Gynecol 2002;187:412–18.

37. National Academy of Sciences. Nutrition during pregnancy. I. Weight gain. II. Nutrient supplements. Washington, DC: National Academy Press, 1990.

38. Ventura SJ et al. Births: final data for 1998. Natl Vital Stat Rep 2000;48:1–100.

39. Carr SR. Effect of maternal hyperglycemia on fetal development, American Dietetic Association Annual Meeting and Exhibition, Kansas City, MO, Oct. 10, 1998.

40. Damm P. Gestational diabetes mellitus and subsequent development of overt diabetes mellitus. Dan Med Bull 1998;45:495–509.

41. Branchtein L et al. Waist circumference and waist-to-hip ratio are related to gestational glucose tolerance. Diabetes Care 1997;20:509–11.

42. Jovanovic L. Aberrations in diabetes metabolism throughout pregnancy, 60th Scientific Sessions of the American Diabetes Association, San Antonio, TX, June 11, 2000.

43. Butte NF. Carbohydrate and lipid metabolism in pregnancy: normal compared with gestational diabetes mellitus. Am J Clin Nutr 2000;71:1256S–61S.

44. Carducci AA, Corrado F, Sobbrio G et al. Glucose tolerance and insulin secretion in pregnancy. Diabetes, Nutrition & Metabolism—Clinical & Experimental 1999;12:264–70.

45. Homko CJ, Sivan E, Reece EA et al. Fuel metabolism during pregnancy. Semin Reprod Endocrinol 1999;17:119–25.

46. Silverman BL et al. Impaired glucose tolerance in adolescent offspring of diabetic mothers. Relationship to fetal hyperinsulinism. Diabetes Care 1995;18:611–7.

47. Lindsay RS et al. Secular trends in birth weight, BMI, and diabetes in the offspring of diabetic mothers. Diabetes Care 2000;23:1249–54.

48. MacNeill S et al. Rates and risk factors for recurrence of gestational diabetes. Diabetes Care 2001;24:659–62.

49. McMahon MJ et al. Gestational diabetes mellitus. Risk factors, obstetric complications and infant outcomes. J Reprod Med 1998;43:372–8.

50. Pole JD, Dodds LA. Maternal outcomes associated with weight change between pregnancies. Can J Public Health 1999;90:233–236.

51. Kjos SL, Buchanan TA. Gestational diabetes mellitus. N Engl J Med 1999;341:1749–56.

52. Coustan DR, Carpenter MW. The diagnosis of gestational diabetes. Diabetes Care 1998;21(suppl)2:B5–8.

53. Crowe SM et al. Oral glucose tolerance test and the preparatory diet. Am J Obstet Gynecol 2000;182:1052–4.

54. Gestational diabetes. http://my.webmd.com, accessed 10/02.

55. Coleman T. Patient centered care of diabetes in general practice. Study failed to measure patient centeredness of GPs' consulting behavior. BMJ1999;318:1621–2.

56. Adams KM, Li H, Nelson RL et al. Sequelae of unrecognized gestational diabetes. Am J Obstet Gynecol 1998;178:1321–32.

57. Mello G, Parretti E, Mecacci F et al. What degree of maternal metabolic control in women with type 1 diabetes is associated with normal body size and proportions in full-term infants? Diabetes Care 2000;23:1494–8.

58. Jovanovic L. Time to reassess the optimal dietary prescription for women with gestational diabetes [editorial]. Am J Clin Nutr 1999;70:3–4.

59. Jovanovic L. Management of diet and exercise in gestational diabetes mellitus. Prenat Neonat Med 1998;3:534–541.

60. Ecker JL et al. Gestational diabetes. N Engl J Med 2000;342:896–7.

61. Pezzarossa A et al. Effects of maternal weight variations and gestational diabetes mellitus on neonatal birth weight. J Diabetes Complications 1996;10:78–83.

62. Merlob P et al. Oral antihyperglycemic agents during pregnancy and lactation: a review. Paediatr Drugs Related Articles 2002:4:755–60.

63. Fagen C et al. Nutrition management in women with gestational diabetes mellitus: a review by ADA's Diabetes Care and Education Dietetic Practice Group. J Am Diet Assoc 1995;95:460–7.

64. McFarland MB et al. Dietary therapy for gestational diabetes: how long is long enough? Obstet Gynecol 1999;93:978–82.

65. Butte NF. Dieting and exercise in overweight, lactating women [editorial]. N Engl J Med 2000;342:502–3.

66. Kalkwarf HJ et al. Dietary fiber intakes and insulin requirements in pregnant women with type 1 diabetes. J Am Diet Assoc 2001;101:305–10.

67. Major CA et al. The effects of carbohydrate restriction in patients with diet-controlled gestational diabetes. Obstet Gynecol 1998;91:600–4.

68. Riccardi G, Rivellese AA. Diabetes: nutrition in prevention and management. Nutr Metab Cardiovasc Dis 1999;9:33–6.

69. Haynes RB. What's new in medicine? Evidence-based medicine hits a home run for diabetes care. WebMD Scientific American Medicine, posted 12/08/03, www.medscape.com.

70. Landon MB. Diabetes mellitus and other endocrine disorders. In: Gabbe SG et al., eds. Obstetrics, normal and problem pregnancies. New York: Churchill Livingstone; 1996: pp 1037–81.

71. Reece EA et al. The role of free radicals and membrane lipids in diabetes-induced congenital malformations. J Soc Gynecol Investig 1998;5:178–87.

72. Raychaudhuri K, Maresh MJ. Glycemic control throughout pregnancy and fetal growth in insulin-dependent diabetes. Obstet Gynecol 2000;95:190–4.

73. Health, United States, 2003. www.cdc.gov/nchs, accessed 10/03.

74. Brown JE, Schloesser PT. Prepregnancy weight status, prenatal weight gain, and the outcome of term twin gestations. Am J Obst Gynecol 1990;162:182–6.

75. Ventura SJ et al. Births: final data for 1999. Vol. 2001: www.cdc.gov/nchs, 2001.

76. Guyer B, Hoyert DL, Martin JA et al. Annual summary of vital statistics—1998. Pediatrics 1999;104:1229–46.

77. Hoskins RE. Zygosity as a risk factor for complications and outcomes of twin pregnancy. Acta Genet Med Gemellol 1995;44:11–23.

78. Keith LG. Multiple pregnancy: epidemiology, gestation and perinatal outcome. New York: Parthenon Publishing Group, 1995.

79. Redline RW. Non-identical twins with a single placenta-disproving dogma in perinatal pathology. N Engl J Med 2003;349:111–14.

80. Imaizumi Y. A comparative study of twinning and triplet rates in 17 countries, 1971–1996. Acta Genet Med Gemellol 1998;47:101–14.

81. Czeizel AE et al. Higher rate of multiple births after periconceptional vitamin supplementation. New Engl J Med 1994;330:1687–8.

82. Maclennan AH. Multiple gestation. Clinical characteristics and management. In: Creasy RK, Resnick R eds. Maternal-fetal medicine: principles and practice. Philadelphia: W. B. Saunders; 1994: pp 589–601.

83. Landy HJ, Keith LG. The vanishing twin: a review. Hum Reprod Update 1998;4:177–83.

84. Brown JE, Carlson M. Nutrition and multifetal pregnancy. J Am Diet Assoc 2000;100:343–8.

85. Kogan MD et al. Trends in twin birth outcomes and prenatal care utilization in the United States, 1981–1997. JAMA 2000;284:335–41.

86. Cohen SB et al. New birth weight nomograms for twin gestation on the basis of accurate gestational age. Am J Obstet Gynecol 1997;177:1101–4.

87. Ellings JM et al. Reduction in very low birth weight deliveries and perinatal mortality in a specialized, multidisciplinary twin clinic. Obstet Gynecol 1993;81:387–91.

88. Dubois S et al. Twin pregnancy: the impact of the Higgins Nutrition Intervention Program on maternal and neonatal outcomes. Am J Clin Nutr 1991;53:1397–403.

89. Luke B et al. Critical periods of maternal weight gain: effect on twin birth weight. Am J Obstet Gynecol 1997;177:1055–62.

90. Konwinski T et al. Maternal pregestational weight and multiple pregnancy duration. Acta Genet Med Gemellol 1973;22:44–7.

91. Brown JE et al. Early pregnancy weight change and newborn size (Abstract), FASEB, Orlando, FL, March 31–April 4, 2000.

92. Zeijdner EE et al. Essential fatty acid status in plasma phospholipids of mother and neonate after multiple pregnancy. Prostaglandins Leukotrienes and Essential Fatty Acids 1997;56:395–401.

93. Fawzi WW et al. Randomized trial of effects of vitamin supplements on pregnancy outcomes and T cell counts in HIV 1-infected women in Tanzania. Lancet 1998;351:1477–82.

94. French M et al. Highly active antiretroviral therapy. Lancet 1998;351:1056–7.

95. Rich KC et al. Maternal and infant factors predicting disease progression in human immunodeficiency virus type 1-infected infants. Women and Infants Transmission Study Group. Pediatrics 2000;105:E8.

96. Fields-Gardner C, Ayoob KT. Position of the American Dietetic Association and Dietitians of Canada: nutrition intervention in the care of persons with human immunodeficiency virus infection. J Am Diet Assoc 2000;100:708–17.

97. Azais-Braesco V, Pascal G. Vitamin A in pregnancy: requirements and safety limits. Am J Clin Nutr 2000;71:1325S–33S.

98. Wunderlich SM. Nutritional assessment and support of HIV-positive pregnant women and their children: perspectives from an industrialized country. Nutrition Today 2000;35:107–112.

99. Bonne OB et al. Delayed detection of pregnancy in patients with anorexia nervosa: two case reports. Int J Eat Disord 1996;20:423–25.

100. Herzog DB et al. Mortality in eating disorders: a descriptive study. Int J Eat Disord 2000;28:20–6.

101. Becker AE et al. Eating disorders. N Engl J Med 1999;340:1092–8.

102. Little L, Lowkes E. J Midwifery Womens Health 2000;45:301–7.

103. Morrill ES et al. Bulimia nervosa during pregnancy: a review. J Am Diet Assoc 2001;102:448–54.

104. Rosso P. Nutrition and metabolism in pregnancy. New York: Oxford University Press; 1990: pp 117–118, 125, 150–151.

105. Chang S-C et al. Am J Clin Nutr 2003;77:1248–54.

106. Scholl TO et al. Maternal growth during pregnancy and the competition for nutrients. Am J Clin Nutr 1994;60:183–8.

107. Scholl TO et al. Leptin and maternal growth during adolescent pregnancy. Am J Clin Nutr 2000;72:1542–7.

108. Lenders CM et al. Curr Opin Pediatr 2000;12:291–6.

109. Lenfant C. Clinical research to clinical practice—lost in translation. N Engl J Med 2003;868–74.

"One of the best things that only you can do is breastfeed your baby for as long as possible. The longer a mom and baby breastfeeds, the greater the benefits are for both mom and baby"

The National Women's Health Information Center, 2002

Chapter 6

Nutrition and Lactation

Chapter Outline

Prepared by **Maureen A. Murtaugh** and **Carolyn Sharbaugh** with **Denise Sofka**

Key Nutrition Concepts

1   Human milk is the best food for newborn infants for the first year of life or longer.

2   Maternal diet does not significantly alter the protein, carbohydrate, fat, and major mineral composition of breast milk, but it does affect the fatty acid profile and the amounts of some vitamins and trace minerals.

3   When maternal diet is inadequate, the quality of the milk is preserved over the quantity for the majority of nutrients.

4   Health care policies and procedures and the knowledge and attitudes of health care providers affect community breastfeeding rates.

Introduction

The benefits of breastfeeding for both mothers and infants are well established. Federal efforts to promote breast-feeding and greater understanding of its advantages have contributed to a resurgence of breastfeeding in the United States since the 1970s. Nevertheless, racial and ethnic disparities in breastfeeding rates persist, and despite the knowledge that the benefits increase with longer duration, there has been little increase in the duration of lactation.

The health care system, workplace, and community can either hinder or facilitate the initiation and continuation of breastfeeding. Health programs can play a significant role in increasing breastfeeding rates. Health care professionals who want to promote breastfeeding should know the physiology of lactation, the composition of human milk, and the benefits to mothers and infants. Helping women achieve the appropriate nutritional status for breastfeeding requires an understanding of energy needs, weight goals, the effects of exercise during breast-feeding, and vitamin and mineral requirements.

Benefits of Breastfeeding

Breastfeeding Benefits for Mothers

Women who breastfeed experience hormonal, physical, and psychosocial benefits.[1] Breastfeeding immediately after birth increases levels of the hormone oxytocin, which stimulates contractions of the uterus, minimizes maternal postpartum blood loss, and helps the uterus to return to nonpregnant size.[2]

After the birth, the return of fertility (through monthly ovulation) is delayed in most women during breastfeeding, particularly with exclusive breastfeeding.[3] This delay in ovulation results in longer intervals between pregnancies. Breastfeeding alone, however, is not as effective as other available birth control methods. Conse-

quently, many health care professionals in the United States do not suggest breastfeeding as an option for birth control.

Many women experience psychological benefits including increased self-confidence and bonding with their infants.[4] Although some women still consider faster return to prepregnancy weight a benefit of breastfeeding, women may lose or gain weight while nursing. The impact of breastfeeding on maternal weight is discussed in more detail later in this chapter. In addition to these short-term benefits, women who nurse have a lower risk of breast and ovarian cancer.[5,6]

Breastfeeding Benefits for Infants

"Human milk is uniquely superior for infant feeding and is species-specific; all substitute feeding options differ markedly from it. The breastfed infant is the reference or normative model against which all alternative feeding methods must be measured with regard to growth, health, development, and all other outcomes."

American Academy of Pediatrics Policy Statement on Breastfeeding, 1997[7]

NUTRITIONAL BENEFITS   The value of the composition of human milk is widely recognized. Companies that make human milk substitutes (HMSs) often use human milk as the standard, recognizing the many unique properties of human milk:

- With its dynamic composition and appropriate balance of nutrients, human milk provides optimal nutrition to the infant.[8,9]

- The balance of nutrients in human milk matches human infant requirements for growth and development closely;[10] no other animal milk or HMS meets infant needs as well.

- Human milk is isosmotic (of similar ion concentration; in this case human milk and plasma are of similar ion concentration) and therefore meets the requirements for infants without other forms of food or water.

- The relatively low protein content of breast milk compared to cow's milk meets the infant's needs without overloading the immature kidneys with nitrogen.

- Whey protein in human milk forms a soft, easily digestible curd.

- Human milk provides generous amounts of lipids in the form of essential fatty acids, saturated fatty acids, medium-chain triglycerides, and cholesterol.

- Long-chain polyunsaturated fatty acids, especially docosahexaenoic acid (DHA), which promotes optimal development of the central nervous system, are present in human milk but not in most HMSs marketed in the United States.

- Minerals in breast milk are largely protein bound and balanced to enhance their availability and meet infant needs with minimal demand on maternal reserves.

IMMUNOLOGICAL BENEFITS   One of the most important realizations about breastfeeding in the last decade is the ability of human milk to protect against infection. Many components of human milk are active against infection.[8,9] Cellular components (*T-* and *B-lymphocytes, neutrophils, macrophages,* and *epithelial cells*) are especially high in *colostrum* but are also present for months in mature human milk in lower concentrations.

Secretory immunoglobulin A, the predominant *immunoglobulin* in human milk, helps to protect the infant's gastrointestinal tract. These proteins in human milk bind iron and vitamin B_{12}, making the nutrients unavailable for the growth of pathogens in the infant's gastrointestinal tract. Such factors are also responsible for the different kinds of intestinal flora (natural bacteria of the gastrointestinal tract) found in breastfed infants.

Growth factors and hormones in human milk, such as insulin, enhance the maturation of the infant's gastrointestinal tract. These substances also help to protect the infant, especially neonates, against viral and bacterial pathogens.

In addition to protecting against viral and bacterial pathogens, breastfeeding enhances the immune response to immunizations including polio, tetanus, diphtheria, and *Haemophilus* influenza. Breastfeeding also enhances the immune response to respiratory syncytial virus (RSV) infection, a common infant respiratory infection.[1] Protection against infection is strongest during the first several months of life for infants who are exclusively breastfed and continues throughout the duration of breastfeeding.

COGNITIVE BENEFITS   Several reports have linked breastfeeding, and especially duration of breastfeeding, with cognitive benefits, assessed by IQ.[11,12] Recognition that the fatty acid composition of milk plays an important role in neuropsychological development bolsters the credibility of psychological or cognitive benefits from breastfeeding. The differences in *cognitive function* are greater in premature infants fed human milk than for term infants.[13] Cognitive development gains increase with the duration of breastfeeding.[11]

REDUCED MORBIDITY   Not surprisingly, in countries with high infant illness (*morbidity*) and death rates (*mortality rates*), poor sanitation, and questionable water supplies, breastfed infants experience reduced rates of illness. Even in the United States, where modern health care systems, safe water, and proper sanitation are commonplace, there is a clear relationship between breastfeeding and reduced rates of illness in infants. In U.S. samples, the incidence of diarrhea is estimated to be 50% lower in exclusively breastfed infants.[14] Internationally, gastrointestinal infection was lower among infants exclusively breastfed for 6 months when compared to those exclusively breastfed for only 3 months.[14,15] Ear infections are 19% lower, and the number of prolonged episodes of ear infection was 80% lower among breastfed infants than among infants fed HMS. In a U.S. population study, breastfed infants experienced 17% less coughing and wheezing, and 29% less vomiting than infants fed HMS.[16]

In addition to the lower rate of acute illnesses in breastfed children, breastfeeding also seems to protect against chronic childhood diseases. Breastfeeding may reduce the risk of celiac disease,[17] inflammatory bowel disease,[18] and neuroblastoma.[19] HMS feeding results in an increase in the risk of allergy (30%) and asthmatic disease (25%).[20]

These reductions in acute and chronic infant illness increase with greater use of human milk.[21] For example, infants who receive some human milk and some HMS are at 60% greater risk of ear infection than those fed exclusively human milk.[21] The risks, particularly for allergy and asthmatic disease, are reduced for the duration of breastfeeding and for months to years after weaning.[20]

Breastfeeding may also play a role in reducing the risk of sudden infant death syndrome (SIDS), but this is still under investigation. Researchers disagree about whether breastfeeding has a primary effect in reducing risk of SIDS. A recent analysis of available studies found that bottle-feeding increases the risk of SIDS, but other factors related to feeding choice may be responsible for this finding.[22]

Considerable attention has been paid to the role of breastfeeding in preventing obesity, but this relationship is still a topic of controversy. Breastfed infants typically are leaner than HMS-fed infants at 1 year of age but have similar activity level and development.[23] A recent review of the topic revealed that a majority of studies have found small reductions in risk of overweight in children greater than 3 years of age who were breastfed rather than given HMS.[24] The effect of breastfeeding on the incidence of overweight

T-lymphocytes   White blood cells that are active in fighting infection may also be called T-cells; the *T* in *T-cell* stands for *thymus*). These cells coordinate the immune system by secreting hormones that act on other cells.

B-lymphocytes   White blood cells that are responsible for producing immunoglobulins.

Neutrophils   A class of white blood cells that are involved in protecting against infection.

Macrophages   A white blood cell that acts mainly through phagocytosis.

Epithelial Cells   Cells that line the surface of the body.

Colostrum   The milk produced in the first 2–3 days after the baby is born. Colostrum is higher in protein and lower in lactose than milk produced after the milk supply is established.

Secretory Immunoglobulin A   One of the proteins found in secretions that protect the body's mucosal surfaces from infections. It may act by reducing the binding of a microorganism with cells lining the digestive tract. It is present in human colostrum but not transferred across the placenta.

Immunoglobulin   A specific protein that is produced by blood cells to fight infection.

Cognitive Function   The process of thinking.

Morbidity   The rate of illnesses in a population.

Mortality Rate   The rate of death.

was greater with longer duration of breastfeeding in some, but not all studies. Several potential mechanisms have been identified for the modest reduction in obesity in children who breastfed and include metabolic programming related to chemical substances in human milk, learned self-regulation of energy intake, and other characteristics of families or parents such as healthy lifestyles. Given the epidemic of obesity in the United States, it is likely that research addressing this important issue will continue.

SOCIOECONOMIC   BENEFITS   OF BREASTFEEDING A decrease in medical care for breastfed infants is the primary socioeconomic benefit of breastfeeding. Medicaid costs for breastfed WIC infants in Colorado were $175 lower than for infants who were fed HMS.[25] Never-breastfed infants required more care for lower respiratory tract illness, otitis media, and gastrointestinal disease than infants breastfed for at least 3 months.[26] Each 1000 never-breastfed infants had 2033 more sick care visits, 212 more days of hospitalization, and 609 more prescriptions. In addition, in one study of two companies with established lactation programs, the incidence of 1-day maternal absenteeism from work due to infant illness was approximately two-thirds lower in breastfeeding women than in nonbreastfeeding women.[27] Thus, companies benefit through lower medical costs and greater employee productivity.

ANALGESIC EFFECTS Breastfeeding seems to work as an analgesic to infants. Breastfeeding during venipucture reduces infant pain equally well as does ingesting 30% glucose solution followed by pacifier use.[28] It appears that the infant must be breastfeeding during the heel prick in order to reduce pain.[29] Breastfeeding may be used to reduce infant discomfort during minor invasive procedures.

Breastfeeding Goals for the United States

"During the twentieth century, infant feeding practices have undergone dramatic changes that reflect shifts in values and attitudes in the U.S. society as a whole. They have tended to occur first among those women at the forefront of changes in dominant social values and among those with resources (whether it is time, energy or money) to permit adoption of new feeding practices."

Institute of Medicine, Subcommittee on Nutrition during Lactation, 1991[30]

In the early 1900s, almost all infants in the United States were breastfed. As safe HMSs became widely available, breastfeeding rates steadily declined, reaching levels below 30% in the 1950s and 1960s and then rising dramatically in the 1970s (Illustration 6.1).[31] In the early 1980s, levels rose above 60% at hospital discharge and then steadily declined until the early 1990s. For the past decade, breastfeeding rates have been steadily increasing among all sociodemographic groups, but they still remain below the Healthy People 2000 breastfeeding goals for the nation.[32] Despite the well-documented advantages of breastfeeding, in 2000 68.4% of all mothers breastfed in the early postpartum period and less than one-third (31.4%) breastfed at 6 months postpartum.[33] Although significant increases in breastfeeding are evident, ethnic and racial disparities in breastfeeding remain wide; the rates of breastfeeding at 6 and 12 months among African American women (21% and 13%, respectively) remain lower than those for white (34% and 18%) or Hispanic women (28% and 18%). Rates of breastfeeding are highest among college-educated women and women aged 30 years and older. The rates of breastfeeding are lowest among women whose infants are at greatest risk of poor health and development—women aged 21 years and under and those with low educational levels.[33,34]

Breastfeeding rates vary across the United States with the highest rates in the Pacific Northwest and Rocky Mountain states and some of the lowest rates in the southern states. Breastfeeding initiation rates are near 82% in the Pacific and Mountain regions of the United States, but in the East South Central region they are only 53%. In five states, (Alaska, Colorado, New Mexico, Utah, and Washington), breastfeeding initiation rates of 75% have been achieved.[33]

In recognition of the health and economic benefits of breastfeeding, national goals for breastfeeding rates have been established and revised since 1980.[32,35] Healthy People 2010 contains wide-ranging national health goals for the new decade, focusing on two major themes: (1) increasing the quality and years of healthy life and (2) eliminating racial and ethnic disparities in health status.[35] In addition to adding specific breastfeeding goals for black or African Americans, and Hispanic or Latina Americans, Healthy People 2010 places increased emphasis on the duration of breastfeeding. By the year 2010, the goal is for at least 75% of American women to breastfeed their infants in the early postpartum period, at least 50% at 6 months, and 25% at 1 year (Table 6.1).[36] Increasing the rates of breastfeeding is a compelling public health goal, particularly among the racial and ethnic groups that are less likely to initiate and sustain breastfeeding throughout the infant's first year.[1]

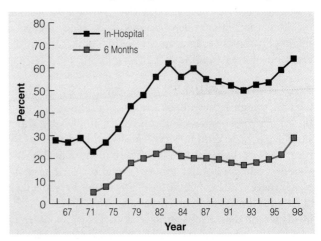

Illustration 6.1 U.S. breastfeeding rates: 1965–1998.[35,36]

Table 6.1 Healthy People 2010: Breastfeeding objectives for the nation[1]

Objective: Increase the proportion of mothers who breastfeed their babies.

	1998 BASELINE	2010 GOAL
In Early Postpartum Period		
All women	64%	75%
Black or African American	45	75
Hispanic or Latina	66	75
White	68	75
At Six Months		
All women	29	50
Black or African American	19	50
Hispanic or Latina	28	50
White	31	50
At One Year		
All women	16	25
Black or African American	9	25
Hispanic or Latina	17	25
White	19	25

SOURCE: U.S. Department of Health and Human Services. Healthy People 2010: Conference Edition—volumes I and II. Washington, DC: U.S. Department of Health and Human Services, Public Health Service, Office of the Assistant Secretary for Health, January 2000.[35]

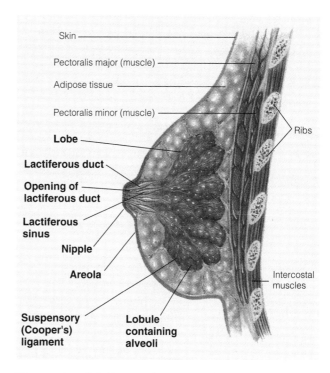

Illustration 6.2 Breast of a lactating female.
This cutaway view shows the mammary glands and ducts.

If the United States is to reach these breastfeeding goals, health professionals must take an active role in promoting and supporting lactation. Becoming a breastfeeding advocate requires a thorough understanding of lactation physiology.

Lactation Physiology

Functional Units of the Mammary Gland

The functional units of the *mammary gland* are the *alveoli* (Illustration 6.2). Each alveolus is composed of a cluster of cells whose job it is to secrete milk (*secretory cells*) with a duct in the center. The ducts are arranged like branches of a tree, with each smaller duct leading to a larger duct. These branchlike ducts lead to the nipple. *Myoepithelial cells* surround the secretory cells. The myopethelial cells can contract under the influence of *oxytocin* and cause milk to be ejected into the ducts. Dilations in the *lactiferous sinuses* behind the nipple allow for storage of milk.

Mammary Gland Development

During puberty, the ovaries mature, and the release of estrogen and progesterone increases (Table 6.2 on the next page). The cyclic release of these two hormones governs pubertal breast development (Illustration 6.3 on the following page). The mammary gland develops its lobular structure (*lobes*) under the cyclic production of progesterone and is usually complete within 12 to 18 months after menarche. As the ductal system matures, cells that can secrete milk develop, and the nipple grows and its pigmentation changes. Fibrous and fatty tissues increase around the ducts.

In pregnancy, the luteal and placental hormones (placental lactogen and chorionic gonadotropin) allow further preparation for breastfeeding (Illustration 6.3). Estrogen stimulates development of the glands that will make milk. Progesterone allows the ducts to elongate and the cells that line them (epithelial cells) to duplicate. Production of milk is possible as early as 16 weeks of gestation.

Mammary Gland The source of milk for offspring, also commonly called the breast. The presence of mammary glands is a characteristic of mammals.

Alveoli Rounded or oblong cavities present in the breast (singular = *alveolus*).

Secretory Cells Cells in the acinus (milk gland) that are responsible for secreting milk components into the ducts.

Myoepithelial Cells Specialized cells that line the alveoli and can contract to cause milk to be secreted into the duct.

Oxytocin A hormone produced during letdown that causes milk to be ejected into the ducts.

Lactiferous Sinuses Larger ducts for storage of milk behind the nipple.

Lobes Rounded structures of the mammary gland.

Lactogenesis Another term for human milk production.

Lactogenesis

Breast-milk production, or *lactogenesis,* occurs in three stages.[37] The first stage, or lactogenesis I, begins during the last trimester of pregnancy; the second and third stages, lactogenesis II and III, occur after birth.

- *Lactogenesis I.* During lactogenesis I, milk begins to form, and the lactose and protein content of milk increases. This stage extends through the first

Table 6.2 Hormones contributing to breast development and lactation

Hormone	Role in Lactation	Stage of Lactation
Estrogen	Ductal growth	Mammary gland differentiation with menstruation
Progesterone	Alveolar development	After onset of menses and during pregnancy
Human growth hormone	Development of terminal end buds	Mammary gland development
Human placental lactogen	Alveolar development	Pregnancy
Prolactin	Alveolar development and milk production	Pregnancy and breastfeeding (from the third trimester of pregnancy to weaning)
Oxytocin	Letdown: ejection of milk from myoepithelial cells	From the onset of milk secretion to weaning

few postpartum days, when suckling is not necessary for initiating milk production.

- *Lactogenesis II.* This stage begins 2–5 days postpartum and is marked by increased blood flow to the mammary gland. Clinically, it is considered the onset of copious milk secretion, or "when milk comes in." Significant changes in both the milk composition and the quantity of milk that can be produced occur over the first 10 days of the baby's life.

- *Lactogenesis III.* This stage begins about 10 days after birth and is the stage in which the milk composition becomes stable.[38]

Hormonal Control of Lactation

Prolactin and oxytocin are necessary for establishing and maintaining a milk supply. *Prolactin* is a hormone that stimulates milk production. Suckling is a major stimulator of prolactin secretion: prolactin levels double with suckling.[8] Stress, sleep, and sexual intercourse also stimulate prolactin levels. To prevent milk production in the last 3 months of pregnancy, prolactin activity is suppressed by a prolactin-inhibiting factor that is released by the hypothalamus. This inhibition of prolactin allows the mother's body to prepare for milk production during pregnancy while preventing milk production until the baby is born. The actual level of prolactin in the blood is not related to the amount of milk made, but prolactin is necessary for milk synthesis to occur.[39]

Oxytocin release is also stimulated by suckling or nipple stimulation. Its main role is in letdown, or the ejection of milk from the milk gland (acinus) into the milk ducts and lactiferous sinuses. Women may experience tingling or sometimes a sharp or shooting sensation that lasts less than a minute and corresponds with contractions in the milk ducts. Oxytocin also acts on the uterus, causing it to contract, seal blood vessels, and shrink in size.

Secretion of Milk

Although the process of milk production is complex, understanding the basic mechanisms of milk secretion is important to understanding how factors such as nutritional status, supplementation, medications, and disease may affect breastfeeding or milk composition. As described by Neville et al.,[40]

Prolactin A hormone that stimulates milk production.

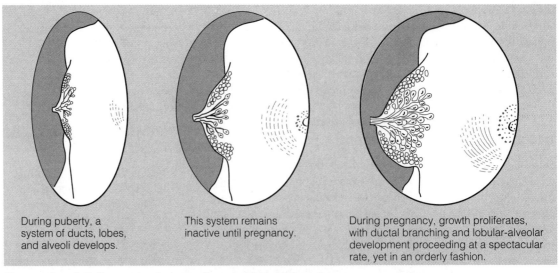

During puberty, a system of ducts, lobes, and alveoli develops.

This system remains inactive until pregnancy.

During pregnancy, growth proliferates, with ductal branching and lobular-alveolar development proceeding at a spectacular rate, yet in an orderly fashion.

Illustration 6.3 Breast development from puberty to lactation.

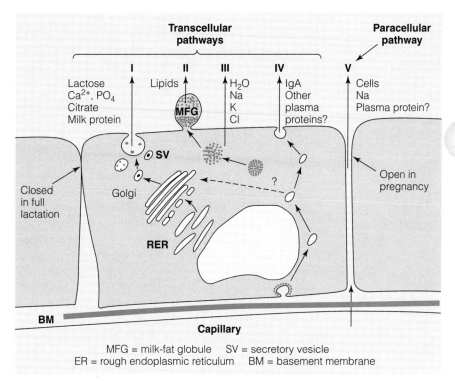

Transcellular pathways | Paracellular pathway

I — Lactose Ca²⁺, PO₄ Citrate Milk protein
II — Lipids — MFG
III — H₂O Na K Cl
IV — IgA Other plasma proteins?
V — Cells Na Plasma protein?

SV

Closed in full lactation

Golgi

RER

BM

Capillary

Open in pregnancy

?

MFG = milk-fat globule   SV = secretory vesicle
ER = rough endoplasmic reticulum   BM = basement membrane

Illustration 6.4 The pathways for secretion of milk components.

SOURCE: From Lactation: Physiology, Nutrition, & Breastfeeding by J. C. Allen and C. Watters, p. 49–102. Copyright © 1983 Kluwer Academics. Reprinted with permission.

the secretory cell in the breast uses five pathways for milk secretion (Illustration 6.4). Briefly, some components, like lactose, are made in the secretory cells and secreted into ducts. Water, sodium, potassium, and chloride are able to pass through alveolar cell membranes in either direction (passive diffusion). Other components are processed within the cells of the breast. Milk fat comes from triglycerides from the mother's blood and from new fatty acids produced in the breast. Fats are made soluble in breast milk by addition of a protein carrier to form milk-fat globules.[41] These milk-fat globules are then secreted into the ducts. Immunoglobulin A and other plasma proteins are captured from the mother's blood and taken into the alveolar cells. These proteins are then secreted into the milk ducts.

The Letdown Reflex

"Nursing moms don't always feel their milk let down. Some moms never notice when their milk ejects, and others have a very strong (pins and needles) feeling as their milk lets down."

Debbie Donovan, Director, ParentsPlace.com

The letdown reflex stimulates milk release from the breasts. An infant suckling at the nipple usually causes the reflex to occur. The stimuli from the infant suckling are passed through nerves to the hypothalamus, which responds by releasing oxytocin from the posterior pituitary gland (Illustration 6.5). The oxytocin causes contraction of the myoepithelial cells surrounding the secretory cells. As a result, milk is ejected through the

ducts to the lactiferous sinus, making it available to the infant. Other stimuli, such as hearing a baby cry, sexual arousal, and thinking about nursing, can also cause letdown, and milk leakage from the breasts.

Breast-Milk Supply and Demand

Can Women Make Enough Milk?

Typical milk production averages approximately 600 ml (240 ml = 1 cup, or 8 ounces) in the first month postpartum, increasing to approximately 750–800 ml per day by 4–5 months postpartum.[30] Milk production can range from 450– to 1200 ml per day in women who are nursing one infant.[42] Infant weight, the caloric density of milk, and the infant's age contribute significantly to the infant's demand for milk. Milk production increases to meet the demands of twins, triplets, or infants and toddlers suckling simultaneously; it can also be increased by pumping the milk.

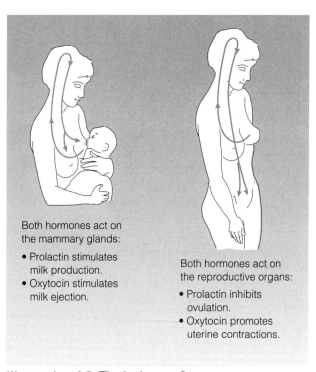

Both hormones act on the mammary glands:
- Prolactin stimulates milk production.
- Oxytocin stimulates milk ejection.

Both hormones act on the reproductive organs:
- Prolactin inhibits ovulation.
- Oxytocin promotes uterine contractions.

Illustration 6.5 The letdown reflex.

An infant suckling at the breast stimulates the pituitary to release the hormones prolactin and oxytocin.

Traditionally, factors such as how vigorously an infant nurses, how much time the infant is at the breast, and how many times he or she nurses in a day were thought to control milk production. We now know that milk synthesis (rate of accumulation of milk in the breast) is related to infant demand.[43] That is, the removal of milk from the breast seems to be the signal to make more milk, and most women are able to increase their milk production to meet infant demand.[44]

An average of 24% of milk is left in the breast after feeding.[45] Thus, the storage capacity of the breast does not seem to be a limit on infant milk intake. The average rate of milk synthesis in a day is only 64% of the highest rate, suggesting that milk synthesis could be increased considerably. Comparisons of milk production between mothers of singletons and twins show that the breasts have the capacity to synthesize much more than a singleton infant usually drinks.[46]

Does the Size of the Breast Limit a Woman's Ability to Nurse Her Infant?

The size of a woman's breast does not determine the amount of milk production tissue (clusters of alveoli containing secretory cells that produce the milk).[47] Much of the variation in breast size is due to the amount of fat in the breast. The size of the breast does limit *storage* because it puts limits on the expansion of the ducts and lactiferous sinuses. Daily milk production is not related to the total milk-storage capacity within the breast, however.[45] This means that a woman with small breasts can produce the same amount of milk as a woman with large breasts, although the latter may be able to feed her infant less frequently to deliver the same volume of milk.

Is Feeding Frequency Related to the Amount of Milk a Woman Can Make?

Feeding frequency is not consistently related to milk production. The rate of milk synthesis is highly variable between breasts and between feedings.[48] The amount of milk produced in 24 hours and the total milk withdrawn in that 24-hour period are highly related, however.[45] Milk synthesis is able to quickly respond to infant demand.

The breast responds to the degree of emptying during a feeding, and this response is a link between maternal milk supply and infant demand. Daly proposed that the breast responds to the infant's need by measuring how completely the infant empties the breast.[43,45,48] For example, if a lot of milk is left in the breast, then milk synthesis will be low to prevent engorgement; if the breast is fully emptied, synthesis will be high to replenish the milk supply.

Exact mechanisms of milk supply and demand are not well understood, but they seem to be related to a protein called feedback inhibitor of lactation (FIL).[48] FIL is an active whey protein that inhibits milk secretion. The protein inhibits all milk components equally according to their concentration in milk. Therefore, this protein seems to affect milk quantity only, not milk composition.

Pumping or Expressing Milk

A woman may need to pump or express milk for many reasons, including maternal or infant illness or separation. Women can express milk manually or by using a hand pump, a commercial electric pump, or a hospital-grade electric pump. A pump that allows a mother to pump both breasts at the same time (10 minutes per session) can save time over single pumping (20 minutes per session).[49] Electric pumps are efficient and may increase prolactin more than hand expression or hand pumping.[49,50] Insufficient milk production is a common problem among women who express milk. Researchers working with women who pumped their breasts report that 8 to 12 or more milk expressions per day were necessary to stimulate adequate production of milk.[49] The optimal number of expressions in a 24-hour period is likely to differ according to how well women empty the breast and the storage capacity of the breast. Women who are able to establish an adequate volume of milk (>500 ml per day) in the first two weeks postpartum are more likely to have enough milk for their infant at 4 to 5 weeks postpartum than are women who express less milk.[49] This recommendation is consistent with the advice to nurse the infant (or pump) early and often to build a good milk supply.

Can Women Breastfeed after Breast Reduction or Augmentation Surgery?

Information regarding breastfeeding rates after breast surgery is scarce, but women generally have more problems after breast reduction surgery than after augmentation surgery. Accumulating evidence suggests that women who undergo breast reduction surgery may be at risk for unsuccessful lactation.[51] The type of surgery appears to be an important determinant of ability to breastfeed. Some experts recommend locating the incision around the lower part of the breast to avoid damage to the ductal system caused by incision in the middle of the breasts. Women with periareolar (around the nipple) incisions experience greater difficulties with breastfeeding.[51,52] Lactation consultants recommend that the surgery date, type, and incision used, as well as prior breastfeeding, be ascertained. Infants should be closely followed to prompt intervention when needed.

Does Silicone from Breast Implants Leach into the Milk?

The American Academy of Pediatrics does not consider silicone implants a reason not to breastfeed.[53] Surprisingly, nearly a million women in the United States have breast implants containing silicone. Early reports introduced concern of esophageal dysfunction in children of women who had silicone implants, but more recent research found no evidence to support such claims. Silicone concentrations in milk from women with implants are not elevated[54] and, in fact, are lower than amounts of silicone found in formula and cow's milk.[55] Although it is possible that immunological responses could cause unfavorable effects, there is no evidence of direct toxicity to the infant from silicone implants.

Human Milk Composition

Human milk is an elegantly designed natural resource. It is the only food needed by the majority of healthy infants for approximately 6 months. Human milk is designed not only to nurture, but also to protect infants from disease. Its composition is changeable over a single feeding, over a day, according to the infant's age or gestation at delivery, with the presence of infection in the breast, with menses, and with maternal nutritional status.

As our ability to measure and identify breast-milk components increases, we recognize that the composition of human milk is complex. Hundreds of components of human milk have been identified, and their nutritive and nonnutritive roles are under investigation. The basic nutrient composition of colostrum and mature milk is provided in Table 6.3. *The Handbook of Milk Composition*[9] and *Breastfeeding: A Guide for the Medical Professional*[8] provide more complete descriptions of human and other milks.

Table 6.3 Human milk composition (per L)[a]		
Milk Component	Early Milk	Mature Milk
Lactose (g)	20–30	67
Total Protein (g)	16	9
Fat%	2	3.5
Calories	—	2730–2940
Retinol (mg)	2	0.3–0.6
Caretenoids (mg)	2	0.3–0.6
Thiamin (ug)	20	200
Riboflavin (ug)		400–600
Niacin (mg)	0.5	1.8–6.0
Vitamin B$_6$ (mg)		0.9–0.31
Pantothenic acid (mg)		2–2.5
Biotin (ug)		5–9
Folate (ug)		80–140
Vitamin B$_{12}$ (ug)		0.5–1.0
Vitamin C (mg)		100
Vitamin D (µg)		0.33
Vitamin E (mg)	8–12	3–8
Vitamin K (µg)	2–5	2–3
Calcium (mg)	250	200–250
Phosphorus (mg)	120–160	120–140
Magnesium (mg)	30–35	30–35
Copper (mg)	0.5–0.8	0.2–0.4
Iron (mg)	0.5–1.0	0.3–0.9
Zinc (mg)	8–12	1–3

[a]Adapted from: Picciano, M.F., *Representative values for constituents of human milk*. Ped Clin N Am 2001;48:1–3.

Colostrum

The first milk, colostrum, is a thick, often yellow fluid produced during lactogenesis II (days 1–3 after infant birth). Infants may drink only 2 to 10 ml (½ to 2 teaspoons) of colostrum per feeding in the first two to three days. Colostrum provides about 58–70 cal/100 ml and is higher in protein, and lower in carbohydrate and fat, than mature milk (produced 2 weeks after infant birth). Secretory immunogolulin A and lactoferrin are the primary proteins in colostrum, but other proteins found in mature milk are not present. The concentration of mononuclear cells (a specific type of white blood cell) from the mother that provides immune protection is highest in colostrum. Colostrum has higher concentrations of sodium, potassium, and chloride than are found in mature milk.

Water

Breast milk is isotonic with maternal plasma. This biological design means that babies do not need water or other fluids to maintain hydration, even in hot climates.[56] As a major component of human milk, water allows suspension of milk sugars, proteins, immunoglobulin A, sodium, potassium, citrate, magnesium, calcium, chloride, and water-soluble vitamins.

Energy

Human milk provides approximately 0.65 cal/ml, although the energy content varies with its fat (and, to a lesser degree, protein and carbohydrate) composition. Breastfed infants consume fewer calories than those fed HMS.[57,58] Whether this difference in energy intake has to do with the composition of human milk, the inability to see the volume of breast milk consumed, or other factors is not known. Infants who are breastfed are thinner for their weight at 8–11 months than HMS-fed infants, but these differences disappear by 12–23 months of age. Few differences are notable by 5 years of age.[59]

Lipids

Lipids are the second largest component of breast milk by concentration (3–5% in mature milk). Lipids provide half of the energy of human milk.[9] Human milk fat is low at the beginning of a feeding, in foremilk, and higher at the end in hindmilk, which follows.

EFFECT OF MATERNAL DIET ON FAT COMPOSITION The fatty acid profile, but not the fat content, of human milk varies with the diet of the mother.[60]

When diets rich in polyunsaturated fats are consumed, more polyunsaturated fatty acids are present in the milk. When very low fat diets with adequate calories from carbohydrate and protein are consumed, more medium-chain fatty acids are synthesized in the breast. When a mother loses weight, the fatty acid profile of her fat stores is reflected in her milk.[61]

DHA   Recent interest in lipids in human milk stems from studies showing developmental advantages provided by docosahexaenoic acid (DHA) and cholesterol.[62] Neither DHA nor cholesterol are available in most HMS. DHA, essential for brain and retinal development, accumulates during the last months of pregnancy. The advantages of human milk seem particularly great for premature infants born before 36–38 weeks. Advantages for term infants have been demonstrated as well. Both maternal diet and supplements can change the level of DHA in human milk.[63] For example, a Norwegian study suggests that DHA supplementation by cod liver oil during pregnancy was associated with higher IQ scores at 4 years of age in breast- versus bottle-fed infants.[12] Cod liver oil contains high levels of vitamin A and vitamin D, so it should be used with caution.

CHOLESTEROL   Cholesterol, an essential component of all cell membranes, is needed for growth and replication of cells. Cholesterol concentration in breast milk ranges from 10–20 mg/dL, and changes with the time of day.[9] Breastfed infants have higher intakes of cholesterol and higher levels of serum cholesterol than infants fed HMS.[64] Early consumption of cholesterol through breast milk appears to be related to lower blood cholesterol levels later in life.[65]

Protein

The protein content of mature human milk is relatively low (0.8–1.0%) compared to other mammalian milks. The concentration of proteins synthesized in the breast is more affected by the infant's age (time since delivery) than by maternal protein intake and maternal serum proteins. Proteins synthesized by the breast are variable because hormones that regulate gene expression and guide protein synthesis change with time.[66] Despite the relatively low concentration, human milk proteins have important nutritive and nonnutritive value. Proteins exhibit a variety of antiviral and antimicrobial effects. Enzymes present in human milk may also protect infants' health by facilitating reactions that prevent inflammation.

CASEIN   Casein is the major type of protein in mature milk from women who deliver either at term or preterm.[67] Casein, calcium phosphate, and other ions such as magnesium and citrate appear as an aggregate and are the source of milk's white appearance.[68] Casein's digestive

products, casein phosphopeptides, keep calcium in soluble form and facilitate its intestinal absorption.

WHEY PROTEINS   Whey proteins are proteins that remain soluble in water after casein is precipitated from milk by acid or enzymes. Whey proteins include serum proteins, enzymes, and immunoglobulins, among others. Several mineral-, hormone-, or vitamin-binding proteins have also been identified as components of whey proteins. These include lactoferrin, which carries iron in a form that is easy to absorb and has bacteriostatic activity. The enzymes present in whey proteins aid in digestion and protecting against bacteria.

NONPROTEIN NITROGEN   Nonprotein nitrogen provides 20–25% of the nitrogen in milk.[68] Urea accounts for 30–50% of nonprotein nitrogen and nucleotides for 20%, depending on the stage of lactation and the diet of the mother. Some of this nitrogen is available for the infant to use for producing nonessential amino acids. Some is used to produce other proteins with biological roles, such as hormones, growth factors, nucleic acids, nucleotides, and carnitine. The role of individual nucleotides in human milk is under investigation; however, nucleotides appear to play important roles in growth and disease resistance.

Milk Carbohydrates

Lactose is the dominant carbohydrate in human milk. Other carbohydrates, including monosaccharides such as glucose, polysaccharides, and protein-bound carbohydrates are also present.[69] Lactose enhances calcium absorption. As the second largest carbohydrate component, polysaccharides contribute calories, stimulate the growth of bifidus bacteria in the gut, and inhibit the growth of *E. coli* and other potentially harmful bacteria.

Fat-Soluble Vitamins

VITAMIN A   The vitamin A content of colostrum is approximately twice that of mature milk. Some of the vitamin A in human milk is in the form of beta-carotene. Its presence is responsible for the characteristic yellow color of colostrum. In mature milk, vitamin A is present at 75 mcg/dL or 280 IU/dL.[70] These levels are adequate to meet infant needs.[71]

VITAMIN D   Vitamin D is present in both lipid and aqueous (water) components of human milk. Most vitamin D is in the form of $25-OH_2$ vitamin D. Vitamin D levels in human milk vary with maternal diet and exposure to sunshine.[72] Maternal exposure to sunlight reportedly increases the vitamin D level tenfold.[73] In response to an increasing incidence of vitamin D–deficiency rickets among infants who are exclusively breastfed, the American

Academy of Pediatrics recommends that breastfed infants be given a supplement of 200 IU of Vitamin D per day beginning in the first 2 months of life.[74]

VITAMIN E   The level of total tocopherols in human milk is related to the milk's fat content. Human milk contains 40 mcg of vitamin E per gram of lipid in the milk.[75] Levels of alpha-tocopherol decrease from colostrum to transitional milk to mature milk, whereas beta- and gamma-tocopherols remain stable throughout each stage of lactation. The level of vitamin E present in human milk is adequate to meet the needs of full-term infants for muscle integrity and resistance of red blood cells to hemolysis (breaking of red blood cells). The levels of vitamin E in preterm milk have been reported to be the same[76] and higher[77] than term milk. In both reports, however, the levels present were not considered adequate to meet the needs of preterm infants.

VITAMIN K   Vitamin K is present in human milk at levels of 2.3 mcg/dL.[78] Approximately 5% of breastfed infants are at risk for vitamin K deficiency. Cases of vitamin K deficiency among exclusively breastfed infants who did not receive vitamin K at birth have been reported. Infants fed either human milk or HMS are vulnerable to hemorrhagic disease of the newborn, however. All infants in the United States receive a vitamin K supplement (0.5–1.0 mg by injection) at birth.[79]

Water-Soluble Vitamins

Water-soluble vitamins in human milk are generally responsive to the vitamin content of the maternal diet and supplements (vitamin C, riboflavin, niacin, B_6, and biotin). Problems with levels of these nutrients in human milk are related to their deficiency in the mother's diet and not to excesses. Clinical problems relating to water-soluble vitamins are rare in infants nursed by mothers with inadequate diets.[8] Vitamin B_6 is the most likely to be deficient in human milk; levels of B_6 in human milk directly reflect maternal intake.[80]

VITAMIN B_{12} and Folic Acid   Vitamin B_{12} and folic acid are bound to whey proteins in human milk; therefore, their content in milk is less influenced by maternal intake than is that of the other water-soluble vitamins. Factors that influence protein secretion (hormones, the infant's age, or time since delivery) are more likely to alter the human milk levels of B_{12} and folate than is dietary intake.[30,81] Infant illness associated with low folate levels in milk has not been reported. Folate levels increase with the duration of lactation despite a decrease in maternal serum and red blood cell folate levels.[81,82] Low levels of B_{12} have been found in the milk of women who follow vegan diets, have latent pernicious anemia (B_{12} deficiency) caused by hypothyroidism, or are generally malnourished.[82,83]

Minerals in Human Milk

The minerals in human milk contribute substantially to its osmolality. ***Monovalent ion*** secretion is managed closely by the alveolar cells, in balance with lactose, to maintain the isosmotic composition of human milk.

> **Monovalent Ion**   An atom with an electrical charge of +1 or −1.

The mineral content of different mammalian milks is related to the growth rate of the offspring. Because human infants grow at a relatively slow rate, the mineral content of human milk is lower than that for milks of other animals whose offspring grow faster. With the exception of magnesium, the concentration of minerals decreases over the first 4 months. This decline during the period of rapid growth is not what one would expect, but infant growth is well supported.[84] The lower mineral concentration of human milk is easier for the kidneys to handle and is considered a significant benefit of human milk.

BIOAVAILABILITY   An important feature of magnesium, calcium, iron, and zinc in human milk is the packaging that makes them highly available for absorption (or bioavailable) to the infant.[85] Packaging minerals so that the infant can use them efficiently also reduces the burden on the mother because less of the mineral is needed in the milk. For example, iron is 49% bioavailable from human milk, but only 10% of the iron in from cow's milk and cow's-milk-based HMS is bioavailable. Exclusively breastfed infants have little risk of iron-deficiency anemia,[86] despite the seemingly low concentration of iron in human milk. One study suggests that infants who are exclusively breastfed for 6½ months are less likely to be anemic than those nursed exclusively for 5½ months.[87]

ZINC   The importance of zinc to human growth is well established. Human-milk zinc is bound to protein and is highly available compared to zinc in cow's milk and cow's-milk-based HMS. Both the zinc intake (per kilogram) and the zinc requirements of infants decline after the first few months.[88] Normally, zinc homoestasis and human-milk zinc levels are maintained even in the face of low maternal zinc intake.[89] Rare cases of zinc deficiency, which appears as intractable diaper rash, occur in exclusively breastfed infants, however.[90] A defect in the mammary gland uptake of zinc has been described as the cause of low milk concentration when maternal serum zinc concentrations are normal.[90] In these cases, infants seem to respond to zinc supplementation.

TRACE MINERALS   Trace minerals (copper, selenium, chromium, manganese, molybdenum, nickel, and fluorine) are present in the human body in small concentrations and are essential for growth and development. Less is known about trace minerals and infant health than about other nutrients. In general, however, the levels of trace minerals in human milk are not altered by the mother's diet or supplement use.

Taste of Human Milk

> "...too full o' th' milk of human kindness to catch the nearest way."
>
> William Shakespeare, *MacBeth,* Act I, Scene V

This line from Shakespeare reflects the centuries-old belief that a breastfeeding woman's diet influences the composition of her milk and has a long-lasting influence on the child. The flavor of human milk is an important taste experience for newborn infants, but it is often ignored in discussions of the benefits of human milk or its composition. Human milk tastes slightly sweet,[91] and it carries the flavors of compounds mothers ingest such as mint, garlic, vanilla, and alcohol.[92]

Infant responses to flavors in milk seem to depend on the length of time since the mother consumed the food, the amount of the flavor that the mother consumed, and the frequency with which the flavor is consumed (new versus repeated exposure). Infants seem more interested in their mothers' milk when flavors are new to them. Researchers found that infants nursed at the breast longer if a flavor (garlic) was new to them than if the mother had taken garlic tablets for several days.[92] Infants who were exposed to carrot-juice flavor in the mother's milk ate less of a carrot-flavored cereal and spent less time feeding at the breast than infants who had not been exposed to the carrot flavor. Thus, exposing infants to a variety of flavors in human milk may contribute to their interest in and consumption of human milk as well as their acceptance of new flavors in solid foods.[93]

The Breastfeeding Infant

Optimal Duration of Breastfeeding

The wide gap between the "best breastfeeding practice" and the norms for breastfeeding in the United States makes study of the optimal duration particularly important for health care professionals. The health benefits to the mother-child pair should be the primary criteria used to determine the optimal duration of breastfeeding, rather than whether the cultural environment makes such duration practical. The American Academy of Pediatrics (AAP)[7] has taken a clear stand on this issue, saying that breastfeeding should continue for a year or longer. The U.S. Surgeon General recommends human-milk feeding exclusively for 6 months, noting further that it is better to breastfeed for 6 months and best to do so for 12 months, with solid foods being introduced at 4–6 months.[1]

Infants who are exclusively breastfed for 6 months experience fewer illnesses from gastrointestinal infection than infants who are given HMS and breast milk at 3 or 4 months of age. Deficits in growth have not been demonstrated among infants from developing or developed countries who are exclusively breastfed for 6 months or longer.[94] Breastfeeding can prevent intestinal blood loss in infants—a factor that should be considered when determining its optimal duration. Infants fed cow's milk before the age of 6 months suffer nutritionally significant losses of iron via intestinal blood loss[95]—an observation that supports the AAP's recommendation.[7] Through 1 year, breastfed infants suffer fewer acute infections than do formula-fed infants, a finding that supports continued breastfeeding beyond the introduction of solids.

Reflexes

Healthy term infants are born with several reflexes that enable them to nourish themselves. Observations show that the sucking reflex is developed by 18 weeks of gestation. By 34 weeks gestation, the suck has adequate pace and rhythm to be nutritive. The gag reflex that prevents the fetus from taking fluids into the lungs is developed by 28 weeks gestation. These reflexes allow term infants to suck and swallow in a coordinated pattern that protects the airways.

Two other reflexes enhance infants' ability to breastfeed. The oral search reflex directs infants to open their mouths wide when near the breast and to thrust their tongues forward. The rooting reflex prompts the infant to turn to the side when stimulated by touch to the lips or cheek. The infant comes forward, opens her or his mouth, and extends the tongue when the center of the upper or lower lip is stimulated.

The presence of these reflexes is important to the success of breastfeeding. Successful nursing also requires appropriate positioning of the infant and adequate maternal letdown and milk production. Appropriate positioning and maternal assessment of infant nursing behaviors must be learned. Support from lactation consultants and other health care professionals trained in lactation may be necessary.

Breastfeeding Positioning

Proper positioning of the infant at the breast is important to breastfeeding success. Mothers need to learn from health professionals experienced in optimal positioning because improper positioning causes pain and possible damage to nipple and breast tissue. The mother may need to use cushions, pillows, or a footstool to be comfortable and positioned well to nurse the infant (see Illustration 6.6). Once the mother is comfortable, she should hold the baby with his mouth directly in front of the nipple and stimulate the oral search reflex by touching the baby's bottom lip with her nipple. The infant will open his mouth wide and should be brought to the breast with the nipple centered in his mouth (Illustration 6.7 on page 157). This process is called latching on. An infant who is properly

a. Cradle hold

b. Football or clutch hold

c. Cross-cuddle hold

Illustration 6.6 Positions for breastfeeding.

latched at the breast has all or most of the areola (the dark pigmented skin around the nipple) in his mouth. If the mother pulls down the infant's lower lip she should see the tongue lying around the areola. The infant's tongue should extend beyond the lower gum line. The baby's nose should be close to the breast with his breathing unrestricted. The mother should hear swallowing, but not smacking, clicking, or slurping. Women who consistently have pain when the infant is sucking should consult a professional trained in lactation.

Identifying Hunger and Satiety

When infants are hungry they begin to bring their hands to their mouths and suck on them, and start moving their heads from side to side with their mouths open. Infants should be fed when these signs of hunger are displayed, rather than waiting for crying, a late sign. Recognizing early hunger behaviors and initiating feeding before the infant becomes very upset can be particularly helpful for mothers and infants who have difficulty nursing.

Nutritive and nonnutritive sucks are different. Feedings begin with nonnutritive sucking. The infant sucks quickly and not particularly rhythmically. Nutritive sucking is slower and more rhythmic as the infant begins to suck and swallow. A mother can hear the infant suck in a quiet room.

Infants should be allowed to nurse as long as they want at one breast. Infants who are fed for shorter periods from both breasts can get larger amounts of foremilk, which can cause diarrhea because of the high lactose content.[96] Allowing the infant to nurse at one breast until satisfied creates a pattern that assures that the infant gets both foremilk and hindmilk. The higher fat content of hindmilk may help in signaling satiety. Infants will stop nursing when full. If they are still hungry, after burping, infants can be offered the other breast.

Feeding Frequency

Stomach emptying after breastfeeding occurs in about 1½ hours. Ten to 12 feedings per day are normal for newborn infants. As infants develop, feeding frequency will depend, in part, on maternal storage. Different feeding patterns can meet infant needs. In one study, infants who did not feed from midnight through early morning consumed more in the other feedings, particularly in the morning. Milk intake and weight gain of these infants in the first 4 months of life were similar to those of infants whose feedings were distributed over 24 hours.[97]

Identifying Breastfeeding Malnutrition

Newborns normally lose up to 7% of their body weight in the first week postpartum. A loss of 10% should trigger an evaluation of milk transfer to the infant by a lactation consultant or other trained professional and provide support needed to maintain breastfeeding. Malnourished infants become sleepy, are nonresponsive, have a weak cry, and wet few diapers. The clinician can use the flowchart in Illustration 6.8 on page 158 to help diagnose failure to thrive in breastfed infants. By the fifth to seventh day postpartum, infants who are getting adequate nourishment have wet diapers approximately six times a day and three to four soft yellowish stools per day.[7] Infants who are slow gainers and not malnourished are alert, bright, responsive, and develop normally.[8] Their urine is pale yellow and dilute while stools are loose and seedy (some small particles are present in the stool). In contrast, infants who are failing to thrive are apathetic, are hard to arouse, and have a weak cry. They have few wet diapers, and their urine is concentrated. Their stools are infrequent.

Mothers of slow-gaining infants, in particular, should be advised to let the infant nurse at one breast until it is empty or the infant stops nursing, rather than

Case Study 6.1

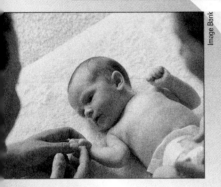

Image Bank

Breastfeeding and Adequate Nourishment

Molly G. is a 24-year-old office manager and part-time aerobics instructor who has delivered vaginally, without complications, a healthy, full-term son, Daniel. With a birthweight of 3200 grams (7 lb), Daniel is the first child for Molly and her husband. Molly is 162 cm (5 ft 4 in) tall with a prepregnancy weight of 56.8 kg (125 lb). She gained 25 kg (55 lb) during her uncomplicated pregnancy and has been a lacto-ovo vegetarian for 5 years. After a 12-hour stay in a birthing center, Molly and her husband bring Daniel home.

At 4 days postpartum Molly, her husband, and her mother-in-law bring the baby to the Health Care Center for his first follow-up visit. At this visit, Daniel weighs 3000 grams (6 lb 6 oz). Molly and her husband are very concerned about whether their son is getting adequate nourishment. They report that Daniel nurses vigorously about every 1½ to 2 hours and never sleeps for more than a couple of hours. Molly says that her milk "came in" on the second postpartum day and that she feels like all she does is nurse. Her nipples are tender, but not uncomfortably sore. She reports Daniel has at least 6 to 8 wet diapers and 2–3 very loose stools each day. She wonders if she has enough milk and worries about how she will ever return to work in 2 months. She also wants to lose the excess weight she gained during the pregnancy and is eager to return to her aerobics classes. Her husband and mother-in-law are supportive, but worry about the baby. The pediatrician's examination of the infant concludes Daniel is a healthy infant with no medical problems.

Questions

1. What factors put Molly at high risk for early termination of breastfeeding?
2. What factors indicate that Daniel is getting adequate nourishment?
3. What concerns do you have about Molly's diet? What advice would you give her about her weight-loss plans and eagerness to return to exercise? Do Molly or Daniel need any vitamin-mineral supplements?
4. If Molly lived in your community, what resources would be available for help and support for breastfeeding mothers?
5. What steps can Molly take to continue successful breastfeeding when she returns to work in 2 months?

(Answers are in Appendix D.)

---

switching breasts after a specific amount of time. This regimen assures that the infant gets hindmilk with its higher fat and calorie content. Lawrence recommends evaluation of slow-gaining infants when the slow-gaining pattern is first recognized (Illustration 6.8).[8]

Infant Supplements

Breastfed infants need few supplements except in special conditions. Breastfed infants should be given 200 IU vitamin D supplements.[74] Fluoride supplements are also recommended. Breastfed infants should not be supplemented with fluoride until 6 months of age. Breastfed infants do not need iron-fortified HMS or supplements,[8] because they rarely experience iron deficiency. The excess iron in HMS might bind with lactoferrin in human milk, resulting in a loss of the lactoferrin's protective activity against anemia.

Tooth Decay

Human milk has infection-fighting components that inhibit the formation of dental caries. Nevertheless, caries can occur in children who are breastfed.[98] Frequent nursing at night after children reach 1 year of age is a risk factor for dental caries. Nevertheless, the prevention of dental caries is not justification for advising early weaning. Rather, the mother should be instructed on the prevention and treatment of early childhood dental caries. All children should be seen by a qualified dentist 6 months after the first tooth erupts, or by 12 months of age. Because mothers are the primary source of the bacteria that cause early childhood caries, mothers or primary caregivers of children should be advised on oral hygiene, diet, fluoride, caries removal, and delay of oral colonization of bacteria-causing caries.

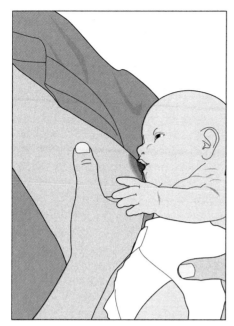

a. Touching the baby's bottom lip with the nipple stimulates the oral search reflex.

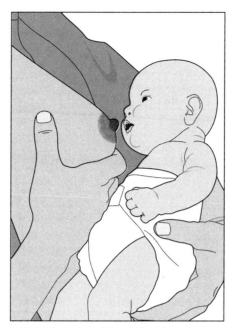

b. The infant opens his mouth wide.

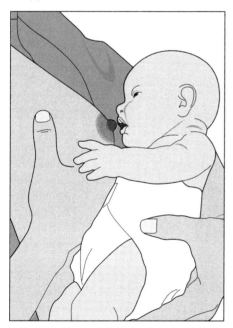

c. The infant is brought to the breast with the nipple centered in his mouth.

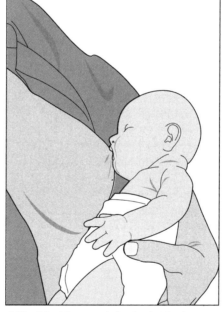

d. The infant is properly latched at the breast and has all of the areola in his mouth.

Illustration 6.7 Attachment.

Breastfed babies have straighter teeth due to the development of a well-rounded dental arch.[99] A well-rounded arch may also help prevent sleep apnea later in life.

Maternal Diet

Both the Food Guide Pyramid[100] and the "Dietary Guidelines for Americans"[101] are appropriate guidelines for breastfeeding women (see Chapter 1). Breastfeeding women should follow the Food Guide Pyramid when choosing their foods. Four servings of dairy products fortified with vitamins A and D each day help to achieve adequate calorie, protein, vitamin D, and calcium intake.

Although it is widely believed that components of the maternal diet are related to infant colic, few well-designed studies have examined this relationship. One retrospective cross-sectional study of the diets of breastfeeding women suggests that cow's milk, onions, cruciferous vegetables, and chocolate are associated with an increased risk of colic in infants.[102] The results may have been biased, however, by the participating women's belief that these foods were related to their infant's colic.

Energy and Nutrient Needs

Until recently, when safe methods of studying energy metabolism became available, determining the energy needs of lactating women was very difficult. As a result, the energy needs of lactation are estimated using the factorial method, which adds the estimated requirements of lactation to the requirements of nonlactating women. The RDIs for normal-weight lactating women assume that the energy spent for milk production is 500 calories per day in the first 6 months and 400 calories afterward.[103] The 2002 RDI is 330 additional calories with 0.8 kg per month weight loss (170 calories per day) per day for the first 6 months and 400 calories per day afterward. It is important to understand that public health policy supports estimating the energy and nutrient needs of breastfeeding women conservatively to ensure adequate nutrition for women and their children. An overestimation of needs is possible using this approach.

Vitamin and mineral intakes that do not meet recommended levels have been reported for lactating

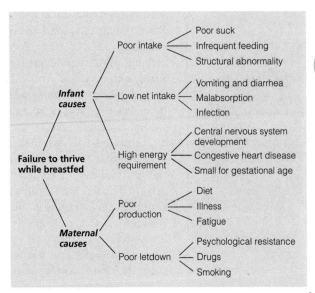

Illustration 6.8 Diagnostic flowchart for failure to thrive.[8]

SOURCE: Reprinted from Breastfeeding: A Guide for the Medical Professions, 5th Ed. by R. A. Lawrence and R. M. Lawrence, copyright © 1999, with permission from Elsevier.

women.[104–107] Intakes of folate, thiamin, vitamin A, calcium, iron, and zinc have been reported to be inadequate. Ten percent of lactating women have thiamin intakes below the recommended levels,[106] whereas fewer than 5% of nonlactating women have intakes below the 1998 DRI. These reports of inadequate intake have not been followed by reports of deficits in the nutritional status of the mother-infant pair, however. Concern over inadequate maternal dietary intake must be carefully balanced against the possibility that women will be discouraged from breastfeeding because they believe their diets are not optimal.

Studies of dietary intake of lactating women in developed countries[108,109] have rarely documented energy intakes that met the recommended levels.[70] In contrast, studies that measure components of energy expenditure and intake[110] or adequacy of milk production and infant growth[42] have simultaneously documented adequate lactation (and infant growth) with dietary intakes below the RDA.[70]

We now understand that women use several mechanisms to meet the needs of lactation. Adjustments in energy intake and energy expenditure must be balanced to meet those needs. Goldberg et al.[111] found that women increased food intake (56% of the energy need for milk production) and decreased physical activity (44% of the energy need for milk production) to meet the increase in energy needs for lactation. Doubly labeled water studies suggest that the components of energy expenditure vary greatly and that measurements of dietary intake can be unreliable.[112] Therefore, a single recommendation for lactating women could never address all of the individual ways that women meet their energy needs. Assessment of adequacy of energy intake of breastfeeding women should be made within the

context of the mother's overall nutritional status, weight changes, and the adequacy of the infant's growth.

Maternal Energy Balance and Milk Composition

The composition of breast milk depends on maternal nutritional status. Protein-calorie malnutrition results in an energy deficit that reduces the volume of milk produced, but usually does not compromise the composition of the milk. Several studies show that milk production is maintained when there is a modest level of negative calorie balance. Animal models first identified a potential threshold effect of energy restriction. Baboons fed 60% of their voluntary intake significantly reduced milk production.[113] Yet baboons fed 80% of their usual intake maintained milk production. Randomized studies such as the one performed on baboons cannot ethically be done with humans. Nevertheless, a series of human studies on weight loss during breastfeeding (discussed below) support a threshold effect of energy limitations on lactation.

Maternal Weight Loss during Breastfeeding

Current RDIs are written assuming a weight loss of 0.8 kg/month. In addition, mechanisms that favor use of maternal fat stores and delivery of nutrients to the breast seem to occur during lactation.[114] Despite these mechanisms that should favor weight loss in breastfeeding women, on average, loss by 12 months postpartum is less than the amount needed to return to prepregnancy weight.[115] Even more surprising, postpartum weight changes (–0.1 kg/mo) are smaller in developing countries than in industrialized nations (–0.8 kg/mo). The failure to return to prepregnancy weight may be due to changes in energy intake, energy expenditure, and fat mobilization that easily meet energy needs.

A number of small studies in the United States have addressed the issue of whether weight loss influences milk production. Strode et al.[116] first observed women who voluntarily reduced their energy intake to 68% of their estimated needs for seven days. No differences in infant intake or milk composition were observed. In the week following the diet, women who consumed fewer than 1500 cal tended to experience a decrease in milk volume. Women who consumed over 1500 cal per day experienced no decrease. Despite its voluntary nature and the lack of milk samples throughout the day, this study provides important support for the idea that modest and/or short-term reductions in energy are not associated with decreases in milk production. In addition, 22 healthy postpartum women who participated in a 10-week weight-loss program, which reduced energy intake by 23%, maintained milk production.[117] The women lost an average of about a pound a week during the 10 weeks. A number of study design limitations, including small numbers and lack of randomization and/or control groups, limit generalizations based on this information, however.

Exercise and Breastfeeding

Two studies[118,119] have examined the effect of increasing energy expenditure on weight and lactation. In a cross-sectional study, vigorous exercise increased energy expenditure, but these women also increased their energy intake, so the calorie deficit was similar in the two groups.[118] There were no significant differences in milk volume between the groups. A later 12-week exercise intervention trial studied women who were 6 weeks postpartum and exclusively breastfeeding to examine the effects of exercise on body composition and energy expenditure during breastfeeding.[119] The women were randomly assigned to two groups. One group followed an aerobic exercise regimen for 45 minutes, 5 days a week. The other group did no exercise. The exercising women increased energy expenditure by 400 cal per day and increased energy intake by about the same amount. Both groups experienced similar weight changes, milk volume, milk composition, and infant weight gain and serum prolactin levels. These studies suggest that lactating women efficiently balance their energy intake to support energy expenditure.

The available evidence suggests that modest energy restriction combined with increases in activity may be effective in helping women to lose weight, while improving their metabolic profile and increasing fat losses. Although only a small number of women have been studied, the consistent lack of effect on milk production (infant intake), milk energy output, immunologic components,[120] and infant growth is encouraging.

Mechanisms responsible for milk production may be modified by exercise. Aerobic activity appears to enhance fatty acid mobilization to meet the needs of milk production, whereas restricted energy intake requires increases in prolactin levels to promote use of dietary fatty acids or to promote mobilization from fat. These prolactin increases explain how milk production can be maintained despite a somewhat negative energy balance.

Vitamin and Mineral Supplements

A 1991 Institute of Medicine report, *Nutrition during Lactation*, stated that well-nourished breastfeeding women do not need routine vitamin or mineral supplementation.[30] Instead, supplementation should target specific nutritional needs of individual women. Supplementation strategies should take into account how nutrients are secreted into human milk and the potential for nutrient-nutrient interactions in mothers and their infants.

Breastfed infants can suffer from deficiencies of vitamins D and B_{12} because human milk can have inadequate amounts of these nutrients even when mothers do not exhibit deficiency symptoms. Exclusively breastfed infants should be given a 200 IU vitamin D supplement to prevent rickets.[74] In North Carolina alone, at least 30 cases of nutritional rickets have been identified in exclusively breastfed infants.[121] B_{12} deficiency has affected infants of women with gastric bypass and vegan diets.[122,123] Women with intestinal disease, such as Crohn's disease, should also be considered at risk and supplemented.

Functional Foods

Concern has been expressed about possible ill effects of high intakes of fortified foods in addition to supplements. Although this is an important issue, studies to date have not identified adverse reactions related to fortified food consumption and RDA levels of nutrients in supplements.

Fluids

There is no evidence that increasing fluid intake will increase milk production or that a short-term fluid deficit results in a decrease in milk production. Fluid demands rise during breastfeeding, however, so women should drink fluids to thirst. Once a mother and her infant have the nursing routine down, she may find it convenient to have something to drink while she nurses. Although many women want to know how many glasses to drink per day, the amount needed varies depending on climate, milk production, body size, and other factors. Therefore, women are advised to drink enough fluids to keep their urine pale yellow. The current RDI for water for lactating women sums the recommendation for nonpregnant and nonlactating women (2.7L for women 19–30) and the water content of the average milk output during the first 6 months (0.78L milk × 87% water = 0.68L) for a total need of 3.4L for women 19–30.[124]

Alternative Diets

Breastfeeding women can follow alternative eating patterns and be well nourished. The goal is to adequately nourish the mother and child, not to force women to use supplements and/or products that are not part of their normal eating patterns. Incorporation of soy products, vegetarian diets of various sorts, and other alternative diet choices can be followed as long as they meet maternal nutritional needs.

Vegans who do not consume dairy products and eggs, however, may need to plan carefully to consume adequate amounts of calories, protein, calcium, vitamin D, vitamin B_{12}, iron, and zinc. Vegetarians' intakes of protein are generally adequate as long as energy intake is adequate. Breastfeeding women who consume no animal products should use plant foods with bioavailable B_{12} from plant sources such as yeasts, seaweed, and fortified soy products. Women who are unable to get adequate B_{12} from foods should take a vitamin B_{12} supplement. The 1991 Institute of Medicine[30] report recommends a multiple vitamin-mineral supplement for vegetarians because human milk may be low in vitamin B_{12} even when mothers do not exhibit deficiency symptoms.

Factors Influencing Breastfeeding Initiation and Duration

All new mothers, both low income and more affluent, need support for breastfeeding. Low-income women, however, may lack the education, support, and confidence to interpret the abundant and pervasive supply of messages on infant feeding practices.[125] Consider the strikingly different context for pregnancy, birth, and parenting for low-income women and their more affluent counterparts:

> Profile of a low-income pregnant woman: "She says she wants to do what is best for her baby and, in fact, knows the breast is best. However, she is afraid breastfeeding will cause her baby to be too 'clingy.' She feels extremely uncomfortable about nursing around family, much less in public. She is certainly not up for the pain she has heard breastfeeding causes. To make things more difficult, in the hospital, she is separated from her baby soon after delivery and is given little assistance for getting started. She is sent home from the hospital with samples of free formula."

> Profile of an affluent pregnant women: "The affluent expectant mother has friends who have breastfed and have helped build her confidence that she can breastfeed successfully. She may have been able to choose her birth setting and select a hospital with knowledgeable staff allowing mother and baby to stay together around the clock. Because she knows there may be bumps in the road getting started, she seeks out support from friends or the doctor after discharge. At home, she has a supportive husband who is proud of her for offering the best for their baby. If she returns to work, she knows she can still breastfeed to keep that special closeness with her baby even after returning to work."

Common barriers to breastfeeding initiation expressed by expectant mothers include the following:

- Embarrassment.
- Time and social constraints, and concerns about loss of freedom (particularly issues for working moms).
- Lack of support from family and friends.
- Lack of confidence.
- Concerns about diet and health practices.
- In adolescents, fear of pain.[125–127]

Additional obstacles to the initiation and continuation of breastfeeding include:

- Insufficient prenatal breastfeeding education.
- Health care provider apathy and misinformation.
- Inadequate health care provider lactation management training.
- Disruptive hospital policies.

- Early hospital discharge.
- Lack of routine follow-up care and postpartum home health visits.
- Maternal employment, especially in the absence of workplace facilities and support for breastfeeding.
- Lack of broad societal support.
- Media portrayal of bottle-feeding as the norm.
- Commercial promotion of infant formula through distribution of hospital discharge packs, coupons for free or discounted formula, and television and general magazine advertising.[7,128–130]

Breastfeeding Promotion, Facilitation, and Support

"Significant steps must be taken to increase breastfeeding rates in the United States and to close the wide racial and ethnic gaps in breastfeeding. This goal can only be achieved by supporting breastfeeding in the family, community, workplace, health care sector, and society."

U.S. Department of Health and Human Services Blueprint for Action
on Breastfeeding, 2000[1]

The support a woman receives from those around her has a direct effect on her capability to breastfeed optimally.[1] The health care system, her workplace, and her community can all work to facilitate the initiation and continuation of breastfeeding.

Role of the Health Care System in Supporting Breastfeeding

Health care providers and facilities can exert tremendous influence over the mother-infant dyad by promoting and modeling optimal breastfeeding practices during prenatal care, at delivery, and after discharge. There is clear evidence for the effectiveness of professional support on the duration of any breastfeeding; a review of quasi trials from 10 countries involving 12 mother-infant pairs showed that all forms of breastfeeding support reduced the risk of ending breastfeeding before 6 months.[131]

Prenatal Breastfeeding Education and Support

Culturally competent prenatal breastfeeding education that is given frequently in person can have a significant positive influence on breastfeeding rates.[126,132] Best Start Social Marketing has developed an effective three-step counseling strategy (Table 6.4) that quickly identifies a woman's particular barriers to choosing breastfeeding and provides targeted education while affirming the woman's ability to breastfeed.[127]

Table 6.4 The Best Start three-step breastfeeding counseling strategy[124]

1. Ask open-ended questions to identify the woman's concerns:
   - Dietitian: "What have you heard about breastfeeding?"
   - Client: "I hear it's best for my baby, but all my friends say it really hurts!"
2. Affirm her feelings by reassuring her that these feelings are normal:
   - Dietitian: "You know, most women worry about whether it will hurt."
3. Educate by explaining how other women like her have dealt with her concerns. Avoid overeducating or giving the impression that breastfeeding is hard to master:
   - Dietitian: "Did you know that it is not supposed to be painful, and if you are having discomfort, there are people who can help make it better?"

Table 6.5 Key teaching points prior to birth[127]

In the hospital or birthing center, the mother should:
- Request early first feeding.
- Practice frequent, exclusive breastfeeding.
- Ask to be taught swallowing indicators.
- Learn indicators of sufficient intake.
- Ask for help if breastfeeding hurts.
- Know sources for help.
- Understand postpartum rest and recovery needs.
- Avoid supplements unless medically indicated.

The Best Start approach avoids questions that force the woman to choose, such as "Are you going to breast-feed or bottle-feed?" and instead uses open-ended questions, such as "What have you heard about breastfeeding?" or "What questions do you have about breastfeeding?" Such questions give the woman an opportunity to begin a dialogue with her provider about the infant feeding decision. By using follow-up probes as necessary, the counselor can identify the woman's specific concerns. Counselors should avoid overwhelming the woman with too much information, which can give the impression that breastfeeding is difficult.[127]

Another effective strategy utilizes peer counselors and peer group discussions with at least one or two women who have successfully breastfed.[133,134] Exposure to mothers nursing their babies increases a woman's level of comfort with breastfeeding and provides a forum for informal discussion with family and friends. Hearing about someone else's personal experience can help a woman realize that other women share her concerns.[125]

Toward the end of pregnancy, women need information on what to expect in the hospital or birthing center and practical tips for initiating breastfeeding.[125] Several key points are shown in Table 6.5. Because fathers,[135–137] grandmothers,[135] *doulas*, friends, and social networks[138] all play a powerful role in infant feeding decisions, these influential people should be included as often as possible in breastfeeding promotion efforts.[125,135,139]

The environment for the delivery of prenatal care can inhibit or facilitate breastfeeding. It should provide positive messages about breastfeeding, such as posters on the walls and magazines and literature in the waiting room that promote breastfeeding; there should be no advertise-

ments or promotions of formula. Nevertheless, patient education materials that include formula advertising, samples, and business reply cards for free formula[140–142] are often available in U.S. and Canadian prenatal care settings, in direct violation of the World Health Organization's (WHO's) International Code on the Marketing of Breast Milk Substitutes (Table 6.6). Women who have been exposed to materials and products from formula companies prenatally are more likely to cease breastfeeding in the first 2 weeks.[141] Use of these materials conveys a subtle message that infant formula is equivalent to breast milk.

Doula An individual who gives psychological encouragement and physical assistance to a mother during pregnancy, birth, and lactation; the doula may be a relative, friend, or neighbor and is usually but not necessarily female.[8]

Table 6.6 World Health Organization's International/UNICEF Code on the marketing of breast milk substitutes

- No advertising of breast-milk substitutes.
- No free samples or supplies.
- No promotion of products through health care facilities.
- No company sales representative to advise mothers.
- No gifts or personal samples to health workers.
- No gifts or pictures idealizing formula feeding, including pictures of infants, on the labels of the infant milk containers.
- Information to health workers should be scientific and factual.
- All information on artificial feeding, including labels, should explain the benefits of breastfeeding and the costs and hazards associated with formula feeding.
- Unsuitable products should not be promoted for babies.
- Manufacturers and distributors should comply with the Code's provisions even if countries have not adopted laws or other measures.

Lactation Consultant   A health care professional who provides education and management to prevent and solve breastfeeding problems and to encourage a social environment that effectively supports the breastfeeding mother-infant dyad. Those who successfully complete the International Board of Lactation Consultant Examiners (IBLCE) certification process are entitled to use IBCLC (International Board Certified Lactation Consultant) after their names (www.iblce.org/).

Le Leche League   An international, nonprofit, nonsectarian organization dedicated to providing education, information, support, and encouragement to women who want to breastfeed. It was founded in 1956 by seven women who had learned about successful breastfeeding while nursing their own babies. Currently, approximately 7100 accredited lay leaders facilitate more than 3000 monthly mother-to-mother breastfeeding support group meetings around the world (www. lalecheleague.org).

Although not all women will choose to breastfeed, the goal of prenatal breastfeeding education is to empower every woman with sufficient knowledge to make an informed decision about how to feed her baby. Some professionals view breastfeeding as a personal choice rather than a public health issue and voice concern that breastfeeding promotional efforts may cause women who choose formula-feeding to feel guilty. Experience indicates that women want information to make the best possible decision for themselves and their infants. Some women who formula-feed report feeling angry about not getting enough breastfeeding information during their pregnancies.[126] In recognition of the benefits of breastfeeding and the important role of health professionals in promoting and supporting it, the leading health and professional organizations in the United States that provide perinatal care have established policies supporting breastfeeding as the preferred infant feeding method.[7,146,147]

Lactation Support in Hospitals and Birthing Centers

Hospital policies and routines can have effects on a woman's critical early experience with breastfeeding that extend far beyond her short stay at the facility.[148] Illustration 6.9 provides examples of hospital practices that influence this pivotal initiation experience. Like prenatal care settings, hospitals should not distribute free samples of infant formula or coupons in their discharge packs because these can have detrimental effects on breastfeeding success, particularly among vulnerable groups such as new mothers and low-income women.[148]

In an effort to promote, protect, and support breastfeeding in hospitals and birthing centers worldwide, the WHO and UNICEF established the Baby Friendly Hospital Initiative in 1992.[149] This initiative focuses on 10 evidence-based components of hospital care that influence breastfeeding success (Table 6.7 on page 164). In early 2003, 38 hospitals and birthing centers in the United States had met all of the criteria in Table 6.8 and were designated as Baby Friendly.

Lactation Support after Discharge

Breastfeeding support is essential in the first few weeks after delivery, when lactation is being established.[1] A study of inner-city Baltimore WIC program partici-

pants[129] provides strong evidence that 7 to 10 days postpartum is the critical window for providing breastfeeding support; 35% of mothers who initiated breastfeeding in the hospital had stopped within 7 to 10 days.

A pediatrician, nurse, or other knowledgeable health care practitioner should see all breastfeeding mothers and their newborns for a home visit or in the office when the newborn is 2 to 4 days old.[7] During the visit, the practitioner should observe the mother breastfeeding to ensure that she is successful, revisit the major concerns she identified during pregnancy, and discuss any new concerns. The practitioner should also arm the mother with information on sources of trained help available in the community, such as *lactation consultants*, peer counselors, the WIC program, and *La Leche League*, should questions or complications arise.[7,125] Follow-up telephone calls, as necessary, provide additional support to mothers who are not fully confident in their ability to breastfeed successfully.[7,125]

The Workplace

The increase in the proportion of women working outside the home that began after World War II has been one of the most significant social and economic trends in modern U.S. history.[151] About 70% of employed mothers with children under 3 years of age work full-time;[1] about one-third of these mothers return to work within 3 months and about two-thirds within 6 months after childbirth.

Barriers to breastfeeding and employment have been recognized by the Surgeon General for at least a decade.[150] These barriers include lack of on-site day care, insufficiently paid maternity leave, rigid work schedules, and employers who lack knowledge about breastfeeding. Current law ensures a woman's right to breastfeed her infant anywhere on federal property that she and her child are authorized to be. Further legislation is still in process aiming to require that women cannot be fired or discriminated against if they breastfeed or express milk during their own lunchtime or break time, to provide a tax credit for employers who provide lacation support services, and to develop minimum standards for breast pump safety.

Planning to return to work full-time does not appear to affect breastfeeding initiation rates substantially.[31] Breastfeeding duration, however, is adversely affected by employment. At 6 months 35% of nonemployed women are still breastfeeding, compared with 33% of mothers working part-time, and only 23% of those employed full-time.[31] Occupation influences duration of breastfeeding. Women in professional occupations breastfeed significantly longer than women in sales, clerical, or technical occupations.[152] Part-time work is more conducive to breastfeeding; the number of hours mothers work per day is inversely associated with the likelihood that the mother will continue to breastfeed.[153]

Studies indicate that women who continue to breastfeed after returning to work miss less time from work because of baby-related illnesses and have shorter

Hospital Practices That Influence Breastfeeding Initiation

	◄—— Strongly Encouraging ——►	◄—— Encouraging ——►	◄—— Discouraging ——►	◄—— Strongly Discouraging ——►
Physical Contact	• Baby put to breast immediately in delivery room • Baby not taken from mother after delivery • Woman helped by staff to suckle baby in recovery room • Rooming-in; staff help with baby care in room, not only in nursery	• Staff sensitivity to cultural norms and expectations of woman	• Scheduled feedings regardless of mother's breastfeeding wishes	• Mother-infant separation at birth • Mother-infant housed on separate floors in postpartum period • Mother separated from baby due to bilirubin problem • No rooming-in policy
Verbal Communication	• Staff initiates discussion re: woman's intention to breastfeed pre- and intrapartum • Staff encourages and reinforces breastfeeding immediately on labor and delivery • Staff discusses use of breast pump and realities of separation from baby, re: breastfeeding	• Appropriate language skills of staff, teaching how to handle breast engorgement and nipple problems • Staff's own skills and comfort re: art of breastfeeding and time to teach woman on one-to-one basis	• Staff instructs woman "to get good night's rest and miss the feed" • Strict times allotted for breastfeeding regardless of mother/baby's feeding "cycle"	• Woman told to "take it easy," "get your rest" . . . impression that breastfeeding is effortful/tiring • Woman told she doesn't "do it right," staff interrupts her efforts, corrects her re: positions, etc.
Nonverbal Communication	• Pictures of woman breastfeeding • Literature on breastfeeding in understandable terms • Staff (doctors as well as nurses) give reinforcement for breastfeeding (respect, smiles, affirmation) • Nurse (or any attendant) making mother comfortable and helping to arrange baby at breast for nursing • Woman sees others breastfeeding in hospital	• Closed-circuit TV show in hospital on breastfeeding	• Pictures of woman bottle-feeding • Staff interrupts her breastfeeding session for lab tests, etc. • Woman doesn't see others breastfeeding	• Woman given infant formula kit and infant food literature • Woman sees official-looking nurses authoritatively caring for babies by bottle-feeding (leads to woman's insecurities re: own capability of care)
Experiential	• If breastfeeding not immediately successful, staff continues to be supportive • Previous success with breastfeeding experience in hospital			• Previous failure with breastfeeding experience in hospital

Illustration 6.9 Hospital practices that influence breastfeeding initiation.[1,141]

Table 6.7 The Baby-Friendly Hospital Initiative: Ten Steps to Successful Breastfeeding[149]

1. Have a written breastfeeding policy that is routinely communicated to all health care staff.

2. Train all health care staff in skills necessary to implement this policy.

3. Inform all pregnant women about the benefits and management of breastfeeding.

4. Help mothers initiate breastfeeding within a half-hour of birth.

5. Show mothers how to breastfeed and how to maintain lactation even if they should be separated from their infants.

6. Give newborn infants no food or drink other than breast milk, unless medically indicated.

7. Practice rooming-in—allow mothers and infants to remain together—24 hours a day.

8. Encourage breastfeeding on demand.

9. Give no artificial teats or pacifiers (also called dummies or soothers) to breastfeeding infants.

10. Foster the establishment of breastfeeding support groups and refer mothers to them on discharge from the hospital and clinic.

Table 6.8 Important elements of worksite lactation support programs[1]

- Prenatal lactation education tailored for working women.
- Corporate policies providing information for all employees on the benefits of breastfeeding and on why their breastfeeding co-workers need support.
- Education for personnel about the services available to support breastfeeding women.
- Adequate breaks, flexible work hours, job sharing, and part-time work.
- Private "Mother's Rooms" for expressing milk in a secure and relaxing environment.
- Access to hospital-grade, autocycling breast pumps at the workplace.
- Small refrigerators for the safe storage of breast milk.
- Subsidization or purchase of individually owned portable breast pumps for employees.
- Access to lactation professional on-site or by phone to give breastfeeding education, counseling, and support during pregnancy, after delivery, and when the mother returns to work.
- Coordination with on-site or near-site child-care programs so the infant can be breastfed during the day.
- Support groups for working mothers with children.

absences when they do miss work than women who do not breastfeed.[25] Worksite programs that support breastfeeding facilitate the continuation of breastfeeding after mothers return to their jobs and offer additional advantages to employers; companies with such programs find that employee morale and loyalty, the company's image as family friendly, recruiting for personnel, and the retention rate of employees after childbirth all improve.[153,154] Companies that have adopted breastfeeding support programs have noted cost savings of $3 per $1 invested in breastfeeding support in addition to lower health care over the first year.[155] Key elements of worksite lactation support programs are presented in Table 6.8.

Mothers planning to return to work have several choices. Breast milk can be expressed into sterile containers during the day, refrigerated or frozen, and then used for subsequent bottle-feedings when they are at work. With on-site child care, a woman can breastfeed during breaks and lunch hours. Another possibility is to train the body to produce milk only when the mother is home during the evenings and at night. To train her body, a woman should omit one feeding at a time during the periods of the day when she will not be feeding or expressing milk. Doing this will help her reduce her milk supply without experiencing engorgement. Then she can gradually wean to the feedings at the appropriate time of the day. This method works because removal of milk is the stimulus for milk production. Generally, at least two feedings per day are needed for women to continue making milk. No evidence indicates that it is necessary to introduce a bottle sooner than 10 days before returning to work.[36] Unless a mother is returning to work immediately after delivery, a bottle should not be introduced before lactation is well established, which is usually at least 4 weeks. Information about hospital-grade breast pumps is readily available on the Internet.

The Community

To increase breastfeeding rates in a community, it is important to identify community attitudes and obstacles to breastfeeding and to solicit support for breastfeeding from community leaders. A multidisciplinary task force with representatives from physicians, hospitals and birthing centers, public health, home visitors, La Leche League, government, industry, school boards, and journalists can be an effective vehicle for assessing community breastfeeding support needs and sponsoring collaborative efforts to overcoming obstacles to breastfeeding.[125] Barriers to breastfeeding may include lack of access to reliable and culturally appropriate sources of information and social support, cultural perception of bottle-feeding as the norm, aggressive marketing of breast-milk substitutes, and laws that prohibit breastfeeding in public.

In the past decade, legislative efforts have been made to protect a woman's right to breastfeed. Legislation is used to protect a woman's right to breastfeed in public, to consider breastfeeding in family law situations, to regulate breast pumps, and to provide incentives to employers who provide breastfeeding support. Legislation addresses issues such as a woman's right to breastfeed in public, express milk at work, and be exempt from jury duty.

To facilitate breastfeeding support advocacy among health professionals in health care facilities and in the community, the National Alliance of Breastfeeding Advocacy (NABA) maintains a searchable database on the Web. The database includes a state-by-state listing of all known breastfeeding coalitions and task forces, American Academy of Pediatrics state chapter residents, International Lactation Consultant Association affiliates, La Leche League International contacts, WIC breastfeeding directors, and Lamaze state contacts, and Title V Maternal and Child Health directors.

Public Food and Nutrition Programs

National Breastfeeding Policy

The U.S. Department of Health and Human Services (DHHS) is the lead federal agency in the effort to promote, protect, and support breastfeeding for U.S. families. Over the past 15 years, the Office of the Surgeon General and the Maternal and Child Health Bureau have highlighted the public health importance of breastfeeding though numerous workshops and publications (Table 6.9). In 2000, the DHHS Blueprint for Action on Breastfeeding outlined a comprehensive framework to increase breastfeeding rates in the United States and to promote optimal breastfeeding practices. The action plan is based on education, training, awareness, support, and research.[1]

In the 1990s, the DHHS, through its Maternal and Child Health Bureau and the Centers for Disease Control, supported the establishment of the U.S. Breastfeeding Committee (USBC) to fulfill one of the goals identified in the *Innocenti Declaration:* Each nation should establish "a multisectoral national breastfeeding committee composed of representatives from relevant government departments, nongovernmental organizations, and health professional associations." The USBC is a collaborative partnership of about 30 major organizations. The committee has worked collaboratively to develop a strategic plan for protecting, promoting, and supporting breastfeeding in the United States[160] and has published online a series of issue papers on the benefits of breastfeeding, breastfeeding and child care, economic benefits of breastfeeding, state breastfeeding legislation, and workplace breastfeeding support (www.usbreastfeeding.org).

Many DHHS agencies have breastfeeding initiatives. The Title V Maternal and Child Health programs of the Health Resources and Services Administration provide substantial support services, training, and research for breast-

Table 6.9 Landmark U.S. breastfeeding policy statements and conferences

- Report of the Surgeon General's Workshop on Breastfeeding and Human Lactation[150]
- Follow-Up Report: Surgeon General's Workshop on Breastfeeding and Human Lactation[156]
- Healthy People 2000 Breastfeeding Goals for the Nation[32]
- DHHS Maternal and Child Health Bureau National Workshop "Call to Action: Better Nutrition for Mothers, Children, and Families"[157]
- Second Follow-Up Report: Surgeon General's Workshop on Breastfeeding and Human Lactation[158]
- 1998 National Breastfeeding Policy Conference, presented by the UCLA Center for Healthier Children, Families and Communities, Breastfeeding Resource Program in cooperation with the U.S. Department of Health and Human Services, Health Resources, and Services Administration, Maternal and Child Health Bureau, and the Centers for Disease Control and Prevention[159]
- Healthy People 2010 Breastfeeding Goals for the Nation[35]
- DHHS Blueprint for Action on Breastfeeding[1]
- U.S. Breastfeeding Committee Strategic Plan

feeding (www.mchb.hrsa.gov). In 2003, the Health and Human Services Office on Women's Health launched a comprehensive 3-year national breastfeeding awareness campaign targeting first-time mothers. In addition, the National Women's Health Center, a project of the Office of Women's Health, has a breastfeeding helpline (1-800-994-WOMEN) and Web site (www.4women.gov) to help mothers with common breastfeeding problems and challenges. The Centers for Disease Control and Prevention plays a major role in supporting breastfeeding nationally through applied research, program evaluation, and surveillance (www.cdc.gov).

USDA WIC Program

WIC (Special Supplemental Nutrition Program for Women, Infants, and Children) is a federal program operated by the U.S. Department of Agriculture (USDA) Food and Nutrition Service in partnership with state and local health departments. Created in 1972, WIC is designed to provide nutrition education, supplementary foods, and referrals for health and social services to economically disadvantaged women who are pregnant, postpartum, or caring for infants and children under age 5.

Innocenti Declaration The Innocenti Declaration on the Protection, Promotion, and Support of Breastfeeding was produced and adopted by participants at the WHO/UNICEF policymakers' meeting on "Breastfeeding in the 1990s: A Global Initiative," held at the Spedale degli Innocenti, in Florence, Italy, on August 1, 1990. The Declaration established exclusive breastfeeding from birth to 4–6 months of age as a global goal for optimal maternal and child health.

In 1989, Congress mandated (Public Law 101-147) that a specific portion of each state's WIC budget allocation to be used exclusively for the promotion and support of breastfeeding among its participants and authorized the use of WIC administrative funds to purchase breastfeeding aids such as breast pumps. Through this legislation, each state has a breastfeeding coordinator and a plan to coordinate operations with local agency programs for breastfeeding promotion. Reauthorization legislation, the Healthy Meals for Healthy Americans Act of 1994 (Public Law 103-448), increased the budget allocation for breastfeeding promotion to $21 for each pregnant and breastfeeding woman. WIC's 1999 budget appropriation increased this amount further, to $23.[53] The USDA hosts semiannual meetings of the U.S. Breastfeeding Consortium to exchange ideas on how the USDA, other federal agencies, and private health interests can work together to promote breastfeeding, especially in the WIC program. More than 25 organizations participate, including health professional associations, breastfeeding advocacy groups, and other federal agencies.

Social Marketing A marketing effort that combines the principles of commercial marketing with health education to promote a socially beneficial idea, practice, or product.[122]

The USDA Food and Nutrition Information Center supports a WIC Works Web site to serve health and nutrition professionals working in the WIC program. The WIC Works site (www.nal.usda.gov/wicworks) includes an e-mail discussion group, links to training materials on breastfeeding promotion, and information on how to share resources and recommendations.

Model Breastfeeding Promotion Programs

WIC National Breastfeeding Promotion Project—Loving Support Makes Breastfeeding Work

Although breastfeeding initiation and duration rates among WIC participants increased in the 1990s, the rates remained considerably lower than among women in higher socioeconomic levels. In 1995, the USDA Food and Nutrition Service entered into a cooperative agreement with Best Start Social Marketing (a nonprofit organization assisting public health, education, and social service organizations with *social marketing* services) to develop a national WIC breastfeeding promotion project to be implemented at the state level. The project has four goals: (1) to encourage WIC participants to begin and to continue breastfeeding, (2) to increase referrals to WIC clinics for breastfeeding support, (3) to increase general public acceptance of and support for breastfeeding, and (4) to provide support and technical assistance to WIC professionals in promoting breastfeeding.[161]

Best Start collected data in 10 pilot states through observations, personal and telephone interviews, and focus groups with WIC participants and individuals who might influence their infant feeding decisions such as mothers, boyfriends and husbands, health care providers, and WIC staff.[162] The researchers looked for motivations and perceived barriers to breastfeeding as well as social network influences on feeding choice.

Results from the research were then used to develop a marketing plan and program material. In contrast to the traditional public health approach of addressing breastfeeding as a medical health decision, the marketing plan repositioned breastfeeding as a way for a family to establish a special relationship with their child from the very onset of its life.[127,162] The campaign slogan, "Loving Support Makes Breastfeeding Work," capitalizes on the concept that everyone is important to a woman's breastfeeding success—her family, friends, doctors, and community. Campaign materials and a counseling program were developed to help mothers work through individual barriers and constraints to breastfeeding.

In 1997, the campaign was implemented at the state level. Participating states could implement all or part of the campaign with technical assistance from the federal level. The states had the opportunity to purchase campaign materials including pamphlets, posters, and radio and television public service announcements that address barriers and encourage breastfeeding. Training was provided on coalition building, utilizing local media, promoting effective breastfeeding counseling strategies and techniques (Table 6.4), and managing peer counseling programs.[127] Since 1997, the Loving Support Makes Breastfeeding Work campaign has expanded to other states; currently, 72 state agencies, Indian tribal organizations, and territories are participating at various levels.[163,164]

A preliminary evaluation of the program's impact in Mississippi found improvement in both breastfeeding rates and attitudes toward, and awareness of, breastfeeding. Prior to initiation of the Loving Support campaign in 1997, Mississippi ranked 50th in the nation in breastfeeding initiation and duration rates. Mississippi's campaign included a public awareness component with television and radio spots, newspaper ads, and billboards in high-traffic areas; a patient and family education program in health departments and WIC clinics; extensive outreach with health providers in the community and 25 hospitals; and community outreach, including training at child care centers and worksites. In 1999, the state moved into 48th place nationally in breastfeeding initiation rates at hospital discharge. Duration rates at 6 months, which climbed from 7.0% to 15.4% from 1998 to 1999, moved Mississippi WIC from 50th to 33rd in the country. Preliminary data from focus groups conducted across the state in 1999, with support from the USDA, revealed that the campaign had a significant impact on WIC program participants' breastfeeding knowledge and attitudes.[163]

Other states are currently assessing their individual campaigns.[162] Based on data from the Ross Breastfeeding Survey,[31] breastfeeding initiation rates in Iowa increased from 57.8% to 65.1% after a year of the campaign. The rates of women still nursing 6 months after birth also increased; before the start of the campaign, 20.4% of women were still breastfeeding after 6 months compared with 32.2% a year after the program's start.[162] Increased breastfeeding support from relatives and friends was also documented from data collected in a mail survey.

Wellstart International

Wellstart International is an independent, nonprofit organization headquartered in San Diego, California, whose mission is "promoting optimal maternal and infant nutrition and health through clinical services and professional education about breastfeeding and lactation management." Wellstart provides education and technical assistance to perinatal health care providers and educators around the world, enabling them to promote maternal and child health in their own settings by supporting breastfeeding. Wellstart faculty and staff offer in-depth clinical and program expertise and both domestic and international experience to hospitals, clinics, and university schools of medicine, nursing, and nutrition, as well as to a wide variety of governmental and nongovernmental health and population agencies. Wellstart is a designated WHO Collaborating Center on Breastfeeding Promotion.

Since 1977 Wellstart has helped health care professionals establish self-sustaining breastfeeding promotion programs worldwide. One example is the National Training and Technical Support Center for Breastfeeding in Cairo, Egypt, established in collaboration with the Egyptian Ministry of Health and Population. The Support Center trains health workers and provides technical support for other related program areas. It is part of the countrywide Technical Support Collaboration involving applied research, evaluation of university curricula, countrywide training of health workers, efforts to achieve Mother-Baby Friendly Hospital status, community outreach, behavior change activities, and monitoring and evaluation. The Support Center also creates documents such as Counseling Guidelines on Infant Feeding for Use in Egypt and the Final Report on the Technical Support Collaboration on Infant Feeding and disseminates them to service providers and community workers.

By combining community outreach activities with the use of information, education, and communication (IEC) materials targeting key behaviors, the Support Center in Cairo has achieved significant behavior changes at the local, regional, and national levels:[164,165]

- *Initiation of breastfeeding.* Behaviors related to initiation of the first breastfeeding improved (p <0.001), with the optimal behavior of initiating breastfeeding within the first hour increasing from 46% to 72% within the demonstration areas.

- *Exclusive breastfeeding.* Exclusive breastfeeding for all age groups increased (p <0.001), with levels increasing from 24% to 74% for the 0–3 month age group, from 8% to 55% for the 4–6 month age group, and from 15.5% to 65% for the combined 0–6 month age group.

- *Infants receiving water and herbal teas.* The use of water or herbal teas within the 0–3 and 4–6 month age groups declined (p <0.001), as would be expected if exclusive breastfeeding was increasing.

- *Bottle and pacifier use.* For infants 0–12 months of age, bottle use decreased from 49% to 12%; pacifier use decreased from 42% to 13.6% (p <0.001).

- *Complementary feeding.* Data related to timely introduction of complementary foods, as well as types of complementary foods received by infants, also showed substantial improvements (p <0.001). The percentage of infants 7–9 months of age receiving no complementary food decreased from 16% to 3%.

On a global level, we must do everything we can

"to increase women's confidence in their ability to breastfeed. Such empowerment involves the removal of constraints and influence that manipulate perceptions and behavior towards breastfeeding, often by subtle and indirect means. Furthermore, obstacles to breastfeeding within the health care system, the workplace, and the community must be eliminated."

These words are from the WHO/UNICEF Innocenti Declaration on the Protection, Promotion, and Support of Breastfeeding, Florence, Italy, 1990. As we have seen, breastfeeding is best for the vast majority of infants and is physiologically possible for the vast majority of women. The challenge is to overcome social and cultural barriers and to provide support systems at the local, national, and international level so that the initiation and duration of breastfeeding will continue to increase.

Resources

Breastfeeding Promotion in Pediatric Office Practices Program
Telephone: (847) 228-5005, extension 4779
Web site: **www.aap.org/visit/brpromo.htm**

Breastfeeding Policy
Web site: **www.aap.org/policy/re9729.html**

Baby-Friendly USA Hospital Initiative
Telephone: (508) 888-8092
Web site: www.aboutus.com/a100/bfusa

Best Start Social Marketing
Telephone: (813) 971-2119
Web site: beststart@beststartinc.org

Best Start, Inc.
Loving Support Campaign materials
Best Start, Inc., Tampa, FL.
Beststart@mindspring.com

Bright Future Lactation Resource Center
Web site: www.bflrc.com

Breastfeeding Legislative Updates
Web site: www.house.gov/maloneyissues/breastfeeding/index.htm

Case Western Reserve University School of Medicine
Online Breastfeeding Basics Course
Web site: www.breastfeedingbasics.org

DHHS HRSA Maternal and Child Health Bureau
Telephone: (202) 784-9770
Web site: www.mchb.hrsa.gov

International Lactation Consultants Association
Telephone: (919) 787-5181
Web site: www. ilca.org

Lactation Education Resources
Web site: www.LERon-line.com

Lactation Study Center, University of Rochester
Web site: www.neonate.pediatrics.rochester.edu

La Leche League
Telephone: (847) 519-7730
Web site: www. lalecheleague.org

National Center for Education in Maternal and Child Health
Telephone: (703) 524-7802
Web site: www.ncemch.org

Support Breastfeeding.com
Web site: www.supportbreastfeeding.com

US Breastfeeding
Web site: www.usbreastfeeding.org

USDA Women, Infants, and Children (WIC) Program
Telephone: (703) 305-2736
Web site: www.fns.usda.gov/wic

UNICEF Baby-Friendly Hospital Initiative
Web site: www.unicef.org

Wellstart International
Telephone: (619) 295-5192
Web site: www.wellstart.org

World Alliance for Breastfeeding
Web site: www. Elogica.com.br/waba

References

1. U.S. Department of Health and Human Services, *HHS blueprint for action on breastfeeding.* Washington, DC: U.S. Department of Health and Human Services, Office on Women's Health, 2000.

2. Heinig, M. J., and K. G. Dewey, *Health effects of breast feeding for mothers: a critical review.* Nutrition Reviews, 1997. 10: 59–73.

3. McNeilly, A. S., *Lactational amenorrhea.* Endocrinol Metab Clin North Am, 1993. 10: 35–56.

4. Kuzela, A. L., C. A. Stifter, and J. W. Worobey, *Breastfeeding and mother-infant interactions.* J Reprod Infant Psychol, 1990. 8: 185–94.

5. Newcomb, P. A., K. M. Egan, and L. Titus-Ernstoff et al., *Lactation in relation to postmenopausal breast cancer.* Am J Epidemiol, 1999. 150(2): 174–82.

6. Rosenblatt, K. A., and D. B. Thomas, *Lactation and the risk of epithelial ovarian cancer. The WHO Collaborative Study of Neoplasia and Steroid Contraceptives.* Int J Epidemiol, 1993. 22(2): 192–7.

7. American Academy of Pediatrics, (AAP), *Breastfeeding and the use of human milk.* Pediatrics, 1997. 100(6): 1035–39.

8. Lawrence, R. A., and R. M. Lawrence, *Breastfeeding: A guide for the medical professional.* 5th ed. St. Louis: Mosby, 1999.

9. Jensen, R. G., ed., *Handbook of Milk Composition.* Academic Press: New York, 1995.

10. Picciano, M. F., *Human milk: nutritional aspects of a dynamic food.* Biol Neonate, 1998. 74(2): 84–93.

11. Rogan, W. J., and B. C. Gladen, *Breastfeeding and cognitive development.* Early Hum Dev, 1993. 31(3): 181–93.

12. Helland, I. B., L. Smith, and K. Saarem et al., *Maternal supplementation with very-long-chain n-3 fatty acids during pregnancy and lactation augments children's IQ at 4 years of age.* Pediatrics, 2003. 111(1): e39–e44.

13. Smith, M. M., M. Durkin, and V. J. Hinton et al., *Influence of breastfeeding on cognitive outcomes at age 6–8 years: follow-up of very low birth weight infants.* Am J Epidemiol, 2003. 158(11): 1075–82.

14. Dewey, K. G., M. J. Heinig, and L. A. Nommsen-Rivers, *Differences in morbidity between breast-fed and formula-fed infants.* J Pediatr, 1995. 126(5 Pt 1): 696–702.

15. Kramer, M. S., T. Guo, and R. W. Platt et al., *Infant growth and health outcomes associated with 3 compared with 6 mo. of exclusive breastfeeding.* Am J Clin Nutr, 2003. 78(2): 291–5.

16. Raisler, J., C. Alexander, and P. O'Campo, *Breastfeeding: A dose-response relationship?* Am J Pub Health, 1999. 89: 25–30.

17. Ivarsson, A., L. A. Persson, and L. Nystrom et al., *Epidemic of coeliac disease in Swedish children.* Acta Paediatr, 2000. 89(2): 165–71.

18. Koletzko, S., P. Sherman, and M. Corey et al., *Role of infant feeding practices in development of Crohn's disease in childhood.* BMJ, 1989. 298(6688): 1617–18.

19. Daniels, J. L., A. F. Olshan, and B. H. Pollock et al., *Breast-feeding and neuroblastoma, USA and Canada.* Cancer Causes Control, 2002. 13(5): 401–5.

20. Oddy, W. H., P. G. Holt, and P. D. Sly et al., *Association between breast feeding and asthma in 6-year-old children: findings of a prospective birth cohort study.* BMJ, 1999. 319(7213): 815–19.

21. Scariati, P. D., L. M. Grummer-Strawn, and S. B. Fein, *A longitudinal analysis of infant morbidity and the extent of breastfeeding in the United States.* Pediatrics, 1997. 99(6): E5.

22. McVea, K. L., P. D. Turner, and

D. K. Peppler, *The role of breastfeeding in sudden infant death syndrome*. J Hum Lact, 2000. 16(1): 13–20.

23. Dewey, K. G., J. M. Peerson, and K. H. Brown et al., *Growth of breast-fed infants deviates from current reference data: A pooled analysis of US, Canadian, and European data sets. World Health Organization Working Group on Infant Growth*. Pediatrics, 1995. 96(3 Pt 1): 495–503.

24. Dewey, K. G., *Is breastfeeding protective against child obesity?* J Hum Lact, 2003. 19(1): 9–18.

25. Splett, P. L., and D. L. Montgomery, *The economic benefits of breastfeeding an infant in the WIC program twelve-month follow-up study*. Washington, DC: Food and Consumer Service, U.S. Department of Agriculture, 1998.

26. Ball, T. M., and A. L. Wright, *Health care costs of formula-feeding in the first year of life*. Pediatrics, 1999. 103(4 Pt 2): 870–6.

27. Cohen, R., M. B. Mrtek, and R. G. Mrtek, *Comparison of maternal absenteeism and infant illness rates among breast-feeding and formula-feeding women in two corporations*. Am J Health Promot, 1995. 10(2): 148–53.

28. Carbajal, R., S. Veerapen, and S. Couderc et al., *Analgesic effect of breast feeding in term neonates: randomised controlled trial*. BMJ, 2003. 326(7379): 13.

29. Ors, R., E. Ozek, and G. Baysoy et al., *Comparison of sucrose and human milk on pain response in newborns*. Eur J Pediatr, 1999. 158(1): 63–6.

30. Institute of Medicine and Subcommittee on Nutrition during Lactation, *Nutrition during lactation*. Washington, DC: National Academy Press, 1991, p. 303.

31. Ross Products Division, *Mothers survey: Updated breastfeeding trend through 1996*. Columbus, OH: Abbot Laboratories, 1998.

32. U.S. Department of Health and Human Services, *Healthy People 2000: National health promotion and disease prevention objectives*. Washington, DC: U.S. Department of Health and Human Services, Public Health Service, Office of the Assistant Secretary for Health.

33. Division of Ross Products, *Ross Mothers Survey*. Columbus, OH: Abbott Laboratories, 2000.

34. *Prevalance of selected maternal behaviors and experiences, pregnancy risk assessment monitoring system (PRAMS), 1999, in Morbidity and Mortality Weekly Report*. Atlanta, GA: Centers for Disease Control, 2002, p. 1–27.

35. U.S. Department of Health and Human Services, *Healthy People 2010: Conference Edition—volumes I and II*. Washington, DC: U.S. Department of Health and Human Services, Public Health Service, Office of the Assistant Secretary for Health, 2000.

36. Ryan, A. S., *The resurgence of breastfeeding in the United States*. Pediatrics, 1997. 99(4): e12.

37. Hartmann, P. E., *Changes in the composition and yield of the mammary secretion of cows during the initiation of lactation*. J Endocrinol, 1973. 59(2): 231–47.

38. Vorherr, H., *Human lactation and breastfeeding in The mammary gland/human lactation/milk synthesis*, B. L. Larson, ed. New York: Academic Press, 1978.

39. Cox, D. B., R. A. Owens, and P. E. Hartmann, *Blood and milk prolactin and the rate of milk synthesis in women*. Exp Physiol, 1996. 81(6): 1007–20.

40. Neville, M. C., *The physiological basis of milk secretion. Part I: Basic physiology*. Ann NY Acad Sci, 1990. 5: 861–68.

41. Jensen, R. G., *The lipids of human milk*. Boca Raton, FL: CRC Press, 1989.

42. Butte, N. F., C. Garza, and E. O. Smith et al., *Human milk intake and growth in exclusively breast-fed infants*. J Pediatr, 1984. 104(2): 187–95.

43. Daly, S. E., and P. E. Hartmann, *Infant demand and milk supply. Part 1: Infant demand and milk production in lactating women*. J Hum Lact, 1995. 11(1): 21–6.

44. Dewey, K. G., and B. Lonnerdal, *Infant self-regulation of breast milk intake*. Acta Paediatr Scand, 1986. 75(6): 893–8.

45. Daly, S. E., R. A. Owens, and P. E. Hartmann, *The short-term synthesis and infant-regulated removal of milk in lactating women*. Exp Physiol, 1993. 78(2): 209–20.

46. Saint, L., P. Maggiore, and P. E. Hartmann, *Yield and nutrient content of milk in eight women breastfeeding twins and one woman breastfeeding triplets*. Br J Nutr, 1986. 56: 49–58.

47. Newton, M., *Human lactation, in Milk: The mammary gland and its secretion*, S. K. Kon, ed. New York: Academic Press, 1961.

48. Daly, S.E.J., and P. E. Hartmann, *Infant demand and milk supply. Part 2: The short-term control of milk synthesis in lactating women*. J Hum Lactation, 1995. 11: 27–37.

49. Hill, P. D., J. C. Aldag, and R.T.J. Chatterton, *Breastfeeding experience and milk weight in lactating mothers pumping for preterm infants*. Birth, 1999. 26(4).

50. Niefert, M., and J. M. Seacat, *Practical aspects of breastfeeding the premature infant*. Perinatal Neonatol, 1988. 12: 24.

51. Souto, G. C., E. R. Giugliani, and C. Giugliani et al., *The impact of breast reduction surgery on breastfeeding performance*. J Hum Lact, 2003. 19(1): 43–9; quiz 66–9, 120.

52. Hurst, N., *Breastfeeding after breast augmentation*. J Hum Lact, 2003. 19(1): 70–1.

53. *Transfer of drugs and other chemicals into human milk*. Pediatrics, 2001. 108(3): 776–89.

54. Berlin, C. M. Jr., *Silicone breast implants and breast-feeding*. Pediatrics, 1994. 94(4 Pt 1): 547–9.

55. Semple, J. L., S. J. Lugowski, and C. J. Baines et al., *Breast milk contamination and silicone implants: preliminary results using silicon as a proxy measurement for silicone*. Plast Reconstr Surg, 1998. 102(2): 528–33.

56. Almroth, S. G., *Water requirements of breastfed infants in a hot climate*. Am J Clin Nutr, 1978. 31: 1154–58.

57. Axelson, I., S. Borulf, and L. Righard et al., *Protein and energy intake during weaning: I. Effects on growth*. Acta Paediatrica Scandinavica, 1987. 76(2): 321–7.

58. Butte, N. F., C. Garza, and C. A. Johnson et al., *Longitudinal changes in milk composition of mothers delivering preterm and term infants*. Early Hum Dev, 1984. 9(2): 153–62.

59. Hediger, M. L., M. D. Overpeck, and W. J. Ruan et al., *Early infant feeding and growth status of US–born infants and children aged 4–71 mo: analyses from the third National Health and Nutrition Examination Survey, 1988–1994*. Am J Clin Nutr, 2000. 72(1): 159–67.

60. Connor, W. E., R. Lowensohn, and L. Hatcher, *Increased docosahexaenoic acid levels in human newborn infants by administration of sardines and fish oil during pregnancy*. Lipids, 1996. 31(Suppl): S183–7.

61. Insull, W., and E. H. Ahrens, *The fatty acids of human milk from mothers on diets taken ad libitum*. Biochem J, 1959. 72: 27.

62. Agostoni, C., E. Riva, and S. Trojan et al., *Docosahexaenoic acid status and developmental quotient of healthy term infants*. Lancet, 1995. 346(8975): 638.

63. Makrides, M., M. A. Neumann, and R. A. Gibson, *Effect of maternal docosahexaenoic acid (DHA) supplementation on breast milk composition*. Eur J Clin Nutr, 1996. 50(6): 352–7.

64. Wong, W. W., D. L. Hachey, and W. Insull et al., *Effect of dietary cholesterol on cholesterol synthesis in breast-fed and formula-fed infants*. J Lipid Res, 1993. 34(8): 1403–11.

65. Owen, C. G., P. H. Whincup, and K. Odoki et al., *Infant feeding and blood cholesterol: a study in adolescents and a systematic review*. Pediatrics, 2002. 110(3): 597–608.

66. Rosen, J. M., W. K. Jones, and J. R. Rodgers et al., *Regulatory sequences involved in the hormonal control of casein gene expression*. Ann N Y Acad Sci, 1986. 464: 87–99.

67. Velona, T., L. Abbiati, and B. Beretta et al., *Protein profiles in breast milk from mothers delivering term and preterm babies*. Pediatr Res, 1999. 45(5 Pt 1): 658–63.

68. Lonnerdal, B., and S. Atkinson, *Nitrogenous components of milk, in Handbook of milk composition*, R. G. Jensen, ed. San Diego, CA: Academic Press, 1995, p. 351–68.

69. Newburg, D. S., and S. H. Neubauer, *Carboyhdrates in milks: analysis, quantities, and signficance., in Handbook of milk composition*, R.G. Jensen, Editor. San Diego, CA: Academic Press, 1995, p. 273–349.

70. Food and Nutrition Board National Research Council, *Recommended Dietary Allowances.* 10th ed. Washington, DC: National Academy Press, 1989.

71. Canfield, L. M., A. R. Giuliano, and E. M. Neilson et al., *Beta-carotene in breast milk and serum is increased after a single beta-carotene dose.* Am J Clin Nutr, 1997. 66(1): 52–61.

72. Rothberg, A. D., J. M. Pettifor, and D. F. Cohen et al., *Maternal-infant vitamin D relationships during breast-feeding.* J Pediatr, 1982. 101(4): 500–3.

73. Ala-Houhala, M., T. Koskinen, and M. T. Parviainen et al., *25-Hydroxyvitamin D and vitamin D in human milk: effects of supplementation and season.* Am J Clin Nutr, 1988. 48(4): 1057–60.

74. Gartner, L. M., and F. R. Greer, *Prevention of rickets and vitamin D deficiency: new guidelines for vitamin D intake.* Pediatrics, 2003. 111(4 Pt 1): 908–10.

75. Lammi-Keefe, C. J., C. R. Jonas, and A. M. Ferris et al., *Vitamin E in plasma and milk of lactating women with insulin-dependent diabetes mellitus.* J Pediatr Gastroenterol Nutr, 1995. 20(3): 305–9.

76. Haug, M., C. Laubach, and M. Burke et al., *Vitamin E in human milk from mothers of preterm and term infants.* J Pediatr Gastroenterol Nutr, 1987. 6(4): 605–9.

77. Chappell, J. E., T. Francis, and M. T. Clandinin, *Vitamin A and E content of human milk at early stages of lactation.* Early Hum Dev, 1985. 11(2): 157–67.

78. Lammi-Keefe, C. J., and R. G. Jensen, *Fat-soluble vitamins in human milk.* Nutr Rev, 1984. 42(11): 365–71.

79. Pediatrics, A.A.P., *Controversies concerning vitamin K.* Pediatrics, 2003. 112: 191–92.

80. Andon, M. B., R. D. Reynolds, and P. B. Moser-Veillon et al., *Dietary intake of total and glycosylated vitamin B-6 and the vitamin B-6 nutritional status of unsupplemented lactating women and their infants.* Am J Clin Nutr, 1989. 50(5): 1050–8.

81. Salmenpera, L., J. Perheentupa, and M. A. Siimes, *Folate nutrition is optimal in exclusively breast-fed infants but inadequate in some of their mothers and in formula-fed infants.* J Pediatr Gastroenterol Nutr, 1986. 5(2): 283–9.

82. Kuhne, T., R. Bubl, and R. Baumgartner, *Maternal vegan diet causing a serious infantile neurological disorder due to vitamin B$_{12}$ deficiency.* Eur J Pediatr, 1991. 150(3): 205.

83. Specker, B. L., A. Black, and L. Allen et al., *Vitamin B-12: low milk concentrations are related to low serum concentrations in vegetarian women and to methylmalonic aciduria in their infants.* Am J Clin Nutr, 1990. 52(6): 1073–6.

84. Butte, N. F., C. Garza, and E. O. Smith et al., *Macro- and trace-mineral intakes of exclusively breast-fed infants.* Am J Clin Nutr, 1987. 45(1): 42–8.

85. Fransson, G. B., and B. Lonnerdal, *Zinc, copper, calcium, and magnesium in human milk.* J Pediatr, 1982. 101(4): 504–8.

86. Duncan, B., R. B. Schifman, and J. J. Corrigan Jr. et al., *Iron and the exclusively breast-fed infant from birth to six months.* J Pediatr Gastroenterol Nutr, 1985. 4(3): 421–5.

87. Pisacane, A., B. DeVizia, and A. Vallente et al., *Iron status in breast-fed infants.* J Pediatr, 1995. 127(3): 429–21.

88. Krebs, N. F., and K. M. Hambidge, *Zinc requirements and zinc intakes of breast-fed infants.* Am J Clin Nutr, 1986. 43(2): 288–92.

89. Sian, L., N. F. Krebs, and J. E. Westcott et al., *Zinc homeostasis during lactation in a population with a low zinc intake.* Am J Clin Nutr, 2002. 75(1): 99–103.

90. Atkinson, S. A., D. Whelan, and R. K. Whyte et al., *Abnormal zinc content in human milk: Risk for development of nutritional zinc deficiency in infants.* Am J Dis Child, 1989. 143(5): 608–11.

91. McDaniel, M. R., E. Barker, and C. L. Lederer, *Sensory characterization of human milk.* J Dairy Sci, 1989. 72(5): 1149–58.

92. Menella, J. A., and G. K. Beauchamp, *Smoking and the flavor of breastmilk.* N Engl J Med, 1998. 339: 149–56.

93. Gerrish, C. J., and J. A. Menella, *Flavor variety enhances food acceptance in formula-fed infants.* Am J Clin Nutr, 2001. 73: 1080–5.

94. Kramer, M. S., and R. Kauma, *Optimal duration of exclusive breastfeeding.* Cochrane Review, 2003(2): 1–63.

95. Ziegler, E. E., S. J. Foman, and S. E. Nelson et al., *Cow milk feeding in infancy: Further observations on blood loss from the gastrointestinal tract.* J Pediatr, 1990. 116: 11–18.

96. Wooldrige, M. S., and C. Fischer, *Colic, "overfeeding" and symptoms of lactose malabsorption in the breast-fed baby.* Lancet, 1988. 2: 382–4.

97. Butte, N. F., Garza, C., Smith, E. O., and Nichols, Human milk intake and growth in exclusively breast-fed infants during the first four months of life. Early Hum Dev, 1985. 12:291–300.

98. Brams, M., and J. Maloney, *"Nursing bottle caries" in breast-fed children.* J Pediatr, 1983. 103: 415–6.

99. Palmer, B., *The influence of breastfeeding on the development of the oral cavity: A commentary.* J Hum Lact, 1998. 14(2): 93–98.

100. U.S. Department of Agriculture, *Food Guide Pyramid.* Washington, DC: U.S. Department of Agriculture and the U.S. Department of Health and Human Services, 1992.

101. U.S. Department of Agriculture, *Nutrition and your health: dietary guidelines for Americans.* Washington, DC: U.S. Department of Agriculture, U.S. Department of Health and Human Services, 1995.

102. Lust, K. D., J. E. Brown, and W. Thomas, *Maternal intake of cruciferous vegetables and other foods and colic symptoms in exclusively breast-fed infants.* J Am Diet Assoc, 1996. 96(1): 46–8.

103. Food and Nutrition Board Institue of Medicine, *Dietary Reference Intakes for energy, carbohydrate, fiber, fat, fatty acids, cholesterol, protein, and amino acids (macronutrients).* Washington, DC: National Academy Press, 2002.

104. Food and Nutrition Board Institute of Medicine, *Dietary Reference Intakes for vitamin c, vitamin e, selenium, and carotenoids.* Washington, DC: National Academy Press, 2002.

105. Food and Nutrition Board Institute of Medicine, *Dietary Reference Intakes for calcium, phosphorus, magnesium, vitamin D, and fluoride.* Washington, DC: National Academy Press, 1999, p. 448.

106. Food and Nutrition Board Institute of Medicine, *Dietary Reference Intakes for thiamin, riboflavin, niacin, vitamin B$_6$, folate, vitamin B$_{12}$, pantothenic acid, biotin, and choline.* Washington, DC: National Academy Press, 1998.

107. Food and Nutrition Board Institute of Medicine, *Dietary Reference Intakes for vitamin A, vitamin K, arsenic, boron, chromium, copper, iodine, iron, manganese, molybdenum, nickel, silicon, vanadium, and zinc.* Washington, DC: National Academy Press, 2001.

108. Butte, N. F., C. Garza, and J. E. Stuff et al., *Effect of maternal diet and body composition on lactational performance.* Am J Clin Nutr, 1984. 39(2): 296–306.

109. Brewer, M. M., M. R. Bates, and L. P. Vannoy, *Postpartum changes in maternal weight and body fat deposits in lactating vs nonlactating women.* Am J Clin Nutr, 1989. 49(2): 259–65.

110. Sadurskis, A., N. Kabir, and J. Wager et al., *Energy metabolism, body composition, and milk production in healthy Swedish women during lactation.* Am J Clin Nutr, 1988. 48: 44–49.

111. Goldberg, G. R., A. M. Prentice, and W. A. Coward et al., *Longitudinal assessment of energy expenditure in pregnancy by the doubly labeled water method.* Am J Clin Nutr, 1993. 57: 494–505.

112. Prentice, A. M., S. D. Poppitt, and G. R. Goldberg et al., *Energy balance in pregnancy and lactation,* in *Nutrient regulation during pregnancy, lactation, and infant growth,* L. H. Allen, J. King, and B. Lonnerda, eds. New York: Plenum Press, 1994.

113. Roberts, S., T. Cole, and W. Coward, *Lactational performance in relation to energy intake in the baboon.* Am J Clin Nutr, 1985. 41: 1270–76.

114. Macnamara, J. P., *Role and regulation of metabolism in adipose tissue during lactation.* J Nutr Biochem, 1995. 6: 120–9.

115. Butte, N. F., and J. M. Hopkinson, *Body composition changes during lactation are highly variable among women.* J Nutr, 1998. 128: 381S–385S.

116. Strode, M. A., K. G. Dewey, and B. Lonnerdal, *Effects of short-term caloric restriction on lactational performance of well-nourished women.* Acta Paediatr Scand, 1986. 75(2): 222–9.

117. Dusdieker, L. B., D. L. Hemingway, and P. J. Stumbo, *Is milk production impaired by dieting during lactation?* Am J Clin Nutr, 1994. 59(4): 833–40.

118. Lovelady, C. A., B. Lonnerdal, and K. G. Dewey, *Lactation performance of exercising women.* Am J Clin Nutr, 1990. 52(1): 103–9.

119. Dewey, K. G., and M. A. McCrory, *Effects of dieting and physical activity on pregnancy and lactation.* Am J Clin Nutr, 1994. 59(2 Suppl): 446S–452S; discussion, 452S–453S.

120. Lovelady, C. A., C. P. Hunter, and C. Geigerman, *Effect of exercise on immunologic factors in breast milk.* Pediatrics, 2003. 111(2): E148–52.

121. Kreiter, S. R., R. P. Schwartz, and H.N.J. Kirkma et al., *Nutritional rickets in African American breast-fed infants.* J Pediatr, 2000. 137(2): 143–5.

122. Wardinsky, T. D., R. G. Montes, and R. L. Friederich et al., *Vitamin B$_{12}$ deficiency associated with low breast-milk vitamin B$_{12}$ concentration in an infant following maternal gastric bypass surgery* [letter]. Arch Pediatr Adolescent Med, 1995. 149(11): 1281–4.

123. Renault, F., P. Verstichel, and J. P. Ploussard et al., *Neuropathy in two cobalamin-deficient breast-fed infants of vegetarian mothers.* Muscle Nerve, 1999. 22(2): 252–4.

124. Food and Nutrition Board, Institute of Medicine. *Dietary Reference Intakes for water potassium, sodium, chloride, and sulfate.* Washington, DC: National Academy Press, 2004.

125. Lazarov, M., and A. Evans, *Breastfeeding—encouraging the best for low-income women.* Zero to Three, 2000(August/September): 15–23.

126. Bryant, C., J. Coreil, and S. L. D'Angelo et al., *A strategy for promoting breastfeeding among economically disadvantaged women and adolescents.* NAACOG's Clin Issu Perinat Women's Health Nurs, 1992. 3(4): 723–30.

127. Bryant, C., and M. Roy, *Best Start's three-step counseling strategy.* Tampa, FL: Best Start, Inc., 1997.

128. American Dietetic Association, *Position of the American Dietetic Association: Promotion of breast-feeding.* J Am Diet Assoc, 1997. 97(6): 662–66.

129. Caulfield, L. E., S. M. Gross, and M. E. Bentley et al., *WIC-based interventions to promote breastfeeding among African American women in Baltimore: Effects on breastfeeding initiation and continuation.* J Hum Lact, 1998. 14(1): 15–22.

130. Freed, G. L., S. J. Clark, and J. Sorenson et al., *National assessment of physicians' breast-feeding knowledge, attitudes, training, and experience.* JAMA, 1995. 273(6): 472–6.

131. Sidorski, J., M. J. Renfrew, and S. Pindoria et al., *Support for breastfeeding mothers.* Cochrane Review, 2003(2).

132. Moreland, J. C., L. Lloyd, and S. B. Braun et al., *A new teaching model to prolong breastfeeding among Latinos.* J Hum Lact, 2000. 16(4): 337–41.

133. Cadwell, K., *Reaching the goals of "Healthy People 2000" regarding breastfeeding.* Clin Perinatol, 1999. 26(2): 527–37.

134. Merewood, M., and B. Phillip, *Peer counselors for breastfeeding mothers in the hospital setting: Trials, training, tributes, and tribulations.* J Hum Lact, 2003. 19(1): 72–76.

135. Bentley, M. E., L. E. Caulfield, and S. M. Gross et al., *Sources of influence on intention to breastfeed among African American women at entry to WIC.* J Hum Lact, 1999. 15(1): 27–34.

136. U.S. Department of Agriculture, F.A.N.S., *Fathers supporting breastfeeding.* Washington, DC: United States Department of Agriculture, 2002.

137. Pavill, B., *Fathers and breastfeeding.* AWHONN Lifelines, 2002. 6(4): 324–31.

138. Barron, S. P., H. W. Lane, and T. E. Hannan et al., *Factors influencing duration of breast feeding among low-income women.* J Am Diet Assoc, 1988. 88(12): 1557–61.

139. Freed, G. L., J. K. Fraley, and R. J. Schanler, *Attitudes of expectant fathers regarding breast-feeding.* Pediatrics, 1992. 90(2 Pt 1): 224–7.

140. Valaitis, R. K., J. D. Sheeshka, and M. F. O'Brien, *Do consumer infant feeding publications and products available in physicians' offices protect, promote, and support breastfeeding?* J Hum Lact, 1997. 13(3): 203–8.

141. Howard, C. R., S. J. Schaffer, and R. A. Lawrence, *Attitudes, practices, and recommendations by obstetricians about infant feeding.* Birth, 1997. 24(4): 240–6.

142. Heiser, B., and M. Walker, *Selling Out Mothers and Babies: Marketing of Breast Milk Substitutes in the USA.* Ellicot City, MD: National Alliance for Breastfeeding Promotion, 2003.

143. World Health Organization. *International code of marketing of breast milk substitutes.* World Health Organization, Geneva, Switzerland, 1981.

144. World Health Organization. *Protecting, promoting, and supporting breastfeeding: the special role of maternity services* (a joint WHO/UNICEF Statement). Geneva, Switzerland: WHO/UNICEF; 1989.

145. Howard, C., F. Howard, and R. Lawrence et al., *Office prenatal formula advertising and its effect on breast-feeding patterns.* Obstet Gynecol, 2000. 95(2): 296–303.

146. American College of Obstetricians and Gynecologists, *Breastfeeding: Maternal and infant aspects,* in ACOG Educational Bulletin. Washington, DC: American College of Obstetricians and Gynecologists, 2000.

147. Nurse-Midwives, A.C.O., *Clinical Practices Statement on Breastfeeding.* Washington, DC: American College of Nurse-Midwives, 1992.

148. Perez-Escamilla, R., E. Pollitt, and B. Lonnerdal et al., *Infant feeding policies in maternity wards and their effect on breastfeeding success: an analytical overview.* Am J Public Health, 1994. 84(1): 89–97.

149. UNICEF. The State of the World's Children. New York: Oxford University Press, 1998.

150. U.S. Department of Health and Human Services, *Report of the Surgeon General's Workshop on Breastfeeding and Human Lactation.* Washington, DC: U.S. Department of Health and Human Services, 1984.

151. Naylor, A. J., *Baby-friendly hospital initiative: Protecting, promoting, and supporting breastfeeding in the twenty-first century.* Pediatr Clin North Am, 2001. 48(2): 475–83.

152. Kurinu, N., P. H. Shiono, S. F. Ezrine, and G. G. Rhodads, *Does maternal employment affect breast-feeding?* Am J Public Health, 1989. 79: 1247–50.

153. Meek, J. Y., *Breastfeeding in the workplace.* Pediatr Clin North Am, 2001. 48(2): 461–74, xvi.

154. Healthy Mothers Healthy Babies Coalition, *Workplace models of excellence 2000: Outstanding programs supporting working women that breastfeed.* Alexandria, VA: National Healthy Mothers, Healthy Babies Coalition, 2000.

155. Committee, United States Breastfeeding, *Workplace brestfeeding suppport* (issue paper). Raleigh, NC: United States Breastfeeding Committee, 2002.

156. U.S. Department of Health and Human Services. *Follow-up report: the surgeon general's workshop on breastfeeding and human lactation.* Washington, DC. U.S. Department of Health and Human Services, Public Health Services, Health Resources and Services Administration, 1985.

157. Sharbaugh, CS. *Call to action: better nutrition for mothers, children, and families.* Washington, DC: National Center for Education in Maternal and Child Health, 1990.

158. Spisak, S. G., *Second follow-up report: the surgeon general's workshop on breastfeeding and human lactation.* Washington, DC: National Center for Education in Maternal and Child Health, 1991.

159. Skisser, W. L., and Thomas, S. *Report of the national breastfeeding policy conference.* U.C.L.A. Center for Healthier Communities, Families, and Children, 1999.

160. United States Breastfeeding Committee, *Breastfeeding in the United States: A national agenda.* Rockville, MD. Washington, DC: U.S. Department of Health and Human Services, Health Resources and

Services Administration, Maternal and Child Health Bureau, 2001.

161. United States Department of Agriculture, *Glickman kicks off USDA campaign to promote National Breastfeeding Week.* USDA Press Release No. 0283.97, August 6, 1997.

162. Andreason, A., *Marketing social change: Changing behavior to promote health, social development, and the environment.* San Francisco: Jossey-Bass, 1995.

163. Khoury, A., *Social Marketing Institute success stories: National WIC breastfeeding promotion project.* http://www.social-marketing.org/success/cs-nationalwic.html. 2001, Mississippi WIC Program.

164. Carothers, C., *Social Marketing Institute success stories: National WIC breast-feeding promotion project.* Tampa, FL: Best Start Social Marketing, 2001.

165. Personal Communication. Amy Khoury, Ph.D. Mississippi WIC Program, 2001.

"The establishment of breastfeeding for at least six months, but optimally for at least a year, as a cultural norm supported by medical, social, and economic practices is a fundamental cornerstone of true wellness promotion."

American Dietetic Association, 2001

Chapter 7

Nutrition and Lactation:

Conditions and Interventions

Chapter Outline

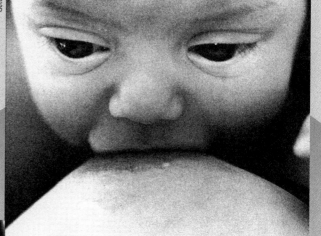

Prepared by **Carolyn Sharbaugh** and **Maureen A. Murtaugh** with **Denise Sofka**

Key Nutrition Concepts

1 Human milk is the preferred food for all premature and sick newborns, with rare exceptions.

2 Breastfeeding women need consistent, informed, and individualized care in the hospital and at home after discharge.

3 It is usually not necessary to discontinue breastfeeding to manage medical problems of the mother or infant; any medical decision to limit a mother's breastfeeding must be justified by the fact that the risk to her baby clearly outweighs the benefits of breastfeeding.

4 Feeding infants early in the postdelivery period whenever possible is important to successful breastfeeding. Early intervention to address questions or problems is equally important for maintaining breastfeeding.

5 Most medications—including over-the-counter as well as prescription drugs, drugs of abuse, alcohol, nicotine, and herbal remedies taken by nursing mothers—are excreted in breast milk.

6 Twins and other multiples can be successfully breast-fed without formula supplementation.

Introduction

The key to successful breastfeeding management is for the mother-infant breastfeeding dyad to receive support and informed, consistent, and individualized care from health care professionals both in the hospital and after discharge. The vast majority of women do not experience significant problems with breastfeeding, and many of the most common problems that do arise can be prevented through prenatal breastfeeding education and a positive, supportive breastfeeding initiation period.

This chapter discusses prevention and treatment of common breastfeeding conditions. Issues related to maternal use of medications, herbal remedies, drugs of abuse, and environmental contaminants are addressed. The chapter presents important considerations for breastfeeding multiples, preterm infants, and infants with medical problems. It provides information on the safe collection and storage of human milk and milk banks. The chapter concludes with case studies providing examples of management of challenging breastfeeding problems and with examples of model programs promoting support for breastfeeding in the health care system.

Common Breastfeeding Conditions

Sore Nipples

Almost all women have some nipple pain when initiating nursing. Some women have pain when the infant first latches on, but pain should not be considered a normal part of breastfeeding.[1] The first important step to prevent pain is proper positioning of the baby on the breast. The areola should be in the baby's mouth and the baby's tongue should be extended and against the lower lip. If a woman is experiencing pain, a lactation consultant or a health care professional well trained in lactation should observe the mother nursing her baby. The lactation consultant can determine whether the pain is simply related to early breastfeeding, or if a problem exists. Sore nipples can result from poor positioning of the infant at the breast, infection (thrush or staphylococcus aureus), pumping with too much suction, or a problem with the infant's suck.[2]

Women can take simple steps to prevent or manage nipple pain. Martin and Krebs[3] recommend that women let their breasts air dry after nursing, rub expressed milk and an all-purpose ointment (not petroleum based) on nipples, and use warm compresses on sore nipples. The common belief that limiting the frequency or duration of feedings will prevent or heal sore nipples is not substantiated by the literature.[4] Using a pump to express milk can help to maintain supply if the nipples are so sore that the mother cannot nurse. However, the suction on the breast pump should be adjusted carefully. High suction can make nipples sore and red.

Letdown Failure

"After she latches on, take deep, long breaths—think yoga, not Lamaze when you nurse. As you exhale, visualize the milk letting down through your breasts into the baby's mouth."

C. Martin and N. F. Krebs[3]

Letdown failure is not common, but because letdown is necessary to successful breastfeeding, it is important to address the matter. Oxytocin nasal spray can be prescribed by a physician for letdown failure. The synthetic oxytocin is sprayed into the nose and stimulates letdown, but it can be used for only a few days to help women get through a tough time. Other methods should be used at the same time to stimulate letdown. Martin and Krebs recommend a number of techniques to help women relax and enhance letdown:[3]

- Play soothing music that the mother can focus on while nursing.
- Have the partner rub his knuckles down her spine.
- Try different nursing positions.
- Get out of the house. Most babies enjoy a walk.
- Arrange for some time alone (a few hours).
- Switch from caffeinated to decaffeinated beverages and water for a few weeks.

Overactive Letdown

Overactive letdown can also be a problem, especially among first-time mothers. When letdown is overactive, milk streams from the breast when feeding begins. Milk may also leak from the other, unused breast. The milk streams quickly and the infant may be overwhelmed by the volume. The infant may choke or gulp to keep up with the flow. When the infant gulps, she takes in air, develops gas pain, and then may be fussy.

To manage overactive letdown, the mother should express milk until the flow slows and then put the infant to the breast.[1] (Expressed milk can be frozen for later use.) Expressing milk also allows infants to get hindmilk and prevents gas and colic that may result from a large volume of relatively low-fat milk with high lactose content.

Engorgement

Engorgement occurs when breasts are overfilled with milk, and it is common in first-time mothers. Engorgement takes place when the supply-and-demand process is not yet established and the milk is abundant. The best way to prevent engorgement is to nurse the infant frequently. (Newborn infants will often nurse every 1 to 2 hours.) If the infant is not available to nurse, expressing milk every few hours will prevent engorgement while helping to build and maintain a milk supply.

The peak time for engorgement varies among women and can occur any time from day 2 through day 14. Women with profound, persistent engorgement are more at risk for breastfeeding difficulties and more likely to wean early.[4] Severe engorgement inhibits milk flow because the swollen tissue is compressing the milk ducts—not because the mother is failing to experience the letdown or milk ejection reflex. Once engorgement occurs, there are several simple treatments to help ease the discomfort. It is important for the mother to express milk until her breasts are no longer hard before putting the infant to breast. This will make it more comfortable for her and easier for her baby to latch on. When an infant is unable to extract milk effectively, expressing by hand or using an electric breast pump can help establish milk flow and soften the breast to make it easier for the infant to attach properly and further extract milk. Women can use analgesics to reduce pain from engorgement. A warm shower, warm compresses with massage before feedings, and expressing milk will help to relieve pressure and trigger milk flow. Application of cold compresses between feedings helps to reduce pain and swelling.[1]

Plugged Duct

Milk stasis, or milk remaining in the ducts, is considered the cause of plugged ducts.[5] Treatment for plugged ducts is gentle massage, warm compresses, and complete emptying of the breast. Women should consider changing nursing positions to facilitate emptying the breast. For example, if the woman is nursing while lying down, she may try a sitting position, or she may switch the position of holding the infant (see Illustration 6.6 in Chapter 6). When plugging occurs repeatedly, a gentle manual massage before nursing often results in the plug being expelled.

Infection

Mastitis is a bacterial infection of the breast most commonly found in breastfeeding women. The incidence in the first 7 weeks postpartum is reported at 2.9%.[4] Some women get mastitis after having cracked or sore nipples, and some get it without any noticeable problem on the surface of the breast, probably from a blood-borne source of infection. Missing a feeding or the infant sleeping through the night may precipitate engorgement, plugged ducts, and then mastitis.[4] Symptoms of mastitis are similar to those of a plugged duct (Table 7.1). In both conditions, there is a painful, enlarged, hard area in the breast, and often an area of redness on the surface of the breast. Cases of mastitis are usually accompanied by a fever and flu-like symptoms.

It is important for the mother to seek early treatment and to continue nursing through mastitis, unless it is too painful. The techniques used to minimize pain from engorgement may also be used for mastitis. Acetaminophen is commonly recommended to help with the pain. The pairing of antibiotics with emptying of the breast to treat bacterial mastitis is important.[6] In a randomized trial, half of 55 women treated only with antibiotics had breast abscess, recurrent mastitis, or symptoms lasting longer than 2 weeks compared to only 2 of 55 who also emptied their breasts (breasts can be emptied by feeding the baby or pumping). Significant delays in seeking treatment for mastitis are associated with the development of abscess and recurrent mastitis.[7] See also Case Study 7.1 on the next page.

Table 7.1 Comparison of symptoms of engorgement, plugged duct, and mastitis

CHARACTERISTICS	ENGORGEMENT	PLUGGED DUCT	MASTITIS
Onset	Gradual, most common early postpartum	Gradual after feedings	Sudden, after 10 days or more postpartum
Site	Both breasts	One breast	Usually one breast
Swelling and Heat	Generalized	May change positions, little or no heat	Red, hot, swollen area on breast
Pain	Generalized	Mild but in a specific location	Intense in a specific location
Fever	No fever	No fever	Fever (101° F)
Other Symptoms	None	None	Flu-like symptoms

SOURCE: Reprinted from Breastfeeding: A Guide for the Medical Professions, 5th Ed. by R. A. Lawrence and R. M. Lawrence, copyright © 1999, with permission from Elsevier.

Case Study 7.1

Chronic Mastitis

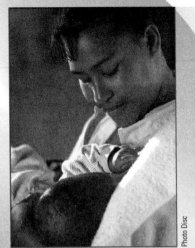

Photo Disc

This was the first and unremarkable pregnancy for 29-year-old Barbara Ann. Barbara Ann has reported experiencing "a little" breast enlargement during her pregnancy.

Her infant is first put to the breast at 2 hours postpartum, and the infant latches well and suckles vigorously. The infant nurses every 2 hours over the first 3 to 4 days postpartum. Barbara Ann's breasts became noticeably fuller during the third postpartum day, and by the fourth postpartum day they are painfully engorged. In addition, Barbara Ann reports painful, burning, cracked nipples. The engorgement makes it difficult for her baby to latch at the breast. The baby becomes irritable, and Barbara Ann experiences a significant amount of pain. A lactation consultant gives Barbara Ann guidelines for engorgement management.

On day 5, the engorgement is still causing discomfort. Barbara Ann's nipples have become more cracked and painful. The lactation consultant notes that the infant's latch has become shallow and tight, probably in an attempt to control the flow of milk. However, the infant shows all the signs of adequate intake, including 10 very wet and 3 soiled diapers during the 24 hours prior to the consultation.

By day 7 postpartum, Barbara Ann has mastitis. She is treated with a 7-day course of dicloxacillin. A lactation consultant assists her in achieving a proper infant latch.

By day 14, Barbara Ann is feeling much better. The mastitis has resolved and her nipples are healing. She still has tenderness during infant feedings and a healing crack on the right side. Her breasts are still uncomfortably full and are occasionally swollen and tender.

At 3 weeks postpartum, Barbara Ann develops an inflamed area on the right breast that remains red and tender despite applying warmth and massage to the area. The lactation consultant helps Barbara Ann to position the infant in a way that allows drainage of the inflamed area. Barbara Ann is treated with dicloxacillin. The crack on the right nipple has improved, but is still not completely healed. Barbara Ann continues to show signs of oversupply, such as breasts feeling uncomfortably full, even after feeding, and excessive milk leakage between feedings.

After 10 days of persistent burning pain in the nipple area, Barbara Ann is treated with fluconazole for a yeast infection. Seven days after starting the fluconazole, a topical nystatin ointment is prescribed for her nipples and an oral suspension for her infant.

At 7 weeks postpartum, Barbara Ann calls the lactation consultant to report another flare-up of bacterial mastitis. Her health care provider prescribes a 10-day course of dicloxacillan. Barbara Ann is still treating her nipples with nystatin ointment. At 8 weeks postpartum her mastitis resolved; her nipple pain is still present, but improving. Barbara Ann is nursing the infant on one side only per feeding and reports that the infant latches better when she is in a more reclined position.

SOURCE: Adapted from: Anonymous. Case management of a breastfeeding mother with persistent oversupply and recurrent breast infections.[108]

Questions

1. Name the causes of engorgement.
2. Name at least two recommendations the lactation consultant might make to decrease engorgement.
3. What measures (other than using the medications prescribed by her physician) should Barbara Ann take to manage the mastitis?
4. How might Barbara Ann decrease the flow rate of her milk?

(Answers are in Appendix D.)

Maternal Medications

"It is equally inappropriate to discontinue breastfeeding when it is not medically necessary to do so as it is to continue breastfeeding while taking contraindicated drugs."

R. A. Lawrence[5]

The single most common medical issue health care providers face in managing breastfeeding patients is maternal medication use.[5] Ninety to 99% of breastfeeding women receive some type of medication during their first week postpartum.[8] Most medications—including over-the-counter as well as prescription drugs—taken by nursing mothers are excreted in breast milk, yet data on drug safety are meager and sometimes conflicting. Two key questions to address in analyzing the risk of an infant's exposure to a drug excreted in breast milk are: How much of the drug is excreted in milk, and at the level of excretion, what is the risk of adverse effects?[9,10] Here are some of the numerous variables to examine when answering these questions:

- Pharmacokinetic properties of the drug
- Time-averaged breast *milk/plasma drug concentration ratio (M/P ratio)* of the drug
- Drug *exposure index*
- Infant's ability to absorb, detoxify, and excrete the drug
- Dose, strength, and duration of dosing
- Infant's age, feeding pattern, total diet, and health[10,11]

Additional considerations are the well-established interethnic and racial differences in drug responsiveness, exposure of the infant to the drug during pregnancy, and whether the drug can be safely given to the infant directly.[5] The ultimate test of drug safety is the measurement of the infant plasma drug concentration and any pharmacodynamic effects on the infant.[10] With so many active variables, carefully controlled studies on large enough samples to validate the results are rare but have increased during the last decade.

Fortunately, numerous resources (Table 7.2) based on a thorough evaluation of available evidence can assist the health care provider and mother in identifying which drugs are safe and which are not. The American Academy of Pediatrics (AAP) Committee on Drugs publishes guidelines for practitioners.[12] The guidelines provide a list of drugs divided into the following seven categories according to risk factors in relationship to breastfeeding:

1. Cytotoxic drugs that may interfere with the cellular metabolism of the nursing infant
2. Drugs of abuse for which adverse effects on the infant during breastfeeding have been reported
3. Radioactive compounds that require temporary cessation of breastfeeding
4. Drugs for which the effect on nursing infants is unknown, but may be of concern
5. Drugs that have been associated with significant effect on some nursing infants and should be given to nursing mothers with caution
6. Maternal medications usually compatible with breastfeeding
7. Food and environmental agents having no effect on breastfeeding

The list, which is updated periodically, includes only those drugs about which there is published information. Other useful monographs and review articles provide additional information on a wide array of medications.[8,9,13–16]

The Breastfeeding and Human Lactation Study Center at the University of Rochester (see Table 7.2) continually updates its database of more than 3000 references on drugs, medications, and contaminants in human milk and is a resource for complex questions on the risks to the breastfed infant. The *Physicians's Desk*

Milk/Plasma Drug Concentration Ratio (M/P Ratio) The ratio of the concentration of drug in milk to the concentration of drug in maternal plasma.[10] Since the ratio varies over time, a time-averaged ratio provides more meaningful information than data obtained at a single time point. It is helpful in understanding the mechanisms of drug transfer and should not be viewed as a predictor of risk to the infant as it is the concentration of the drug in milk, and not the M/P ratio, is critical to the calculation of infant dose and assessment of risk.[8,10]

Exposure Index The average infant milk intake per kilogram body weight per day × (the milk to plasma ratio divided by the rate of drug clearance) × 100. It is indicative of the amount of the drug in the breast milk that the infant ingests and is expressed as a percentage of the therapeutic (or equivalent) dose for the infant.[9]

Table 7.2 Resources on drugs, medications, and contaminants in human milk

- American Academy of Pediatrics, Committee on Drugs. The transfer of drugs and other chemicals into human milk (RE9403). *Pediatrics* 2001; 108:776.[12]
- G. G. Briggs, R. K. Freeman, and S. J. Yaffe. *Drugs in Pregnancy and Lactation*, 6th ed. Baltimore, MD: Williams and Wilkins, 2001.[16]
- T. W. Hale. *Medications and mother's milk*. 10th ed. Amarillo, TX: Pharmasoft Medical Publishing; 2002.[15] Dr. Hale will answer questions from health professionals posed to the Pharmasoft Web site: http://www.iBreastfeeding.com.
- M. Blumenthal, W. Busse, and A. Goldberg et al. (eds.). *The Complete German Commission E Monographs: Therapeutic Guide to Herbal Medicines*. Boston: Integrative Medicine Communications; 1998.[25]
- The Breastfeeding and Human Lactation Center, University of Rochester. This service is available for complex medication questions (9:30 a.m. to 4:00 p.m. EST, Monday to Friday at (585) 275-0088).

Reference (PDR) is not a good source for information about drugs and breastfeeding, because the information is derived directly from pharmaceutical companies whose first concern is avoiding liability. When there are no studies that prove beyond a doubt that a drug is safe for nursing mothers, the drug companies must advise against its use while breastfeeding—even if what is known about the drug suggests that there is little cause for concern.

Only a few medications are thought to be contraindications to breastfeeding: antineoplastic agents, radioactive isotopes, drugs of abuse, and drugs that suppress lactation.[11,12] Fortunately, safer alternative medications can be recommended as a substitute for most other drugs with known adverse effects on infants. Specific knowledge about a medication's safety during breastfeeding will allow proper treatment and avoid unnecessary maternal anxiety and undue risk.[13]

Many women have questions on the safety of oral contraceptive use during lactation. Preparations containing 2.5 mg or less of a 19-norprogestogen and 50 mcg or less of ethinyl estradiol or 100 mcg or less of mestranol present no problem for the mother or infant.[5] Milk yield can be decreased with larger doses. Progestin-only implants are also safe and effective during lactation.[17] Implants that deliver orally active steroids should be used only after 6 weeks postpartum to avoid transferring of steroids to the newborn.[17]

If a drug or surgery is elective, a mother may be able to delay it until the baby is weaned. If a breastfeeding mother needs a specific medication, and the hazards to the infant are minimal, she should be instructed to take the medication after breastfeeding—at the lowest effective dose, and for the shortest duration.[11,14] Other important steps can be taken to further minimize the effects (Table 7.3). It is also sometimes possible to choose alternative routes for administration of a medication to reduce exposure. For example, prescribing an inhalant instead of a drug taken by mouth, or a topical application rather than oral dosing, reduces infant exposure. If a drug is to be taken for diagnostic testing (such as a radioactive agent) a mother may need to withhold breastfeeding for a short period of time, pumping and discarding her milk. Discontinuing breastfeeding due to maternal medications is a last resort; but it may be necessary for the health and well-being of the mother, for example, if she needs chemotherapy or radioactive treatment. Any decision to limit a mother's breastfeeding must be justified by the fact that the risk to her baby clearly outweighs the benefits of breastfeeding.

Herbal Remedies

Numerous herbs have been used in folk and traditional systems of healing to affect the flow of milk (Table 7.4), or to treat mastitis, infant colic, and thrush.[18–20] However, scientific information about herb use during lactation, particularly recent studies, is sparse. The limited

---

Table 7.3 Minimizing the effect of maternal medication[5]

1. Avoid long-acting forms: Accumulation in the infant is a genuine concern because the infant may have more difficulty excreting a long-acting form of a drug, which usually requires detoxification in the liver.

2. Schedule doses carefully: Check usual absorption rates and peak blood levels of the drug, and schedule the doses so that the least amount possible gets into the milk. To minimize milk levels of most drugs, the safest time for a mother to take the drug is usually immediately after her infant nurses.

3. Evaluate the infant: Watch for any unusual signs or symptoms, such as changes in feeding pattern or sleeping habits, fussiness, or rash.

4. Choose the drug that produces the least amount in the milk.

---

pertinent safety data are based on traditional use, animal studies, and knowledge of the pharmacologic activities of the product's constituents. Medicinal herbs should be viewed as drugs, with evaluation of both their pharmacological and toxicological potential.[16,21]

A mother may perceive herbs as natural and, therefore, safe and even preferable to conventional beverages, over-the-counter medicines, or prescription drugs. However, the risks of using some herbal remedies may outweigh the potential benefits. Many herbs are far from benign and many are contraindicated during lactation (Table 7.5). Because little is known about the amount secreted in human milk or the effects on preterm or term infants, herbs that are central nervous system stimulants,

---

Table 7.4 Herbs traditionally used to affect milk production

Herbs That Promote Milk Flow
Caraway
Celery root and seed
Chaste tree berry
Fennel
Fenugreek
Goat's rue
Raspberry
Rauwolfia
Verbena

Herbs That Reduce Milk Flow
Castor bean
Jasmine flower
Sage

SOURCE: From Hardy, M. L., "Herbs of special interest to women. Journal of the American Pharmaceutical Association, 40:2234. Copyright © 2000. Used by permission.

cathartic laxatives, hepatotoxic, carcinogenic, cytotoxic, or mutagenic, or that contain potentially toxic essential oils, are not recommended during lactation.[18]

The toxic effects of herbs are often not the fault of the herb itself, but are caused by products containing misidentified plants or contaminants such as heavy metals, synthetic drugs, microbial toxins, and toxic botanicals.[4,22,23] *Medicinal herbs* in the United States are regulated as dietary supplements and are not tested for safety or efficacy.

Herbal teas that are safe for the infant and mother during lactation are presented in Table 7.6 on the next page. Lawrence[5] recommends using only herbal teas "that are prepared carefully, using only herbs for essence (e.g., Celestial Seasonings brand tea) and avoiding heavy doses of herbs with active principles." Careful attention should also be given to preparation, avoiding long steeping times.

Some culinary herbs may lead to problems when used extensively. In lactation and herbal texts, sage has a folk reputation for lowering milk supply,[18] as do parsley and peppermint, especially if the oil is taken by mouth in large doses.[23] Consumed on occasion, however, in small amounts as part of a reasonably varied diet, peppermint, parsley, sage, and other culinary herbs currently have no documented negative effect on lactation.

Although various herbal gels, ointments, or creams are often suggested for use on the nipples, any substance applied to the breast or nipples could easily be ingested by the nursing child. The use of herbal oils is not recommended.[23] In one infant, severe breathing difficulties were documented after the mother used menthol, a significant component of peppermint oil, on her nipples.[24]

Medicinal Herbs   Plants used to prevent or remedy illness.[21]

Although health care practitioners may wish otherwise, some mothers may refuse prescription drugs and insist on using herbal alternatives. If a mother is consuming a large amount of any herbal product, its contents should be checked. Important information to obtain on the product includes its name, a list of all ingredients, the names of the plants or other components (include the plants' Latin names if possible), details of the preparation, and the amount consumed.[23] Reliable sources of herbal

---

Table 7.5
Medicinal herbs considered not appropriate for use during pregnancy or lactation[22,28]

Agnus castus	Cottonroot	Heliotropium	Poplar
Aloes	Crotalaria	Hops	Prickly ash
Angelica	Darniana	Horehound, black	Pulsatilla
Apricot kernel	Devil's claw	Horehound, white	Queen's delight
Aristolchia	Dogbane	Horsetail	Ragwort
Asafoetida	Dong quai	Hydrocotlye	Red clover
Avens	Echinacea	Jamaica dogwood	Roman chaparral
Blue flag	Ephedra (ma huang)	Juniper	Sassafras
Bogbean	Eucalyptus	Liferoot	Senna
Boldo	Eupatorium	Lobelia	Shepherd's purse
Bonese	Euphorbia	Male fern	Skullcap
Borage	Fennel	Mandrake	Skunk cabbage
Broom	Feverfew	Mate	Squill
Buchu	Foxglove	Meadowsweet	St. John's Wort
Buckthorn	Frangula	Melilot	Stephania
Burdock	Fucus	Mistletoe	Tansy
Calamus	Gentian	Motherwort	Tonka bean
Calendula	German chamomile	Myrrh	Uva-ursi
Cascara	Germander	Nettle	Valerian root
Cayenne pepper	Ginseng, eleuthero	Osha	Vervain
Chamomile	Ginseng, panax	Passionflower	Wild carrot
Cohosh, black	Golden seal	Pennyroyal	Willow
Cohosh, blue	Ground ivy	Petasites	Wormwood
Coltsfoot	Groundsel	Plantain	Yarrow
Comfrey	Guarana	Pleurisy root	Yellow dock
Cornsilk	Hawthorne	Pokeroot	Yohimbine

Note: Exclusion from this list should not be considered a recommendation for safety.

Table 7.6 Herbal teas considered safe during lactation[5]

TEA	ORIGIN/USE
Chicory	Root/caffeine-free coffee substitute
Orange spice	Mixture/flavoring
Peppermint	Leaves/flavoring
Raspberry	Fruit/flavoring
Red bush tea	Leaves, fine twigs/beverage
Rose hips	Fruits/vitamin C

SOURCE: Reprinted from Breastfeeding: A Guide for the Medical Professions, 5th Ed. by R. A. Lawrence and R. M. Lawrence, copyright © 1999, with permission from Elsevier.

information[15,25–27] (see Table 7.2) or the regional poison control center may be able to identify potentially harmful pharmacological and toxicological ingredients.

In balancing the risks and benefits in a given situation, consideration should be given to the benefits of continued breastfeeding to the baby and the mother. It is also important to consider the varied nature of lactation: newborns face different risks than older babies or toddlers because of immaturity; infants consume varying amounts of human milk; mothers may be looking forward to many months of lactation yet need or desire the benefits of medicinal herbs. A few of the widely used herbs in the United States are discussed next.

Allergy Hypersensitivity to a physical or chemical agent.

Specific Herbs Used in the United States

ECHINACEA Echinacea is used for the common cold and to enhance the immune system. Insufficient reliable data are available on its entry into breast milk and effects on the infant. Gastrointestinal distress in some women has been reported. Consumption of echinacea during lactation is not recommended.[28]

GINSENG ROOT Ginseng root, widely believed to increase capacity for mental work and physical activity and to reduce stress, contains dozens of steroid-like glycosides, sterols, coumarins, flavonoids, and polysaccharides.[5] It is reported to have estrogen-like effects in some women, with mastalgia common with extended use and mammary nodularity also reported. Although there has been considerable animal experimentation with ginseng root, human data are not extensive. The lack of standardized preparations, information on dosage, and accurate recording of side effects is a problem. Because of the reported breast effects and occasional reports of vaginal bleeding, the use of ginseng during lactation may not be advisable.[4,5]

ST. JOHN'S WORT St. John's wort, widely used in the United States and Europe as a mood stabilizer and antide-pressant, has the potential to suppress lactation. One of its active ingredients, hypericin, inhibits dopamine beta-hydroxylase. This could lead to increased dopamine, increased prolactin inhibitory factor, and suppression of prolactin. No clinical studies have investigated the effect of this herb on lactation. There is little conclusive evidence on the transfer of Saint John's wort into breast milk or its effects on the newborn.[19] Antidepressants with an established safety profile during breastfeeding may be preferable alternatives.

EPHEDRA (MA HUANG) This nervous system stimulant is typically found in herbal diet aids. It is contraindicated during lactation because it may accelerate heart rate, raise blood pressure, and cause infant colic.[28]

FENUGREEK Used in teas, poultices, ointments, syrups, and as an ingredient in East Indian cooking, fenugreek is considered a galactogogue (milk production stimulant). However, there is no scientific evidence to back this claim. This herb is derived from a plant in the same family as peanuts and chickpeas and has potential for *allergy* in sensitive infants. The usual dosage, 2 to 3 capsules three times daily, is considered compatible with breastfeeding.[4]

CABBAGE LEAVES Cabbage leaves (either cool or at room temperature) have been reported to reduce breast discomfort and swelling, although it is not known how the effects are mediated.[29] In recent randomized trials, cabbage leaves and gel packs were equally effective in the treatment of engorgement, as were cabbage extract and a placebo cream.[30]

Alcohol and Other Drugs

"Avoid prescribing or proscribing it [alcohol] and . . . assist the mother in appropriately adjusting her alcohol consumption in both timing and volume."

R. A. Lawrence[5]

Alcohol

The harmful effects of alcohol consumption during pregnancy are well documented, and drinking during pregnancy is clearly not recommended. Recommendations on alcohol consumption during lactation, however, are less clear-cut and are controversial. Alcohol consumed by the mother passes quickly into her breast milk, and the effects on the breastfeeding baby are directly related to the amount the mother consumes.[31,32]

The level of alcohol in breast milk matches the maternal plasma levels at the time of the infant feeding. Peak maternal plasma and breast-milk levels are reached 30 to 60 minutes after alcohol consumption and at 60 to 90 minutes when taken with food.[5] As the alcohol clears from a mother's blood, it clears from her milk. It takes a 120-pound

Table 7.7 Alcohol and breastfeeding: time (hr:min) until zero level in milk is reached for women at different body weights[32]

MATERNAL BODY WEIGHT		ALCOHOL REMAINS IN BREAST MILK (hr:min)		
lbs.	(kg)			
100	(45.4)	2:42	5:25	8:08
120	(54.4)	2:30	5:00	7:30
140	(63.5)	2:19	4:38	6:58
160	(72.6)	2:10	4:20	6:30
180	(81.6)	2:01	4:03	6:05

SOURCE: Adapted from E. Ho and A. Collantes et al., Alcohol and breast feeding: Calculation of time to zero level in milk. *Biol Neonate* 2001;80:219–22.

woman about 2 to 3 hours to eliminate from her body the alcohol in 1 serving of beer or wine (Table 7.7).[32] The common practice of pumping the breasts and then discarding the milk immediately after drinking alcohol does not hasten the disappearance of alcohol from the milk, because the newly produced milk still will contain alcohol as long as the mother has measurable blood alcohol levels.[31]

In many cultures folklore passed down for generations encourages the use of alcohol as a galactologue that facilitates milk letdown and rectifies milk insufficiency as well as sedating and calming the fussy infant.[31] In contrast to this folklore, there is evidence of a negative dose-related impact of alcohol on maternal oxytocin levels and the milk-letdown reflex. At least a partial decrease in milk letdown is seen at alcohol levels of 1.0 to 1.5 g/kg, and at higher levels all women have a partial to complete block in milk ejection.[5] Maternal alcohol consumption affects the odor of breast milk and the volume consumed by the infant. Breastfed infants consumed, on average, 20% less breast milk during the 3 to 4 hours following their mother's consumption of an alcoholic beverage (0.3 g ethanol/kg).[31,33] Compensatory increases in intake were then observed during the 8 to 16 hours after exposure when mothers refrained from drinking.[33]

Recent studies on the impact of maternal alcohol ingestion during lactation on infant sleep patterns and psychomotor development have raised concerns about regular consumption of alcohol while lactating. In one study, 11 of 13 breastfed infants had a reduction of more than 40% in active sleep after consuming their mother's expressed breast milk flavored with alcohol (32 mg) on one testing day and expressed breast milk alone on the other.[34] All infants spent significantly less total time sleeping after consumption of the breast milk with alcohol (56.8 minutes with alcohol compared to 78.2 minutes without). This reduction was due in part to a decrease in the amount of time the infants spent in rapid eye movement (REM) sleep.[31] The investigators concluded that short-term exposure to small amounts of alcohol in breast milk produces distinctive changes in the infant's sleep-wake patterning.

In a study of 400 infants born to members of a health insurance plan, no differences in the infant's cognitive development scores were found at 1 year of age between infants whose mothers consumed alcohol while nursing and those that did not. However, Psychomotor Development Index scores at 1 year of age were slightly lower among infants who were exposed to alcohol through breast milk than among those who were not exposed.[35] For example, if a mother consumed 2 drinks daily while nursing, the score decreased by nearly 0.5 standard deviation. The association between maternal drinking and delayed motor development persisted even after the investigators controlled for more than 100 potentially confounding variables, such as maternal tobacco, marijuana, and heavy caffeine use during pregnancy and the first 3 months after delivery.[36]

Because current research does not show that occasional use (1 to 2 drinks) of alcohol is harmful to the baby, La Leche League continues to support the opinion that the occasional use of alcohol in limited amounts is compatible with breastfeeding.[37] The American Academy of Pediatrics places alcohol in the category "Maternal Medication Usually Compatible with Breastfeeding."[12] The Institute of Medicine Subcommittee on Nutrition during Lactation recommends that lactating women should be advised that if alcohol is consumed, intake should be limited to "no more than 0.5 grams of alcohol per kilogram of maternal body weight per day."[38] For a 60-kilogram (132-pound) woman, 0.5 grams of alcohol per kilogram of body weight corresponds to approximately 2 to 2.5 ounces of liquor, 8 ounces of table wine, or 2 cans of beer.[11] There is concern that prohibiting alcohol may be too restrictive, especially when the research does not support any serious impact on the baby when a mother has an occasional drink. Many feel that nursing mothers are already placed under too many restrictions and may be discouraged from initiating or continuing to breastfeed because they feel they will face too many limitations.

If a mother does choose to have a drink or two, she can wait for the alcohol to clear her system before nursing, according to the times given in Table 7.7. She can plan ahead and have alcohol-free expressed milk stored for the occasion. If she becomes engorged, she can pump her breasts as a means of comfort, and discard her alcohol-containing milk. Drinking water, resting, or pumping and discarding breast milk will not hasten the removal of alcohol from the milk, because the alcohol content of milk matches the maternal plasma alcohol levels.

Nicotine (Smoking Cigarettes)

It is not ideal to smoke and breastfeed, but it is worse to smoke and not breastfeed. Well-documented data provide clear evidence that children of smoking mothers do better if breastfed in regard to general health, respiratory illness, and risk of sudden infant death syndrome (SIDS)[5] than if they are bottle-fed. Unfortunately, women who smoke cigarettes are less likely to breastfeed than nonsmokers, are less likely to seek help with breastfeeding difficulties than nonsmokers, and are at increased risk for stopping breastfeeding by 3 months.[39] Although lower milk output has been reported among smoking mothers, it is unknown which components of cigarette smoke are responsible for the reduced milk production, and several studies provide evidence that smoking does not necessarily hinder breastfeeding.[39] There is also substantial epidemiological evidence that social and behavioral factors, and not physiological factors, are largely responsible for the lower rates of breastfeeding found among smokers.[39] Some women believe that smoking is a barrier to breastfeeding; they do not believe they could, or should, adhere to the kinds of healthy practices they think are required of mothers to breastfeed.

Nicotine levels in breast milk of women who smoke are 1.5–3.0 times higher than the levels in the mother's blood,[39,12] and the mean 24-hour nicotine concentrations in breast milk rise as cigarette consumption increases. There is no evidence to document whether this amount of nicotine presents a health risk to the nursing infant; and because breastfeeding and smoking may be less detrimental to the child than bottle-feeding and smoking are, the AAP Committee on Drugs removed nicotine (and thus smoking) from its 2001 list of drugs of abuse with adverse effects on the infant during breastfeeding.[12]

Dahlstrom and colleagues[40] estimated that the dose of nicotine in breastfeeding infants was 1 mcg per kilogram per feeding, based on data on nicotine concentrations in breast milk within 30 minutes after smoking. Women who smoke 10 to 20 cigarettes per day have 0.4 to 0.5 mg of nicotine/L in their milk.[41] The total daily systematic infant dose from breast milk is estimated at less than 6 or 10 mcg/kg/day, or 50 times less than the exposure of a 70-kg (about 160-lb) adult smoking 20 cigarettes per day or using a 21-mg nicotine patch.[41] With gradual intake over a day's time, the infant can metabolize nicotine in the liver and excrete the chemical in the kidneys. Numerous studies of nicotine and cotinine concentrations in the nursing mother and her infant confirm that although bottle-fed infants born to smoking mothers and raised in a smoking environment have significant levels of nicotine and metabolites in their urine, breastfed infants have higher levels.[5]

Women should be counseled not to smoke while nursing or in the infant's presence. Mothers who are not willing to stop smoking should cut down, consider low-nicotine cigarettes, and delay feedings as long as possible after smoking. The half-life of nicotine is 95 minutes.

When used as directed, smoking cessation aids that replace nicotine do not appear to pose any more problems for the breastfeeding infant than maternal smoking does.[41] Because transdermal nicotine (nicotine patch) provides a steady level of nicotine in plasma and thus in breast milk, the mother cannot control the level of nicotine in breast milk except by changing the strength of the patch. Mothers who use nicotine replacement therapy intermittently (gum, nasal spray, or inhalation) might minimize the nicotine in their milk by prolonging the duration between nicotine administration and breastfeeding.[41]

Marijuana

Delta-9-tetrahydrocannabinol (THC), the active ingredient in marijuana, transfers to and concentrates in breast milk and is absorbed and metabolized by the nursing infant. There is evidence from animal studies of structural changes in the brain cells of newborn animals nursed by mothers whose milk contained THC. Impairment of deoxyribonucleic acid (DNA) and ribonucleic acid (RNA) formation and proteins essential for proper growth and development has been described.[5] In one study following breastfeeding mothers and their infants for 12 months, marijuana exposure via the mother's milk during the first month postpartum appeared to be associated with a decrease in infant motor development at 1 year of age.[42] Concerns about marijuana use during lactation include the amount of the drug the infant ingests while nursing and the amount inhaled from the environment. The possible effect on DNA and RNA metabolism should discourage any maternal use, especially because brain cell development is still taking place in the first months of life. The American Academy of Pediatrics (AAP) classifies THC as a drug of abuse that is contraindicated during lactation.[12]

Caffeine

Although caffeine ingestion is a frequent concern of breastfeeding mothers, moderate intake causes no problems for most breastfeeding mothers and babies. A dose of caffeine equivalent to a cup of coffee results in breast-milk levels of 1% of the level in maternal plasma and, consequently, low levels in the infant.[5] However, because the infant's ability to metabolize caffeine does not fully develop until 3 to 4 months of age, caffeine does accumulate in the infant. Cases of caffeine excess in breastfed infants have been documented.[43] Symptoms, which include infants being wakeful, hyperactive, and fussy, did not require hospitalization and disappeared over a week's time after caffeine was removed from the maternal diet. No long-term effects of caffeine exposure during lactation have been documented.[44]

Whereas most breastfed infants can tolerate a maternal caffeine intake equivalent to 5 or fewer 5-ounce cups of coffee per day, or less than 750 ml per day, some babies may be more sensitive than others. If a mother suspects her baby is reacting to caffeine, she may try avoiding caffeine from all sources (coffee, tea, soft drinks, over-the-counter medications, chocolate) for 2 to 3 weeks.[37]

Other Drugs of Abuse

Amphetamines, cocaine, heroin, and phencyclidine hydrochloride (angel dust, PCP) are classified by the AAP Committee on Drugs[12] as drugs of abuse that are contraindicated during lactation. The AAP guidelines strongly state that these compounds and all other drugs of abuse are hazardous not only to the nursing infant but also to the mother's physical and emotional health. In addition to their adverse pharmacological effects on the mother and infant, street drugs lack standardization and may be contaminated with other active ingredients, bacteria, heavy metals, or pesticides.[11]

Environmental Exposures

Environmental pollutants in human milk have been studied since the 1950s, when the pesticide DDT was first detected in breast milk. When a mother is exposed to chemicals in the workplace or at home, she is likely to absorb and distribute these chemicals to breast milk to some extent. The presence of low levels of environmental chemicals, such as those in Table 7.8, in most human-milk samples is well documented, but the levels are generally low. Furthermore, the significance of the presence of these contaminants on the well-being of the mother and the infant is not well defined;[45] there is little conclusive data on factors related to infant exposure via breastfeeding, particularly on those with a time-dependent nature.[46] There are no established "normal" or "abnormal" levels in breast milk for clinical interpretation, and breast milk is not routinely tested for environmental exposures.[47]

Unless the mother has a high-level of occupational exposures, extreme dietary exposures (e.g., from fish in contaminated waters), or unusual residential exposures to hazardous or toxic chemicals, breastfeeding remains overwhelmingly the preferred choice compared with breast-milk substitutes.[45,48] The World Health Organization,[49] the American Academy of Pediatrics,[50] the U.S. Department of Health and Human Services,[47] and other major health organizations overwhelmingly support the importance of breastfeeding even in a contaminated world. The benefits of breastfeeding, which include high levels of antioxidants, may prove to be essential to compensate for and outweigh the risk of toxic effects from the environment.

> **Hyperbilirubinemia** Elevated blood levels of bilirubin, a yellow pigment that is a by-product of the breakdown of fetal hemoglobin.

Currently the focus of scientific concerns is being directed toward removing potentially toxic substances from the environment and on establishment of a U.S. Breast Milk Monitoring Program to track trends in exposure levels over time.[46] Encouragingly, data from several countries show a decline in the level of DDT and dioxin metabolites in human milk.[46] However, data from the United States are limited and inconclusive.

Women should be advised about how to reduce exposures that may affect breast-milk quality rather than abandoning breastfeeding for artificial methods. Women should avoid eating swordfish, king mackerel, tile fish, and shark or freshwater fish from waters reported as contaminated by local health agencies. Women should also limit exposure to chemicals such as pesticides and solvents found in paints, non-water-based glues, furniture strippers, nail polish, and gas fumes.

Neonatal Jaundice and Kernicterus

> "The AAP discourages the interruption of breastfeeding in healthy term newborns and encourages continued and frequent breastfeeding (at least eight to ten times every 24 hours)."
>
> American Academy of Pediatrics[56]

Neonatal jaundice is a yellow discoloring of the skin caused by too much bilirubin in the blood (*hyperbilirubinemia*). It is a common and usually benign condition that resolves on its own or with minimal intervention. At least 60 to 70% of term infants will become visibly jaundiced with their serum bilirubin levels exceeding 5 to 7 mg/dl (85 to 199 mol/L).[51] If hyperbilirubinemia does not resolve and becomes sufficiently severe, the elevated bilirubin levels can cause permanent neurological damage.[52]

Table 7.8 Environmental pollutants that may be found in human milk

POLLUTANT	POTENTIAL HEALTH EFFECT
• DDT, DDE	Estrogenic, antiandrogenic activity
• PCB/PCDF	Ectodermal defects, developmental delay
• TCDD (Dioxin)	Choloracne
• Chlordane	Neurotoxicity
• Heptachlor	Neurotoxicity
• Hexachlorobenzine	Hypotonia, seizures, rash
• Volatile organic compounds	
Tetracholorethylene	Hepatotoxicity
Trichlorethylene	Hepatotoxicity
Halothane	Hepatotocicity
Carbon disulfide	Neurotoxicity
• Lead	Renal, central nervous system effects
• Mercury	Central nervous system effects
• Brominated flame retardants	Thyroid disorders, brain development

SOURCE: Adapted from U.S. Department of Health and Human Services, *HHS Blueprint for Action on Breastfeeding.* 2000.

In recent years, the overall incidence of infant jaundice has risen,[53] and hyperbilirubinemia is the most frequent cause for hospital readmission during the first 2 weeks of life in the United States.[54] More infants are becoming jaundiced, and their jaundice is more severe. The Joint Commission on Accreditation of Healthcare Organizations (JCAHO),[55] the Centers for Disease Control, and the American Academy of Pediatrics[56] have all noted the increasing rates and the need for prevention, early detection, and prompt treatment. Risk factors for the development of severe hyperbilirubinemia have been identified (Table 7.9). The higher rates of breast-feeding, in conjunction with shorter postpartum hospital stays, is the leading explanation for the higher prevalence of neonatal jaundice.[51]

It is important for all health professionals to understand the causes of, risk factors for, and early signs of hyperbilirubinemia. Preventing toxicity from excessive jaundice and protecting and ensuring successful breast-feeding require an understanding of the normal and abnormal patterns and mechanisms of jaundice in the newborn period, particularly the mechanisms related to human-milk intake.[57]

Bilirubin Metabolism

Bilirubin is a by-product of the normal physiologic degradation of heme. Most heme in the neonate is derived from fetal erythrocytes. Because higher levels of hemoglobin are necessary in utero to carry the oxygen delivered to the fetus by the placenta, the normal full-term infant has a hematocrit of 50% to 60%. As soon as the infant is born and begins to breathe room air, the need for high levels of hemoglobin is gone, and excess erythrocytes are destroyed. The released hemoglobin is broken down by the reticuloendothelial system; bilirubin, an insoluble by-product of the breakdown of hemoglobin, is released into

the circulation bound to albumin or another transport protein. The insoluble form of bilirubin is removed from circulation by the liver, which conjugates bilirubin to a water-soluble form and excretes it via the bile to the stool. The balance between liver cell uptake of bilirubin, the rate of bilirubin production, and the rate of bilirubin resorption through the intestines determines the total serum bilirubin (TSB) level.

Before birth, the maternal liver is responsible for metabolism and clearance of fetal bilirubin. After birth, unique developmental factors that control the production, conjugation, and excretion of bilirubin predispose the neonate to hyperbilirubinemia.[51,58]

- Bilirubin production in the neonate is double that of an adult due to breakdown of fetal erythrocytes.
- Uptake of insoluble bilirubin by the liver is limited because of a reduction in the concentration of ligandin, a bilirubin-binding protein in the liver cell.
- Conjugation to a water-soluble form is limited in the liver because of deficient activity of uridine diphosphoglucuronosyl transferase (UDPGT), a liver enzyme responsible for bilirubin conjugation.
- Excretion of bilirubin is delayed because of an enzyme present in the intestine of the newborn, beta-glucuronidase, that converts conjugated bilirubin back into its unconjugated state, which is reabsorbed.

Physiologic versus Pathologic Newborn Jaundice

After the first 24 hours, rising bilirubin levels in healthy term infants are reflective of the physiological breakdown of fetal hemoglobin, the increased resorption of the bilirubin from the intestines, and the limited ability of the newborn's immature liver to process large amounts of bilirubin as effectively as a mature liver. Neonates tend to produce more bilirubin than they can eliminate. Prematurity magnifies this imbalance. Retention of unconjugated bilirubin by the newborn is known as normal newborn jaundice, or physiologic jaundice of the newborn.[57] Excessive bilirubin is deposited in various tissues, including the skin, muscles, and mucous membranes of the body, causing the skin to take on a yellowish color. In healthy newborns, this condition is temporary and usually resolves within a few days without treatment. In the typical newborn population, the bilirubin level rises steadily in the first 3 to 4 days of life, peaking around the fifth day, and then declining. Bilirubin levels of healthy preterm infants peak later (day 6 to day 7) and take longer to resolve. Bilirubin levels in physiologic jaundice are usually less than 12 mg of bilirubin per dL of blood of infants of white or black mothers, and average 10 to 14 mg in infants of Asian ancestry, including Chinese, Japanese, and Korean, and Native Americans.[58]

Table 7.9 Risk factors for severe hyperbilirubinemia[55,60]

- Jaundice in the first 24 hours of life
- Jaundice before discharge
- Previous jaundiced sibling
- Preterm birth (35–38 weeks gestation)
- Exclusive breastfeeding
- East Asian race
- Bruising, cephalohematoma, vacuum extraction birth
- Maternal age > 25 years
- Male gender
- Macrosomic infant of diabetic mother
- Oxytocin used in labor
- ABO incompatibility

In contrast to physiologic jaundice of the newborn, pathologic jaundice begins earlier (sometimes before 24 hours of age), rises faster, and lasts longer. A TSB greater than 8 mg/dL in the first 24 hours should be investigated for pathologic origin. Causes of pathologic jaundice include the following:[51]

- Hemolytic disease (immune disorders, Rh isoimunization, ABO or minor blood type incompatibility)
- Erythrocyte disorders (glucose-6-phosphate-dehydrogenase deficiency, hereditary spherocytosis)
- Extravasation of blood (cephalohematoma, subgaleal hemorrhage, bruising)
- Inborn errors of metabolism/conjugation defects (galactosemia, Crigler-Najjar syndrome types 1 and II, Gilbert's syndrome, Lucey-Driscoll syndrome)
- Hypothyroidism
- Polycythemia
- Macrosomic infant of diabetic mother
- Intestinal obstruction; delayed passage of meconium
- Sepsis

In most cases of pathological jaundice, frequent breast-feeding (10 to 12 times every 24 hours) can continue during diagnosis and treatment.[56] An advantage of colostrum and mature human milk is stimulation of bowel movements, speeding elimination of bilirubin.[5] However, in jaundice caused by galactosemia, breastfeeding is contraindicated.

Because bilirubin is a cell toxin, concern arises when TSB elevates to levels with the potential to cause permanent damage. The brain and brain cells, if destroyed by bilirubin deposits, do not regenerate.[59] *Bilirubin encephalopathy,* or *kernicterus,* has a mortality rate of 50%, and survivors usually are burdened with severe problems including cerebral palsy, hearing loss, paralysis of upward gaze, and intellectual and other handicaps.[60] Although full-scale kernicterus is rare, there has been an increase in reported cases in the last 15 years.[52] There is also concern that mild effects of bilirubin on the brain may be manifested clinically in later life with symptoms such as incoordination, hypertonicity, and mental retardation, or perhaps learning disabilities.[5,59]

Hyperbilirubinemia and Breastfeeding

Jaundice in the breastfed infant has been divided into types, early or late, based on the age of onset (Table 7.10). It is important to differentiate between the two to establish effective prevention and treatment. Early onset of elevated unconjugated bilirubin unexplained by other pathologic factors is associated with inadequate feeding and is called "breastfeeding jaundice" or more precisely, "breast-non-feeding jaundice."[57] Late onset (after day 5), prolonged elevation of unconjugated bilirubin associated with the ingestion of breast milk is called breast-milk jaundice.[57,61]

BREAST-NON-FEEDING JAUNDICE Differences in bilirubin levels between adequately breastfed infants and formula-fed infants have not been found to be significant.[57] However, infants who nurse infrequently or inefficiently ingest fewer calories and lose more weight (more than 8% of birthweight) than formula-fed infants are at risk for elevated bilirubin levels. Suboptimal breastfeeding can delay passage of meconium and reduce fecal weight, increasing the enterohepatic circulation of bilirubin. In addition, reduced milk intake produces a state of partial starvation in the infant, which further increases the intestinal absorption of bilirubin.[61] Delay in initiation of breastfeeding beyond the first hour of life, and administration of water to infants either before initiation of breastfeeding or in addition to breastfeeding, significantly reduce the frequency of breastfeeding and increase bilirubin concentrations. Excessive hyperbilirubinemia causes lethargy and poor feeding in some infants,

> **Kernicterus or Bilirubin Encephalopathy** The end result of very high untreated bilirubin levels. Excessive bilirubin in the system is deposited in the brain, causing toxicity to the basal ganglia and various brain-stem nuclei.[60]

Table 7.10 Comparison of early and late jaundice associated with hyperbilrubinemia while breastfeeding

	Early Jaundice Occurs 2–5 Days of Age	Late Jaundice Occurs 5–10 Days of Age
Onset	Transient: 10 days	Persist >1 month
Incidence	More common with first child Approximately 60% of newborns in the United States become clinically jaundiced	All children of a given mother
Feeding	Infrequent feeds at breast Receiving water or dextrose water	Milk volume not a problem No supplements
Stools	Stools delayed and infrequent	Normal stooling
Total Serum Bilirubin Treatment	Peaks ≤15 mg/dL None or phototherapy	May be >20 mg/dL Phototherapy Discontinue breastfeeding temporarily (4 to 12 hours) Rarely, exchange transfusion
Associations	Low Apgar scores, water or dextrose water supplement, prematurity	None identified

SOURCE: Reprinted from Breastfeeding: A Guide for the Medical Professions, 5th Ed. by R. A. Lawrence and R. M. Lawrence, copyright © 1999, with permission from Elsevier.

further reducing breastfeeding frequency, duration, and milk production, and promoting a vicious cycle of increasing bilirubin levels.

It is now well established that early onset breastfeeding jaundice results from reduced volume of milk transfer to the infant, limiting caloric intake and producing a state of partial starvation and weight loss—the equivalent of the adult disorder known as starvation jaundice. Lawrence outlines treatment guidelines (Table 7.11) aimed at treating the actual cause; that is, failed breastfeeding or inadequate stooling or underfeeding.[5] The goal is to evaluate the breastfeeding for frequency, length of suckling, and apparent supply of milk, and then adjust the breastfeeding to solve the problem.[61] If stooling is an issue, the infant should be stimulated to stool. If starvation is the problem, the infant should receive temporary supplemental feeding by cup or bottle while the milk supply is being increased by better breastfeeding techniques.

Early discharge of infants from the hospital at less than 72 hours of life has raised concerns about the ability to evaluate breastfeeding and the opportunity to evaluate infants for jaundice.[55] Formal observation of breastfeeding with evaluation of effectiveness of breastfeeding and milk transfer at regular intervals throughout the first days of life can identify breastfeeding problems sufficiently early to ensure correction of problems. The AAP strongly recommends that all breastfed infants be evaluated by a trained observer within 2 or 3 days after discharge from the hospital.[62] Particular attention must be paid to infants who are <38 weeks gestation.[60]

Table 7.11 Management outline for early jaundice while breastfeeding

1. Monitor all infants for initial stooling. Stimulate stool if no stool in 24 hours.
2. Initiate breastfeeding early and frequently. Frequent, short feeding is more effective than infrequent, prolonged feeding, although total time may be the same.
3. Discourage water, dextrose water, or formula supplements.
4. Monitor weight, voidings, and stooling in association with breastfeeding pattern.
5. When bilirubin level approaches 15 mg/dL, stimulate stooling, augment feeds, stimulate breast milk production with pumping, and use phototherapy if this aggressive approach fails and bilirubin exceeds 20 mg/dL.
6. Be aware that no evidence suggests early jaundice is associated with "an abnormality" of the breast milk, so withdrawing breast milk as a trial is only indicated if jaundice persists longer than 6 days or rises above 20 mg/dL or the mother has a history of a previously affected infant.

SOURCE: Reprinted from Breastfeeding: A Guide for the Medical Professions, 5th Ed. by R. A. Lawrence and R. M. Lawrence, copyright © 1999, with permission from Elsevier.

BREAST-MILK JAUNDICE SYNDROME In contrast to physiological newborn jaundice, which peaks in the third day and then begins to drop, breast-milk jaundice syndrome becomes apparent after the third day, and bilirubin levels may peak any time from the seventh to the tenth day, with untreated cases being reported to peak as late as the fifteenth day.[5] In breast-milk jaundice syndrome, no correlation exists with weight loss or gain, and stools are normal. Initially breast-milk jaundice syndrome was thought to be an unusual and distinct type of newborn jaundice affecting 1% of all breastfed neonates. More recent research reports demonstrated that at least one-third of all breastfed infants are clinically jaundiced in the third week of life, and that two-thirds have significant unconjugated hyperbilirubinemia in the third week, in contrast to the absence of hyperbilirubinemia in the third week in full-term artificially fed infants.[61] What was once believed to be a clinical disorder is now recognized as a normally occurring extension of physiologic jaundice of the newborn.[60,61] However, at this time there is insufficient evidence to support the popular theory that breast-milk jaundice may provide protective effects for newborns by the antioxidant effects of bilirubin, compensating for the relative deficiency of endogenous antioxidants in newborns.[57]

The undisputed cause of breast-milk jaundice syndrome is unresolved. It is believed to be caused by a combination of factors: a substance in most mothers' milk that increases intestinal absorption of bilirubin and individual variations in the infant's ability to process bilirubin.[5,57] To treat severe elevations in unconjugated bilirubin in breast-milk jaundice syndrome, the AAP Hyperbilirubinemia Management Guidelines (Table 7.12) are applied with the goal of lowering TSB bilirubin levels substantially over 20 mg/dL promptly.[5,56,60] To establish the diagnosis of breast-milk jaundice firmly when the bilirubin level is above 16 mg/dL for more than 24 hours, a short, temporary interruption of breastfeeding (12 to 24 hours) while monitoring bilirubin levels is recommended.[5,57]

The belief that severe hyperbilirubinemia in breastfed infants cannot result in kernicterus is erroneous and dangerous; 98% of the 105 cases of kernicterus in the U.S. Kernicterus Registry were breastfed infants.[52] Kernicterus, while rare, can develop in otherwise healthy full-term breastfed newborns or breastfed infants with sepsis. Breastfed infants need to be followed closely, supported effectively, and evaluated appropriately in order to avoid rare cases of severe hyperbilirubinemia and kernicterus.[52]

INTERRELATIONSHIPS BETWEEN BREAST-NON-FEEDING JAUNDICE AND BREAST-MILK JAUNDICE SYNDROME Although breast-non-feeding jaundice and breast-milk jaundice are two separate entities, they can have an interactive effect on each other. Infants with breast-milk jaundice who manifest higher levels of bilirubin in the second and third weeks of life, often over 15 mg/dl, have been noted to have had relatively high serum bilirubin concentrations during the first

3 to 5 days of life due either to breast-non-feeding jaundice, hemolysis, or unknown etiology.[61] Gartner postulates that these early, elevated bilirubin levels may produce an enlarged bilirubin pool. Then the ingestion of mature milk and a consequent enhancement of the enterohepatic circulation may enlarge the pool even further.[61]

Treatment for Jaundice

Table 7.12 shows the American Academy of Pediatrics guidelines for the management of hyperbilirubinemia in healthy term newborns.[56] Phototherapy involves placing the newborn under special fluorescent lights that, like sunlight, assist in removing jaundice from the skin. The light is absorbed by the bilirubin, changing it to a water-soluble product, which can then be eliminated without having to be conjugated by the liver.

Historically, treatment for jaundice in American hospitals involved phototherapy and discontinuing breastfeeding either permanently or until the bilirubin levels were acceptable. In addition, many health professionals believed that newborn infants would become dehydrated if they were not supplemented with water or formula during the first days of breastfeeding. Recent research shows these practices are counterproductive.[52,57] The benefits of early and frequent breastfeeding in the first days of life to prevention of hyperbilirubinemia through maintaining hydration and stimulating the passage of stool are now well documented. The passage of stool in the newborn is important because there are 450 mg of bilirubin in the intestinal tract *meconium* of the average newborn.[5] To avoid reabsorption of bilirubin from the gut into the serum, passing meconium in the stool is critical. The current AAP hyperbilirubinemia management guidelines state: "The AAP discourages the interruption of breastfeeding in healthy term newborns and encourages continued and frequent breastfeeding (at least 8 to 10 times every 24 hours). Supplementing nursing with water or dextrose water does not lower the bilirubin level in jaundiced, healthy breastfeeding infants."[56,60]

Information for Parents

Health professionals should convey a balanced approach when communicating with parents about jaundice. Parents need to know that most breastfed infants will become jaundiced and that the overwhelming majority of cases will be benign. Only a small fraction of these infants are at risk for developing extreme hyperbilirubinemia and kernicterus. However, parents need to fully understand the serious consequences of extremely elevated bilirubin levels and should have their infants evaluated by a health professional if jaundice develops. Health professionals need to understand that feelings of guilt are common among mothers of jaundiced infants because many mothers feel that they caused the jaundice by breastfeeding.[63,64] By providing accurate information and encouragement to breastfeed, health professionals have great influence on whether a mother continues breastfeeding after her experience with neonatal jaundice.

Meconium Dark green mucilaginous material in the intestine of the full-term fetus.

Breastfeeding Multiples

The benefits of breastfeeding to mother and infant are multiplied with twins and higher-order multiples, who often are born at risk.[65] History and numerous case reports[66–68] offer ample evidence that an individual mother can provide adequate nourishment for more than one infant. In seventeenth-century France, wet nurses in foundling homes fed from three to six infants, who were

Table 7.12 Management of hyperbilirubinemia in the healthy term newborn

INFANT AGE (hours)	TOTAL SERUM BILIRUBIN (TSB) LEVEL, mg/dL (pmol/L)			
	Consider Phototherapy*	Phototherapy	Exchange Transfusion If Intensive Phototherapy Fails**	Exchange Transfusion and Intensive Phototherapy
=≤24***	......	....	......	....
25–48	≥12 (170)	≥15(260)	≥20 (340)	≥25 (430)
49–72	≥15 (260)	≥18 (310)	≥25 (430	≥30 (510)
>72	≥17 (290)	≥20 (340)	≥25 (430)	≥30 (510)

*Phototherapy at these TSB levels is a clinical option, meaning that the intervention is available and may be used on the bases of individual clinical judgment.
**Intensive phototherapy should produce a decline of TSB of 1 to 2 mg/dL within 4 to 6 hours and the TSB should continue to fall and remain below the threshold level for exchange transfusion. If this does not occur, it is considered a failure of phototherapy.
***Term infants who are clinically jaundiced at =≤24 hours old are not considered healthy and require further evaluation.
SOURCE: Adapted from American Academy of Pediatrics Provisional Committee for Quality Improvement and Subcommittee on Hyperbilirubinemia. Practice parameter: Management of hyperbilirubinemia in the healthy term newborn. Pediatrics 94(4):558–565. Copyright © 1994. Reprinted by permission of the American Academy of Pediatrics.

often of differing ages with different daily requirements.[5,69] Breastfeeding initiation rates of nearly 70% have been reported by surveys of members of Mothers of Super Twins (MOST), Parents of Multiple Births Association (POMBA) of Canada, and Double Talk, a newsletter for parents of multiples. Some mothers of triplets and quadruplets have fully breastfed their babies.[66]

Frequency and effectiveness of breastfeeding are the keys to building a plentiful milk supply. The more often a baby nurses, the more milk there will be.[5,65] Mothers who exclusively breastfeed twins or triplets can produce 2 to 3 kg/day, although this involves nursing an average of 15 or more times per day.[70] The main obstacle to nursing multiples is not usually the milk supply, but time and fatigue of the mother. Parents of twins and higher-order multiples need support in four major areas: organization, feeding, individualization, and stress management.[65]

> **Food Allergy (Hypersensitivity)** Abnormal or exaggerated immunologic response, usually immunoglobulin E (IgE) mediated, to a specific food protein.

Mothers of twins or higher-order multiples often face special challenges in the intrapartum establishment of lactation.[66] Breastfeeding initiation may take place in the neonatal intensive care unit, usually because of prematurity and low birthweight. Mothers may be coping with the effects of a more physically demanding pregnancy and birth or complications of pregnancy. Mothers may experience exaggerated postpartum sleep deprivation related to round-the-clock care of two or more newborns, concern for sick newborns, or staggered infant discharge, which results in time divided between infants at home and in the hospital. In addition, every aspect of breastfeeding management is affected by the dynamics that multiple newborns create.[5,66]

Health care professionals can help the mothers of multiples face the many challenges of breastfeeding by offering consistent, informed, individualized care and support in the hospital and after discharge.[71] Knowledgeable care providers can help parents distinguish between multiples-specific issues, normal variations in individual infant's breastfeeding abilities and patterns, and actual breastfeeding problems. Mothers need information on when and how to initiate simultaneous feedings, practical tips for managing nighttime nursing and fatigue, and how to assure themselves that their babies are getting ample nourishment.[66] Parents need to be informed of resources for parenting multiples and for receiving support for breastfeeding in their communities (Table 7.13).

A well-defined plan for the lactating woman's health care, including screening for nutritional problems and dietary guidance, is also important.[38] Mothers should be encouraged to drink to satisfy their thirst, to eat nutritious foods, and to sleep when the babies sleep. As in singleton nursing, women nursing multiples should be encouraged to obtain their nutrients from a well-balanced, varied diet rather than from vitamin-mineral supplements.

Table 7.13 Resources for the parents of multiples

- www.lalecheleague.org/bfmultiple.html
- www.tripletconnection.org/oldroot/brstfeed.html
- K. K. Granada, *Mothering Multiples: Breastfeeding and Caring for Twins and More!!!* Rev. ed. Schaumburg, IL: La Leche League International, 1999.

Infant Allergies

Protection from allergies one of the most important benefits of breastfeeding. There is accumulating evidence that exclusive breastfeeding throughout the first 6 months of life can protect against the development of asthma and other allergic diseases.[72-74] Eliminating major food allergens from the diets of infants at high risk for atopic disease, and from the diets of their lactating mothers, can delay or prevent some food allergy and atopic dermatitis.[5,50,75,76] Several mechanisms (Table 7.14) are thought to contribute to the protective effect of breastfeeding.

The development of infant *food allergy (hypersensitivity)* is influenced by genetic risk for allergy, duration of breastfeeding, time for introduction of other foods, maternal cigarette smoking during pregnancy and parental smoking, air pollution, exposure to infectious disease, and maternal diet and immune systems.[72]

Common pediatric food allergens include:

- Cow's milk
- Wheat
- Eggs
- Peanuts
- Soybeans
- Tree nuts (e.g., almonds, Brazil nuts, walnuts, hazelnuts)

Infants with a positive family history of allergies should be exclusively breastfed for 6 months, with continuance of breastfeeding for as long as possible.[5,50,72] Breastfeeding mothers with a family history of allergies also should be counseled to avoid common allergens in their own diets.

Table 7.14 Possible reasons for allergy-preventive effects of breastfeeding[72,76]

- Low content of allergens
- Transfer of maternal immunity
- Long-chain fatty acids and IGA in breast milk protect against inflammation and infections
- Regulation of infant immunity
- Influence on gut microbial flora

Several studies have documented the presence of food allergens, particularly milk and peanut protein, in breast milk sufficient to induce food reactions in infants.[77] The American Academy of Pediatrics specifically suggests restricting milk, egg, fish, peanuts, and tree nuts in the maternal diet if symptoms of food allergy are noted in the infant.[78] If there is no history of allergy to a specific food in the mother's or father's family, avoiding a food because it is a potential allergen is an unnecessary precaution. Only if there is a family history of allergies, or if a baby shows allergic symptoms, should a mother consider avoiding specific foods. If a mother avoids certain foods, she must be careful to ensure that her diet remains nutritionally adequate.[77]

Food Intolerance

Although infants may have sensitivities to certain foods, there is no scientific basis for the concern about gassy foods, such as cabbage or legumes, causing gas in the breastfed baby. In mothers, the normal intestinal flora produces gas from the action on fiber in the intestinal tract. Neither the fiber nor the gas is absorbed from the intestinal tract, and neither enters the mother's milk. Likewise, the acid content of the maternal diet does not affect the breast milk, because it does not change the pH of the maternal plasma.

Characteristic essential oils in foods such as garlic and spices may pass into the milk, and an occasional infant objects to their presence. Studies by Mennella and Beauchamp confirm that the diet of a lactating woman alters the sensory qualities of her milk.[79,80] Extensive clinical experience also suggests that some infants are sensitive to certain foods in the mother's diet. According to Lawrence, garlic, onions, cabbage, turnips, broccoli, beans, rhubarb, apricots, or prunes may be bothersome to some infants, making them colicky for 24 hours.[5] In the summer, a heavy diet of melon, peaches, and other fresh fruits may cause colic and diarrhea in the infant. Red pepper has been reported to cause dermatitis in the breastfed infant within an hour of milk ingestion.[81] Contrary to popular belief, chocolate rarely causes problems and can be consumed in moderation without causing colic, diarrhea, or constipation in most infants.[5]

If a mother suspects that her baby reacts to a specific food, it may be helpful for her to keep a record of foods eaten, along with notes on the baby's symptoms or behavior. If highly allergic or sensitive, infants may react to foods their mothers have eaten within minutes, although symptoms generally show up between 4 and 24 hours after exposure. Symptoms will improve in most infants after the offending food has been removed from the mother's diet for 5 to 7 days, but it may take 2 weeks to totally eliminate all traces of the offending substance from both the mother and baby.[82]

Near-Term Infants

Infants born near 37 weeks may have subtle immaturity that makes establishing breastfeeding difficult and places the infant at risk for insufficient milk intake, hypoglycemia, jaundice, and poor weight gain.[83] Infants may have cardiorespiratory instability, especially in an upright position; poor temperature control; lower glycogen and fat stores to prevent hypoglycemia; and immature immune systems; and their suck-swallow coordination may be poorly coordinated and result in poor latch-on and milk transfer.

The main emphasis in postpartum care should be on building and maintaining the milk supply and feeding the infant. If an infant is not sucking vigorously, mothers of near-term infants should pump after each feeding attempt, or at least every 3 hours, to build the milk supply. A specific feeding plan, including waking a sleepy infant every 2–4 hours, is recommended to avoid the near-term breastfeeding cascade (Illustration 7.1). Mothers and their near-term infants should have a feeding assessment by a trained lactation professional who can provide intervention for positioning, potential suck problems, and other breastfeeding issues. Weighing an infant before and after feedings may be useful to determine milk transfer. Discharge planning should include follow-up care, home visits, and lactation counseling as needed. Close follow-up should be continued until the infant is gaining weight and the mother is comfortable.[83]

Human Milk and Preterm Infants

"Human milk is the preferred feeding for all infants, including premature and sick newborns, with rare exceptions. The ultimate decision on feeding of the infant is the mother's. Pediatricians should provide parents with complete, current information on the benefits and methods of breastfeeding to ensure that the feeding decision is a fully informed one. When direct breastfeeding is not possible, expressed human milk, fortified when necessary for the premature infant, should be provided."

AAP[62]

The benefits of breastfeeding may be most visible among preterm infants, who are born immature and without adequate stores of nutrients. The nutritional benefits include ease of protein digestion, fat absorption, and improved lactose digestion.[62,84] The known health and developmental benefits include better visual acuity,[85] greater motor and mental development at 1.5 years of age,[86] greater verbal intelligence quotient at 7–8 years of age,[87] and a lower incidence of serious infectious disease—including necrotizing enterocolitis and sepsis—even among infants who also receive some human-milk substitutes.[86–89] Lower weight and length gains and

Case Study 7.2

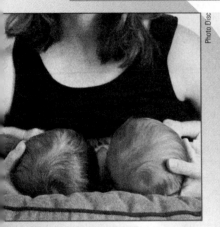

Photo Disc

Breastfeeding Premature Infants

At 30 weeks gestation, 35-year-old Stacey delivers twin boys: baby Andrew is 2 pounds, 9 ounces and 13.5 inches long: baby Mark is 1 pound, 13 ounces and 14 inches long. Stacey had a difficult pregnancy that included severe nausea and vomiting, heartburn, preeclampsia, and preterm labor.

Stacey is very committed to breastfeeding and was able to use an electronic breast pump approximately 18 hours after delivery. She pumps successfully every 3–4 hours or 7–8 times daily in order to establish her milk supply. At the end of the first week, Stacey is pumping about 14–16 oz daily and by 3½ weeks she is pumping 2–2.5 oz per breast at each pumping. Stacey wakes at night to pump when she is engorged. She has placed the pumping equipment by the bed and become adept at pumping, getting out of bed only to put the milk in the refrigerator.

At 2 weeks postpartum, Stacey experiences a plugged duct and has difficulty emptying the right breast for 2 days.

The twin boys have suffered the usual preterm difficulties with breathing, apnea, and bradycardia. Initially the twins are tube-fed. As their condition improves, baby Andrew is first put to breast 3 weeks after his birth, and baby Mark several weeks later. Baby Mark has more difficulty learning to latch on and suck and is growing more slowly than is his brother. Multiple interventions are used to achieve breastfeeding success. On advice from a lactation consultant, a nipple shield helps baby Mark latch on. In response to slow weight gain in baby Mark, the lactation consultant recommends that the baby receive hindmilk, which is often higher in fat and calories.

Mark and Andrew are released from the hospital a day after their due date. Mom continues using the nipple shield for several weeks with Mark, trying without the nipple shield every few days. After 3 weeks at home, baby Mark is able to latch on without the nipple shield. The twins are fed human milk and up to 3 bottles of premature infant formula for the first 2 months at home. The babies take feedings equally well from bottle or breast.

Questions

1. Describe the appropriate steps that a mother of a premature infant should take to establish a good milk supply by pumping.
2. What are the causes of plugged ducts?
3. What are the treatments for plugged ducts?
4. Describe how a mother would go about feeding a baby only hindmilk.

(Answers are in Appendix D.)

---

poorer bone mineralization have been reported for preterm infants fed human milk.[90,91]

The composition of milk from women who deliver preterm infants is higher in protein, slightly lower in lactose, and higher in energy content (58–70 cal/oz) compared to the milk of women who deliver full-term infants (approximately 62 cal/oz). Although preterm milk has higher calcium and phosphorus than term milk, study of human milk in the 1980s found that calcium absorption from human milk did not meet the level of fetal growth that would have been expected. Lawrence and Lawrence[5] suggest that human milk is not adequate to meet the needs for fat-soluble vitamins (vitamins A, D, E and K) and vitamin C if the infant receives protein supplements. Concerns about bone mineralization and growth of infants fed unfortified human milk led to fortification of preterm human milk to prevent development of rickets, and to increase the calorie and nutrient content, as recommended by the American Academy of Pediatrics.[62] Infants who receive fortified human milk do not need additional sup-

Illustration 7.1 Near-term infant breastfeeding cascade.

Near-Term Breastfeeding Cascade

SOURCE: Reprinted with permission from N. E. Wight, "Breastfeeding the borderline (near-term) preterm infant," in Pediatrics Annals, 32:329–336. Copyright © 2003 Slack Inc.

plementation unless a specific nutritional problem is identified. The health benefits, particularly reduction in sepsis and necrotizing enterocolitis, outweigh the slightly lower rate of weight gain and length gain observed among preterm infants fed fortified human milk compared to those who receive a human-milk substitute (HMS) for premature infants.[90]

Early feeding seems to be important for preterm infants (when medically appropriate). Early feeding may be important to the ability to digest[84] and to the development of the infant's digestive system.[92] Often, milk must be expressed and stored for preterm infants. Establishing a milk supply early seems to be important to the mother's ability to maintain a milk supply that will meet her infant's demands after several weeks (see Chapter 6, "Can Women Make Enough Milk?"). A woman who is pumping less than 750 mL of milk by 2 weeks may need additional support to establish a milk supply that will meet her infant's needs beyond the first month. See Case Study 7.2 on the facing page.

Medical Contraindications to Breastfeeding

Few medical problems in the mother or baby are absolute contraindications to breastfeeding (Table 7.15 on the next page). Very few infectious pathogens pose a risk to the newborn that outweighs the potential benefits of breastfeeding.[11,93,94] Even infants with metabolic disorders such as phenylketonuria can continue to breastfeed in combination with a specialized formula to meet calorie and protein needs. When mothers or infants have medical or other problems that cause a poor suck or other feeding prob-

lems, early identification and appropriate support from a lactation consultant are necessary for successful breastfeeding. In some cases, pumping milk may be required to maintain a supply of milk until problems with the infant's suck can be addressed and corrected. Whenever a medical situation presents a potential risk for breastfeeding infants, the theoretical risk must be carefully measured against the projected benefits of breastfeeding.[94]

Breastfeeding and HIV Infection

The transmission of HIV type 1 (HIV-1) from mother to child through breastfeeding is well documented and causes one-third of all pediatric HIV infections.[95] Reports of transmission rates range between 5 and 20%, with rates as high as 15 to 53% when the mother acquires the HIV infection just before or during the breastfeeding period.[94] Factors contributing to these variable rates include strain of HIV, maternal illness, immune status and viral load, duration of breastfeeding (timing of transmission), primary infection of the mother during the breastfeeding period, exclusive breastfeeding versus mixed feeding, mastitis, and the availability of antiretroviral therapy.[94,96]

In developed countries, where safe and affordable breast-milk substitutes are available, HIV-infected women should be counseled strongly not to breastfeed.[97] The U.S. Department of Health and Human Services's Blueprint for Action on Breastfeeding states that "HIV-infected women in the United States should not breastfeed or provide their breast milk for the nutrition of their own or other infants because of the risk of HIV transmission to the child."[47]

The choice for women with HIV in developing countries is not so clear. In most cases, breast-milk substitutes are not affordable to families or to government-sponsored public health programs, and they pose a serious health risk to infants both with and without HIV.[96] Women and infants in developing countries who are most at risk for HIV also face poor water quality and sanitation and are at high risk for diarrheal diseases and other infections. Boiling of water for decontamination presents another obstacle because obtaining fuel to boil water is either difficult or expensive. A recent WHO-sponsored meta-analysis documented a sixfold increase in mortality due to diarrheal disease in the first 6 months of life for infants who were not breastfed, when compared with breastfed infants, and twice the risk of pneumonia deaths.[98] In certain populations, the benefits of breastfeeding may outweigh the risks of HIV transmission. Available data on the impact of breastfeeding on the mortality of HIV-infected mothers are insufficient and contradictory.[93]

Originally, the World Health Organization (WHO) recommended that in developing countries or areas where the risk of infant mortality from infection is great, breastfeeding

Table 7.15 Summary of medical contraindications to breastfeeding in the United States

PROBLEM	BREASTFEEDING	CONDITION
Infectious Diseases		
Acute infectious disease	Yes	Respiratory, reproductive, gastrointestinal infection
HIV	No	HIV positive in developed countries
Active tuberculosis	Yes	After mother has received 2 or more weeks of treatment
Hepatitis		
A	Yes	As soon as mother receives gamma globulin
B	Yes	After infant receives HBIG, first dose of hepatitis B vaccine should be given before hospital discharge
C	Yes	If no co-infections (e.g., HIV)
Venereal warts	Yes	
Herpes viruses		
Cytomegalovirus	Yes	
Herpes simplex	Yes	Except if lesion on breast
Varicella-zoster (chickenpox)	Yes	As soon as mother becomes noninfectious
Epstein-Barr	Yes	
Toxoplasmosis	Yes	
Mastitis	Yes	
Lyme disease	Yes	As soon as mother initiates treatment
HTLV-I	No	
Over-the-Counter/Prescription Drugs and Street Drugs		
Antimetabolites	No	
Radiopharmaceuticals		
Diagnostic dose	Yes	After radioactive compound has cleared mother's plasma
Therapeutic dose	No	
Drugs of abuse	No	Exceptions: cigarettes, alcohol
Other medications	Yes	Drug-by-drug assessment
Environmental Contaminants		
Herbicides	Usually	Exposure unlikely (except workers heavily exposed to dioxins)
Pesticides		
DDT, DDE	Usually	Exposure unlikely
PCBx, PBBs	Usually	Levels in milk very low
Cyclodiene pesticides	Usually	Exposure unlikely
Heavy metals		
Lead	Yes	Unless maternal level >40 mg/dL
Mercury	Yes	Unless mother symptomatic and levels measurable in breast milk
Cadmium	Usually	Exposure unlikely
Radionuclides	Yes	Risk greater to bottle-fed infants
Metabolic Disorders		
Galactosemia	No	
Pheynylketonuria	Yes	Human milk supplemented with phenylalanine-free formula

SOURCE: Reprinted from Breastfeeding: A Guide for the Medical Professions, 5th Ed. by R. A. Lawrence and R. M. Lawrence, copyright © 1999, with permission from Elsevier.

is recommended even if the mother has *AIDS*.[99] This policy was clarified at a meeting of the WHO, the Children's Charity UNICEF, and UNAIDS in Geneva in October 2000. The policy now states: "When replacement feeding is acceptable, feasible, affordable, sustainable, and safe, avoidance of all breastfeeding by HIV infected mothers is recommended. Otherwise, exclusive breastfeeding is recommended during the first months of life."[97]

Recent clinical trials in sub-Saharan African populations suggest that targeted breast-milk replacement and early weaning strategies could be combined with effective antiretroviral to decrease total mother-to-infant HIV-1 transmission and promote child survival.[100] Insufficient data are available concerning the treatment of breast milk by freezing

AIDS = acquired immunodeficiency syndrome.

or rapid, brief heating to inactivate HIV-1.[94] Pasteurization (65°C for 30 minutes) will usually inactivate HIV.[93-93]

Detailed instructions on counseling HIV-infected women are available.[101] All women should be encouraged to know their HIV status and seek early prenatal care. Women need to be aware of the risks of HIV transmission during pregnancy and lactation.

Where the risk of mortality from other infections is not great, mothers with HIV should be counseled—if it is feasible—to obtain milk from a breast-milk bank, where screened mothers donate milk and the milk is pasteurized.[93,102] The possibility of providing the mother's own treated, expressed breast milk to the baby at risk of HIV infection via breastfeeding is an alternative that needs to be fully explored.

Human-Milk Collection and Storage

"Human milk is the most appropriate food for infants, and is also used as medical therapy for older children and adults with certain medical conditions. Human milk has a long history and proven track record both as nutrition and therapy."

Human Milk Banking Association of North America

The appropriate collection and storage of human milk is important whether the milk is for the mother's own infant or to be donated. All of the collection containers and tubing used should be cleaned by dishwasher or sterilized by boiling. Hand pumps, electric handheld pumps, hospital-grade electric breast pumps, and manual expression can be used to extract the milk.

Milk that is used for a mother's own baby is safe at room temperature for 4 hours and can be refrigerated at 0 to 4°C (32 to 39°F) for up to 8 days without significant bacterial growth.[103,104] (Refrigeration inhibits bacterial growth, whereas freezing does not.) At home, milk can be frozen for 3 months in a self-defrosting refrigerator and up to 12 months in a freezer that maintains a stable temperature at 0°C.

Milk Banking

Human milk banks provide human milk to infants who cannot be breastfed by their mothers. Premature and sick infants are most likely to receive banked milk. A woman can donate milk once, or on a continuing basis if her supply exceeds the demands of her infant. There is a long history of providing human milk to infants by persons other than the biological mother.[105] Wet nurses were the main source of human milk until the early 1900s for infants not fed by their biological mothers. Milk banks began in Europe and followed in the United States. Some neonatal intensive care units had informal milk banks until the 1980s. As a result of the human immunodeficiency virus,

the resurgence of tuberculosis, and risks related to donors who might abuse drugs, human-milk banks are now scarce in North America but due to recognition of the importance of human milk, demand is increasing.

Human-milk donors are chosen by their health profiles. Women are carefully screened before they can donate extra milk to milk banks. Milk banks that belong to the Milk Banking Association of North America require telephone screening, a written health and lifestyle history, and verification of the health of the mother and baby by the health care provider of each. Blood samples are tested for hepatitis B, hepatitis C, HIV, human T-cell lymphotropic virus (HTLV), and syphilis by the milk bank. Women are not accepted if they are acutely ill, have had a blood transfusion or an organ transplant within a year, drink more than 2 ounces of liquor daily, regularly use medications or megavitamins, smoke, or use street drugs. Additionally, women who eat no animal products must take vitamin supplements with B_{12} to be eligible to donate.

Human milk is carefully pasteurized to kill any potential pathogens while preserving the nutrients and active immune properties of the milk. The North American Human Milk Banking Association communicates closely with the Food and Drug Administration to follow guidelines for use of human tissues and fluids. Human milk for milk banks is stored frozen to preserve the immunologic and nutritional components. Rigid plastic (polypropylene) containers are recommended for keeping the milk composition stable. White blood cells stick to glass, but not to plastic containers.[106]

A prescription from a physician or a hospital is needed to order milk for an infant from one of the North American Milk Banking Association milk banks. Costs are approximately $3.25 per ounce before shipping charges, significantly more than the cost of human-milk substitutes.[105] Some insurance companies and Medicaid programs cover the fees when it is demonstrated that donor milk is the most appropriate therapy for a specific patient.

Model Programs

Loving Support for a Bright Future Breastfeeding Support Kits

The Maternal and Child Health Bureau (MCHB), working with Best Start Social Marketing, the U.S. Department of Agriculture, and many other organizations, developed breastfeeding support kits for physicians and other health providers. The kits were developed in response to a need identified in a national survey of physicians' knowledge and attitudes about breastfeeding.[107] The survey revealed that physicians were not prepared with sufficient information and training to provide support and counseling to breastfeeding mothers.

Family physicians, pediatricians, and obstetricians are in a position to offer encouragement and advice to potential and current breastfeeding mothers. In fact, research has shown that accurate, appropriate counseling and encouragement by physicians can improve rates of breastfeeding initiation and duration.[107,108] The Loving Support for a Bright Future Breastfeeding Kits are designed to help health care providers develop an optimal environment for breastfeeding promotion within their practices or clinic settings. The kits contain helpful, easy-to-use references, such as a pocket guide for breastfeeding management, a resource guide, consumer materials, and community-based activities. The physician and health provider kits are distributed through professional organizations including the American Academy of Pediatrics, the American College of Obstetricians and Gynecologists, the American Academy of Family Physicians, Association of Women's Health, Obstetric and Neonatal Nurses, and the American College of Nurse Midwives. The kits are available through Best Start Social Marketing.

Breastfeeding Promotion in Physicians' Office Practices

The American Academy of Pediatrics (AAP) receives funding support from the MCHB, U.S. DHHS, and the AAP Friends of the Children Fund for Breastfeeding Promotion in Physicians' Office Practices (BPPOP), an innovative program designed to boost breastfeeding promotion and support in underserved populations. Initiated in 1997, BPPOP's original mission was to improve the ability of AAP members to support new mothers and their breastfeeding infants and to encourage pediatricians to collaborate with others to develop breastfeeding promotion programs. Pediatricians enrolling in the program received a resource kit of educational materials—including the Loving Support Kits described earlier—and other strategies to more effectively promote, support, and manage breastfeeding with all families in their practices. In addition, pediatricians were provided technical assistance by telephone and e-mail from AAP staff regarding breastfeeding concerns, and were encouraged to participate in community and regional collaborative action groups.

Getty

After over 700 pediatricians nationwide joined the program, BPPOP was expanded in 2002 to include obstetricians, family physicians, and other health care providers and to specifically target office practices working with racially and ethnically diverse populations. A speaker's kit and materials targeting underserved populations were added to the resource kit, along with the newest strategies and opportunities for multidisciplinary networking for community breastfeeding promotion. Physicians joining the program complete a self-assessment questionnaire at the beginning and end of the program and measure the impact of breastfeeding promotion efforts by tracking the breastfeeding initiation and duration rates within their practices.

The Rush Mothers' Milk Club

The Rush Mothers' Milk Club at Rush Presbyterian—St. Luke's Medical Center has four main components: (1) expressing milk for the club members' babies, (2) providing skin-to-skin care and suckling at the empty breast as practice for the newborn, (3) feeding babies at the breast as soon as they are able to suck and swallow effectively, and (4) nursery staff helping the mother prepare for breastfeeding after discharge (www.rush.edu/patients/children/publications/notes/preemies.html). The club serves as a place for mothers to discuss their goals and concerns about breastfeeding. Mothers learn the value of their milk to their high-risk babies from the Special Care Nursery staff and from the Rush Mothers' Milk Club. The mothers learn to measure the amount of fat and calories in their own milk; they also learn to capture the highest-calorie portion of milk, which is usually produced during the last 10 minutes of pumping. To create a bond between the mother and baby, mothers are encouraged to use breast pumps at the baby's bedside. Family members and friends are also encouraged to participate in the weekly Mothers' Milk Club meetings to learn the importance of breastfeeding to the high-risk infant.

The success of the Rush Mothers' Milk Club is measured by its breastfeeding initiation rates. Between 95 and 97% of all mothers who deliver high-risk infants at Rush Presbyterian—St. Luke's Medical Center begin to nurse, compared with national rates for high-risk infants of only 30 to 40%. Some low-income mothers who delivered babies below 1500 grams were able to provide enough milk for 95% of their baby's feedings in the first 3 days and 87% over the first 60 days. This initiation rate is approximately twice the national initiation rate for premature infants. Many low-income women report that they decided to provide breast milk to their babies only after a health care provider or other mother discussed the benefits of human milk with them.

Resources

Best Start Social Marketing
> *Web site:* www.Bestartinc.org

The Human Milk Banking Association of North America, Inc.
> Represents all of the North American human-milk banks that collect, pasteurize, and distribute donated mother's milk. *Web site:* www.hmbana.org

International Lactation Consultants Association
> *Web site:* www.ilca.org

La Leche League
> *Web site:* www. Lalecheleague.org

C. Martin and N. F. Krebs
> *The Nursing Mother's Problem Solver.* New York: Fireside Publishing, 2000.

See Chapter 6 for additional resources.

References

1. Wight NE. Management of common breastfeeding issues. Pediatr Clin North Am 2001;48:321–44.

2. Tait P. Nipple pain in breastfeeding women: causes, treatment, and prevention strategies. J Midwifery Women's Health 2000;45:212–15.

3. Martin C, Krebs NF. The nursing mother's problem solver. New York: Simon and Schuster; 2000.

4. Smith JW, Tully MR. Midwifery management of breastfeeding: using the evidence. J Midwifery Women's Health 2001;46:423–38.

5. Lawrence RA, Lawrence RM. Breastfeeding: A guide for the medical profession. St. Louis: Mosby; 1999.

6. Thomsen AC, Espersen T, Maigaard S. Course and treatment of milk stasis, noninfectious inflammation of the breast, and infectious mastitis in nursing women. Am J Obstet Gynecol 1984;149:492–5.

7. Matheson I, Aursnes I, Horgen M, Aabo O, Melby K. Bacteriological findings and clinical symptoms in relation to clinical outcome in puerperal mastitis. Acta Obstet Gynecol Scand 1988;67:723–6.

8. Chung AM, Reed MD, Blumer JL. Antibiotics and breast-feeding: a critical review of the literature. Paediatr Drugs 2002;4:817–37.

9. Ito S. Drug therapy for breast-feeding women. N Engl J Med 2000;343:118–26.

10. Begg EJ, Duffull SB, Hackett LP, Ilett KF. Studying drugs in human milk: time to unify the approach. J Hum Lact 2002;18:323–32.

11. Howard CR, Lawrence RA. Xenobiotics and breastfeeding. Pediatr Clin North Am 2001;48:485–504.

12. Transfer of drugs and other chemicals into human milk. Pediatrics 2001;108:776–89.

13. Hansen WF, Peacock AE, Yankowitz J. Safe prescribing practices in pregnancy and lactation. J Midwifery Women's Health 2002;47:409–21.

14. Nice FJ, Snyder JL, Kotansky BC. Breastfeeding and over-the-counter medications. J Hum Lact 2000;16:319–31.

15. Hale TW. Medications and mothers milk (10th ed.). Amarillo, TX: Pharmasoft Medical Publishing; 2002.

16. Briggs GG, Freeman RE, Yaffee SJ. Drugs in pregnancy and lactation (6th ed.). Baltimore: Williams & Wilkins; 2001.

17. Diaz S. Contraceptive implants and lactation. Contraception 2002;65:39–46.

18. Hardy ML. Herbs of special interest to women. J Am Pharm Assoc (Wash.) 2000;40:234–42.

19. Kopec K. Herbal medications and breastfeeding. J Hum Lact 1999;15:157–61.

20. Belew C. Herbs and the childbearing woman. Guidelines for midwives. J Nurse Midwifery 1999;44:231–52.

21. Humphrey SL, McKenna DJ. Herbs and breastfeeding. Breastfeeding Abstracts 1997;17:11–12.

22. Foote J, Rengers B. Medicinal herb use during pregnancy and lactation. The Perinatal Nutrition Report 1998;5:1–2.

23. Humphrey S. Sage advice on herbs and breastfeeding. LEAVEN 1998;34:43–47.

24. Leung A, Foster S. Encyclopedia of common natural ingredients used in foods, drugs, and cosmetics. New York: Wiley; 1996.

25. Blumenthal M, Busse W, Goldberg A et al. (eds.). The complete German Commission E monographs: therapeutic guide to herbal medicines. Boston: Integrative Medicine Communications; 1998.

26. Cunningham E, Hansen K. Question of the month: Where can I get information on evaluating herbal supplements? J Am Diet Assoc 1999;99:1240.

27. The American Botanical Council. Herbal Information. American Botanical Council; 2003.

28. Cartwright, MM. Herbal use during pregnancy and lactation: a need for caution. The Digest 2001;(Summer):1–3. American

Dietetic Association Public Health/ Community Nutrition Practice Group.

29. Roberts KL, Reiter M, Schuster D. A comparison of chilled and room temperature cabbage leaves in treating breast engorgement. J Hum Lact 1995;11:191–4.

30. Snowden HM, Renfrew MJ, Woolridge MW. Treatments for breast engorgement during lactation. Cochrane Database Syst Rev 2003;CD000046.

31. Mennella J. Alcohol's effect on lactation. Alcohol Res Health 2001;25:230–4.

32. Ho E, Collantes A, Kapur BM, Moretti M, Koren G. Alcohol and breast feeding: calculation of time to zero level in milk. Biol Neonate 2001;80:219–22.

33. Mennella JA. Regulation of milk intake after exposure to alcohol in mothers' milk. Alcohol Clin Exp Res 2001;25:590–3.

34. Mennella JA, Gerrish CJ. Effects of exposure to alcohol in mother's milk on infant sleep. Pediatrics 1998;101:E2.

35. Little RE, Anderson KW, Ervin CH, Worthington-Roberts B, Clarren SK. Maternal alcohol use during breast-feeding and infant mental and motor development at one year. N Engl J Med 1989;321:425–30.

36. Little RE. Maternal use of alcohol and breastfed infants. New England Journal of Medicine 1990;322:339.

37. Mohrbacher N, Stock J. The breastfeeding answer book. La Leche League International; 2003.

38. Subcommittee on Nutrition during Lactation, Committee on Nutritional Status during Pregnancy and Lactation. Food and Nutrition Board Institute of Medicine, National Academy of Sciences. Nutrition during lactation: summary, conclusions, and recommendations. Washington DC: National Academy Press; 1991.

39. Amir LH, Donath SM. Does maternal smoking have a negative physiological effect on breastfeeding? The epidemiological evidence. Birth 2002;29:112–23.

40. Dahlstrom A, Lundell B, Curvall M, Thapper L. Nicotine and cotinine

concentrations in the nursing mother and her infant. Acta Paediatr Scand. 1990;79:142–7.

41. Dempsey DA, Benowitz NL. Risks and benefits of nicotine to aid smoking cessation in pregnancy. Drug Saf 2001;24:277–322.

42. Astley SJ, Little RE. Maternal marijuana use during lactation and infant development at one year. Neurotoxicol Teratol 1990;12:161–8.

43. Rivera-Calimlim L. The significance of drugs in breast milk. Pharmacokinetic considerations. Clin Perinatol. 1987;14:51–70.

44. Nehlig A, Debry G. Consequences on the newborn of chronic maternal consumption of coffee during gestation and lactation: a review. J Am Coll Nutr 1994;13:6–21.

45. Schreiber JS. Parents worried about breast milk contamination. What is best for baby? Pediatr Clin North Am 2001;48: 1113–27, viii.

46. LaKind JS, Berlin CM, Naiman DQ. Infant exposure to chemicals in breast milk in the United States: what we need to learn from a breast milk monitoring program. Environ Health Perspect 2001;109:75–88.

47. U.S. Department of Health and Human Services. HHS Blueprint for action on breastfeeding. 2000. Department of Health and Human Services, Office of Women's Health.

48. World Health Organization. World Health Organization: report of the WHO working group on the assessment of health risks for human infants from exposure to PCDDs, PCDFs, PCBs. Chemosphere 1998;57:1627–43.

49. Pronczuk J, Akre J, Moy G, Vallenas C. Global perspectives in breast milk contamination: infectious and toxic hazards. Environ Health Perspect 2002;110:A349–A351.

50. Committee on Nutrition, American Academy of Pediatrics. Pediatric Nutrition Handbook. Elk Grove Village, IL: American Academy of Pediatrics; 2003.

51. Stokowski LA. Early recognition of neonatal jaundice and kernicterus. Adv Neonatal Care 2002;2:101–14.

52. Gourley GR. Breast-feeding, neonatal jaundice and kernicterus. Semin Neonatol 2002;7:135–41.

53. American Academy of Pediatrics. Kernicterus in full-term infants—United States. Morbidity and Mortality Monthly Report 2001;108:763–5.

54. Geiger AM, Petitti DB, Yao JF. Rehospitalisation for neonatal jaundice: risk factors and outcomes. Paediatr Perinat Epidemiol 2001;15:352–8.

55. Joint Commission on Accreditation of Healthcare Organizations. Kernicterus threatens healthy newborns. Sentinel Event Alert, April 2001(18).

56. American Academy of Pediatrics Provisional Committee for Quality Improvement and Subcommittee on Hyperbilirubinemia. Practice parameter: management of hyperbilirubinemia in the healthy term newborn. Pediatrics 1994;558–65.

57. Gartner LM, Herschel M. Jaundice and breastfeeding. Pediatr Clin North Am 2001;48:389–99.

58. Reiser D. Hyperbilirubinemia. AWHONN Lifelines 2001;5:55–61.

59. Hansen TWR., Bratlid D. Bilirubin and brain toxicity. Acta Paediatr Scand 1986;75:513.

60. Neonatal jaundice and kernicterus. Pediatrics 2001;108:763–5.

61. Gartner LM. Breastfeeding and jaundice. J Perinatol 2001;21(suppl 1):S25–S29.

62. American Academy of Pediatrics Work Group on Breastfeeding. Breastfeeding and the use of human milk. Pediatrics 1997;100: 1035–9.

63. Willis SK, Hannon PR, Scrimshaw SC. The impact of the maternal experience with a jaundiced newborn on the breastfeeding relationship. J Fam Pract 2002;51:465.

64. Hannon PR, Willis SK, Scrimshaw SC. Persistence of maternal concerns surrounding neonatal jaundice: an exploratory study. Arch Pediatr Adolesc Med 2001;155: 1357–63.

65. Flidel-Rimon O, Shinwell ES. Breastfeeding multiples. Semin Neonatol 2002; 7:231–9.

66. Gromada KK, Spangler AK. Breastfeeding twins and higher-order multiples. J Obstet Gynecol Neonatal Nurs 1998; 27:441–9.

67. Gromada KK. Breastfeeding more than one: multiples and tandem breastfeeding. NAACOGS Clin Issu Perinat Women's Health Nurs 1992;3:656–66.

68. Auer C, Gromada KK. A case report of breastfeeding quadruplets: factors perceived as affecting breastfeeding. J Hum Lact 1998; 14:135–41.

69. Palmer G. The politics of breastfeeding. 2nd, 32. London: Pandora Press/Harper Collins; 1993.

70. Saint L, Maggiore P, Hartmann PE. Yield and nutrient content of milk in eight women breast-feeding twins and one woman breast-feeding triplets. Br J Nutr 1986;56: 49–58.

71. Hattori R, Hattori H. Breastfeeding twins: guidelines for success. Birth 1999; 26:37–42.

72. O'Connell EJ. Pediatric allergy: a brief review of risk factors associated with developing allergic disease in childhood. Ann Allergy Asthma Immunol 2003;90:53–8.

73. Gdalevich M, Mimouni D, David M, Mimouni M. Breast-feeding and the onset of atopic dermatitis in childhood: a systematic review and meta-analysis of prospective studies. J Am Acad Dermatol 2001; 45:520–7.

74. Gdalevich M, Mimouni D, Mimouni M. Breast-feeding and the risk of bronchial asthma in childhood: a systematic review with meta-analysis of prospective studies. J Pediatr 2001;139:261–6.

75. Chandra RK. Food allergy and nutri-

tion in early life: implications for later health. Proc Nutr Soc 2000;59:273–7.

76. Bjorksten B, Kjellman NI. Does breastfeeding prevent food allergy? Allergy Proc 1991;12:233–7.

77. Mofidi S. Nutritional management of pediatric food hypersensitivity. Pediatrics 2003;111:1645–53.

78. Einarson A, Lawrimore T, Brand P, Gallo M, Rotatone C, Koren G. Attitudes and practices of physicians and naturopaths toward herbal products, including use during pregnancy and lactation. Can J Clin Pharmacol 2000;7:45–9.

79. Mennella JA, Beauchamp GK. The effects of repeated exposure to garlic-flavored milk on the nursling's behavior. Pediatr Res 1993;34:805–8.

80. Mennella JA, Beauchamp GK. Maternal diet alters the sensory qualities of human milk and the nursling's behavior. Pediatrics 1991;88:737–44.

81. Cooper RL, Cooper MM. Red pepper-induced dermatitis in breast-fed infants. Dermatology 1996;193:61–2.

82. Zeretzke K. Allergies and the breastfeeding family. New Beginnings 1998;15(4):100.

83. Wight NE. Breastfeeding the borderline (near-term) preterm infant. Pediatr Ann 2003;32:329–36.

84. Shulman RJ, Schanler RJ, Lau C, Heitkemper M, Ou CN, Smith EO. Early feeding, feeding tolerance, and lactase activity in preterm infants. J Pediatr 1998;133: 645–9.

85. Gibson RA, Makrides M. Polyunsaturated fatty acids and infant visual development: a critical appraisal of randomized clinical trials. Lipids 1999;34:179–84.

86. Lucas A, Cole TJ. Breast milk and neonatal necrotising enterocolitis. Lancet 1990;336:1519–23.

87. Lucas A, Morley R, Cole TJ. Randomised trial of early diet in preterm babies and later intelligence quotient. BMJ 1998;317:1481–7.

88. Hylander MA, Strobino DM, Dhanireddy R. Human milk feedings and infection among very low birth weight infants. Pediatrics 1998;102:E38.

89. Schanler RJ, Shulman RJ, Lau C, Smith EO, Heitkemper MM. Feeding strategies for premature infants: randomized trial of gastrointestinal priming and tube-feeding method. Pediatrics 1999;103:434–9.

90. Schanler RJ, Shulman RJ, Lau C. Feeding strategies for premature infants: beneficial outcomes of feeding fortified human milk versus preterm formula. Pediatrics 1999;103:1150–7.

91. Schanler RJ, Abrams SA. Postnatal attainment of intrauterine macromineral accretion rates in low birth weight infants fed fortified human milk. J Pediatr 1995;126:441–7.

92. Schanler RJ. Overview: the clinical perspective. J Nutr 2000;130:417S–9S.

93. Jones CA. Maternal transmission of infectious pathogens in breast milk. J Paediatr Child Health 2001;37:576–82.

94. Lawrence RM, Lawrence RA. Given the benefits of breastfeeding, what contraindications exist? Pediatr Clin North Am 2001;48:235–51.

95. Humphrey J, Iliff P. Is breast not best? Feeding babies born to HIV-positive mothers: bringing balance to a complex issue. Nutr Rev 2001;59:119–27.

96. Jackson DJ, Chopra M, Witten C, Sengwana MJ. HIV and infant feeding: issues in developed and developing countries. J Obstet Gynecol Neonatal Nurs 2003;32: 117–27.

97. WHO Technical Consultation on Behalf of the UNFPA/UNICEF/WHO/ UNAIDS Inter-Agency Task Team on Mother-to-Child Transmission of HIV. New data on the prevention of mother-to-child transmission of HIV and their policy implications: conclusions and recommendations. Geneva, Switzerland: World Health Organization; Oct. 13, 2000.

98. Effect of breastfeeding on infant and child mortality due to infectious diseases in less developed countries: a pooled analysis. WHO Collaborative Study Team on the Role of Breastfeeding on the Prevention of Infant Mortality. Lancet 2000;355:451–5.

99. World Health Organization. HIV and Infant Feeding. A review of HIV transmission through breastfeeding. Geneva, Switzerland: World Health Organization; 1998.

100. Nolan ML, Greenberg AE, Fowler MG. A review of clinical trials to prevent mother-to-child HIV-1 transmission in Africa and inform rational intervention strategies. AIDS 2002;16:1991–9.

101. UNAIDS, UNICEF World Health Organization. HIV and infant feeding: guidelines for health care managers and supervisors. Geneva, Switzerland: World Health Organization; 1998.

102. Tully DB, Jones F, Tully MR. Donor milk: what's in it and what's not. J Hum Lact 2001;17:152–5.

103. Pardou A, Serruys E, Mascart-Lemone F et al. Human milk banking: influence of storage processes and of bacterial contamination on some milk constituents. Bio Neonat 1994;65:302–9.

104. Tully MR. Recommendations for handling mother's own milk. J Hum Lact 2000;16:149–51.

105. Jones F. History of North American donor milk banking: one hundred years of progress. J Hum Lact 2003;19:313–18.

106. Paxson CI Jr, Cress CC. Survival of human milk leukocytes. J Pediatr 1979; 94:61–4.

107. Freed GL, Clark SJ, Sorenson J, Lohr JA, Cefalo R, Curtis P. National assessment of physicians' breast-feeding knowledge, attitudes, training, and experience. JAMA 1995;273:472–6.

108. Schanler RJ, O'Connor KG, Lawrence RA. Pediatricians' practices and attitudes regarding breastfeeding promotion. Pediatrics 1999;103:E35.

"A babe is fed with
milk and praise."
Charles Lamb, *The First Tooth*

Chapter 8

Infant Nutrition

Chapter Outline

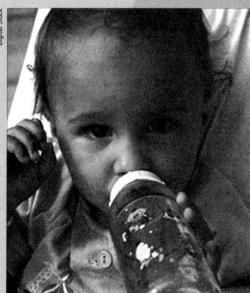

Prepared by **Janet Sugarman Isaacs**

Key Nutrition Concepts

1   The dynamic growth experienced in infancy is the most rapid of any age. Inadequate nutrition in infancy, however, leads to consequences that may be lifelong, harming both future growth and future development.

2   Progression in feeding skills expresses important developmental steps in infancy that signal growth and nutrition status.

3   Nutrient requirements of term newborns have to be modified for preterm infants. Knowing the needs of sick and small newborns results in greater understanding of the complex nutritional needs of all newborns and infants.

4   Changing feeding practices, such as the care of infants outside the home and the early introduction of foods, markedly affects nutritional status of infants.

Introduction

This chapter is about healthy *full-term infants* born at 37 weeks of gestation or later, and healthy *preterm infants* born at 34 weeks or later. These newborns are expected to have typical growth and development. The term *normal* is not used much in this chapter, because its opposite, *abnormal,* is an emotionally laden term, particularly when describing babies to their parents. *Typical* is used in place of *normal* when possible.

This chapter discusses how nutrition is an important contributor to the complex development of infants. Both biological and environmental factors interact during infant growth and development. Models about the interaction of biological and environmental factors are often incomplete. They are not always adequate for describing complex interactions, such as how mealtime stimulates language development and how food preferences develop during infancy. The complexity of infant development contributes to our individuality later on.

Full-Term Infants   Infants born between 37 and 42 weeks of gestation.

Preterm Infants   Infants born at or before 37 weeks of gestation.

Infant Mortality   Death that occurs within the first year of life.

Assessing Newborn Health

Birthweight as an Outcome

The weight of a newborn is a key measure of health status during pregnancy. The average gestation for a full-term infant is 40 weeks, with a range from 37 to 42 weeks. Full-term infants usually weigh 2500–3800 grams (5.5 to 8.5 pounds) and are 47–54 centimeters (18.5–21.5 inches) in length.[1] There were over 4 million births in the United States in 2002, and 88% of these newborns were full term.[2] Infants with normal birthweights are less likely to require intensive care and are usually healthy in the long run. Conversely, preterm births, regardless of birthweight, are those born at 37 weeks or less. Preterm means incomplete development has taken place.[3]

Infant Mortality

Worldwide *infant mortality* rates rank the United States at 27th place, far worse than many other countries.[3] The reasons for the high rate of mortality are many, but the prevalence of low birthweight is a major factor. In 2002, 7.8% of live births in the United States were low birthweight or less than 2500 grams (5.5 pounds).[2] The three leading causes of infant mortality in 2002 were congenital malformations, preterm births, and sudden infant death syndrome.[3] The higher incidence of infant mortality, low birthweight, and preterm births in African American infants is of particular concern. Table 8.1 shows infant deaths in the United States based on race of the mother.[3] The basis for racial disparity and preterm birth is a major focus in federal initiatives to combat infant deaths in the United States. Despite efforts to lower neonatal and infant deaths, rates are still too high.

Combating Infant Mortality

Efforts to improve newborn health are under way on many levels. In the United States, improved access to specialized care for mothers and infants has been credited in part for a decline in the infant mortality rate.[2] This is a multifaceted problem, however, affected by:

- Social and economic status of the families and women
- Access to health care
- Medical interventions
- Teenage pregnancy rates
- Availability of abortion services
- Failure to prevent preterm and low-birthweight births

Resources have been concentrated on the proportion of newborns identified at risk. Some of these resources are

Table 8.1 United States infant mortality rates by race of the mother[3]

	DEATHS PER 1000 LIVE BIRTHS
All newborns	6.9
White	5.7
Black	13.5
Pacific Islander	9.0
Chinese	3.5
American Indian or Alaska Native	8.3
Hispanic/Mexican	5.4

major payers of health services, such as Medicaid and the Child Health Initiatives Program (CHIP).[4] The following concepts underscore the commitment of resources for infants:

- Recognition that birthweight is important for long-term health outcomes.
- Prevention and treatment of complications for at-risk infants are investments for the future.

The emphasis on prevention is seen in various programs. The Early Periodic Screening, Detection, and Treatment Program (*EPSDT*) is a major source of preventive and routine care for infants in low-income families. Immunizations during infancy are another example of a prevention approach.

Nutrition is included in some national prevention programs. The Special Supplemental Food Program for Women, Infants, and Children (WIC) and the Centers for Disease Control (CDC) collaborate to track infant growth as a part of the Nutrition Surveillance Program.[5] The "Bright Futures" program promotes and improves the health, education, and well-being of infants and children. Nutrition is one component of the program's guidelines about common issues and concerns. Bright Futures is an example of a comprehensive approach to health supervision by a collaboration of government and professional groups.[6]

Standard Newborn Growth Assessment

Newborn health status is assessed by various indicators of growth and development taken right after birth. Indicators include birthweight, length, and head circumference for gestational age. The designation "small for gestational age" (SGA), also called "small for dates," *"intrauterine growth retardation (IUGR),"* or "intrauterine growth restriction" means the newborn's birthweight falls below the 10th percentile of weight for gestational age.[7] Infants above the 90th percentile are considered large for gestational age (LGA). Those in between are appropriate for gestational age (AGA).

Infant Development

Monitoring infants' nutritional status requires an understanding of their overall development. Full-term newborns have a wider range of abilities than previously recognized; they hear and move in response to familiar sounds, such as the mother's voice.[9] Newborns demonstrate four states of arousal, ranging from sleeping to fully alert, and responsiveness differs in part on their state of arousal.[9] Recognizing the state of arousal is a part of nursing successfully.

Organs and systems developed during gestation continue to increase in size and complexity during infancy. The newborn's central nervous system is immature; that is, the neurons in the brain are less organized compared to those of the older infant. As a result, the newborn gives inconsistent or subtle cues of hunger and other needs, compared to the cues given later. The fact that newborns can *root, suckle,* and coordinate swallowing and breathing within hours of birth shows that feeding is directed by reflexes and the central nervous system.[9,10] Newborn *reflexes* are protective for them. These reflexes fade as they are replaced by purposeful movements during the first few months of life.[9] Table 8.2 on the next page lists major reflexes in newborns.[9–11]

Motor Development

Motor development reflects an infant's ability to control voluntary muscle movement. There are several models for describing infant development, but none provide a complete description and explanation of the rapid advances in motor skills achieved during infancy.[9,11] Illustration 8.1 on page 203 depicts motor development during the first 15 months.[7] It is a great source of pride for parents when their baby first rolls over or sits up. The development of muscle control is top down, meaning head control is the start and last comes lower legs.[10] Muscle development is also from central to peripheral, meaning the infant learns to control the shoulder and arm muscles before muscles in the hands.[10,12] Motor development influences both the ability of the infant to feed and the amount of calories expended in the activity.[13] An example of how motor development affects feeding is the ability to sit in a high chair. Only when an infant has achieved the motor development of head control and sitting balance and certain reflexes have disappeared can oral feeding with a spoon be achieved.[12] The development of motor skills slowly increases the caloric needs of infants over time because increased activity requires more energy.[13] Infants who crawl expend more calories in physical activity than do younger infants who cannot roll over.

EPSDT   The Early Periodic Screening, Detection, and Treatment Program is a part of Medicaid and provides routine checkups for low-income families.

Intrauterine Growth Retardation (IUGR)   Fetal undergrowth from any cause, resulting in a disproportionality in weight, length, or weight-for-length percentiles for gestational age. Sometimes called *intrauterine growth restriction.*

Reflex   An automatic (unlearned) response that is triggered by a specific stimulus.

Rooting Reflex   Action that occurs if one cheek is touched, resulting in the infant's head turning toward that cheek and the infant opening his mouth.

Suckle   A reflexive movement of the tongue moving forward and backward; earliest feeding skill.

Critical Periods

The concept of critical period is based on a fixed time period during which certain behaviors emerge. This theory of development suggests that there is a time period, or window of development, during which certain skills must be learned in order for subsequent learning to occur.[11] A critical period for the development of oral

Table 8.2 Major reflexes found in newborns

Name	Response	Significance
Babinski	A baby's toes fan out when the sole of the foot is stroked from heel to toe.	Perhaps a remnant of evolution
Blink	A baby's eyes close in response to bright light or loud noise.	Protects the eyes
Moro	A baby throws its arms out and then inward (as if embracing) in response to loud noise or when its head falls.	May help a baby cling to its mother
Palmar	A baby grasps an object placed in the palm of its hand.	Percursor to voluntary grasping
Rooting	When a baby's cheek is stroked, it turns its head toward the cheek that was stroked and opens its mouth.	Helps a baby find the nipple
Stepping	A baby who is held upright by an adult and is then moved forward begins to step rhythmically.	Precursor to voluntary walking
Sucking	A baby sucks when an object is placed in its mouth.	Permits feeding
Withdrawal	A baby withdraws its foot when the sole is pricked with a pin.	Protects a baby from unpleasant stimulation

SOURCE: Robert V. Kail and John C. Cavanaugh, *Human Development: A Lifespan View.* 2nd ed. Belmont, CA: Wadsworth/Thomson Learning, 2000, p. 99.[9]

feeding may explain some later feeding problems in infancy.[12] In the typical healthy newborn, the mouth is a source of pleasure and exploring, an important form of early learning. When a newborn has a prolonged period of respiratory support, for example, the baby may not associate mouth sensations with pleasure, but rather with discomfort. Under such circumstances the critical period for associating mouth sensations with pleasure and exploration may have been missed. After discharge, such an infant may be a reluctant feeder and have difficulty learning to enjoy food from a spoon.

Cognitive Development

The concept of biological and environmental systems interacting is seen in Illustration 8.2 on page 204 showing *sensorimotor* development.[9] These skills influence feeding in important ways. For example, the stage during which infants are very sensitive to food texture is also when their speech skills are emerging.[9] The interaction between the environment and stimulating the senses in the developing brain is now seen as structuring the nervous system long term.[14] The latest research suggests that access to adequate calories and protein may not be sufficient for maximizing brain maturation if the social and emotional growth of the infant is not stimulated simultaneously.[14] Cognitive development is also subjected to genetic controls, which turn genes on and off in different time frames and sites within the body.[15] Specific vitamins being needed at spe-

Sensorimotor An early learning system in which the infant's senses and motor skills provide input to the central nervous system.

Gastroesophageal Reflux (GER) Movement of the stomach contents backward into the esophagus, due to stomach muscle contractions. The condition may require treatment depending on its duration and degree. Also known as *gastro-esophageal reflux disease (GERD)*.

cific time frames of development is an example of the interaction of genetics and environment.

Digestive System Development

"Now good digestion wait on appetite, and health on both."

William Shakespeare, *Macbeth*

A healthy digestive system is necessary for successful feeding. Parents may worry about gastrointestinal problems, in part because of misinformation about nutrition in infancy.[16] For example, an infant with soft, loose stools may be considered by the parents to have diarrhea if they do not know that this is typical for breastfed infants.[6] Another common example is parents being concerned that their infant's gastrointestinal discomfort may interfere with weight gain, even though growth usually progresses well. It takes up to 6 months for the infant gastrointestinal tract to mature, and the time required varies considerably from one individual to another.[6,7]

During the third trimester the fetus swallows amniotic fluid, and this stimulates the lining of the intestine to grow and mature.[7] At birth, the healthy newborn's digestive system is sufficiently mature to digest fats, protein, and simple sugars and to absorb fats and amino acids. Although healthy newborns do not have the same levels of digestive enzymes or rate of stomach emptying as older infants, the gut is functional at birth.[17]

After birth and through the early infancy period, the coordination of peristalsis within the gastrointestinal tract improves. Maturation of peristalsis and rate of passage are associated with some forms of gastrointestinal discomfort in infants.[15] Infants often have conditions that reflect the immaturity of the gut, such as colic, *gastroesophageal reflux (GER)*, unexplained diarrhea, and constipation.[6,17] Such conditions

Illustration 8.1 Gross motor skills.

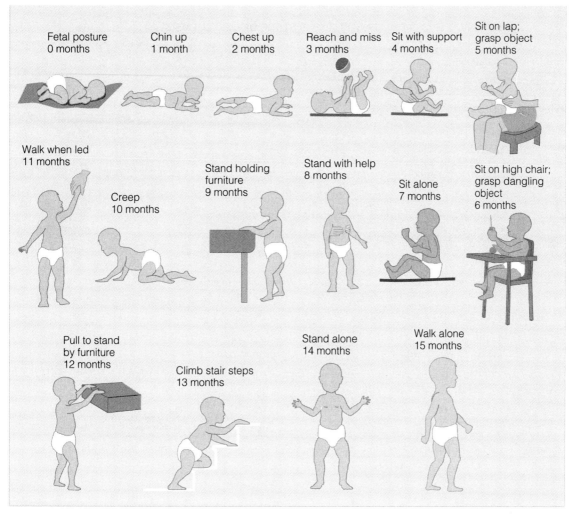

Fetal posture
0 months

Chin up
1 month

Chest up
2 months

Reach and miss
3 months

Sit with support
4 months

Sit on lap;
grasp object
5 months

Walk when led
11 months

Creep
10 months

Stand holding
furniture
9 months

Stand with help
8 months

Sit alone
7 months

Sit on high chair;
grasp dangling
object
6 months

Pull to stand
by furniture
12 months

Climb stair steps
13 months

Stand alone
14 months

Walk alone
15 months

Based on Shirley, 1931, and Bayley, 1969.

SOURCE: Robert V. Kail and John C. Cavanaugh, *Human Development: A Lifespan View*, 2nd ed. Belmont, CA: Wadsworth/Thomson Learning, 2000, p. 99.[9]

do not interfere with the ability of the intestinal villa to absorb nutrients, and typically do not hinder growth. Other factors influence the rate of food passage through the colon and the gastrointestinal discomfort seen in infants. These include:

- *Osmolarity* of foods or liquids (which affects how much water is in the intestine)
- Colon bacterial flora
- Water and fluid balance in the body[7]

Parenting

> "A babe in the house is a well-spring of pleasure."
>
> Martin Farquhar Tupper, *On Education*

Even though the newborn can breast- or bottle-feed after birth, skills of new parents develop slowly. The parents' ability to recognize and respond to infant cues of hunger and satiety improve over time. New parents have to learn the temperament of their infants. Temperament has a biological basis and includes the infant's style or patterns of behavior.[9] Temperament includes the infant's emotional reactions to new situations, activity level, and sociability. The fit between the infant's temperament and that of the parents can increase or decrease feeding problems. For example, new parents may take a while to recognize that the newborn is more comfortable nursing at one noise level in the home than at another. Infants who are 6 months of age or older are better able to let parents know their needs and temperament. Conflicts in temperament may escalate as time progresses; they may become a factor in failure to thrive or other growth and feeding problems in later years.[11]

Osmolarity Measure of the number of particles in a solution, which predicts the tendency of the particles to move from high to low concentration. Osmolarity is a factor in many systems, such as in fluid and electrolyte balance.

Illustration 8.2 Sensorimotor stage of development.

SUBSTAGES DURING THE SENSORIMOTOR STAGE OF DEVELOPMENT

Substage	Age (months)	Accomplishment	Example
1	0–1	Reflexes become coordinated.	Sucking a nipple
2	1–4	Primary circular reactions appear an infant's first learned reactions to the world.	Thumb sucking
3	4–8	Secondary circular reactions emerge, allowing infants to explore the world of objects.	Shaking a toy to hear a rattle
4	8–12	Means–end sequencing of schemes is seen, marking the onset of intentional behavior.	Moving an obstacle to reach a toy
5	12–18	Tertiary circular reactions develop, allowing children to experiment.	Shaking different toys to hear the sounds they make
6	18–24	Symbolic processing is revealed in language, gestures, and pretend play.	Eating pretend food with a pretend fork

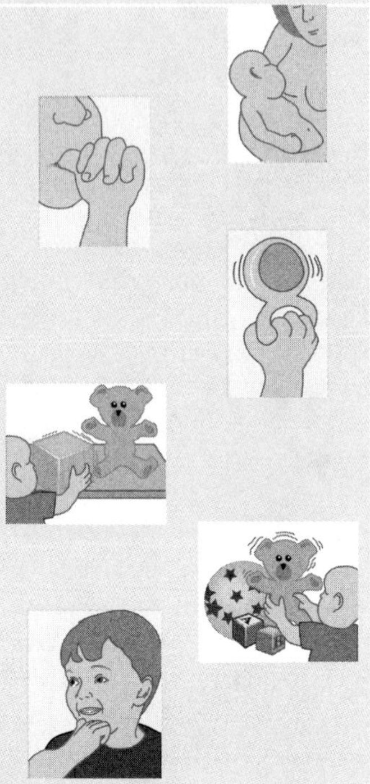

SOURCE: Robert V. Kail and John C. Cavanaugh, *Human Development: A Lifespan View,* 2nd ed. Belmont, CA: Wadsworth/Thomson Learning, 2000, p. 128.[9]

Energy and Nutrient Needs

The Dietary Guidelines for Americans address the needs of children aged 2 and older, and not infants. Recommendations for infants are largely based on information from the Dietary Reference Intakes (DRI) and professional groups such as the National Academy of Medicine and the American Academy of Pediatrics.[13,18–20]

Caloric Needs

The caloric needs of typical infants are higher per pound of body weight than at any other time of life. The range in caloric requirements for individual infants is broad, ranging from 80 to 120 cal per kg (2.2 lb) body weight.[13] The average caloric need of infants in the first 6 months of life is 108 cal per kg body weight, based on growth in breast-fed infants.[13] From 6 to 12 months of age, the average caloric need is 98 cal/kg.[13] Factors that account for the range of caloric needs of infants include the following:

- Weight
- Growth rate
- Sleep/wake cycle
- Temperature and climate
- Physical activity
- Metabolic response to food
- Health status and recovery from illness

Current recommendations for infants of 108 and 98 cal are considered about 15% too high, based on new study results.[13] However, too few energy expenditure studies have been done on infants to reach a consensus about changing energy requirements.

Protein Needs

Recommended protein intake from birth up to 6 months averages 2.2 g of protein per kg of body weight, and from 6 to 12 months the need is for 1.6 g of protein per kg.[13] Protein needs of individual infants vary with the same factors listed for calorie needs. Protein needs are influenced more directly by body composition than calorie needs are, because metabolically active muscles require more protein for maintenance.

Most young infants who breastfeed or consume the recommended amounts of infant formula meet protein needs without added foods. Infants may exceed their protein needs based on the Recommended Daily Allowance (RDA) when they consume more formula than recommended for age and when protein sources such as baby cereal are added to infant formula. Inadequate or excessive protein intake can result for infants who are offered formula that is not made correctly, such as when less or more water is used than appropriate in preparation. Essential amino acids required by healthy infants are constant across the first year of life.

Fats

There is no specific recommended intake level of fats for infants. Fat restriction is not recommended. Breast milk provides 55% of its calories from fat, and this percentage reflects an adequate intake of fat by infants.[17] The main source of fat in most infant diets is breast milk or formula. Cholesterol intake should not be limited in infancy, because infants have a high need for it and its related metabolites in gonad and brain development.[22] The percentage of fat in the diet drops after the infant has baby foods on a spoon added, since most baby foods are low in fat. Infants need fat, which is a concentrated source of calories, to support their high need for calories. Fat requirements in infancy are complicated by the differences in digestion and transport of fats based on fatty acid chain length.[17] *Short-* and *medium-chain fats* such as those in breast milk are more readily utilized than *long-chain-fats,* such as in some infant formulas. Long-chain fatty acids are the most common type in food, but they are more difficult for young infants to utilize. Examples of long-chain fatty acids are C16–C18 and include palmitic (C16:0), stearic (C18:1), and linoleic (C18:2) acids.[23]

Infants use fats to supply energy to the liver, brain, and muscles including the heart. The fact that infants have high caloric needs compared to those for older children means that infants use fats more regularly for generating energy. Young infants cannot tolerate fasting for long, because it quickly uses up both carbohydrate and fat energy sources. This effect of fasting explains in part why infants cannot sleep through the night. In rare cases, some infants cannot metabolize fats due to a genetic condition that blocks specific enzymes needed to generate energy. Such infants may get sick suddenly; a few have been identified with this rare condition of fat metabolism only after dying of what appeared to be sudden infant death syndrome (SIDS).[24]

Fats in food provide the two essential fatty acids, linoleic and alpha-linolenic acid. Essential fatty acids are substrates for hormones, steroids, endocrine, and neuroactive compounds in the developing brain.[23] The long-chain polyunsaturated fatty acids docosahexaenoic acid (DHA) and eicosapentaenoic acid (EPA) are derived from an essential fatty acid. DHA and EPA have been studied in eye and brain development and function.[23] Higher levels of these long-chain fatty acids were documented in breast-fed compared to formula-fed infants, resulting in concern about essential fatty acids and neurological development. Starting in 2002, major baby formula manufacturers added DHA with much fanfare.[21] Examples of products with these added are Enfamil Lipil or Similac Advance. The Food and Drug Administration did not endorse specific health benefits, but decided further study was needed and is monitoring the products' tolerance.[25]

Metabolic Rate, Calories, Fats, and Protein—How Do They All Tie Together?

The metabolic rate of infants is the highest of any period after birth.[7] The higher rate is primarily related to infants' rapid growth rate and the high proportion of infant weight that is made up of muscle. The usual body fuel for metabolism is glucose. When sufficient glucose is available, growth is likely to proceed. When glucose from carbohydrates is limited, amino acids will be converted into glucose and used for energy and therefore are made unavailable for growth. The conversion of amino acids into glucose is a more dynamic process in infants as compared to adults. The breakdown of amino acids for use as energy occurs during illness in adults, but can occur daily in fast-growing infants. Circulating amino acids in the blood from ingested foods will be used for glucose production, and if these are not sufficient, the body will release amino acids from muscles. This process of breaking down body protein to generate energy is known as catabolism. If catabolism goes on too long, it will slow or stop growth in infants. The precise site of all this metabolic activity is inside organs such as the liver, and in *mitochondria* within cells. If carbohydrates are not provided in sufficient amounts, growth will plateau because ingested protein and fats will be used for meeting energy needs.

> **Short-Chain Fats**   Carbon molecules that provide fatty acids less than 6 carbons long, as products of energy generation from fat breakdown inside cells. Short-chain fatty acids are not usually found in foods.
>
> **Medium-Chain Fats**   Carbon molecules that provide fatty acids with 6–10 carbons, again not typically found in foods.
>
> **Long-Chain Fats**   Carbon molecules that provide fatty acids with 12 or more carbons, which are commonly found in foods.
>
> **Mitochondria**   Intracellular unit in which fatty acid breakdown takes place and many enzyme systems for energy production inside cells are regulated.

Other Nutrients and Nonnutrients

FLUORIDE   The DRI for fluoride is 0.1 mg daily for infants less than 6 months of age, and 0.5 mg daily for 7- to 12-month-olds.[26] Fluoride is incorporated into the enamel of forming teeth, including those not yet erupted. Dental caries in early childhood are more frequent if an infant does not meet the DRI for fluoride. If an infant has more fluoride than recommended, tooth discoloration may result later. Community water fluoridation is safe for breastfeeding women and for infants.[6] Most infants who live in areas with fluoridated water do not require another source of fluoride. Fluoride is low in breast milk.[26] In areas in which fluoridated water is not available, prescribed fluoride is recommended for breastfed infants. If families routinely purchase bottled water, they should select water that has fluoride added.[6]

VITAMIN D   Vitamin D, or preformed forms of vitamin D such as cholecalciferol, is required for bone mineralization with calcium.[26] It is not supplied by human milk in sufficient amounts, but it is added to formulas. DRI for infants up to 12 months is 5 mcg (200 IU) each day.[6,26] Sunlight may be a sufficient source for meeting the vitamin D requirement for infants. Infants who are taken outside in clothing that allows direct exposure of the skin on the face and arms to sunlight 10–15 minutes per day increase their vitamin D status. Parents are, however, encouraged to apply ultraviolet-light skin-protective lotions to their children's exposed skin when outside. The high levels of skin-protective creams block vitamin D formation. Infants with dark skin and those living in areas with limited sunny days and covered skin produce little vitamin D. The identification of babies with rickets has resulted in stronger recommendations for breastfed infants from the American Academy of Pediatrics (AAP).[19] Breastfed infants should receive vitamin D supplements beginning at 2 months of age and until they begin taking at least 17 ounces daily of vitamin D–fortified milk.[19] Multivitamin supplements, containing 200 international units of vitamin D, are available as over-the-counter liquid drops or tablets.

SODIUM   Sodium is a major component of extracellular fluid and an important regulator of fluid balance. Estimated minimum requirements for sodium are 120 mg for 0- to 5-month-olds and 200 mg for 6- to 12-month-olds.[18] Breast milk's content of sodium was used as the basis for setting the sodium requirements for infants, and infant formula is supplemented with sodium to match the amount in breast milk. Typical infants do not have difficulty maintaining body fluids and electrolytes, even though they may not show thirst as a separate signal from hunger. Young infants do not sweat as much as older children, so losses from sweating are not major losses for infants. Illnesses such as diarrhea or vomiting cause the loss of sodium and water and increase the risk of dehydration. Infants do not need salt added to foods to maintain adequate sodium intake.

FIBER   Although there are dietary fiber recommendations for toddlers and children, there are no dietary fiber recommendations for infants.[27] Commercial and home-made baby foods are generally not significant sources of dietary fiber, because preparation methods reduce dietary fiber.[28] However, fiber-containing foods such as fruits, vegetables, and grains are appropriate foods for older infants.

LEAD   Although lead is not a nutrient, it can be associated with iron and calcium status during infancy. Elevated blood lead levels can be toxic to the developing brain, interfere with calcium and iron absorption, and bring about slowed growth and shorter stature.[29,30] Infants may inadvertently be exposed to environmental sources of lead. Lead may be a contaminant in water from lead pipes, particularly if the house was built before 1950. Older homes may contain lead-based paints that taste sweet to infants. Screening for lead poisoning is recommended starting at 9 to 12 months of age.[6,30] If siblings have been found to have lead poisoning, screening for lead may be started at 6 months. Infancy is not the peak age for lead poisoning, but infants can be exposed if their parents work with lead-containing products. For example, if the father is a truck driver who uses leaded gasoline, his work clothes may have lead dust on them. If these clothes are mixed with the rest of the household laundry, or children play in the laundry room where lead dust has settled, an infant in the home may become exposed.

Physical Growth Assessment

Tracking growth in length and in weight helps identify health problems early, preventing or minimizing slowing of the growth rate. Parents understand that a sign of health is growth of their babies. By the time children reach school age, most families have a wall in their homes that proudly displays many height measurements over time. Healthy newborns double their birthweight by age 4–6 months and triple it by 1 year.[1] Growth reflects nutritional adequacy, health status, and economic and other environmental influences on the family. There is a wide range of growth attainment considered normal, however, and healthy babies may follow different patterns of growth. Often, healthy infants have short periods when their weight gain is slower or faster than at other times. Slight variations in growth rate can result from illness, teething, inappropriate feeding position, or family disruption. The overall growth pattern is important, and each assessment is compared to the whole picture. Table 8.3 shows typical growth rates during the first year of life.

Accurate assessment of growth and interpretation of growth rates are important components of health care for infants (Illustration 8.3). Accuracy requires calibrated scales, a recumbent-length measurement board with an attached right-angled headpiece, and a nonstretch tape for measurement of head circumference. Table 8.4 on page 208 shows how to avoid common errors that interfere with accurate measurements. Makers of measuring equipment recommend checking their accuracy periodically, such as once a month. Calibration of measuring equipment is carried out by using standard weights (or lengths) to confirm accuracy and precision over the range that the equipment measures.

Table 8.3 Typical gains in weight and height for age in infancy[31]

AGE	WEIGHT GAIN grams	WEIGHT GAIN grams (pounds)	LENGTH GAIN mm	LENGTH GAIN mm (inches)
	Per Day	Per Month	Per Day	Per Month
0–3 months	20–30	600–900 (1.3–2)	1	30 (1.2)
3–6 months	15–21	450–630 (1–1.4)	0.68	20 (0.8)
6–12 months	10–13	300–390 (0.7–0.9)	0.47	14 (0.6)

Standard techniques should be used to measure growth; these require practice and consistency. Equipment needed to measure infant growth is different from equipment for measuring children and adults. The scale bed must be long enough to allow the infant to lie down. Length is measured with the infant lying down with a head and foot board at right angles to the firm surface. Positioning the baby quickly and carefully is a skill needed for accurate measurement of recumbent length. In weighing, clothing, hair ornaments, and how much the baby jiggles the scale are examples of factors that could add error to measurements.

Interpretation of Growth Data

The National Center for Chronic Disease Prevention and Health Promotion (CDC) 2000 infant growth charts are based on five national surveys and represent a larger sample of infants than that used for previous growth charts.[1] The infant growth charts are based on infants weighed nude on calibrated scales. The charts take into account differences in the growth patterns for formula-fed compared to breastfed infants, regardless of race or ethnic background.[1] Various growth charts are provided in Appendix A. Growth charts for 0- to 36-month-olds consist of a prepared graph for each gender, showing:

- Weight for age
- Length for age
- Weight for length
- Head circumference for age

The more times a baby has been measured and the growth plotted, the more likely the growth trend will be clear, in spite of minor errors. Measures over time can identify a change in rate of weight or length gain and the need for intervention. Growth is so fast during infancy that it may be easier to determine growth problems during infancy than later. Every month in infancy there is an increase in both weight and length, which is not expected in older children. Warning signs of growth difficulties are lack of weight or height gain; plateau in weight, length, or head circumference for more than 1 month; or a drop in weight without regain within a few weeks.[7] Head circumference increases as a result of brain growth. If head circumference is not increasing typically, neither weight nor height increases are likely to track on standard growth percentiles. In the rare circumstance that an infant has a rapid increase in head circumference, this is not a sign of

Illustrations 8.3 Infant being measured on length board and scale.

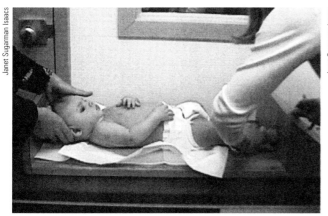

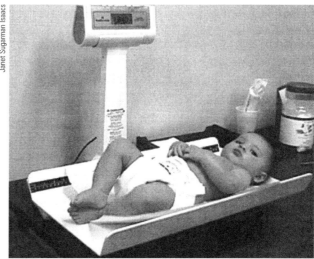

Table 8.4 Measuring growth accurately in infants

To Avoid Measurement Errors

- Use measuring equipment that was calibrated recently.
- Confirm that the scale is on zero before starting.
- Make sure the infant is not holding or wearing anything that adds weight or length.
- Confirm the position of the infant for length measurements:

  Head position—the infant's eyes are looking straight up and the head is in midline, touching the head board.

  Neither hips nor knees are bent.

  Heel is measured with foot flat against the foot board.

- Head circumference measure is at the widest part of the head.

To Avoid Growth Plotting Errors

- Calculate the age accurately in months after confirming the date of birth.
- Confirm plotting on the metric scale if kilograms were measured, not the pound scale.
- Confirm that the plotted weight and length are marked well enough to read easily without being so large as to change percentiles.

Table 8.5 Typical daily volumes for young infants not being breastfed

Age of Infant	Typical Intake of Formula per Day (24 hours)
Birth to 1 month	16–20 fl oz per day, 8–12 feedings/day, 1–2.5 fl oz per feeding
1 to 2 months	18–26 fl oz per day, 8–10 feedings/day, 2–4 fl oz per feeding
2 to 3 months	22–30 fl oz per day, 6–8 feedings/day, 3–5 fl oz per feeding
3 to 4 months	24–32 fl oz per day, 4–6 feedings/day, 4–8 fl oz per feeding

good nutrition or normal growth, but it may signal a condition that requires immediate attention to protect brain development. The rate of weight gain during infancy is not necessarily predictive of future growth patterns after infancy, nor a risk for long-term overweight, compared to the weight-gain pattern later in childhood.[31]

Feeding in Early Infancy

"Food is the first enjoyment of life."

Lin Yutan, *The Importance of Living*

Breast Milk and Formula

The American Academy of Pediatrics and the American Dietetic Association recommend exclusive breastfeeding for the first 6 months of life and continuation of breastfeeding for the second 6 months as optimum nutrition in infancy.[32,33] Infants who are born preterm benefit from breastfeeding, too. Encouragement of breastfeeding right after birth, before the mother's milk supply is available, is an example of a birthing practice that is endorsed.[34] Other recommended practices include teaching safe handling and storage of expressed human milk.[34] The nutrient composition of breast milk is presented in Chapter 6. For young infants less than 4–6 months old, no other liquids or foods are recommended in addition to breast milk

and formula.[6] Recommendations for formula intake of young infants are shown in Table 8.5.

The growth rate and health status of an infant are better indicators of the adequacy of the baby's intake than is the volume of breast milk or formula. Infant formulas for full-term newborns are typically 20 cal/fl oz when prepared as directed. Formula for infants who were born prematurely provide 22 or 24 cal/fl oz. Some health providers recommend further increasing the caloric density of formula for some preterm infants, but such recommendations are not appropriate for most infants. Most infants can be quite flexible in accepting formula, lukewarm or cold, or changes in formula brands.

Table 8.6 gives an overview of the composition of commercially available infant formulas compared to breast milk.[35] Some formulas have been developed for common conditions of healthy infants, such as gastroesophageal reflux (GER) or frequent diarrhea. The specialty formula market appears to be growing, such as follow-up and hypoallergenic formulas. Formula manufacturers are exploring possible additions such as the trace essential micronutrients found in breast milk. Selenium and nucleotides for preterm infants are examples of recent formula additives.[25]

Cow's Milk during Infancy

The American Academy of Pediatrics recommends that whole cow's milk, skim milk, and reduced-fat milks not be used in infancy.[36] Iron-deficiency anemia has been linked to early introduction of whole cow's milk. Low iron availability may come about as a result of gastrointestinal blood loss, low absorption of other minerals (calcium and phosphorus), or the lack of other iron-rich foods in the diet.[37] Studies on infants who were 7.5 months of age confirmed earlier findings that blood loss with whole cow's milk is more likely if the infant had been breastfed earlier rather

Table 8.6 How infant formulas are modified compared to breast milk

MACRONUTRIENTS	BREAST MILK	COW'S-MILK-BASED FORMULA	SOYBEAN-BASED FORMULA
Protein	7% of calories	9–12%	11–13%
Carbohydrates	38% of calories	41–43%	39–45%
Fats	55% of calories	48–50%	45–49%

OTHER WAYS INFANT FORMULAS ARE MODIFIED COMPARED TO BREAST MILK

What Is Modified	How It Is Modified	Examples from Two Major Manufacturers
Calorie level	Increase in calories from 20 calories/fl oz to 22 or 24 calories/fl oz (for preterm infants).	EnfaCare Lipil is 22 calories/fl oz. Similac with Iron 24 is 24 calories/fl oz.
Form of protein	Protein is broken down to short amino acid fragments (hydrolyzed protein) or into single amino acids. Source of protein changed.	Similac Neosure Advance has amino acids. Enfamil Nutramigen has hydrolyzed milk protein. Prosobee has hydrolyzed soy protein in place of milk-based protein.
Type of sugar	Lactose is replaced by other sugars, such as sucrose or glucose polymers from various carbohydrate sources.	Enfamil LactoFree has lactose replaced by corn syrup solids (which provides glucose). Prosobee has carbohydrates from corn syrup solids. Neither has sucrose or lactose.
Type of fat	Long-chain fatty acids partially replaced with medium-chain fatty acids (MCT) and source of fat changed.	Pregestimil has about half of the long-chain fats, replaced by a mixture of vegetable oils. Enfamil Nutramigen has no MCT oil, but has vegetable oils in place of animal-based fats.
Allergy/intolerance	Replacement of milk-based protein with protein from soybeans or replacement of whole proteins with amino acid fragments or single amino acids.	Similac Isomil and Enfamil Prosobee have milk protein replaced by soy protein.
Micronutrients	Increased calcium and phosphorus concentration for preterm infants. Decreased minerals related to renal function. Added essential fatty acids (see above). Lower supplemental iron.	Enfamil PrematureLipil Similac PM 60/40 is modified in calcium, phosphorus, and is low in iron. Similac Special Care Advance 24 is a low-iron formula sold only to hospitals for preterm infants. Enfamil Low Iron and Similac Low Iron have lower levels of iron than the standard formula.
Thickness	Added rice or fiber for gastrointestinal problems.	Similac Isomil D.F. (DF = diarrhea free) for short-term use; it has added fiber from soy. Enfamil A.R. has added rice.
Age of infant	Target age 0–12 months	Similac Isomil Advance
	Target age 9–24 months	Similac Isomil 2

than fed infant formula.[37] In 1997, 39% of infants were fed whole cow's milk at or before age 12 months.[37] The high cost of infant formulas may result in families selecting cow's milk for older infants who are not breastfed.

Development of Infant Feeding Skills

Infants are born with reflexes that prepare them to feed successfully. As noted earlier, these reflexes include rooting, mouthing, head turning, gagging, swallowing, and coordi-nating breathing and swallowing.[9] Infants are also born with food-intake regulation mechanisms that adjust over time with development of the infant.[38] In early infancy, self-regulation of feeding is mediated by the pleasure of the sensation of fullness. Inherent preferences are in place for a sweet taste, which is also a pleasurable sensation.[39] After the first 4–6 weeks, reflexes fade and infants learn to purposely signal wants and needs. However, it is not until much later—about age 3—that children can verbalize that they are hungry. In between the reflexes fading and the child speaking, appetite and food intake are regulated by biological and environmental factors interacting with one another.

Table 8.7 Development of infant feeding skills

Chronological Age	Developmental Milestone	Feeding Skills
Birth to 1 month	Vision is blurry; hears clearly. Head is oversized for muscle strength of the neck and upper body.	Suckling and sucking reflexes. Frequent feedings of 8–12 per 24 hr. Only thin liquids tolerated.
1–3 months	Cannot separate movement of tongue from head movements. Head control emerges. Smiles and laughs. Puts hands together.	Volume increases up to 6–8 fl oz per feeding, so number of feedings per day drops to 4–8 per 24 hr. Sucking pattern allows thin liquids to be easily swallowed. Learns to recognize bottle (if bottle-fed).
4–6 months	Able to move tongue from side to side. Working on sitting balance with stable sitting emerging. Drooling is uncontrolled. Disappearance of newborn reflexes allows more voluntary movements. Teething and eruption of upper and lower central incisors.	Interest in munching, biting, and new tastes. Cannot easily swallow lumpy foods, but pureed foods swallowed. 6–8 fl oz per feeding and 4–5 feedings per day (may be variable if breastfeeding). Holds bottle (if bottle-fed).
7–9 months	Hand use emerges, with pincer grasp and ability to release. Stable independent sitting. Crawling on hands and knees. Starting to use sounds, may say "mama" and "dada."	Self-feeding with hands emerges. Munching and biting emerges. Indicates hunger and fullness clearly. Prefers bottle, but little loss from a held, open cup.
10–12 months	Can pull to stand, standing alone emerges. Enjoys making sounds as if words. Can pick up small objects, such as a raisin. Can bang toys together with two hands. Has consistent routines about bedtime, diaper changing. Usually does not drool anymore.	Likes self-feeding with hands. Spoon self-feeding emerges. Drinks from an open cup as well as from a bottle. Uses upper and lower lip to clear food off a spoon. Enjoys chopped or easily chewed food or foods with lumps. Sitting position for eating. Enjoys table foods even if some baby foods still used.

Table 8.7 shows infant developmental milestones and readiness for feeding skills.[9,10,12] The interaction of biology and environment prevails here, too. For example, depression in a caregiver may be an underestimated variable in the development of infant feeding. Maternal depression may bring about a lower level of interaction between the parent and infant during feeding, reducing the number or volume of feedings and increasing the risk of slower weight gain.[40] Media influences and changes in social practices also affect how babies are fed. Examples are cultural and ethnic perceptions of breastfeeding and the availability of quality child care for infants.

Several models help assess readiness for a breastfed infant to begin eating from a spoon at around 4–6 months. The developmental model is based on looking for signs of readiness, such as being able to move the tongue from side to side without moving the head.[10,12] The infant must be able to keep her head upright and sit with little support before initiating spoon-feeding. Models based on

chronological age as well as those based on cues from the infant are considered outdated. Most infants adapt to a variety of feeding regimens, and various feeding practices can be healthy for them. The parents' ability to read the infant's cues of hunger, satiation, tiredness, and discomfort influence feeding skill progression. The cues infants give may include:

- Watching the food being opened in anticipation of eating
- Tight fists or reaching for the spoon as a sign of hunger
- Showing irritation if the feeding pace is too slow or if the feeder temporarily stops
- Starting to play with the food or spoon as the infant begins to get full
- Slowing the pace of eating, or turning away from food when they want to end the meal
- Stopping eating or spitting out food when they have had enough to eat

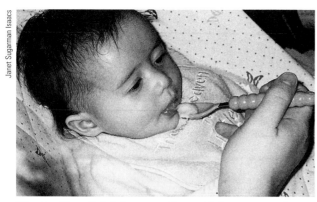

Illustration 8.4 Infant reaching with her tongue for a spoon.

Infants relate positive and pleasurable attributes to satisfaction of their hunger as part of a successful feeding experience. If there are long episodes of pain from gastroesophageal reflux or constipation, these can become the basis for later feeding problems as the association of eating and pleasure is replaced by an association of eating and discomfort.[17] An infant who makes the association between eating and discomfort is likely to be seen as an irritable baby. This may set up a cycle of the infant being difficult to calm and the parent being frustrated. If this cycle is not replaced by the more positive association of eating and pleasure, the feeding difficulty in infancy may later be characterized by pickiness, food refusals, and difficult mealtime behavior in an older child. A negative association of pain and eating may persist, and appear as a behavioral problem at mealtimes.

Introduction of Solid Foods

Infants begin with food offered on a spoon in a small portion size of 1–2 tablespoons for a meal, with one or two meals per day. The purpose of offering food on a spoon to infants at 4–6 months is to stimulate mouth muscle development, and less so for nutritional needs, which are met from breast milk. Watching a baby learn how to eat from a spoon is fun for new parents. If the baby has achieved the developmental milestones in Table 8.7, it may take him only a few days of practice to start spoon-feeding and to learn to consume 1 tablespoon of semisoft food as a meal. Spoon-feeding is really two new experiences for a baby: a spoon is not soft and warm in the mouth like the breast, and whatever food is selected does not feel the same as breast milk on the tongue. At first the baby tries to suck food on a spoon like it was a liquid, so some food will come out of the mouth.

Babies respond strongly to new tastes or smells, regardless of the first food. Introducing a baby to food on a spoon includes these recommendations (in addition to those in the following section on infant feeding position):

- Time the first spoon-feeding experiences for when the baby is not overly tired or hungry, but active and playful.

- Offer a small spoon with a shallow bowl. The temperature of the spoon may have to be considered if it can conduct hot or cold readily.

- Give the baby time to open his or her mouth and extend the tongue toward the food. If the baby cannot extend the tongue farther out than the lower lip, the baby is not ready for spoon-feeding.

- Place the bowl of the spoon on the tongue with slight downward pressure toward the front of the mouth. Touching the back of the tongue may elicit a gag response.

- The spoon should be almost level. It is not a good practice to scrape the food off the spoon with the baby's gums by tilting the spoon handle up too high. The baby's chin should be slightly down to protect the airway.

- The pace of eating should be based on watching for the baby to swallow. Rushing will increase the risk of choking and of the infant having an unpleasant experience.

- First meals may be small in volume—only 5 or 6 baby spoons—and last about 10 minutes, based on the baby's interest.

After mastering the new skill of eating from a spoon, babies quickly teach their parents how to feed them by indicating the rate of eating they like. Common mistakes happen if the person feeding the baby does not read the signs that the baby is giving.

The Importance of Infant Feeding Position

Positioning infants for feeding with a bottle and for eating from a spoon are important because improper positioning is associated with choking, discomfort while eating, and ear infections.[12] The semi-upright feeding position as exemplified in car seats or infant carriers is recommended for the first few months.[10] Unsafe feeding positions, such as propping a bottle or placing the baby on a pillow, increase the risk for choking and overfeeding. The recommended sleeping position for young infants is lying on the back without elevating the head on a pillow. This position is not recommended for feeding, which reinforces why feeding an infant in bed with a bottle is not generally recommended.[6,10]

Spoon-feeding also has a recommended infant feeding position. The infant can better control his mouth and head in a seated position with good support for the back and feet. The person offering the spoon should sit directly in front of the infant and make eye contact without requiring the baby to turn his head.[10,12] A high chair is an appropriate feeding chair when the infant can sit without assistance. The infant should be kept in a sitting position by use of a seat belt so that the hips and legs are at 90 degrees. This position assists the infant's balance and digestion. If the infant is sliding out from under the tray

Case Study 8.1

Photo Disc

Baby Samantha Will Not Eat

Samantha is a healthy 8-month-old girl who lives with her mother Kathy, her father, and her older sister, who is almost 3 years old. Both parents now work full-time, and both children attend day care full-time. Kathy nursed Samantha exclusively before she returned to work, and built up a supply of frozen breast milk. She nurses her twice per day now, early in the morning and before Samantha goes to sleep. Samantha gets breast milk offered in bottles at day care. Samantha is reported by the day care staff to be a good baby. However, when Kathy picks her up after work, Samantha wants to be held and will not sit in her highchair or eat dinner. She cries if she is not held. Samantha's sister wants to eat as soon at they get home. Kathy has so much to do at home after work, she finds it difficult to hold Samantha at such a busy time. Kathy thinks that Samantha must be hungry, and that she would be less irritable if she ate her dinner.

Questions

1. What signs is baby Samantha giving to show that she needs comforting rather than hunger?
2. How might Kathy change her routine to give baby Samantha more attention and meet the needs of her older daughter?
3. At 8 months, is Samantha too young to overeat out of emotional needs?
4. Should Kathy stop or continue breastfeeding to improve Samantha's eating?

(Answers are in Appendix D.)

---

of the high chair with the hips forward, the stomach is under more pressure and spitting up is more likely.

Some apparently healthy infants show resistance to learning feeding skills or react to the introduction of foods in an unusual manner. These problems in early feeding experiences are sometimes warning signs of more general health or developmental difficulties. They may signal emerging problems that cannot be diagnosed until later. Families who call attention to early feeding problems may assist their infants in the long run by having the problem recognized earlier. For example, some infants who start and stop feeding frequently, but then do not feed for several minutes in a row, may later be discovered to have heart problems. The coordination of eating and breathing may have been the basis for the starting and stopping. Some infants are very reluctant feeders and are later diagnosed with a milk protein intolerance. Case Study 8.1 describes a baby who refuses to eat when her mother thinks she must be hungry.

> **Weaning** Discontinuation of breastfeeding or bottle-feeding and substitution of food for breast milk or infant formula.

Preparing for Drinking from a Cup

The process of *weaning* starts in infancy, and usually is completed in toddlerhood. The recommended age for weaning the infant from the breast or from a bottle to drinking from a cup is from 12 to 24 months.[12] Breastfed infants may make the transition to drinking from a cup without ever having any liquids in a bottle. If breastfeeding is continued as recommended for the first year of life, introducing a cup for water and for juices after 6 months is recommended, near the time that foods are offered on a spoon. By the time weaning from breastfeeding is planned, the 1-year-old infant will be skilled enough at drinking from a cup to meet her fluid needs without a bottle. Infants who are not exclusively breastfed, or are breastfed for less than 12 months, need to have additional fluids offered in a bottle because their ability to meet their fluid needs by drinking from a cup are not sufficiently developed. Developmental readiness for a cup begins at 6–8 months.[12] Eight-month-old infants enjoy trying to mimic drinking from open cups that they see in the home. However, the ability to elevate the tongue and control the liquid emerges later, at closer to a year. The 10- to 12-month-old infant enjoys drinking from a held cup and trying to hold his own cup, even though the breast or bottle is the main feeding method. Infants are likely to decrease total intake of calories from breast milk or infant formula if served in a cup, because they are less efficient in the mouth skills needed. At first the typical portion size of fluid from a cup is 1 to 2 ounces. The infant who is weaned too soon may plateau in weight because of decreased total calorie intake. The drop in total fluids consumed may result in constipation. Changing from a bottle to a covered sippy cup with a small spout is not the same developmental step as weaning to an open cup.[12] The mouth skills needed in controlling liquids with the tongue are more advanced with an open cup. The skills learned in drinking from an open cup also encourage speech development.

Food Texture and Development

"They say fingers were made before forks and hands before knives."

Jonathan Swift

Weaning is not complete until the caloric intake from breast milk is provided from foods and beverages. Infants advance from swallowing only fluids to pureed soupy foods at 4–6 months.[6,12] Before that they can move liquids only from the front to the back of the mouth. The mouth is exquisitely sensitive to texture. If food with soft lumps is presented too soon, it causes an unpleasant sensation of choking. When infants are 4–6 months old, they can move their tongues from side to side. At 6–8 months, they are ready for foods with a lumpy but soft texture to elicit munching and jaw movements.[12] These movements simulate chewing. By 8–10 months, infants are able to chew and swallow soft mashed foods without choking. It is important to offer infants foods that do not require much chewing, because infants do not develop mature chewing skills until they are toddlers.

First Foods

The first food generally recommended for infants at 4–6 months is baby cereal, such as iron-fortified cereal mixed with water or breast milk. Rice cereal is a common first food because it is easily digested and *hypoallergenic.* When to add baby cereal or other food to an infant's diet is determined not only by developmental milestones as recommended, but by other reasons, such as these:

- Some parents add baby foods because they think this will make the baby sleep longer. This practice is neither recommended nor effective for most infants. This common belief may result in introduction of baby cereal before the infant has developed the skills to eat from a spoon.

- Some families are instructed by pediatricians to add dry rice cereal as part of the treatment for gastrointestinal problems, because it tends to thicken infant formula.

Fruits and vegetables, such as pears, applesauce, or carrots, are also sometimes first foods for infants. What are considered healthy first foods for infants varies in different cultures and ethnic groups. Regardless of what foods are offered first, the timing and spacing of new foods can be used to identify any negative reactions. Com-

mon recommendations for parents of 6-month-olds are to add only one new food at a time and to offer it over 2 or 3 days. There are specific recommendations regarding the timing and spacing of foods known to trigger food allergies in families with histories of this problem (discussed later in the "Food Allergies and Intolerances" section).

Commercial baby foods are not required by infants. Parents can prepare baby foods at home using a blender or food processor, or by mashing with a fork. Care must be taken, however, to provide a soupy texture and to avoid contamination of home-prepared baby food by bacteria on food or from unsanitary storage methods. The nutrient content of home-prepared baby foods can vary widely depending on how they are prepared and stored. Adding salt and sugar to home-prepared baby foods are examples of variables that can decrease nutritional quality. The advantage of home-prepared baby foods is that a wider variety of foods may be introduced that are likely to be a part of the diet later. Additionally, money not spent on commercial baby foods may be significant savings for some families.

Commercial baby foods are commonly selected because of their sanitation and convenience. Several are listed in Table 8.8 on the next page.[28] Moreover, families who pack food for day care or who travel with infants find commercial baby foods convenient. Parents have a lot of choices to make in selecting baby foods. Selection should be based on the nutritional needs of the infant, not on what is available in local stores and the eating habits of the adult shoppers. Examples of baby foods that may reflect shopper's selections rather than baby needs are fruits with added tapioca or baby food desserts and snack foods. They are not recommended for most infants. Jar sizes of baby foods are based on industry standards, not necessarily recommended portion sizes. Portion sizes for infants should be based on appetite. Finishing an opened jar of baby food may encourage overeating if parents do not pay attention to signs from the infant.

Hypoallergenic Foods or products that have a low risk of promoting food or other allergies.

Many foods eaten by other family members are appropriate foods for infants who are 9–12 months of age. Examples are regular applesauce, yogurt, soft cooked green beans, mashed potatoes, cooked hot cereal, and Cheerios.

Inappropriate and Unsafe Food Choices

New parents may inadvertently select foods for infants based on their own likes and dislikes, rather than on the infant's needs. Such choices are problematic when they increase the risk of choking. Here are some examples of unsafe foods for infants:[6,10,12]

- Popcorn
- Peanuts
- Raisins, whole grapes

Table 8.8 Commercial baby foods[28]

Type	Portion Sizes	Target Age/Skills
Single-ingredient pureed fruits and vegetables, 25–70 cal/jar	71 g (2.5 oz) jar	4–6 months, introduced on a spoon, with portion size of 2–3 tbs/meal
Powdered cereals, 60 cal/serving	Dry: ½ oz or 4 tbs/serving, mixed with water, juice, breast milk, formula, or other liquids	4–8 months
Jarred cereal mixed with fruits, 90 cal/jar	"Wet:" 113 g (4 oz.) jar	4–8 months
Juices, fruit, and vegetables, 60–100 cal/bottle with added vitamin C, calcium	4 and 6 fl oz bottles	4–10 months
Fruits and vegetables, pureed textures, added ingredients, such as tapioca and mixtures	113 g (4 oz) jar, two jars per meal	4–8 months, no munching or food intolerances
Meat mixtures, containing 3–4 g protein/jar, 50–70 cal/jar	113 g (4 oz) jar, one jar per meal	6–9 months, no munching or food intolerances
Desserts, 0–2 g protein/jar and 80–100 cal/jar	113 g (4 oz) jar	6–9 months, no munching
Fruits and vegetables with mixed textures, 0–4 g protein/jar, 70–160 cal/jar	170 gm (6 oz) jar, one jar or 12 tbs/meal	9–12 months, side-to-side tongue movements and munching
Meat-based dinners, with mixed textures, 3–5 g protein/jar and 90–130 cal/jar	170 g (6 oz) jar, one jar or 12 tbs/meal	9–12 months, side-to-side tongue movements and munching
Meats with textures, 10–11 g protein/jar and 70–100 cal/jar	71 g (2.5 oz) jars	9–12 months, side-to-side tongue movements and munching
High-texture baby foods Diced fruits, vegetables, and meats Dinners have pieces as lumps	Dices are 71 or 128 g jars (2.5 or 4.5 oz) Dinners are 170 g (6 oz) jars	Finger foods (to be picked up), requires self-feeding with hands and spoon, chewing
Baked products, such as zwieback toast and biter biscuits	One zwieback toast is 7 gm One biter biscuit is 11 gm	10–12 months, requires biting and munching, limited chewing

- Uncut stringy meats
- Gum and gummy-textured candies
- Hard candy, jelly beans
- Hot dog pieces
- Hard raw fruits or vegetables, such as apples, green beans

Some foods present a choking risk for infants because of their lower chewing skills. Under-chewed pieces of food can obstruct the infant's airway because voluntary coughing and clearing the throat are skills not yet learned.[10] Moreover, the infant may not be able to clear food from the roof of the mouth. A sticky food such as peanut butter against the hard palate may fall to the back of the mouth and present a risk of choking. Foods that do not easily stay together, such as potato chips, also can cause choking. A chip breaks apart in the mouth, but little pieces may stay crunchy. Small pieces may present to the back of the mouth before the infant can use the tongue to move the pieces to the sides and initiate munching.

Water

Breast milk or formula generally provides adequate water for healthy infants for the first 4–6 months.[6,13] Infant drooling generally does not increase the need for water. In hot, humid climates, infants have increased needs for water; but water should not replace breast milk or formula. Added water can be used to meet fluid needs, but not caloric and nutrient needs. All forms of fluids contribute to meeting the infant's water needs. Often parents are reluctant to say they offered their infants sips from their own glasses containing soft drinks or drinks containing caffeine or alcohol. This may be important information to include in a food intake record, especially if the contents are not recommended for babies. The replacement of an infant formula with a less nutritionally rich alternative such as juice, "sports drinks," cola, or tea has been found to be a contributor to lower-quality diets for infants.[41]

Water needs of infants are a concern because dehydration is such a common response to illness in infancy.

The infant has limited ability to signal thirst, especially when sick. Vomiting and diarrhea result in dehydration more rapidly in infants than in older children, with symptoms that are more difficult to interpret.[42,43] Replacement of electrolytes has been the basis for a variety of over-the-counter fluid replacement products, such as Pedialyte, "sports drinks," and Gatorade, that are marketed to parents. These products contain some glucose (dextrose) along with sodium, potassium, and water. The amount of glucose provides significantly lower calories than do breast milk or formula—usually 3 cal per fl oz compared to 20 cal per fl oz. Such liquids can be overused, and they may result in weight loss even in healthy infants. Juice is not needed to meet the fluid needs of infants. The American Academy of Pediatrics recommends that juices need not be introduced into the diet before 6 months of age, and never at bedtime.[41] There is a recommendation for juice after age 1, but not before. If juice is offered to an infant over 6 months, it is recommended to be offered in a cup and not in a bottle. Infancy may be the time in which a habit of excessive juice intake starts, so limiting juice volume is a way to avoid problems later.

How Much Food Is Enough for Infants?

Parents report of infant feeding behavior changes as the infant/parent interaction matures from early to late infancy. During early infancy while the sleep/wake cycle of the infant is irregular, it is common for new parents to interpret all discomfort as signs of hunger. The infant's ability to self-calm is a developmental step that plays out differently with different temperaments and parenting styles.[9] The infant who is quite sensitive to what is happening around him is likely to be viewed as irritable and hungry if he cries frequently. In contrast, the infant who sleeps through usual household noises and is less reactive to her immediate environment is likely to be offered food fewer times per day. As a result of different responses to temperament, a pattern of excessive or inadequate food and formula intakes may result.

The following is an example of excessive intake for a 3-month-old not being breastfed. Total formula intake: 33 fl oz (seven bottles per day, ranging from 3 to 5 fl oz per bottle), offered at 7:30 a.m., 11:00 a.m., 12:45 p.m., 2:30 p.m., 5:30 p.m., 8:45 p.m., and 11:30 p.m. One jar of baby food applesauce, fed on a spoon at 9:00 a.m. This is overfeeding because excessive formula is being offered, along with premature offering of spooned food. The frequency of the bottles being offered suggest that the parents are interpreting the baby's signs of discomfort as hunger when she may have other needs, such as for being held, changed, or calmed by movement or touch. Overfeeding is less likely with breastfeeding.

In the first few months, the oral need to suck is easily confused with hunger by new parents. The typical forward and backward tongue movements of the infant's first attempts to eat from a spoon may seem to be a rejection of food by new parents.[12] The infant appears to spit out the food, but this is a sign of learning to swallow and not necessarily a taste preference. The same food that appears to be rejected will be accepted as the infant learns to move the food from the front to the back of the mouth.[12] It may appear to a parent that the infant does not like a food if he appears to choke. This choking response is more likely based on the position of the spoon on the tongue.[10,12] The mouth is very sensitive, particularly toward the back. If the bowl of the spoon is too far back, it will cause a gagging reaction, regardless of the taste of the food.

How Infants Learn Food Preferences

Infants learn food preferences largely based on their experiences with food. Breastfed infants may be exposed to a wider variety of tastes within breast milk than are infants offered formula.[44] The different foods that the breastfeeding mother eats may result in some flavor compounds being transmitted to the nursing infant. Studies on infants in the age range of 4–7 months showed that acceptance of new foods was more rapid than acceptance of new foods after the first year of life.[40] In the 1920s and 1930s, pediatrician Clara Davis conducted studies on the self-selection of food by weaning infants.[45] Dr. Davis was able to demonstrate that older infants are able to select and consume amounts of food needed to sustain normal growth.[45] Her classic studies were interpreted by later generations as supporting the concept that infants and children will instinctively select a well-balanced diet. However, these studies were subject to misperception because careful attention was not paid to the original methods, in which only nutritious and unsweetened foods were available.[46]

Food preferences of infants are largely learned, but genetic predisposition toward sweet tastes and against bitter foods may modify food preferences. Food preferences developed in infancy set the stage for lifelong food habits. The development of trust and security for an infant are crucial, but this need not be linked to overfeeding the infant.[14] Recognizing an infant's specific needs and responding to them appropriately is important. If offered only a limited variety of foods with little interaction during the meal, infants may learn to refuse to eat as a method of gaining attention. For example, 10-month-old infants enjoy throwing food on the floor just to have someone bring them more. They may enjoy the sound of banging a spoon on a high chair more than eating. Infants who learn to get attention by not eating are likely to manipulate the behavior of adults even more successfully as toddlers.

Nutrition Guidance

Nutrition guidance materials have been developed for parents from many sources, such as the WIC program, makers of infant foods, and professionals such as those in the

Table 8.9 Infant feeding recommendations

Topics	Nutrition Education Sample Content
Appropriate use of infant formula (if not breastfeeding)	Mixing instructions for diluting concentrated formula, keeping formula sanitary by refrigeration, and monitoring how long offered formula is left out. Feeding positions for the infant and bottle, and burping the infant during feedings.
Baby food and sanitation	Serving sizes for infants of different ages, refrigeration and sanitation for opened jars of baby food, problems from mixing different baby foods.
Preparing baby foods at home	Avoiding spices, salt, and pepper in baby foods. Using safe food-handling techniques in preparing and storing servings.
Prevention of dental caries	Recommendations for bedtime and naptime to avoid sugary liquids pooling in the mouth. Identifying liquids that may promote dental caries.
Feeding position	Feeding positions for starting food on a spoon. How to tell if the infant's high chair is safe for feeding.
Signs of hunger and fullness	Identifying early signs compared to later signs of hunger. How infants of various ages communicate at mealtime. Reinforcing and rewarding infant signs of hunger and fullness.
Preventing accidents and injury	Checking temperature of baby foods and liquids. Use of appropriate car seats and safety belts.
Spitting up—when to be concerned	Typical feeding behaviors in young infants. Signs of overfeeding. Spitting up and signs of illness. Discussing signs and symptoms with health providers.

Bright Future in Practice initiative.[6,28,35] The need for nutrition education was demonstrated in a study of mothers and pregnant women, which showed that misunderstanding about infant nutrition was common.[16] Infant feeding recommendations from nutrition education materials are sampled in Table 8.9.

Infants and Exercise

The exercise and fitness benefits for adults do not apply to infants. Providing a stimulating environment is recommended, so infants can explore and move as a part of their developmental milestones. The American Academy on Pediatrics Committee on Sports Medicine policy statement recommends that structured infant exercise programs should not be promoted as being therapeutically beneficial for healthy infants.[47] Infants do not have the strength or reflexes to protect themselves, and their bones are more easily broken than those of older children and adults.

Failure to Thrive (FTT) Condition of inadequate weight or height gain thought to result from a caloric deficit, whether or not the cause can be identified as a health problem.

Organic Failure to Thrive Inadequate weight or height gain resulting from a health problem, such as iron-deficiency anemia or a cardiac or genetic disease.

Supplements for Infants

Specific supplements are recommended for breastfed infants in the United States and Canada, under certain circumstances:

- Fluoride supplements are recommended if the family lives in a place that does not provide fluoridated water.

- If breast milk is the only form of nutrition after 4–6 months, fluoride is recommended.[32]
- Elemental iron (at 3 mg per kg body weight of the infant) may be prescribed if the mother was anemic during pregnancy.[6]
- Vitamin B_{12} may be prescribed if the mother is a vegan.[6,48]
- Vitamin D supplements may be needed if the infant is not exposed to adequate amounts of sunlight.

Supplements may also be prescribed for infants who were born early at low birthweight. They may need vitamin A and E and iron, due to low stores of these nutrients usually accumulated late in pregnancy. A liquid multivitamin and mineral with fluoride is a common prescription for the healthy premature baby, regardless of breastfeeding status.[6]

Common Nutritional Problems and Concerns

Common nutritional problems during infancy are failure to thrive, colic, iron-deficiency anemia, constipation, dental caries, and food allergies. Parents often overestimate the association between eating and these common health problems. Parents should discuss their concerns with the infant's health care providers.

Failure to Thrive

Failure to thrive (FTT) is a diagnosis that can be made during infancy or later. Various terms are used to refine FTT, such as *organic* (meaning a diagnosed medical illness is the basis), *nonorganic* (meaning not based on a medical diagnosis), and mixed type.[49] Although growth failure may be brought about by a variety of medical and social conditions, FTT is primarily used to describe conditions in which a calorie deficit is suspected.[8,50] FTT is an emotionally loaded diagnosis for parents, because the term implies that someone failed. Examples of nonorganic or environmental factors are maternal depression, mental illness, alcohol or drug abuse in the home, feeding delegated to siblings or others unable to respond to the infant, and overdilution of the formula. The relationship of FTT to poverty has been well documented.[51] Examples of organic reasons for FTT commonly found are untreated GER, chronic illnesses such as ear infection or respiratory illness, and *developmental disabilities.* (The connection between FTT and developmental disabilities is further discussed in Chapter 9).[40,42] If there is a medical basis for expecting that the infant will not fit standard growth projections, FTT is not an appropriate term to use. For example, growth for an infant born with IUGR should be based on this medical history and related testing near birth. If this infant at 11 months of age is taken to a new health care setting, FTT may be suspected because of the infant's small size—unless the true cause, IUGR, is revealed.

Table 8.10 provides an example of an assessment that can be used to determine if FTT is present. The assessment of FTT depends on tracking growth. Once FTT is suspected, review of medical records often indicates that growth measurements have been taken in a variety of health care settings with different equipment and personnel, at times when the infant was both well and sick. These records may produce an irregular growth pattern that is difficult to interpret.

Nutrition Intervention for Failure to Thrive

Failure to thrive may be a basis for referral to a registered dietitian. Correction of FTT usually is not as simple as just feeding the baby, but increasing caloric and protein intake is the first step.[49,50] The registered dietitian's role is to assess the growth and nutritional adequacy, establish a care plan, and provide follow-up as part of a team approach. She may work with other specialists concerning medical or psychological aspects. Nutrition interventions may establish caloric and protein intake goals and a feeding schedule to assure adequate nutrition is being provided.[35] Other interventions may include:

> **Nonorganic Failure to Thrive** Inadequate weight or height gain without an identifiable biological cause, so that an environmental cause is suspected.
>
> **Developmental Disabilities** General term used to group specific diagnoses together that limit daily living and functioning and occur before age 21.
>
> **Colic** A condition marked by a sudden onset of irritability, fussiness, or crying in a young infant between 2 weeks and 3 months of age who is otherwise growing and healthy.

- Gaining agreement from the caregivers about how and when intake and weight monitoring will be done
- Enrolling the infant in an early intervention program in the local area
- Arranging for transportation or solving other barriers to follow-up care
- Assessing social supports to assure a constant supply of food and formula (if used)
- Assisting the family in advocating for the infant within the health care delivery system, such as locating a local pediatrician and getting prescriptions filled

FTT is one reason that social service agencies become involved with families. Most new parents handle stress without hurting their infants, but a few react in ways that result in infants presenting with FTT or worse. After investigation, FTT may be determined to be a form of child abuse, as a result of neglect. Some infants diagnosed with FTT become at risk for child abuse and need foster care if the home situation does not improve.

Colic

Colic is the sudden onset of irritability, fussiness, or crying in a young infant.[52] Parents usually think that the infant has abdominal pain. Episodes may have a pattern of onset at the same time of day, for about the same duration every day, with all symptoms disappearing by the third or fourth month. The association of colic symptoms, gastrointestinal upset, and infant feeding practices has been studied, but no definitive cause has been shown.[52]

Table 8.10 Complete nutritional assessment of an infant to rule out failure to thrive

Components:

- Review birth records with attention to weight, length, head circumference, fetal or maternal risk factors such as rate of weight gain during pregnancy, newborn screening results, Apgar scores, and physical exams after birth.
- Review and interpret growth records from all available sources such as primary care, WIC, and emergency room visits.
- Interpret current growth measurements of the infant, including an indication of body composition, such as fat skinfold measurements.
- Review of family structure, education, and social supports with attention to access to food and formula (if not breastfed).
- Analyze and interpret current food and fluid intakes as reported by the primary caregiver(s).
- Review the complete physical assessment and medical history to rule out a biological basis for FTT.
- Observe infant feeding by the primary caregiver and interpret parent-child interactions, feeding duration, and the feeding skills of the infant.

The response to colic is often to change baby formulas if the infant is not breastfeeding, although the change in formulas usually does not change the pattern of colic. Recommendations to relieve colic may include rocking, swaddling, bathing, or other ways of calming the infant, positioning the baby well for eating, or burping to relieve gas.[6,35] One theory about why infants have colic points to the mother's diet while breastfeeding, particularly her consumption of milk or specific foods such as onions. Identifying the origin of colic requires more research.

Iron-Deficiency Anemia

Iron deficiency in infants is less common than iron deficiency in toddlers. Iron reserves in full-term infants reflect the prenatal iron stores of the mother.[53] Women with iron-deficiency anemia during pregnancy pass on less iron to the fetus, a condition that may increase risk of anemia during infancy. Infants who have iron deficiency may be exposed to other risk factors to their overall development, such as low birthweight, elevated lead levels, or generalized undernutrition.[53] Family income at or below the poverty level is also a risk for iron deficiency.[53,54] Research in infants who have long-term and severe iron-deficiency anemia suggests inadequate iron contributes to long-term learning delays from its role in central nervous system development.[15,30] Treatment of iron-deficiency anemia in infancy is generally by prescribed oral elemental iron administered as a liquid.[6,35]

Breastfed infants may be prescribed elemental oral iron and also receive iron through iron-fortified baby cereal after 4–6 months of age. For infants who are not breastfed, a usual source of iron for formula-fed babies is iron-fortified infant formula. Iron from this source improves iron status measured during the first year and is well accepted.[53] In a randomized study of healthy infants, those who received iron-fortified infant formula had by the end of the first year significantly improved biochemical measures of iron status compared to those who received infant formula without added iron. However, there were no differences in the developmental scores of the two groups by 15 months of age.[54]

The level of iron in iron-fortified formula has been 15 mg per liter, or 11.5 mg per quart, based on the RDA of 6 mg of iron for infants up to 6 months and 10 mg for infants from 6 to 12 months.[18] New infant formulas with a lower level of supplementation are also marketed in part as a result of gastrointestinal side effects that have been attributed to iron added to formula.[43] The "low-iron" formula has 8 mg iron per liter, or 4.5 mg of iron per quart.[35] Some manufacturers are not recommending the low-iron formula beyond 4 months of age, because it would not meet the RDA for iron.

Diarrhea and Constipation

Diarrhea and constipation may be attributed to dietary components such as breast milk or use of an iron supplement. Parents think that diarrhea and constipation are related to the infant's diet and want to change the diet or feeding plan to lower gastrointestinal upset. In fact, diarrhea can result from viral and bacterial infections, food intolerance, or changes in fluid intake.[44] Typically, young infants have more stools per day than do older infants, and have them soon after oral intake.[55] The number of stools varies widely from two per day to six per day, decreasing as the infant matures.[55] Parents of breastfed infants generally do not have concerns about constipation, because the infant's stool is generally soft. Infants fed soybean-based infant formulas may have more constipation than those fed a cow's-milk-based formula. Recommendations for avoiding constipation are to assure that the infant is getting sufficient fluids and to avoid medications unless prescribed for the infant. Some parents use prune or other juices that have a laxative effect for an older infant, but there is a risk of creating a fluid imbalance and subsequent diarrhea.[43] Foods with high dietary fiber are generally not recommended for infants with constipation, because many sources such as whole-wheat crackers or apples with peels present a choking hazard and are not recommended for infants.

The cause of diarrhea during infancy may or may not be identifiable. Diarrhea in an infant may become a serious problem if the infant becomes dehydrated or less responsive.[43] Most infants have one or two days of loose stools without weight loss or signs of illness, such as after getting immunizations. General recommendations are to continue to feed the usual diet during diarrhea.[6,35] Breast milk does not cause diarrhea. During a bout of diarrhea, continuing adequate intake of fluids such as breast milk or infant formula is generally sufficient to prevent dehydration.

Prevention of Baby-Bottle Caries and Ear Infections

Baby-bottle caries are found in children older than 1 year, but are initiated by feeding practices during infancy. Infants have high oral needs, which means they love to suck, and to explore by putting things in their mouths. They derive comfort from sucking and may relax or fall asleep while sucking. The use of a bottle containing formula, juice, or other high-carbohydrate foods to calm a baby enough to sleep, however, may set her up for dental caries.[56] During sleep the infant swallows less, allowing the contents of the bottle to pool in the mouth. These pools of formula or juice create a rich environment for the bacteria that cause tooth decay to proliferate, increasing the risk for tooth decay.

Risk for ear infections is also correlated with excessive use of a baby bottle as a bedtime practice, as a result of the feeding position.[56] The shorter and more vertical tubes in the ears of infants are under different pressure during the process of sucking from a bottle.[35] If the infant is feeding by lying down while drinking, the liquid does not fully drain from the ear tubes. The buildup of liquid in the tubes increases the risk of ear infections. In a study of over 200 infants, pacifiers and bottle-feeding were correlated with greater prevalence of ear infections.[56] Infants who were breastfed did not have as high a rate of ear infections.[56]

Here are some good feeding practices to limit baby-bottle caries and ear infections related to baby bottles:

- Limit the use of a bottle as part of a bedtime ritual.
- Offer juices in a cup, not a bottle.
- Put only water in a bottle if offered for sleep.
- Examine and clean emerging baby teeth to prevent caries from developing.

Food Allergies and Intolerances

The prevalence of true food allergies is higher in younger than in older children. About 6–8% of children under four years of age have allergies that started in infancy.[57,58] An infant may develop a food allergy to the protein in a cow's-milk-based formula over time. Often such a problem follows a gastrointestinal illness. When the infant is well, protein is broken down during digestion so that absorption in the small intestine is of groups of two or three amino acids linked together. After an illness, small patches of irritated or inflamed intestinal lining may allow protein fragments of larger lengths of amino acids to be absorbed. Such fragments are hypothesized to trigger a reaction as if a foreign protein had invaded, setting up a local immune or inflammatory response.[58] This absorption of intact protein fragments is the basis for allergic reactions. When this happens with cow's-milk protein, it is likely that soy-based formulas will also cause the same allergic reaction.

The most common allergic reactions are respiratory and skin symptoms, such as wheezing or skin rashes.[57] Food allergies are confirmed by specific laboratory tests after infancy.[57] True allergies can present as an array of reactions building up over time, so that it may take several years for the initiating cause to be identified.

Food intolerances are frequently suspected in infants. Families may consider skin rashes, upper airway congestion, diarrhea, and other forms of gastrointestinal upset to be food allergies, but often they are not.[57] As in the case of food allergies, infants are not usually given all the tests for food intolerance, but instead are treated for symptoms. The infant with suspected protein intolerance may be changed to a specialized formula composed of *hydrolyzed protein*.[58] Because the protein of a hydrolyzed formula is already broken down, it does not trigger the same response as intact protein fragments do. A family with a known allergy or intolerance may lower the risk of the allergy occurring in their infant by breastfeeding, and by postponing into the second or third year introduction of allergy-causing foods such as wheat, eggs, and peanut butter. It is important for families not to overly restrict such foods thought to cause allergies from an infant's diet unless required. If many foods are being avoided, there may be consequences such as decreasing the nutritional adequacy of the diet and reinforcing behaviors of rejecting foods and limiting variety. Allergy and intolerance symptoms are more common in response to nonfood items such as grasses and dust, so many different sources of symptoms must be considered.

Lactose Intolerance

Lactose intolerance is a food intolerance in infancy characterized by cramps, nausea, and pain, and by alternating diarrhea and constipation. Infants who are breastfeeding may develop lactose intolerance, because breast milk has lactose in it.[43,35] Gastrointestinal infections may temporarily cause lactose intolerance because the irritated area of the intestine interferes with lactase production.[42] The ability to digest lactose generally returns shortly after the illness subsides. *Lactose* is in all dairy products, so it is in cow's-milk-based infant formulas.[43] Lactose-free infant formulas are soybean based or based on lactose-free cow's milk. Lactose intolerance is less common during infancy than at older ages in groups that are susceptible to it. An infant who was fed a lactose-free formula is likely to be able to eat dairy products later. Because dairy products are such an important source of calcium, introducing foods with low lactose is recommended for older infants who appeared to be lactose intolerant when younger.

> **Hydrolyzed Protein Formula** Formula that contains enzymatically digested protein, or single amino acids, rather than protein as it naturally occurs in foods.
>
> **Lactose** A form of sugar or carbohydrate composed of galactose and glucose.

Cross-Cultural Considerations

Commercial baby foods reflect the bias of the dominant American culture. There is no ethnic diversity in baby foods—no collards or Mexican beans. Many successful avenues to nourish a healthy infant are available, and room ought to be made for different cultural patterns in the development of feeding practices. Some cultural practices are clearly unsafe and must be discouraged, such as a mother prechewing meats for a baby. Cultural practices that support the development of competence in parents can be encouraged, even if not part of the dominant culture. Examples of practices that may reflect cultural choices are swaddling an infant, or having an infant sleep in the parent's bed or in a room with a certain temperature. Practices based on family traditions may be forms of social support for new parents. Only if new parents have not considered the safety of the infant or have little knowledge of other, safer alternatives should cultural practices be discouraged. For example, it may be a cultural practice to offer meats to adults but not to infants. Parents should be informed that older infants may safely eat meats that are cut up or soft cooked to avoid causing choking. Some cultures consider meat-based soups as infant foods.

Cultural considerations may affect the family's willingness to participate in assistance programs such as WIC or early intervention programs. The dignity of the family unit, including extended relatives, has to be considered when counseling families about infant feeding practices. Food-based cultural patterns may be part of a religious tradition, so sensitivity to the family unit would include recognizing and understanding such beliefs.

Vegetarian Diets

Studies have found that infants who receive well-planned vegetarian diets grow normally.[59] The most restrictive diets, vegan and macrobiotic diets, have been associated with slower growth rates in infancy, particularly if infants do not receive enough breast milk.[48,59] Breastfed vegan infants should receive supplements containing vitamin D, vitamin B_{12}, iron, and possibly iron and zinc.[59] The composition of the breast milk from vegan mothers may differ in small ways from standard breast milk.[48] An example is in the ratio of types of fat, although the total fat in breast milk from vegan mothers is the same. Impacts of these differences on the health of infants are generally not known.

Galactosemia A rare genetic condition of carbohydrate metabolism in which a blocked or inactive enzyme does not allow breakdown of galactose, causing serious illness in infancy.

Hypothyroidism Condition in which thyroid hormone is not produced in sufficient quantities, interfering with growth and mental development if untreated in infants.

Vegetarian diets range from adequate to inadequate, depending on the degree to which the diet is restricted, just as diets for omnivores range from adequate to inadequate.[59] Either food sources, such as fortified infant cereals and soymilk, or supplements can be used to assure adequate vitamin and mineral intake. Vegetarian families who avoid all products of animal origin, including milk and eggs, require carefully selected fortified foods or a higher degree of supplementation for their infants.[59] Periodic assessments of dietary intake, growth, and health status can be used to monitor the infant fed a vegetarian diet. Vegetarian infants have similar risk for developing food allergies from soy products, wheat, and nuts as do other infants.

Nutrition Intervention for Risk Reduction

The Early Head Start program is an example of a federal program that is focused on preventing and reducing risks to infant development.[6] The Early Head Start program was developed to work with infants and their families, especially new families at risk due to drug abuse, infants with disabilities, or teenage mothers. Nutrition services are among a wide range of services typically offered in an Early Head Start program. Other services may include home-based early childhood education, case management, and mental health support services, as well as health and socialization services. The Early Head Start program assists families in coordinating WIC participation with food stamps, routine well-baby visits, and day care, as needed.

Model Program: Newborn Screening and Expanded Newborn Screening

In the United States and many other countries, all newborns are screened for rare conditions that may cause disability or death. Such screening, from a small dried blood spot, was initiated in the 1960s after early treatment of phenylketonuria (PKU) was shown to prevent later mental retardation in young children.[24] Most states screen for three to six different conditions, such as PKU, *galactosemia, hypothyroidism,* and sickle-cell disease.[24] New technology has resulted in expanded newborn screening, so some states are now testing for as many as 30 different conditions from the same dried-blood test.[24] Many of the disorders that can be detected by expanded newborn screening are treated by diet. Dietary treatment avoids the substance that has a metabolic block and replaces other dietary components that are usually provided in the foods that are avoided.[24] Expanded infant screening for genetic disorders is likely to continue to expand nutrition knowledge overall.

WIC

WIC serves almost 47% of all infants born in the United States.[60] WIC Program participation is related to improved infant feeding practices, increased rates of breastfeeding, and improved iron status of infants. The WIC program also functions as a gateway program to other health services for low-income families with infants, including immunization services, medical care, and social services.

Resources

American Academy of Pediatrics

Reliable and credible sources of pediatric medical expertise in position papers, with sections for consumers and health care providers.

Web site: www.pediatrics.org

American Dietetic Association

Consumer and provider information includes child nutrition and health and access to credible resources.

Web site: www.eatright.org

Food Allergy Network

Reliable and scientifically based information for families with diagnosed allergies. This site routinely includes recipes.

Web site: **www.foodallergy.org**

Health/Infants

Consumer and marketing information about infant growth and development, based on information available to mass media.

Web site: www.cnn.com/health/indepth.health/infants

National Center for Growth Statistics (source of growth charts)

Site for obtaining growth charts and guidelines for their use.

Web site: www.cdc.gov/nchs

Nutrition/WIC (part of United States Department of Agriculture)

Nutrition education materials and information for low-income families.

Web site: www.nal.usda.gov/fnic/Wicdbase

References

1. National Center for Health Statistics: NCHS growth curves for children 0–19 years. U.S. Vital and Health Statistics, Health Resources Administration. Washington, DC: U.S. Government Printing Office; 2000.

2. U.S. Vital and Health Statistics. Births: preliminary data for 2002. National Vital Statistics Report 2003:51;1–20.

3. U.S. Vital and Health Statistics. Infant mortality statistics from the 2000 period linked birth/infant death data set. National Vital Statistics Report 2002:50;1–26.

4. United States Department of Health and Human Services Centers for Medicare and Medicaid Services; www.cms.hhs.gov/schip, accessed 11/15/2003.

5. United States Department of Health and Human Services National Center for Chronic Disease Prevention and Health Promotion; http://www.cdc.gov/nccdphp/dnpa/pednss.htm, accessed 11/15/2003.

6. Story M, Holt K, Sofka D, eds. Bright Futures in practice: nutrition. National Center for Education and Child Health, Arlington, VA 2000:25–56, 266–70.

7. Beaver PK. Gestational age and birth weight. In Lowdermilk DL, Perry SE, and Bobak, IM, eds. Maternity and women's health care. 6th ed. St. Louis, MO: Mosby; 1997: 1015–42.

8. DC Board of Dietetics and Nutrition Movements for District of Columbia Children 2003;1:2.

9. Kail RV, Cavanaugh JC. Human development: a lifespan view. 2nd ed. Belmont, CA: Wadsworth/Thomson Learning; 2000: 83–121.

10. Cech D, Martin ST. Functional movement development across the life span. Philadelphia: W. B. Saunders; 1995: 76–86.

11. Santrock JW. Life-span development. 7th ed. New York, NY: McGraw Hill; 1999: 124–9.

12. Colangelo CA. Biomechanical frame of reference. In: Kramer P, Hinojosa J. Frames of reference for pediatric occupational therapy. Philadelphia: Williams and Wilkins; 1993: 233–56.

13. Institute of Medicine Food and Nutrition Board. Dietary Reference Intakes for energy, carbohydrate, fiber, fat, fatty acids, cholesterol, protein, and amino acids. Washington, DC: National Academy Press; 2002.

14. Ramey CT, Ramey SL. Right from birth, building your child's foundation for life. New York: Goddard Press; 1999.

15. Batshaw ML. Chromosomes and heredity. In: Batshaw, ML ed. Children with disabilities. 5th ed. Baltimore, MD: Paul H. Brookes; 2002: 3–26, 770.

16. Hobbie C, Baker S, Bayerl C. Parental understanding of basic infant nutrition: misinformed feeding choices. J Pediatr Health Care 2000;14:26–31.

17. Anderson DM. Nutrition for the low-birth weight infant. In: Mahan LK, Escott-Stump S, eds. Krause's food, nutrition and diet therapy. 10th ed. Philadelphia: W.B. Saunders; 2000: 214–38.

18. Trumbo P, Yates AA, Schlicker SA et al. Dietary reference intakes; vitamin A, vitamin K, arsenic, boron, chromium, copper, iodine, iron, manganese, molybdenum, nickel, silicon, vanadium and zinc. J Am Diet Assoc 2001;101:294–301.

19. American Academy of Pediatrics. Prevention of rickets and vitamin D deficiency: new guidelines for vitamin D intake. Pediatrics 2003;111:908–10.

20. Report of the Scientific Committee on Food on the Revision of Essential Requirements of Infant Formulae and Follow-Up Formulae, European Commission Health and Consumer Protection Directorate-General. Adopted April 4, 2003; Brussels, Belgium.

21. Retsinas, G. The marketing of a superbaby formula. New York Times Money and Business/Financial desk 6/1/2003.

22. Elias ER, Irons MB, Hurley AD et al. Clinical effects of cholesterol supplementation in six patients with Smith-Lemli-Opitz syndrome. Am J Med Genetics 1997;68: 305–10.

23. Birch, EE et al: A randomized controlled trial of long-chain polyunsaturated fatty acid supplementation of formula in term infants after weaning at 6 wks of age. American Journal of Clinical Nutrition 2002;75: 570–80.

24. Batshaw ML, Tuchman M. PKU and other inborn errors of metabolism. In: Batshaw, ML ed. Children with disabilities. 5th ed. Baltimore, MD: Paul H. Brookes; 2002: 333–45.

25. Carver JD. Advances in nutritional modifications of infant formulas. Am J Clin Nutr 2003;77(suppl):1550S-4S.

26. Yates AA, Schlicker SA, Suitor CW. Dietary reference intakes; the new basis for recommendations for calcium and related nutrients, B vitamins, and choline. J Am Diet Assoc 1998;98:699–706.

27. Hampl JS, Betts NM, Benes BA. The "age+5" rule: comparisons of dietary fiber intake among 1–4-year-old children. J Am Diet Assoc 1998;98:1418–23.

28. Gerber Products Company. Nutrient values. Fremont, MI; 2000.

29. Ballew C, Khan LJ, Kaufman R et al. Blood lead concentration and children's anthropometric dimensions in the Third National Health and Nutrition Examination

Study (NHANES III), 1988–1994. J Pediatr 1999;134:623–30.

30. McLaren DS. Clinical manifestations of human vitamin and mineral disorders: a resume. In: Shils ME et al., eds. Modern nutrition in health and disease. 9th ed. Baltimore, MD: Williams and Wilkins; 1999: 485–503.

31. Dietz WH. Health consequences of obesity in youth: childhood predictors of adult disease. Pediatrics 1998;101:518S–25S.

32. American Academy of Pediatrics Work Group on Breastfeeding. Breastfeeding and the use of human milk. Pediatrics 1997;100:1035–39.

33. American Dietetic Association. Position of the American Dietetic Association: promotion and support of breast-feeding. J Am Diet Assoc 1997;97:662.

34. Trachtenbarg DE, Golemon TB. Care of the premature infant: part 1. Monitoring growth and development. Amer Family Physician 1998;57;2123–30.

35. American Dietetic Association. Pediatric manual of clinical dietetics. N Nevin-Folino, ed., 2003; appendix 1.

36. American Academy of Pediatrics. Committee on Nutrition (RE9251). Policy Statement. The use of whole cow's milk in infancy. Pediatrics 1992;89:1105–09.

37. Ziegler EE, Jiang T, Romero E et al. Cow's milk and intestinal blood loss in late infancy. J Pediatrics 1999;135:720–6.

38. Birch LL, Gunder L, Grimm-Thomas K et al. Infants' consumption of a new food enhances acceptance of similar foods. Appetite 1998;30:283–95.

39. Capaldi ED, Powley TL, eds. Taste, experience, and feeding. Washington, DC: American Psychological Association; 1990: 75–93.

40. Cadzow SP, Armstrong KL, Fraser JA. Stressed parents with infants: reassessing physical abuse risk factors. Child Abuse & Neglect 1999;23:845–53.

41. American Academy of Pediatrics. The use and misuse of fruit juice in pediatrics. Pediatrics 2001;107:1210–13.

42. Sandritter T. Gastroesophageal reflux disease in infants and children. J Pediatr Health Care 2003;17(4):198–205.

43. Baldassano RN, Cochran WJ. Diarrhea. In: Altschuler SM, Liacouras CA, eds. Clinical pediatric gastroenterology. Philadelphia, PA: Harcourt Brace and Co.; 1998: 9–18.

44. Mennella JA, Beauchamg GK. Maternal diet alters the sensory qualities of human milk and the nurslings' behavior. Pediatrics 1991;88:737–44.

45. Davis CM. Self-selection of diet by newly weaned infants: an experimental study. Am. J Diseases of Children 1928;36:651–79.

46. Story M, Brown JE. Do young children instinctively know what to eat? New England J Med 1998;103–6.

47. American Academy of Pediatrics. Policy Statement (RE8132). Infant exercise programs. Pediatrics 1988;82:800. Reaffirmed 11/94.

48. Mangels AR, Messina V. Considerations in planning vegan diets: infants. J Am Diet Assoc 2001;101:670–7.

49. Khoshoo V, Reifen R. Use of energy-dense formula for treating infants with non-organic failure to thrive. Eur J Clin Nutr (England). 2002;56(9):921–4.

50. Goldsmith E. Nonorganic failure to thrive. MCN Am J Matern Child Nurs. 2001; 26(4):221.

51. Needell B, Barth RP. Infants entering foster care compared to other infants using

birth status. Child Abuse & Neglect 1998;22:1179–87.

52. Metcalf T, Irons TG, Sher L, et al. Simethicone in the treatment of infant colic: a randomized, placebo-controlled multicenter trial. Pediatrics 1994;94:29–34.

53. Recommendations to prevent and control iron deficiency in the United States. MNWR April 3, 1998;47.

54. Moffatt MEK, Longstaffe S, Besant J et al. Prevention of iron deficiency and psychomotor decline in high-risk infants through use of iron-fortified infant formula: a randomized clinical trial. J. Pediatics 1994;125:527–34.

55. Wenner WJ. Constipation and encopresis. In: Altschuler SM, Liacouras CA, eds. Clinical pediatric gastroenterology. Philadelphia, PA: Harcourt Brace and Co.; 1998: 165–8.

56. Jackson JM, Mourino AP. Pacifier use and otitis media in infants twelve months of age or younger. Pediatric Dentistry 1999;21:255–60.

57. Sicherer SH, Morrow EH, Sampson HA. Dose-response in double-blind, placebo-controlled oral food challenges in children with atopic dermatitis. J Allergy and Clinical Immunology 2000;105:582–6.

58. Falci, KJ, Gombas KL, Elliot EL. Food allergen awareness: an FDA priority. Food Safety Magazine 2001:Feb–March.

59. Sanders T. Vegetarian diets and children. Pediatric Clinics of North America 1995;42(4):955–65.

60. U.S. Department of Agriculture WIC Program; www.usda.fns, accessed 10/30/2003.

"Man eats to live,
he does not live to eat."
—Abraham ibn Ezra, Spanish poet
and scientist (1092–1161)

Chapter 9

Infant Nutrition:

Conditions and Interventions

Chapter Outline

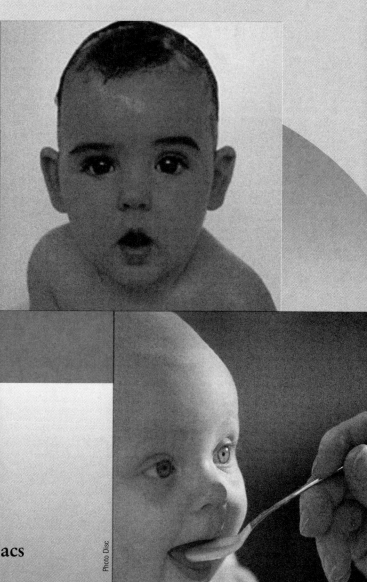

Prepared by **Janet Sugarman Isaacs**

Key Nutrition Concepts

1 Infants who are born preterm or who are sick early in life often require nutritional assessment and interventions that ensure that they are meeting their nutritional needs for growth and development.

2 Early nutrition services and other interventions can improve long-term health and growth among infants born with a variety of conditions.

3 The number of infants requiring specialized nutrition and health care is increasing due to the improved survival rates of small and sick newborns.

Introduction

Most infants are born healthy and then grow and develop in the usual manner. This chapter addresses the nutritional needs of infants who have health problems before or shortly after birth and are at risk for health or developmental difficulties. Within the first year of life, most infants have minor illnesses that do not interfere with growth and development. However, infants who were sick or small as neonates are likely to have conditions that may change the course of growth or development. *Children with special health care needs* is a broad term that includes the infants discussed in this chapter. As in Chapter 8, this chapter models sensitive communication with families by avoiding the word *normal* and, by implication, *abnormal* when referring to infants with special health care needs. Similarly, the designation *normal growth* is replaced by the phrase *standard growth* when referring to the CDC growth charts. Language such as "She is below normal on the growth charts" is replaced with "family-friendly" language, such as "She is the weight of a typical 6-month-old." Most families use the word *premature* (or *prematurity*) comfortably, but *preterm birth* is the conventional usage in maternity care. Both terms refer to infants born before 37 weeks of gestation, and both are used in this chapter.

Children with Special Health Care Needs A federal category of services for infants, children, and adolescents with, or at risk for physical or developmental disability, or with a chronic medical condition caused by or associated with genetic/metabolic disorders, birth defects, prematurity, trauma, infection, or perinatal exposure to drugs.

Low-Birthweight Infant (LBW) An infants weighing <2500 g or <5 lb 8 oz at birth.

Very Low Birthweight Infant (VLBW) An infant weighing <1500 g or <3 lb 5 oz at birth.

Extremely Low Birthweight Infant (ELBW) An infant weighing <1000 g or 2 lb 3 oz at birth.

Neonatal Death Death that occurs in the period from the day of birth through the first 28 days of life.

Perinatal Death Death occurring at or after 20 weeks of gestation and through the first 28 days of life.

Down Syndrome Condition in which three copies of chromosome 21 occur, resulting in lower muscle strength, lower intelligence, and greater risk for overweight.

Seizures Condition in which electrical nerve transmission in the brain is disrupted, resulting in periods of loss of function that vary in severity.

Infants at Risk

The overall U.S. infant mortality rate decreased 45% between 1980 and 2002.[1] It appears, however, that the health care system has been more successful at saving ill infants than in preventing preterm birth, low birthweight, or chronic conditions. The number of infants requiring nutritional services is increasing in large measure because of advances in neonatal intensive care. Small preterm infants who did not survive in the past are now being "saved." These are *low-birthweight infants, very low birthweight infants,* and *extremely low birthweight infants* who require the most intensive resources to support life and account for many *neonatal* and *perinatal deaths.* The smallest living newborns, who weigh 501–600 grams (1 pound 2 ounces to 1 pound 5 ounces), have a 31% chance of survival at birth.[2] This birthweight range corresponds to about 23 weeks of gestation in the second trimester of pregnancy. Infants with birthweights of 901–1000 g (2 lb to 2 lb 3 oz) are in the 29-week range of gestation and have an 88% change of survival.[2] Infants with genetic disorders, malformations, or birth complications have also benefited from advances in treatment and are less likely to die in infancy. However, they are much more likely to have chronic conditions, with increased need for medical, nutritional, and educational services later.

Regardless of what condition is involved, these nutrition questions are likely:

- How is the baby growing?
- Is the diet providing all required nutrients?
- How is the infant being fed?

In-depth nutrition assessments make sure nutrition is not limiting an infant's growth and development. Such assessments are needed by three main groups of infants:

- Infants born before 34 weeks of gestation. Preterm infants are born at less than 37 weeks of gestation, but generally only those born before 34 weeks of gestation have higher nutritional risks.
- Infants born with consequences of abnormal development during pregnancy, such as infants born with heart malformations as a result of the heart not forming correctly or exposure to toxins during gestation. Exposure to alcohol during gestation may interfere with brain formation and result in permanent changes in brain function. This second category includes infants with genetic syndromes, such as *Down syndrome.*
- Infants at risk for chronic health problems. Risks may come from the treatment needed to save their lives, or from the home environment that the baby enters. Examples of conditions that increase risks are *seizures* or cocaine withdrawal symptoms. Long-term consequences, such as later learning problems, may not be known for years.

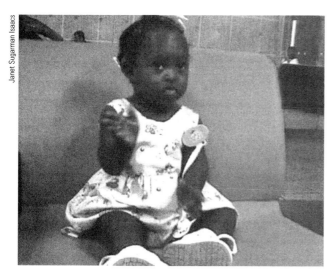

Illustration 9.1 Infant girl with Down syndrome, after her heart surgery.

Families of Infants with Special Health Care Needs

Every parent's wish is that his or her baby will be healthy. When parents find out their newborn has medical problems, they grieve for the loss of the perfect child of their dreams. The emotional impact of having a sick newborn overwhelms many parents, and providers of services for these families must be sensitive to parents' emotional needs. Coping styles of various family members vary, even if they are well prepared. It may take over a year for some family members to understand how the baby is doing and to adjust to the special needs of the child. For conditions with long-term consequences, the first year may not be long enough for parents to see that their infant is developing differently from other babies.

Energy and Nutrient Needs

Nutrient requirements for infants with health conditions are based on the recommendations for healthy infants.[3] Specific nutrients may be adjusted higher or lower based on the health condition involved. Adjusting the diet to changing conditions and close monitoring of growth and development may result in changing recommendations quite frequently in the first year. Scientific frontiers of medicine, genetics, nutrition, and technology interact in caring for sick infants. Nutrition requirements are not known for every condition, and individuals respond at their own pace of growth and development, so many nutrition recommendations are based on the best judgment under the circumstances.

Energy Needs

For infants with special health care requirements, caloric needs may be the same, less, or more than the RDA for infants (108 calories per kilogram body weight).[3] Some conditions in newborns that have caloric needs based on the RDA are cleft lip and palate or phenylketonuria (PKU). The more common situation is that caloric needs are increased. Caloric needs of sick infants can be estimated with measurements such as *indirect calorimetry*.[4] Machines that measure indirect calorimetry are often available in intensive care nurseries. Estimated energy needs can change, however, with medications, activity, health conditions, and growth. Such estimates show that sick and small infants vary in their caloric needs more widely than expected, and that caloric deficits in preterm infants may be more common than previously known.[5] Extra calories are needed in circumstances such as the following:

> **Indirect Calorimetry** Measurement of energy requirements based on oxygen consumption and carbon dioxide

- Infections
- Fever
- Difficulty breathing
- Temperature regulation
- Recovery from surgery and complications

Infants who are born preterm at less than 34 weeks of gestation particularly need higher energy intakes. The American Academy of Pediatrics suggests that premature infants need 120 cal/kg.[3] Intakes for recovering premature babies may be even higher, and the range of caloric need can be wide.[5,6] The European Society for Gastroenterology and Nutrition gives a caloric intake range from 95–165 cal/kg.[7] The amount of calories needed to gain 15 grams per day is recommended.[5] Infants born with VLBW or ELBW may still be weak and have difficulty feeding when they come home. Their higher calorie needs may be more difficult to meet given the small volume that they can consume. Over time, the recovering infant may increase intake so that even 180 cal/kg per day may be consumed.

Some infants need less energy than the RDA. These are infants born with smaller muscles or lower activity as a result of the inability to move certain muscles. An infant who has Down syndrome or one with repaired spina bifida needs fewer calories than the RDA of 108 cal/kg body weight.[3] Too many calories would interfere with her efforts to crawl; her weak muscles would have more body weight to move.

Protein Requirements

As noted for energy needs, protein requirements of infants with special health care needs may be higher, lower, or the same as other infants. Protein requirements based on the RDA of 2.2 grams of protein per kilogram body weight are recommended if the condition does not affect growth or digestion.[3] Protein recommendations are sufficient if total calories are high enough to meet energy needs.[8] The concept of protein sparing is important in fast-growing infants. If enough energy is available from glucose generated from foods containing fats and carbohydrates, then the amino acids generated from protein-containing foods are spared; that is, the amino acids are available for growth. If glucose from foods is not sufficient for meeting energy needs, however, amino acids from digestion of protein foods will be used to meet energy needs, and therefore less will be available for growth. When total caloric intake is low, protein-rich foods become an energy source. In this circumstance, providing the RDA for protein may be inadequate and result in slow growth. With preterm infants a sign of inadequate protein may be slow head growth, which is an indicator of brain growth.[9]

> **MCT Oil** A liquid form of dietary fat used to boost calories; composed of medium-chain triglycerides.
>
> **Catch-Up Growth** Period of time shortly after a slow growth period when the rate of weight and height gains is likely to be faster than expected for age and gender.
>
> **Hypocalcemia** Condition in which body pools of calcium are unbalanced, and low levels are measured in blood as a part of a generalized reaction to illnesses.

Conditions that could slow growth may require higher protein levels than the RDA for protein for infants. Higher protein recommendations are common in early infancy for conditions such as recovery from surgery or LBW. Protein intakes of 3.0–3.5 g/kg are appropriate for premature and recovering infants.[8] For recovering from some complications of ELBW, high protein intakes, as much as 4 g/kg, appear safe with adequate fluids, and without kidney problems.[8] The importance of protein to growing neonates is hard to overemphasize. Protein deficits in preterm infants have been shown even when high protein is provided, depending on how small and sick they are and how soon after birth the infant is being fed.[5]

Protein recommendations lower than the RDA are unusual in infancy. Infants with lower muscle activity as a result of smaller-sized muscles generally need lower protein. One example is Down syndrome. Conditions that lower physical activity and movement are often not identified until the infant is old enough to be moving around, closer to the end of the first year. During infancy, muscle tone is known to change over time, so muscle coordination and movement problems are usually confirmed later.

FORM OF PROTEIN Many illnesses interfere with the functioning of the gastrointestinal tract and digestion, even though newborns are born with intact enzymes for protein digestion. Protein and fat digestion depend on liver and pancreatic enzymes for intestinal absorption. However, many conditions associated with preterm birth and illness stress the liver and reduce its ability to function, causing changes in protein and fat digestion. Sick infants may require forms of protein in which amino acids are in short chains, such as in hydrolyzed protein, or single amino acids.[9] Other examples of infants needing protein that has been broken down are those with metabolic disorders.[10] Total protein may be limited, and partially replaced by mixtures of specific amino acids. For infants with PKU, meats and dairy product intake have to be limited because they contain too much of the amino acid phenylalanine.

Fats

Infants need a high-fat diet compared to older people, because fats provide energy. Up to 55% of calories from fats may be recommended.[9] The need for calories provided by fats is especially important in sick or recovering infants. Low-fat diets are generally not recommended for infants. Conditions that require limiting fats for infants are uncommon. One example is very sick infants who require heart-lung bypass machines as a part of major surgery. Such infants are given low-fat diets after age 2, without a specific level of fat restriction.[8] Fats are more difficult to absorb for infants with VLBW or ELBW because they require pancreatic and liver enzymes.[9] These enzyme systems may also be impaired in sick infants. Naturally occurring long-chain fats in breast milk may be supplemented with medium-chain fatty acids for sick infants. Medium-chain triglycerides do not require bile for absorption, so they are preferred.[9] Making sufficient bile for digesting long-chain fats requires healthy livers, like those in full-term infants, that are not likely in preterm births. **MCT oil** can be added to ensure calories from fats are available. Additionally, the essential fatty acids—alpha-linolenic and linoleic acid, as well as docosahexaenoic acid (DHA) and arachidonic acid (AA)—are provided in breast milk, human-milk fortifier, or special formulas.[11]

Vitamins and Minerals

DRIs for vitamins and minerals are appropriate for many infants with health conditions because recommendations are set with a safety margin.[12] However, DRIs are based on growth of typical infants, not those in which **catch-up growth** is required.[5] Vitamin and mineral requirements are affected by various health conditions, particularly those involving digestion. Prescribed medications may increase the turnover of specific vitamins.[13] Some infants with special health care needs have restrictions in volume consumed or activity that increase or decrease needs for specific vitamins or minerals. For example, limited volume of liquids may rule out vitamin-rich juices in the diet of infants with breathing problems.

High-potency vitamin and mineral supplements are usually prescribed for sick or recovering infants. Calcium is a potentially limiting nutrient in sick infants because calcium imbalance and **hypocalcemia** occur with a variety of conditions.[9] Iron, B_{12}, vitamin D, and fluoride are limiting in some specific situations.[8,13] Even after infants who were LBW or VLBW are eating well, vitamin and mineral requirements may be higher than the RDA, depending on specific health conditions. After preterm birth and discharge to the home, deficiencies in copper, zinc, and vitamin D are rare; but they may be checked by blood tests.[14] The early signs of rickets as seen by X ray are considered a sign of needing more vitamin D, above the 400 IU daily recommendation for term and preterm infants.[14]

Human-milk fortifiers are used to boost calories as well as to provide additional vitamins and minerals in some infants in neonatal intensive care units.[14] In order to meet the higher requirements of specific vitamins and minerals for VLBW infants, such products can be added to breast milk. Such products are used under specific conditions, such as when an infant can only tolerate a low volume per feeding. They are intended to bridge the gap between breast milk and the extra needs of a VLBW infant.[8] Major ingredients are vitamins A, D, and C, and minerals such as calcium, phosphorus, sodium, and chloride.[9] Iron is not provided in human-milk fortifiers, and is prescribed as needed.[9]

Some infant formula products also provide vitamins and minerals in concentrated amounts in formulas for premature infants. Such formulas are supplemented with extra calcium, phosphorus, copper, and zinc compared to standard formulas.[8] (The high levels of vitamins and minerals in premature infant formulas are shown in Table 9.2 later in this chapter). For some conditions, vitamins are used not only as dietary components, but as pharmaceuticals. An example is a condition in which vitamin B_{12} is injected as a part of therapy for a rare genetic disorder of protein metabolism.[10]

Growth

Growth in infancy is usually a reassuring sign that sufficient calories and nutrients are provided. The Centers for Disease Control's 2000 growth charts are a good starting point for monitoring of growth in infants with health risks.[15] The first goal of nutritional care is to maintain growth for age and gender. Later, this approach may be modified if there is a growth pattern typical for a specific condition that is identified after the first year. A steady accretion of weight or height is a sign of adequate growth, even if gains are not at the typical rate. Plateaus in weight or height, or weight gain followed by weight loss, are signs of inadequate growth. Growth may be assessed reliably using each infant as her own control regardless of health conditions. As noted in Chapter 8, the methods of assessing growth require consistency and accuracy in order to make sure growth is interpreted correctly. Errors such as confusing pounds and kilograms in plotting interfere with interpretation no matter what growth chart is being used.

Usually, providing sufficient calories and nutrients results in good growth, but not in all cases. Sometimes slow growth is a symptom of an underlying condition, rather than a sign of inadequate nutrition. For example, infants who are born with genetic forms of kidney disease are short even when they consume adequate diets during the first year. Refinements in the usual methods and interpretation of growth are needed in conditions known to influence growth and development. These include:

- Using growth charts for specific diagnoses, such as Down syndrome growth charts.[16] (A list of specialty growth charts is included in Chapter 11.)

- Biochemical indicators of tissue stores of nutrients such as iron or protein, and of electrolytes such as potassium and sodium.

- Indicators of body composition, such as body fat measurement. These can be used to show caloric intake is not limiting growth because fat stores are adequate.

- Special attention to indicators of brain growth, such as measuring head circumference, may be helpful to explain short stature or other unusual growth patterns.

- Using treatment guidelines or published protocols, including disorder-specific weight-gain graphs and recommendations instead of standard growth charts.

- Medications that change weight gain, appetite, or body composition. Side effects of medications can explain rapid changes in weight.

Growth in Preterm Infants

Growth charts developed by the CDC can be used to assess growth progress of preterm infants with birthweights over 2500 grams.[15] The body composition of infants born preterm is not the same as that of term infants, in part because these infants have missed part of the third trimester, when fat is added rapidly.[11] In fact, body fat buildup is a late sign of recovery from preterm delivery. Body composition at various gestational ages is used to adjust growth expectations based on age. Treatment of the infant's medical condition may also affect growth expectations; for instance, fluid accumulation may artificially increase weight. As a result of such considerations, VLBW and ELBW infants are not represented by the standard growth charts.[15] Growth for newborns with birthweights between 501 and 1501 grams can be tracked by the Neonatal Research Network Growth Observational Study Research Network, sponsored by the National Institutes of Health, National Institute of Child Health and Human Development (NICHD). Its Web site (www.Neonatal.rit.org) has a VLBW Postnatal Growth Chart component that projects growth from whatever birthweight, birth length, or birth head circumference is entered. The management of preterm birth is so rapidly changing that recently published growth charts may not reflect growth patterns achieved with current practices.[13]

The most commonly used growth charts intended for premature infants are called the "IHDP Growth Charts"; these are based on the *Infant Health and Development Program (IHDP)*.[17] This large research program has created four charts, two for girls and two for boys, at two different birthweights. One set, called "LBW Premature," is for infants with birthweights in the range of 1501–2500 grams; the other is "VLBW Premature," and is for use

> **Infant Health and Development Program (IHDP)** Growth charts with percentiles for VLBW (<1500-g birthweight) and LBW (<2500-g birthweight).

with infants who weighed 1500 grams or less and were less than 38 weeks of gestation.[17] Each chart has:

- Head circumference for gestation-adjusted age
- Weight for gestation-adjusted age
- Length for gestation-adjusted age

These growth charts for preterm infants can be used into toddlerhood, although standard growth charts often replace them if the child is growing well.

CORRECTION FOR GESTATIONAL AGE

Gestation-adjusted age is calculated by subtracting gestational age at birth from 40 weeks (the length of a full-term pregnancy). The resulting number of weeks is divided by four to obtain months. The result in months is then subtracted from the infant's current age. For example, if an infant was born at 30 weeks gestation, she is 10 weeks early. This equals 2.5 months preterm. When she is 3 months old, her gestation-adjusted age is 2 weeks, or 0.5 months. This age of 0.5 months would be used in plotting her growth on the IHDP growth chart as part of assessing her growth and development.

Does Intrauterine Growth Predict Growth Outside?

"Apples don't fall from a pear tree."

French saying

The answer is yes, no, and maybe! Fetal monitoring during pregnancy and in-depth knowledge about the development of various organ systems provides a clear pattern of growth at various gestational ages. However, many factors during and after pregnancy are known to affect growth rate. In summary, these are:

- Intrauterine environment, particularly the adequacy of the placenta in delivering nutrients; the presence of toxins such as viruses, alcohol, or maternal medications; or the depletion of a needed substance, such as folic acid.
- Fetal-origin errors in cell migration or formation of organs, whether or not a cause is known. Various nutrients, such as vitamin A, have been implicated in such errors.
- Unknown factors that cause preterm birth, such as environmental toxins in air pollution.

Knowledge gained from studies of gene-nutrient interactions will identify additional genetic and nutritional factors influencing fetal development and improve preventative and treatment strategies.

Research is emerging that shows some infants were born prematurely due to conditions originating during the intrauterine period.[6] As discussed in Chapter 8, SGA and intrauterine growth retardation are terms used to describe infants who are smaller than expected at birth. SGA is the more general term because it is based on the population of infants of the same gestational age.[6] Both predict higher medical risks and need for close growth monitoring.

If the intrauterine insult was early in gestation, body weight, length, and head size (brain size) are affected.[9,13] There has been a change in the number and size of fetal cells. The abnormal fetal growth pattern may persist despite adequate medical and nutrition support after birth. Examples of conditions causing early insults are infants born after cocaine and alcohol exposure, which are associated with IUGR and preterm birth.[18] Later exposure in the second or early part of the third trimester may result in preservation of head size and body length, but low weight. Some genetic conditions characterized by small size are not diagnosed until childhood, but have IUGR noted in the medical history.[16]

Intrauterine growth may not predict growth for some infants whose birth removes them from adverse exposure within the intrauterine environment. Examples of this situation are maternal uncontrolled diabetes, smoking, phenylketonuria, or maternal seizures treated by medications. In such cases the rate of growth after birth may be improved and normalized during the first year of life.[9] In general, the earlier the exposure to the toxin, the worse the effects on later growth, but there is quite a bit of variability.[18] Sometimes marijuana, alcohol, tobacco, and crack/cocaine have been used at various times during pregnancy, so growth impacts may be based on amount, timing, and interactions of toxins. The most important risks for later growth may be neonates born with smaller head circumferences.

A large category that changes the rate of intrauterine growth is early medical treatment. If the intrauterine growth was fine, preterm birth or its complications may slow or plateau growth early in infancy. This may mask whether or not typical growth goals are appropriate all during infancy. Various studies have measured whether infants recover from early growth problems.[19–21] A group of ELBW infants who were followed into adolescence and did not have major disabilities were shorter in stature and lower in weight than others of the same age and gender, suggesting growth is affected long term.[20] Another study of infants who were short for age and were provided nutritional supplements for 2 years found that the infants were still short for age at 11 and 12 years.[21] Whether or not adequate nutrition is provided early in life of course depends on how *adequate nutrition* is defined. Advocates of increasing current nutritional recommendations for preterm infants expect growth problems could be lessened if higher levels of nutrition are provided, regardless of other factors.[5]

Conditions that affect growth may be time limited. Catch-up growth may be seen during recovery in many infants, resulting in changing growth interpretation. Surgery may delay growth for a short time, when required for heart conditions. Growth may slow and then catch up with respiratory illnesses that resolve with medications. Usually increased access to adequate nutrition improves growth. Only close monitoring over time may show signs of catch-up growth early, such as an increase in fat stores or length.

The amount of time needed for catch-up growth for premature infants differs based on gestational age at birth and subsequent complications. Clinical conventions are to provide 1 year for catch-up growth for infants born 32 weeks or later, and 3 years for catch-up growth for VLBW or ELBW infants. It is likely such infants will have growth percentiles in the lower end of the normal range during childhood.

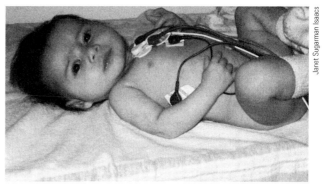

Illustration 9.2 Baby girl wearing a heart monitor at home.

Interpretation of Growth

Hospital discharge after preterm birth may be based on a pattern of weight gain, such as 20–30 grams per day.[13,14] Strong emphasis is placed on growth as a sign of improving health in monitoring small and sick infants after discharge, but complications make this difficult to achieve. An example of a common diagnosis in preterm babies and difficulty in interpretation of growth is the lung condition bronchopulmonary dysplasia (BPD). Rates of growth among infants with BPD are different than those in full-term infants and preterm infants who do not have BPD. While infants with BPD are recovering, the growth pattern of these infants is affected for the entire first year of life.[19] The reasons for the slower growth are higher nutritional requirements, changes in endocrine and pulmonary systems, and perhaps interaction among these systems.

Growth rate changes are closely associated with the frequency of illness, hospitalizations, and medical history.[13,14] Conditions acquired as a result of preterm birth that make growth difficult to interpret include

- Symptoms related to intestinal absorption that can temporarily or permanently change nutrient requirements.
- *Microcephaly* (small head size) or *macrocephaly* (large head size) compared to other growth indicators may be a sign that growth may be affected as a result of neurological consequences. Both large and small head size in infancy can affect muscle mass, body composition, and subsequent growth.
- Variable rates of recovery and growth are seen for many infants. Infants are as different from each other as the rest of us, but these differences are hard to see soon after birth.

Common Nutritional Problems

Infants who are small or sick near birth may have major growing and feeding problems. Infancy is such a vulnerable time of life that most health conditions occurring this

early interfere with growth and development. Over time, most of these problems resolve, although some become chronic and a few result in death. Nutrition plays an important part in preventing illness, maintaining health, and treating conditions in infancy. Nutrition tends to become more important over time to maintain growth if conditions are chronic (Illustration 9.2).[22]

Table 9.1 on the next page shows nutrition problems in infants with special health care needs. Nutrition assessment documents these concerns, and nutrition services are provided based on the assessment.

Nutrition Risks to Development

Many health conditions change the infant's rate of development. *Developmental delay* describes the interaction of a chronic condition with development. The terms *children with special health care needs* and *developmental delay* are general terms used to allow nutritional, medical, and developmental services to be provided for infants.

Developmental delay is used to describe a wide range of symptoms that reflect slow development. Symptoms that relate to nutrition are common. These symptoms include infants who are growing slower than expected for age, or have difficulty in feeding, such as refusing food from a spoon by 8 months of age. An example is a 2-month-old girl who does not breastfeed for more than a few minutes per side. At first this may appear to be a problem of breastfeeding position or frequency. By 4 months, weight gain is slower than expected, so now growth concerns and feeding concerns are interacting; it is not clear if they are separate or related problems. These concerns are sufficient to request an evaluation for eligibility for intervention services. Several months later, after various services have been put in place, it may still be

Microcephaly Small head size for age and gender as measured by centimeters (or inches) of head circumference.

Macrocephaly Large head size for age and gender as measured by centimeters (or inches) of head circumference.

Developmental Delay Conditions represented by at least a 25% delay by standard evaluation in one or more areas of development, such as gross or fine motor, cognitive, communication, social, or emotional development.

Table 9.1 Nutrition concerns in infants with special health care needs

Growth	Slow rate of weight gain
	Fast rate of weight gain
	Slow rate of gain in length
	Disproportionate rate of weight to height gain
	Unusual growth pattern with plateau in weight or length gain
	Altered body composition that decreases or increases muscle size or activity
	Altered brain size that decreases or increases muscle size or activity
	Altered size of organs or skeleton, such as an enlarged liver or shortened leg length
Nutritional adequacy	Calorie needs are higher or lower
	Nutrient requirements higher or lower overall
	Specific nutrients, such as protein or sodium, are required in higher or lower amounts
	Vitamins, minerals, or cofactors (such as carnitine) are required in higher or lower amounts
Feeding	Disruption of the delivery of nutrients as a result of:
	• Structure or functioning of the mouth or oral cavity
	• Structure or functioning of the gastrointestinal tract, including diarrhea, vomiting, and constipation
	• Appetite suppression by constipation or medications
	• Disrupted interaction of the infant with the parent, such as infant cues being so subtle that parent responses are delayed
	• Posture or position promotes or interferes during meal times
	• Timing of nursing, meals, and snacks throughout the day
	• Inappropriate food choices or methods of preparation
	• Interruptions in adequate shelter for feeding and sleeping
	• Instructions were unclear or too complicated for the parent to follow

unclear if these nutrition problems are from development, a health condition such as a heart murmur, or the interaction of both. In any case, the girl fits the category of a child with a special health care need. This allows the family to benefit from nutritional, medical, and developmental interventions, without requiring a specific diagnosis. Infants generally are not old enough to have a specific diagnosis related to development, such as mental retardation or *autism*.

Down syndrome is an example of a condition in which developmental delay is noted in infancy. Down syndrome prevalence is about 9 cases per 10,000 live births.[23] Nutrition concerns with infants who have Down syndrome are feeding difficulties related to weak muscles in the face, and overall; high risk of overweight; and constipation. Heart and intestinal conditions are more common in infants with Down syndrome, so their nutrient needs may be increased if surgery is required. This is also an example of a chronic condition in which nutrition problems such as overweight increase over time if prevention and maintaining health are not addressed. Growth requires close monitoring to identify and prevent overweight starting in infancy.

Autism Condition of deficits in communication and social interaction with onset generally before age 3 years, in which mealtime behavior and eating problems occur along with other behavioral and sensory problems.

Infants with Down syndrome love to suck and have things in their mouths so much that it is easy to overfeed them. Development of movement occurs at a slower rate, with lower physical activity, which also can contribute to overweight. Giving parents their own copy of Down syndrome growth charts for infants is recommended after the diagnosis is confirmed.[16] These charts may be helpful in recognizing typical growth and preventing overweight early. These special growth charts are available from places that serve children with Down syndrome, such as developmental or genetics clinics in major medical centers.[16]

Not all children with developmental delay in infancy have developmental disabilities later. For example, an infant with breathing problems may be slower to grow and to crawl as a result of his higher caloric needs during the first year of life. Such an infant may show developmental delay in motor skills, but by age 3 he will have improved in overall health and have caught up to others in motor skills. He would not have a developmental disability. Other examples are infants from high-risk pregnancies, such as those born large for gestational age as a result of poorly controlled gestational diabetes. Many infants require short stays in intensive care units for glucose regulation; some may have long-term risks for their development. Some infants with developmental delays continue to have slower development over time. After infancy, when standard testing and evaluation can be performed, the term *developmental delay* may be replaced with a more specific type of medical or developmental diagnosis.

Severe Preterm Birth and Nutrition

The yearly incidence of VLBW in the United States is approximately 60,000 infants—about the same as the population of a small city, such as Iowa City, Iowa. Infants with birthweights near 1500 grams (3 pounds 4 ounces) have gestational ages from 28–32 weeks and a survival rate of almost 90%.[24] Each infant requires immediate intensive care hospitalization; and if they survive, these infants continue to have high nutritional needs throughout infancy. ELBW infants weigh under 1500 grams and have gestational ages ranging from 23 to 28 weeks. Despite advances in the care of such infants, disability such as delayed development is a common outcome of ELBW.[25] Some outcome studies are demonstrating lifelong consequences of low birthweights, such as impacting later employment as adults.[26]

Nutrition problems resulting from VLBW and ELBW preterm births are addressed as they present. The initial problem after birth is that the newborn cannot nurse like a full-term infant, and most require respiratory support to breathe. Getting adequate calories and nutrients into the preterm infant requires **nutrition support,** usually first **parenteral feeding** and then **enteral feeding** methods.[9,27] Feeding problems of preterm infants are discussed later in this chapter. All newborns have high metabolic rates, and they will use fat stores and protein in tissues and muscles to meet glucose needs if consumed calories and nutrients are not sufficient. This happens sooner in infants than in adults.[4,5] Providing sufficient calories and nutrients to meet requirements and preserve ingested protein and calories for growth is the goal, but it may be difficult and take more time than expected in sick and recovering infants.

How Sick Babies Are Fed

Gastrointestinal upset is a response to many conditions in newborns, whether the intestines are the initial problem or not. VLBW, ELBW, and sick infants are especially vulnerable to problems related to the gastrointestinal tract. Such problems directly affect how calories and nutrients are provided, as well as the composition of the diet. For example, if a newborn gets an infection, an early sign may be inflammation of the intestine. As a response, the method of feeding the infant has to be adjusted. Inflamed or damaged areas may slow or interrupt typical intestinal muscle movements, resulting in signs of increasing illness.[9,13] Blood loss from the intestines is a sign of **necrotizing enterocolitis (NEC),** a serious condition in the neonate. When this occurs, oral feeding is stopped and replaced by parenteral nutrition.

Many gastrointestinal conditions interfere with infant feeding, such as gastroesophogeal reflux, constipation, spitting up, and vomiting. In small and sick newborns, these gastrointestinal conditions may represent slow or uncoordinated movements of the intestinal muscles.[9] These conditions do not rule out enteral feeding, which stimulates the intestines and keeps them healthy. Feeding methods are selected based on the length of time before it is expected the baby can nurse or feed without help. Gavage feedings may be used. These are slow feedings sent from the mouth or nose into the stomach though a tube. Infants who are too weak to breastfeed may be offered the comfort of the breast or pacifier along with gavage feeding. *Oral-gastric (OG) feeding* is also used. Other enteral methods are *transpyloric feeding (TP), gastrostomy feeding,* and *jejunostomy feeding.*[9] These methods are used when nutrition support is expected to be needed for several months.[22]

FOOD SAFETY Preterm babies with immature immunological systems are prone to infection, so every effort is made to assure their feedings do not become contaminated. The rate of feeding preterm infants is often much slower than that for full-term infants, and formula or breast milk is at room temperature for a longer time. Contamination of feeding equipment increases with time; consequently, hospitals have policies requiring them to change the feedings often, such as every four hours.[28–30]

What to Feed Preterm Infants

Breast milk is the recommended source of nutrition for preterm infants. Colostrum and breast milk are produced even when the mother delivers very early.[31] Preterm human milk has increased protein content compared to term milk.[31] Hospital protocols and policies for having mothers pump and freeze breast milk for later use by their preterm infants are highly recommended.[32] Staff training to encourage new mothers at home to rest enough and pump enough to stimulate breast-milk production is also recommended.[33]

Nutrition Support Provision of nutrients by methods other than eating regular foods or drinking regular beverages, such as directly accessing the stomach by tube or placing nutrients into the bloodstream.

Parenteral Feeding Delivery of nutrients directly to the bloodstream.

Enteral Feeding Method of delivering nutrients directly to the digestive system, in contrast to methods that bypass the digestive system.

Necrotizing Enterocolitis (NEC) Condition with inflammation or damage to a section of the intestine, with a grading from mild to severe.

Oral-Gastric (OG) Feeding A form of enteral nutrition support for delivering nutrition by tube placement from the mouth to the stomach.

Transpyloric Feeding (TP) Form of enteral nutrition support for delivering nutrition by tube placement from the nose or mouth into the upper part of the small intestine.

Gastrostomy Feeding Form of enteral nutrition support for delivering nutrition by tube placement directly into the stomach, bypassing the mouth through a surgical procedure that creates an opening through the abdominal wall and stomach.

Jejunostomy Feeding Form of enteral nutrition support for delivering nutrition by tube placement directly into the upper part of the small intestine.

Barriers to breastfeeding small and sick newborns are partially based on their abilities, on how sick they are, and on the care system. Promoting breast milk for preterm infants is recommended as hospital policy by the American Academy of Pediatrics, but hospitals differ in practices.[31] Medical conditions in the infant may undermine breastfeeding. Infants are generally able to nurse successfully at about 37 weeks of gestation. Prior to that age, they may benefit from being put to the breast to stimulate non-nutritive sucking, or sucking that does not deliver milk to be swallowed. There are a few conditions in which human milk is unsafe for preterm or sick infants. Breastfeeding is contraindicated when breast milk contains harmful medications, street drugs, viruses, or other infective agents, or when the infant has a specific type of gastrointestinal tract malformation or inborn errors of metabolism.[31]

Depending on the infant's birthweight and health status, breast milk may be insufficient in nutrients unless supplemented by human-milk fortifier and/or other sources of calories, such as MCT oil. If not fed modified breast milk or nursing, the infant's source of nutrition may be cow's-milk- or soybean-based formulas.[32] Whey as the predominant form of protein from cow's milk is recommended because its amino acids profile is closer to that of human milk.[8]

Infant formulas for preterm infants are available for home use after hospital discharge, if breast milk is not available. They provide the higher calories and nutrient levels that small infants need compared to term infants.[34] Standard formula that is 20 calories per fluid ounce can also be used for preterm infants, modified in a manner similar to breast milk to boost calories and nutrients. High-calorie formulas, such as 28 cal/fl oz, may be appropriate for some infants. Such high-calorie formulas are not routinely used because they have high osmolarity, which may affect fluid and electrolyte balance. Table 9.2 shows a comparison of premature and standard formulas.[34] If the infant easily fatigues or is too weak to suck enough volume, 22 or 24 cal/fl oz formulas may be recommended.[9] The sources to add extra calories and nutrients are selected based on the infant's gastrointestinal tolerance and volume requirements. They may include MCT oil; polycose; rice baby cereal; and rarely, human-milk fortifier. Routine nutritional assessment of the infant's growth tracks the diet's effectiveness in providing adequate nutrients and calories.

Preterm Infants and Feeding

VLBW or ELBW infants usually progress at their own rate of development regarding feeding skills. The goal is the same as for all infants—to achieve good nutritional status, as indicated by growth and feeding skills progression. Most families enjoy feeding their infants and experience few long-term feeding problems. Infants who were born preterm, however, may be hard to feed. There are several reasons that an infant may be difficult to feed:[32]

- Fatigue. Low levels of arousal of weak or sick infants may lessen feeding duration.

- Low tolerance of volume. Abdominal distention due to feeding may result in changes in breathing and heart rate, so that the infant stops feeding.

- "Disorganized feeding" may result from the infant having experienced defensive and unpleasant reactions to feedings or procedures, so anything coming to the mouth causes a stress reaction rather than a pleasurable reaction.[9]

Regardless of the associated conditions, certain feeding characteristics of preterm infants are distinct from those of term infants (Table 9.3). Most recovering infants improve in their feeding abilities with time. Anxiety decreases as parents become more comfortable with caring for their infant at home. The underlying reflexes that associate pleasure with feeding reemerge, and the interac-

Table 9.2 Selected nutrient composition of term and preterm formulas

Nutrients	20 Cal/Fl Oz	22 Cal/Fl Oz	24 Cal/Fl Oz
Protein	2.1 g	2.8 g	3 g
Linoleic acid	860 mg	950 mg	1060 mg
Vitamin A	300 IU	450 IU	1250 IU
Vitamin D	60 IU	80 IU	270 IU
Vitamin E	2 IU	4 IU	6.3 IU
Thiamin (B_1)	80 mcg	200 mcg	200 mcg
Riboflavin(B_2)	140 mcg	200 mcg	300 mcg
Vitamin B_6	60 mcg	100 mcg	150 mcg
VitaminB_{12}	0.3 mcg	0.3 mcg	0.25 mcg
Niacin	1000 mcg	2000 mcg	4000 mcg
Folic acid	16 mcg	26 mcg	35 mcg
Pantothenic acid	500 mcg	850 mcg	1200 mcg
Biotin	3 mcg	6 mcg	4 mcg
Vitamin C	12 mg	16 mg	20 mg
Inositol	6 mg	30 mg	17 mg
Calcium	78 mg	120 mg	165 mg
Copper	75 mcg	120 mcg	125 mcg

Table 9.3 Preterm and term infant feeding differences

Preterm Infant	Term Infant
Central nervous system does not signal hunger	Signals hunger; has supportive newborn feeding reflexes
Unstable feeding position, such as a forward head position	Stable and facilitating feeding position from newborn reflexes
Oral hypersensitivity	Readily accepts food by mouth

Table 9.4 Example diet for a VLBW infant at 8 months of age, with a corrected age of 4.5 months (weight 4.5 kg 5–25% on VLBW Premature Boys growth chart)

Food and Formula	Feeding Instructions
Five feedings per day of formula, each 5 fl oz, High-calorie formula (24 cal/fl oz) with 2 tb rice cereal added	Provide support for semi-reclining feeding position during and up to 30 minutes after feeding
Medications for stomach added to 2 fl oz of 50% diluted apple juice two times per day	Encourage use of pacifier for comfort between feedings
Two meals of pureed baby foods fed with spoon, total intake one 2-oz jar	Offer bottles every 3 hours except overnight if no signs of hunger before then
Liquid vitamin/mineral supplement	Keep scheduled appointments for WIC; weigh in at MD office, bringing diet log book

tion of the infant with the feeder improves. There is a lot of room for hope in the feeding process after discharge.

Major advances in understanding the nutritional needs of preterm babies have come about from working with smaller and smaller infants. Table 9.4 presents an example of a typical diet for a premature baby, who was not breast-fed and had a 3-month hospital stay after birth. The infant has gastroesophageal reflux and prescribed medications that are included in his feeding instructions. This example shows that providing an adequate diet is an important part of the infant's growth and development and his recovery from preterm birth complications (see Case Study 9.1 on the next page).

Infants with Congenital Anomalies and Chronic Illness

Infants who are not preterm but require neonatal intensive care may be at risk for chronic illness. About half of babies in neonatal intensive care units have normal birth-weights and experience lower mortality than LBW infants. They tend to have a higher rate of *congenital anomalies* (22%) and often require rehospitalization.[35] These infants need more nutrition services than typical infants because their growth and feeding development requires close monitoring and intervention.

Congenital anomalies are recorded in the Birth Defects Monitoring Program (BDMP).[36] The Centers for Disease Control publishes prevalence data based on states and hospitals that participate voluntarily in surveillance programs. The United States keeps prevalence data not just for infants but also for children of various ages. *Infant mortality attributable to birth defects (IMBD)* accounts for approximately 2 deaths per 1000 live births.[36] Whereas infant mortality is decreasing in the United States, the proportion of infants who die from IMBD is increasing.[36] The major types of birth defects associated with death are first, heart malformations and

next, central nervous system defects.[36] Examples of central nervous system congenital anomalies are spina bifida and *anencephaly*. Since folic acid has been added as a supplement in grains and flours, rates of spina bifida and related conditions have declined 20–30%.[37]

Infants with congenital anomalies, genetic syndromes, and malformations fit in the category of children with special health care needs, so they are eligible for a wide range of medical, nutritional, and educational services to maximize their growth and development.[25] All babies born with such conditions have risks to maintaining good nutrition status as they receive treatment. The nutrition concerns are growth, adequacy of the diet in providing required nutrients, and feeding development, as shown in Table 9.1. Nutrition services range from temporary to long term, and they are as diverse as the many different types of conditions involved. Several examples of disorders with minor and major nutritional consequences follow.

Major nutritional impacts are exemplified by disorders that involve the gastrointestinal tract. Infants with *diaphragmatic hernia* or *tracheoesophageal atresia* cannot safely eat by mouth and require nutrition support and several surgeries during infancy.[32,36] Diaphragmatic hernia occurs in 1 in 4000 live births due to failure of the diaphragm to form completely. Tracheoesophageal atresia occurs in 1 in 4500 live births due to an error in development of the trachea. These examples of conditions treated by neonatal surgery and intensive care have been credited with lower infant mortality for such congenital

Congenital Anomaly Condition evident in a newborn that is diagnosed at or near birth, usually as a genetic or chronic condition, such as spina bifida or cleft lip and palate.

Infant Mortality Attributable to Birth Defects (IMBD) Category used in tracking infant deaths in which specific diagnoses have a high mortality.

Anencephaly Condition initiated early in gestation of the central nervous system in which the brain is not formed correctly, resulting in neonatal death.

Diaphragmatic Hernia Displacement of the intestines up into the lung area due to incomplete formation of the diaphragm in utero.

Tracheoesophageal Atresia Incomplete connection between the esophagus and the stomach in utero, resulting in a shortened esophagus.

Case Study 9.1

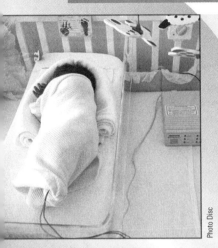

Photo Disc

Premature Birth in an At-Risk Family

A baby named Eric is born at 30 weeks of gestation, appropriate for gestational age, at 1.4 kilograms (3 pounds). Like his mother, he tests positive for cocaine. Eric receives routine intensive care services, which include ruling out sepsis, and is given head ultrasound studies. By 33 weeks he is being fed OG and has only transient respiratory difficulties. Prior to discharge at 37 weeks, he appears to be developing normally. He drinks 22 fl oz formula/day at the rate of about 1.5–2 fl oz/feeding, with 10 feedings per day. He is placed in foster care with an experienced foster mother who has two older children.

Eric's custody is reconsidered when his biological mother expresses interest when he is about 9 months of age. Shortly after that, his mother's parental rights are terminated based on criminal charges. His foster mother reports that he has been a colicky baby, and he has had at least three ear infections during his first year of life. He is enrolled in an *early intervention program* based on his prematurity and intrauterine drug exposure. Eric's initial developmental testing was within normal limits at 6 months.

Eric's foster mother expresses concern about his intervals of periodic crying, during which he accepts neither soothing nor a bottle. Eric is diagnosed with gastroesophageal reflux (GER) and slow gastric emptying, and he is treated medically starting at 8 months. He is slow to accept foods on a spoon, with gagging and spitting up. His growth on the IHDP LBW chart is near the 50th percentile for weight and height. His head circumference is at the 5th percentile. Eric does not sit up unassisted until 8.5 months, but he turns over from stomach to back and vice versa easily. He is sent to a genetic specialist because he appears to have some facial features consistent with fetal alcohol syndrome, such as low-set ears, wide nasal bridge, and thin upper lip. The diagnosis is not confirmed because he is too young, but reported as a possibility to be reevaluated after he is older and can be given developmental tests. If this diagnosis were confirmed, standard growth charts would not fit Eric, because short stature and low weight are part of the diagnosis, even when adequate nutrition is provided. In infancy Eric's growth was within normal limits after correction for prematurity, with the same trend of a lower head circumference.

Eric's foster mother expresses an interest in adoption when he is almost 1 year old. She is pleased that he needs no medication and is growing well. His early intervention services are continued based on his at-risk status, because no specific 25% delay has been documented at 1 year. (Later, at 34 months of age, he is diagnosed with mixed developmental delay based on cognitive and speech delays.) Eric is adopted by the foster family.

Questions

1. Did Eric's early birth account for his slow growth later?
2. Did nutrition affect when Eric's developmental delay probably started? Note that it was diagnosed at about age 3.
3. What are the signs that Eric can outgrow his problems?

(Answers are in Appendix D.)

Early Intervention Program
Educational intervention for the development of children from birth up to 3 years of age.

---

anomalies.[36] Both conditions change the motility of the gastrointestinal tract, so sufficient calories, nutrients to maintain growth, and oral feeding are important parts of the treatment plan.[32] Such infants miss the windows of development when oral feeding is pleasurable and may have residual feeding problems, such as disliking eating by mouth, well into early childhood. Financing such intensive health care and maintaining the child's normal social and emotional development are major issues for the family. The families of such infants have many specialty health care providers, and their complex financial and emotional reactions can also influence the infant's development.

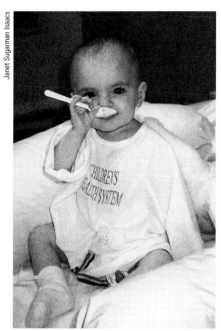

Illustration 9.3 Infant in hospital being encouraged to eat with a spoon.

Both of these conditions eventually result in children being able to eat like everyone else (Illustration 9.3).

Common examples of congenital anomalies are those for infants with *cleft lip and palate.* Major feeding difficulties occur before and after corrective surgeries, which sometimes interfere with growth in infancy and early childhood.[38,39] Assistance in feeding by registered dietitians as part of a team approach is needed, because hearing, speech, and language problems are associated with cleft lip and palate. Positions for eating are adapted, and use of special feeding devices for the infant's cleft are needed. Cleft lip and palate may occur alone or as a part of various rare genetic conditions, so growth and feeding problems should also be assessed after corrective surgery.[38]

Infants with Genetic Disorders

Infants diagnosed with genetic disorders near birth are a small subset of infants with congenital anomalies or chronic conditions. They also fit in the category of infants with special health care needs. The expanded use of prenatal genetic tests has the consequence that some families know ahead of time that the baby has a specific condition at birth. When the condition is treated by a special diet, nutritional therapy may begin right after birth. The mother can plan ahead to breastfeed by pumping and freezing her milk. The newborn is fed the appropriate diet while confirmatory testing proceeds. Once the newborn's testing confirms the prenatal test results, therapy can set the amounts of breast milk to be mixed with a special formula at home.

The number of genetic disorders that can be identified in newborns is increasing rapidly, particularly through expanded infant screening programs. The nutritional implication of expanded genetic screening and diagnosis is that more newborns require special diets immediately. Infants with rare genetic conditions such as galactosemia or *maple syrup urine disease* need the diet started within days of birth—waiting even one week can result in irreversible brain damage or death.[10] The infants who are identified in newborn screening with metabolic or genetic conditions usually require special formulas. Genetic centers or inborn errors of metabolism clinics are notified when a newborn screening result needs follow-up. Immediate action is taken to locate the family and confirm the diagnosis. In such circumstances, newborn screening results in early diagnosis and avoids a costly stay in the hospital intensive care unit. For example, an infant with galactosemia, if not picked up by the initial abnormal screening result, is likely to have been hospitalized with possible sepsis or liver problems by the time a second screen is to be collected. If the baby with galactosemia receives supportive measures, such as sugar solutions, and then soy-based infant formulas without galactose, recovery is usually rapid. Some of the disorders picked up by newborn screening do not make the baby sick in early infancy, but later. It is difficult for parents of a healthy-appearing newborn to be told that a special diet is needed to prevent illness later. An example of a condition that can be picked up by newborn screening before illness presents is cystic fibrosis.

This concept of increased use of genetic tests is exemplified by the condition called *DiGeorge syndrome.* This test may be ordered for any infant with a heart defect. DiGeorge sydrome is a condition in which a small piece of chromosome 22 is deleted.[40] Recent incidence estimates place it at 1 in every 4000 births.[40] This makes DiGeorge relatively common for a genetic condition, more common than PKU or cystic fibrosis, and second only to Down syndrome as a cause of mental retardation. Infants with DiGeorge syndrome may have a wide range of conditions affecting the heart, immune system, and calcium balance, and later speech and learning problems. Only when the genetic probe became available was it understood that three separate disorders involved the same deletion. As a result, the incidence of DiGeorge syndrome was underreported before the probe was available. Nutrition services may be required based on short stature, heart malformations, heart surgery, and resulting feeding problems (see Case Study 9.2 on the next page).

> **Cleft Lip and Palate**   Condition in which the upper lip and roof of the mouth are not formed completely and are surgically corrected, resulting in feeding, speaking, and hearing difficulties in childhood.
>
> **Maple Syrup Urine Disease**   Rare genetic condition of protein metabolism in which breakdown by-products build up in blood and urine, causing coma and death if untreated.
>
> **DiGeorge Syndrome**   Condition in which chromosome 22 has a small deletion, resulting in a wide range of heart, speech, and learning difficulties.

Case Study 9.2

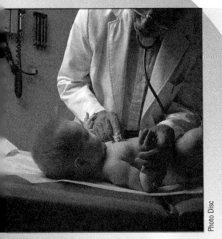

Photo Disc

Noah's Cardiac and Genetic Condition

A baby named Noah requires open heart surgery, which results in a diagnosis of DiGeorge syndrome. His mother is successful at expressing her milk and maintaining her breast-milk supply by using the breast pump provided by the medical center during the hospitalization. Pumping and freezing the breast milk for the baby is important to the family. Noah is a reluctant eater by breast or bottle, with intake usually only 1 or 2 ounces per feeding, when allowed to feed. He is too weak to feed throughout most of the hospitalization.

At discharge the family is referred to a local early intervention program, WIC, Supplemental Social Insurance (SSI), and the state program for children with special needs. Specialty clinic and local follow-up appointments are made. Feeding difficulties concern the family, and both a lactation consultant and a registered dietitian are involved at home. Noah nurses frequently, but briefly, due to fatigue. Growth is slower than expected, but the cardiologist does not think Noah's slow weight gain is a result of the cardiac problem.

Noah tolerates only small volumes, even after his family implements the recommendations from the lactation consultant. The consultant recommends offering breast milk in a bottle so that its caloric density can be increased by added rice cereal and MCT oil. The family perceives this recommendation as undermining the mother's effort to breastfeed, and they are unwilling to try it. Noah likes to nurse, but for such a short time that the richer hindmilk may not be available. Over time, the family offers food on a spoon and continues nursing. Weight and height gain do not fit the expected rate, but appear to be fairly consistent.

The family increasingly enjoys parenting, and think Noah is a beautiful infant. They have not contacted WIC, nor the early intervention program. They regard the baby as getting better after surgery. They consider his small size a result of heart surgery. Many people they meet assume he is a premature baby, but his small size is not as much a matter of concern to his parents as it is to health providers.

This case example demonstrates inadequate growth rate, feeding problems, and questionable adequacy of the diet. The genetic syndrome, heart condition, nutritional inadequacy of the baby's diet, and the impact of stress and coping on the mother's milk supply all could explain his slow growth. Growth expectations are unclear because there is no growth chart for this genetic syndrome, nor for infants with cardiac anomalies. The standard growth chart is the only one available, but it may not be appropriate to predict future growth.

Important parts of Noah's growth assessment include that his fat stores are good, showing he has access to enough calories, and that his head circumference percentile is low, which suggests that his brain is not growing at the typical rate. This could be due to neurological damage during surgery or afterward, or to the underlying genetic syndrome. It is probably not due to inadequate nutrition in early infancy, because the body tries to preserve brain growth, but there is no way to rule that out. Noah's feeding problems are subtle signs of his developmental delay, although this is clear only in hindsight.

Questions

1. How does breastfeeding benefit Noah?
2. How do you know what is good growth in such a case, when the standard growth chart may not fit?
3. Why doesn't the family want to have WIC or early intervention services, although they are eligible?

(Answers are in Appendix D.)

Feeding Problems

Infants who were born preterm or have chronic health problems tend to be more irritable and less able to signal their wants and needs compared to healthy infants. Feeding difficulties are reported in 40–45% of families with VLBW infants.[32] Children with developmental disabilities have more frequent feeding problems, as high as 70%, that may or may not be identified in infancy.[6] Table 9.5 gives signs of feeding difficulties that can be caught early.

By the time feeding problems require interventions to prevent further growth and developmental problems, families and infants may be frustrated from their feeding experiences. Infants who are difficult to feed are at risk for failure to thrive (FTT), child abuse, and neglect.[22] Infant feeding guidelines for term infants are appropriate for many preterm infants if they were healthy and had gestational ages such as 35 weeks. Preterm infants who were VLBW or ELBW need infant feeding guidelines based on their adjusted gestational age. As an example, the recommendation for adding food on a spoon at 4–6 months would be adjusted to 6–8 months for an infant who was born at 32 weeks of gestation. Even with this adjustment, feeding problems are common because preterm infants may be extra sensitive. The emphasis on weight gain and catch-up growth inadvertently results in overfeeding and signs of gastrointestinal discomfort, such as spitting up. Some preterm infants by late infancy have learned to get attention by devices such as refusing to eat, dropping food off the high chair, or throwing a cup.

Nutrition Interventions

When feeding problems are identified in infancy, interventions are required to assure growth and development. Interventions may include any or all of these:

- Assess growth more frequently or in more depth, such as by measuring body fat stores to identify a change in rate of weight or length gain. This would include head growth measurement.
- Monitor the infant's intake of all liquids and foods by a diet analysis to document that enough calories and nutrients are being consumed. The infant's intake may be variable due to illness, congestion, or medications that lower the appetite.
- Change the frequency or volume of feedings as needed to meet calorie and nutrient needs.
- Adjust the timing of nursing, snacks, or meals as needed to fit medication or sleeping schedules.
- Assess the infant's feeding position and support as needed. This may be important if the infant cannot sit without support.
- Change the diet composition to improve the nutrient density, so that the infant has to expend less effort to meet energy or nutrient needs.
- Provide parent education or support services as needed, so that the feeding environment is positive and low in stress.
- Observe the interaction of the infant and mother (or whoever is routinely feeding the infant) at home or in a developmental program to make sure

---

Table 9.5 Signs of feeding problems in infants

In Early Infancy (under 6 months of age)

- Baby has a weak suck and cannot make a seal on the nipple; breast milk or formula runs out of the mouth on whatever side is lower, with obvious fatigue after a few minutes of sucking.
- Baby appears to be hungry all the time due to low volume consumed per feeding, and/or time between feedings does not appear to increase from one month to the next.
- Extended feeding times are seen, with the baby napping during the feeding despite efforts to keep the baby interested in the feeding.
- The mother is not sure that the baby is swallowing, although she is appearing to suck.

In Later Infancy (over 6 months of age)

- The baby cannot maintain good head control while being fed from a spoon.
- The baby resists spoon-feeding by not opening her mouth when food is offered.
- The baby drinks from a bottle but does not accept baby foods after trying repeatedly.
- The baby resists anything in the mouth except a bottle, breast, nipple, or pacifier.
- The baby does not explore the mouth with fingers or try to mouth toys.
- The baby resists lumpy and textured foods; may turn face away or push food away.
- The baby does not give signs to the parents that clearly indicate hunger or fullness.

Table 9.6 Examples of infant formula for special needs

Condition	Example of Special Infant Formula
Pulmonary problems such as bronchopulmonary dysplasia or cardiac defect	Breast milk or standard infant formula with polycose and MCT oil to provide 28 cal/fl oz (high calories in a low volume)
Phenylketonuria (genetic disorder of protein metabolism)	Mixture of amino acids, carbohydrates, fats, vitamins, and minerals without the amino acid phenylalanine
Maple syrup urine disease (genetic disorder of protein metabolism)	Mixture of amino acids, carbohydrates, fats, vitamins, and minerals without the amino acids leucine, isoleucine, and valine
VLBW infant who required surgery after necrotizing enterocolitis	Mixture of amino acids, carbohydrates, fats, vitamins, and minerals
Gastroesophageal reflux and swallowing problem	Standard infant formula with baby rice cereal (increased thickness is to lower risk of choking and vomiting)
Chronic renal failure (hereditary kidney disease)	Concentrated natural protein, fats, and carbohydrates providing 40 cal/fl oz

that signs of hunger and comfort result in a positive feeding experience for the pair.

- Adjust routine nutrition guidelines to the developmental abilities of the infant even if different from the chronological age or gestation-corrected age.

Often, attempts to improve the feeding experience are successful in meeting the infant's calorie and nutrient needs. However, when calorie and nutrient needs are higher than usual, additional steps are needed to make sure that the diet is enriched. Table 9.6 shows some of the special formulas that may be used by infants who have feeding problems or chronic conditions that increase their nutrient requirements.

Nutrition Services

Infants who were born preterm or with special health care needs have access to more nutrition services than other infants. The following programs are sources of nutrition services or finances to pay for nutrition services:[41]

- Federal disability programs
- Individuals with Disabilities Education Act (IDEA), Part C
- Early Head Start
- WIC
- State funding from the Maternal and Child Health (MCH) Block Grant

Infants with disabling conditions are eligible for Supplemental Social Insurance (SSI), a federal program within the Social Security Administration.[41] SSI provides the family with a disability check and access to health insurance if its income meets federal guidelines.

Nutrition services are part of educational programs in IDEA, including services for children 0–2 years of age.[47] Early Head Start programs enroll infants with special health care needs. Infants with a nutrition-related diagnosis such as PKU or a cardiac problem would be eligible. Early Head Start staff are trained to feed infants special diets.

Each state has to designate a portion of the federal MCH Block Grant for children with special health care needs.[41] Services differ from state to state, but all states provide care for infants with chronic conditions. Following are some examples of how nutrition services are provided:

- Specialty clinic services, such as having a nutrition consultant attend a cystic fibrosis clinic
- Contractual services for providing special formulas or therapy for groups of patients who need more nutrition care than usually provided
- Visiting at schools or programs to conduct nutrition assessments or coordinate follow-up recommendations, such as making sure that a specific diet is being offered or that mealtime behavior is being monitored
- Transporting teams of specialists to rural or isolated areas for direct care
- Development and distribution of nutrition education materials for staff training

Every state also has a program funded by MCH to identify and advocate for children with special needs, such as the Developmental Disabilities Council.[41]

Resources

Emory University Pediatrics Department

This Web site includes information on preterm births and risks for providers and parents, with sections on nutrition in vari ous health conditions.

Web site: www.emory.edu/PEDS

National Association of Developmental Disabilities Councils

This Web site includes public policy and advocacy resources for providers and parents all over the United States concerning various disabilities, including services for infants.

Web site: www.naddc.org

National Early Childhood Technical Assistance Center

This Web site provides publications and other resources about programs for infants and children. Staff training materials are also available for those who are working with infants and young children.

Web site: www.nectus.unc.edu

Neonatology on the Web

This Web site provides resources for health care professionals, including practice guidelines and consensus statements.

Web site: www.neonatology.org

Preemieparents.com

This Web site lists sites for recommended reading and parent resources.

Web site: www. Preemieparents.com

Tufts Nutrition and Health

This site (newsletter) has extensive information for parents about credible nutrition information sources as well as a section on special diets and resources.

Web site: www.healthletter.tufts.edu

United Cerebral Palsy Associations

This Web site provides links to service sites in the United States.

Web site: www.ucpa.org

References

1. Trends in infant mortality attributable to birth defects, United States 1980–2002. MMWR, accessed 8/15/03.

2. Cooper TR, Bereth CL, Adams JM et al. Actuarial survival in the premature infant less than 30 weeks gestation. Pediatrics 1998;101;975–8.

3. Institute of Medicine Food and Nutrition Board. Dietary Reference Intakes for energy, carbohydrate, fiber, fat, fatty acids, cholesterol, protein, and amino acids. National Academy Press; 2002.

4. Stallings VA. Resting energy expenditure. In: Altschuler SM, Liacouras CA, eds. Clinical pediatric gastroenterology. Philadelphia: Harcourt Brace; 1998: 607–11.

5. Embleton NE, Pang N, Cooke RJ. Postnatal malnutrition and growth retardation: an inevitable consequence of current recommendations in preterm infants? Pediatrics 2001:107, 270–3.

6. Hill JB, Haffner WHJ. Growth before birth. In: Batshaw ML, ed. Children with disabilities. 5th ed. Baltimore, MD: Paul H. Brookes; 2002: 43–53.

7. Committee on Nutrition of the Preterm Infant. European Society of Paediatic Gastroenterology and Nutrition in Nutrition and feeding of the preterm infant. Oxford, UK: Blackwell Scientific Publications; 1987.

8. Hall RT, Carroll RE. Infant feeding. Pediatrics in Review 2000;21:191–200.

9. Rais-Bahrami K, Short BL, Batshaw ML. Premature and small-for-dates infants. In: Batshaw ML, ed. Children with disabilities. 5th ed. Baltimore, MD: Paul H. Brookes; 2002: 85–103.

10. Batshaw ML, Tuchman M. PKU and other inborn errors of metabolism. In:

Batshaw ML, ed. Children with disabilities. 5th ed. Baltimore, MD: Paul H. Brookes; 2002: 333–45.

11. Birch, EE et al. A randomized controlled trial of long-chain polyunsaturated fatty acid supplementation of formula in term infants after weaning at 6 wks of age. American Journal of Clinical Nutrition 2002;75: 570–80.

12. Trumbo P, Yates AA, Schlicker SA et al. Dietary reference intakes; vitamin A, vitamin K, arsenic, boron, chromium, copper, iodine, iron, manganese, molybdenum, nickel, silicon, vanadium, and zinc. J Am Diet Assoc 2001;101:294–301.

13. Ehrenkranz RA, Younes N, Lemons JA, Fanaroff AA et al. Longitudinal growth of hospitalized very low birth weight infants. Pediatrics, 1999;104:280–9.

14. York J Devoe, M. Health issues in survivors of prematurity. South Med J 2002; 95(9):969–76.

15. National Center for Health Statistics: NCHS growth curves for children 0–19 years. U.S. Vital and Health Statistics, Health Resources Administration. Washington, DC: U.S. Government Printing Office; 2000.

16. Growth references: third trimester to adulthood. 2nd ed. Clinton SC: Greenwood Genetic Center; 1998.

17. The Infant Health and Development Program: enhancing the outcomes of low-birth-weight, premature infants. JAMA 1990;263(22):3035–42.

18. Eyler FD, Behnke M, Conlon M et al. Birth outcome from a prospective, matched study of prenatal crack/cocaine use: I. Interactive and dose effects on health and growth. Pediatrics 1998;101:229–37.

19. Huysman WA, de Ridder M, de Bruin NC et al. Growth and body composition in preterm infants with bronchopulmonary dysplasia. Arch Dis Child Fetal Neonatal Ed 2003; 88:F46–F51.

20. Peralta-Carcelen M, Jackson DA, Goran MI et al. Growth of adolescents who were born at extremely low birth weight without major disability. Pediatrics 2000;136: 633–40.

21. Walker SP, Grantham-Mcgregor SM, Powell CA et al. Effects of growth restriction in early childhood on growth, IQ, and cognition at age 11 to 12 years and the benefits of nutritional supplementation and psychosocial stimulation. J Pediatrics 2000; 137:36–41.

22. Eicher PS. Feeding. In: Batshaw, ML ed. Children with disabilities. 5th ed. Baltimore, MD: Paul H. Brookes;2002:549–66.

23. Down syndrome prevalence at birth— United States 1983–1990. MMWR August 26, 1994:43(33):617–22.

24. U.S. Vital and Health Statistics. Births: preliminary data for 2002. National Vital Statistics Report 2003:51;1–20.

25. Wood, NS, Marlow N, Costeloe K et al. Neurological and developmental disability after extremely preterm birth. N Eng J of Med 2000;343:378–84.

26. Strauss R. Adult functional outcome of those born small for gestational age: twenty-six-year follow-up of the 1970 British birth cohort. JAMA 2000;283:625–32.

27. Kilbride HW, Bendorf K, Wheeler R. Total parenteral nutrition. In: Merenstein GB, Gardner SL, eds. Handbook of neonatal intensive care. 4th ed. St. Louis: Mosby; 1998: 300–16.

28. Matlow A, Wray R, Goldman C, Streitenberger L et al. Microbial contamination of enteral feed administration sets in a pediatric institution. Amer J Infect Contr 2003;51:49–53.

29. American Dietetic Association. Infant feedings: guidelines for preparation of formula and breast milk in health care facilities. 2003.

30. Mudd SH, Braverman N, Pomper M, Tezcan K et al. Infantile hypermethioninemia and hyperhomocysteinemia due to high methionine intake: a diagnostic trap. Molecular Genetics and Metabolism 2003; 79:6–16.

31. American Academy on Pediatrics. Breastfeeding and the use of human milk. (RE9729) Pediatrics 1997;100:1035–9.

32. Gardner S, Snell BJ, Lawrence RA. Breastfeeding the neonate with special needs. In: Merenstein GB, Gardner S. Handbook of neonatal intensive care. 4th ed. St. Louis: Mosby; 1998: 333–66.

33. Elliott S, Reimer C. Postdischarge telephone follow-up program for breastfeeding preterm infants discharged from a special care nursery. Neonat Net 1998;17:41–5.

34. Mead-Johnson Nutritionals Web site; available at www.meadjohnson.Com/ products/index.html/, accessed 9/5/03.

35. Gray JE, McCormick MC, Richardson DK, et al. Normal birth weight intensive care unit survivors: outcome assessment. Pediatrics 1996;97:832–8.

36. Trends in infant mortality attributable to birth defects, United States 1980-1995. MMWR September 25, 1998;47(37):773–8.

37. Centers for Disease and Prevention. Folic acid and prevention of spina bifida and anencephaly MMWR 2002;51(RR13):1–3.

38. Lee J, Nunn J, Wright C. Height and weight achievement in cleft lip and palate. Arch Dis Child 1997;76:70–72.

39. Milerad J, Larson O, Hagberg C et al. Associated malformations in infants with cleft lip and palate: a prospective, population-based study. Pediatrics 1997; 100:180–6.

40. Moss EM, Batshaw ML, Solot CB et al. Psychoeducational profile of the 22q11.2 microdeletion: a complex pattern. J Pediatrics 1999;134:193–8.

41. General Information about disabilities. National Information Center for Children and Youth with Disabilities; available at www.nichcy.org/general.htm, accessed 9/5/03.

"Enough is as good
as a feast."
Proverbs

Chapter 10

Toddler and Preschooler Nutrition

Chapter Outline

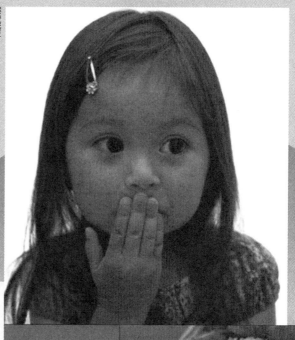

Photo Disc

Photo Disc

Prepared by **Nancy H. Wooldridge**

brand X pictures

Key Nutrition Concepts

1 Children continue to grow and develop physically, cognitively, and emotionally during the toddler and preschool years, adding many new skills rapidly with time.

2 Learning to enjoy new foods and developing feeding skills are important components of this period of increasing independence and exploration.

3 Children have an innate ability to self-regulate food intake. Parents and caretakers need to provide children nutritious foods and let children decide how much to eat.

4 Parents and caretakers have tremendous influence on children's development of appropriate eating, physical activity, and other health behaviors and habits formed during the toddler and preschool years. These lessons are mainly transferred by example.

Introduction

This chapter describes the growth and development of toddlers and preschool-age children and their relationships to nutrition and the establishment of eating patterns. Growth during the toddler and preschool years is slower than in infancy but steady. This slowing of *growth velocity* is reflected in a decreased appetite; yet young children need adequate calories and nutrients to meet their nutritional needs. The eating and health habits established at this early stage of life may affect food habits and subsequent health later in life. The development of new skills and increasing independence mark the toddler and preschool stages. Learning about and accepting new foods, developing feeding skills, and establishing healthy food preferences and eating habits are important aspects of this stage of development.

Growth Velocity The rate of growth over time.

Toddlers Children between the ages of 1 and 3 years.

Gross Motor Skills Development and use of large muscle groups as exhibited by walking alone, running, walking up stairs, riding a tricycle, hopping, and skipping.

Fine Motor Skills Development and use of smaller muscle groups demonstrated by stacking objects, scribbling, and copying a circle or square.

Preschool-Age Children Children between the ages of 3 and 5 years, who are not yet attending kindergarten.

Iron-Deficiency Anemia A condition often marked by low hemoglobin level. It is characterized by the signs of iron deficiency plus paleness, exhaustion, and a rapid heart rate.

Definitions of the Life-Cycle Stage

Toddlers are generally defined as children between the ages of 1 and 3 years. This stage of development is characterized by a rapid increase in *gross* and *fine motor skills* with subsequent increases in independence, exploration of the environment, and language skills. *Preschool-age children* are between 3 and 5 years of age. Characteristics of children at this stage of development include increasing autonomy; experiencing broader social circumstances, such as attending preschool or staying with friends and relatives; increasing language skills; and expanding their ability to control behavior.

Importance of Nutrition

Adequate intake of energy and nutrients is necessary for toddlers and preschool-age children to achieve their full growth and developmental potential. Undernutrition during these years impairs children's cognitive development as well as their ability to explore their environments.[1] Long-term effects of undernutrition, such as failure to thrive and cognitive impairment, may be prevented or reduced with adequate nutrition and environmental support.

Tracking Toddler and Preschooler Health

There are approximately 72 million children under 18 years of age in the United States today.[2] Of these, 7% live in extreme poverty, defined as households with incomes below 50% of the poverty level.[2] Twelve percent of children have no health insurance. When evaluating young children's nutritional status and offering nutrition education to parents, it is important to consider children's home environments. Establishing healthy eating habits may not be high on a family's priority list when the home environment is one of poverty and food insecurity.

Disparities in nutrition status indicators in this age group exist among races. For example:

- Eight percent of low-income children under age 5 years are growth retarded, but up to 15% of low-income African American children are.[3]
- Seventeen percent of Mexican American children and 10% of African American children aged 1 to 2 years have *iron-deficiency anemia* compared to 8% of white children.[3] (See page 252 for a definition of iron-deficiency anemia.)

Healthy People 2010

Healthy People 2010, objectives for the nation for improvements in health status by the year 2010, includes a number of objectives that directly relate to toddlers and preschoolers.[3] These are listed in Table 10.1.

Normal Growth and Development

An infant's birthweight triples in the first 12 months of life, but growth velocity slows thereafter until the adolescent growth spurt. On average, toddlers gain 8 ounces

Table 10.1 Healthy People 2010 objectives related to toddlers and preschool-age children[3]

Objective 19-4	Reduce growth retardation among low-income children under age 5 years from 8% to 5%.
Objective 19-5	Increase the proportion of persons aged 2 years and older who consume at least two daily servings of fruit from 28% to 75%.
Objective 19-6	Increase the proportion of persons aged 2 years and older who consume at least three daily servings of vegetables, with at least one-third being dark green or deep yellow vegetables, from 3% to 50%.
Objective 19-7	Increase the proportion of persons aged 2 years and older who consume at least six daily servings of grain products, with at least three being whole grains, from 7% to 50%.
Objective 19-8	Increase the proportion of persons aged 2 years and older who consume less than 10% of calories from saturated fat from 36% to 75%.
Objective 19-9	Increase the proportion of persons aged 2 years and older who consume no more than 30% of calories from fat from 33% to 75%.
Objective 19-10	Increase the proportion of persons aged 2 years and older who consume 2400 mg or less of sodium daily from 21% to 65%.
Objective 19-11	Increase the proportion of persons aged 2 years and older who meet dietary recommendations for calcium from 46% to 75%.
Objective 19-12	Reduce iron deficiency among young children and females of childbearing age from 4% and 11% to 1% and 7%.

(0.23 kg)/month and 0.4 in (1 cm) of height/month, while preschoolers gain 4.4 pounds (2 kg) and 2.75 in (7 cm) per year.[4] This decrease in rate of growth is accompanied by a reduced appetite and food intake in toddlers and preschoolers. A common complaint of parents of children this age is that their children have much less appetite and a lower interest in food or eating compared to their appetite and food intake during infancy. Parents need to be reassured that a decrease in appetite is part of normal growth and development for children in this age group.

In monitoring a child's physical growth, it is important for children to be accurately weighed and measured at periodic intervals. Toddlers less than 2 years of age should be weighed without clothing or a diaper. The *recumbent length* of toddlers should be measured on a length board with a fixed head board and movable foot board. Proper measurement of recumbent length requires two adults— one at the child's head, making sure the crown of the head is placed firmly against the head board, and the other making sure that the child's legs are fully extended and placing the foot board at the child's heels. Proper positioning for measurement of a child's recumbent length is shown in Illustration 10.1. Preschool-age children should be weighed and measured without shoes and in lightweight clothing. Calibrated scales should be used, and a height board should be used for measuring *stature*. Illustrations 10.2 and 10.3 on the next page further demonstrate the proper techniques for weighing and measuring young children. It is important that both weight and height be plotted on the appropriate growth charts, such as the 2000 CDC growth charts discussed next.

The 2000 CDC Growth Charts

A full set of the "CDC Growth Charts: United States" can be found in Appendix A.[5] Illustration 10.4 on page 245 shows an example of one of the charts, which depicts the growth of a healthy child. Charts are gender specific and are available for birth to 36 months and 2 to 20 years. With these growth charts, the health care professional can plot and monitor weight-for-age, length- or stature-for-age, head circumference-for-age, weight-for-length, weight-for-

> **Recumbent Length** Measurement of length while the child is lying down. Recumbent length is used to measure toddlers less than 24 months of age, and those between 24 and 36 months who are unable to stand unassisted.
>
> **Stature** Standing height.

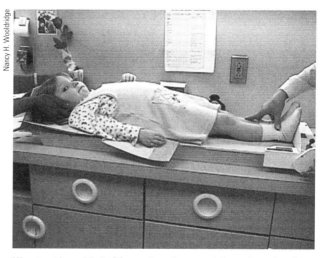

Illustration 10.1 Measuring the recumbent length of a toddler is a two-person job!

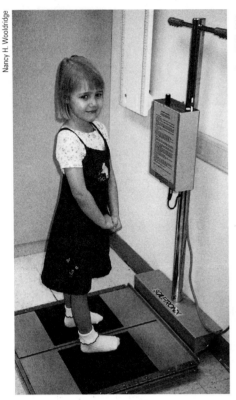

Illustration 10.2 Young child being weighed.

Illustration 10.3 Measuring the stature of a preschool-age child.

stature, and body mass index (BMI)-for-age. There is overlap between the two sets of growth charts for children between 24 and 36 months of age. If the child's recumbent length is measured, then the birth- to 36-month growth chart is the appropriate one to use. If the child over 2 years of age is measured standing, the 2- to 20-year-old growth chart is the correct choice. Children's growth usually "tracks" within a fairly steady percentile range. It is important to monitor a child's growth over time and to identify any deviations in growth. It is the pattern of growth that is important to assess rather than any one single measurement. A weight measurement without a length or stature measurement doesn't indicate how appropriate the weight is for the child's length or stature.

Body mass index, or BMI, provides a guideline for assessing underweight and overweight in children and adults. Body mass index is predictive of body fat for children over 2 years of age. A BMI of 85th percentile or greater but less than the 95th percentile indicates risk of being overweight, and a BMI of 95th percentile or greater indicates overweight.[8] A BMI less than 5th percentile indicates underweight. BMI fluctuates throughout childhood. BMI increases in infancy; it decreases during preschool years, hitting its lowest point at approximately 4 to 6 years of age; and it then increases into adulthood. Because of this normal fluctuation of BMI, the only way

Body Mass Index An index that correlates with total body fat content or percent body fat, and is an acceptable measure of adiposity or body fatness in children and adults.[6,7] It is calculated by dividing weight in kilograms by the square of height in meters (kg/m²).

to know if a child's BMI is within a normal range is to plot BMI-for-age on the appropriate growth curve. In pediatrics, the goal is to strive for a BMI-for-age in the normal range and not a specific BMI range, as is the goal for adults.

Growth charts visually aid parents by demonstrating the expected slowing of growth velocity during the toddler and preschool stages of development. Although the curves for weight-for-age and length- or stature-for-age continue to increase during the toddler and preschool years, the slope of the curve is not as steep as during the first year of life.

Common Problems with Measuring and Plotting Growth Data

Growth in young children that is measured or plotted incorrectly can lead to errors in health status assessment. Standard procedures should be followed; calibrated and appropriate equipment should be used; and plotting should be double-checked, including checking the age of the child, to avoid such errors. Choosing the appropriate growth chart based on how the child was measured (recumbent length versus stature) and on the child's gender is important as well as using the updated growth charts.

Physiological and Cognitive Development

Toddlers

An explosion in the development of new skills happens during the toddler years. Most children begin to walk independently at about their first birthday. At first the walking is more like a "toddle," with a wide-based gait.[4] After practicing for several months, the toddler achieves greater steadiness and soon will be able to stop, turn, and stoop without falling over. Gross motor skills, such as sitting on a small chair and climbing on furniture, develop rapidly at this age; and with practice, great improvements in balance and agility take place. At about 15 months, children can crawl up stairs; by about 18 months, they can run stiffly. Most toddlers can walk up and down stairs one step at a time by 24 months, and jump in place. At about 30 months, children have advanced to going up stairs by alternating their feet. By 36 months of age, children are ready for tricycles.

Illustration 10.4 Birth to 36 months: girls length-for-age and weight-for-age percentiles[5]

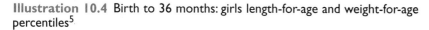

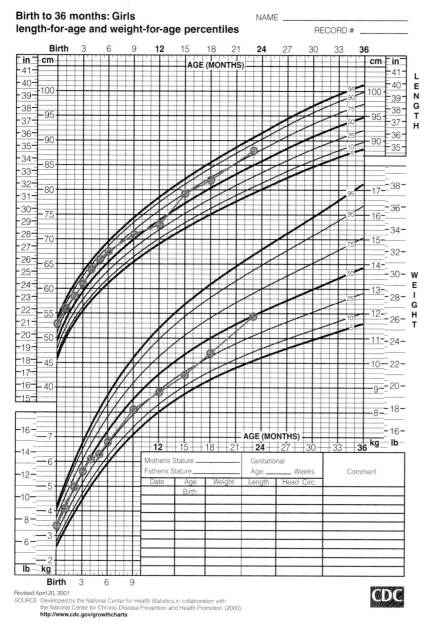

Birth to 36 months: Girls
length-for-age and weight-for-age percentiles

NAME _____

RECORD # _____

Mother's Stature _____
Father's Stature _____

Gestational
Age: _____ Weeks

Comment

Date	Age	Weight	Length	Head Circ.
Birth				

Revised April 20, 2001.
SOURCE: Developed by the National Center for Health Statistics in collaboration with
the National Center for Chronic Disease Prevention and Health Promotion (2000).
http://www.cdc.gov/growthcharts

found independence becomes very important to them. Toddlers now have the power to control the distance between themselves and their parents. *Nelson's Textbook of Pediatrics*[4] describes how toddlers often "orbit" around their parents, like planets, moving away, looking back, moving farther, and then returning.

From a socialization standpoint, the child moves from being primarily self-centered to being more interactive. The toddler now possesses the ability to explore the environment and to develop new relationships. Fears of certain situations, such as separation, darkness, loud sounds, wind, rain, and lightning commonly emerge during this period as the child learns to deal with changes in the environment. Children develop rituals in their daily activities in an attempt to deal with these fears.

Social development also involves imitating others, such as parents, caretakers, siblings, and peers, during this time. The child in this stage begins to learn about the family's cultural customs, including those related to meals and food.

Dramatic development of language skills occurs from 18 to 24 months. Once a child realizes that words can stand for things, his vocabulary erupts from 10 to 15 words at 18 months to 100 or more words at 2 years of age. The toddler soon begins combining words to make simple sentences. By 36 months, the child uses three-word sentences.[4]

Children become increasingly mobile and independent with improvements in gross motor skills. Toddlers are fascinated with these newfound skills, showing a readiness to put these skills into practice and to develop new skills. However, toddlers have no sense of dangerous situations. At this age, children are especially vulnerable to accidental injuries and ingestion of harmful substances. In fact, the leading cause of death among young children is unintentional injuries.[3] Parents and caregivers have to constantly watch over toddlers, preferably in environments made "child safe."

COGNITIVE DEVELOPMENT IN TODDLERS

With the toddler's newly acquired physical skills, exploring the environment accelerates, and exerting their new-

An important social change for toddlers is increased determination to express their own will. This expression often comes in the form of negativism and the beginning of temper tantrums, which give this stage of development its label of "the terrible twos." With an increase in motor development coupled with an increasing quest for independence, the toddler tries to do more and more things, pushing her capabilities to the limit. Thus the toddler can become easily frustrated and negative. The child seeks more independence and at the same time needs the parents and caretakers for security and reassurance. Toddler behavior uncannily parallels the same type of behavior commonly seen in adolescents!

Illustration 10.5 Toddler enjoying mealtime!

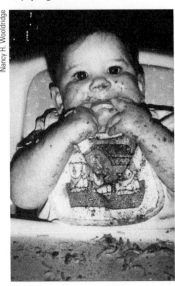

Nancy H. Wooldridge

DEVELOPMENT OF FEEDING SKILLS IN TODDLERS

Many babies begin to wean from the bottle at about 9 to 10 months of age, when their solid food intake increases and they learn to drink from a cup.[9] Parents need to pay attention to cues of readiness for weaning, such as disinterest in breastfeeding or bottle-feeding. The time it takes to wean is variable and depends on both the child and the mother. Weaning will be easier for those babies who adapt well to change. Weaning is a sign of the toddler's growing independence and is usually complete by 12 to 14 months of age, although the age varies from child to child.

Gross and fine motor development during the toddler years enhances children's ability to chew foods of different textures and to self-feed. Between 12 and 18 months, toddlers are able to move the tongue from side to side (or laterally) and learn to chew food with rotary, rather than just up-and-down movements. Toddlers can now handle chopped or soft table food.

At about 12 months, children have a refined pincer grasp that enables them to pick up small objects, such as cooked peas and carrots, and put them into their mouths. Children will be able to use a spoon around this age, but not very well. At 18 to 24 months, toddlers are able to use the tongue to clean the lips and have well-developed rotary chew movements. Now the toddler can handle meats, raw fruits and vegetables, and multiple textures of food.

A strong need for independence in self-feeding emerges during the toddler age. "I do it!" and "No, no, no!" are commonly heard phrases in households where toddlers reside. As toddlers busily practice their newly found skills, they become easily distracted. Parents need to realize that their toddler's sometimes-fierce independence is part of normal growth and development and represents an ongoing process of separation from dependency on the parents and caretakers.

Increasing fine motor and visual motor coordination skills allow toddlers to use cups and spoons more effectively. Although toddlers' skill with a spoon increases during the second year, they prefer to eat with their hands. Initial attempts at self-feeding are inevitably messy, as Illustration 10.5 depicts, but represent an important stage of development. It is important that parents and caretakers keep distractions during mealtimes, such as television, to a minimum, and allow their toddlers to practice self-feeding skills and to experience new foods and textures. The child derives pleasure in self-feeding and exploring new tastes. Learning to self-feed allows the child to develop mastery of an important part of everyday life.

Adult supervision of eating is imperative due to the high risk of choking on foods at this age. Toddlers should always be seated during meals and snacks, preferably in a high chair or booster seat with the family, and not allowed to "eat on the run." Foods that may cause choking, such as hard candy, popcorn, nuts, whole grapes, and hot dogs, should not be served to children less than 2 years of age.[9]

FEEDING BEHAVIORS OF TODDLERS

The toddler's need for rituals, a hallmark of this stage of development, may be linked to the development of food jags. Many toddlers demonstrate strong food preferences and dislikes. They can go through prolonged periods of refusing a particular food or foods they previously liked. The intensity of the refusal or the negative attitude toward a particular food will be influenced by the child's temperament (see the "Temperament Differences" section on page 249). To circumvent food jags, parents can serve new foods along with familiar foods. New foods are better accepted if they are served when the child is hungry, and if she sees other members of the family eating these foods. Eventually, toddlers' natural curiosity will get the best of them. Toddlers are great imitators, which includes imitating the eating behavior of others.

Mealtime is an opportunity for toddlers to practice newly acquired language and social skills and to develop a positive self-image. It is not the time for battles over food or "force feedings." Establishing the habit of eating breakfast is an important part of healthy eating behaviors. Family mealtime provides an opportunity for parents and caretakers to model healthy eating behaviors for the young child.

APPETITE AND FOOD INTAKE IN TODDLERS

Parents need to be reminded that toddlers naturally have a decreased interest in food because of slowing growth, and a corresponding decrease in appetite. Besides, with all of their newfound gross and fine motor skills, they have places to go and new environments to explore! It is a part of normal growth and development for toddlers to have a decreased interest in food and to be easily distracted at mealtime.

Toddlers need toddler-sized portions. One rule of thumb for serving size is 1 tablespoon of food per year of age. So, a serving for a 2-year-old child would be about 2

Developing Appropriate Feeding Behaviors

Lindsey, a 24-month-old little girl, lives with her parents. She stays at a child care center during the week while both of her parents are at work. On the weekends, her parents enjoy their time with Lindsey, although a lot of their time is spent running errands and catching up on household chores. Partly to appease Lindsey, her parents allow her to have as much of her favorite beverage, apple juice, from a "sippy" cup as she wants between meals. Lindsey also has free access to snacks such as crackers, slices of cheese, and cookies. When the family sits down to have a meal together, Lindsey plays with her food and usually doesn't eat much. She tells her parents that she doesn't like the food being served and wants "something else." She soon becomes fussy and wants to get down from her booster seat. In trying to keep her at the table with them, her parents turn on the television or play Lindsey's favorite cartoons on the VCR. If that does not quiet Lindsey, her mother offers to prepare another food item of Lindsey's choice. Mealtime has become an unpleasant experience for the family.

Questions

1. Identify some of the inappropriate eating habits that Lindsey's parents have allowed her to develop.
2. Considering her stage of development, what advice would you give Lindsey's parents in their attempts to increase the number of foods that she will eat?
3. In what types of food-preparation activities would it be appropriate for Lindsey's parents to have her participate? Why is this important?
4. What suggestions do you have for snack-food items for Lindsey?
5. Would you advise Lindsey's parents to give her a daily multivitamin supplement?
6. What advice would you give the family regarding physical activity?

(Answers are in Appendix D.)

---

tablespoons. It is better to give the child a small portion and allow him to ask for more than to serve large portions. Parents often overestimate portion sizes needed by the young child, which may contribute to labeling the child as a "picky" eater. Because toddlers can't eat a large amount of food at one time, snacks are vital in meeting the child's nutritional needs. It is important that toddlers not be allowed to "graze" throughout the day on sweetened beverages and foods such as cookies and chips. These foods can "kill" their limited appetite for basic foods at meal and snack times. In considering the toddler's need for rituals and limit setting, parents and caretakers need to establish regular but flexible meal and snack times, allowing enough time between meals and snacks for the toddler to get hungry.

Preschool-Age Children

Preschool-age children continue to expand their gross and fine motor capabilities. At age 4, the child can hop, jump on one foot, and climb well. The child can ride a tricycle, or a bicycle with training wheels, and can throw a ball overhand.[4]

COGNITIVE DEVELOPMENT OF PRESCHOOL-AGE CHILDREN Magical thinking and egocentrism characterize the preschool period.[4] Egocentrism does not mean that the child is selfish, but rather that the child is not able to accept another's point of view. The child is beginning to interact with a widening circle of adults and peers. During the preschool years, children gradually move from primarily relying on external behavioral limits, such as those demanded by parents and caregivers, to learning to limit behavior internally. This transition is a prerequisite to functioning in a school classroom.[4] Also during this time, children's play starts to become more cooperative, such as building a tower of blocks together. Toward the end of the preschool years, children move to more organized group play, such as playing tag or "house."

Control is a central issue for preschool children. They will test their parents' limits and still resort to temper

Table 10.2 Meal-preparation activities for young children[10]

2-year-olds

Wipe tabletops	Snap green beans
Scrub vegetables	Wash salad greens
Tear lettuce or greens	Play with utensils
Break cauliflower	Bring ingredients from one place to another

3-year-olds

Wrap potatoes for baking	Shake liquids in covered container
Knead and shape yeast dough	Spread soft spreads
Pour liquids	Place things in trash
Mix ingredients	

4-year-olds

Peel oranges or hard-cooked eggs	Mash bananas using fork
Move hands to form round shape	Set table
Cut parsley or green onions with dull scissors	

5- to 6-year olds

Measure ingredients	Use an eggbeater
Cut with blunt knife	

tantrums to get their way. Temper tantrums generally peak between the ages of 2 and 4 years.[4] The child's challenge is to separate, and the parent's challenge is to appropriately set limits and at the same time to let go, another parallel with adolescence. Parents need to strike an appropriate balance for setting limits. Too tightly controlled limits can undermine the child's sense of initiative and cause them to act out, whereas loose limits can cause the child to feel anxious and that no one is in control.

Language develops rapidly during the preschool years and is an important indicator of both cognitive and emotional development. Between ages 2 and 5, children's vocabularies increase from 50 to 100 words to more than 2000 words, and their language progresses from two- to three-word sentences to complete sentences.[4]

DEVELOPMENT OF FEEDING SKILLS IN PRESCHOOL-AGE CHILDREN

The preschool-age child can use a fork and a spoon and uses a cup well. Cutting and spreading with a knife may need some refinement. Children should be seated comfortably at the table for all meals and snacks. Eating is not as messy a process during the preschool years as it was during toddlerhood. Spills still do occur, but they are not intentional. Foods that cause choking in young children should be modified to make them safer, such as cutting grapes in half lengthwise and cutting hot dogs in quarters lengthwise and then cutting into small bites. Adult supervision during mealtime is still important.

FEEDING BEHAVIORS OF PRESCHOOL-AGE CHILDREN

As during the toddler years, parents of preschool-age children need to be reminded that the child's rate of growth continues to be relatively slow, with a relatively small appetite and food intake. Growth occurs in "spurts" during childhood. Appetite and food intake increase in advance of a growth spurt, causing children to add some weight that will be used for the upcoming spurt in height. Therefore, the appetite of a preschool-age child can be quite variable.

Preschool-age children want to be helpful and to please their parents and caretakers. This characteristic makes the preschool years a good time to teach children about foods, food selection, and preparation by involving them in simple food-related activities. For instance, outings to a farmers' market can introduce children to a variety of fresh vegetables and fruit. Allowing children to be involved in meal-related activities such as those listed in Table 10.2 can be quite instructive.[10] Families of preschool-age children need to continue to be encouraged to eat together, like the family in Illustration 10.6.

INNATE ABILITY TO CONTROL ENERGY INTAKE

An important principle of nutrition for young children, and one with direct application to child feeding, is children's ability to self-regulate food intake. If allowed to decide when to eat and when to stop eating without outside interference, children eat as much as they need.[11,12] Children have an innate ability to adjust their caloric intake to meet caloric needs. The preschool-age child's intake may fluctuate widely from meal to meal and day to day. But over a week's time, the young child's intake remains relatively stable.[13] Parents who try to interfere with the child's ability to self-regulate intake by forcing the child to "clean her plate" or using food as a reward are asking the child to overeat or undereat.

Although children can self-regulate caloric intake, no inborn mechanisms direct them to select and consume a well-balanced diet.[14] Children learn healthful eating habits.[15] Parents give up some control over what their preschool child eats if the child spends more time away from home in a child care center or with extended family members. Preschool children continue to learn about food and food habits by observing their parents, caretakers, peers, and siblings, and they begin to be influenced by what they see on television. Their own food habits and food preferences are established at this time.

APPETITE AND FOOD INTAKE OF PRESCHOOLERS

Parents of preschool age children often describe their children's appetite as being "picky." One reason a child may want the same foods all of the time is because familiar foods may be comforting to the child. Another reason is that the child may be trying to exert control over this aspect of her life. The child's eating and food selection can easily become a battleground between parent

Illustration 10.6 Sharing family meals is an important aspect of development in young children.

and child; this scenario should be avoided. Some practical suggestions for parents and caretakers of children this age include serving child-sized portions and serving the food in an attractive way. Young children often do not like their foods to touch or to be mixed together, such as in casseroles or salads. They typically do not like strongly flavored vegetables and other foods, or spicy foods at this young age. Just as with toddlers, parents of preschool-age children should not allow their children to eat and drink indiscriminately between meals and snacks. This behavior often "kills" the appetite at mealtime. Children should not be forced to stay at the table until they have eaten a certain amount of food as determined by the parent.

Temperament Differences

> "Better is a dinner of herbs where love is, than a fatted ox and hatred with it."
>
> Proverbs 15:17

Temperament is defined as the behavioral style of the child, or the "how" of behavior. This definition was derived from analysis of data from the New York Longitudinal Study, begun in 1956 by Chess and Thomas.[16] These investigators defined three temperamental clusters—the "easy" child (about 40% of children), the "difficult" child (10%), and the "slow-to-warm-up" child (15%). The remaining children, classified as "intermediate-low" or "intermediate-high," demonstrated a mixture of behaviors but gravitated to one end of the spectrum.[16]

Children's temperaments affect feeding and mealtime behavior. The "easy" child is regular in function, adapts easily to regular schedules, and tries and accepts new foods readily. The "difficult" child, on the other hand, is characterized by irregularity in function and slow adaptability. This child is more reluctant to accept new foods and can be negative about them. The "slow-to-warm-up" child exhibits slow adaptability and negative responses to many new foods with mild intensity. With repeated exposures to new foods, this child can learn to accept them over time with limited complaining.[16]

The "goodness of fit" between the temperaments of the child and the parent or caretaker can influence feeding and eating experiences.[16] A mismatch can result in conflict over eating and food. Parents and caretakers need to be aware of the child's temperament when attempting to meet nutritional needs. The difficult or slow-to-warm-up child may pose special challenges that need to be addressed by gradually exposing the child to new foods and not hurrying him to accept them.[16]

Food-Preference Development, Appetite, and Satiety

Food-preference development and regulation of food intake have been studied extensively by Leann Birch and associates.[17,18] It is clear that children's food preferences do determine what foods they consume. Children naturally prefer sweet and slightly salty tastes and generally reject sour and bitter tastes. These preferences appear to be unlearned and present in the newborn period. Children eat foods that are familiar to them, a fact that emphasizes the importance environment plays in the development of food preferences. Children tend to reject new foods but may learn to accept a new food with repeated exposures to it. It may, however, take eight to ten exposures to a new food before it is accepted. Children who are raised in an environment where all members of the family eat a variety of foods are more likely to eat a variety of foods. One study showed that 5-year-old girls' fruit and vegetable intakes were related to their parents' fruit and vegetable intakes.[19]

Children also appear to have preferences for foods that are energy dense due to high levels of sugar and fat.[17,18] This preference may develop because children associate eating energy-dense foods with pleasant feelings of satiety, or because these types of foods may be associated with special social occasions such as birthday parties. The context in which foods are offered to a child influences the child's food preferences. Foods served on a limited basis but used as a reward become highly desirable. Restricting a young child's access to a palatable food may actually promote the desirability and intake of that food.[20] Coercing or forcing children to eat foods can have a long-term negative impact on their preference for these foods.[17,18, 19]

APPETITE, SATIETY   Children's energy intake regulation has been studied by giving children *preloads* of foods or beverages of varying energy content, followed by self-selected meals. In one such study, children aged 3 to 5 years were given either a low-energy preload beverage made with aspartame (Nutrasweet), a low-calorie sugar substitute, or a high-energy preload beverage made with sucrose. Fat and protein content of the preloads did not differ. Children were then allowed to self-select their lunches. Children who had the low-calorie beverage before lunch consumed more calories at lunch, while those who had the higher-calorie beverage consumed fewer calories. These results indicate that young children are able to adjust caloric intake based on caloric need.[18,21] Similar studies were conducted in 2- to 5-year-olds using foods with dietary fat or olestra, a nonenergy fat substitute. Results indicate that children compensated for the lower level of calories in food when olestra was substituted for dietary fat.

> **Preloads**   Beverages or foods such as yogurt in which the energy/macronutrient content has been varied by the use of various carbohydrate and fat sources. The preload is given before a meal or snack and subsequent intake is monitored. This study design has been employed by Birch and Fisher in their studies of appetite, satiety, and food preferences in young children.[18]

The preloading protocol just described was also used to study children's responsiveness to caloric content of foods in the presence or absence of common feeding advice from adults. In one group, teachers were trained to minimize their control over how much the children ate. In the other group, teachers were trained to focus the children on external factors to control their intake, such as rewarding the children for finishing the portions served to them or encouraging them to eat because "it was time to eat." Results of this investigation show that when the adults focused the children on external cues for eating, children lost their ability to regulate food intake based on calories. It appears children's innate ability to regulate caloric intake can be altered by child-feeding practices that focus on external cues rather than the child's own hunger and satiety signals.[21]

The effects of portion size on children's intakes were compared between classes of 3- and 5-year-old children. The children were served either a small, medium, or large portion of macaroni and cheese along with standard amounts of other foods in their usual lunchtime setting. Analysis of amount of food eaten showed that portion size did not affect the younger children's intakes; their intakes remained constant despite the amount of food served to them. In contrast, the 5-year-old children's intakes increased significantly with the larger portion sizes. The researchers conclude that by 5 years of age, children are influenced by the size of portions served to them, another external factor that influences intake.[22] These investigators also raise the question as to what effect large portion sizes have on overeating and, consequently, on the development of childhood overweight.

Another study of 5-year-old girls and their parents looked at the effects of parents' restriction of palatable foods on their children's consumption of these foods. After a self-selected standard lunch, these 5-year-old girls were given free access to snack foods, such as ice cream, potato chips, fruit-chew candy, and chocolate bars. The daughters of parents who reported restricting access to snack foods indicated to the investigators that they ate "too much" of the snack foods and also reported negative emotions about eating the snack foods. Parents' restriction of foods actually promoted the consumption of these foods by their young daughters and, of even more concern, the daughters reported feeling bad about eating these "forbidden" foods.[23] A related study found lowered self-concept in 5-year-old girls with high body weight.[24] Daughters of parents who restricted access to food and expressed concern about their daughter's weight status tend to have negative self-evaluations.[21] Mothers in particular seem to have the most influence over their young daughters' beliefs about food and dieting.[25]

Satter describes the optimal "feeding relationship" as one in which parents and caretakers are responsible for what children are offered to eat and the environment in which the food is served, while children are responsible for how much they eat or even whether they eat at a particular meal or snack. According to Satter, if this feeding relationship is respected, then feeding and potential weight problems can be prevented.[26] Parenting includes influencing what is served to children and the environment in which it is served, at home and in child care settings as well.

What implications does all this research have on child-feeding practices? Based on the results of these studies, it appears that by late preschool age, children are more respon-

Table 10.3 Practical applications of child-feeding research[18]

- Parents should respond appropriately to children's hunger and satiety signals.
- Parents should focus on the long-term goal of developing healthy self-controls of eating in children, and look beyond their concerns regarding composition and quantity of foods children consume or fears that children may eat too much and become overweight.
- Parents should not attempt to control children's food intakes by attaching contingencies ("No dessert until you finish your rutabagas") and coercive practices ("Clean your plate; children in Bangladesh are starving").
- Parents should be cautioned not to severely restrict "junk foods," foods high in fat and sugar, as that may make these foods even more desirable to the child.
- Parental influence should be positively focused on the child developing food preferences and selection patterns of a variety of foods consistent with a healthy diet. Parental modeling of eating a varied diet at family mealtime will have a strong influence on children.
- Children have an unlearned preference for sweet and slightly salty tastes; they tend to dislike bitter, sour, and spicy foods.
- Children tend to be wary of new foods and tastes, and it may take repeated exposures to new foods before they are accepted.
- Children need to be given appropriate child-sized servings of food.
- Child-feeding experiences should take place in secure, happy, and positive environments with or under adult supervision.
- Children should never be forced to eat anything.

sive to external cues than to their innate ability to self-regulate intake. Table 10.3 sums up the practical applications of Birch's work.[18] The importance of appropriate parenting skills in helping children learn to self-regulate food intake and possibly avoid problems with obesity is echoed by a panel of obesity experts.[27] Birch's research also reinforces the important role that parents and caretakers play in modeling healthy eating behaviors for young children.

Energy and Nutrient Needs

Dietary Reference Intakes (DRIs) were developed from 1997 through 2004. Information about the various DRI publications can be found on the National Academy Press Web site (see the "Resources" at the end of this chapter). Published DRI tables are provided on the inside front cover of this book.

Energy Needs

DRIs have been established for the energy needs of young children.[28] The formula for Estimated Energy Requirements (EER) for children aged 13–35 months is (89 × weight of infant [kg] −100) + 20 (kcal for energy deposition). For example, a healthy 24-month-old girl who weighs 12 kg would have an EER of (89 × 12 kg −100) + 20 = 988 kilocalories. Beginning at age 3, the DRI equations for estimating energy requirements are based on a child's gender, age, height, weight, and physical activity level (PAL). Categories of activity are defined in terms of walking equivalence. Table 10.4 on the next page depicts estimated energy requirements for reference boys and girls at selected ages. Energy needs of toddlers and preschool-age children reflect the slowing of the growth velocity of children in this age group.

Protein

The DRIs for protein for the toddler and preschool age groups can be found in Table 10.5 on the following page.[28] The recommended level of protein intake is generally reached or exceeded in children consuming the usual American diet as well as vegetarian diets. Adequate energy intake to meet an individual child's needs has a protein-sparing effect; that is, with adequate energy intake, protein is used for growth and tissue repair rather than for energy. Ingestion of high-quality protein, such as milk and other animal products, lowers the amount of total protein needed in the diet to provide the essential amino acids.

Vitamins and Minerals

Dietary Reference Intakes (DRIs) for vitamins and minerals have been established for the toddler and preschool-age child. Analyses of data from National Health and Nutrition Examination Surveys (NHANES I, II, and III) and the Continuing Survey of Food Intake by Individuals (CSFII) indicate that children's average intakes of most nutrients meet or exceed the recommendations.[29,30] Most children from birth to 5 years are meeting the targeted levels of consumption of most nutrients, except for iron, calcium, and zinc. The DRIs for these key nutrients are listed in Table 10.6 on the next page.[31,32]

Dietary Reference Intakes (DRIs) Quantitative estimates of nutrient intakes, used as reference values for assessing the diets of healthy people. DRIs include Recommended Dietary Allowances (RDAs), Adequate Intakes (AI), Tolerable Upper Intake Level (UL), and Estimated Average Requirement (EAR).

Common Nutrition Problems

Iron-Deficiency Anemia

Iron deficiency and iron-deficiency anemia are prevalent nutrition problems among young children in the United States. A rapid growth rate coupled with frequently

Table 10.4 Estimated energy requirements for reference boys and girls at selected ages and varying physical activity levels (PAL)[28]

AGE/GENDER	REFERENCE WEIGHT (kg [lbs])	REFERENCE HEIGHT (m [in])	SEDENTARY PAL kcal/d	LOW ACTIVE PAL kcal/d	ACTIVE PAL kcal/d	VERY ACTIVE PAL kcal/d
3-year-old boy	14.3 (31.5)	0.95 (37.4)	1162	1324	1485	1683
4-year-old boy	16.2 (35.7)	1.02 (40.2)	1215	1390	1566	1783
5-year-old boy	18.4 (40.5)	1.09 (42.9)	1275	1466	1658	1894
3-year-old girl	13.9 (30.6)	0.94 (37.0)	1080	1243	1395	1649
4-year-old girl	15.8 (34.8)	1.01 (39.8)	1133	1310	1475	1750
5-year-old girl	17.9 (39.4)	1.08 (42.5)	1189	1379	1557	1854

SOURCE: From the Institute of Medicine, Food & Nutrition Board.

inadequate intake of dietary iron places toddlers, especially 9- to 18-month-olds, at the highest risk for iron deficiency.[33] According to the NHANES, 1999–2000, 7% of toddlers aged 1 to 2 years and 4% of 6- to 11-year-old children are iron deficient;[34] the full impact of this nutrition problem is profound. Iron-deficiency anemia in young children appears to cause long-term delays in cognitive development and behavioral disturbances.[1,33]

Table 10.7 depicts the progressing signs of iron deficiency. Iron deficiency can be defined as absent bone marrow iron stores, an increase in hemoglobin concentration of <1.0 g/dL after treatment with iron, or other abnormal lab values, such as serum ferritin concentration.[33] Iron-deficiency anemia is defined as less than the 5th percentile of the distribution of *hemoglobin* concentration or *hematocrit* in a healthy reference population. Age- and gender-specific cutoff values for anemia are derived from NHANES III data. For children 1 to 2 years of age, the diagnosis of anemia would be made if the hemoglobin concentration is <11.0 g/dL and hematocrit <32.9%. For children aged 2 to 5 years, a hemoglobin value <11.1 g/dL or hematocrit <33.0% is diagnostic of iron-deficiency anemia.

Hemoglobin A protein that is the oxygen-carrying component of red blood cells. A decrease in hemoglobin concentration in red blood cells is a late indicator of iron deficiency.

Hematocrit An indicator of the proportion of whole blood occupied by red blood cells. A decrease in hematocrit is a late indicator of iron deficiency.

Anemia A reduction below normal in the number of red blood cells per cubic mm in the quantity of hemoglobin, or in the volume of packed red cells per 100 ml of blood (hematocrit). This reduction occurs when the balance between blood loss and blood production is disturbed.

Not all anemias are due to iron deficiency. Other causes of *anemia* include other nutritional deficiencies such as insufficient folate or vitamin B_{12}, chronic inflammation, or recent or current infection.[33]

One Healthy People 2010 objective is to reduce iron deficiency to 5% in children aged 1 to 2 years and to 1% in children aged 3 to 4 years.[3] Part of reaching this goal will mean reducing or eliminating disparities in iron deficiency by race and family income level. The prevalence of iron deficiency is higher in African Americans than in white children, and it is highest in Mexican American children (17% of children aged 1 to 2 years).[3] Children of families with incomes less than 130% of the poverty threshold have a higher incidence of iron deficiency than those with a higher income (12% versus 7%).

PREVENTING IRON DEFICIENCY The Centers for Disease Control has published recommendations for preventing iron deficiency in the United States.[33] It is recommended that children 1 to 5 years of age drink no more than 24 ounces of cow's milk, goat's milk, or soymilk each day because of the low iron content of these milks. Larger intakes may displace high-iron foods. For detecting iron deficiency, it is recommended that children at high risk for iron deficiency, such as low-income children and migrant and recently arrived refugee children, be tested for iron deficiency between the ages of 9 and 12 months, 6 months later, and then annually from ages 2 to 5 years. For children who are not at high risk for iron deficiency, selective

Table 10.5 Dietary Reference Intakes for protein[28]

Age	RDA g/kg/d
1–3-year-olds	1.1 g/kg/d or 13 g/day*
4–8-year-olds	.95 g/kg/d or 19 g/day*

SOURCE: From the Institute of Medicine, Food & Nutrition Board.

Table 10.6 Dietary Reference Intakes for key nutrients for toddlers and preschoolers[31,32]

AGE	RECOMMENDED DIETARY ALLOWANCES		ADEQUATE INTAKE
	Iron (mg/d)	Zinc (mg/d)	Calcium (mg/d)
1–3 years	7	3	500
4–8 years	10	5	800

Table 10.7 Progression of iron deficiency

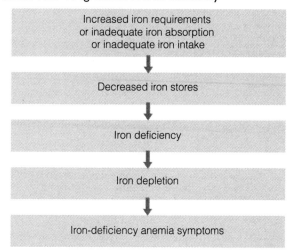

Increased iron requirements
or inadequate iron absorption
or inadequate iron intake

↓

Decreased iron stores

↓

Iron deficiency

↓

Iron depletion

↓

Iron-deficiency anemia symptoms

screening of children at risk only is recommended. Children at risk include those who have a low-iron diet, consume more than 24 ounces of milk per day, have a limited access to food because of poverty or neglect, and who have special health care needs, such as an inborn error of metabolism or chronic illness.

NUTRITION INTERVENTION FOR IRON-DEFICIENCY ANEMIA   Treatment of iron-deficiency anemia includes supplementation with iron drops at a dose of 3 mg/kg per day, counseling of parents or caretakers about diets that prevent iron deficiency, and repeat screening in 4 weeks. An increase of >1 g/dL in hemoglobin concentration, or >3% in hematocrit, within 4 weeks of initiation of treatment confirms the diagnosis of iron deficiency. If the anemia is responsive to treatment, dietary counseling should be reinforced and the iron treatment should be continued for 2 months. At that time, the hemoglobin and hematocrit should be rechecked, and the child should be reassessed in 6 months. If the hemoglobin and hematocrit do not increase after 4 weeks of iron treatment, further diagnostic tests are needed. Iron status will not improve with iron supplements if the cause of the anemia is not directly related to a need for iron.[33]

Dental Caries

Approximately one in five children aged 2 to 4 years has decay in the primary or permanent teeth.[3] A primary cause of dental decay is habitual use of a bottle with milk or fruit juice at bedtime or throughout the day. Prolonged exposure of the teeth to these fluids can produce **baby-bottle tooth decay.** Upper front teeth are most severely affected by decay, which is where fluids pool when toddlers fall asleep while drinking from a bottle. Toddlers with baby-bottle tooth decay are at increased risk for caries in the permanent teeth.[35] The incidence of baby-bottle tooth decay is highest among Hispanic, American

Indian, and Alaska Native children, and among children whose parents have less than a high school education.[3]

Food sources of carbohydrates such as milk and fruit juice can have direct effects on dental caries development because *streptococcus mutans,* the main type of bacteria that cause tooth decay, uses carbohydrates for food. Bacteria present in the mouth excrete acid that causes the tooth decay.[35] Consequently, the more often and longer teeth are exposed to carbohydrates, the more the environment in the mouth is conducive to the development of tooth decay. Foods containing carbohydrates that stick to the surface of the teeth, such as sticky candy such as caramel, are strong caries promoters. Rinsing the mouth with water or brushing teeth to get rid of the carbohydrate stuck to teeth reduces caries formation. Young children allowed to graze, or indiscriminately eat or drink throughout the day, likely expose their teeth to carbohydrates longer, which encourages bacteria proliferation and tooth decay. Crunchy foods such as carrot sticks and apple slices, when age appropriate, are good choices for snacks because they are less likely to promote tooth decay than sticky candies are.

FLUORIDE   Children need a source of fluoride in the diet. If the water supply is not adequately fluoridated, then a fluoride supplement is recommended. The American Dental Association, the American Academy of Pediatrics, and the American Academy of Pediatric Dentistry have devised a fluoride supplementation schedule, which is based on the child's age and the fluoride content of the local water supply.[36] Children aged 6 months to 3 years need 0.25 mg of fluoride per day if their local water supply has <0.3 ppm of fluoride. Children 3 to 6 years of age need 0.5 mg fluoride per day if their water supply has <0.3 ppm, but only 0.25 mg fluoride per day if the local water has 0.3 to 0.6 ppm of fluoride.[36] Excessive fluoride supplementation, consumption of toothpaste with fluoride, and natural water supplies high in fluoride can cause *fluorosis.*

Although otherwise harmless, fluorosis produces permanent staining of the tooth enamel, particularly for permanent teeth.[35] Due to the risk of fluorosis, fluoride supplements are available only by prescription. Few foods contain much fluoride, but fluoridated water used in beverages and food preparation does provide fluoride.

> **Baby-Bottle Tooth Decay**   Dental caries in young children caused by being put to bed with a bottle or allowed to suck from a bottle for extended periods of time. Also called *baby- or nursing-bottle dental caries.*
>
> **Fluorosis**   Permanent white or brownish staining of the enamel of teeth caused by excessive ingestion of fluoride before teeth have erupted.[35]

Constipation

Constipation, or hard and dry stools associated with painful bowel movements, is a common problem of young children. Sometimes "stool holding" develops when the child does not completely empty the rectum, which can lead to chronic overdistension so that eventually the child is retaining a large fecal mass.[37] Then having a bowel movement can become painful to the child, which leads to more stool holding, and a vicious cycle ensues. A health care provider should manage the treatment of stool holding.

Diets providing adequate total fiber for age (see fiber recommendations on page 260) guard against constipation. Some of the best food sources of fiber for toddlers and preschoolers are whole grain breads and cereals, legumes, and fruits and vegetables appropriate for age. Too much fiber should be avoided, however. Young children easily develop diarrhea from high amounts. High-fiber foods may displace other energy-dense foods and may decrease the bioavailability of some minerals, such as iron and calcium.

Lead Poisoning

Approximately 2.2% of children 1 to 5 years of age have high blood lead levels, or those exceeding 10 mcg/dL.[38] According to the latest surveillance data, the number of children with high blood lead levels continues to decline throughout the United States.[3,38] Young children are particularly at risk for developing high levels of lead because they are growing so rapidly, and in exploring their environment, they enjoy putting things into their mouths. Depending on the children's surroundings, some of these objects may be high in lead. There are racial and ethnic disparities in children with high lead levels, with higher rates found in African American and Mexican American children than in white children living in comparable housing.[39]

High blood lead levels affect the functioning of many tissues in the body, including the brain, blood, and kidneys. Low-level exposure to lead is associated with decreases in IQ and behavioral problems, and elevated blood lead levels may decrease growth in young children.[40] Damage caused by lead exposure may begin during pregnancy as lead is transported across the placenta to the fetus.

Children living in housing built before 1950 are at increased risk of high lead levels because lead-based paint may have been used on these houses. Lead-based paint chips taste sweet, tempting children to consume them. As the age of housing decreases nationally, so does the incidence of high lead levels in children.[3] Lead can enter the food supply through lead-soldered water pipes, contaminated water supplies, and from certain canned goods from other countries that contain lead-soldered seals. Nonfood items containing lead include dirt, lead weights, and other objects. Lead is also found in ceramic glazes, pewter, and in some folk remedies. One study of a small group of toddlers from 18 to 36 months of age, who lived in an urban area with potentially high lead levels, found a correlation between the amount of lead wiped from the children's hands and the level of lead in their food.[41]

Food Security Access at all times to a sufficient supply of safe, nutritious foods.

These results emphasize the importance of hand washing when preparing and consuming food.

Iron deficiency in preschool-age children increases the risk of elevated lead levels.[42] With iron deficiency and a reduction in iron availability, binding receptors for iron are made available for lead transport. Also, with iron deficiency, iron uptake receptors in the small intestine are available for lead absorption. Adequate dietary calcium intake appears to protect against high blood lead levels by decreasing absorption of lead.

The CDC published guidelines for screening children for lead poisoning in 1997.[43] The American Academy of Pediatrics endorsed these guidelines in a policy statement published in 1998.[44] Lead screening should be considered at 9 to 12 months of age and again at around 24 months of age. Local and state health departments have established recommendations for universal or targeted screening. Universal screening is recommended for communities with more than 27% of the housing built before 1950 or where data is insufficient to know the prevalence of elevated blood lead levels. Targeted screening is recommended in communities where less than 27% of the housing was built before 1950 or where less than 12% of children have blood lead levels greater than 10 mcg/dl.[9,38,43,44] Besides the age of the housing, other risk factors for high blood lead levels include living in poverty and having a sibling or playmate who has had high blood lead levels. Most states require lead screeening of children who receive public assistance services such as Medicaid and Supplemental Nutrition Program for Women, Infants, and Children (WIC).[9,38,43,44]

To summarize, eliminating sources of lead in the child's environment is the most important step toward eliminating elevated blood lead levels in children. In addition, preventing iron deficiency and promoting a well-balanced diet, which includes good sources of calcium, help to prevent this problem in young children.

Food Security

One of the Healthy People 2010 objectives is to increase *food security* among U.S. households to 94% from a baseline of 88%.[3] In 1998, nearly 10 million people, more than one-third of them children, lived in households in which at least some members experienced hunger due to the lack of food.[45] According to the Household Food Security Report, the level of households having uncertain food availability has remained relatively constant since 1995. Food insecurity is more likely to exist among American Indian or Alaska Native, African American, and Hispanic or Latino people as compared to whites; among households with children, particularly those headed by single women; and in lower-income-level households (<130% of poverty threshold).

Food security is particularly important for young children because of their high nutrient needs for growth and development. Young children are a vulnerable group because they must depend on their parents and caretakers to supply them with adequate access to food. It appears that children who are hungry and have multiple experiences with food insufficiency are more likely to exhibit behavioral, emotional, and academic problems as compared to other children who do not experience hunger repeatedly.[46]

Food Safety

Young children are especially vulnerable to food poisoning because they can become ill from smaller doses of organisms. Key foodborne pathogens include *Campylobacter* species and *Salmonella* species, which are the most frequently reported foodborne illnesses in the United States, and the emerging pathogens, *E. coli* O157:H7 and *Listeria monocytogenes*.[3] The highest rate of *Campylobacter* infections is seen in children under age 1.[3] *Campylobacter* is transmitted by handling raw poultry, eating undercooked poultry, drinking raw milk or nonchlorinated water, or handling infected animal or human feces.[47] The most common cause of salmonella poisoning is consumption of foods containing undercooked or raw eggs, such as raw cookie dough containing eggs. Children less than 10 years of age account for a disproportionate percent of cases of *E. coli*. It is a serious disease that can cause bloody diarrhea and *hemolytic uremic syndrome (HUS)*. Outbreaks of *E. coli* have been associated with ingestion of contaminated, undercooked hamburger meat, unpasteurized apple cider and juice, and unpasteurized milk. Employing proper food storage and preparation techniques at home, in child care centers, and in retail food establishments is essential for decreasing the incidence of foodborne illnesses in young children. Contamination of food products can occur at any point along the way from production to consumption. Therefore, risk reduction should be targeted at various steps in food processing. One major food safety education program, called FightBAC, has issued four food-safety practice messages:[3,48]

- Clean: Wash hands and surfaces often.
- Separate: Don't cross-contaminate.
- Cook: Cook to proper temperatures.
- Chill: Refrigerate promptly.

One of the U.S. Dietary Guidelines is "Keep food safe to eat," which helps to emphasize the importance of food safety in American life today.[49] The Dietary Guidelines offer additional strategies for keeping food safe. Child care workers as well as family members and other caretakers of children need to be well educated in food safety issues.

The U.S. Environmental Protection Agency is in the process of evaluating all existing standards for pesticides by 2006.[3] A major objective of this evaluation is to ensure that the current levels of pesticides in the food supply and drinking water are safe for young children.

Prevention of Nutrition-Related Disorders

The prevalence of *overweight* and *obesity* among children, adolescents, and adults in the United States has increased and represents a major public health problem. High-energy, high-fat diets coupled with sedentary lifestyles are thought to be major contributors to the increase in weight. Cardiovascular disease, a major cause of death and morbidity in the United States today, is also thought to be influenced by diets and sedentary lifestyles. Food habits, preferences, and behaviors established during the toddler and preschool ages logically influence not only dietary habits later in life but also subsequent health status. However, this logical deduction has not yet been substantiated by scientific research in this age group, which is sorely needed.[52] In the meantime, families are encouraged to adopt dietary and exercise patterns that promote a healthy lifestyle.

Overweight and Obesity in Toddlers and Preschoolers

According to the NHANES 1999–2000 data, 10.4% of children ages 2 to 5 years are overweight, defined as having a BMI greater than or equal to the 95th percentile. This is up from 7.2% in the NHANES III survey (1988–1994). Another 20.6% of children ages 2 to 5 years are at risk for overweight with BMIs greater than or equal to the 85th percentile but less than the 95th percentile. Data collected on infants from birth through 23 months indicated that 11.4% had weight-for-length percentiles greater than or equal to the 95th percentile.[53] No significant differences among ethnic groups were seen at these young ages. Increases in the prevalence of overweight are seen in children of all ages and racial and ethnic groups, but changes are greatest for older preschool-age children.[54]

During the preschool years, a decrease in body mass index (BMI), or weight-for-height squared (wt[kg]/ht[m]^2), is a normal part of growth and development. BMI usually reaches its lowest point at approximately 4 to 6 years of age and then increases gradually in the period called *adiposity rebound* or *BMI rebound.*

Hemolytic Uremic Syndrome (HUS) A serious, sometimes fatal complication associated with illness caused by *E. coli* O157:H7, which occurs primarily in children under the age of 10 years. HUS is characterized by renal failure, **hemolytic anemia,** and a severe decrease in **platelet** count.[3]

Hemolytic Anemia Anemia caused by shortened survival of mature red blood cells and inability of the bone marrow to compensate for the decreased life span.

Platelets A component of the blood that plays an important role in blood coagulation.

Overweight Body mass index at or above the 95th percentile.

Obesity BMI-for-age greater than the 95th percentile with excess fat stores as evidenced by increased triceps skinfold measurements above the 85th percentile.[50,51]

Adiposity or BMI Rebound A normal increase in body mass index that occurs after BMI declines and reaches its lowest point at 4 to 6 years of age.[56]

Early adiposity rebound in children increases the risk of adult obesity.[55] An annual increase in BMI of three to four units may reflect a rapid increase in body fat for most children.[27]

Obesity is a multifaceted problem that is difficult to treat, making prevention the preferred approach. A committee of obesity experts has recommended that children with a BMI greater than or equal to the 85th percentile on the CDC growth charts with complications, such as hypertension or gallbladder disease, or with a BMI greater than or equal to the 95th percentile, be evaluated and possibly treated for obesity.[27] Children less than 2 years of age who fall into this category should be evaluated by a pediatric obesity specialist. Obtaining a *triceps skinfold* measurement and comparing it to standards will further validate that the child has excess fat stores rather than increased lean muscle mass. Standards for triceps skinfold measurements include those derived from data of the National Health and Nutrition Examination Survey (NHANES) I, which are age and gender specific.[57]

Maintaining weight while gaining height can be the best treatment for obese children between the ages of 2 and 7. This approach allows the obese child to "grow into his or her height" and to lower BMI. However, if the child is already exhibiting secondary complications of obesity, such as mild hypertension or high cholesterol or triglyceride levels, gradual weight loss may be indicated.[27] Reducing weight at this young age is tricky because sufficient nutrients must be provided for children to reach their full height potential and to remain healthy.

Triceps Skinfold   A measurement of a double layer of skin and fat tissue on the back of the upper arm. It is an index of body fatness and measured by skinfold calipers. The measurement is taken on the back of the arm midway between the shoulder and the elbow.

Heart Disease   The leading cause of death and a common cause of illness and disability in the United States. Coronary heart disease, the principal form of heart disease, is caused by buildup of cholesterol deposits in the coronary arteries that feed the heart.

LDL Cholesterol   Low-density lipoprotein cholesterol, the lipid most associated with atherosclerotic disease. Diets high in saturated fat, trans fatty acids, and dietary cholesterol have been shown to increase LDL-cholesterol levels.

Familial Hyperlipidemia   A condition that runs in families and results in high levels of serum cholesterol and other lipids.

Trans Fatty Acids   Fatty acids that have unusual shapes resulting from the hydrogenation of polyunsaturated fatty acids. Trans fatty acids also occur naturally in small amounts in foods such as dairy products and beef.

Atherosclerosis   A type of hardening of the arteries in which cholesterol is deposited in the arteries. These deposits narrow the coronary arteries and may reduce blood flow to the heart.

Prevention and Treatment of Overweight and Obesity

Prevention is the best approach for overweight and obesity. Parenting techniques, such as finding reasons to praise the child's behavior, but never using food as a reward, foster the development of healthy eating behaviors in children and help children to self-regulate food intake. (See "Food-Preference Development, Appetite, and Satiety," earlier in this chapter.) When weight control for a young child is warranted, the expert committee on obesity evaluation and treatment recommends a general approach, including family education and involvement.[27] Examples of behavior changes, which a family can incorporate into its lifestyle, include increasing physical activity, offering nutrient-dense and not calorie-dense snacks, and focusing on behavior changes rather than weight changes.

Nutrition and Prevention of Cardiovascular Disease in Toddlers and Preschoolers

Heart disease is the number one cause of death in the United States today.[3] A leading risk factor for cardiovascular disease, which includes diseases of the heart and blood vessels, is elevated levels of *LDL cholesterol.* Children with *familial hyperlipidemias* and obese children can have high levels of LDL cholesterol. High intakes of saturated fat, *trans fatty acids,* and, to a lesser extent, dietary cholesterol elevate LDL cholesterol levels in children and adults alike. In relation to health promotion and disease prevention, the level and type of fat in children's diets will affect their risks of developing cardiovascular disease as adults.[52] Fatty streaks, which can be precursors to the buildup of fat deposits in blood vessels, have been found in the arteries of young children. Some experts believe that these streaks can represent the beginning of *atherosclerosis* and cardiovascular disease, but the evidence is not conclusive.[58]

According to the new DRIs,[28] the Acceptable Macronutrient Distribution Ranges (AMDRs) for fat are 30 to 40 percent for children 1 to 3 years and 25 to 35 percent for children 4 to 18 years. No specific recommendations for total fat per day in the diets of young children have been made. Adequate intake levels for the essential fatty acids, linoleic acid and alpha-linolenic acid, have been determined.[28]

Dietary recommendations are different for children who are at increased risk of developing premature cardiovascular disease because a parent has heart disease or because of familial hyperlipidemias. These children need periodic screening for blood cholesterol levels and close follow-up. If LDL-cholesterol levels are high, restriction of total calories from saturated fat to less than 7% and of dietary cholesterol to no more than 200 mg per day is recommended. These children need to be closely monitored by a physician and registered dietitian.[58]

Vitamin and Mineral Supplements

Children who consume a variety of basic foods can meet all of their nutrient needs without vitamin or mineral supplements. Eating a diet of a variety of foods is the preferred way to get needed nutrients because foods contain many other substances, such as phytochemicals, in addition to nutrients that benefit health.

The American Academy of Pediatrics recommends vitamin and mineral supplementation for children who are at high risk of developing or have one or more nutrient deficiencies.[36] Children at risk of nutrient deficiency include:

- Children from deprived families or who suffer from abuse or neglect
- Children with anorexia or poor appetite and poor eating habits
- Children who consume a "fad diet" or only a few types of food
- Children who consume a vegetarian diet without dairy products

Despite these recommendations, data from the NHANES III indicate that children 1 to 5 years of age are major users of supplements.[59] Approximately one in two 3-year-olds in the United States are given a vitamin and mineral supplement by their parents.[60] Characteristics of mothers who give their children supplements include non-Hispanic white, older, more years of education, married, have health insurance, receive care from a private health care provider, have greater household income, and took supplements themselves during pregnancy. Children most likely to receive a supplement are those at low risk of developing nutrient deficiencies; children who would most likely benefit from supplements are less likely to receive them.

If given to children, vitamin and mineral supplement doses should not exceed the DRI for age. Parents and caretakers should be warned against giving high amounts of vitamins and minerals to children, particularly vitamins A (retinol) and D. The *Tolerable Upper Intake Levels* shown in the DRI tables should serve as a guide to excessive levels of nutrient intake from fortified foods and supplements.

Herbal Supplements

The use of herbal remedies for various disorders is increasing in the United States today, as is the use of complementary and alternative medicine practices in general. Parents and caretakers who take herbs are likely to give these products to their children. Few definitive studies exist on the effectiveness of these substances in preventing disease and promoting health in adults, much less in children. Despite the lack of scientific evidence, anecdotal reports of benefits abound. However, some reports have linked herbal preparations to adverse effects.[61] Information on herb use should be obtained during the nutrition assessment of a child to rule out herbs as a source of health problems. At the current time, there is no regulation of or consistency in the composition of the products; this can lead to uncertain results. Children given various herbs are the "test subjects" in these uncontrolled studies. Parents should be advised of the potential risks of herbal therapies and the need for closely monitoring the child if they choose

to give the child herbs. The National Institutes of Health's (NIH) National Center for Complementary and Alternative Medicine (NCCAM) Web site, which is listed in the "Resources" section at the end of this chapter, provides reports on the known safety and effectiveness of various herbal remedies and alternative medical practices.

Dietary and Physical Activity Recommendations

"Children ages 2 to 11 years should achieve optimal physical and cognitive development, attain a healthy weight, enjoy food, and reduce the risk of chronic disease through appropriate eating habits and participation in regular physical activity."[62]

The American Dietetic Association

Taking into consideration the energy and nutrient needs of young children and the common nutritional problems and concerns of this age group, it is easy to understand the importance that underlies dietary recommendations for toddlers and preschoolers. A primary recommendation is that young children eat a variety of foods. This recommendation is more easily achieved if healthful food preferences and eating habits are acquired during the early years. Food preferences in conjunction with food availability form the foundation of the child's diet. Limited food selection, therefore, will influence the adequacy of the child's diet by decreasing variety. Parents and caretakers cannot expect a child to "do as I say, but not as I do." Nutrition education aimed at the adults in the child's life becomes as important, if not more so, than nutrition education directed to the child.

> **Tolerable Upper Intake Levels** Highest level of daily nutrient intake that is likely to pose no risk of adverse health effects to almost all individuals in the general population; gives levels of intake that may result in adverse effects if exceeded on a regular basis.

Dietary recommendations have been developed and disseminated by the federal government and professional organizations. Two sets of guidelines for young children's diets are available: the Dietary Guidelines for Americans and the Food Guide Pyramid.[10,49] Recommendations for caloric and nutrient intake are represented in the DRIs.

Dietary Guidelines

Dietary Guidelines for Americans, 2000 (discussed in Chapter 1), emphasize that children be offered a variety of foods including grain products, vegetables and fruits, and low-fat dairy products.[49] The guidelines also recommend that beans, lean meats, and poultry be added as appropriate for the child. Foods high in fat and sugar, such as candy, cookies, and cakes, should be limited in children's diets. The Dietary Guidelines also emphasize the importance of parents modeling this type of diet for their children or "to do as I say and as I do." The guidelines also emphasize the importance of physical activity. Parents are advised to encourage their children to engage in at least 60 minutes per day of vigorous physical activity and to limit the time spent in sedentary activities, such as TV watching and computer game playing, that replace physical activity.[49] It is important that parents model for their children a lifestyle that includes a varied diet and regular physical activity.

Food Guide Pyramid

The USDA has developed a Food Guide Pyramid for young children (Illustration 10.7) targeted at children aged 2 to 6 years.[10] The guide encourages children to consume a variety of foods, with grains, fruits, and vegetables forming the base of the diet and with foods high in fat and sweets being consumed in limited amounts.

Foods shown in the guide represent single-serving portions of foods commonly eaten by young children. To make the guide "child friendly," the food groups have shorter names, and the number of servings for each group is a single number rather than a range. To emphasize the importance of physical activity, illustrations of children being active are depicted around the pyramid. In the booklet, "Tips for Using the Food Guide Pyramid for Young Children 2 to 6 Years Old," parents are advised to

give 2- to 3-year-old children the same number of servings as recommended for 4- to 6-year-old children, but to give about two-thirds of a serving for the younger children. Children 2 to 6 years of age need the same number of servings from the milk group each day—2 servings. The booklet also includes age-appropriate meal-preparation activities for young children and some simple recipes children can help prepare. A sample meal and snack plan based on the "Food Guide Pyramid for Young Children" can be found in Table 10.8. Illustration 10.8 on page 260 depicts a chart that parents can utilize in evaluating their children's intake as compared to the Food Guide Pyramid. The source for this booklet for parents is given in the "Resources" section at the end of this chapter.

Recommendations for Intake of Iron, Fiber, Fat, and Calcium

Adequate iron intake is necessary to prevent iron deficiency and iron-deficiency anemia in toddlers and preschoolers. Appropriate fiber intake is needed to prevent constipation and may provide long-term disease prevention. Fat is an important source of calories, essential fatty acids, and fat-soluble vitamins in young children's diets. Adequate calcium intake is important for children to achieve peak bone mass, and yet about 20% of children do not meet the DRIs for calcium.[3]

IRON Adequate iron intake is important in this age group to prevent iron deficiency. A list of good sources of dietary iron can be found in Chapter 1, page 25, Table 1.12. Meats, which are good sources of iron, can be ground or chopped to make them easier for toddlers to chew. Fortified breakfast cereals and dried beans and peas are also good sources of iron.

"Toddler" milks, or iron-fortified commercial formulas for toddlers, are available. Healthy children who consume a variety of foods, and whose milk intake is less than 24 ounces daily, obtain adequate iron without these special products. Other commercial beverages being marketed to parents include formulas that were originally designed for children with illnesses or who had to be fed complete nutrition through a feeding tube. Such special products are expensive, and they are unnecessary for healthy children. It would be better for parents of healthy children to spend their food dollars on a variety of healthy foods rather than on these special products.

FIBER Ample fiber intake has been associated with the prevention of heart disease, certain cancers, diabetes, and hypertension in adults. Whether fiber helps prevent these problems as young children become adults is not known, but it is clear that fiber in a child's diet helps prevent constipation and is part of a healthy diet. Too much fiber in a child's diet can be detrimental because high-fiber diets have the potential of reducing the energy density of the

Illustration 10.7 Food Guide Pyramid for Young Children.[10]

Table 10.8 One day's sample meals and snacks for 4- to 6-year-old children.[10] (Offer 2- to 3-year-olds the same variety but smaller portions.)

	GRAIN	VEGGIE	FRUIT	MILK	MEAT
Breakfast					
100% fruit juice, ¾ c			1		
Toast, 1 slice	1				
Fortified cereal, 1 oz	1				
Milk, ½ c				½	
Mid-Morning Snack					
Graham crackers, 2 squares	1				
Milk, ½ c				½	
Lunch					
Meat, poultry, or fish, 2 oz					2 oz
Macaroni, ½ c	1				
Vegetable, ½ c		1			
Fruit, ½ c			1		
Milk, ½ c				½	
Mid-Afternoon Snack					
Whole grain crackers, 5	1				
Peanut butter, 1 tb					½ oz
Cold water, ½ c					
Dinner					
Meat, poultry, or fish, 2½ oz					2½ oz
Potato, 1 medium		1			
Broccoli, ½ c		1			
Cornbread, 1 small piece	1				
Milk, ½ c				½	
Total Food Group Servings	6	3	2	2	5 oz

diet, which could affect growth.[63] High-fiber diets could also affect the bioavailability of some minerals, such as iron and calcium.

The new recommendations for total fiber intake based on the DRIs can be found in Table 10.9 on the next page.[28] Total fiber is the sum of dietary fiber and functional fiber. Earlier recommendations were based on dietary fiber alone. Including fruits, vegetables, and whole grain breads and cereal products in the diet can increase the dietary fiber intake of children. Those children who meet the fiber intake recommendation consume more high- and low-fiber breads and cereals, fruits, vegetables, legumes, nuts, and seeds than those who do not. Children with adequate fiber intake tend to have lower intakes of fat and cholesterol, and higher intakes of fiber, vitamins A and E, folate, magnesium, and iron than do those children who have low dietary fiber intakes.[64]

FAT An appropriate amount of fat in a young child's diet can be achieved by employing the principles of the Dietary Guidelines and the Food Guide Pyramid that promote a diet of whole grain breads and cereals, beans and peas, fruits and vegetables, low-fat dairy products after 2 years of age, and lean meats.[10,49] Foods high in fat are used sparingly, especially foods high in saturated fat and trans fatty acids. However, an appropriate amount of dietary fat is necessary to meet children's needs for calories, essential fatty acids, and fat-soluble vitamins. As discussed in Chapter 1, good sources of linoleic acid, an essential fatty acid, are peanut, canola, corn, safflower, and other vegetable oils. Flaxseed, soy, and canola oils, as well as fish, are good sources of alpha-linolenic acid, another essential fatty acid.

It is important to include sources of fat-soluble vitamins in the diets of young children. Good sources of vitamin A include whole eggs and dairy products. Sources of vitamin D include exposure to sunlight and vitamin D–fortified milk. Corn, soybean, and safflower oils are excellent sources of vitamin E. Vitamin K is widely distributed in both animal and plant foods.

CALCIUM Adequate calcium intake in childhood affects peak bone mass. A high peak bone mass is thought to be protective against osteoporosis and fractures later in

PLAN FOR YOUR YOUNG CHILD... The Pyramid Way

Use this chart to get an idea of the foods your child eats over a week. Pencil in the foods eaten each day and pencil in the corresponding triangular shape. (For example, if a slice of toast is eaten at breakfast, write in "toast" and fill in one Grain group pyramid.) The number of pyramids shown for each food group is the number of servings to be eaten each day. At the end of the week, if you see only a few blank pyramids...keep up the good work. If you notice several blank pyramids, offer foods from the missing food groups in the days to come.

	SUNDAY	MONDAY	TUESDAY	WEDNESDAY	THURSDAY	FRIDAY	SATURDAY
Milk	△△	△△	△△	△△	△△	△△	△△
Meat	△△	△△	△△	△△	△△	△△	△△
Vegetable	△△△	△△△	△△△	△△△	△△△	△△△	△△△
Fruit	△△	△△	△△	△△	△△	△△	△△
Grain	△△△	△△△	△△△	△△△	△△△	△△△	△△△
	△△△	△△△	△△△	△△△	△△△	△△△	△△△
Breakfast							
Snack							
Lunch							
Snack							
Dinner							

Illustration 10.8 Plan for Your Young Child ... The Pyramid Way.[10]

life.[65] Yet many children do not consume adequate calcium. An estimated 21% of children 2 to 8 years of age consume less than their DRI for calcium.[3] The recommendations for daily calcium intake in the DRI table are 500 mg/day for children aged 1 to 3 years and 800 mg/day for children aged 4 to 8 years. An important aspect of adequate calcium intake in toddlers and preschoolers is the development of eating patterns that will lead to adequate calcium intake later in childhood.[65]

Dietary sources of calcium are listed in Chapter 1, on page 24, Table 1.12. Dairy products are good sources of calcium, as are canned fish with soft bones such as sardines, dark-green leafy vegetables such as kale and bok choy, tofu made with calcium, and calcium-fortified foods and beverages such as calcium-fortified orange juice. Nonfat and low-fat dairy products are low in saturated fat while still providing a good source of calcium.

Fluids

Healthy toddlers and preschoolers will consume enough fluid through beverages, foods, and sips and glasses of water to meet their needs. Fluid requirements increase with fever, vomiting, diarrhea, and when children are in hot, dry, or humid environments.

Consumption of milk has decreased among young children since the late 1970s, but consumption of carbonated soft drinks has increased by about the same amount. Since that time, consumption of noncitrus juices has increased almost threefold.[66] According to food consumption surveys, young children consume large amounts of sweetened beverages—including fruit juice, soft drinks, and sweetened iced tea—to the detriment of the overall nutritional balance of the diet and oral health. Approximately 50% of children aged 2 to 5 years

Table 10.9 Adequate Intake of total fiber for children[28]	
AGE	**TOTAL DAILY FIBER**
1–3 years old	19 grams/day
4–8 years old	25 grams/day

Table 10.10 Mean daily intakes of the Food Guide Pyramid for children in the 1989–1991 CSFII[68]

Gender/Age	Number in Study	Grain	Vegetable	Fruit	Dairy	Meat
Recommended servings		6–11	3–5	2–4	2–3	5–7 oz
Males 2–5 years	429	5.2	1.9	1.5	2.0	2.8
Females 2–5 years	416	4.9	2.0	1.5	1.9	2.7

consume soft drinks.[67] Children with high consumption of regular soft drinks (more than 9 ounces per day) consume more calories and less milk and fruit juice than those with lower consumptions of regular soft drinks. Water is a good but underused thirst quencher for toddlers and preschoolers, as long as milk (2 cups) and fruit juice consumption (1 cup) is part of the regular diet. Parents and caretakers can offer children water to drink between meals and snacks.

Recommended versus Actual Food Intake

Several national surveys have examined food and nutrient intakes of young children. Comparing the Household Food Consumption Survey to the Food Guide Pyramid reveals that the mean food group intakes for children aged 2 to 5 years were below minimum recommendations for all food groups except the dairy group, for which intakes were at or near recommendations (see Table 10.10).[68] In this study cohort, about 70% of the children did not meet the recommended servings for fruits, grains, meats, and dairy; and about 64% did not meet the recommendations for vegetables. Only about 1% of children met the recommendations for food intake as outlined by the Food Guide Pyramid. Children who met none of the recommendations had nutrient intakes well below the 1989 ***Recommended Dietary Allowances (RDAs)*** for vitamin B$_6$, calcium, iron, and zinc and had a low fiber intake.[69]

According to the Food and Nutrient Intakes by Children report,[66] mean energy intakes of young children met the RDAs for energy. Mean percentages of total energy from carbohydrate, protein, total fat, saturated fat, and cholesterol intake in the diets of toddlers and preschoolers are shown in Table 10.11.

The total fat intake of about 32% of total calories is within the target range for this age child. Three- to 5-year-olds have average sodium intakes of 2600 mg per day, which slightly exceeds the sodium recommendation of 2400 mg per day.

In general, young children consume more than enough protein and fat. Mean vitamin and mineral intakes of young children exceeded the RDAs except for vitamin E and zinc.[66] In a longitudinal study of the nutrient and food intakes of preschool children aged 24 to 60 months, mean intakes of zinc, folic acid, and vitamins D and E were consistently below the recommended levels.[70] Low intakes of zinc, vitamin E, and iron were found in toddlers aged 12 to 18 months, a time of dietary transition.[71] The means for nutrient intakes often hide problems at the extremes, however. They fail to indicate the percentage of children with low nutrient intakes less than 66% of the recommended levels, and children with high nutrient intakes that exceed the Tolerable Upper Intake Levels.

According to the USDA report on Food and Nutrient Intakes by Children,[66] 89% of 1- to 2-year-old children ate snacks, and for these children snacks contributed 24% of calories and 23% of total fat intake for the day. Similarly, 88% of 3- to 5-year-old children consumed snacks, which contributed 22% of total energy and 20% of total fat. These data indicate the important contribution of snacks to children's overall nutrient intakes.

Children's portion sizes have remained constant over recent years except for meat portions, which have decreased. This stability in portion sizes of young children over time reinforces the hypothesis that young children are capable of self-regulating energy intake. Portion sizes are positively related to both body weight percentiles and energy intake. It seems that young children self-regulate energy intake by adjusting portion size.[72]

Recommended Dietary Allowances (RDAs) The average daily dietary intake levels sufficient to meet the nutrient requirements of nearly all (97% to 98%) healthy individuals in a population group. RDAs serve as goals for individuals.

Cross-Cultural Considerations

When working with families from various cultures, it is important to learn as much as possible about the culture's

Table 10.11 Mean percentages of total calories from carbohydrate, protein, total fat, and saturated fatty acids, and cholesterol intake[66]

Age	Carbohydrate (%)	Protein (%)	Total Fat (%)	Saturated Fatty Acids (%)	Cholesterol (mg/d)
1–2 years	54	15	32	13	189
3–5 years	55	14	32	12	197

food-related beliefs and practices. Ask the parents and caretakers about their experiences with food, including foods used for special occasions. It is also helpful to know whether foods are used for home remedies or to promote certain aspects of health. Cultural beliefs influence many child-feeding practices, such as what foods are best for young children, which cause digestive upsets, or which help relieve illnesses. Perhaps one of the best-known examples is the use of chicken soup to cure what ails you! It is important for the health care provider to build on cultural practices and to reinforce those positive practices while attempting to effect change in those practices that could be harmful to the young child.

A series of booklets entitled *Ethnic and Regional Food Practices* is available from the American Dietetic Association. This series addresses food practices, customs, and holiday foods of various ethnic groups. The booklets provide examples for incorporating traditional foods of various groups of people into dietary recommendations. For example, peanut or polyunsaturated oils are recommended to Chinese Americans for stir-frying instead of the more traditional use of lard or chicken fat. Ordering information for this series can be found in the "Resources" section at the end of this chapter.

Vegetarian Diets

Young children can grow and develop normally on vegetarian or vegan diets, as long as their dietary patterns are intelligently planned. Vegetarian diets are rich in fruits, vegetables, and whole grains, the consumption of which is encouraged for the general population. However, young children in particular need some energy-dense foods to reduce the total amount of food required. The amount of vegetarian foods needed to meet nutrient needs may be more food than young children can eat. Young children need to eat several times a day to meet their energy needs because their stomachs cannot hold a lot of food at one time.

Children who are fed *vegan* and *macrobiotic diets* tend to have lower rates of growth, although still within the normal ranges, during the first 5 years of life compared to chil-

Vegan Diet The most restrictive of vegetarian diets, allowing only plant foods.

Macrobiotic Diet This diet falls between semivegetarian and vegan diets and includes foods such as brown rice and other grains, vegetables, fish, dried beans, spices, and fruits.

dren given a mixed diet.[73] Strict vegan diets, which exclude all foods of animal origin, may be deficient in vitamins B_{12} and D, zinc, and omega-3 fatty acids, and they may also be low in calcium, unless fortified foods are consumed. Protein needs are usually met if the diet is adequate in energy and a variety of foods are included.[74] Children on vegan diets should receive vitamin B_{12} supplements or consume fortified breakfast cereals, textured soy protein, or soymilk fortified with vitamin B_{12}. The vitamin B_{12} status of children following vegetarian and vegan diets should be monitored on a regular basis because vitamin B_{12} deficiency may cause vitamin B_{12} deficiency anemia. Iron-

deficiency anemia is an infrequent problem among children consuming a vegetarian diet.

Vitamin D adequacy can be met by diet or by sun exposure. Good sources of vitamin D for children include fortified soymilk, fortified breakfast cereals, and fortified margarines. Zinc is found in foods of animal origin. Plant sources of zinc include legumes, nuts, and whole grains. Vegetable products are also lacking in omega-3 fatty acids. Therefore, including a source of these fatty acids, such as canola or soybean oil, is advisable.[73]

Foods containing phytates, such as unrefined cereals, may interfere with calcium absorption. Thus, if the child's diet contains a lot of unrefined cereals, higher calcium intakes may be needed.[73] Good sources of calcium for children on strict vegetarian diets include fortified soymilk, calcium-fortified orange juice, tofu processed with calcium, and certain vegetables, such as broccoli and kale.[9] Supplements may be necessary for some children with inadequate intakes that are not remedied by dietary means.

Guidelines recommended for vegetarian diets for young children have been developed and are given here:[9]

- Allow the child to eat several times a day (i.e., three meals and two to three snacks).
- Avoid serving the child bran and an excessive amount of bulky foods, such as bran muffins and raw fruits and vegetables.
- Include in the diet some sources of energy-dense foods such as cheese and avocado.
- Include enough fat (at least 30% of total calories) and a source of omega-3 fatty acids, such as canola or soybean oil.
- Include sources of vitamin B_{12}, vitamin D, and calcium in the diet, or supplement if required.

Tables 10.12 and 10.13 (on the next page) provide suggested daily food guides for lacto-ovo vegetarians and vegans.[9]

Table 10.12 Suggested Daily Food Guide for lacto-ovo vegetarians at various intake levels[9]

| Food Groups | SERVINGS PER DAY, BY AGE AND DAILY CALORIC INTAKE | |
	1–2 Years (1300 kcal)	3–6 Years (1800 kcal)
Breads, grains, cereal	5 servings	5 servings
Legumes	½	1
Vegetables	2	3
Fruits	3	3
Nuts, seeds	½	½
Milk, yogurt, cheese	3	3
Eggs (limit 3/week)	½	½
Fats, oils (added)	2	4
Sugar (added teaspoons)	3	4

SOURCE: Data from Bright Future in Practice: Nutrition by Mary Story, et al., (Eds.), p. 165. © 2000 National Center for Education in Maternal and Child Health.

Table 10.13 Suggested Daily Food Guide for vegan children at various intake levels[9]

| | SERVINGS PER DAY, BY AGE AND DAILY CALORIC INTAKE | |
Food Groups	1–2 Years (1300 kcal)	3–6 Years (1800 kcal)
Breads, grains, cereal	5	6
Legumes	½	1
Vegetables, dark-green leafy	1	1
Vegetables, other	1	2
Fruits	2	4
Nuts, seeds	½	½
Milk alternatives	3	3
Fats, oils (added)	2	4
Sugar (added teaspoons)	3	4

SOURCE: Data from Bright Future in Practice: Nutrition by Mary Story, et al., (Eds.), p. 165. © 2000 National Center for Education in Maternal and Child Health.

Child Care Nutrition Standards

"All child-care programs should achieve recommended standards for meeting children's nutrition and nutrition education needs in a safe, sanitary, supportive environment that promotes healthy growth and development."[75]

The American Dietetic Association

An estimated 23 million children in the United States require child care while their parents work, making foods children eat away from home a major contribution to their overall intake. Nearly half of 3- and 4-year-olds are enrolled in nursery school, preschool, or kindergarten.[2] Young children eat away from home often: 37% of 1- to 2-year-olds, and 52% of 3- to 5-year-olds, eat away from home daily. The most common places for away-from-home meals of children 5 years of age and younger are fast-food restaurants, day care centers, and friends' houses.[66]

Nutrition standards for child care services exist, and specify minimum requirements for amounts and types of foods to include in meals and snacks, as well as food service safety procedures.[75,76] These standards also address nutrition learning experiences and education for children, staff, and parents as well as the physical and emotional environment in which meals and snacks are served. It is recommended that children in part-day programs (4 to 7 hours per day) receive food that provides at least one-third of their daily calorie and nutrient needs in at least one meal and two snacks or two meals and one snack. A child in a full-day program (8 hours or more) should receive foods that meet one-half to two-thirds of the child's daily needs based on the DRIs in at least two meals and two snacks or three snacks and one meal. Food should be offered at intervals of not less than 2 hours and not more than 3 hours and should be consistent with the Dietary Guidelines for Americans.[49]

Physical Activity Recommendations

Inactivity is thought to be a major contributor to the increasing prevalence of obesity. The Dietary Guidelines for Americans, 2000, recommend that young children engage in play activity for at least 60 minutes every day.[49] Some of the suggested activities include

- Playing tag
- Riding a tricycle or bicycle
- Walking, skipping, or running

Parents are encouraged to set a good example for their children by being physically active themselves. Parents are also encouraged to limit the time that they allow their children to watch television and play computer games.

Nutrition Intervention for Risk Reduction

Model Program

"Bright Futures is a vision, a philosophy, a set of expert guidelines, and a practical developmental approach to providing health supervision for children of all ages, from birth through adolescence."[77]

Bright Futures in Practice: Nutrition is an example of a model program for nutrition intervention for risk reduction.[9] This guide is a component of the larger project *Bright Futures Guidelines for Health Supervision of Infants, Children, and Adolescents.*[77]

The purpose of *Bright Futures* is to foster trusting relationships among the child, health professionals, the family, and the community to promote optimal health for the child.[77] *Bright Futures* guidelines are developmentally based and address the physical, mental, cognitive, and social development of infants, children, and adolescents and their families.

Many different user-friendly materials and tools are available from this program to assist in implementing the guidelines. In addition to the *Bright Futures Guidelines for Health Supervision*, implementation guides have been published not only for oral health, general nutrition, physical activity, and mental health, but also for families. *Bright Futures in Practice: Nutrition* is based on three critical principles:[9]

1. Nutrition must be integrated into the lives of infants, children, adolescents, and families.
2. Good nutrition requires balance.
3. An element of joy enhances nutrition, health, and well-being.

The program is based on the premise that optimal nutrition for children be approached from the standpoint of development of the child and put in the context of the environment in which the child lives.[9] It emphasizes the development of healthy eating and physical activity behaviors. Nutrition supervision guidelines are given for each age group; and within each broad age group, interview questions, screening and assessment, and nutrition counseling topics are provided. The program lists desired outcomes for the child and discusses the role of the family, as well as frequently asked questions. For example, by utilizing the guidelines, health care providers will be able to provide anticipatory guidance to parents of toddlers for the proper advancements of their diets based on growth and development. The implementation guide also addresses special topics related to pediatric nutrition, including oral health, vegetarian eating practices, iron-deficiency anemia, and obesity. Useful information is included in the appendix, such as nutrition questionnaires for the various age groups. *Bright Futures in Practice: Nutrition* is a valuable resource for anyone who is interested in promoting healthy eating and physical activity behaviors in children. Ordering information for *Bright Futures* materials is available at its Web site, which is listed in the "Resources" section at the end of this chapter.

Public Food and Nutrition Programs

Young children and their families can benefit from a number of federally sponsored food and nutrition programs. Four example programs are presented here.

WIC

The Special Supplemental Nutrition Program for Women, Infants, and Children,[78] previously described in Chapter 8, is administered by the Food and Nutrition Service of the U.S. Department of Agriculture (USDA). It is one of the most successful federally funded nutrition programs in the United States. Participation in WIC services improves the growth, iron status, and the quality of dietary intake of nutritionally at-risk infants and children up to age 5 years.[79] The WIC Program is cost effective in that each dollar invested in the program saves up to $3 in health care.[78]

As in infancy, to be eligible for WIC services children must live in a low-income household at 185% or less of the federal poverty level, and be at nutrition risk. *Nutrition risk* means that a child has a medical or dietary-based condition which places the child at increased risk. Such conditions include iron-deficiency anemia, underweight, overweight, a chronic illness such as cystic fibrosis, or an inadequate dietary intake.[78]

Children receive nutrition assistance, education, and follow-up services by specially trained registered dietitians and nutritionists. Vouchers for food items such as milk, juice, eggs, cheese, peanut butter, and fortified cereals are given to eligible families. These vouchers are exchanged for the food items at authorized retailers.

WIC's Farmers' Market Nutrition Program

The Farmers' Market Nutrition Program is a special seasonal program for WIC participants. This program provides vouchers for the purchase of locally grown produce at farmers' markets. The program is designed to help low-income families increase their consumption of fresh fruits and vegetables.

Head Start and Early Head Start

Administered by the U.S. Department of Health and Human Services, Head Start and Early Head Start are comprehensive child development programs, serving children from birth to 5 years of age, pregnant women, and their families. Nearly 1 million U.S. children participate in this program. The overall goal is to increase the readiness for school of children from low-income families. A range of individualized, culturally appropriate services are provided through Head Start and related agencies working in education and early childhood development; medical, dental, and mental health services; nutrition services; and parent education.[80] About three in four Head Start families have incomes less than $12,000 annually. More specific information about Early Head Start can be found in Chapter 9.

Food Stamps

The Food Stamp Program, administered by the USDA, is designed to help adults in low-income households buy food. The monetary amount of food vouchers provided to an eligible household depends on the number of people in it, and on the income of the household. Income eligibility criteria for this and a number of other federal programs can be found at the USDA's Food and Nutrition Service Web site: www. fns.usda.gov/fsp. The average monthly amount of benefits received through the Food Stamp Program in 2002 was $79.68 per person or $185.71 per household, enough to help families and individuals pay for a portion of the food they need. Each state must develop a food stamp nutrition education plan based on federal guidance.[81] Participation in the Food Stamp Program is associated with increased intakes of a number of nutrients.[79]

Resources

American Academy of Pediatrics
    The American Academy of Pediatrics Web site contains consumer information on current news topics affecting the pediatric

population. Policy statements can be found on this page, and consumer publications can be ordered through the Bookstore.
Web site: **www.aap.org**

American Dietetic Association

Besides information for members of the American Dietetic Association, this Web site contains information for consumers on various topics of interest. There is a "Daily Tip" and a feature article, plus consumer information on topics such as food safety and healthy lifestyles.

Web site: www.eatright.org

Green M, Palfrey JS (eds.)

Bright Futures: Guidelines for Health Supervision of Infants, Children, and Adolescents. 2nd ed. Arlington, VA: National Center for Education in Maternal and Child Health, 2000. This *Bright Futures* publication contains developmentally based guidelines for health supervision and addresses the physical, mental, cognitive, and social developments of infants, children, adolescents, and their families.

Bright Futures

The *Bright Futures* publications are developmentally based guidelines for health supervision. They address the physical, mental, cognitive, and social development of infants, children, and adolescents and their families. Three implementation guides have also been published to date addressing oral health, general nutrition, and physical activity.

Web site: www. brightfutures.org. *Bright Futures* materials may also be ordered from National Maternal and Child Health Clearinghouse, 2070 Chain Bridge Road, Suite 450, Vienna, VA 22182-2536, www.nmchc.org, (703) 356-1964.

Centers for Disease Control and Prevention, National Center for Health Statistics

CDC growth charts: United States, May 30, 2000. The growth charts were released in 2000. Some of the individual charts have been modified since then. But when you pull up the Web site, it reads "2000 CDC Growth Charts: United States." The CDC Web site provides background information on the recent growth chart revisions. Also, individual growth charts can be downloaded and printed from this Web site.

Web site: www.cdc.gov/growthcharts

Diabetes Care and Education, Dietetic Practice Group of the American Dietetic Association

Ethnic and Regional Food Practices, A Series. Chicago, IL: American Dietetic Association, 1994–1999.

This series of booklets addresses food practices, customs, and holiday foods of various ethnic groups. Booklets describing the following ethnic groups are available: Alaska Native, Chinese American, Filipino American, Hmong American, Jewish, and Navajo.

Graves DE, Suitor CW

Celebrating Diversity: Approaching Families through Their Food. 2nd ed. Arlington, VA: National Center for Education in Maternal and Child Health, 1998.

The purpose of this publication is to assist health professionals in learning to communicate effectively with a diverse clientele. Topics covered in the book include using food to create com-mon ground, changing food patterns, examining how food choices are made, communicating with clients and families, and working within the community.

National Academy Press, Dietary Reference Intakes

The Web site of the National Academy Press, publisher for the National Academies, contains descriptions of available books. Over 2000 books are available online free of charge. The current Dietary Reference Intakes are also available at this Web site and can be read online free of charge.

Web site: www.nap.edu

National Center for Complementary and Alternative Medicine, National Institutes of Health

The National Center for Complementary and Alternative Medicine is dedicated to science-based information on complementary and alternative healing practices. This Web site provides information for consumers and practitioners as well as information about related news and events.

Web site: www.nccam.nih.gov/nccam

National Network for Child Care

This Web site, hosted by Iowa State University Extension, provides articles, resources, and links on a variety of topics of interest to professionals and families who care for children and youth. Topics include child development, nutrition, and health and safety.

Web site: www.nncc.org

Partnership for Food Safety Education

This partnership was formed in 1997 for the purpose of educating the public about safe food handling to reduce foodborne illnesses. This Web site promotes the partnership's four food safety practices to the educator, consumer, and the media.

Web site: www.fightbac.org

Patrick K, Spear BA, Holt K, Sofka D

Bright Futures in Practice: Physical Activity. Arlington, VA: National Center for Education in Maternal and Child Health, 2001.

This colorful and user-friendly spiral-bound book is one of the implementation guides of the *Bright Futures* publications. Developmentally appropriate activities are presented for infants, children, and adolescents. Special issues and concerns are addressed, such as physical activity for the child with a chronic condition such as asthma or diabetes mellitus.

USDA, Center for Nutrition Policy and Promotion

Tips for Using the Food Guide Pyramid for Young Children 2 to 6 Years Old. Government Printing Office, 202/512-1800, Stock Number 001-00004665-9.

This Web site provides information on USDA materials including the Dietary Guidelines for Americans and the Food Guide Pyramid. This particular document can be found by clicking on "Food Guide Pyramid for Young Children."

Web site: www.usda.gov/cnpp

References

1. Center on Hunger, Poverty and Nutrition Policy. Statement on the link between nutrition and cognitive development in children. Tufts University, School of Nutrition Science and Policy; 1998.

2. Kids count data book. Baltimore: The Annie E. Casey Foundation; 2003.

3. U.S. Department of Health and Human Services. Healthy People 2010 (conference edition, in two volumes). Washington, DC: U.S. Department of Health and Human Services, January 2000.

4. Behrman RE, Kliegman RM, Jenson

HB, eds. Nelson's textbook of pediatrics, 17th ed. Philadelphia: WB Saunders; 2004.

5. Centers for Disease Control and Prevention, National Center for Health Statistics. CDC growth charts: United States. Available at www.cdc.gov/nchs/about/major/nhanes/growthcharts/charts.htm, accessed 4/19/04.

6. Pietrobelli A, Myles FA, Alison DB et al. Body mass index as a measure of adiposity among children and adolescents: a validation study. J Pediatr 1998;132:204–10.

7. Dietz WH, Robinson TN. Use of the body mass index (BMI) as a measure of overweight in children and adolescents. J Pediatr 1998;132:191–3.

8. Troiano RP, Flegal KM. Overweight children and adolescents: description, epidemiology, and demographics. Pediatrics 1998;101:497–504.

9. Story M, Holt K, Sofka D, eds. Bright futures in practice: nutrition. Arlington, VA: National Center for Education in Maternal and Child Health; 2000.

10. USDA, Center for Nutrition Policy and Promotion. Tips for using the food guide pyramid for young children 2 to 6 years old. Washington, DC: U.S. Government Printing Office; phone (202) 512-1800, Stock Number 001-00004665-9. Available at www.usda.gov/cnpp.

11. Satter E. The feeding relationship: problems and interventions. J Pediatr 1990;117:181–9.

12. Birch LL. Children's food acceptance patterns. Nutr Today 1996;31:234–40.

13. Birch LL, Johnson SL, Andresen G et al. The variability of young children's energy intake. N Eng J Med 1991;324:232–5.

14. Story M, Brown JE. Do young children instinctively know what to eat? The studies of Clara Davis revisited. New Eng J Med 1987;316:103–6.

15. Van den Bree MBM et al. Genetic and environmental influences on eating patterns of twins ages >50 years. Am J Clin Nutr 1999;70:456–65.

16. Chess S, Thomas A. Dynamics of individual behavioral development. In: Levine MD, Carey WB, Crocker AC, eds. Developmental-behavioral pediatrics. Philadelphia: WB Saunders; 1992: 84–94.

17. Birch LL, Fisher JO. Development of eating behaviors among children and adolescents. Pediatrics 1998;101:539–49 (S).

18. Birch LL, Fisher JA. Appetite and eating behavior in children. Pediatr Clinic N Amer 1995;42:931–53.

19. Fisher JO, Mitchell DC, Smiciklas-Wright H, Birch LL. Parental influences on young girls' fruit and vegetable, micronutrient, and fat intakes. J Amer Diet Assoc 2002;102:58–64.

20. Fisher JO, Birch LL. Restricting access to palatable foods affects children's behavioral response, food selection, and intake. Am J Clin Nutr 1999;69:1264–72.

21. Birch LL, Fisher JO. Food intake regulation in children, fat and sugar substitutes and intake. Ann NY Acad Sci 1997;819:194–220.

22. Rolls BJ, Engell D, Birch LL. Serving portion size influences 5-year-old but not 3-year-old children's food intake. J Amer Diet Assoc 2000;100:232–4.

23. Fisher JO, Birch LL. Parents' restrictive feeding practices are associated with young girls' negative self-evaluation of eating. J Amer Diet Assoc 2000;100:1341–6.

24. Davison KK, Birch LL. Weight status, parent reaction, and self-concept in five-year-old girls. Pediatrics 2001;107:46–53.

25. Abramovitz BA, Birch LL. Five-year-old girls' ideas about dieting are predicted by their mothers' dieting. J Amer Diet Assoc 2000;100:1157–63.

26. Satter E. Feeding dynamics: helping children to eat well. J Pediatr Health Care 1995;9:178–84.

27. Barlow SE, Dietz WH. Obesity evaluation and treatment: expert committee recommendations. Pediatrics [serial online] 1998;102:E3. Available at www.pediatrics.aappublications.org/cgi/content/full/102/3/e29, accessed 4/19/04.

28. Institute of Medicine, Food and Nutrition Board. Dietary Reference Intakes for energy, carbohydrate, fiber, fat, protein, and amino acids. Washington, DC: National Academy Press; 2003.

29. Kennedy E, Goldberg J. What are American children eating? Implications for public policy. Nutr Rev 1995;53:111–26.

30. Kennedy E, Powell R. Changing eating patterns of American children: a view from 1996. J Amer Coll of Nutr 1997;16:524–9.

31. Institute of Medicine, Food and Nutrition Board. Dietary reference intakes for calcium, phosphorus, magnesium, vitamin D, and fluoride. Washington, DC: National Academy Press; 1997.

32. Institute of Medicine, Food and Nutrition Board. Dietary reference intakes for vitamin A, vitamin K, arsenic, boron, chromium, copper, iodine, iron, manganese, molybdenum, nickel, silicon, vanadium, and zinc. Washington, DC: National Academy Press; 2001.

33. Centers for Disease Control and Prevention. Recommendations to prevent and control iron deficiency in the United States. MMWR April 3, 1998;47(RR-03).

34. Centers for Disease Control and Prevention. Iron deficiency in the United States, 1999–2000. MMWR 2002; 51:897–9.

35. Casamassimo P. Bright futures in practice: oral health. Arlington, VA: National Center for Education in Maternal and Child Health; 1996.

36. American Academy of Pediatrics, Committee on Nutrition. Pediatric nutrition handbook. Elk Grove Village, IL: American Academy of Pediatrics; 1998.

37. McClung HJ, Boyne L, Heitlinger L. Constipation and dietary fiber intake in children. Pediatrics 1995;96:999–1001 (S).

38. Meyer PA, Pivetz T, Dignam TA et al. Surveillance for elevated blood lead levels among children—United States, 1997–2001. In: Surveillance summaries, September 12, 2003. MMWR 2003;52(SS-10):1–21.

39. Centers for Disease Control, National Center for Environmental Health. Childhood lead poisoning. Available at www.cdc.gov/nceh/lead/factsheets/childhoodlead.htm, accessed 10/3/03.

40. Ballew C, Khan LK, Kaufmann R et al. Blood lead concentration and children's anthropometric dimensions in the Third National Health and Nutrition Examination Survey (NHANES III), 1988–1994. J Pediatr 1999;134:623–30.

41. Stanek K, Manton W, Angle C et al. Lead consumption of 18- to 36-month-old children as determined from duplicate diet collections: nutrient intakes, blood lead levels, and effects on growth. J Amer Diet Assoc 1998;98:155–8.

42. Mushak P, Crocetti AF. Lead and nutrition. Nutr Today 1996;31:12–17.

43. Centers for Disease Control and Prevention. Screening young children for lead poisoning: guidance for state and local public health officials. Atlanta, GA: Centers for Disease Control and Prevention, 1997. Also available at www.cdc.gov/nceh/programs/lead/guide/1997/guide97.htm.

44. American Academy of Pediatrics. Screening for elevated blood lead levels. Pediatrics 1998;101:1072–8.

45. Household food security in the United States 1995–1998, advance report summary. Available at www.fns.usda.gov/oane/MENU/Published/FSP/FILES/ fsecsum.htm.

46. Kleinman RE, Murphy JM, Little M et al. Hunger in children in the United States: potential behavioral and emotional correlates. Pediatrics [serial online] 1998; 101:E3. Available at www. pediatrics.aappublications.org/cgi/content/full/101/1/e3, accessed 4/19/04.

47. National Institute of Allergy and Infectious Diseases, National Institutes of Health, U.S. Department of Health and Human Services. Foodborne diseases. Available at www.niaid.nih.gov/factsheets/foodbornedis.htm.

48. Partnership for Food Safety education. FightBAC™. Available at www.fightbac.org.

49. U.S. Department of Agriculture and U.S. Department of Health and Human Services. Dietary guidelines for Americans, 2000. 5th ed. Home and Garden Bulletin No. 232. Washington, DC: U.S. Government Printing Office.

50. Must A, Dallal GE, Dietz WH. Reference data for obesity: 85th and 95th percentiles of body mass index (wt/ht²) and triceps skinfold thickness. Am J Clin Nutr 1991;53:839–46.

51. Dwyer JT, Stone EJ, Yang M. Prevalence

of marked overweight and obesity in a multiethnic pediatric population: findings from the Child and Adolescent Trial for Cardiovascular Health (CATCH) study. J Amer Diet Assoc 2000;100:1149–56.

52. Milner JA, Allison RG. The role of dietary fat in child nutrition and development: summary of an ASNS workshop. J Nutr 1999;129:2094–105.

53. Ogden CL, Flegal KM, Carroll MD, Johnson CL. Prevalence and trends in overweight among U.S. children and adolescents, 1999–2000. JAMA 2002; 288:1728–32.

54. Mei Z, Scanlon KS, Grummer-Strawn LM et al. Increasing prevalence of overweight among U.S. low-income preschool children: the Centers for Disease Control and Prevention pediatric nutrition surveillance. Pediatrics [serial online] 1998;101: E12. Available at www. pediatrics .aappublications.org/cgi/content/ full/101/1/e12, accessed 4/19/04.

55. Whitaker RC, Pepe MS, Wright JA et al. Early adiposity rebound and the risk of adult obesity. Pediatrics [serial online] 1998;101:E5. Available at www.pediatrics .aappublications.org/cgi/content/full/101/3/e5, accessed 4/19/04.

56. Dietz WH, Gortmaker SL. Preventing obesity in children and adolescents. Annu Rev Public Health 2001;22:337–53.

57. Frisancho AR. New norms of upper limb fat and muscle areas for assessment of nutritional status. Am J Clin Nutr 1981;34:2540–5.

58. National Cholesterol Education Program. Report of the expert panel on blood cholesterol in children and adolescents. Bethesda, MD: National Cholesterol Education Program; 1991.

59. Vital & Health Statistics Series 11: Data from the National Health Survey. 1999;244:1–14.

60. Yu SM, Kogan MD, Gergen P. Vitamin-mineral supplement use among preschool children in the United States. Pediatrics [serial online] 1997;100:e4. Available at www.pediatrics.aappublications.org/cgi/content/full/100/5/e4, accessed 4/19/04.

61. Buck ML, Michael RS. Talking with families about herbal products. J Pediatr 2000;136:673–8.

62. American Dietetic Association. Position of the American Dietetic Association: dietary guidance for healthy children ages 2 to 11 years. J Amer Diet Assoc 2004;104: 660–77.

63. A summary of conference recommendations on dietary fiber in childhood: Conference on Dietary Fiber in Childhood, New York, May 24, 1994. Pediatrics 1995;96: 1023–8.

64. Hampl JS, Betts NM, Benes BA. The age+5 rule: comparisons of dietary fiber intake among 4- to 10-year-old children. J Amer Diet Assoc 1998;98:1418–23.

65. American Academy of Pediatrics, Committee on Nutrition. Calcium requirements of infants, children, and adolescents. Pediatrics 1999;104:1152–7.

66. U.S. Department of Agriculture, Agricultural Research Service. 1999. Food and nutrient intakes by children 1994–96, 1998. ARS Food Surveys Research Group; available on the "Products" page at www.barc.usda.gov/bhnrc/foodsurvey/home.htm, accessed 4/19/04.

67. Harnack L, Stang, J, Story M. Soft drink consumption among U.S. children and adolescents: nutritional consequences. J Amer Diet Assoc 1999;99:436–41.

68. Munoz KA, Krebs-Smith SM et al. Food intakes of US children and adolescents compared with recommendations. Pediatrics 1997;100:323–9.

69. National Research Council. Recommended dietary allowances, 10th ed. Washington, DC: National Academy Press; 1989.

70. Skinner JD, Carruth BR, Houck KS et al. Longitudinal study of nutrient and food intakes of white preschool children aged 24 to 60 months. J Amer Diet Assoc 1999;99: 1514–21.

71. Picciano MF, Smiciklas-Wright H, Birch L et al. Nutritional guidance is needed during dietary transition in early childhood. Pediatrics 2000;106:109–14.

72. McConaby KL, Smiciklas-Wright H, Birch LL et al. Food portions are positively related to energy intake and body weight in early childhood. J Pediatr 2002;140:340–7.

73. Sanders TAB. Vegetarian diets and children. Pediatr Clinic N Amer 1995;42: 955–65.

74. American Dietetic Association. Position of the American Dietetic Association and Dietetians of Canada: Vegetarian diets. J Amer Diet Assoc 2003;103:748–65.

75. American Dietetic Association. Position of the American Dietetic Association: nutrition standards for child-care programs. J Amer Diet Assoc 1999;99:981–6.

76. American Public Health Association, American Academy of Pediatrics. Caring for our children, national health and safety performance standards: guidelines for out-of-home child care programs; 1992.

77. Green M, Palfrey JS, eds. Bright futures: guidelines for health supervision of infants, children, and adolescents, 2nd ed. Arlington, VA: National Center for Education in Maternal and Child Health; 2000.

78. U.S. Department of Agriculture. WIC. Available at www.usda.gov.

79. Rose D, Habicht JP, Devaney B. Household participation in the Food Stamp and WIC Programs increases the nutrient intakes of preschool children. J Nutr 1998; 128: 548–55.

80. U.S. Department of Health and Human Services. Head Start. Available at www .dhhs.gov.

81. U.S. Department of Agriculture. Food Stamps Program. Available at www.fns .usda.gov/fsp.

"Nothing in life is to be
feared, it is only to
be understood."
Marie Curie

Chapter 11

Toddler and Preschooler Nutrition:
Conditions and Interventions

Chapter Outline

- Introduction
- Who Are Children with Special Health
  Care Needs?
- Nutrition Needs of Toddlers and Preschoolers
  with Chronic Conditions
- Growth Assessment
- Feeding Problems
- Nutrition-Related Conditions
- Food Allergies and Intolerance
- Dietary Supplements and Herbal Remedies
- Sources of Nutrition Services

Prepared by **Janet Sugarman Isaacs**

Key Nutrition Concepts

1 Nutrition problems in young children with special health care needs are underweight, overweight, feeding difficulties, and higher nutrient needs as a result of chronic health conditions.

2 Feeding difficulties in preschoolers and toddlers appear as food refusals, picky appetites, and concerns about growth.

3 Nutrition services for toddlers and preschoolers with chronic health problems are provided in various settings, including schools and other educational programs and specialty clinics.

4 Toddlers and preschoolers at risk for chronic conditions have the same nutritional problems, concerns, and needs as other children.

Introduction

Most toddlers and preschoolers are healthy and develop as expected. This chapter discusses children who do not fit the typical pattern, *children with special health care needs* associated with a *chronic condition* or disability, or children who are at risk. Sometimes no diagnosis has been made, and yet parents, health care providers, or preschool teachers have a nagging feeling that something is not right about how the child is growing and developing. This chapter covers nutrition needs and services for young children with food allergies, breathing or *pulmonary* problems, feeding and growth problems, developmental delays, and those at risk for needing nutrition support.

Children with Special Health Care Needs A general term for infants and children with, or at risk for, physical or developmental disabilities, or chronic medical conditions from genetic or metabolic disorders, birth defects, premature births, trauma, infection, or prenatal exposure to drugs.

Chronic Condition Disorder of health or development that is the usual state for an individual and unlikely to change, although secondary conditions may result over time.

Pulmonary Related to the lungs and their movement of air for exchange of carbon dioxide and oxygen.

Early Intervention Services Federally mandated evaluation and therapy services for children in the age range from birth to 3 years under the Individuals with Disabilities Education Act.

Who Are Children with Special Health Care Needs?

The child who does not see, hear, or walk is easily recognized as having a chronic condition. It can be difficult and expensive, however, to identify some other children with special health care needs. Criteria for labeling chronic conditions in children vary from state to state. More than 40 different federal definitions describe the term *disability*.[1] Criteria used for identifying disabilities in adults do not fit children, because the criteria are related to a person's ability to work or perform household chores. *Chronic condition* and *disability* mean the same thing in referring to toddlers and preschoolers. Prevalence estimates for disabilities range from 5% to 31% of children.[1,2] Whatever the number, nutrition problems are common in children with disabilities. Up to 90% of children with disabilities have some type of nutritional problem.[3]

Toddlers and preschoolers with chronic illnesses are entitled to the same services as older people with chronic illnesses are, with additional help. They are covered by the Americans with Disabilities Act, the Social Security Disability Program, Supplemental Social Security Insurance (SSI) Program, and services for families without health insurance coverage.[4] Additional help comes from educational regulations ensuring that all children with disabilities have a free, appropriate public education. Nutrition services are funded within education regulations in the Individuals with Disabilities Education Act (IDEA).[4] Most children start school at age 5 or 6 years, but children at risk or who have special needs may attend well before that, as soon as the need is identified. The sooner special educational, nutritional, and health care interventions are started, the better for the overall development of the child. Parents of a typical child choose and pay for day care or a preschool program. For a child with a special health care needs, day care or educational programs are selected based on nutrition and other types of therapy provided by state and federal resources. Nutrition services can be provided to young children within special educational programs and services both as preschoolers (3 to 5 years old) and from birth up to 3 years old.[4,5] Services have to be culturally appropriate for various ethnic groups, reflecting food preferences, religious beliefs, and sensitivity to dress and language; otherwise, they are likely to be rejected.

Eligibility for services does not require a specific diagnosis. *Early intervention services* are based on the following:[4,5]

- Developmental delays in one or more of the following areas: cognitive, physical, language and speech, psychosocial, or self-help skills
- A physical or mental condition with a high probability of delay, such as Down syndrome
- At risk medically or environmentally for substantial developmental delay if services are not provided

A number of chronic conditions are suspected, but not obvious, in the first year of life. The diagnosis often becomes clear in the toddler and preschool years, however. Standardized developmental screening, evaluation, and testing for these ages show more reliability than they do for infants. Parents who were told about possible disabilities during infancy move beyond coping by denial or disbelief and are willing to seek out services. The tendency is to resist labeling a young child with a diagnosis, so some suspected conditions are not confirmed until school age if the delay in diagnosis will not harm the child.

Nutrition Needs of Toddlers and Preschoolers with Chronic Conditions

Toddlers or preschoolers with chronic health conditions are at risk for the same nutrition-related problems and concerns as other children.[6] Consequently, every attempt should be made to meet their overall nutritional needs and to assure normal growth and development. The DRIs for toddlers and preschoolers provide a good starting point for setting protein, vitamin, and mineral needs for children with chronic conditions.[7] (DRI tables are located on the inside front cover of this text.) The recommendations for typical children concerning dietary fiber, prevention of lead poisoning, and iron-deficiency anemia apply to children at risk or already diagnosed with special health care needs.[5] However, some specific conditions require adjustments to the general guidelines. The DRI for dietary fiber for children may be too low for some conditions and too high for others.[8] Children with sickle-cell disease have more specific blood iron and lead testing than the usual guidelines. Iron-rich foods to increase their iron stores may not be appropriate when iron also comes with blood transfusions. Consequently, nutritional needs must be customized to the child.[6,8]

Chronic conditions may result in poor appetites, although there are increased caloric needs.[8] Table 11.1 gives examples of conditions in which caloric needs may be high or low.[6,8] Each child must be assessed to confirm caloric needs. A child may have an interval of needing additional calories based on the course of the chronic condition. Changes in caloric needs may explain why both underweight or overweight are more common in children with chronic conditions than in other children.[6] Overweight and obesity are common in Down syndrome and spina bifida in part because lower caloric needs are due to low muscle mass, lower mobility, and short stature.[6] Overall health status is worsened by excessive body weight, so matching caloric intake to needs is important no matter how difficult. Underweight results in part from the chronic illness and its treatment. Children with chronic illnesses may be more likely to experience weight loss with any illness. Underweight children with a chronic condition may or may not benefit from food choices for weight gain. In underweight children, it is inappropriate to make some of the usual recommendations, such as reducing fat intake.[5]

Recommendations regarding food intake, vitamin and mineral supplementation, and mealtime behaviors also should be customized to the individual child. Children who are frequently sick or have low energy levels and appetites may dislike eating foods that are hard to chew or take a long time to eat. Some food-intake problems related to chronic illness may result from the children's behavior. It is age appropriate for children to express their food likes and dislikes, insist on their independence, and go on food jags. It can be difficult but important to distinguish between food-intake problems related to the chronic condition and those related to "growing up" in the toddler and preschool years.

Cystic Fibrosis Condition in which a genetically changed chromosome 7 interferes with all the exocrine functions in the body, but particularly pulmonary complications, causing chronic illness.

Diplegia Condition in which the part of the brain controlling movement of the legs is damaged, interfering with muscle control and ambulation.

Pediatric AIDS Acquired immunodeficiency syndrome in which infection-fighting abilities of the body are destroyed by a virus.

Prader-Willi Syndrome Condition in which partial deletion of chromosome 15 interferes with control of appetite, muscle development, and cognition.

Growth Assessment

Most toddlers and preschoolers with chronic conditions are provided an assessment of nutritional status as a first step in determining whether more intensive levels of nutrition services are needed. The need for nutrition services is identified by answers to these sorts of questions:

- Is the child's growth on track?
- Is his or her diet adequate?
- Are the child's feeding or eating skills appropriate for the child's age?
- Does the diagnosis affect nutritional needs?

A variety of nutrition screening tools exist for assessing the nutritional status of children with chronic conditions.[6,8] Such tools are useful for children at risk as well as those already diagnosed with conditions such as asthma, HIV infection, allergies, and cerebral palsy. After assessment, nutrition intervention services provide methods to improve nutritional status. Several conditions that require nutrition intervention services include failure to thrive, celiac disease, breathing problems, and muscle coordination problems.

Children with special health care needs often have conditions that affect growth even when adequate nutrients are provided. In such cases the 2000 CDC growth

---

Table 11.1 Chronic conditions generally associated with high and low caloric needs

Higher Caloric Need Conditions	Lower Caloric Need Conditions
Cystic fibrosis	Down syndrome
Renal disease	Spina bifida
Ambulatory children with *diplegia*	Nonambulatory children with diplegia
Pediatric AIDS	*Prader-Willi syndrome*
Bronchopulmonary dysplasia (BPD)	Nonambulatory children with short stature

charts require interpretation based on the child's previous growth pattern.[9] If a thin and small-appearing child has adequate fat stores, adding calories may be harmful; it is important to recognize the growth pattern as healthy for that child. Trying to add calories in such a case may promote overweight in the form of excess fat stores. Growth patterns in children with special health care needs are also affected by some prescribed medications, particularly those that change body composition, such as steroids and growth hormone.[11,12]

Specific growth charts developed for chronic conditions, when available, are preferred. For children up to age 38 months who are born low birthweight (LBW) or very low birthweight (VLBW), the Infant Health and Development Program (IHDP) growth percentile charts are appropriate.[10] Correction for prematurity, as discussed in Chapter 9, makes the charts useful for preterm babies as well as toddlers. For a child born 3 months early, for example, the IHDP growth charts could be used at a chronological age of 41 months. Plotting on both the 2000 CDC growth charts and the IHDP chart documents catch-up growth. When catch-up growth happens, the child's growth pattern crosses channels on the growth chart, for weight as well as for length.

> **Rett Syndrome** Condition in which a genetic change on the X chromosome results in severe neurological delays, causing children to be short, thin appearing, and unable to talk.
>
> **Meningitis** Viral or bacterial infection in the central nervous system that is likely to cause a range of long-term consequences in infancy, such as mental retardation, blindness, and hearing loss.

Special health care providers commonly use a special head growth chart.[13] It provides head circumference percentiles from birth to age 18 years and is used to determine whether head growth falls within normal limits or indicates a neurological condition, such as *Rett syndrome*.[13] Rett syndrome is a rare disorder, characterized by a reduced rate of head growth beginning in the toddler years (see Illustration 11.1).[14] Later, over time, the rate of weight and height accretion slows in girls with Rett syndrome.[14] Decreased rate of head growth in toddlers and preschools may be indicative of problems from infancy, such as prematurity, or consequences of infection such as *meningitis*. Some clinics plot head circumference on the back of the CDC 2000 growth chart (up to age 3 years) as well as on the special head circumference growth chart.

Feeding Problems

Children with special health care needs have many of the same developmental feeding issues as other children, such as using food to control their parents' behavior at mealtime and going on food jags. Feeding problems that are part of underlying health conditions may emerge in the toddler and preschool years on top of the usual feeding difficulties and require extra attention. Some feeding problems during the toddler years are typical in children who are later diagnosed with a chronic condition. Examples of such conditions include gastroesophageal reflux, asthma (pulmonary problems in general), developmental delay, cerebral palsy, attention deficit hyperactivity disorder, and autism.[3,8] These children as toddlers tend to display signs of feeding problems, such as low interest in eating, long mealtimes (over 30 minutes), preferring liquids over solids, and food refusals. Children at risk for developmental delay often prove more difficult to feed as toddlers and preschoolers.[3] The child may drink liquids excessively, or eat foods usually preferred by younger children (see Case Study 11.1, page 274). Recognizing that the child needs to be treated as younger than current chronological age may be a necessary step. Offering the child food textures that she can eat successfully within a monotonous diet, or continuing to offer a bottle, may be appropriate choices in these circumstances.

Table 11.2 shows an example of the likes and dislikes of a 2.5-year-old child. The child likes only a few foods

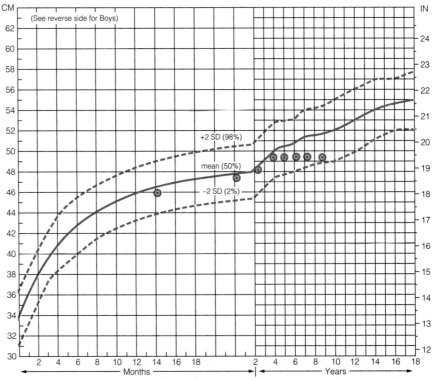

Head Circumference GIRLS

Illustration 11.1 Nellhaus head circumference growth chart plotted for girl with Rett syndrome.

Table 11.2 Food choices of a 2.5-year-old child with suspected developmental delay

Likes	Dislikes
3 packets instant Cream of Wheat with added sugar and margarine (refused offered apple slices)	Hamburger meats, or any other kind of meat
Macaroni and cheese (refused offered sandwich with lettuce and bologna)	Green beans or any kind of vegetables
	Vegetable soup
Banana, with peel removed (refused other cut-up fruits offered)	Salads of all kinds
Pudding, only chocolate	Casseroles or any mixtures of foods
Cheese puffs (refused corn chips)	Milk, and milk with any flavoring added
Juices of all kinds, in a sippy cup	

---

that are not especially nutritious. Usual recommendations are to add variety to the child's diet and to assure intake of meats, milk, and vegetables. This recommendation is appropriate for a typical child, but this child's eating pattern suggests a feeding problem. The soft textures and mild tastes of preferred foods characterize a child closer to 1 year of age, and the foods the child dislikes require higher oral skills to eat. An evaluation of the child's overall level of functioning will likely indicate a developmental delay of the child's feeding skills.

Behavioral Feeding Problems

"Every mouth prefers its own soup."

Sephardic saying

Mealtime behavioral problems and food refusals are common in children with behavioral and attention disorders. These concerns often bring parents to nutrition experts for solutions. Behavioral disorders that affect nutritional status are autism and attention-focusing problems, such as *attention deficit hyperactivity disorder (ADHD)*. ADHD may be suspected during the preschool years, but it is primarily treated during school years.[15] (ADHD is discussed further in Chapter 13.) Table 11.3 shows the intake of a 2-year-old with a feeding problem resulting in a self-restricted diet typical of autism. The child refuses to eat many foods and is rigid in what he will eat. He does not respond to feeling hungry, as other 2-year-old children do. When he is not given foods he likes, he refuses to eat at all and has temper tantrums during which he can injure himself. His self-restricted diet is a part of the condition, which affects how he senses everything in his environment. He prefers to drink rather than eat foods, so a high proportion of his total calories come from one type of drink. Interventions to improve the diet for this child may include a complete vitamin and mineral supplement, and adding one new food by offering it many times (15–20 times) over one or two months. Nutrition interventions should be incorporated into this child's overall treatment plan, provided within a special education program. See Case Study 11.1 on the next page.

Excessive Fluid Intake

The food intake noted in Table 11.3 highlights a common issue related to excessive fluids. Many young children prefer to drink rather than eat solid foods, especially when they are not feeling well. Families of chronically ill children tend to offer juices and lower-nutrient beverages in an effort to achieve growth when eating is difficult. The American Academy of Pediatrics recommendation to limit juice intake to 4 to 6 fluid ounces per day for ages 1–6 years applies to all children.[16] Calcium-fortified juices may be appropriate if other sources of calcium are limited, but these juices can also be overconsumed. In a child who already has gastrointestinal problems, it may not be clear if problems are caused by excessive juice intake.[16] Excess juice resulting in a pattern of low milk intake has been documented to result in smaller stature and lower bone density.[17] In young children who may be less active due to chronic conditions, the negative impact may be larger.

Attention Deficit Hyperactivity Disorder (ADHD) Condition characterized by low impulse control and short attention span, with and without a high level of overall activity.

Feeding Problems and Food Safety

Toddlers and preschoolers with chronic conditions are at greater risk for food-contamination problems. Some

---

Table 11.3 Dietary intake of 2-year-old child with suspected autism

Dry Fruit Loops cereal
10 fl oz calcium-supplemented orange juice drink
Chicken fingers from a specific fast-food restaurant
French fries
10 fl oz calcium-supplemented orange juice drink
Waverly crackers
Pringles potato chips
10 fl oz calcium-supplemented orange juice drink
oatmeal cake
10 fl oz calcium-supplemented orange juice drink

Case Study 11.1

Photo Disc

A Picky Eater

Greg is a well-groomed boy almost 3 years old. He has been growing as expected, but he does not talk. He can walk and move about well, but he prefers to play alone. Favorite foods are juices in his sippy cup, which he likes to carry around; macaroni and cheese; white bread without crusts; mashed potato; Honeycomb cereal; and crackers. Greg cries and throws food that he does not like, such as hamburgers, fruits, most vegetables, and any food combinations. He will periodically eat cheese pizza, scrambled eggs, and applesauce.

His mother tries talking to the pediatrician about his picky appetite, but the pediatrician reassures her that Greg will eat when he is hungry and not to worry. Greg's mother is frustrated that he is so difficult to take out to eat because of the tantrums he throws in restaurants and friends' homes. He sometimes eats a large portion of a food he likes. Most of the time, he is satisfied just drinking juices all day from his sippy cup, and he is rarely interested in eating when others eat. He is able to eat with a spoon, but he does not like to touch foods with his hands. Greg has been referred for speech therapy, but his therapy does not address his eating. His medical history shows that he was born full-term and has had three ear infections but no major illnesses.

Nutrition assessment shows that Greg is consuming adequate calories at 1350 calories/day, or 85 calories per kilogram. His diet is excessive in vitamin C and B vitamins, with adequate protein at the RDA for his age. His sources of protein are mainly his starchy foods of bread, crackers, and dry cereal.

Questions

1. What are the signs that Greg's feeding problem may be related to his speech?
2. Because Greg is growing well and meeting his calorie needs, why not just wait for him to mature to accept other foods?
3. Was his pediatrician wrong to say that Greg will eat when he is hungry?

(Answers are in Appendix D.)

Neuromuscular Disorders Conditions of the nervous system characterized by difficulty with voluntary or involuntary control of muscle movement.

feeding problems result in prolonged needs for soft, easy-to-eat food textures well past the age when baby foods are eaten. Fork-mashing or blending foods may invite bacterial contamination or spoilage over time.[6] Similarly, complete nutritional supplements and formulas are subject to contamination, particularly in the tubing used to give them. How often tubing and devices to deliver formulas are changed can be a food-safety issue. Some families, aware of the high cost of such devices, tend to use them longer than recommended.[18] Microbial contamination of powdered dry formulas has been found, so pasturized canned liquid formulas are recommended.[18]

Feeding Problems from Disabilities Involving Neuromuscular Control

Children who have feeding problems related to muscle control of swallowing or control of the mouth or upper body may choke or cough while eating or refuse foods that require chewing.[3] These types of feeding problems result from conditions such as cerebral palsy or other *neuromuscular disorders* and genetic disorders such as Down syndrome. These signs of feeding and swallowing problems in toddlers or preschoolers generally appear more severe than the reactions of infants who are learning how to munch and chew foods.[8] The decrease in appetite expected in toddlers and preschoolers may be pronounced in children who find eating difficult and unpleasant. These feeding problems may require further study to make sure eating is safe for a child, and not related to frequent illness such as bronchitis or pneumonia.

A child with *hypotonia* or *hypertonia* in the upper body may experience difficulty sitting for a meal and self-feeding with a spoon.[8] If these feeding problems are not resolved by providing therapy in early intervention programs or schools, children are likely to resist eating over time. They may then need a form of nutrition support, such as placement of a *gastrostomy.*

Nutrition-Related Conditions

Failure to Thrive

Failure to thrive (FTT) is a condition in which a caloric deficit is suspected.[19] FTT has a slightly different basis in toddlers and preschoolers, who may have grown adequately during the first year. Their decrease in growth rate occurs at the age when appetite typically decreases and control issues at mealtime are expected, making identifying the cause of FTT more difficult. Generally FTT is suspected when a child's growth declines more than two growth percentiles, placing him near or below the lowest percentile in weight-for-age, weight-for-length, and/or BMI. FTT may result from a complex interplay of medical and environmental factors, such as the following:[4,6]

- Digestive problems such as gastrointestinal reflux or celiac disease
- Asthma or breathing problems
- Neurological conditions such as seizures
- Pediatric AIDS

Children who have chronic illnesses or were born preterm have a higher risk of FTT as a result of abuse or *medical neglect.*[19] They have greater needs than other children do, and they may be more irritable and demanding, which places them at risk. Often a specific nutrient or group of nutrients are suspected of being inadequate in the diet of children with FTT, when the more appropriate emphasis should be placed on energy and protein. Copper and zinc in the blood of toddlers with failure to thrive was reported to be the same as age-matched controls, although protein intake was lower.[20]

Recovery from FTT can include catch-up growth, which is an acceleration in growth rate for age.[3] If calories are provided at a higher level than for a typical child of the same age, catch-up growth is likely (see Illustration 11.2 on the following page). The length of time needed for catch-up growth varies, but some weight gain should be documented within a few weeks. For example, recovery from FTT for one 3-year-old was a gain of 6 pounds—more weight than typical to gain in 1 year—within the first 3 months of living in a new home.

Toddler Diarrhea and Celiac Disease

Toddlers are likely to develop diarrhea. The condition is called toddler diarrhea, in which otherwise healthy growing children have diarrhea so often that their parents bring them for a checkup.[3] Testing shows no intestinal damage and normal blood levels, without FTT or weight loss. The dietary culprit is likely to be excessive intake of juices that contain sucrose or sorbitol. The diarrhea results from excess water being pulled into the intestine, so limiting juices intake may be recommended.[16]

Celiac disease occurs in people who are sensitive to gluten, a component of wheat, rye, and barley. It has a prevalence of 1 in 3000 people within certain ethnic groups, such as those of Middle Eastern or Irish ancestry.[21] Symptoms of diarrhea and other digestive problems usually develop by 2 years of age. Confirmation of the condition is based on testing blood for the antibodies to gluten. Dietary management requires complete restriction of any foods with gluten. This list includes everything made with flour, such as bread and pasta, as well as foods with wheat, barley, or rye as an additive.[21] The allowed foods include rice, soy, corn, and potato flours. Meats, fruits, and vegetables are not restricted, but many processed foods use wheat flour for thickening. After instituting dietary restrictions, the intestinal damage heals and the digestive symptoms disappear. The parents of preschoolers with celiac disease learn to be expert readers of food labels because intestinal damage recurs if gluten is eaten by mistake.

> **Hypertonia** Condition characterized by high muscle tone, stiffness, or spasticity.
>
> **Hypotonia** Condition characterized by low muscle tone, floppiness, or muscle weakness.
>
> **Gastrostomy** Form of enteral nutrition support for delivering nutrition by tube directly into the stomach, bypassing the mouth through a surgical procedure that creates an opening through the abdominal wall and stomach.
>
> **Medical Neglect** Failure of parent or caretaker to seek, obtain, and follow through with a complete diagnostic study or medical, dental, or mental health treatment for a health problem, symptom, or condition that, if untreated, could become severe enough to present a danger to the child.

Autism

The toddler and preschool years are when behavioral signs of what may later be found to be autism are noted by families. No scientifically proven diet is now recommended for prevention or treatment of autism.[22] A gluten-free and casein-free diet is well known to families who educate themselves on the Internet and through autism support resources. Many products are marketed for the diet. Milk substitutes to avoid casein may or may not meet the child's need for calcium, vitamin D, protein, or other nutrients. Parents may choose to restrict gluten in the child with autism who has been found not to have celiac disease. Until the major studies now under way about autism release their findings, dietary recommendations for autism are the same as for any other child of the same age who has feeding problems.

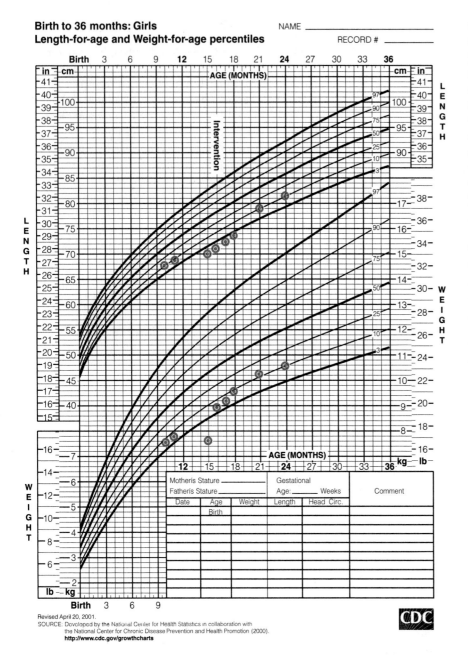

Birth to 36 months: Girls
Length-for-age and Weight-for-age percentiles

NAME _____

RECORD # _____

Revised April 20, 2001.
SOURCE: Developed by the National Center for Health Statistics in collaboration with the National Center for Chronic Disease Prevention and Health Promotion (2000).
http://www.cdc.gov/growthcharts

Illustration 11.2 Growth chart for a girl with failure to thrive before and after intervention.

Muscle Coordination Problems and Cerebral Palsy

Toddlers and preschoolers at risk for or confirmed with cerebral palsy need nutrition assessments that include body composition indexes, such as fat stores.[6,23] Nutrition interventions are then based on these findings, and they may include encouraging weight gain if body fat stores are low. If the child appears thin as a result of small muscle size and not low fat stores, weight gain is not needed. Growth tracking may be based

Spastic Quadriplegia A form of cerebral palsy in which brain damage interferes with voluntary muscle control in both arms and legs.

on spastic quadriplegia growth charts for some toddlers or preschoolers.[24] Part of the growth assessment for a preschooler with cerebral palsy may include an estimate of caloric needs for activity, which may be higher or lower than expected. A girl may expend higher energy in her efforts to coordinate walking while receiving physical therapy 3 days per week at school. Her activity may be lower if she is in a wheelchair most of the time.

Feeding assessment for a child with severe cerebral palsy (also called *spastic quadriplegia*) may be necessary as part of the overall nutritional assessment.[8] The assessment may include an observation of eating to determine any restrictions in the type of foods that the child can eat, and whether coordinating muscles for chewing, swallowing, and/or using a spoon or fork are working well. Table 11.4 on page 278 provides a food-intake record for a 4-year-old girl with spastic quadriplegia who does not walk and is receiving nutritional services for weight gain. Her meal pattern was adjusted because she tires easily while eating. She does not like to eat too much at a time, and refuses to be fed by another person (which is appropriate for her age). She can chew foods such as fresh apple, but then is too tired to eat something else. She eats a larger portion if the food is soft and does not require her to work so hard. She has gained weight at a slow rate, and her fat stores are low. The first plan is to use regular foods that are easy for her to eat to meet her nutritional needs, including cooked rather than fresh vegetables and fruits, and to avoid hard-to-chew foods, such as roast beef or corn on the cob. If she does not gain weight by eating foods such as those suggested in Table 11.4, she may need nutritional supplementation to assure her nutritional needs are met within her feeding limitations. See Case Study 11.2.

Pulmonary Problems

Breathing conditions are examples of common problems in children with special health care needs with major nutritional consequences. Breathing problems increase nutri-

Early Intervention Services for a Boy At Risk for Nutrition Support

Robert is 2.3 years old and in an early intervention program. He is eligible based on his preterm birth at 30 weeks gestation. His premature birth was related to exposure to an intrauterine infection. All in the family agree that he is small, but their main concern is that he is difficult to feed. He cries and refuses to eat when offered meals, even those with his favorite foods.

The registered dietitian who consults at the early intervention program meets with the family, assesses Robert, and reviews his medical records. Nutrition services are first planned to boost calories to stimulate weight gain. Observing Robert being fed by his mother is part of the nutrition services. Other therapists at the early intervention center are involved in making sure that Robert is positioned well to eat, so that he is sitting up without extra effort. The nutritionist and occupational therapist are concerned that Robert is choking so easily, and they talk to the family about contacting his pediatrician. They send a fax to the pediatrician's office recommending tests to study Robert's swallowing.

Robert does not attend the early intervention program for the next 3 weeks. The tests demonstrate that he is aspirating some of his liquids into his lungs, so oral feeding is unsafe. He requires a gastrostomy for feeding and is hospitalized for surgery. His parents learn how to feed him through the gastrostomy.

When Robert returns to the early intervention program, his pediatrician asks the early intervention staff to monitor his weight and to reinforce the discharge feeding instructions with the family. Nutrition services provided in the early intervention program are changed from working on Robert's oral feeding to monitoring and documenting his growth as adjustments are made in his gastrostomy feeding schedule. Over the next 6 months, Robert gains weight. He starts being more interactive with the staff at the early intervention center and makes some developmental progress in his walking and speaking. He is still a small child, but his improved nutritional status is confirmed by his adequate body fat measurements. His ability to return to eating by mouth will be reassessed later in the year.

Questions

1. What are the signs that Robert needs gastrostomy feeding?
2. Could the gatrostomy placement have been prevented if Robert had gained weight?
3. Can Robert enjoy life if he cannot eat?

(Answers are in Appendix D.)

Photo Disc

---

tional needs, lower interest in eating, and can slow growth rate.[25] (See Case Study 11.2). Infants who were born preterm are especially likely as toddlers to have breathing problems. Up to 80% of 1000-gram infants can develop chronic lung disease.[25] Examples of pulmonary diseases or chronic lung disease are ***bronchopulmonary dysplasia (BPD)*** and ***asthma***. Asthma is self-reported in 58 of every 1000 children under 5 years of age.[26] Asthma results in more emergency room visits for children under 5 years—at 121 visits per 10,000 people—than it does in older children with asthma.[26] Asthma does not necessarily require nutrition services, but some children have asthma as a result of food allergies.[27] Toddlers and preschoolers with BPD have a positive long-term prognosis because new lung tissue can grow until about 8 years of age.[25]

Toddlers and preschoolers with serious breathing problems generally need extra caloric intake due to the

Bronchopulmonary Dysplasia (BPD) Condition in which the underdeveloped lungs in a preterm infant are damaged so that breathing requires extra effort.

Asthma Condition in which the lungs are unable to exchange air due to lack of expansion of air sacs. It can result in a chronic illness and sometimes unconsciousness and death if not treated.

extra energy expended in breathing. Increased *work of breathing (WOB)* occurs with different pulmonary conditions and generally leads to low interest in feeding, partially as a result of tiredness.[26] Feeding difficulties have several causes in a toddler treated for BPD:[25]

- The normal progression of feeding skills is interrupted.
- Medications and their side effects contribute to high nutrition needs.
- Interrupted sleep and fatigue make hunger and fullness cues harder to interpret.

By the preschool years, the impact of BPD on slowing the rate of weight gain is usually clear. Exposure to common respiratory illnesses, which are minor in typical children, can require a trip back to the hospital for some children with BPD. Increased frequency of infections adds another limitation to catch-up growth. Neither the CDC growth chart nor the IHDP preterm growth chart may be helpful in predicting the child's growth pattern, but periods of good health are usually accompanied by a increase in weight gain and appetite.

Dietary recommendations for toddlers with BPD are similar to those for children with weakness (see Table 11.4). Small, frequent meals with foods that are concentrated sources of calories are needed. Easy-to-eat foods may still be recommended, so that fatigue from meals is low. If the toddler with breathing problems does not gain weight as a result of dietary recommendations such as those in Table 11.4, the next step will be to add complete nutritional supplements to meet the higher caloric needs. The supplements, such as Pediasure, are also a source of vitamins and minerals.

Developmental Delay and Evaluations

Developmental delay may be suspected when specific nutrients are consumed in inadequate or excessive amounts. Iron deficiency and lead toxicity are risk factors for developmental problems.[5,6] Developmental evaluations are recommended for young children who have been sick for a long time and isolated from other children. Standardized testing aids in finding a definitive diagnosis and appropriate educational programs. Developmental delay is a specific diagnosis that may be replaced by *mental retardation* when the child is 6 or 7 years old.[4] Changes in growth rate are typical in children with developmental delay.[23] Short stature is common and part of the unusual growth pattern that often prompts referrals for genetic testing.[28] The evaluation of growth from a genetic expert may include more in-depth analyses, such as measurements of hand and foot size and bone age.[28] Genetic syndromes also can be associated with unusually fast growth. Soto's syndrome is a rare disorder in which the child is tall and large, but has delayed development.[28]

Work of Breathing (WOB) A common term used to express extra respiratory effort in a variety of pulmonary conditions.

Mental Retardation Substantially below average intelligence and problems in adapting to the environment, which emerge before age 18 years.

Anaphylaxis Sudden onset of a reaction with mild to severe symptoms, including a decrease in ability to breathe, which may be severe enough to cause a coma.

Table 11.4 Meal pattern and recommended foods for an underweight girl with feeding problems as a result of weakness

Meal Pattern: Small Frequent Meals and Snacks to Prevent Tiredness at Meals	Recommended Foods That Are Easy to Chew, with Small Portions
Breakfast at home	**Breakfast:** ½ c oatmeal with added soft fruit, margarine, and brown sugar
Mid-morning snack	
Lunch (at preschool)	**Snacks:** 1 slice deli meat with 6 fl oz whole milk with Carnation Instant Breakfast added
Afternoon snack (at preschool)	½ c soft-cooked sliced apples with added margarine
After-school snack	cake-type cookie (frosting allowed)
Dinner	
Bedtime snack	**Dinner:** ½ c mashed potato with added margarine
	3 tb meat loaf
	3 tb soft-cooked carrots with added margarine
	Bedtime snack: chocolate cake with frosting and 4 fl oz whole milk

Food Allergies and Intolerance

True food allergies are estimated to be present in 2% to 8% of children.[27] Food allergies are usually identified in toddlers and preschoolers because allergy testing in infancy is not useful due to the incomplete development of the immune system. True food allergies can result in life-threatening episodes of *anaphylaxis*.[27] Examples of food allergies that may result in anaphylaxis for some children include the following:[27]

- Milk
- Eggs
- Wheat
- Peanuts
- Walnuts
- Soy
- Fish

Cow's milk protein allergy rarely persists into the toddler and preschool years. However, when cow's milk protein allergy does persist, symptoms in the toddler and preschool years may appear as more general allergy symptoms, such as asthma or skin rashes.[29] A high incidence of other food allergies are present in a child with confirmed cow's milk protein allergy, with, for example, 35% reacting also to oranges or 47% reacting also to soymilk.

Strict and complete avoidance of the food that causes the allergy is required. This abstinence includes all settings, such as eating nothing prepared at bake sales when the food ingredients are unknown. If the preschool child is on an extensively restricted diet, the quality of the diet may not meet all her nutritional needs. Such restrictions are also likely to result in mealtime behavioral problems. The parents may become overprotective, or the child may quickly learn to use restricted foods to get a parent's concern and attention. Diagnosed food allergies can greatly affect the family. For children at risk for anaphylaxis, parents and caregivers should be given instruction in emergency lifesaving procedures and use of an injectible form of epinephrine.[27]

Dietary Supplements and Herbal Remedies

> "It is better to take food into the mouth than to take worries into the heart."
>
> Yiddish saying

Families who are concerned that something may be wrong with their young children may be attracted to health and nutritional claims targeted and packaged for adults. The family that is having difficulty finding effective treatment for a child is most at risk for inappropriate or ineffective alternative products. Parent coalitions and advocacy groups are excellent sources of networking for families, but they can also be sources of nutritional claims for products and dietary regimens that have no scientific testing behind them. Down syndrome, for example, is a disorder for which nutritional supplementation has been marketed to parents. No specific nutrients, combinations of nutrients, or herbal remedies have been shown to improve the intellectual functioning of individuals with Down syndrome.[30,31] The National Down Syndrome Society cautions parents about the ineffectiveness of nutrient and herbal supplements to discourage their use, but interest continues.[30,32] What is really being marketed is hope, which families always want and need.

Constipation remedies are examples of over-the-counter products used often for children with special health care needs. Constipation is a common condition in children with various neuromuscular conditions in which muscles are weak.[6] Parents tend to try over-the-counter remedies, dietary methods, and home remedies for consti-

pation management. The effectiveness of dietary fiber may be low when muscle weakness is an underlying problem.[3] Both overtreatment and undertreatment can get the child in trouble by worsening the constipation problem. A young child recently died as a result of poisoning by a laxative product administered at a higher dose than recommended.[33] Effective prescription medications for constipation management are available, but the family has to bring the problem to the attention of the health care provider. Encouraging the family to discuss the problem with the physician before trying over-the-counter products is important for many children.

Sources of Nutrition Services

Infants and toddlers who have chronic conditions are served by a variety of resources. Registered dietitians who have training in pediatrics are qualified to provide services to toddlers and preschool children with chronic conditions. Programs in which nutrition care may be accessed include the following:[34]

- State programs for children with special health care needs
- Early intervention programs (age 0 up to 36 months)
- Early childhood education programs (IDEA, ages 3–5 years)
- Head Start; regular program or special needs category (ages 3–5 years)
- Early Head Start; regular program or special needs category (0 up to 36 months)
- WIC
- Low birthweight follow-up programs
- Child care feeding programs (USDA)

These programs are described in Chapters 8 and 10. Efforts to increase program accessibility come from state and federal governmental offices, toll-free outreach services, and Web sites. Specific outreach programs to locate toddlers and preschoolers at risk are funded in each state, under names such as "Child Find."[4] Because every child at risk is eligible for a screening, contacting a neighborhood public school is a good starting place to locate services, even if the child is not old enough to attend the school.

Resources

Federal Interagency Coordinating Council Site for Families with Children with Disabilities
Identifies, by state and city, resources for finding local intervention programs.
Web site: **www.fed-icc.org**

Federation for Children with Special Needs
Includes support for families of children with special needs.
Web site: **www.fcsn.org**

Food Allergy

A credible source of recommendations for preventing food allergy reactions; the newsletter provides recipes to avoid foods that cause reactions.
Web site: **www.foodallergy.org**

National Information Center for Children and Youth with Disabilities

A useful site for parents and providers who are looking for intervention services for children with special needs. It is targeted mainly toward educational programs.
Web site: **www.NICHCY.org**

National Organization for Rare Diseases (NORD)

A credible source for parents and providers of information and resources about rare "orphan" diseases.
Web site: **www.rarediseases.org**

Quackwatch

Includes information on dietary supplements and products that are claimed to benefit health and nutrition.
Web site: **www.quackwatch.com**

References

1. Westbrook LE, Silver EJ, Stein, REK. Implications for estimates of disability in children: a comparison of definitional components. Pediatrics 1998;101:1025–30.

2. Newacheck PW, Taylor WR. Childhood chronic illness, prevalence, severity, and impact. Amer J of Public Health 1992; 82:364–71.

3. Staiano A. Food refusal in toddlers with chronic diseases. J Ped Gastro Nutr 2003; 37:225–27.

4. National Information Center for Children and Youth with Disabilities. General information about disabilities. Available at www.nichcy.org/general.htm, accessed 8/19/03.

5. Story M, Holt K, Sofka D, eds. Bright futures in practice: nutrition. Arlington, VA: National Center for Education in Maternal and Child Health; 2000: 266–70.

6. Beker LT, Farber AF, Yanni CC. Nutrition and children with disabilities. In: Batshaw ML, ed. Children with disabilities. 5th ed. Baltimore, MD: Paul H. Brookes; 2002: 141–64.

7. Trumbo P, Yates AA, Schlicker SA, Poos M. Dietary reference intakes: vitamin A, vitamin K, arsenic, boron, chromium, copper, iodine, iron, manganese, molybdenum, nickel, silicon, vanadium, and zinc. J Amer Diet Assoc 2001;101:294–301.

8. Eicher PS. Feeding. In: Batshaw, ML ed. Children with disabilities. 5th ed. Baltimore, MD: Paul H. Brookes; 2002: 549–66.

9. National Center for Health Statistics. NCHS growth curves for children 0–19 years. U.S. Vital and Health Statistics, Health Resources Administration. Washington, DC: U.S. Government Printing Office; 2000.

10. The Infant Health and Development Program: enhancing the outcomes of low-birthweight, premature infants. JAMA 1990;263(22):3035–42.

11. Anneren G, Gustafsson J, Sara VR, Tuvemo T. Normalized growth velocity in children with Down's syndrome during growth hormone therapy. J Intel Disabil Res 1993;37l:381–7.

12. Wollmann HA, Schultz U, Grauer ML, Ranke MB. Reference values for height and weight in Prader-Willi syndrome based on 315 patients. Eur J Pediatr 1998;157: 634–42.

13. Nellhaus, G. Composite international and interracial graphs. Pediatrics 1968;41:106–14.

14. Isaacs JS, Murdock M, Lane J, Percy AK. Eating difficulties in girls with Rett syndrome compared with other developmental disabilities. J Amer Diet Assoc 2003:103(2); 224–30.

15. American Academy of Pediatrics clinical practice guideline: diagnosis and evaluation of the child with attention deficit/hyperactivity disorder. Pediatrics 2000;105:1158–70.

16. American Academy of Pediatrics. The use and misuse of fruit juice in pediatrics. Pediatrics 2001;107:1210–13.

17. Black RE, Williams SM, Jones IE, Goulding A. Children who avoid drinking cow milk have low dietary calcium intakes and poor bone health. Am J Clin Nutr 2002: 76:675–80.

18. Bott, L. Contamination of gastrostomy feeding systems in children in a home-based enteral nutrition program. J Pediatr Gastroentol 2001;33:266–70.

19. Orelove FP, Hollahan DJ, Myles KT. Maltreatment of children with disabilities: training needs for a collaborative response. Child Abuse & Neglect 2000;24:185–94.

20. Berkovitch M, Heyman E, Afriat R, Matz-Khromchenko I et al. Copper and zinc blood levels among children with nonorganic failure to thrive. J Clin Nutr 2003; Apr;22(2):183–6.

21. Jennings JSR, Howdle PD. New developments in celiac disease. Curr Opin Gastroenterol 2003;19:118–29.

22. Cunningham E, Marcason W. Is there any research to support a gluten- and casein-free diet for a child that is diagnosed with autism? J Amer Diet Assoc 2001;101:222.

23. Minns RA. Neurological disorders. In: Kelnar CJH, Savage MO et al., eds. Growth disorders. Chapman and Hall; 1998: 447–70.

24. Krick J, Murphy-Miller P, Zeger S, Wright, E. Pattern of growth in children with cerebral palsy. J Amer Diet Assoc 1996;96:680–5.

25. Wooldridge, N. Pulmomary diseases. In: Samour PQ, Kelm KK, and Lang CE. Handbook of pediatric nutrition, 2nd ed. Gaithersburg, MD: Aspen Publishers: 1999: 315–54.

26. Surveillance for Asthma—United States, 1980–1999. MMWR 2002;51:SS1–10.

27. Falci KJ, Gombas KL, Elliot EL. Food allergen awareness: an FDA priority. Food Safety Magazine 2001:Feb.–March.

28. Batshaw ML. Chromosomes and heredity. In: Batshaw, ML ed. Children with disabilities. 5th ed. Baltimore, MD: Paul H. Brookes; 2002: 3–26, 770.

29. Bishop JM, Hill DJ, Hosking CS. Natural history of cow milk allergy: clinical outcome. J Pediatrics 1990;116:862–7.

30. Position statement on vitamin-related therapies. National Down Syndrome Society; 1997. www.ndss.org/content.cfm, accessed 4/04.

31. Trissler RJ. Folic acid and Down syndrome. J Amer Diet Assoc 2000;159.

32. Statement on nutritional supplements and piracetam for children with Down Syndrome. American College of Medical Genetics; 1996. www.acmg.net/resources/policies/pol-006.asp, accessed 4/04.

33. McGuire J, Kulkarni M, Baden H. Fatal hypermagnesemia in a child treated with megavitamin/megamineral therapy. J Pediatr 2000;105:318.

34. Maternal and Child Health Bureau, Health and Human Services. Available at http://mchb.hrsa.gov/html/drte.html, accessed 8/19/03.

"Men are but children of
a larger growth."
John Dryden

Chapter 12

Child and Preadolescent Nutrition

Chapter Outline

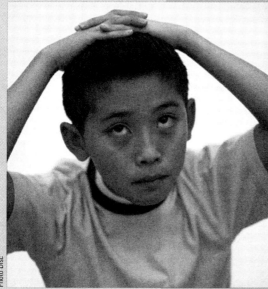

Prepared by **Nancy H. Wooldridge**

Key Nutrition Concepts

1 Children continue to grow and develop physically, cognitively, and emotionally during the middle childhood and preadolescent years in preparation for the physical and emotional changes of adolescence.

2 Children continue to develop eating and physical activity behaviors that affect their current and future states of health.

3 Although children's families continue to exert the most influence over their eating and physical activity habits, external influences, such as teachers, coaches, peers, and the media, begin to have more impact on children's health habits.

4 With increasing independence, children begin to eat more meals and snacks away from home and need to be equipped to make good food choices.

Introduction

This chapter focuses on the growth and development of school-age and preadolescent children and their relationships to nutritional status. Children continue to grow physically at a steady rate during this period, but development from a cognitive, emotional, and social standpoint is tremendous. This period in a child's life is preparation for the physical and emotional demands of the adolescent growth spurt. Having family members, teachers, and others in their lives who model healthy eating and physical activity behaviors will better equip children for making good choices during adolescence and later in life.

Middle Childhood   Children between the ages of 5 and 10 years; also referred to as school-age.

Preadolescence   The stage of development immediately preceding adolescence; 9 to 11 years of age for girls and 10 to 12 years of age for boys.

Working-Poor Families   Families where at least one parent worked 50 or more weeks a year and the family income was below the poverty level.

High-Poverty Neighborhoods   Neighborhoods where 40% or more of the people are living in poverty.

Definitions of the Life-Cycle Stage

Middle childhood is a term that generally describes children between the ages of 5 and 10 years. This stage of growth and development is also referred to as school-age, and the two terms are used interchangeably in this chapter. *Preadolescence* is generally defined as ages 9 to 11 years for girls and ages 10 to 12 years for boys. School-age is also used to describe preadolescence.

Importance of Nutrition

Adequate nutrition continues to play an important role during the school-age years in assuring that children reach their full potential for growth, development, and health. Nutrition problems can still occur during this age, such as iron-deficiency anemia, undernutrition, and dental caries. Regarding weight, both ends of the spectrum are seen during this age. The prevalence of obesity is increasing, but the beginnings of eating disorders can also be detected in some school-age and preadolescent children. Therefore, adequate nutrition and the establishment of healthy eating behaviors can help to prevent immediate health problems as well as promote a healthy lifestyle, which in turn may reduce the risk of the child developing a chronic condition, such as obesity, type 2 diabetes, and/or cardiovascular disease later in life.[1] Adequate nutrition, especially eating breakfast, has been associated with improved academic performance in school and reduced tardiness and absences.[2] Meeting energy and nutrient needs, addressing common nutrition problems, and preventing nutrition-related disorders while establishing healthy eating and physical activity habits will be discussed later in this chapter.

Tracking Child and Preadolescent Health

Not all families benefited from the economic boom of the 1990s. Statistics indicate that the conditions under which many children are growing up are alarming. Approximately 22% of U.S. children live in poverty, which is one of the highest poverty rates in the world.[3] Twelve percent of children had no health insurance in 2000.[3] Additional statistics include:

- 17.6 million children had no parent in the household who had a full-time, year-round job in 2000.
- 6.9 million children lived in *working-poor families*.
- 28% of families with children were headed by a single parent in 2000.
- 4% of children lived with their fathers only in 2001.
- 22% of children lived with their mothers only in 2001.
- 4% of children lived with neither parent in 2001.
- 23% of children lived in *high-poverty neighborhoods*.[3]

The environment in which a child lives affects the child's health and education. In 2000, 36% of fourth-grade students did not achieve the expected level of knowledge about science.[3] Lack of transportation is a significant limitation for many families. In 2001, 7% of all children younger than 18 years lived in a family that did not own a car or other vehicle.[3] In the discussions that follow regarding nutrition during childhood, the recommendations must always be considered in the context of the individual child's environment.

Disparities in nutrition status indicators exist among different races and ethnic groups. For example:

- African American and Mexican American girls have higher average body mass index (BMI)-for-age than do Caucasian girls.

- Overweight and obesity are more prevalent among African Americans and Hispanics of both genders than whites.
- Minorities have higher percentages of total calories from dietary fat.[4]

Healthy People 2010

A number of objectives in the Healthy People 2010 document are specific to children's health and well-being. Table 12.1 lists the specific Healthy People 2010 objectives that are pertinent to a discussion of middle childhood and preadolescence.

Normal Growth and Development

During the school-age years, a child's growth is steady; but the growth velocity is not as great as it was during infancy or as great as it will be during adolescence. The average annual growth during the school years is 7 pounds (3–3.5 kg) in weight and 2.5 inches (6 cm) in height.[5] Children of this age continue to have spurts of growth that usually coincide with periods of increased appetite and intake. During periods of slower growth, the child's appetite and intake will decrease. Parents should not be overly concerned with this variability in appetite and intake in their school-age children.

Periodic monitoring of growth continues to be important in order to identify any deviations in the child's growth pattern. Children should continue to be weighed on calibrated scales without shoes and in lightweight clothing. The child's stature or standing height should be measured without shoes and utilizing a height board (see Illustration 10.3 in Chapter 10). A height board consists of a nonstretchable tape on a flat surface like a wall with a moveable right-angle head board. The child's heels should be up against the wall or flat surface; and the child should be instructed to stand tall, looking straight ahead with arms by the sides, during the measurement. Both weight and height should be plotted on the appropriate 2000 CDC growth charts, discussed next.

The 2000 CDC Growth Charts

The "CDC Growth Charts: United States," found in Appendix A, are excellent tools for monitoring the growth of a child.[6] The growth charts, which are pertinent to the school-age child, are weight-for-age, stature-for-age, and body mass index (BMI)-for-age for boys and girls. The growth charts are based on data from cycles 2 and 3 of the National Health and Examination Survey (NHES) and the National Health and Nutrition Examination Surveys (NHANES) I, II, and III. However, weight data for children greater than 6 years of age who participated in NHANES III were not included, because there was a known higher

Table 12.1 Healthy People 2010 objectives related to school-age children[4]

Objective 19-3	Reduce the proportion of children and adolescents who are overweight or obese from 11% to 5%.
Objective 19-5	Increase the proportion of persons aged 2 years and older who consume at least two daily servings of fruit from 28% to 75%.
Objective 19-6	Increase the proportion of persons aged 2 years and older who consume at least three daily servings of vegetables, with at least one-third being dark green or deep yellow vegetables, from 3% to 50%.
Objective 19-7	Increase the proportion of persons aged 2 years and older who consume at least six daily servings of grain products, with at least three being whole grains, from 7% to 50%.
Objective 19-8	Increase the proportion of persons aged 2 years and older who consume less than 10% of calories from saturated fat from 36% to 75%.
Objective 19-9	Increase the proportion of persons aged 2 years and older who consume no more than 30% of calories from fat from 33% to 75%.
Objective 19-10	Increase the proportion of persons aged 2 years and older who consume 2400 mg or less of sodium daily from 21% to 65%.
Objective 19-11	Increase the proportion of persons aged 2 years and older who meet dietary recommendations for calcium from 46% to 75%.
Objective 19-15	Increase the proportion of children and adolescents aged 6 to 19 years whose intake of meals and snacks at schools contributes proportionally to good overall dietary quality (developmental objective).
Objective 22-8	Increase the proportion of the nation's public and private schools that require daily physical education for all students from 17% to 25%.
Objective 22-11	Increase the proportion of children and adolescents who view television 2 or fewer hours per day from 60% to 75%.
Objective 22-14	Increase the proportion of trips made by walking from 28% to 50%.
Objective 22-15	Increase the proportion of trips made by bicycling from 2.2% to 5.0%.

Illustration 12.1 Age 2 to 20 years: Girls' stature-for-age and weight-for-age percentiles.[6]

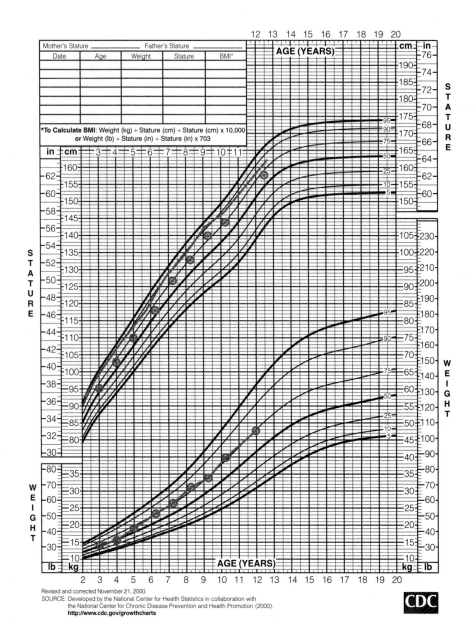

Revised and corrected November 21, 2000.
SOURCE: Developed by the National Center for Health Statistics in collaboration with the National Center for Chronic Disease Prevention and Health Promotion (2000).
http://www.cdc.gov/growthcharts

prevalence of overweight for these ages. Incorporating this data into the growth charts would reflect an unhealthy standard.[7] Gender-specific BMI-for-age greater than or equal to the 95th percentile defines overweight, and BMI-for-age values greater than or equal to the 85th but less than the 95th percentiles identify children at risk for becoming overweight.

Illustrations 12.1 and 12.2 depict the growth of a healthy child. A chart for weight-for-stature up to a height of 122 cm (or 48 in.) is also available for the younger school-age child. As with the toddler and preschooler, it is the child's pattern of growth over time that is important rather than any single measurement. The tracking of BMI-for-age is an important screening tool for overweight as well as undernutrition. Making sure to use the correct age of the child when plotting on the growth charts, and using the most current growth curves, will help to avoid errors.

Physiological and Cognitive Development of School-Age Children

Physiological Development

During middle childhood, muscular strength, motor coordination, and stamina increase progressively.[5] Children are able to perform more complex pattern movements, therefore affording them opportunities to participate in activities such as dance, sports, gymnastics, and other physical activities.

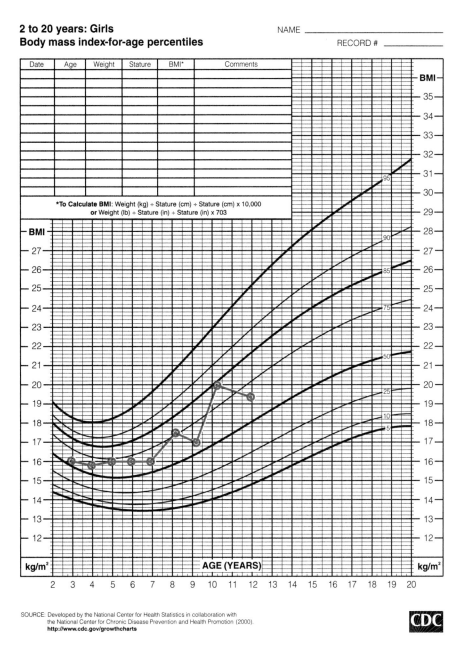

2 to 20 years: Girls
Body mass index-for-age percentiles

NAME _____

RECORD # _____

To Calculate BMI: Weight (kg) ÷ Stature (cm) ÷ Stature (cm) x 10,000 or Weight (lb) ÷ Stature (in) ÷ Stature (in) x 703

SOURCE: Developed by the National Center for Health Statistics in collaboration with the National Center for Chronic Disease Prevention and Health Promotion (2000).
http://www.cdc.gov/growthcharts

Illustration 12.2 Age 2 to 20 years: Girls' body mass index-for-age percentiles.[6]

During the early childhood years, percent body fat reaches a minimum of 16% in females and 13% in males. Percent body fat then increases in preparation for the adolescent growth spurt. This increase in percent body fat, which usually occurs on average at 6.0–6.3 years of age, is called *adiposity rebound* or *BMI rebound* and is reflected in the BMI-for-age growth charts.[8] The increase in percent body fat with puberty is earlier and greater in females than in males (19% for females versus 14% for males). During middle childhood, boys have more lean body mass per centimeter of height than girls. These differences in body composition become more pronounced during adolescence.[1] It is important to understand that BMI is not constant throughout childhood. Plotting BMI-for-age on the growth charts is the only way to know if a child's BMI is outside the normal range for his age. The goal for children is not to strive for a certain range of BMI values, as it is in adults, but rather to have a BMI-for-age percentile within the normal range.

With the increase in body fat, preadolescents, especially girls, may be concerned that they are becoming overweight. Parents need to be aware that an increase in body fat during this stage is part of normal growth and development. Parents need to be able to reassure the child that these changes are most likely not permanent; parents also need to be careful not to reinforce a preoccupation with weight and size. Boys may become concerned about developing muscle mass and need to understand that they will not be able to increase their muscle mass until middle adolescence (see Chapter 14).[1]

Cognitive Development

The major developmental achievement during middle childhood is self-efficacy, the knowledge of what to do and the ability to do it. During the school-age years, children move from a preoperational period of development to one of "concrete operations."[5] This stage is characterized by being able to focus on several aspects of a situation at the same time; being able to have more rational cause-effect reasoning; being able to classify, reclassify, and generalize; and a decrease in egocentrism, which allows the child the ability to see another's point of view. Schoolwork becomes increasingly complex as the child gets older. School-age children also enjoy playing strategy games, displaying growing cognitive and language development.

During this stage, the child is developing a sense of self. Children become increasingly independent and are learning their roles in the family, at school, and in the community.[1] Peer relationships become increasingly important, and children begin to separate from their own families by spending the night at a friend's or relative's house. More and more time is spent watching television and playing video games. Older children may be able to walk or ride a bicycle to a neighborhood store and purchase snack items. Thus, influences outside the home environment play an increasing role in all aspects of the child's life.

Development of Feeding Skills

With increased motor coordination, school-age children develop increased feeding skills. During childhood, the child masters the use of eating utensils, can be involved in simple food preparation, and can be assigned chores related to mealtime such as setting the table. By performing these tasks, the child learns to contribute to the family, which boosts developing self-esteem. Complexities of the tasks can be increased as the child grows older. At the same time, the child can be learning about different foods, simple food preparation, and some basic nutrition facts.

EATING BEHAVIORS Parents and older siblings continue to have the most influence on a child's attitudes toward food and food choices during middle childhood and preadolescence. The eating behaviors and cultural food practices and preferences of parents will influence the child's food likes and dislikes. The feeding relationship between parent and child, as described in Chapter 10, still applies to the school-age child. Parents are responsible for the food environment in the home, what foods are available, and when they are served. The child is responsible for how much she eats.[9] Parents need to continue to be positive role models for their children in displaying healthy eating behaviors. They also need to provide the necessary guidance so that the child will be able to make healthy food choices when she is away from home.

Families should try to eat meals together. When children are involved in school-related activities, eating together is often difficult for families to achieve because of the family members' hectic schedules. But eating together as a family should be encouraged as a goal, allowing time for conversation (see Illustration 12.3). Excessive reprimanding and arguments should be avoided during mealtime.

One study of 9- to 14-year-old children of participants of the Nurses' Health Study II found a positive rela-

Illustration 12.3 A family enjoying mealtime together.

tionship between families eating dinner together and the overall quality of the children's diets.[10] Children who ate dinner with their families had higher energy intakes as well as higher intakes of nutrients such as fiber, calcium, folate, iron, and vitamins B_6, B_{12}, C, and E. These children also reported eating more fruits and vegetables, eating less fried foods when away from home, and drinking fewer soft drinks. The percentage of children who reported eating family dinner decreased with the age of the child. So, a higher percentage of 9-year-olds than 14-year-olds ate dinner with the family, indicating that family dinner becomes more of a challenge as children get older.

School-age children spend more and more time away from home, which is an important part of normal growth and development. Peer influence becomes greater as the child's world expands beyond the family. The increased peer influence extends to attitudes toward foods and food choices. Children may suddenly request a new food or refuse a previous favorite food, based on recommendations from a peer.

Teachers and coaches have an increasing influence on the child's attitudes toward food and eating behaviors. Nutrition should be part of the health curriculum, and what is learned in the classroom should be reinforced by foods available in the school cafeteria. Vending machines present in school as a source of extra funding can also either reinforce good nutrition with appropriate choices or be a source of high-fat, high-sugar foods and beverages.

In their expanding world, children come under the influence of the media. Children want to try foods they see advertised on television. One study that analyzed the commercials aired during Saturday morning television programming found that 56.5% of all advertisements were for food.[11] Of these, 43.6% were classified in the fats, oils, and sweets food group, which is just the opposite of recommendations of the U.S. Department of Agriculture's Food Guide Pyramid (see Chapter 1).[11,12] Fast-food establishments, with their playgrounds and give-aways, are also attractive to children.

Snacks continue to contribute significantly to a child's daily intake. During middle childhood, children cannot consume large amounts of food at one time and therefore need snacks to meet their nutrient needs. Many children prepare their own breakfasts or after-school snacks. These children need to have a variety of foods available to them, be equipped with nutrition education for making their own food choices, and have some age-appropriate knowledge and skill in food preparation—assuming, of course, that the family has adequate access to food.

BODY IMAGE AND EXCESSIVE DIETING Food-preference development, appetite, and satiety in young children were thoroughly discussed in Chapter 10. Researchers have described the innate ability of young children to internally control their energy intake and their responsiveness to energy density. The internal controls can be altered by external factors such as child-feeding practices. Studies in 9- to 10-year-old children found that these older children were not as responsive to energy density as were young preschool-age children.[13] External factors such as the time of day, the presence of other people, and the availability of good food begin to override the internal controls of hunger and satiety as children get older.

Birch and associates, who have performed extensive research in the area of the development of food preferences and appetite control in children, have also examined the relationships among children's adiposity, child-feeding practices, and children's responsiveness to energy density.[14] These researchers found that children of parents who imposed authoritarian controls on their children's eating were less likely to be responsive to energy density. In other words, these children were not able to listen to internal cues in energy regulation. In girls, regulation of energy intake was inversely related to their adiposity. Heavier girls were less likely to be able to regulate their intake based on internal cues. Parents who had difficulty controlling their own intakes seemed to impose more restrictions on their children. A study of mothers and their 5-year-old daughters found that this transfer of "restrictive" eating practices may begin as early as the preschool age.[15] The more the mother is concerned with her own weight and with the risk of her daughter becoming overweight, the more likely she is to employ restrictive child-feeding practices. These researchers hypothesize that chronic dieting and dietary restraints, which are commonly seen in adolescent girls and young women, may have their beginnings in the early regulation of energy intake and may be related to the amount of parental control exerted over the child's eating.[14]

Young girls seem to have a preoccupation with weight and size at an early age. With the normal increase in BMI or body fatness in preadolescence, many girls and their mothers may interpret this phenomenon of normal growth and development as a sign that the child is developing a weight problem. By imposing controls and restrictions over their daughters' intakes, mothers may actually be promoting the intake of the forbidden or restricted foods.[16] Similar results were found in 5-year-old girls whose parents restricted palatable snack foods.[17] Not only did parental restriction promote the consumption of these forbidden foods by the young girls, but these children reported feeling bad about eating these foods. Early "dieting" may actually be a risk factor for the development of obesity.[18] Dieting, which imposes restrictions, is similar to controlling child-feeding practices, which restrict children's intake. Both methods ignore internal cues of hunger and satiety. Not only do these types of child-feeding practices contribute to the onset of obesity and possibly a nutritionally inferior diet, but they may also be contributing to the beginnings of eating disorders. Eating disorders are discussed in more detail in Chapter 14.

Many studies of ethnic differences in body image and body size preferences have been conducted, hypothesizing that there are ethnic and gender differences in these parameters. The research conducted to date is not conclusive. A recent study of men and women of four ethnicities/races—black, Hispanic, Asian, and white—was conducted to examine body image and body size assessments, while controlling for age, body weights, and education level as a measure of socioeconomic status.[19] This study found that ethnicity alone did not influence the preference for body shapes or tolerance for obesity. In working with individual families, however, it is important to try to assess their health beliefs and their preference for body size, which may affect their readiness for nutrition counseling.

Energy and Nutrient Needs of School-Age Children

New Dietary Reference Intakes (DRIs) have recently been developed (1997–2004). DRI tables are provided on the inside front cover of this book. Children need a variety of foods that provide enough energy, protein, carbohydrate, fat, vitamins, and minerals for optimal growth and development.[1]

Energy Needs

Energy needs of school-age children reflect the slow but steady growth rate during this stage of development. Energy needs of an individual child are dependent on the child's activity level and body size. Equations for estimating energy requirements have been developed as part of the Dietary Reference Intakes, based on a child's gender, age, height, weight, and physical activity level (PAL).[20] Estimated energy expenditure (EER) has been defined as total energy expenditure plus kilocalories for energy deposition. Categories of activity are defined in terms of walking equivalence. For example, an 8-year-old girl who weighs 56.4 pounds (25.6 kg) and is 50.4 inches (128 cm) tall will require 1360 kilocalories per day if she is sedentary, 1593 kcal/day if low active, 1810 kcal/day if active, and 2173 kcal/day if very active. Energy allowances based on body weight are lower for school-age children than for toddlers and preschoolers. The decrease in the energy requirement per kilogram of body weight is a reflection of slowing growth rate.

Protein

Based on the new DRIs, the recommended protein intake for school-age children is 0.95 gram of protein/kg body weight/day for 4- to 13-year-old girls and boys.[20] Like younger children, school-age children can meet this recommendation by consuming diets that follow the Food Guide Pyramid recommendations for children.[12] Vegetarian diets are also appropriate for school-age children if they provide sufficient energy, complementary protein foods, a variety of foods, and adequate levels of intake of

vitamins and minerals.[1] By meeting an individual child's energy needs, protein is spared for tissue repair and growth.

Vitamins and Minerals

Dietary Reference Intakes (DRIs) for vitamins and minerals have been established for the school-age and preadolescent child. Analysis of data from NHANES I, II, and III and the Continuing Survey of Food Intake by Individuals (CSFII) indicates that children's mean intakes of most nutrients meet or exceed the recommendations. Still, certain subsets of children do not meet their needs for key nutrients such as iron and zinc, which are important for growth; and calcium, needed to achieve peak bone mass.[21,22] According to NHANES III data, calcium intakes are declining in 6- to 11-year-old children.[4] Dietary Reference Intakes for these key nutrients are listed in Table 12.2.[23,24]

Common Nutrition Problems

During the last century, common nutrition problems have shifted from problems of nutrient deficiencies to problems of excess nutrition, such as energy, fat, and salt. During middle childhood, some children still experience problems such as iron-deficiency anemia and dental caries, especially with easy accessibility to high-sugar foods. These nutrition problems are addressed here, followed by a thorough discussion of prevention of nutrition-related disorders.

Iron Deficiency

Iron deficiency is not as common in middle childhood as it is in the toddler age group. According to the NHANES 1999–2000 survey data, 4% of 6- to 11-year-old children were found to be iron deficient, compared to 7% of toddlers.[25] Although the prevalence of iron-deficiency anemia is decreasing, these rates are still above the 2010 national health objectives.

Age- and gender-specific cutoff values for anemia are based on the 5th percentile of hemoglobin and hematocrit for age from NHANES III. For children from 5 to under

Table 12.2 Dietary Reference Intakes for key nutrients for school-age children[23,24]		
RECOMMENDED DIETARY ALLOWANCES		**ADEQUATE INTAKE**
AGE	Iron, mg/d Zinc, mg/d	Calcium, mg/d
4–8 years	10 5	800
9–13 years	8 8	1300

8 years of age, the diagnosis of anemia is made if the hemoglobin concentration is <11.5 g/dL and hematocrit <34.5%. For children from 8 to under 12 years of age, a hemoglobin value <11.9 g/dL or hematocrit <35.4% is diagnostic of iron-deficiency anemia.[25,26]

Dental Caries

Approximately one in two children aged 6 to 8 years has decay in his primary or permanent teeth.[4] The amount of time that children's teeth are exposed to carbohydrates influences the risk of dental caries or tooth decay. (See the explanation of the carogenic process in Chapter 10.) Complex carbohydrates such as fruits, vegetables, and grains are better choices than simple sugars, such as soft drinks and candy, in relation to oral health and nutrition. Sticky carbohydrate-containing foods, such as raisins and gummy candy, are strong caries promoters. Fats and proteins may have a protective effect on enamel. So choosing snacks that are combinations of carbohydrates, proteins, and fats may decrease the risk of developing dental caries. Having regular meal and snack times versus continually snacking throughout the day is also beneficial. Rinsing the mouth after eating—or better yet, brushing the teeth regularly—also decreases the development of cavities.[27] It is important that the school-age child continue to have a source of fluoride, either from the water supply or through supplementation. Details about fluoride supplementation were reported in Chapter 10.

During middle childhood, children lose their primary or baby teeth and begin to get their permanent teeth. If several teeth are missing, children may experience difficulty in chewing some foods, such as meat. Also, the orthodontic appliances commonly worn by school-age children may interfere with the child's ability to eat certain foods. Modifying food, such as by chopping meat or slicing fresh fruit, can help.[1]

Prevention of Nutrition-Related Disorders in School-Age Children

The prevalence of overweight among children is increasing at an alarming rate. Increases in prevalence of overweight are present in the adult population and in populations of other countries, indicating that social and environmental factors may be having an impact. Despite the increase in the prevalence of overweight, analysis of dietary data from NHANES I, II, and III indicates no corresponding increase in energy intake among children over the years. This finding suggests that physical inactivity may be a significant contributing factor to the increased prevalence of overweight.[28] The problem of increasing overweight in the United States needs to be addressed from a public health perspective.[7] Furthermore, children

who are overweight are at increased risk for developing risk factors for chronic conditions, such as cardiovascular disease and type 2 diabetes mellitus.[29]

Overweight and Obesity in School-Age Children

PREVALENCE   According to the NHANES 1999–2000 data, approximately 15% of children aged 6 through 11 years are overweight, with BMIs-for-age greater than or equal to the 95th percentile; an additional 30% are considered to be at risk for overweight, with BMIs greater than or equal to the 85th but less than the 95th percentiles.[30] These percentages have increased significantly since the NHANES III study (1988–1994), when 11% of children aged 6 through 11 years were overweight and 14% were considered to be at risk for overweight. The proportion of children aged 6 through 11 years who are at risk for becoming overweight ranges from about 23% for non-Hispanic white females to approximately 43% for Mexican American males.[30] When the NHANES III data were controlled for socioeconomic status and age, BMI levels were significantly higher for black and Mexican American girls than for white girls.[31] These ethnic differences among girls first become apparent at 6 to 9 years of age. Black girls and boys had greater increases in BMI across age groups than whites did, resulting in greater differences in BMI in the older age groups. The prevalence of overweight and obesity among participants of the Child and Adolescent Trial for Cardiovascular Health (CATCH) is comparable to the NHANES III population data, with a higher prevalence among blacks and Hispanics than whites for both genders.[32]

Bone Age   Bone maturation; correlates well with stage of pubertal development.

The prevalence of overweight among children has increased over time, with the increases occurring between NHANES III (1988–1994) and NHANES 1999–2000 being similar to those occurring between NHANES II (1976–1980) and NHANES III (1988–1994).[30] In fact, an intrasurvey increase of about 2 to 6 percentage points occurred for most of the gender, age, and racial-ethnic groups during the 6 years of the NHANES III survey.[7] As mentioned in the description of the growth charts, BMI data for children older than 6 years of age were not included in the revised growth charts because of the known increased prevalence of overweight for these ages in NHANES III. Inclusion of this data in the revised growth charts would reflect a heavier population and would not be a healthy standard. Further analysis shows that the heaviest, or obese, children are getting heavier.

CHARACTERISTICS OF OVERWEIGHT CHILDREN   Overweight children are usually taller, have advanced *bone ages,* and experience sexual maturity at an earlier age than their nonoverweight peers. From a psychosocial standpoint, overweight children look older than they are, and often adults expect them to behave as if they were older. Health consequences of obesity, such as hyperlipidemia, higher concentrations of liver enzymes, hypertension, and abnormal glucose tolerance, occur with increased frequency in obese children as compared to children of normal weight.[33] Analysis of data from the Bogalusa Heart Study, a community-based study of adverse risk factors in early life in a biracial population, confirms an increase in chronic disease risk factors with increasing BMI-for-age.[29] Increasing insulin levels show the strongest association with increasing BMI-for-age. Additionally, overweight children are more likely to have more than one chronic disease risk factor.

> **Subscapular Skinfold Thickness** A skinfold measurement that can be used with other skinfold measurements to estimate percent body fat; the measurement is taken with skinfold calipers just below the inner angle of the scapula, or shoulder blade.

Type 2 diabetes mellitus, typically considered to be a disease of adults, is increasing in children and adolescents in the United States today, with up to 85% of affected children being either overweight or obese at diagnosis.[34] According to the recommendations of a panel of experts in diabetes in children, any child who is overweight—defined by this group as having a BMI above the 85th percentile—and who has other risk factors should be monitored for type 2 diabetes beginning at age 10 or at puberty. Other risk factors include a family history of type 2 diabetes; belonging to certain racial and ethnic groups including African American, Hispanic American, Asian and South Pacific Islander, and Native American; and having signs of insulin resistance.[34]

It is still unclear what effect an early onset of obesity in childhood has on the risk of adult morbidity and mortality.[33] But consequences of obesity and the precursors of adult disease do occur in obese children. More studies have been performed on the relationship between obesity during adolescence and the risks of obesity in adulthood than have been performed on the relationship between childhood obesity and obesity and Type 2 diabetes in adulthood (see Chapter 14).[35]

PREDICTORS OF CHILDHOOD OBESITY Dietz describes critical periods in childhood for the development of obesity: gestation and early infancy, the period of adiposity rebound, and adolescence.[8] "Adiposity rebound" (or rebound in BMI) is the normal increase in body mass index, which occurs after BMI declines and reaches its lowest point, at about 4 to 6 years of age, and is reflected in the BMI-for-age growth chart. Studies suggest that the age at which adiposity rebound occurs may have a significant effect on the amount of body fat that the child will have during adolescence and into adulthood. Early adiposity rebound is defined as beginning before 5.5 years of age, while the average age of adiposity rebound is 6.0–6.3 years. Adiposity rebound after age 7 is considered late. Studies have shown that adolescents and adults who as children had an early adiposity rebound have higher BMI and *subscapular skinfold thicknesses* than those subjects who had an average or late adiposity rebound. Three possible mechanisms explain the relationship between adiposity rebound and subsequent obesity.[36] The period of adiposity rebound may be when children are beginning to express learned behaviors related to food intake and activity. Early adiposity rebound may be related to infants who were exposed to gestational diabetes during fetal development and consequently have large birthweights. Although more study is needed, the conclusion is that preventive efforts need to focus on these developmental stages.[8]

Another predictor of childhood obesity is the child's home environment. Children from birth to 8 years were followed over a 6-year period as part of the National Longitudinal Survey of Youth.[37] The associations between the home environment and socioeconomic factors and the development of childhood obesity were examined. Maternal obesity was found to be the most significant predictor of childhood obesity, followed by low family income and lower cognitive stimulation.

Parental obesity is associated with an increased risk of obesity in children.[38] In one study, parental obesity doubled the risk of adult obesity for both obese and nonobese children less than 10 years of age. An analysis of data from NHANES III indicated a higher percentage of overweight youth who had one obese parent as compared to those children who had no obese parent. The percentage of overweight youth increased further if both parents were obese.[39] The connection between parental obesity and obesity in children is likely due to genetic as well as environmental factors.[38]

EFFECTS OF TELEVISION VIEWING TIME ON THE INCIDENCE OF OVERWEIGHT Analysis of data collected during cycles II and III of the National Health Examination Survey (NHES) revealed significant associations between the time spent watching television and the prevalence of obesity in children and adolescents.[40] A dose-response relationship was detected. For each additional hour of television viewed in the 12- to 17-year-old group, the prevalence of obesity increased by 2%.

A strong dose-response relationship between TV viewing time and the prevalence of overweight was found in the National Longitudinal Survey of Labor Market Experience, Youth Cohort (NLSY). This study consists of a nationally representative sample of youths aged 10 to 15 years.[41] The odds of having a BMI above the 85th percentile for age and gender are significantly greater for those youths who view more than 5 hours of television per day as compared to those who watch 2 or fewer hours of television daily. These odds remain the same when adjustments are made for confounding variables such as previous overweight of the child, maternal overweight, socioeconomic status, household structure, ethnicity, and child aptitude test scores. Approximately 33% of the youth report watching more than 5 hours of television per day, while only 11% watch 2 or fewer hours of daily television, which is a Healthy People 2010 objective.[4,41]

According to NHANES III data, children aged 11 through 13 years have the highest rates of daily television viewing.[42] Children, both males and females, who watch 4 or more hours of television daily have greater body fat and BMI than those who watch less television.[42,43] A school-based intervention program aimed at reducing third- and fourth-grade children's television, videotape, and video game use was shown to be effective in reducing television viewing and meals eaten in front of the television. In addition, decreases in body mass index, triceps skinfold, waist circumference, and *waist-to-hip ratio* were seen.[44]

The proposed mechanisms by which television viewing contributes to obesity include reduced energy expenditure by displacing physical activity and increased dietary intake by eating during viewing or as a result of food advertising.[44] Analysis of NHANES III data showed a positive correlation between intake and number of hours of television watched.[43] One study found that energy expenditure during television viewing was actually significantly lower than *resting energy expenditure* in 15 obese children and 16 normal-weight children who ranged in age from 8 to 12 years.[45] Based on these findings, it is hypothesized that television viewing does contribute to the prevalence of obesity, and that treatment for childhood obesity should include a reduction in the number of hours spent watching television and videos and playing video and computer games.

One of the Healthy People 2010 objectives is to increase the proportion of children and adolescents who view television 2 or fewer hours per day from 60% to 75%. Related data analyzed by race and ethnicity, gender, and family income level are depicted in Table 12.3.[4]

Table 12.3 Percentage of children and adolescents viewing television 2 or fewer hours per day by race/ethnicity, gender, and family income level[4]

Children and Adolescents Aged 8 to 16 Years, 1988–1994	Television 2 or Fewer Hours per Day
Race and Ethnicity	
Mexican American	53%
Black or African American	42
White	65
Gender	
Female	64
Male	54
Family Income Level	
Poor	53
Near poor	54
Middle/high income	64

Prevention of Overweight and Obesity

"An ounce of prevention is worth a pound of cure."

Recognizing the increase in the prevalence of childhood overweight and its associated chronic health problems, the American Academy of Pediatrics has issued a policy statement on the prevention of pediatric overweight and obesity.[46] The policy statement advocates for (1) early recognition, using BMI for age as a screening tool, and providing anticipatory and appropriate guidance regarding healthy eating and physical activity; and (2) advocacy for opportunities for physical activity, improvements in foods available to children, research, and third-party reimbursement for treatment of overweight.[46]

TREATMENT OF OVERWEIGHT AND OBESITY

Recommendations of an expert committee on obesity evaluation and treatment for children were described in Chapter 10.[47] The committee recommends that children and adolescents with BMIs greater than or equal to the 95th percentile for age and gender should have an in-depth medical assessment. For children older than 7 years of age, prolonged weight maintenance is an appropriate goal if their BMI is greater than or equal to the 85th but less than the 95th percentile and if they do not have any secondary complications of obesity. However, for children whose BMI is in the range just described and who have a nonacute secondary complication of obesity, such as mild hypertension or hyperlipidemia—or whose BMI is at or above the 95th percentile—weight loss is recommended. Reducing sedentary behaviors and increasing physical activity are important components of obesity treatment for children.[48] Rather than focusing on attaining an ideal body weight, treatment should focus on changing unhealthy lifestyle behaviors and maintaining healthy behaviors.[47] See Case Study 12.1 on the next page.

Some potential consequences of a weight-loss program in childhood are a slowing of linear growth and the beginnings of eating disorders. To reduce the risks associated with weight loss in childhood, the program must ensure nutritional adequacy of the diet, a nonjudgmental approach, and attention to the child's emotional state.[47]

> **Waist-to-Hip Ratio** The ratio of the waist circumference, measured at its narrowest, and the hip circumference, measured where it is widest. This ratio is an easy way to measure body fat distribution, with a higher ratio indicative of an abdominal fat pattern. A high waist-to-hip ratio is associated with a high risk of chronic disease.
>
> **Resting Energy Expenditure** The amount of energy needed by the body in a state of rest.

Nutrition and Prevention of Cardiovascular Disease in School-Age Children

In the new DRIs, no recommendations for total grams of fat per day in the diet of children have been made.[20] Studies have shown that as long as enough energy is provided for growth, no effect of fat intake on growth has been

Case Study 12.1

Pediatric Overweight

Photo Disc

Seven-year-old Timothy's mother takes him to his pediatrician for a checkup. His weight of 77 pounds, plotted at the 95th percentile, and his height of 52 inches are between the 75th and 90th percentiles for his age. His body mass index of 20 kg/m² is plotted at just over the 95th percentile for his age. His growth percentiles have been increasing over the last several years.

Timothy's mother expresses concern to the pediatrician about her son's weight. His older and younger brothers are both thinner than Timothy. Timothy's mother is obese, but his father is a normal weight for height. Timothy is in the second grade. He rides the school bus to and from school. He participates in the School Lunch Program at his school, but his parents gave him extra money in case he wants to buy some additional à la carte food items from the cafeteria or items from the vending machines. After school, Timothy and his brothers stay in their home with a babysitter until one of their parents returns home from work. Timothy usually watches TV or plays video games after school. His parents leave snack foods—chips, cookies, and sodas—in the house for their sons to have after school. His mother usually prepares their evening meal, which consists of a meat, starch, vegetables, and a dessert item. After dinner, Timothy does his homework and then usually watches more TV with his parents. He usually has a dish of ice cream before going to bed.

Questions

1. What is your assessment of Timothy's body size based on his weight-for-age, height-for-age, and BMI-for-age percentiles?
2. What suggestions do you have for Timothy's parents about improving his eating habits?
3. What suggestions do you have for Timothy's parents for increasing his physical activity level?
4. Is it significant that Timothy's mother also has a weight problem?

(Answers are in Appendix D.)

---

found. In addition, the evidence is still insufficient to be able to define the optimal fat intake for promoting growth while also preventing obesity and other chronic diseases. According to the DRIs, the Acceptable Macronutrient Distribution Range (AMDR) for fat is 25 to 35% of energy for children 4 to 18 years of age.[20]

The new DRIs do stress the importance of including sources of linoleic acid (omega-6 fatty acid) and alpha-linolenic acid (omega-3 fatty acid). Adequate intake levels for these essential fatty acids have been determined and can be found in Table 12.4. Sources of linoleic acid include vegetable oils, seeds, nuts, and whole grain breads and cereals. Fish, as well as flaxseed, soy, and canola oils, are good sources of alpha-linolenic acid. (See Chapter 1 for a complete list of food sources of linoleic acid and alpha-linolenic acid.) A diet that emphasizes fruits, vegetables, low-fat dairy products, whole grain breads and cereals, nuts, seeds, fish, and lean meats is recommended

for promoting nutrition and preventing cardiovascular disease in school-age children.[20]

The American Heart Association and the American Academy of Pediatrics have jointly issued guidelines for cardiovascular health promotion in all children and adolescents.[49,50] For children over 2 years of age, these guide-

Table 12.4 Adequate intake* of linoleic acid and alpha-linolenic acid[20]

GENDER AND AGE	LINOLEIC ACID, g/day	ALPHA-LINOLENIC ACID, g/day
Children 4–8 years	10	0.9
Boys 9–13 years	12	1.2
Girls 9–13 years	10	1.0

*See Chapter 1 for a list of food sources.
SOURCE: From the Institute of Medicine, Food & Nutrition Board.

lines recommend limiting foods high in satuarated fats (<10 percent of total calories per day), cholesterol (<300 mg per day), and trans fatty acids.[49,50]

As discussed in Chapter 10, children with hyperlipidemias require further dietary restrictions to help control LDL cholesterol. For these children, restricting saturated fat to less than 7% of total calories and cholesterol to 200 mg per day is recommended.[49–51] Restricting trans fatty acids to a level as low as possible is also important. Increasing soluble fiber intake, emphasizing weight management and physical activity, and follow-up by a registered dietitian are also treatment recommendations.[50]

Dietary Supplements

"For most healthy children who eat a variety of foods, experts generally agree that dietary supplements beyond a daily multivitamin and mineral supplement are not necessary, let alone safe or effective."[52]

Children who are healthy and consume a diet of a variety of foods do not require a vitamin and mineral supplement to meet their nutrient needs. The American Academy of Pediatrics recommends vitamin and mineral supplementation for children who are at high risk of developing, or have one or more, nutrient deficiencies.[53] (See Chapter 10 for a list of children at risk for nutrient deficiency.)

If vitamin and mineral supplements are given to school-age children, the supplement should not exceed the Dietary Reference Intakes for age. Parents should be warned against giving amounts of vitamins and minerals that exceed the Tolerable Upper Intake levels designated in the DRI tables.

It is not clear to what extent herbal supplements are given to school-age children. Herbal supplements are used in some cultures as home remedies. It is important to obtain this information from parents and caretakers as part of the child's health history. The use of herbal supplements, botanicals, and vitamin/mineral supplements may be a more prevalent practice by parents of children with special health care needs (see Chapters 11 and 13).

Dietary Recommendations

The basic dietary recommendation for school-age and preadolescent children is to eat a diet of a variety of foods, which is why it remains so important throughout these school years for children to have a variety of foods available to them. The available food environment will affect children's food choices. Parents and other adult role models need to continue to model appropriate eating behaviors for children.

Dietary recommendations, as outlined by the USDA in the Dietary Guidelines for Americans and the Food Guide Pyramid, apply to school-age children as well as to other segments of the population.[12,54] Professional organizations such as the American Dietetic Association have also published positions on dietary guidance for healthy children, supporting the federal guidelines.[55]

Recommendations for Intake of Iron, Fiber, Fat, and Calcium

Adequate iron nutrition is still important during middle childhood and preadolescence to prevent iron-deficiency anemia and its consequences. According to food consumption surveys, children are not eating the recommended amounts of fiber in their diets. Children are exceeding the recommendations of total calories from fat and saturated fat. Calcium requirements increase during the preadolescent years, but calcium intake decreases with age.

IRON Although iron deficiency is not as prevalent during the school-age years as it was during the toddler and preschool-age years, adequate intake of iron is still important. The inclusion of iron-rich foods—such as meats, fortified breakfast cereals, and dry beans and peas—in children's diets is important. A good vitamin C source, such as orange juice, will enhance the absorption of iron. (See Chapter 1 for a more complete list of high-iron foods.)

> **Total Fiber** Sum of dietary fiber and functional fiber.
>
> **Dietary Fiber** Complex carbohydrates and *lignins* naturally occurring and found mainly in the plant cell wall. Dietary fiber cannot be broken down by human digestive enzymes.
>
> **Lignin** Noncarbohydrate polymer that contributes to dietary fiber.
>
> **Functional Fiber** Nondigestible carbohydrates including plant, animal, or commercially produced sources, that have beneficial effects in humans.

FIBER As reported in Chapter 10, many health effects of fiber intake have been identified—including prevention of chronic disease in adulthood, such as heart disease, certain cancers, diabetes, and hypertension. The new recommendations for *total fiber* intake based on the DRIs can be found in Table 12.5. Total fiber is the sum of *dietary fiber* and *functional fiber*. Earlier reommendations were based on dietary fiber.[20]

To increase the dietary fiber in children's diets, parents and caretakers can begin by increasing the amount of fresh fruits and vegetables and whole grain breads and cereals being offered. High-fiber fruits, such as apples with peels, have about 3 grams per serving, while fruit juices are low in fiber. High-fiber vegetables, such as broccoli, have about 2.5 grams per serving. Whole grain breads, cereals, and brown rice have about 2.5 grams per serving. High-fiber cereals, such as bran flakes and raisin bran, have about 8 to 10 grams per serving. Served alone, these high-fiber cereals may not be well accepted by young children; but they can be mixed with other cereals or used

Table 12.5 Adequate intake of total fiber[20]

GENDER AND AGE	TOTAL FIBER, g/day
Children 4–8 years	25
Boys 9–13 years	31
Girls 9–13 years	26

SOURCE: From the Institute of Medicine, Food & Nutrition Board.

in recipes for food items such as muffins. Dried beans and peas are also excellent sources of fiber, providing 4 to 7 grams of fiber per half-cup serving.[56]

FAT   Food intakes that follow recommendations of the Dietary Guidelines for Americans and the Food Guide Pyramid provide an appropriate amount of fat for school-age and preadolescent children.[12,54] Healthy diets include whole grain breads and cereals, beans and peas, fruits and vegetables, low-fat dairy products, and lean meats, fish, and poultry. Foods high in fat, especially those high in saturated fat and trans fatty acids, should be kept to a minimum. However, an appropriate amount of dietary fat is necessary to meet children's needs for calories, essential fatty acids, and fat-soluble vitamins.

CALCIUM   The recommendations for adequate daily intakes of calcium are 800 milligrams for children aged 4 to 8 years and 1300 mg for children 9 through 18 years.[23] The higher recommendation for older children reflects the fact that most bone formation occurs during puberty. Adequate calcium intake during this time is necessary to achieve peak bone formation, which may prevent osteoporosis later in life.[57]

Good sources of calcium are listed in Chapter 1. It is difficult to meet the higher recommendations of calcium without the inclusion of dairy products, preferably low-fat dairy products. For those individuals who are lactose intolerant, lactose-reduced dairy products are available. Calcium-fortified foods such as fruit juice and soymilk are also available for children such as those on a vegan diet. For children whose calcium intake is inadequate, calcium supplements need to be given under the guidance of a physician or registered dietitian.

Fluids

It is of particular importance for school-age children to drink enough fluids to prevent dehydration during periods of exercise and during participation in sports, because children are at risk for dehydration and heat-related stress. Preadolescent children need to be more careful about staying hydrated than do adults and adolescents, for several reasons.[58] Children sweat less, and they get hotter during exercise. Some sports, such as football and hockey, require special protective gear that may prevent the body from cooling off. Children should never deprive themselves of food or water in order to meet a certain weight category, such as in wrestling.

Adults who are supervising children's physical activities need to make sure that children drink fluids before, during, and after exercise. The thirst mechanism may not work as well during exercise, and children may not realize that they need fluids. Cold water is the best fluid for children. However, children may be more likely to drink extra fluids if they are flavored. Sports drinks, which contain 4 to 8% carbohydrate and diluted fruit juice, are appropriate for children. Children should not be given soft drinks or undiluted juice, because the high carbohydrate load is too high to be hydrating and could cause stomach cramps, nausea, and diarrhea.[58]

SOFT DRINKS   Approximately 32% of school-age children consume up to 8.9 ounces of soft drinks per day, and 32% consume greater than or equal to 9 ounces per day; only 36% are nonconsumers of soft drinks.[59] School-age children consume more soft drinks than preschool-age children do, but not as much as adolescents—indicating an increase in consumption with age. Energy intake increases with increased consumption of nondiet soft drinks. Children with high consumption of regular soft drinks (more than 9 ounces per day) consume less milk and fruit juice than do those with lower consumptions of regular soft drinks. According to analysis of NHANES III data, overweight children have a higher proportion of their energy intake from soft drinks than nonoverweight children do.[28] A study of 548 ethnically diverse school-age children, average age 11.7 years, showed that BMI and the frequency of high BMI greater than 95th percentile increased along with increased consumption of sugar-sweetened beverages.[60] These investigators controlled for anthropometric, demographic, dietary, and lifestyle variables. Soft drinks can contribute significantly to children's overall calorie intake, while contributing little to the overall nutritional value of their diets and displacing more nutritious foods. Soft drinks can also contribute to a child's caffeine intake, providing 35–50 mg caffeine per 12-ounce serving.[61] Diet soft drinks do not provide sugar, and the aspartame content of diet sodas does not appear to pose a risk to healthy children.[62] Soft drinks in excess are not recommended for school-aged children because they provide empty calories and promote tooth decay.

Recommended versus Actual Food Intake

Comparing children's food intake data to the Food Guide Pyramid reveals that the mean food group intakes for children aged 2 to 19 years are below minimum recommendations for all food groups except for the dairy group, which shows intakes are at or near recommendations (see Table 12.6).[12,63] When the data are analyzed by racial/

Table 12.6 Mean daily intakes of the Food Guide Pyramid for children in the 1989–1991 CSFII[63]

Gender and Age	Number in Study	Grain	Vegetable	Fruit	Dairy	Meat
Recommendations (servings)		6–11	3–5	2–4	2–3	5–7 oz
Males 6–11 years	599	5.7	2.4	1.3	2.2	3.8
Females 6–11 years	573	5.5	2.4	1.4	2.1	3.7

Table 12.7 Mean percentages of food energy from carbohydrate, protein, total fat, saturated fatty acids, and cholesterol intake of 6- to 11-year-old children[65]

Gender and Age	Carbohydrate, %	Protein, %	Total Fat, %	Saturated Fatty Acids, %	Cholesterol, mg/d
Males:					
6–9 years	55	14	33	12	225
6–11 years	55	14	33	12	232
Females:					
6–9 years	55	14	32	12	190
6–11 years	55	14	33	12	199

ethnic groups, blacks have a significantly higher number of servings of meat than do whites or Hispanics. Whites have significantly higher intakes of dairy than do blacks or Hispanics. Fruit consumption is lower for children living in households where the family income is less than 131% of the poverty index. Dairy consumption tends to go up with increasing income.

Another measure of diet quality is provided by the Healthy Eating Index (HEI—available at www.usda.gov/cnpp), which is part of the Continuing Survey of Food Intake by Individuals.[64] The HEI measures the degree to which a person's diet conforms to the Food Guide Pyramid;[12] measures total fat and saturated fat as a percentage of total energy, total cholesterol, and sodium intake; and indicates the variety of foods in a person's diet. According to the most recently available HEI data, only 8% of children aged 7 to 12 years of age had a good diet, while 79% needed to improve their diets and 13% had a poor diet. The average HEI score for children aged 7–12 years was 64.1 out of 100, indicating that their diets needed improvement.[64]

The composition of children's diets, based on data from the Food and Nutrient Intakes by Children report, can be found in Table 12.7.[65] This table depicts the mean percentages of total energy from carbohydrate, protein, total fat, and saturated fatty acids, and cholesterol intake for 6- to 9-year-old and 6- to 11-year-old males and females. Fat and saturated fat intakes have decreased slightly since the NHANES III survey, which found a total fat intake of 33.8% and a saturated fat intake of 12.6% of total calories among 6- to 11-year-old boys and a total fat intake of 33.6% and a saturated fat intake of 12.3% among 6- to 11-year-old girls.[28] According to these data, both boys and girls are exceeding the recommendation for total calories from saturated fat of less than 10%. Cholesterol intake is well below the recommendation of 300 mg per day. Analysis of NHANES III data shows that the percentage of energy from fat is higher for black and Mexican American girls and black boys than for white girls and boys.[31] These differences are seen by 6 to 9 years of age in black and Mexican American girls and by 10 to 13 years of age in black boys.

As in the toddler and preschool-age group, mean vitamin and mineral intakes for school-age children exceeded the recommendations except for vitamin E and zinc.[65,66] Table 12.8 depicts a further analysis of children's diets in relation to dietary fiber, sodium, and caffeine intake. School-age children's caffeine intake has risen dramatically for both males and females from an average daily intake of 12.7 milligrams during the preschool years. Higher caffeine consumption is seen in the 6- to 11-year-old group versus the 6- to 9-year-old group, indicating an increased consumption of caffeine with age. This coincides with an increase in soft-drink consumption.

Tables 12.9 and 12.10 on the next page depict the mean percentages of nutrient intake contributed by foods eaten as snacks and foods eaten away from home for 1 day. These figures further illustrate the important contributions that snacks and meals eaten away from home make to the total daily food intake of children, especially in relation to total calories, total fat, and saturated fat. Analysis of food consumption data indicates that snacking among children has increased over the years, and the contribution of snacks to energy intake has increased from 20% in 1977 to 25% in 1996.[67] When eating away from home, school-age children most often eat at the school cafeteria, followed by someone else's house, and fast-food restaurants.[65]

According to the NHANES III data, children aged 6 to 11 years obtain about 20% of their total energy intake from beverages, with milk, soft drinks, and juice drinks being the largest contributors.[28] Drinking whole milk contributes significantly to children's saturated fat intake.

Table 12.8 Mean dietary fiber, sodium, and caffeine intake of 6- to 11-year-old children[65]

Gender and Age	Dietary Fiber, g	Sodium, mg	Caffeine, mg
Males:			
6–9 years	13	3195	23
6–11 years	14	3264	25
Females:			
6–9 years	12	2764	19
6–11 years	12	2839	23

Table 12.9 Mean percentages of nutrient intake contributed by foods eaten as snacks for 1 day[65]

Gender and Age	Individuals Eating Snacks, %	Food Energy, %	Total Fat, %	Saturated Fatty Acids, %
Males:				
6–9 years	83	21	19	19
6–11years	83	21	20	20
Females:				
6–9 years	84	21	20	20
6–11 years	82	20	19	19

Cross-Cultural Considerations

Healthy People 2010 has as one of its major goals the elimination of health disparities among different segments of the population.[4] The reasons for the health disparities are complex but may include genetic variations, environmental factors, and health behaviors, including diet. Access to community-based, culturally competent, linguistically appropriate preventive health care is needed to eliminate these disparities.[4]

A unique characteristic of every ethnic group in America is its culturally based foods and food habits.[68] As discussed earlier, children learn food habits within the context of the family's culture. It is important for a health care professional to try to learn as much as possible about the foods and diets of the ethnic groups served, including where food is purchased and how it is prepared. The next step is to evaluate the diet within the context of the culture. Which foods or food habits have positive health benefits and should be encouraged? Which food behaviors have harmful effects on health and should be limited or modified? For example, in working with the Latino population, it is important to first of all establish the country of origin. Latino immigrants may be from Mexico, Central America, South America, or the Carribean Islands. Food habits are unique for each of these ethnic groups. For example, Central Americans eat a lot of legumes, rice, and corn. Fruits and vegetables are also included in the diet. So, these dietary practices form the basis of a healthy diet. However, lard is the most commonly used fat. Encouraging Central Americans to use a vegetable oil instead of lard is an example of modifying a food practice to make it healthier.

As mentioned in Chapter 10, a series of booklets entitled *Ethnic and Regional Food Practices* is available from the American Dietetic Association. These booklets provide examples for incorporating traditional foods of various ethnic groups into dietary recommendations. Ordering information for this series can be found in the Resources section at the end of this chapter.

Vegetarian Diets

Young children who are consuming vegetarian diets are usually following their parents' eating practices. Preadolescents, on the other hand, may choose to follow a vegetarian diet independently of the family, motivated by concerns about animal welfare, ecology, and the environment.[1] A vegetarian diet is a socially acceptable way to reduce caloric intake and may be adopted by adolescents with eating disorders (see Chapter 14). A Vegetarian Food Guide Pyramid with suggested number of servings from different food groups has been developed.[69] Key nutrients in planning vegetarian diets for children include adequate calories, protein, calcium, zinc, iron, omega-3 fatty acids, vitamin B_{12}, riboflavin, and vitamin D.[70] Suggested daily food guides for lacto-ovo vegetarians and for vegans can be found in Tables 12.11 and 12.12.[1]

Table 12.10 Mean percentage of nutrient intake contributed by foods obtained and eaten away from home for 1 day[65]

Gender and Age	Individuals Eating Snacks, %	Food Energy, %	Total Fat, %	Saturated Fatty Acids, %
Males:				
6–9 years	65	26	27	28
6–11 years	65	26	28	28
Females:				
6–9 years	64	29	30	31
6–11 years	66	30	31	32

Table 12.11 Suggested daily food guide for lacto-ovo vegetarians at various intake levels[1]

| Food Groups | SERVINGS PER DAY, BY AGE AND DAILY CALORIC INTAKE | |
	7–10 Years (2000 kcal)	11+ Years (2200–2800 kcal)
Breads, grains, cereal	6	9–11
Legumes	1	2–3
Vegetables	3–5	4–5
Fruits	4	4
Nuts, seeds	1	1
Milk, yogurt, cheese	3	3
Eggs (limit 3/week)	½	½
Fats, oils (added)	4	4–6
Sugar (added teaspoons)	6	6–9

SOURCE: Data from Bright Future in Practice: Nutrition by Mary Story, et al., (Eds.), p. 165. © 2000 National Center for Education in Maternal and Child Health.

Physical Activity Recommendations

Physical activity has many proven health benefits, including prevention of coronary heart disease. Physical activity is one of the health behaviors that is important to establish in childhood to increase the chances of a physically active lifestyle that will continue into adolescence and adulthood. With the increased prevalence of childhood obesity, increasing physical activity and decreasing sedentary behaviors become important factors in controlling childhood overweight.[43]

Recommendations versus Actual Activity

It is recommended that children engage in at least 60 minutes of physical activity every day.[54] Parents are encouraged to:

- Set a good example by being physically active themselves and joining their children in physical activity.
- Encourage children to be physically active at home, at school, and with friends.
- Limit television watching, computer games, and other inactive forms of play by alternating with periods of physical activity.

Additionally, physical activity and daily physical education should be encouraged at schools and during after-school care programs. But presently, only about 17% of middle and junior high schools require daily physical activity for all students.[4] Healthy People 2010 objectives include increasing the proportion of trips that school-age children make by walking and by bicycling. Currently, only about 28% of children and adolescents ages 5 to 15 years take walking trips to school of less than 1 mile, and only about 2% of children and adolescents ages 5 to 15 years take bicycle trips to school of less than 2 miles. To meet the Healthy People 2010 goals, communities need to assure safe places for children to walk and ride their bicycles. Bicycle safety measures, such as wearing a helmet, need to be employed. Communities can also offer youth sports and recreation programs that are developmentally appropriate and fun for all young people. To achieve such a community environment, partnerships need to be established among federal, state, and local governments, nongovernment organizations, and private entities. The Centers for Disease Control and Prevention (CDC) have proposed strategies for promoting physical activity for children in family, school, and community settings.[71]

Determinants of Physical Activity

It is important to understand children's physical activity patterns and determinants of physical activity, so that vulnerable groups can be identified and appropriate intervention programs designed. Potential determinants of physical activity behaviors among children include physiological, environmental, psychological, social, and demographic factors.[72] Childhood physical activity has been difficult to assess and to track into adulthood. Many studies have identified correlates of physical activity behavior rather than predictors. The determinants of childhood physical activity are probably multidimensional and interrelated.

Table 12.12 Suggested daily food guide for vegan children and adolescents at various intake levels[1]

| Food Groups | SERVINGS PER DAY, BY AGE AND DAILY CALORIC INTAKE | |
	7–10 Years (2000 kcal)	11+ Years (2200–2800 kcal)
Breads, grains, cereal	7	10–12
Legumes	1	2–3
Vegetables, dark-green leafy	1	1
Vegetables, other	2	3–4
Fruits	5	4–6
Nuts, seeds	1	1
Milk alternatives (fortified with calcium and vitamins B_{12} and D)	3	3
Fats, oils (added)	4	4–6
Sugar (added teaspoons)	6	6–9

SOURCE: Data from Bright Future in Practice: Nutrition by Mary Story, et al., (Eds.), p. 165. © 2000 National Center for Education in Maternal and Child Health.

More work needs to be done in this area, but here are some generalities resulting from existing studies:

- Girls are less active than boys.
- Physical activity decreases with age.
- Seasonal and climate differences are seen in children's activity levels.
- Physical education in schools has decreased.

School and neighborhood safety is an important issue in promoting physical activity. In addition, parents have direct and indirect effects on children's physical activity levels.

Organized Sports

Many school-age and preadolescent children participate in organized sports activities, through schools or other community organizations. An analysis of NHANES III data indicates that children who participate in team sports and exercise programs are less likely to be overweight as compared to nonparticipants.[39] The American Academy of Pediatrics (AAP) recommends that children who are involved in sports be encouraged to participate in a variety of activities. The proper use of safety equipment such as helmets, pads, mouth guards, and goggles should be encouraged. The AAP warns against intensive, specialized training for children. The AAP's recommendations include:

- Regularly monitoring the child athlete's physical condition and development
- Preventing stress or overuse injuries, with the child's physician and coach working together
- Identifying and addressing eating disorders
- Instructing families, coaches, and child athletes about recognizing and preventing heat injury[73]

Recommendations for sports participation by children stress the importance of proper hydration. As noted earlier, children are at greater risk for dehydration and heat-related stresses than are adults.

Nutrition Intervention for Risk Reduction

"It is the position of the American Dietetic Association, the Society for Nutrition Education, and the American School Food Service Association that comprehensive nutrition services must be provided to all of the nation's preschool through grade twelve students. These nutrition services shall be integrated with a coordinated, comprehensive school health program and implemented through a school nutrition policy."[74]

Nutrition Education

Eating a healthy diet and participating in physical activity are important components of a healthy lifestyle that may prevent chronic disease in childhood and into adolescence and adulthood. School age is a prime time for learning about healthy lifestyles and incorporating them into daily behaviors. Schools can provide an appropriate environment for nutrition education and learning healthy lifestyle behaviors. Nutrition education studies have been conducted in school settings as well as outside of schools. Some of these programs have been knowledge-based nutrition education programs, with the focus on improving the knowledge, skills, and attitudes of children in regard to food and nutrition issues.[75] Other nutrition education programs have been more behaviorally focused, emphasizing disease risk reduction as well as enhancing health. The CDC has published "Guidelines for School Health Programs to Promote Lifelong Healthy Eating"; see Table 12.13.[76]

Table 12.13 Recommendations for school health programs promoting healthy eating[76]

Recommendation 1. Policy: Adopt a coordinated school nutrition policy that promotes healthy eating through classroom lessons and a supportive school environment.

Recommendation 2. Curriculum for nutrition education: Implement nutrition education from preschool through secondary school as part of a sequential, comprehensive school health education curriculum designed to help students adopt healthy eating behaviors.

Recommendation 3. Instruction for students: Provide nutrition education through developmentally appropriate, culturally relevant, fun, participatory activities that involve social learning strategies.

Recommendation 4. Integration of school food service and nutrition education: Coordinate school food service with nutrition education and with other components of the comprehensive school health program to reinforce messages on healthy eating.

Recommendation 5. Training for school staff: Provide staff involved in nutrition education with adequate preservice and ongoing in-service training that focuses on teaching strategies for behavioral change.

Recommendation 6. Family and community involvement: Involve family members and the community in supporting and reinforcing nutrition education.

Recommendation 7. Program evaluation: Regularly evaluate the effectiveness of the school health program in promoting healthy eating, and change the program as appropriate to increase its effectiveness.

Nutrition Integrity in Schools

"The school and community have a shared responsibility to provide all students with access to high-quality foods and nutrition services. . . . Educational goals, including the nutrition goals of the National School Lunch Program and School Breakfast Program, should be supported and extended through school district policies that create an overall school environment with learning experiences that enable students to develop lifelong, healthful eating habits."[77]

The American Dietetic Association

Nutrition integrity in schools is defined as ensuring that all foods available to children in schools are consistent with the U.S. Dietary Guidelines for Americans and the Dietary Reference Intakes.[4,54] School nutrition programs are vital to reinforcing healthy eating habits in school-age children. Sound nutrition policies need the support of the community and school environments, and must involve students in order to be successful. Preparing community leaders for involvement in policy development is one of the nutrition integrity core concepts.[77] Training food service personnel, teachers, administrators, and parents is an integral part of this process. The school environment must be one that supports healthy eating and exercise patterns. Foods sold from vending machines and snack bars often do not support healthy eating and may undermine sound nutrition programs. However, in some schools, vending machine proceeds are important sources of revenue for underfunded schools. Some schools have *pouring rights* contracts with soft-drink companies and receive a percentage of the profits.[74] Many schools sell à la carte items in addition to standard school lunches to increase revenue. One study showed that seventh-grade students in schools with à la carte items ate more fat and fewer fruits and vegetables than did students at schools without an à la carte program.[78] In the same study, for each snack vending machine, students' mean intake of fruit servings declined by 11%. It is against USDA regulations to sell *competitive foods* of minimal nutritional value. However, it is not against USDA regulations to sell these foods to students at times other than mealtimes or in other areas of the school, outside of food service areas. Adequate time allotted for meals is another important component of a sound nutrition program. Students can be involved in a nutrition advisory council, providing feedback about menu preferences and meal environment, and serving as a communication link with other students.

The School Health Index (SHI) for Physical Activity, Healthy Eating, and a Tobacco-Free Lifestyle is a self-assessment and planning tool for schools, which is offered by the National Center for Chronic Disease Prevention and Health Promotion, Centers for Disease Control and Prevention.[79] The SHI helps schools:

- Identify strengths and weaknesses in health promotion policies and strategies

- Develop an action plan for improving student health
- Involve all stakeholders including teachers, parents, students, and the community in improving school policies and programs[79]

The SHI has eight different modules for elementary schools and middle and high schools in the self-assessment, which correspond to the eight components of a coordinated school health program as depicted in Illustration 12.4. Each module consists of a score card, a questionnaire with guidance for arriving at a score, planning questions, and recommentations for implementation. Illustration 12.5 on the following page is an example of items on the Nutrition Services score card.

Model Programs

The 5 a Day for Better Health program models public-private partnership to enhance community nutrition education and to effect change in eating behaviors. In 1991, the National Cancer Institute (NCI), acknowledging the strong association of increased fruit and vegetable consumption with decreased risk of certain cancers, launched the 5 a Day for Better Health program. NCI partnered with the Produce for Better Health Foundation, a nonprofit organization

> **Pouring Rights** Contracts between schools and soft-drink companies whereby the schools receive a percentage of the profits of soft-drink sales in exchange for the school offering only that soft-drink company's products on the school campus.
>
> **Competitive Foods** Foods sold to children, in food service areas during mealtimes, that compete with the federal meal programs.

Illustration 12.4 Eight Components of a Coordinated School Health Program.[79]

School Heath Index — Elementary School

Module 4: Nutrition Services

Score Card (photocopy before using)

Instructions

1. Carefully read and discuss the Module 4 Questionnaire, which contains questions and scoring descriptions for each item listed on this Score Card.
2. Circle the most appropriate score for each item.
3. After all questions have been scored, calculate the overall Module Score and complete the Module 4 Planning Questions located at the end of this module.

		Fully in place	Partially in place	Under Development	Not in place
4.1	Breakfast and lunch programs	3	2	1	0
4.2	Variety of foods in school meals	3	2	1	0
4.3	Low-fat and skim milk available	3	2	1	0
4.4	Meals include appealing, low-fat items	3	2	1	0
4.5	A la carte offerings include appealing, low-fat items	3	2	1	0
4.6	Sites outside the cafeteria include appealing, low-fat items	3	2	1	0
4.7	Food purchasing and preparation practices to reduce fat content	3	2	1	0
4.8	Promote healthy cafeteria selections	3	2	1	0
4.9	Clean, safe pleasant cafeteria	3	2	1	0
4.10	Preparedness for food emergencies	3	2	1	0
4.11	Collaboration between food service staff and teachers	3	2	1	0
4.12	Degree and certification for food service manager	3	2	1	0
4.13	Professional development for food service manager	3	2	1	0

Column Totals: For each column, add up the numbers that are circled and enter the sum in this row

Total points: Add the four sums above and enter the total to the right.

Module Score = (Total points/39) x 100   %

Illustration 12.5 Nutrition Services Score Card.[79]

that represents the produce industry, for this campaign. Industry participants included supermarkets, suppliers, commodity groups, and food service operations.[80]

The 5 a Day program includes retail, media, community, and research components. Supermarkets provide information to consumers at the retail level. The NCI and Produce for Better Health Program work together to develop a media campaign. At the community level, health, educational, agricultural, and voluntary agencies work together, sometimes forming coalitions, to reach consumers at the local level. For the research component, NCI funded nine studies to develop, implement, and evaluate interventions in specific communities to increase the consumption of fruits and vegetables. The results of one of these funded studies, based in elementary schools, are presented here. It is a good example of what can be accomplished through nutrition education and community involvement of parents, teachers, school food service personnel, and industry.

Model Program: High 5 Alabama

The purpose of this study was to evaluate the effectiveness of a school-based dietary intervention program in increasing the fruit and vegetable consumption of fourth-graders.[81] Twenty-eight elementary schools in the Birmingham, Alabama, metropolitan area were paired within three school districts based on ethnic composition and the proportion of students receiving free or reduced-price meals through the National School Lunch Program. One school in each pair was randomly assigned to an intervention group or a usual care control group. Assessments were completed at baseline (at the end of third grade), after year 1 (at the end of fourth grade), and after year 2 (at the end of fifth grade).

The intervention consisted of three components: classroom, parent, and food service. The classroom component of the intervention included 14 lessons, taught biweekly by trained curriculum coordinators with assistance from the regular classroom teachers. The parent component consisted of an overview during a kickoff meeting and completion of seven homework assignments by the parent and the child. Parents were also asked to encourage and support behavior change in their children. The food service component consisted of food service managers and workers receiving half-day training by High 5 nutritionists in purchasing, preparing, and promoting fruit and vegetables within the High 5 guidelines. Data analyzed included 24-hour recalls from the students, cafeteria observations, psychosocial measures, and parent measures.[81]

Results indicate that mean daily consumption of fruit and vegetables was higher at Year 1 follow-up (3.96 vs 2.28) and Year 2 follow-up (3.2 vs. 2.21) for the intervention group as compared to the controls. At Year 1 follow-up, the mean daily consumption of fruits and vegetables was higher for the intervention parents as compared to control parents, but no difference was found at Year 2 follow-up. The intervention was found to be effective in subsamples suggesting that the program can be used with boys and girls; African American and European Americans; low-, middle-, and higher-income families; and parents of low, medium, and high educational levels. The intervention was found to be effective in changing the fruit and vegetable consumption of fourth-grade students. Future studies are recommended to enhance the effectiveness of the intervention in changing parents' consumption patterns and to test the effectiveness of the intervention when delivered by the regular classroom teachers.

Public Food and Nutrition Programs

"It is the position of the American Dietetic Association that all children and adolescents regardless of age, gender, socioeconomic status, racial, ethnic, or linguistic diversity, or health status should have access to food and nutrition programs that ensure the availability of a safe and adequate food supply that promotes optimal physical, cognitive, and social growth and development. Appropriate food and nutrition programs include food assistance and meal programs, nutrition education initiatives, nutrition screening and assessment followed by appropriate nutrition intervention, and anticipatory guidance to promote optimal nutrition status."[82]

The American Dietetic Association

Child nutrition programs, which have had a federal legislative basis since 1946, contribute significantly to the food intake of school-age children. The purpose of the child nutrition programs is to provide nutritious meals to all children. These programs can also reinforce nutrition education, which takes place in the classroom. Increasing the proportion of children and adolescents aged 6 to 19 years whose overall dietary quality is enhanced by meals and snacks at schools is addressed in one of the Healthy People 2010 objectives.[4] Child nutrition programs include the National School Lunch Program, School Breakfast Program, Child and Adult Care Food Program, Summer Food Service Program, Special Milk Program, Commodity Assistance for Child Nutrition Programs, Special Supplemental Food Program for Women, Infants, and Children, Nutrition Education and Training Program, and the National Food Service Management Institute.[83] Descriptions of several of these programs follow.

The National School Lunch Program

The federal government provides financial assistance to schools participating in the National School Lunch Program (NSLP) through cash reimbursements for all lunches served, with additional cash for lunches served to needy children, and through commodities.[83,84] Schools must meet five major requirements in order to participate in the NSLP:

1. Lunches must be based on nutritional standards.
2. Children who are unable to pay for lunches must receive lunches for free or at a reduced price, with no discriminating between paying and nonpaying children.
3. The programs must operate on a nonprofit basis.
4. The programs must be accountable.
5. Schools must participate in the *commodity program.*

School lunches must provide one-third of the Dietary Reference Intakes for the age/grade group of children being served for energy, protein, calcium, iron, vitamin A, and vitaminc C and must be consistent with the most recent version of the U.S. Dietary Guidelines for Americans when analyzed over a week's time.[54,83,84] Special emphasis is placed on serving a variety of foods and having menus that contain a variety of fruits and vegetables, low-fat dairy products, and lean meats. In addition, food school service personnel must make food safety a priority. These programs must also meet the needs of children with disabilities and special health care needs (see Chapter 13). Although not federally mandated, schools are encouraged to allow adequate time for children to eat their lunches. Schools receive payments from the federal government based on the number of meals served by category—paid, free, or reduced price.

Commodity Program A USDA program in which food products are sent to schools for use in the child nutrition programs. Commodities are usually acquired for farm price support and surplus removal reasons.[79]

Schools participating in the National School Lunch Program can choose one of the following four menu-planning approaches to plan school lunches:[85]

1. *Traditional Food-Based Menu–Planning Approach*
   Schools using this approach must plan menus with specific component and quantity requirements by offering five food items from four food components (meat/meat alternate, vegetables and/or fruits, grains/breads, and milk). Minimum portion sizes are established by ages and grade groups. Table 12.14 on the next page shows the meal pattern for lunches.

2. *Enhanced Food-Based Menu–Planning Approach*
   A variation of the traditional food-based menu-planning approach, this approach increases calories from low-fat food sources and increases the weekly servings of vegetables and fruits and grains/breads, in order to meet the Dietary Guidelines. The five food components are retained.

Table 12.14 Traditional food-based menu–planning approach—meal pattern for lunches[85]

Food Components and Food Items	Minimum Quantities Ages 9 and Older Grades 4–12	Recommended Quantities Ages 12 and Older Grades 7–12	SAMPLE MENU Minimum Quantities for Ages 9 and Older
Milk (as a beverage)	8 fl oz	8 fl oz	8 fl oz low-fat milk
Meat or Meat Alternate (quantity of the edible portion as served)			
Lean meat, poultry, or fish	2 oz	3 oz	2-oz hamburger patty
Alternate Protein Products	2 oz	3 oz	
Cheese	2 oz	3 oz	
Large egg	1	½ c	
Cooked dry beans or peas	½ c	¾ c	
Peanut butter or other nut or seed butters	4 tbs	6 tbs	
Yogurt, plain or flavored, unsweetened or sweetened	8 oz or 1 c	12 oz or 1½ c	
The following may be used to meet no more than 50% of the requirement and must be used in combination with any of the above: Peanuts, soynuts, tree nuts, or seeds, as listed in program guidance, or an equivalent quantity of any combination of the above meat/meat alternate (1 oz of nuts/seeds = 1 oz of cooked lean meat, poultry, or fish)	1 oz = 50%	1½ oz = 50%	
Vegetable or Fruit 2 or more servings of vegetables, fruits, or both	¾ c	¾ c	Lettuce, tomato Carrot/raisin salad Fresh apple
Grains/Breads (servings per week) Must be enriched or whole grain. One serving is 1 slice of bread or an equivalent serving of biscuits, rolls, etc., or ½ cup of cooked rice, macaroni, noodles, other pasta products, or cereal grains.	8 servings per week (For the purposes of this table, a week = 5 days.) minimum of 1 serving per day	10 servings per week Minimum of 1 serving per day	Enriched hamburger bun Condiments

3. *Nutrient Standard Menu–Planning Approach* This is a computer-based menu-planning, system that analyzes the specific nutrient content of menus, using approved computer software. This system allows more flexibility in planning, at the same time assuring that the nutrient standards are being met.

4. *Assisted Nutrient Standard Menu–Planning Approach* This variation of the nutrient standard menu–planning approach is for schools that do not have the necessary technical resources to conduct the nutrient analyses. This approach allows schools to have an outside source plan and analyze the menus based on the school's needs and preferences.

An additional provision of the National School Lunch Program allows states and school districts to use any rea-

sonable approach to menu planning as long as the method is reviewed and approved by the state agency or USDA.

School Breakfast Program

First authorized as a pilot program in 1966, the School Breakfast Program is a voluntary federal program. Many state legislatures have mandated breakfast programs for their districts, especially in schools serving needy populations.[84] In general, the NSLP rules also apply to the School Breakfast Program. School breakfasts must provide one-fourth of the Dietary Reference Intakes for the children being served, based on age or grade groups, and comply with the U.S. Dietary Guidelines for Americans when analyzed over a week's time. Table 12.15 shows the traditional, food-based meal pattern for breakfast. It is a special chal-

Table 12.15 Traditional food-based menu planning approach—meal pattern for breakfasts[85]

Food Components and Food Items	Grades K–12	SAMPLE MENU
Milk (fluid)		
As a beverage, on cereal, or both	8 fl oz	8 fl oz low-fat milk
Juice/Fruit/Vegetable		
Fruit and/or vegetable; or full-strength fruit juice or vegetable juice	½ c	½ c orange juice
Select 1 serving from each of the following components, 2 from one component, or an equivalent combination.		
Grains/Breads		
Whole-grain or enriched bread	1 slice	1 slice enriched toast, butter, jelly
Whole-grain or enriched biscuit, roll, muffin, etc.	1 serving	
Whole-grain, enriched or fortified cereal	¾ c or 1 oz	¾ c raisin bran cereal
Meat or Meat Alternates	1 oz	
Meat/poultry or fish	1 oz	
Alternate protein products	1 oz	
Cheese	1 oz	
Large egg	½	
Peanut butter or other nut or seed butters	2 tbs	
Cooked dry beans and peas	4 tbs	
Nuts and/or seeds (as listed in program guidance)	1 oz	
Yogurt, plain or flavored, unsweetened or sweetened	4 oz or ½ c	

lenge for schools to allow enough time for school breakfasts before school when most of the participating children arrive at about the same time. Currently, universal breakfast programs in elementary schools are being tested as pilot programs.

Summer Food Service Program

The Summer Food Service Program provides meals to children from needy areas when school is not in session. Schools, local government agencies, or other public and private nonprofit agencies operate these programs. The federal government provides financial assistance to these programs for providing meals in areas where 50% or more of the participating children are from families whose incomes are lower than 185% of the poverty level.[84] The Summer Food Service Program is an important source of food for many children from food-insecure families.

Team Nutrition

Team Nutrition is a program of the USDA's Food and Nutrition Service. The program is aimed at improving children's lifelong eating and physical activity habits through application of information in the Dietary Guidelines for Americans and the Food Guide Pyramid.[12,54,86] Team Nutrition is a partnership of public and private organizations, including private sector companies, nonprofit organizations, and advocacy groups which are interested in improving the health of the nation's children. Team Nutrition is an excellent example of a program that addresses the establishment of healthy eating and physical activity patterns for children on multiple fronts.

Team Nutrition operates through three behavior-oriented strategies:

- Provide training and technical assistance for Child Nutrition food service professionals to help them serve meals that meet nutrition standards while tasting and looking good.

- Provide integrated nutrition education for children and their parents with the goal of establishing healthy food and physical activity choices as part of a healthy lifestyle.

- Provide support for healthy eating and physical activity by involving community partners, including school administrators and other school and community partners.

Six communication channels are utilized: (1) food service initiatives; (2) classroom activities; (3) school-wide events; (4) home activities; (5) community programs and events; and (6) media events and coverage.[86] Schools are recruited to become Team Nutrition schools with the benefit of receiving a resource kit of materials. Additional information is available on the Team Nutrition Web site at www.fns.usda.gov/tn. The site includes activities for educators, parents, and students.[86]

Resources

The American Dietetic Association, Diabetes Care and Education Dietetic Practice Group

Ethnic and Regional Food Practices A Series. Chicago: The American Dietetic Association, 1994–1999.
This series of booklets addresses food practices, customs, and holiday foods of various ethnic groups.

Bright Futures

The Bright Futures publications are developmentally based guidelines for health supervision, addressing the physical, mental, cognitive, and social development of infants, children, and adolescents and their families. Bright Futures materials may also be ordered from National Maternal and Child Health Clearinghouse, 2070 Chain Bridge Road, Suite 450, Vienna, VA 22182-2536, www.nmchc.org, (703) 356-1964.
Web site: www.brightfutures.org

The Center for Health and Health Care in Schools

This center presents sample "best practices" school policies on healthy eating; it also provides questions for parents and interested community members to ask in assessing school nutrition policies.
Web site: www.healthinschools.org

Centers for Disease Control and Prevention, National Center for Health Statistics

The CDC Web site provides background information on the recent growth chart revisions. Also, individual growth charts can be downloaded and printed from this Web page.
Web site: www.cdc.gov/nchs/about/major/nhanes/growthcharts/charts.htm.

Child Nutrition Programs

The Child Nutrition Program Web site provides information on all of its programs including the National School Lunch Program, School Breakfast Program, Special Milk Program, Summer Food Service Program, and Child and Adult Care Food Program.
Web site: www.fns.udsa.gov/cnd

Coalition for Healthy and Active Kids

This coalition's mission is the education of parents, children, schools, and communities about the importance of physical fitness and nutrition education in trying to reverse the childhood obesity trend.
Web site: www.chaausa.org

Dietary Reference Intakes: National Academy Press

Over 2,000 books are available online at this site, free of charge, including the current Dietary Reference Intakes.
Web site: www.nap.edu

Federal Trade Commission, Bureau of Consumer Protection

See *"Promotions for Kids:" Dietary Supplements Leave Sour Taste.* This article explores what is known about the use of dietary supplements in the pediatric population and includes a list of pointers for parents.
Web site: www.ftc.gov/bcp/conline/features/ kidsupp.htm

Fit, Healthy, and Ready to Learn: A School Health Policy Guide

This in-depth guide includes policies to encourage physical activity, encourage healthy eating, and discourage tobacco use.

The policy to encourage healthy eating addresses nutrition education, the food service program, and other food choices at school.
Web site: www.nasbe.org/healthyschools/fithealthy.mgi

Graves DE, Suitor CW

Celebrating Diversity: Approaching Families through Their Food. 2nd ed. Arlington, VA: National Center for Education in Maternal and Child Health, 1998. The purpose of this publication is to assist health professionals in learning to communicate effectively with a diverse clientele.

Green M, Palfrey JS. (eds.)

Bright Futures: Guidelines for Health Supervision of Infants, Children, and Adolescents. 2nd ed. Arlington, VA: National Center for Education in Maternal and Child Health, 2000. This publication addresses the physical, mental, cognitive, and social development of infants, children, and adolescents and their families.

Healthy Edge 2000

This continuing education program of the American School Food Service Association provides training for school food service personnel and nutrition professionals in nutrition education and in offering healthy meals that are appealing to children while meeting their nutritional needs.
Web site: www.asfsa.org/continuinged/profdev/he2k

Healthy Habits for Healthy Kids

A joint project of the American Dietetic Association and WellPoint, this program provides guidance to health professionals in counseling families about childhood overweight.
Web site: www.eatright.org

Kidnetic

This colorful and interactive Web site is geared to teach older school-age children about healthy eating and exercise habits. It is supported by the International Food Information Council (IFIC) Foundation.
Web site: www.kidnetic.com

Kids Count

This Web site provides data on critical issues affecting at-risk children and their families. The publication *Kids Count,* by the Annie E. Casey Foundation, is also available online at this Web address.
Web site: www.kidscount.org

KidsHealth

This Web site, supported by the Nemours Foundation, has sections for parents, kids (older school age), and teens. It addresses all types of health concerns, including healthy eating and exercise.
Web site: www.KidsHealth.org

Kittler PG, Sucher KP.

Food and Culture. 3rd ed. Belmont, CA: Wadsworth/Thomson Learning, 2001. This book covers culturally related food and nutrition topics. The traditional food habits of key ethnic, religious, and regional groups are comprehensively reviewed. Information on traditional health beliefs and practices is also discussed.

National Center for Complementary and Alternative Medicine, National Institutes of Health

This Web site provides science-based information for consumers and practitioners as well as information about related news and events.

Web site: **www.nccam.nih.gov/nccam**

Patrick K, Spear BA, Holt K, and Sofka D

Bright Futures in Practice: Physical Activity. Arlington, VA: National Center for Education in Maternal and Child Health, 2001. Developmentally appropriate activities are presented for infants, children, and adolescents.

Promoting Better Health for Young People through Physical Activity and Sports

Report to the President from the Secretary of Health and Human Services and the Secretary of Education. This report outlines 10 strategies to promote health through lifelong participation in physical activity and sports.

Web site: **www.cdc.gov/nccdphp/dash/ presphysactrpt**

Team Nutrition

This Web site provides information on the USDA's Team Nutrition program, which is designed to help implement the Dietary Guidelines in Child Nutrition Programs.

Web site: **www.fns.usda.gov/tn**

Using the Food Guide Pyramid: A Resource for Nutrition Educators

This Web site from the USDA's Food, Nutrition, and Consumer Services, the Center for Nutrition Policy and Promotion, provides extensive nutrition education ideas for using the Food Guide Pyramid.

Web site: **www.usda.gov/cnpp/pyramid.gif**

References

1. Story M, Holt K, Sofka D, eds. Bright futures in practice: nutrition. Arlington, VA: National Center for Education in Maternal and Child Health; 2000.

2. Meyer AF, Sampson AE, Weitzman M et al. School breakfast program and school performance. Am J Dis Child 1989;143:1234–9.

3. Kids count data book. Baltimore, MD: The Annie E. Casey Foundation; 2003.

4. U.S. Department of Health and Human Services. Healthy People 2010 (Conference Edition, in two volumes). Washington, DC; January 2000.

5. Behrman RE, Kliegman RM, Jenson HB, eds. Nelson textbook of pediatrics, 17th ed. Philadelphia: WB Saunders; 2004.

6. Centers for Disease Control and Prevention, National Center for Health Statistics. CDC growth charts: United States. Available at www.cdc.gov/nchs/about/major/nhanes/growthcharts/ charts.htm, accessed 4/19/04.

7. Troiano RP, Flegal KM. Overweight children and adolescents: description, epidemiology, and demographics. Pediatrics 1998;101:497–504.

8. Dietz WH. Critical periods in childhood for the development of obesity. Am J Clin Nutr 1994;59:955–9.

9. Satter E. Feeding dynamics: helping children to eat well. J Pediatr Health Care 1995;9:178–84.

10. Gillman MW, Rifas-Shiman SL, Frazier AL et al. Family dinner and diet quality among older children and adolescents. Arch Fam Med 2000;9:235–40.

11. Kotz K, Story M. Food advertisements during children's Saturday morning television programming: are they consistent with dietary recommendations? J Amer Diet Assoc 1994;94:1296–1300.

12. U.S. Department of Agriculture. Food Guide Pyramid: a guide to daily food choice. Washington, DC: USDA, Human Nutrition Information Service; 1992. Home and Garden Bulletin No. 252.

13. Birch LL, Fisher JO. Food intake regulation in children, fat and sugar substitutes and intake. Ann NY Acad of Sci 1997;819: 194–220.

14. Birch LL, Fisher JO. Appetite and eating behavior in children. Pediatr Clinic N Amer 1995;42:931–53.

15. Birch LL, Fisher JO. Mothers childfeeding practices influence daughters eating and weight. Amer J Clin Nutr 2000;71: 1054–61.

16. Birch LL. Psychological influences on the childhood diet. J Nutr 1998;128:407S–10S.

17. Fisher JO, Birch LL. Parents' restrictive feeding practices are associated with young girls' negative self-evaluation of eating. J Amer Diet Assoc 2000;100:1341–6.

18. Birch LL, Fisher JO. Development of eating behaviors among children and adolescents. Pediatrics 1998;101:539–49(S).

19. Cachelin FM, Rebeck RM, Chung GH, Pelayo E. Does ethnicity influence body-size preference? A comparison of body image and body size. Obesity Research 2002; 10:158–66.

20. Institute of Medicine, Food and Nutrition Board. Dietary Reference Intakes for energy, carbohydrate, fiber, fat, protein, and amino acids. Washington, DC: National Academy Press; 2002.

21. Kennedy E, Goldberg J. What are American children eating? Implications for public policy. Nutr Rev 1995;53:111–26.

22. Kennedy E, Powell R. Changing eating patterns of American children: a view from 1996. J Amer Coll Nutr 1997;16:524–9.

23. Institute of Medicine, Food and Nutrition Board. Dietary Reference Intakes for calcium, phosphorus, magnesium, vitamin D, and fluoride. Washington, DC: National Academy Press; 1997.

24. Institute of Medicine, Food and Nutrition Board. Dietary Reference Intakes for vitamin A, vitamin K, arsenic, boron, chromium, copper, iodine, iron, manganese, molybdenum, nickel, silicon, vanadium, and zinc. Washington, DC: National Academy Press; 2001.

25. Centers for Disease Control and Prevention. Iron deficiency. United States, 1999–2000. MMWR 2002;51:897–9.

26. Centers for Disease Control and Prevention. Recommendations to prevent and control iron deficiency in the United States. MMWR April 3, 1998;47:RR-03.

27. Casamassimo P. Bright futures in practice: oral health. Arlington, VA: National Center for Education in Maternal and Child Health; 1996.

28. Troiano RP, Briefel RR, Carroll MD et al. Energy and fat intakes of children and adolescents in the United States: data from the National Health and Nutrition Examination Surveys. Am J Clin Nutr 2000;72 (suppl):1343S–53S.

29. Freedman DS, Dietz WH, Srinivasan SR, Berenson GS. The relation of overweight to cardiovascular risk factors among children and adolescents: the Bogalusa Heart Study. Pediatrics 1999;103:1175–82.

30. Ogden CL, Flegal KM, Carroll MD, Johnson CL. Prevalence and trends in overweight among U.S. children and adolescents, 1999–2000. JAMA 2002;288:1728–32.

31. Winkleby MA, Robinson TH, Sundquist J, Kraemer HC. Ethnic variation in cardiovascular disease risk factors among children and young adults: findings from the Third National Health and Nutrition Examination Survey, 1988–1994. JAMA 1999;281: 1006–13.

32. Dwyer JT, Stone EJ, Yang M. Prevalence of marked overweight and obesity in a multiethnic pediatric population: findings from the Child and Adolescent Trial for Cardiovascular Health (CATCH) study. J Amer Diet Assoc 2000;100:1149–56.

33. Dietz WH. Health consequences of obesity in youth: childhood predictors of adult disease. Pediatrics 1998;101:518S–25S.

34. American Diabetes Association. Type 2 diabetes in children and adolescents. Pediatrics 2000;105:671–80.

35. Dietz WH. Childhood weight affects adult morbidity and mortality. J Nutr 1998; 128:411S–14S.

36. Dietz WH, Gortmaker SL. Preventing obesity in children and adolescents. Ann Rev Public Health 2001;22:337–53.

37. Strauss RS, Knight J. Influence of the home environment on the development of obesity in children. Pediatrics [serial online] 1999;103:e85. Available at www .pediatrics.aapublications.org/cgi/content/full/103/1/e85, accessed 4/19/04.

38. Whitaker RC, Wright JA, Pepe MS et al. Predicting obesity in young adulthood from childhood and parental obesity. New Eng J Med 1997;337:869–73.

39. Dowda M, Ainsworth BE, Addy CL. Environmental influences, physical activity, and weight status in 8- to 16-year-olds. Arch Pediatr Adolesc Med 2001;155:711–7.

40. Dietz WH, Gortmaker SL. Do we fatten our children at the television set? Obesity and television viewing in children and adolescents. Pediatrics 1985;75:807–12.

41. Gortmaker SL, Must A, Sobol A et al. Television viewing as a cause of increasing obesity among children in the United States, 1986–1990. Arch Pediatr Adolesc Med 1996;150:356–62.

42. Andersen RE, Crespo CJ, Bartlett SJ et al. Relationship of physical activity and television watching with body weight and level of fatness among children: results from the Third National Health and Nutrition Examination Survey. JAMA 1998;279:938–42.

43. Crespo CJ, Smit E, Troiano RP et al. Television watching, energy intake, and obesity in U.S. children. Arch Pediatr Adolesc Med 2001;155:360–5.

44. Robinson TN. Reducing children's television viewing to prevent obesity: a randomized controlled trial. JAMA 1999;282:1561–67.

45. Klesges RC, Shelton ML, Klesges LM. Effects of television on metabolic rate: potential implications for childhood obesity. Pediatrics 1993;91:281–6.

46. American Academy of Pediatrics. Committee on Nutrition. Prevention of pediatric overweight and obesity. Pediatrics 2003; 112:424–30.

47. Barlow SE, Dietz WH. Obesity evaluation and treatment: expert committee recommendations. Pediatrics [serial online] 1998; 102:e29. Available at www.pediatrics. aapublications.org/cgi/content/full/102/3/e29, accessed 4/19/04.

48. Epstein LH, Paluch RA, Gordy CC, Dorn J. Decreasing sedentary behaviors in treating pediatric obesity. Arch Pediatr Adolesc Med 2000;154:220–6.

49. Kavey RW, Daniels SR, Lauer RM et al. American Heart Association guidelines for primary prevention of atherosclerotic cardiovascular disease beginning in childhood. Circulation 2003;107:1562–66.

50. Kavey RW, Daniels SR, Lauer RM et al. American Heart Association guidelines for primary prevention of atherosclerotic cardiovascular disease beginning in childhood. J Pediatr 2003;142:368–72.

51. National Cholesterol Education Program. Report of the expert panel on blood cholesterol in children and adolescents. Bethesda, MD: National Cholesterol Education Program; 1991.

52. Federal update. J Amer Diet Assoc 2000;100:877.

53. American Academy of Pediatrics, Committee on Nutrition. Pediatric nutrition handbook. Elk Grove Village, IL: American Academy of Pediatrics; 1998.

54. U.S. Department of Agriculture and U.S. Department of Health and Human Services. Dietary guidelines for Americans, 2000, 5th ed. Home and Garden Bulletin No. 232.

55. American Dietetic Association. Position of the American Dietetic Association: dietary guidance for healthy children aged 2 to 11 years. J Amer Diet Assoc 1999;99:93–101.

56. Dwyer JT. Dietary fiber for children: how much? Pediatrics 1995;96:1019S–22S.

57. American Academy of Pediatrics, Committee on Nutrition. Calcium requirements of infants, children, and adolescents. Pediatrics 1999;104:1152–7.

58. Jennings DS, Steen SN. Play hard eat right, a parents' guide to sports nutrition for children. Minneapolis, MN: Chronimed Publishing; 1995.

59. Harnack L, Stang J, Story M. Soft drink consumption among U.S. children and adolescents: nutritional consequences. J Amer Diet Assoc 1999;99:436–41.

60. Ludwig DS, Peterson KE, Gortmaker SL. Relation between consumption of sugar-sweetened drinks and childhood obesity: a prospective, observational analysis. Lancet 2001;357:505–8.

61. Pennington JAT. Bowes and Church's food values of portions commonly used. 17th ed. Philadelphia: Lippincott-Raven Publishers, 1998.

62. American Dietetic Association. Position of the American Dietetic Association: use of nutritive and nonnutritive sweeteners. J Amer Diet Assoc 1998;98:580–7.

63. Munoz KA, Krebs-Smith SM, Ballard-Barbash R, Cleveland LE. Food intakes of U.S. children and adolescents compared with recommendations. Pediatrics 1997;100:323–9.

64. Federal Interagency Forum on Child and Family Statistics. America's children: key national indicators of well-being 2003. Federal Interagency Forum on Child and Family Statistics, Washington, DC: U.S. Government Printing Office.

65. U.S. Department of Agriculture, Agricultural Research Service. 1999 Food and nutrient intakes by children 1994–1996, 1998. ARS Food Surveys Research Group. Available on "products" page at www.barc.usda.gov/bhnrc/foodsurvey/home.htm, accessed 4/19/04.

66. National Research Council. Recommended dietary allowances, 10th ed. Washington, DC: National Academy Press; 1989.

67. Jahns L, Siega-Riz AM, Popkins BM. The increasing prevalence of snacking among U.S. children from 1977 to 1996. J Pediatr 2001;138:493–8.

68. Kittler PG, Sucher KP. Food and culture. 3rd ed. Belmont, CA: Wadsworth/Thomson Learning; 2001.

69. Messina V, Melina V, Mangels AR. A new food guide for North American vegetarians. J Amer Diet Assoc 2003;103:771–5.

70. Messina V, Mangels AR. Considerations in planning vegan diets: children. J Amer Diet Assoc 2001;101:661–9.

71. Department of Health and Human Services and Department of Education. Promoting better health for young people through physical activity and sports: a report to the president from the secretary of Health and Human Services and the secretary of Education. Silver Spring, MD: Centers for Disease Control and Prevention; 2000. Available at www.cdc. gov/nccdphp/dash/presphysactrpt, accessed 4/19/04.

72. Kohl HW, Hobbs KE. Development of physical activity behaviors among children and adolescents. Pediatrics 1998;101:549–54.

73. American Academy of Pediatrics, Committee on Sports Medicine and Fitness. Intensive training and sports specialization in young athletes. Pediatrics 2000;106:154–7.

74. American Dietetic Association. Position of ADA, SNE, and ASFSA: nutrition services—an essential component of comprehensive school health programs. J Amer Diet Assoc 2003;103:505–14.

75. Lytle L. Nutrition education for school-aged children. J Nutr Ed 1995;27:298–311.

76. Centers for Disease Control. Guidelines for school health programs to promote lifelong healthy eating. J Sch Health 1997;67:9–26.

77. American Dietetic Association. Position of the American Dietetic Association: local support for nutrition integrity in schools. J Amer Diet Assoc 2000;100:108–11.

78. Kubik MY, Lytle LA, Hannan PJ et al. The association of the school food environment with dietary behaviors of young adolescents. Am J Pub Health 2003;93:1168–72.

79. Centers for Disease Control and Prevention. School Health Index for physical activity, healthy eating and a tobacco-free lifestyle: a self-assessment and planning guide. Elementary school version. Atlanta, GA 2002. Available at www.cdc.gov/nccdphp/dash/SHI, accessed 4/19/04.

80. Havas S, Heimendingeer J, Reynolds K. 5 a day for better health: a new research initiative. J Amer Diet Assoc 1994;94:32–6.

81. Reynolds KD, Franklin FA, Binkley D et al. Increasing the fruit and vegetable consumption of fourth-graders: results from the High 5 project. Prev Med 2000;30:309–19.

82. American Dietetic Association. Position of the American Dietetic Association: child and adolescent food and nutrition programs. J Amer Diet Assoc 2003;103:887–93.

83. 7CFR Parts 210 and 220. Child Nutrition Programs; School Meal Initiative for Healthy Children; Final Regulation. June 1995.

84. Martin J. Overview of federal child nutrition legislation. In: Martin J, Conklin MT, eds. Managing child nutrition programs: leadership for excellence. Gaithersburg, MD: Aspen Publishers; 1999.

85. U.S. Department of Agriculture, Food and Nutrition Service. FNS-303, 1998; also available online at www.fns.usda.gov/fns, accessed 4/19/04.

86. U.S. Department of Agriculture, Food and Nutrition Service. Team Nutrition. Available at www.fns.usda.gov/tn, accessed 4/19/04.

"Every sickness begins in the stomach."
Yemenite saying

Chapter 13

Child and Preadolescent Nutrition:

Conditions and Interventions

Chapter Outline

Prepared by Janet Sugarman Isaacs

Key Nutrition Concepts

1   Children are children first, even if they have conditions that affect their growth and nutritional requirements.

2   Common nutrition problems in children with chronic conditions are underweight and overweight, and difficulties in eating enough to meet higher nutrient requirements.

3   Children who have special health conditions receive more intensive nutrition services in schools and health care settings than other children do.

4   Family meal patterns and routines affect nutrition for children, so providing adequate support for families improves the nutritional status of children with chronic health conditions.

Introduction

Nutrition services need to be part of the goal to help a child reach his or her full potential. This chapter discusses the nutrition needs of children with chronic conditions, such as cystic fibrosis, diabetes mellitus, cerebral palsy, phenylketonuria (PKU), and behavioral disorders. Nutrition recommendations are based on those for children generally, but they may be modified by the condition and its consequences on growth, nutrient requirements, and/or eating abilities. Other factors such as activity that increase or decrease caloric needs are also discussed. Expanding school and community resources include nutrition services for children with special health care needs or those with developmental disabilities. Advocates for those with disabilities prefer "people-first language," which this chapter models. This convention names the person first and then the condition. An example is "a girl with *Down syndrome*," rather than the "Down's girl." Advocates for those with disabilities also prefer the word *disabilities* rather than *handicapped*. (The word *handicap* comes from the practice of using a cap to beg.)

"Children Are Children First"—What Does That Mean?

Children with special health care needs are children first, even if their conditions change their nutrition, medical, and social needs. Children are expected to become more independent in making food choices, assisting with meal preparation, and participating at mealtime with other family members. These same expectations are appropriate for children with special health care needs. For example, the child with spina bifida should be encouraged to make a salad or set the table. Modifications may be needed to help the child be successful, such as storing plates and utensils on low shelves, or lowering counter heights to accommodate a wheelchair at a kitchen sink. The cognitive developmental gains of childhood and participation in meal preparation and mealtime are the same for children with chronic conditions.

This concept has been acknowledged in schools through federal legislation in the Individuals with Disabilities Education Act (IDEA). This law requires the least restrictive environment and is resulting in inclusive settings for more children with disabilities.[1] This concept of inclusion has major ramifications for how children receive all types of services, such as schools providing alternative foods as required in the main cafeteria with all the other children. As a result of inclusion, children in wheelchairs or with Down syndrome spend time in regular classrooms with others of the same age. Nutritional problems related to food refusals, mealtime behavior, or special diets are being addressed in the neighborhood school in the regular classroom as often as possible.

This same concept of treating the child as a child first is recommended at home too. A special diet is part of an overall treatment plan that incorporates normal developmental steps of childhood. Children with diabetes or PKU do not benefit from being treated in a special manner at mealtime. As soon as possible, children are taught to take responsibility for making food choices consistent with their diet plans. Consistency and structure in the home support a child's normal development. This structure includes regular meal and snack times, and the child accepting increasing levels of responsibility in preparing foods. These approaches lower the chance of the child being overprotected or manipulating adults because of her illness or its treatment.

General nutrition guidelines for children, school nutrition educational materials, and nutrition prevention strategies may or may not be applicable to children with specific conditions. Many nutrition education curricula appropriately target the goals of preventing overweight, lowering fat, and increasing fruit and vegetable intake. These curricula provide appropriate education for the most part to children with conditions such as diabetes and Down syndrome. Conditions in which slow weight gain and underweight are common, such as cerebral palsy, may not fit such curricula.[2] Nutrition education may not address such children's need for high-fat foods as part of high caloric needs. Another example of how general nutrition recommendations may not fit all children concerns the widely used Food Guide Pyramid.[3] A child with PKU cannot ever have protein-rich meats or dairy products.[4] When this pyramid is discussed and encouraged, the child with PKU may feel isolated and confused about whether his diet is healthy. When possible, children with special health care needs are encouraged to participate in school nutrition education programs with modifications as needed.

"Train a child in the way he should go, and when he is old, he will not depart from it."

Proverbs

Nutritional Requirements of Children with Special Health Care Needs

Caloric and protein needs are lower on a body weight basis in childhood than during the preadolescence, toddler, and preschool years.[5] Children with special health care needs have a wide range of nutritional requirements and more variability than other children based on these factors:[6,7]

- Low caloric intake may be appropriate with small muscle size.
- High protein is needed with high protein losses, such as skin breakdown.
- High fluid volume is needed with frequent losses from vomiting or diarrhea.
- High fiber may be needed for chronic constipation management.
- Long-term use of prescribed medications may increase or decrease vitamin or mineral requirements, or change the balance of vitamins and minerals needed as a result of medication side effects.
- Routine illness is more likely to result in hospitalization or resurgence of symptoms of the underlying disorder.

Energy Needs

Children with special health care needs may need more, less, or the same caloric intakes as other children of the same age. Energy needs are amazingly complex in children, let alone those with special needs. Under ideal conditions the caloric needs are measured by indirect calorimetry, but usually they are estimated using standard calculations that cannot take into account the specific conditions involved.[8] Machines that measure indirect calorimetry are becoming available, but they still can give only estimates of energy needs at rest, without considerations of energy needed for activity and growth. Conditions that slow growth or decrease muscle size generally result in lower caloric needs.[6,9,10] Caloric needs in a child with Prader-Willi syndrome may be only 66% of the caloric needs of a child of the same age and gender.[6]

Other factors that change energy requirements are related to activity level and frequency of illnesses.[5,6] Children with a chronic condition are encouraged to participate in age-appropriate sports activities. Conditions in which activity may be especially beneficial include diabetes and mild cerebral palsy. The level of activity of children with chronic conditions may be higher or lower than

activity in other children. Children who are very active may appear thin as a result of low caloric intake. Children with autism and attention deficit hyperactivity disorder are generally more active than other children, and/or they may sleep less.[11,12] Such a range in level of activity is addressed as a part of a thorough nutrition assessment. Questions such as "Is the child receiving physical therapy one or three times per week?" and "How much time does the child use a walker compared to a wheelchair?" are examples of how activity can be assessed in determining caloric needs.

Protein Needs

Protein needs also can be higher, lower, or the same as those for other children, based on the condition. Healing burns and cystic fibrosis are examples of disorders with high protein needs—at 150% of the RDA.[13] Conditions such as PKU and other protein-based inborn errors of metabolism require greatly reduced amounts of natural protein in the diet.[4] Children with diabetes mellitus do not have modified protein needs.[14] The importance of protein for wound healing and for maintaining a healthy immune system makes protein requirements key for various conditions with frequent illnesses or surgeries. For example, a child with cerebral palsy who is scheduled to have hip surgery would have protein needs evaluated in a complete nutritional assessment. Higher protein may be recommended for wound healing and for skin breakdown while in a cast after surgery.

Other Nutrients

DRIs are good starting places to assess the need for vitamins and minerals in chronic conditions.[15] (DRIs are listed inside the back cover of this textbook.) As in all children, if the diet provides sufficient foods to meet the needs for protein, fats, and carbohydrates, it is likely the vitamin and mineral needs are also met. However, children with chronic conditions may have more difficulty meeting the DRI for vitamins and minerals as a result of these considerations:[15,16]

- Eating or feeding problems may restrict intake of foods requiring chewing, such as meats, so that certain minerals may be low in the diet.
- Prescribed medications and their side effects can increase turnover for specific nutrients, raising the recommended amount needed.
- Food refusals are common with recurrent illness, so total intake may be more variable day to day than in other children of the same age.
- Treatment of the condition necessitates specific dietary restrictions, so that vitamins and minerals usually provided in restricted foods have to be supplemented.

Nutrients such as calcium that are low in the general population of children are also problem nutrients for children with chronic conditions.[17] The American Academy of Pediatrics statement on calcium applies to children with chronic conditions. This statement recommends good-quality food sources of calcium for all children. Food sources of calcium may avoid lead contamination that has been reported in a variety of over-the-counter calcium supplements.[18] Taking high levels of supplemental calcium does not clearly benefit children with chronic conditions.[17]

Growth Assessment

The Centers for Disease Control (CDC) 2000 growth charts are a good starting place for assessing the growth of any child. Identifying children at risk for overweight and preventing long-term cardiovascular risks are important purposes underlying growth assessments of children. Such concerns may or may not apply to children with chronic conditions, but a nutrition assessment can tailor nutrition goals to specific conditions. Families dealing with conditions that shorten life, such as cystic fibrosis, would not benefit from including overweight and its long-term risks as part of the child's nutrition assessment.

> **Cerebral Spinal Atrophy** Condition in which muscle control declines over time as a result of nerve loss, causing death in childhood.
>
> **Secondary Condition** Common consequence of a condition, which may or may not be preventable over time.
>
> **Seizures** Condition in which electrical nerve transmission in the brain is disrupted, resulting in periods of loss of function that vary in severity.
>
> **Scoliosis** Condition in which the vertebral bones in the back show a side-to-side curve, resulting in a shorter stature than expected if the back were straight.
>
> **Neuromuscular** Term pertaining to the central nervous system's control of muscle coordination and movement.

Families of children with severe disabilities who have wheelchairs may be apprehensive about growth as a goal for the child. They may have long-term concerns about caring for the child at home. Families may not want the child to grow at the usual rate when activities such as lifting her from the bathtub or out of a wheelchair may become more difficult. Also, children with rare degenerative conditions such as *cerebral spinal atrophy* have such major decreases in muscle size that growth may not occur.[10]

Most children with chronic conditions do grow, and assessing growth is an important component of nutrition services. If the child's condition is known to change the rate of weight or height gain, either slowing or accelerating it, the following signs need attention regardless of what growth chart is used:

- A plateau in weight
- A pattern of gain and then weight loss
- Not regaining weight lost during an illness
- A pattern of unexplained and unintentional weight gain

Growth Assessment and Interpretation in Children with Chronic Conditions

Factors that affect growth assessment and interpretation in childhood may not have been detectable earlier in younger children. These factors are the age of onset of the condition, *secondary conditions,* and activity. The child's age when the condition started may influence whether CDC growth charts are applicable. Early onset is more likely than later onset to affect growth in conditions such as *seizures.*[10] If the seizures started in middle childhood, the standard growth chart may be appropriate because the child's growth pattern is already established. Onset of seizures in the neonatal period may reflect more severe brain damage, which markedly slows growth rate. Then the child's own growth record over time would be the best indicator of future growth.

Toddlers and preschoolers with cerebral palsy usually do not develop secondary conditions until childhood or later. *Scoliosis* is a secondary condition that interferes with accurate measurement of stature.[6] It may develop as a result of muscle incoordination and weakness in some forms of cerebral palsy in preadolescence. If a child with cerebral palsy has stature measurements that plateau or decline, it may be a result of cerebral palsy, scoliosis, lack of adequate nutritional intake, or a combination of these three factors.[10] Nutritional interventions cannot prevent scoliosis, although nutritional consequences of its treatment may arise. Children may be provided custom-fitted back braces, so weight gain means the brace needs to be replaced. Children with scoliosis braces also may become less active because the brace restricts some types of movement. If scoliosis surgery is performed, the child may become slightly taller immediately, again showing that stature measurements have to be interpreted with care.

Body Composition and Growth

Children with special health care needs may or may not be typically proportioned in muscle size, bone structure, and fat stores. Some children in good nutritional status may plot at or below the lowest percentile on a standard growth chart for height.[19] In fact, low-percentile heights are usual for a child with Down syndrome if growth is plotted on the CDC chart rather than the special growth chart for Down syndrome.[20] Short stature, low muscle tone, and low weight compared to age-matched peers are not attributed to caloric intake. They characterize the natural consequences of the *neuromuscular* changes within Down syndrome. Similarly, a child with low muscle size could have a low weight and short stature. It would be unfair to assume that the child's diet is inadequate because of the low weight. A thorough assessment that includes body composition is necessary. For example, a thin-appearing child needs to have body fat stores measured before diet recommendations are made. If body fat stores are fine, adding calories is more likely to contribute to overweight.

Children with small muscle size will have lower weights than those with regular-sized muscles.[8,10] Conditions with altered muscle size may be described using terms such as *hypotonia* or *hypertonia*. Examples include cerebral palsy, Down syndrome, and spina bifida.[10,20] Not all muscles are affected. For example, some children with spina bifida have larger muscle size in the upper body and smaller muscle size in the lower body. Variation in size of muscles may make growth interpretation more difficult. Any assessment must address risks for overweight, such as body mass index (BMI) and adiposity rebound. Standard interpretation may suggest a risk of overweight, but it may not accurately reflect that short stature is part of the child's condition. By standard interpretation, every child with Down syndrome or spina bifida could be overweight. (See Case Study 13.1, page 312). For now, no established BMI tables cover specific conditions or the appropriate time for adiposity rebound.

Measuring body fat is another indicator of body composition. Skinfold fat measurements and their interpretation have to be based on consistent and repeatable standard methods.[21,22] Measuring fat stores in children is not like measuring fat stores in adults, because of the changes in body composition that come with age and growth. Calculated formulas and methods for body composition for children are not the same as for adults.[8,23] Estimates of body composition for children with chronic conditions may be based on smaller sample sizes than for other children, but still such information is helpful. Identified low fat stores trigger recommendations to boost calories.

In-depth growth assessment may include head circumference measurement for all ages, with plotting and interpretation based on the Nellhaus head circumference growth chart, as discussed in Chapter 9.[24] Head circumference is important because children with unusually small heads have smaller brains, a characteristic associated with short stature. Even with adequate diet and no documented eating problems, children with various genetic disorders tend to be shorter than age-matched peers.[20,25]

SPECIAL GROWTH CHARTS   Special growth charts have been published for a variety of genetic conditions.[20] Table 13.1 includes examples of these special growth charts. The number of children reported in such growth charts is not as large, nor as representative as the CDC 2000 growth charts. Special growth charts are revised often, based on new information emerging about the natural course of rare conditions. Some special growth charts are based on only the most severe forms of the condition, such as for children living in residential care. Many chronic conditions do not have special growth charts because they present with a wide range of severity. Conditions without a

Table 13.1 Examples of specialty growth charts[2,20]

Conditions with Special Growth Charts	Comment
Achrondroplasia,	Form of dwarfism
Down syndrome, Trisomy13, Trisomy 18	Short stature, variable weight
Fragile X syndrome	Short stature, primary in males
Prader-Willi syndrome	Short stature, overweight
Rubinstein-Tabyi syndrome	Short stature
Sickle-cell disease	Short stature
Turner syndrome	Short stature
Spastic quadriplegia	Short stature, low weight
Marfan syndrome	Tall stature

specialty growth chart, which may or may not match the standard growth charts, include the following:

- *Juvenile rheumatoid arthritis*
- Cystic fibrosis
- Rett syndrome
- Spina bifida
- Seizures
- Diabetes

Nutrition Recommendations

Children with chronic conditions require nutrition assessments to determine whether they are meeting their nutrient and caloric needs, whether eating problems such as food refusals or mealtime behavior are interfering with meeting nutritional needs, and whether growth is on target for their age and gender. Then nutrition interventions are provided based on the assessment. The goal is

Juvenile Rheumatoid Arthritis Condition in which joints become enlarged and painful as a result of the immune system; generally occurs in children or

for the child to maintain good nutritional status, and to prevent nutrition-related problems from being superimposed on the primary condition. (See Case Study 13.1 on the following page.)

Children with special health care needs benefit from the same nutritional recommendations as other children do, particularly in general areas such as dietary fiber or appropriate use of soft drinks. However, children with special health care needs may require particular formulas and nutrition support not needed for most children. Most children develop feeding skills during the toddler and preschool years; by childhood, abilities and/or disabilities that limit self-feeding and using utensils may require more aggressive

Case Study 13.1

Photo Disc

Adjusting Caloric Intake for a Child with Spina Bifida

Sam is a third-grader in regular classes at his public school. He uses a wheelchair all the time and can transfer from his wheelchair to a chair by himself. He is on a toileting schedule at school with the assistance of a nurse. He participates in modified physical education as part of his physical therapy treatment. He likes to eat with his friends at school. His mother tries to make him cut back at the evening meal and has stopped buying some of his favorite snacks. He is mad at his mother because he likes his snacks after school when he is bored.

Nutrition assessment from Sam's last visit at the spina bifida clinic at the local hospital showed that he was overweight by measuring his fat stores. Because he cannot stand, his stature was estimated by measuring his length lying down and comparing it with his last length measurement. Standard methods could not be used to measure him, which limits the interpretation of his growth using the CDC growth chart. The chart showed Sam at the 75% percentile in weight for his age, which is not overweight for his age. His rate of weight gain of 8 pounds per year, typical for a boy his age, is too fast for his low level of activity. His estimated calorie needs are 1100 per day due to low activity and short stature, or about two-thirds of the caloric needs of others his age. Sam says he does not care about his size or being overweight. His mother is quite concerned that she would not be able to assist him if he fell or needed to be lifted.

Recommendations: The nutritionist at the clinic completes a school lunch prescription to reduce Sam's caloric intake from 650 calories to 350 calories per lunch. His meal pattern is adjusted to two meals (breakfast and lunch) and two snacks per day at home, which better fits his low caloric needs. Sam is allowed to choose his favorite snack foods to replace his evening meal. Giving him choices about his snacks increases Sam's sense of being in control and lowers the instances of his expressing anger at his mother about snack foods. The clinic nutritionist calls the school to review Sam's level of activity and confirm that the lunch changes are being implemented. The physical therapist at school has found after-school swimming lessons and recommends them to Sam's mother as a way to increase his activity and socialization. To motivate Sam to pay attention to his eating and weight gain, his teacher and mother set up a monthly nonfood reward for him if he does not gain any weight. The effectiveness of the plan to cut Sam's caloric intake and increase his activity will be assessed at his next clinic visit, when he will be weighed and have his fat stores measured.

Questions

1. Sam does not care about his size or being overweight, so why is a diet plan necessary?
2. What are the risks from Sam's weight, since he is only at the 75th percentile for his age on the standard growth chart?
3. When he goes through puberty, will Sam grow taller and thus be able to eat more calories each day?

(Answers are in Appendix D.)

support. Nutritional supports common for children are enteral supplements, when oral feeding of regular foods is insufficient in quality or amount to maintain health and to assure growth is not being limited. Table 13.2 provides a list of commonly used complete nutritional supplements and examples of their use. Children under 10 years of age are generally provided a pediatric formula, but adult formulas may be used for children.

Methods of Meeting Nutritional Requirements

Children with special health care needs who cannot meet their nutrient requirements from regular foods may receive complete nutritional supplements in addition to meals or for partial or complete replacement for meals. The first choice is that required supplements are drunk or eaten in the usual way. If this method does not work out, complete nutritional supplements can be administered by placement of a feeding *gastrostomy*.[6] Gastrostomy feeding may be required in children with kidney diseases, some forms of cancer, and severe forms of cerebral palsy and cystic fibrosis.[26–28] Many families experience difficulty accepting a gastrostomy for meeting nutritional requirements because feeding is such an important aspect of parenting.[29] Aside from emotional aspects, insurance coverage and financial questions of paying for formulas fed by gastrostomies are major concerns for some families.

Children fed by gastrostomy can have many different schedules, such as eating orally during school and being fed by gastrostomy overnight. Table 13.3 gives an example of a feeding plan that includes gastrostomy feeding and oral feeding. If medications are required, they can be given through the gastrostomy also. For example, for children with pediatric AIDS who require many medications during the day, compliance with taking the drugs improved after gastrostomies were placed. The parents spend less time trying to administer the medications to their children, and some children's health improved as a result of taking all of the required medications. Another example is a child who cannot safely drink liquids as a result of cerebral palsy. The child could have fluids given by gastrostomy, but eat solids foods by mouth. Children with gastrostomies can swim, bathe, and do any activities they could do before the gastrostomy was placed.

Most formula fed by gastrostomy can be consumed as beverages. Even regular foods can be blended in a recipe for gastrostomy feeding for some children. Such "home brews" for gastrostomy feeding have to be carefully monitored because they are a rich medium for bacterial contamination. Part of the decision-making process about use of special formulas and gastrostomy feeding often hinges on

Table 13.2 Examples of nutritional supplements and formula for children[26]

Formula	Comments
Pediatric versions of complete nutritional supplements, such as Pediasure	Generally recommended for children under 10 years of age; can be used for gastrostomy or oral nutrition support.
Adult complete nutritional supplements, such as Ensure	Generally 1 calorie per milliliter is recommended for children.
Enrichment of beverages, such as Carnation Instant Breakfast added to milk	Requires that milk is tolerated.
Predigested formula with amino acids and medium-chain fatty acids, such as Peptamen Junior	For conditions in which intestinal absorption may be impaired.
Special formulas for inborn errors of metabolism (PKU), such as Phenex-2	Usually a powder that is mixed as a beverage, but other forms such as bars and capsules are available.
High-calorie booster for cystic fibrosis, such as Scandishake	Generally 2.5 calories per milliliter to concentrate calories in small volume.

Table 13.3 Example of a feeding and eating schedule for an 8-year-old who eats by mouth and by gastrostomy

Daily Schedule	Comments
6:30 a.m. Night feeding pump turned off 7:15 a.m. Breakfast: refused 8:00 a.m. Bus to school 11:30 a.m. School lunch offered and about half is eaten: ½ chicken sandwich, all of french fries, with ⅓ pint of whole-milk 3:30 p.m. After-school snack at home of 4 oz pudding cup, two plain cookies, and 4 fl oz orange drink 6:30 p.m. Evening meal at home: ½ cup mashed potato, 6 fl oz whole milk, refused vegetable and meat 8:30 p.m. Bedtime	Overnight feeding by gastrostomy runs from 9:30 p.m. until 6:30 a.m., providing about 3 fl oz per hour, so no hunger in the morning is common. Child has slow eating pace and is easily distracted by school lunchroom sounds. Mealtime behavior at home includes many attempts to leave the table, with prompting to eat from parents. Parents hook up night feeding pump while the child is sleeping.

prior rejection of other feeding methods. When possible, gastrostomy feeding is planned as a temporary measure, with a return to oral eating later. For example, a child who has a gastrostomy because of a kidney condition may have a kidney transplant that allows removal of the gastrostomy after recovery.

Other nutrition supplements fed by gastrostomy have specific components that are unusual in beverages because they have such a strange taste. For example, formulas that contain individual amino acids generally are accepted only by those who have had them from infancy, as in the formulas for PKU. If a child required a new formula with amino acids, gastrostomy feeding would be more successful than oral feeding in most cases.

VITAMIN AND MINERAL SUPPLEMENTS FOR CHRONIC CONDITIONS   Children's complete vitamin and mineral supplements are recommended for a variety of chronic conditions to assure that the DRIs for essential nutrients are provided. However most over-the-counter supplements are in the form of chewable tablets, so children who cannot chew well may require a liquid form of vitamins and minerals. The composition of vitamin and mineral supplements may be important because some have added ingredients not recommended for certain chronic conditions. Examples are vitamin and mineral brands with added carbohydrates, which are not allowed on a *ketogenic diet,* or those made with an artificial sweetener containing phenylalanine not recommended for children with PKU. (The ketogenic diet and PKU are discussed later in this chapter.)

The underlying diagnosis can make specific nutrients so important in the diet that they may be prescribed as pharmaceuticals. Cystic fibrosis treatment (discussed later in this chapter) requires fat-soluble vitamin supplements due to poor intestinal absorption of these nutrients. Vitamins A, D, E, and K are needed in cystic fibrosis. Vitamin B_{12} injections are needed for some protein-based inborn errors of metabolism.[4,13] Vitamin C may be prescribed above the DRI for some children with spina bifida who have frequent bladder infections.[6] The high dose of vitamin C functions as a medication rather than as a nutrient in this instance.

Use of excessive levels of vitamins and minerals can be risky, especially for underweight children. For example, a child may be counseled to add Carnation Instant Breakfast, while another provider adds a complete nutritional supplement and is unaware that the child takes a chewable children's vitamin/mineral tablet too. Determination of the total intake of supplemented nutrients is part of a nutrition assessment. All medications, prescribed and over the counter, are identified within the nutrition and medical care plans.

Children with chronic conditions that limit activity or require medications that affect bone growth need special attention regarding calcium and vitamin D requirements.[17] *Galactosemia,* in which dairy products are eliminated from the diet, and cerebral palsy are two examples of conditions affecting calcium. Some children with these conditions are like older women with *osteoporosis* in that their calcium may move out of bones faster than it goes in. Providing additional calcium, phosphorous, and vitamin D may be recommended. Selecting calcium supplements that can be taken for years or decades by children raises concerns about the lead content found in some calcium supplements.[18]

Fluids

Guidelines for fluids for all children are appropriate. Particular considerations for children with special health care needs are high fluid losses—for example, from drooling as a result of cerebral palsy—or behaviors that result in low fluid intakes.[6] Because constipation is common in children with neuromuscular disorders, adequate fluids are often stressed as part of a bowel management program. Children with limited ability to talk may have more difficulty indicating thirst. Many chronic health conditions carry higher risks for dehydration due to side effects of prescribed medications. A chronic condition generally does not change the fluid requirement when the child is well.

Eating and Feeding Problems in Children with Special Health Care Needs

Eating and feeding problems are diagnosed when children have difficulty accepting foods, chewing them safely, or ingesting sufficient foods and beverages to meet their nutritional requirements. About 70% of children with developmental delays have feeding difficulties, independent of whether neuromuscular problems have been identified.[6] Examples of these feeding problems include the following situations:

- Self-feeding skills are lower than the child's chronological age, requiring assistance and supervision to assure adequate intake.
- Meals take so long or so much food is lost in the process of eating that the actual food intake is too low.
- The condition requires adjustment in the timing of meals and snacks at home and at school.

In children who do not have developmental delay, the impact of chronic conditions on eating may include behavioral problems at mealtimes, conflicts about control over

Ketogenic Diet   High-fat, low-carbohydrate meal plan in which ketones are made from metabolic pathways used in converting fat as a source of energy.

Galactosemia   A rare genetic condition of carbohydrate metabolism in which a blocked or inactive enzyme does not allow breakdown of galactose. It can cause serious illness if not identified and treated soon after birth.

Osteoporosis   Condition in which low bone density or weak bone structure leads to an increased risk of bone fracture.

food choices, and variability in appetite. Families of children with chronic conditions may focus on mealtimes and foods as methods of coping with their concerns about the child's future. For example, families may be overprotective and restrict a child from eating at friends' homes, when such activities may be appropriate for social development.

Specific Disorders

CYSTIC FIBROSIS Cystic fibrosis (CF) is one of the most common lethal genetic disorders, with an incidence of one in 1500–2000 live births.[13] It is highest among Caucasians, with an incidence of one in 17,000 live births for African Americans. The CF gene is located on the long arm of chromosome 7, and it has many different genetic versions. The most common genetic mutation characterizes 67% of the cases. CF affects all the exocrine functions in the body, with lung complications often causing death. Its major nutrition-related consequence is malabsorption of various nutrients due to the lack of pancreatic enzymes. This can result in a slower rate of weight and height gain, and higher energy needs due to chronic lung infections. Children with cystic fibrosis are likely to develop malnutrition as the condition progresses. Intensive nutrition interventions may be required to meet higher caloric needs.[26,27]

Nutrition interventions for CF include monitoring growth, assessing dietary intake, and increasing calories and protein by two to four times the usual recommendations to compensate for malabsorption. Every time a child with CF eats a meal or snack, he must take pills containing enzymes. Frequent eating and large calorie-dense meals are encouraged. Gastrostomy feeding at night to boost calories is sometimes required. Vitamin and mineral supplementation, particularly fat-soluble vitamins, is a part of daily management. Children with CF are at risk for developing diabetes because the pancreas is a target organ of CF damage.[13] In recent years children diagnosed with CF have achieved longer life expectancies, and many have survived into the young adult years.

Nutrition experts working with children who have CF struggle to balance the children's high nutrient needs and frequent illnesses. Many children with CF have slow growth and are lower in weight and shorter than expected. Even with nutrition support, decline in pulmonary function over time continues. Some children with CF have lung transplants if they meet strict eligibility requirements. CF is on the leading edge of gene therapy research, giving hope to families with young children.

DIABETES MELLITUS Diabetes mellitus is a disorder of *insulin* regulation in which dietary management is crucial. The hormone insulin may be underproduced or mistimed in its release, and/or peripheral tissues may become insensitive to it.[30,31] Both type 1 and type 2 diabetes mellitus are increasing in children, for reasons that are not well understood.[31] Diabetes incidence estimates vary widely in children based on age and ethnicity. Children of Pima Indians in Arizona have a high incidence of type 2 diabetes, at 22 per 1000 children as compared to 7.2 cases per 100,000 in children seen in Ohio.[14] In children, diabetes mellitus type 1 is more common than type 2. Type 1 diabetes is related to immune function and results in virtually no insulin production. For type 2 diabetes, which is more common in older children, teens, and adults, some insulin may be produced in the body. Because it is closely associated with overweight, type 2 diabetes may partially be managed by weight loss.[30,31] Consequences of poorly controlled diabetes in children are the same as in adults—risk of heart, eye, and kidney damage and premature death. Major changes in diabetes management for adults are based on 1994 guidelines that recommend increased monitoring of blood glucose and diet flexibility.[14] These concepts are also appropriate for children with diabetes, although some specific blood-monitoring levels for adults are not appropriate for children. Children with diabetes type 1 are more likely to have both high and low blood sugars, not just high blood sugars.

> **Insulin** Hormone usually produced in the pancreas to regulate movement of glucose from the bloodstream into cells within organs and muscles.
>
> **Postictal State** Time after a seizure of altered consciousness; appears like a deep sleep.

Treatment for diabetes is regulation of the timing and composition of meals and exercise, along with insulin injections or medications.[14] A third-grader with type 1 diabetes is likely to require an insulin injection once or twice per day, with oversight and modification of school breakfast, school lunch, and snack time based on physical activity in school and after-school activities. If the child is invited to a birthday party, the timing of meals and snacks can be adjusted to allow the child to attend the party and eat most of the foods there. Common colds, or foods a child refuses to eat, can cause wide variation in blood sugar, contributing to irritability, sleepiness, or difficulty with schoolwork.

In the summertime, many localities organize summer camps for children with diabetes; diet education and controlled access to a diabetic diet are provided along with the usual camp activities. Such disease-specific camps are good for breaking the social isolation that children experience when they feel they are the only ones required to follow special diets.

SEIZURES Seizures are uncontrolled electrical disturbances in the brain. Epilepsy and seizures are the same disorder. Seizures in children are a relatively common condition, with an incidence of 3.5 per 1000 children.[32,33] Seizure activity has a range of outward signs, from uncontrollable jerking of the whole body to mild blinking. Currently, no known nutrients bring on seizures. Children who have seizures are usually treated by medications that prevent seizures. After some types of seizures, the child may have a period of semiconsciousness called a *postictal state* and appear to be sleeping, but she is difficult to wake.[32,33] Feeding or eating during the postictal state is not recommended, because the child may choke. Some children have long enough postictal states to miss meals. Then adding other eating times is needed to make up for the lost calories and nutrients.

When seizures are controlled by medications, growth usually continues at the rate typical for that child. Dietary consequences of controlled seizures are primarily related to drug-nutrient side effects, such as change in hunger or sleepiness. Some drugs should be taken without food, and others may be offered with snacks or meals. Most drugs have to be taken on a strict schedule and are not stopped without medical supervision.

Some children have uncontrolled seizures that may cause further brain damage over time. For reasons that remain unknown, seizures decrease when brain metabolism is switched from the usual fuel, glucose, to **ketones** from fat metabolism.[33] Some specialty clinics administer the **ketogenic diet** for uncontrolled seizures. The ketogenic diet severely limits carbohydrates and increases calories from fat. An example intake on a ketogenic diet is given in Table 13.4. The diet is adequate in calories and protein. Vitamins and minerals have to be added as supplements because the allowed food sources of carbohydrates are not sufficient to meet vitamin and mineral requirements. The ketogenic diet may allow seizure medications to be reduced or eliminated over time. However, many difficulties, such as measuring

Ketones Small two-carbon chemicals generated by breakdown of fatty acids for energy.

Ketogenic Diet High-fat, carbohydrate meal plan in which ketones are made from metabolic pathways used to convert fat to energy.

growth, blood glucose, and ketones in urine, lie in monitoring the body's reaction to such severe carbohydrate restriction. Growth during the time on a ketogenic diet may be different from that seen in the child's previous pattern. Some children improve in both weight and height when seizure activity declines. The ketogenic diet is so high in fat that some children gain weight faster than expected. The diet is generally recommended for 2 years, if it shows demonstrated effectiveness.

CEREBRAL PALSY   Cerebral palsy (CP) is one of the most common conditions in children with severe disabilities (Illustration 13.1). Overall incidence of cerebral palsy is about 1.4–2.4 per 1000 children.[34] *Cerebral palsy* is a general term well understood by the public; it covers a broad range of conditions resulting from brain damage. Causes of CP all involve damage to the brain early in life, either before or after birth. The initial site of brain damage does not progress, but progression of secondary effects occurs over time. Secondary effects may include contractures, scoliosis, gastroesophageal reflux, and constipation.[6,10] Many children with CP have constipation because coordinated muscle movements are part of bowel emptying, including the muscles in and over the intestines. Muscle coordination problems most easily seen in movements of the arms and legs may occur in muscles all over the body, including the abdominal muscles that assist in bowel evacuation.[34]

The form of cerebral palsy that presents the most nutrition problems is spastic quadriplegia, involving all limbs.[2] Most children with spastic quadriplegia appear thin, but this appearance may be a result of brain damage or muscle size. Children with cerebral palsy often have

---

Table 13.4 Sample day's menu for a ketogenic diet for a 7-year-old treated for seizures; portion sizes are prescribed, and parents are taught to measure foods

Breakfast
Beverage: heavy whipping cream (4 fl oz)
Scrambled egg with bacon and mushrooms

Lunch
Heavy whipping cream (4 fl oz)
Hard-boiled egg mixed with mayonnaise
2 tb green beans with butter

After-school snack
Carbohydrate-free multivitamin and mineral pill
Sugar-free popsicle
Diet soda

Dinner
Heavy whipping cream (4 fl oz)
Black olives (3)
Sugar-free gelatin topped with whipped cream
Slice of full-fat ham
Slice of tomato

Snack
Walnuts (3)
Diet soda

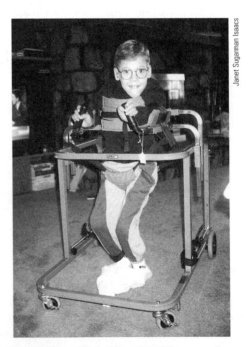

Janet Sugarman Isaacs

Illustration 13.1 Boy with CP in a walker.

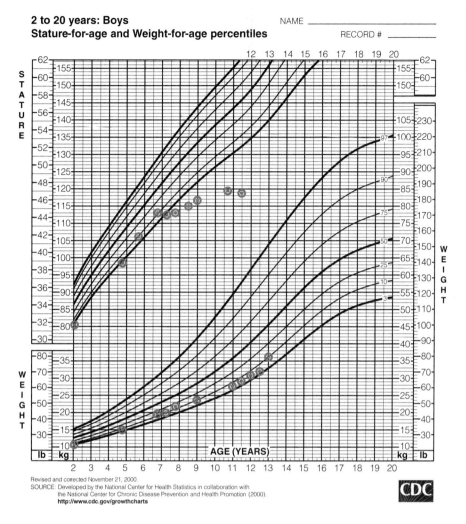

2 to 20 years: Boys
Stature-for-age and Weight-for-age percentiles

NAME _____

RECORD # _____

Illustration 13.2 Growth chart for gastrostomy feeding for a boy with spastic quadriplegia and scoliosis.

Revised and corrected November 21, 2000.
SOURCE: Developed by the National Center for Health Statistics in collaboration with the National Center for Chronic Disease Prevention and Health Promotion (2000).
http://www.cdc.gov/growthcharts

CDC

other forms of brain damage as well: 39–44% have mental retardation; 26–36% have seizures, 14–18% have severe visual impairment. Causes of CP are unknown for more than one-third of affected children, and the condition may or may not be related to preterm birth. About half the time after a preterm birth, a basis for CP can be identified from the perinatal period.[34] The prevalence of CP in children born with very low birthweight has been increasing, but this trend may be a consequence of the overall increase in survivors of severe prematurity.[6] Children with CP can enjoy many activities, attend school, and later contribute to society. Persons with CP display a wide range of abilities. As indicated in the growth chart in Illustration 13.2, children with spastic quadriplegia grow; but their growth is slower than others, with or without gastrostomy feeding.[2]

Nutritional consequences of spastic quadriplegia are slow weight gain and other growth concerns, difficulty with feeding and eating, and changes in body composition. No specific vitamins or minerals are known to correct CP. Problem nutrients are likely to be those related to bone density, calcium, and vitamin D, or nutrients needed in higher amounts as a result of medication side effects. Rec-

ommendations for caloric needs are difficult to determine, even with an in-depth growth assessment. Children with small or weak muscles have lower caloric needs because they are less active as a result of little voluntary muscle control. In contrast, types of CP characterized by increased uncontrolled movement require extra calories as a result of a higher activity level. *Athetosis* is an example of this less common form of CP, in which increased energy needs have been documented.[35] Altered body composition affects many aspects of the child's nutrition and eating abilities.[7] Eating or feeding problems may appear in the forms of spilling food, long mealtimes, fatigue at mealtime, and/or requiring assistance to eat. Difficulty in controlling muscles such as those in the neck and back, used in head position and sitting, and those in the jaw, tongue, and lips and used in swallowing, may contribute to feeding and eating problems.[34]

Athetosis Uncontrolled movements of the large muscle groups as a result of damage to the central nervous system.

Nutrition experts who provide services for children with CP monitor their growth, and then make recommendations for food choices that fit the children's abilities for eating, for nutritional supplements if food and beverages are not providing sufficient nutrients, and for nutrition support if needed. Nutrition interventions may include the following:

- Stimulating oral feeding
- Promoting healthy eating at school
- Adjusting menus and timing of meals and snacks at home or school for meeting nutrient needs from foods that minimize fatigue during meals
- Assessing and adjusting the child's diet over time
- Using adapted self-feeding utensils or other types of feeding equipment

PHENYLKETONURIA (PKU) PKU is an inborn error of protein metabolism with a prevalence of 1 in 12,000 live births.[4] The only treatment is lifelong dietary management, in which more than 80% of protein intake from foods and beverages is replaced by protein in which the amino acid phenylalanine has been removed. The enzyme that uses phenylalanine as a substrate is either not working at all, or only partially active, in a person with PKU. Treatment reduces intake of this amino acid to the minimum amount needed as an essential amino acid. This strategy limits toxic breakdown products of accumulated phenylalanine, which the body has difficulty clearing. How excess phenylalanine causes mental retardation is not known. The PKU diet is required throughout life (Illustration 13.3). If foods with protein are consumed in too high amounts, PKU slowly becomes a degenerative disease affecting the brain at whatever age the treatment is stopped. A woman with PKU has to continue strict dietary adherence because high levels of phenylalanine affect every pregnancy, even if the infant did not inherit PKU.

When their diets are managed correctly, children with PKU appear to be eating meals providing less food than the meals of other children. The diet is adequate in all vitamins, minerals, protein, fats, and calories, but more nutrients are in liquid rather than solid forms. Foods to be avoided completely are protein-rich foods such as meats, eggs, regular dairy products, peanuts, and soybeans in all forms. Allowed natural sources of protein are limited amounts of regular crackers, potato chips, rice, and potatoes. Many fruits and vegetables are encouraged, if offered without added sources of protein. Some foods that are high in fats and sugars and generally low in natural protein, such as fried vegetables or candy canes, are safe for children with PKU. Illustration 13.4 shows a food pyramid for PKU with the base of the pyramid the special medical food or formula that replaces the protein in foods.

The phenylalanine-deficient protein is generally served as a liquid, called a medical food or formula. The vitamins and minerals required to meet the RDA are in the phenylalanine-deficient protein powder. If the child does not drink enough of the PKU formula, foods that the child eats to meet her vitamin, mineral, and calorie needs will elevate the blood phenylalanine. Table 13.5 is a dietary recall from a child in good control of PKU. The phenylalanine-deficient protein also is available as bars and pills. It can be expensive to buy substitute low-protein alternative foods, such as low-protein pizza crusts, low-protein cheese, and low-protein baking mixes. Successful compliance requires use of low-protein foods to allow variety in the diet, such as low-protein pasta.

Illustration 13.3 This girl does not appear to have a chronic illness, but she has PKU.

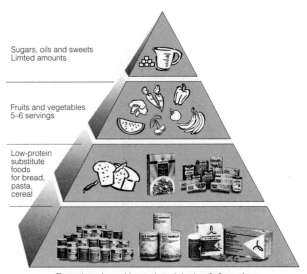

Sugars, oils and sweets
Limted amounts

Fruits and vegetables
5–6 servings

Low-protein substitute foods for bread, pasta, cereal

Formula or bar without phenylalanine 2–3 servings

Illustration 13.4

SOURCE: © 2000 SHS North America. Reprinted with permission.

Table 13.5 Dietary recall for a 5-year-old child with well-controlled PKU

Breakfast

2 slices low-protein bread with jelly and margarine

6 cut-up orange pieces

8 fl oz PKU formula

Lunch

½ c fruit cocktail in heavy syrup

1 c tossed salad (lettuce, tomato, celery, cucumber only) with 2 tb ranch dressing

17 french fries with ketchup

6 fl oz apple juice

After-school snack

½ c microwave popcorn

8 fl oz PKU formula

Dinner

pickle spears (dill, 3 wedges)

1 c low-protein imitation rice containing 1.5 tb margarine

½ c grilled onions, green peppers, and mushrooms (on rice)

1 c canned peaches in heavy syrup

8 fl oz PKU formula

Snack

Skittles candy (small snack size)

4 fl oz apple juice

ATTENTION DEFICIT HYPERACTIVITY DISORDER

Attention deficit disorder and attention deficit hyperactivity (ADHD) disorder are the most common *neurobehavioral* conditions in children. The incidence of ADHD is estimated at 3% to 5% of school-age children.[36,37] Children suspected of having ADHD may have a chaotic meal and snack pattern and the inability to stay seated for a meal. They may be given fewer opportunities to use kitchen appliances and get their own snacks due to impulsiveness. Theories about specific foods or nutrients causing ADHD have not been proven scientifically, but high interest in nutrition as a cause and treatment continues. The sale of herbal medicines and nutritional supplements to families with children with ADHD is common. One large survey found that 64% of children with ADHD had tried at least one type of alternative therapy and that 13% of children took some sort of multivitamin supplement.[38] Megavitamins were found ineffective in treating ADHD in a double-blind crossover trial, but such information has not stopped claims made for them or marketing and sales.[38] Often health care providers, including nutritionists, are not told about the use of these supplements by families.

Recommended procedures to confirm ADHD include at least two sites completing observation checklists about behavior. Treatment with the following two approaches has been most effective:

1. A structured behavioral approach that may also include mental health counseling and support, such as parenting classes

2. A prescribed *psychostimulant* medication; examples are Ritalin or Adderal

Nutritional concerns in ADHD include medication side effects that decrease appetite, maintaining growth while being medicated, and mealtime behavior. Low appetite as a result of treatment of ADHD is quite variable; it depends on the timing of the medications compared to meals, dosage, and medication schedule and on how long the child has had to adjust to the medication. Less interference with appetite and growth is likely if the child does not take the medication during school holidays.

Regardless of the child's dosage schedule, ADHD medication peak activity is aimed for school hours, which includes school lunch. Nutrition interventions for children on psychostimulant medications call for timing meals and snacks around the medication's action peaks. For example, adding a large bedtime snack when the medication's effects are low is a typical recommendation. Monitoring weight and height carefully over time helps identify growth plateaus. Education for the school's lunchroom supervisors and teachers may be helpful to deal with food refusals and mealtime behavior for the child with ADHD.[37]

PEDIATRIC HIV Most children with HIV were infected at around the time of birth. HIV in children under age 13 is classified differently than it is in adults, by age-based categories as well as level of immunosuppression from the virus. Only in the last few years have affected infants lived long enough to benefit from the combination highly active antiretroviral therapy that became available in the mid-1990.[39] As experience has accumulated with these potent medications, growth of children with low viral loads has improved, and opportunistic infections have occurred at a lower rate.

Nutrition is an important component of HIV management. Failure to thrive is common in infected infants regardless of drug treatment, due to the dampening effect of antiretroviral therapy on appetite and food intake. For children too young to be in charge of their own medication and eating schedules, educating the family and arranging support may be part of the nutrition intervention. Other nutrition concerns include food-related infection control measures, assuring access to complete nutritional supplements, and referrals to food banks. If weight-gain and medication compliance problems are unresolved, gastrostomy placement for medications and supplemental feedings may be needed.[39]

Working with children with HIV is complicated and demanding, and dietary approaches have to be customized to the behavioral and developmental realities of each child. For example, an 11-year-old girl is being treated for HIV and its related illnesses. Her diet prescription

> **Neurobehavioral** Pertains to control of behavior by the nervous system.
>
> **Psychostimulant** Classification of medication that acts on the brain to improve mental or emotional behavior.

includes a high-protein/high-calorie diet, with one complete vitamin/mineral supplement daily and three meals and three snacks. Her family members are to check her weight at home weekly and call in if they observe weight loss and low appetite. The girl takes four kinds of HIV-related medications, totaling 17 capsules per day. Two medications have no food-related restrictions, one is best taken with food at two different times per day, and one is best taken on an empty stomach (30 minutes before a meal or 2 hours after a meal). She also gets an injection every other week to strengthen her blood counts.

Dietary Supplements and Herbal Remedies

Families with children in a lengthy process of diagnosis—where the diagnosis does not lead to a definite treatment and when expense, insurance coverage, and administrative problems tax their ability to cope—are more likely to seek alternative therapies. Some of these alternatives have questionable effectiveness and perhaps are even harmful. No herbal remedies or nutritional supplements have been found effective to prevent or treat the conditions covered in this chapter; however, nutritional claims abound for various chronic conditions. Families hear from one another about micronutrients—such as magnesium, zinc, and B_6—sold with various combinations of amino acids for Down syndrome and autism.[40] Restrictive diets, such as avoiding dairy products or gluten, have been researched for one condition and then extrapolated for another. Sports drinks and high-protein products marketed for athletes may attract families with children who have difficulty gaining weight.

Strategies to counter unscientific nutritional claims for various products include the following:

- Recognize the benefits of support for families, such as advocacy groups.
- Improve communication with health care providers, so that families ask more questions about nutrition claims of alternative treatments.
- To give them some control over decision making for their children, give families reliable information, such as scientific literature or fact sheets, without endorsing any claim.

Sources of Nutrition Services

Children with chronic conditions that interfere with their ability to function may be eligible for Supplemental Social Insurance (SSI). Low-income families are eligible for SSI depending on the child's condition. Examples of conditions usually qualifying for SSI are chromosomal disorders; mental retardation; and severe forms of seizures,

cerebral palsy, and CF. A child with treated PKU is generally not eligible for SSI because treatments prevent decline in learning abilities. Also, the Americans with Disabilities Act applies to all ages. It requires, for example, that school cafeteria lines accommodate wheelchairs.

USDA Child Nutrition Program

The U.S. Department of Agriculture Child Nutrition Program, as described in previous chapters, requires that school breakfast and lunch menus be modified for children with diagnosis-specific diets or changes in the texture of foods. Parents who want their children to participate in the Child Nutrition Program cannot be charged an additional fee for providing a special diet for the child. A registered dietitian or another health provider completes a prescription ordering special breakfasts or lunches. Examples of diet prescription orders are a reduced-calorie school lunch and breakfast, a pureed diet, or a nutrient-modified diet, such as a PKU diet (see Case Study 13.2, page 321). If families do not want to participate in the Child Nutrition Program and prefer to pack lunches, they can change their decision any time during the school year. Formulas administered by gastrostomy are not required to be supplied by the Child Nutrition Program.

Maternal and Child Health Block Program of the U.S. Department of Health and Human Services (HHS)

Every state has a designated portion of federal funding for children with special health care needs.[41,42] A wide range of services can be provided based on state planning as reported back to HHS. Nutrition services may be in specialty clinics or county health departments; or they can be contracted with for providing care, assuring resources such as formulas, foods, and nutrition education. Nutrition experts work with children in various settings, including schools, early intervention programs, homes, clinics, and facilities. Also, a program in every state identifies and advocates for children with special needs. An example is the Developmental Disabilities Council.

Public School Regulations: 504 Accommodation and IDEA

Two sets of regulations guide how schools provide nutrition services in addition to the Child Nutrition Program. Nutrition services in schools are generally more available to younger children than they are to older children. Children in regular education have different access to services than do children eligible for special educational services. See Case Study 13.2.

504 ACCOMMODATION Children with special health care needs can have nutrition requirements met

Case Study 13.2

Dealing with Food Allergies in School Settings

Judy is to start regular kindergarten now. When she was 2 years old, she was diagnosed with a peanut food allergy after many episodes of asthma and hives. Her health has improved with avoidance of peanuts in all forms as a preschooler. The family has carefully watched what she eats. However, at age 4 she had an episode of breathing difficulty that required an emergency room visit. This incident makes the family quite concerned about Judy's eating at school. She is generally not allowed to go to friends' homes to play; friends come to her house so the family can watch out for her. She has been instructed not to take any food from anyone. She has not been in day care or preschool, so starting school is a big step for the family.

Judy's mother meets with the school staff to discuss plans for Judy at school to avoid exposure to peanuts in any form. The family does not want to participate in the school lunch program. Her mother plans to pack Judy a lunch from home, although most children eat food provided at school. Judy's mother obtains a letter from her pediatrician about the peanut allergy and gives copies to the school nurse, teacher, and principal. With the school staff, Judy's mother discusses snack time for Judy and her eating at the cafeteria, which periodically serves food cooked in peanut oil, or food containing peanuts. The snack at kindergarten is provided by parents, based on a rotation schedule. It is usually milk or juice with cookies or fruit. A plan is put in place for the teacher to check the snack foods and offer a replacement snack provided by Judy's mother if she is unsure whether that day's snack contains peanuts.

The school working-group has written out a 504 accommodation plan for Judy's peanut allergy. It includes making sure that tables where children are eating peanut-containing foods are washed well, and posting signs in the cafeteria with Judy's picture to make sure she does not inadvertently get peanut-containing food from another child or in a food activity.

After Judy has been in school for 1 month, the family meets with the school group. In that month, two episodes have resulted in what may have been hives, and her family is worried that Judy is not adjusting well. At snack time Judy does not recognize some of the foods; she refuses to eat a snack most days. She appears hungry after school at home. Her mother says she would like to send a snack to school that she knows her daughter will eat, and she wants to attend school during snack time to make sure Judy is not being teased.

Questions

1. Why is it the school's job to check for peanuts when other parents are sending snacks?
2. The parents seem overprotective. Can the teacher transfer Judy to another classroom?
3. What are the chances that Judy will outgrow the peanut allergy by the next year?

(Answers are in Appendix D.)

---

through the provisions of the 504 Accommodation of the Rehabilitation Act of 1973.[1] Examples of children who may need 504 accommodations are those with diabetes, cystic fibrosis, or arthritis. A written plan, set in place, may provide an additional snack during the school day or additional time to eat during school lunch. The 504 accommo-

dation procedures are appropriate for documenting that a child with PKU needs an alternative food during a classroom birthday or Halloween party, or when food activities take place in the classroom. This procedure gives parents a means of making sure that all are aware of required dietary restrictions. It is not the responsibility of the child with

PKU to inform new staff or substitutes about her need for a special diet. As a result of implementing the 504 accommodation, a child with PKU is not left out of food activities at school.

INDIVIDUALS WITH DISABILITIES EDUCATION ACT

Children eligible for special education are covered by regulations within the Individuals with Disabilities Education Act (IDEA).[1] It requires each child to have an individualized education plan (IEP) that may include nutrition-related goals and objectives as needed. The school staff must involve the parent in developing the IEP. For some diagnoses, it would be appropriate for a nutritionist to attend the IEP meeting to make sure the teacher, teacher's aide, and other staff understand what the child needs. Nutrition, eating, and feeding problems may be a part of that plan, and apply to food offered in the classroom as well as that served in the regular school cafeteria. An example of IEP goals and objectives can be found in Table 13.6. For this particular plan, the child's education includes learning to eat by mouth with prompting and assistance. Nutritional supplements may be purchased as part of an education intervention called for in the child's IEP.

Nutrition Intervention Model Program

The Maternal and Child Health Bureau (MCH) is a part of the Department of Health and Human Services and funds nutrition services for chronically ill children.[41,42] MCH develops and promotes model programs by funding competitive grants that emphasize training health care providers, including nutrition experts. Model programs that are targeted for children with special health care needs are necessary because most health care providers are not comfortable caring for children with rare conditions. Families of children with special health care needs often have difficulty locating nutrition experts and other providers who are familiar with their complex nutritional, medical, and educational needs. Training programs vary in length from short, intensive courses to year-long traineeships. Topics vary from nutrition for infants receiving intensive care services to nutrition problems of adolescence, such as warning signs of anorexia nervosa. Examples of such federal grant programs are the Pediatric Pulmonary Centers and Leadership Education in Neurodevelopmental Disabilities.[42] Funding nutrition faculty is one of the provisions of these grants.

Table 13.6 Example of nutrition objectives in an individualized education plan for an 8-year-old boy with limited oral feeding skills

1. In three of five trials JR will hold food on the spoon as he moves it to his mouth without hand-over-hand assistance of his aides during three meals per week.

2. JR will point to what he wants to eat with his left hand three trials after two prompts per meal three days each week.

3. JR will cooperate in having his gastrostomy site checked at feedings by pulling up his shirt three days in a row of each week.

Resources

Ability On-Line Support Network
Connects young people with disabilities and chronic illnesses to peers and mentors with and without disabilities.
Web site: www.ablelink.org

American Diabetes Association
Allows searches about diabetes in children around the world and provides professional and consumer publications.
Web site: www.diabetes.org

Cystic Fibrosis Foundation
This large organization has information about research, services, and policy related to cystic fibrosis, including nutritional products and recommendations.
Web site: www.cff.org

Exceptional Parent
This magazine is an excellent resource for a wide variety of conditions, and includes tips for parents on how to work with care providers and educators.
Web site: www.eparent.com

Ketogenic Diet
This Web site is maintained by the Packard Children's Hospital at Stanford, which provides credible information for parents and providers about the ketogenic diet and places that use it.
Web site: www.stanford.edu/group/ketodiet

National Down Syndrome Society
Includes information about its policy regarding nutritional products and directs parents to local resources for working with schools.
Web site: www.ndss.org

References

1. General Information about disabilities. National Information Center for Children and Youth with Disabilities. Washington, DC: U.S. Government Printing Office. Available at www.Nichcy.org/general.htm, accessed 2/1/01.

2. Krick J, Murphy-Miller P, Zeger S, Wright, E. Pattern of growth in children with cerebral palsy. J Am Diet Assoc 1996; 96:680–685.

3. U.S. Department of Agriculture, Human Nutrition Information Service. Food Guide Pyramid; 1992; Home and Garden Bulletin No. 572.

4. Batshaw ML, Tuchman M. PKU and other inborn errors of metabolism. In: Batshaw ML, ed. Children with disabilities. 5th ed. Baltimore, MD: Paul H. Brookes; 2002: 333–45.

5. Institute of Medicine Food and Nutrition Board. Dietary reference intakes for energy, carbohydrate, fiber, fat, fatty acids, cholesterol, protein, and amino acids 2002.

6. Blackman, JA. Medical aspects of developmental disabilities in children birth to three. Gaithersburg, MD: Aspen Publications; 1997: 44–54, 88–90.

7. Eicher PS. Feeding. In: Batshaw ML, ed. Children with disabilities. 5th ed. Baltimore, MD: Paul H. Brookes; 2002: 549–66.

8. Stallings VA. Resting energy expenditure. In: Altschuler SM, Liacouras CA. Clinical pediatric gastroenterology. Philadelphia: Harcourt Brace; 1998: 607–11.

9. L'Alleman D, Eiholzer U, Schlump M et al. Cardiovascular risk factors improve during three years of growth hormone therapy in Prader-Willi syndrome. Eur J Pediatr 2000;159:836–42.

10. Minns, RA. Neurological disorders. In: Kelnar CJH et al., eds. Growth disorders. London: Chapman and Hall Medical; Philadelphia, PA, USA; 1998: 447–70.

11. Schertz M, Adesman AR, Alfieri NE, Bienkowski RS. Predictors of weight loss in children with attention deficit hyperactivity disorder treated with stimulant medication. Pediatrics 1996;98:763–9.

12. Berney TP. Autism—an evolving concept. British J of Psychiatry 2000;176:20–5.

13. Wooldridge N. Pulmonary diseases. In: Samour PQ, Kelm KK, Lang CE. Handbook of pediatric nutrition, 2nd ed. Gaithersburg, MD: Aspen Publishers; 1999: 315–54.

14. Chalmers KH. Diabetes. In: Samour PQ, Kelm KK, and Lang CE. Handbook of pediatric nutrition, 2nd ed. Gaithersburg, MD; Aspen Publishers: 1999: 425–51.

15. Trumbo P, Yates AA, Schlicker SA, Poos M. Dietary reference intakes; vitamin A, vitamin K, arsenic, boron, chromium, copper, iodine, iron, manganese, molybdenum, nickel, silicon, vanadium, and zinc. J Am Diet Assoc 2001;101:294–301.

16. American Academy of Pediatrics. Alternative routes of drug administration—advantages and disadvantages (subject review). Pediatrics 1997;100:143–52.

17. American Academy of Pediatrics. Calcium requirements of infants, children, and adolescents (RE9904). Pediatrics 1999; 1152–7.

18. Ross EA, Szabo NJ, Tabbett IR. Lead content of calcium supplements. J Am Med Assoc 2000;284:1425–9.

19. National Center for Health Statistics. NCHS growth curves for children 0–19 years. U.S. Vital and Health Statistics, Health Resources Administration. Washington, DC: U.S. Government Printing Office; 2000.

20. Growth references: third trimester to adulthood, 2nd ed. Clinton, SC: Greenwood Genetic Center; 1998.

21. Frisancho AR. Triceps skinfold and upper arm muscle size norms for assessment of nutritional status. Am J Clin Nutr 1974;27:1052–7.

22. Fomon S, Haschke F, Ziegler E, Nelson SE. Body composition of reference children from birth to age 10 years. American J Clin Nutr 1982;35:1169–75.

23. Goran MI, Driscoll P, Johnson R et al. Cross-calibration of body-composition techniques against dual-energy x-ray absorptiometry in young children. Am J Clin Nutr 1996;63:299–305.

24. Nellhaus G. Composite international and interracial graphs. Pediatrics 1968; 41:106–14.

25. Batshaw ML. Chromosomes and heredity. In: Batshaw ML, ed. Children with disabilities. 5th ed. Baltimore, MD: Paul H. Brookes; 2002: 3–26, 770.

26. Steinkamp G, Von der Hardt H. Improvement of nutritional status and lung function after long-term nocturnal gastrostomy feedings in cystic fibrosis. J Pediatrics 1994;121:244–9.

27. Walkowiak J, Przyslawski J. Five-year prospective analysis of dietary intake and clinical status in malnourished cystic fibrosis patients. J Hum Nutr Diet. 2003;16:225–31.

28. Sullivan PB. Gastrostomy feeding in the disabled child: when is the antireflux procedure required? Arch Dis Child. 1999;81: 463–4.

29. Shingadia D, Viani RM, Yogev R et al. Gastrostomy tube insertion for improvement of adherence to highly active antiretroviral therapy in pediatric patients with human immunodeficiency virus. Pediatrics 2000;105:1–5. Available at www .pediatrics.org/cgi/content/full/105/6/e80, accessed 2/1/01.

30. American Diabetes Association. Type 2 diabetes in children and adolescents. Pediatrics 2000;671–80.

31. American Diabetes Association. Nutrition recommendations and principles for people with diabetes mellitus. Diabetes Care 1998;21:532–5.

32. Weinstein S. Epilepsy. In: Batshaw ML, ed. Children with disabilities. 5th ed. Baltimore, MD: Paul H. Brookes; 2002: 493–523.

33. Tallian KB, Nahata MC, Tsao C-Y. Role of the ketogenic diet in children with intractable seizures. Ann Pharmacotherapy 1998;32:349–56.

34. Pellegrino L. Cerebral palsy. In: Batshaw ML, ed. Children with disabilities. 5th ed. Baltimore, MD: Paul H. Brookes; 2002: 443–66.

35. Johnson RK, Goran MI, Ferrara MS, Poehlman ET. Athetosis increases resting metabolic rate in adults with cerebral palsy. J. Amer Diet Assoc 1996;96:145–8.

36. American Academy of Pediatrics Clinical Practice. Guideline: diagnosis and evaluation of the child with attention deficit/hyperactivity disorder. Pediatrics 2000;105:1158–70.

37. Isaacs, JS, Watkins, A Hodgens, JB Zachor DA. Nutrition and ADHD: implications for school lunch. Topics in Clin Nutr 2002;17:27–39.

38. Cala S, Crismon ML, Baumgartner J. A survey of herbal use in children with attention-deficit-hyperactivity disorder or depression. Pharmacotherapy 2003;23: 222–30.

39. Spiegel HML, Bondwit AM. HIV infection in children. In: Batshaw ML, ed. Children with disabilities. 5th ed. Baltimore, MD: Paul H. Brookes; 2002: 123–39.

40. Position statement on vitamin-related therapies. National Down Syndrome Society; 1997.

41. Story M, Holt K, Sofka D, eds. Bright futures in practice: nutrition. Arlington, VA: National Center for Education and Child Health; 2000: 266–70.

42. Maternal and Child Health Bureau, Health and Human Services. Available at http://mchb.hrsa.gov/html/drte.html, accessed 4/5/04.

> "The willingness of adolescents to try out new behaviors creates a unique opportunity for nutrition education and health promotion. Adolescence is an especially important time in the life cycle for nutrition education since dietary habits adopted during this period are likely to persist into adulthood."
>
> Jamie Stang

Adolescent Nutrition

Chapter Outline

- Introduction
- Nutritional Needs in a Time of Change
- Normal Physical Growth and Development
- Normal Psychosocial Development
- Health and Eating-Related Behaviors during Adolescence
- Energy and Nutrient Requirements of Adolescents
- Nutrition Screening, Assessment, and Interventions
- Physical Activity and Sports
- Promoting Healthy Eating and Physical Activity Behaviors

Prepared by **Jamie Stang** and **Mary Story**

Key Nutrition Concepts

1   Nutrition needs should be determined by the degree of sexual maturation and biological maturity (biological age) instead of by chronological age.

2   Unhealthy eating behaviors common among adolescents include frequent dieting, meal skipping, use of unhealthy dieting practices, and frequent consumption of foods high in fat and sugar, such as fast foods, soft drinks, and savory snacks.

3   Concrete thinking and abstract reasoning abilities do not develop fully until late adolescence or early adulthood; therefore, education efforts need to be highly specific and based on concrete principles.

Introduction

Adolescence is defined as the period of life between 11 and 21 years of age. It is a time of profound biological, emotional, social, and cognitive changes during which a child develops into an adult. Physical, emotional, and cognitive maturity is accomplished during adolescence (Illustration 14.1). Many adults view adolescence as a tumultuous, irrational phase that children must go through. However, this view does disservice to its developmental importance. The tasks of adolescence, not unlike those experienced during the toddler years, include the development of a personal identity and a unique value system separate from parents and other family members, a struggle for personal independence accompanied by the need for economic and emo-

> **Puberty**   The time frame during which the body matures from that of a child to that of a young adult.

tional family support, and the adjustment to a new body that has changed in shape, size, and physiological capacity. When the seemingly irrational behaviors of adolescents are reframed as essential endeavors and viewed in light of these developmental tasks, adolescence can and should be viewed as a unique, positive, and integral part of human development.

Nutritional Needs in a Time of Change

The biological, psychosocial, and cognitive changes associated with adolescence have direct effects on nutritional status. The dramatic physical growth and development experienced by adolescents significantly increases their needs for energy, protein, vitamins, and minerals. However, the struggle for independence that characterizes adolescent psychosocial development often leads to the development of health-compromising eating behaviors, such as excessive dieting, meal skipping, use of unconventional nutritional and nonnutritional supplements, and the adoption of fad diets. These disparate situations create a great challenge for health care professionals. The challenging behaviors of adolescents can become opportunities for change at a time during which adult health behaviors are being formed. The search for personal identity and independence among adolescents can lead to positive, health-enhancing behaviors such as adoption of healthful eating practices, participation in competitive and noncompetitive physical activities, and an overall interest in developing a healthy lifestyle. These interests and behaviors provide a good foundation on which nutrition education can build.

This chapter provides an overview of normal biological and psychosocial growth and development among adolescents and how these experiences affect the nutrient needs and eating behaviors of teens. Common concerns related to adolescent nutrition and effective methods for educating and counseling teens are also discussed.

Normal Physical Growth and Development

Early adolescence encompasses the occurrence of *puberty*, the physical transformation of a child into a young adult. The biological changes that occur during puberty include sexual maturation, increases in height and weight, accumulation of skeletal mass, and changes in body composition. Even though the sequence of these events during puberty is consistent among adolescents, the age of onset, duration, and tempo of these events vary a great deal among and within individuals. Thus, the physical appearance of adolescents of the same chronological age covers a wide range. These variations directly affect the nutrition requirements of adolescents. A 14-year-old male who has

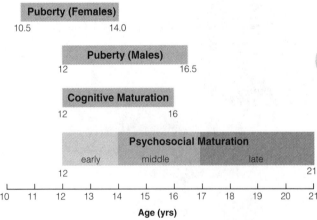

Illustration 14.1 Average ages of pubertal, cognitive, and psychosocial maturation.

SOURCE: From "Adolescent Growth and Development" by R. L. Johnson, in *Adolescent Medicine* by A. Hoffman and D. Greynaus. Copyright © 1988 McGraw-Hill Companies. Reprinted with permission.

already experienced rapid linear growth and muscular development will have noticeably different energy and nutrient needs than a 14-year-old male peer who has not yet entered puberty. For this reason, sexual maturation (or biological age) should be used to assess biological growth and development and the individual nutritional needs of adolescents rather than chronological age.

Sexual maturation rating (SMR), also known as "Tanner Stages," is a scale of *secondary sexual characteristics* that allows health professionals to assess the degree of pubertal maturation among adolescents, regardless of chronological age (Table 14.1). SMR is based on breast development and the appearance of pubic hair among females, and testicular and penile development and the appearance of pubic hair among males. SMR stage 1 corresponds with prepubertal growth and development, while stages 2 through 5 denote the occurrence of puberty. At SMR stage 5, sexual maturation has concluded. Sexual maturation correlates highly with linear growth, changes in weight and body composition, and hormonal changes.[1]

The onset of *menses* and changes in height relative to the development of secondary sexual characteristics that occur in females during puberty are shown in Illustration 14.2 on the next page. Among females, the first signs of puberty are the development of breast buds and sparse, fine pubic hair occurring on average between 8 and 13 years of age (SMR stage 2). *Menarche* occurs 2 to 4 years after the initial development of breast buds and pubic hair, most commonly during SMR stage 4. The average age of menarche is 12.4 years, but menarche can occur as early as 9 or 10 years or as late as 17 years of age. Menarche may be delayed in highly competitive athletes or in girls who severely restrict their caloric intake to limit body fat.

Ethnic and racial differences are evident in the initiation of sexual maturation among females. Recent research suggests that African American girls experience puberty earlier than their Caucasian peers.[2] By 8 years of age, 48% of African American girls had reached SMR stage 2, compared to only 15% of Caucasian females. The average age of initial breast development is approximately 8.8 years for African American girls and 9.9 years for Caucasian females, while pubic hair growth begins at 8.7 years in African American females and about 2 years later in Caucasian girls. Menarche, however, occurs at about the same time: 12.2 years for African American and 12.8 for Caucasian adolescents. These findings suggest that the average length of puberty may be somewhat longer for African American females than for Caucasian females.

> **Secondary Sexual Characteristics** Physiological changes that signal puberty, including enlargement of the testes, penis, and breasts and the development of pubic and facial hair.
>
> **Menses** The process of menstruation.
>
> **Menarche** The occurrence of the first menstrual cycle.

Table 14.1. Sexual maturity rating for girls and boys

Girls

Stage	Breast Development	Pubic Hair Growth
1	Prepubertal; nipple elevation only	Prepubertal; no pubic hair
2	Small, raised breast bud	Sparse growth of hair along labia
3	General enlargement of raising of breast and areola	Pigmentation, coarsening and curling, with an increase in amount
4	Further enlargement with projection of areola and nipple as secondary mound	Hair resembles adult type, but not spread to medial thighs
5	Mature, adult contour, with areola in same contour as breast, and only nipple projecting	Adult type and quantity, spread to medial thighs

Boys

Stage	Genital Development	Pubic Hair Growth
1	Prepubertal; no change in size or proportion of testes, scrotum, and penis from early childhood	Prepubertal; no pubic hair
2	Enlargement of scrotum and testes; reddening and change in texture in skin of scrotum; little or no penis enlargement	Sparse growth of hair at base of penis
3	Increase first in length, then width of penis; growth of testes and scrotum	Darkening, coarsening and curling, increase in amount
4	Enlargement of penis with growth in breadth and development of glands; further growth of testes and scrotum, darkening of scrotal skin	Hair resembles adult type, but not spread to medial thighs
5	Adult size and shape genitalia	Adult type and quantity, spread to medial thighs

SOURCE: From J. M. Tanner, Growth at Adolescence. Copyright © 1962 Blackwell Publishers. Reprinted with permission.

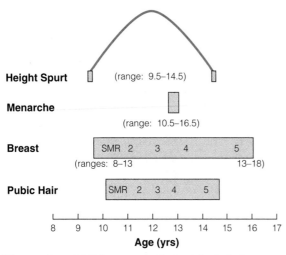

Illustration 14.2 Sequence of physiological changes during puberty in females.

SOURCE: From J. M. Tanner, Growth at Adolescence. Copyright © 1962 Blackwell Publishers. Reprinted with permission.

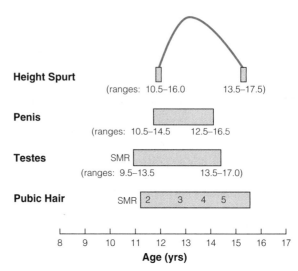

Illustration 14.3 Sequence of physiological changes during puberty in males.

SOURCE: From J. M. Tanner, Growth at Adolescence. Copyright © 1962 Blackwell Publishers. Reprinted with permission.

Onset of the linear growth spurt occurs most commonly during SMR stage 2 in females, beginning between the ages of 9.5 and 14.5 years in most females (Illustration 14.2). Peak velocity in linear growth occurs during the end of SMR stage 2 and during SMR stage 3, approximately 6 to 12 months prior to menarche. As much as 15% to 25% of final adult height will be gained during puberty, with an average increase in height of 9.8 inches (25 cm).[3] During the peak of the adolescent growth spurt, females gain approximately 3.5 inches (8–9 cm) a year. The linear growth spurt lasts 24 to 26 months, ceasing by age 16 in most females. Some adolescent females experience small increments of growth past age 19 years, however. Linear growth may be delayed or slowed among females who severely restrict their caloric intake.

Enlargement of the *testes* and change in *scrotal* coloring are most often the first signs of puberty among males (Illustration 14.3), which occurs between the ages of 10.5 and 14.5 years with 11.6 years being the average age. The development of pubic hair is also common during SMR stage 2. Testicular enlargement begins between the ages of 9.5 and 13.5 years in males (SMR 2 to 3) and concludes between the ages of 12.7 and 17 years. The average age of "spermarche" is approximately 14 years among males. Clearly, males show a great deal of variation in chronological age at which sexual maturation takes place.

On average, peak velocity of linear growth among males occurs during SMR stage 4, coinciding with or just following testicular development and the appearance of faint facial hair. The peak velocity of linear growth occurs at 14.4 years of age, on average. At the peak of the growth spurt, adolescent males will increase their height by 2.8 to 4.8 inches (7 to 12 cm) a year. Linear growth will continue

Testes One of the two male reproductive glands located in the scrotum.

throughout adolescence, at a progressively slower rate, ceasing about 21 years of age.

Changes in Weight, Body Composition, and Skeletal Mass

As much as 50% of ideal adult body weight is gained during adolescence. Among females, peak weight gain follows the linear growth spurt by 3 to 6 months. During the peak velocity of weight change, which occurs at an average age of 12.5 years, girls will gain approximately 18.3 pounds (8.3 kg) per year.[3] Weight gain slows around the time of menarche, but will continue into late adolescence. Adolescent females may gain as much as 14 pounds (6.3 kg) during the latter half of adolescence. Peak accumulation of muscle mass occurs around or just after the onset of menses.

Body composition changes dramatically among females during puberty, with average lean body mass falling from 80% to 74% of body weight while average body fat increases from 16% to 27% at full maturity. Females experience a 44% increase in lean body mass and a 120% increase in body fat during puberty.[4] Adolescent females gain approximately 2.5 pounds (1.14 kg) of body fat mass each year during puberty. Adolescent body fat levels peak among females between the ages of 15 to 16 years. Research by Frisch suggests that a level of 17% body fat is required for menarche to occur and that 25% body fat is required for the development and maintenance of regular ovulatory cycles.[5] Normal changes in body fat mass can be mediated by excessive physical activity and/or severe caloric restriction.

Even though the accumulation of body fat by females is obviously a normal and physiologically necessary process, adolescent females often view it negatively. Weight dissatisfaction is common among adolescent females during and immediately following puberty, leading to potentially health-compromising behaviors such as excessive caloric restriction, chronic dieting, use of diet pills or laxatives, and, in some cases, the development of body image distortions and eating disorders.

Among males, peak weight gain coincides with the timing of peak linear growth and peak muscle mass accumulation.[3] During peak weight gain, adolescent males gain an average of 20 pounds (9 kg) per year. Body fat decreases in males during adolescence, resulting in an average of approximately 12% by the end of puberty.

Almost half of adult peak bone mass is accrued during adolescence. By age 18, more than 90% of adult skeletal mass has been formed.[1] A variety of factors contribute to the accretion of bone mass, including genetics, hormonal changes, weight-bearing exercise, cigarette smoking, consumption of alcohol, and dietary intake of calcium, vitamin D, protein, phosphorus, boron, and iron. Because bone is comprised largely of calcium, phosphorus, and protein and because a great deal of bone mass is accrued during adolescence, adequate intakes of these nutrients are critical to support optimal bone growth and development.

Normal Psychosocial Development

During adolescence, an individual develops a sense of personal identity, a moral and ethical value system, feelings of self-esteem or self-worth, and a vision of occupational aspirations. Psychosocial development is most readily understood when it is divided into three periods: early adolescence (11 to 14 years), middle adolescence (15 to 17 years), and late adolescence (18 to 21 years). Each period of psychosocial development is marked by the mastery of new emotional, cognitive, and social skills (Table 14.2).

During early adolescence, individuals begin to experience dramatic biological changes related to puberty. The development of body image and an increased awareness of sexuality are central psychosocial tasks during this period of adolescence. The dramatic changes in body shape and size can cause a great deal of ambivalence among adolescents, leading to the development of poor body image and eating disturbances if not addressed by family members or health care professionals.

Peer influence is very strong during early adolescence. Young teens, conscious of their physical appearance and social behaviors, strive to "fit in" with their peer group. The need to fit in can affect nutritional intake among adolescents. Focus groups conducted with adolescent females divide food into two main categories: junk food and healthy food.[6] Consumption of junk food was associated with friends, fun, weight gain, and guilt, while consumption of healthy food was associated with family, family meals, and home life. Clearly, teens express their ability and willingness to fit in with a group of peers by adopting food preferences and making food choices based on peer influences and by refuting family preferences and choices.

The wide chronological age range during which pubertal growth and development begins and proceeds can become a major source of personal dissatisfaction for many adolescents. Males considered to be "late bloomers" often feel inferior to their peers who mature earlier and may resort to the use of anabolic steroids and other supplements in an effort to increase linear growth and muscle development. Females who mature early have been found to have more eating problems and poorer body image than their later-developing peers.[7,8] They are also more likely to initiate "grown-up" behaviors such as smoking, drinking alcohol, and engaging in sexual intercourse at an early age.[9] Education of young adolescents on normal variations in tempo and timing of growth and development can help to facilitate the development of a positive self-image and body image and may reduce the likelihood of early initiation of health-compromising behaviors.

Cognitively, early adolescence is a time dominated by concrete thinking, egocentrism, and impulsive behavior. Abstract reasoning abilities are not yet developed to a

Table 14.2 Psychosocial processes and the substages of adolescent development

Substage	Emotionally Related	Cognitively Related	Socially Related
Early adolescence	Adjustment to a new body image; adaptation to emerging sexuality	Concrete thinking; early moral concepts	Strong peer effect
Middle adolescence	Establishment of emotional separation from parents	Emergence of abstract thinking; expansion of verbal abilities and conventional morality; adjustment to increased school demands	Increased health risk behavior; heterosexual peer interests; early vocational plans
Late adolescence	Establishment of a personal sense of identity; further separation from parents	Development of abstract, complex thinking; emergence of postconventional morality	Increased impulse control; emerging social autonomy; establishment of vocational capability

Source: From G. M. Ingersoll, "Psychological and Social Development," In Textbook of Adolescent Medicine. Copyright © 1992 Saunders. Reprinted with permission.

great extent in most adolescents, limiting their ability to understand complex health and nutrition issues. Young adolescents also lack the ability to see how their current behavior can affect their future health status or health-related behaviors.

Middle adolescence marks the development of emotional and social independence from family, especially parents. Conflicts over personal issues, including eating and physical activity behaviors, are heightened during mid-adolescence. Peer groups become more influential and their influence on food choices peaks. Physical growth and development are mostly completed during this stage. Body image issues are still of concern, especially among males who are late to mature and females. Peer acceptance remains important, and the initiation of and participation in health-compromising behaviors often occurs during this stage of development. Adolescents may believe they are invincible during this stage.

The emergence of abstract reasoning skills occurs rapidly during middle adolescence; however, these skills may not be applied to all areas of life. Adolescents will revert to concrete thinking skills if they feel overwhelmed or experience psychosocial stress. Teens begin to understand the relationship between current health-related behaviors and future health, even though their need to "fit in" may supplant this understanding.

Late adolescence is characterized by the development of a personal identity and individual moral beliefs. Physical growth and development is largely concluded, and body image issues are less prevalent. Older teens become more confident in their ability to handle increasingly sophisticated social situations, which is accompanied by reductions in impulsive behaviors and peer pressure. Adolescents become even less economically and emotionally dependent upon parents. Relationships with one individual become more influential than the need to fit in with a group of peers. Personal choice emerges.

Abstract thinking capabilities are realized during late adolescence, which assists teens in developing a sense of future goals and interests. Adolescents are now able to understand the perspectives of others and can fully perceive future consequences associated with current behaviors. This capability is especially important among adolescent females who plan to have children or who become pregnant.

Health and Eating-Related Behaviors during Adolescence

Eating patterns and behaviors of adolescents are affected by many factors, including peer influences, parental modeling, food availability, food preferences, cost, convenience, personal and cultural beliefs, mass media, and body image. Illustration 14.4 presents a conceptual model of the many factors that influence eating behaviors of ado-

Illustration 14.4 Conceptual model for factors influencing eating behavior of adolescents.

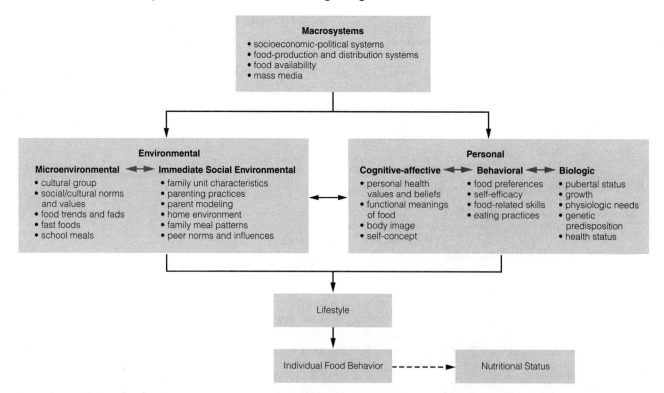

SOURCE: From M. Story and I. Alton, "Becoming a Woman: Nutrition in Adolescence," in D. A. Krummel and P. M. Kris-Etherton (Eds.), in Nutrition in Women's Health. Copyright © 1996 Aspen Publishers. Reprinted with permission.

lescents. The model depicts three interacting levels of influence on adolescent eating behaviors: personal or individual, environmental, and macrosystems. Personal factors that influence eating behavior include attitudes, beliefs, food preferences, *self-efficacy,* and biological changes. Environmental factors include the immediate social environment such as family, friends and peer networks, and other factors such as school, fast-food outlets, and social and cultural norms. Macrosystem factors, which include food availability, food production and distribution systems, and mass media and advertising, play a more distal and indirect role in determining food behaviors, yet can exert a powerful influence on eating behaviors. To improve the eating patterns of youth, nutrition interventions should be aimed at each of the three levels of influence.

Eating habits of adolescents are not static; they fluctuate throughout adolescence, in relation to psychosocial and cognitive development. Longitudinal data of adolescent females suggest that even though body weight percentiles track throughout adolescence, little consistency guides the intakes of energy, nutrients, vitamins, and minerals from early to late adolescence.[10] Health professionals must therefore refrain from jumping to conclusions about the dietary habits of adolescents (even if they have been evaluated for nutritional status at an earlier age) and take the time to assess an individual's current dietary intake.

Adolescents lead busy lives. Many are involved in extracurricular sports or academic activities, others are employed, and quite a few must care for younger children in a family for part of the day. These activities, combined with the increased need for social and peer contact and approval, and increasing academic demands as they proceed through school, leave little time for adolescents to sit down to eat a meal. Snacking and meal skipping are commonplace among adolescents.

Almost all adolescents consume at least one snack per day, with a range of one to seven.[11,12] One study of adolescents found that during an average week, males ate 18.2 meals and 10.9 snacks, while females ate 16.9 meals and 9.9 snacks.[6] Snacks account for 25% to 33% of daily energy intakes among adolescents.[12] National data suggest that the proportion of calories and nutrients from foods consumed as snacks has risen during the past decade.[13] Unfortunately, the food choices made by adolescents while snacking tend to be high in sugar, sodium, and fat while relatively low in vitamins and minerals. Soft drinks, the most commonly chosen snacks for adolescent females, account for about 6% of total caloric intake.[14] This trend causes significant concern because high consumption of soft drinks increases the risk for bone fractures over an individual's lifetime.[15]

The occurrence of meal skipping increases as adolescents mature. Breakfast is the most commonly skipped meal; only 29% of adolescent females tend to eat breakfast daily.[11] Skipping breakfast can dramatically decrease intakes of energy, protein, fiber, calcium, and folate due to the absence of breakfast cereal or other nutrient-dense foods commonly consumed at breakfast. Almost one-quarter of adolescents skip lunch.[16,17] As with breakfast, skipping lunch reduces intakes of energy, protein, and other nutrients. Adolescents who skip meals should be counseled on convenient, portable, and healthy food choices that can be taken with them and eaten as meals or snacks.

> **Self-Efficacy** The ability to make effective decisions and to take responsible action based upon one's own needs and desires.

As adolescents mature, they spend less time with family and more time with their peer group. Eating away from home becomes prevalent: female adolescents, for example, eat almost one-third of their meals away from home, and the average teen eats at a fast-food restaurant twice a week. Fast-food visits account for 31% of all food eaten away from home, and make up 83% of adolescent visits to restaurants.[16,18] Fast-food restaurants and food courts are favorite eating places of teens for several reasons:

- They offer a social setting with an informal, comfortable atmosphere for adolescents.
- Fast foods are relatively inexpensive and offer socially acceptable choices.
- Fast foods can be eaten outside the restaurant, fitting into the busy schedules of adolescents.
- Service is fast and the limited offerings allow for quick decision making.
- Fast-food restaurants employ many adolescents, increasing the social value of these restaurants.

Eating at fast-food restaurants has a direct bearing on the nutritional status of adolescents. Many fast foods are high in fat and low in fiber and nutrients. However, specific choices increase the nutrient content of fast-food meals and decrease the fat content. Adolescents can be counseled to ask for juice or milk instead of soft drinks, order small sandwiches instead of larger choices, choose a salad as a side dish instead of fries, order grilled items as opposed to fried sandwiches, and avoid "supersizing" meals, even if this seems to offer a better economic deal.

Vegetarian Diets

The term *vegetarian* is used quite broadly and can consist of many different eating patterns. Table 14.3 on the following page lists the most common vegetarian diet patterns along with the foods most commonly excluded. Among low-literacy populations, *vegetarian* may be thought to mean that a person eats vegetables. Therefore, health professionals should ask adolescents to define what type of vegetarian diet they consume and elicit a complete list of the foods being avoided.

The prevalence of vegetarianism among adolescents is small—approximately 1% of adolescents report consuming a vegetarian diet.[19] Adolescents adopt vegetarian eating

Table 14.3 Types of vegetarian diets and food excluded

Type of Vegetarian Diet	Foods Excluded
Semi- or partial vegetarian	Red meat
Lacto-ovo vegetarian	Meat, poultry, fish, seafood
Lactovegetarian	Meat, poultry, fish, seafood, eggs
Vegan (total vegetarian)	Meat, poultry, fish, seafood, eggs, dairy products (may exclude honey)
Macrobiotic	Meat, poultry, eggs, dairy, seafood, fish (fish may be included in the diets of some macrobiotic vegetarians)

SOURCE: Reprinted with permission from E. Haddad and P. Johnston, "Vegetarian Diets and Pregnant Teens," in M. Story and J. Strang (Eds.), Nutrition and the Pregnant Adolescent: A Practical Reference Guide. Minneapolis, MN: Center for Leadership, Education, and Training in Maternal and Child Nutrition, University of Minnesota, 2000.

Table 14.4 Suggested daily food guide for lacto-ovo vegetarians and vegan adolescents at various intake levels

	SERVINGS PER DAY, BY DAILY CALORIC INTAKE	
FOOD GROUPS	Lacto-Ovo Vegetarians 11+ years (2200–2800 kcal)	Vegans 11+ years (2200–2800 kcal)
Breads, grains, cereal	9–11	10–12
Legumes	2–3	2–5
Vegetables	4–5	3–5
Fruits	4	4–5
Nuts, seeds	1	4–6
Milk, yogurt, cheese	3	0
Eggs (limit 3/week)	½	0–1
Fats, oils (added)	4–6	4–6
Sugar (added teaspoons)	6–9	6–9

SOURCE: Data used with permission from E. H. Haddad, "Development of a Vegetarian Food Guide," in American Journal of Clinical Nutrition, 1994:59:307–16; and M. Story, et al. (Eds.), Bright Futures in Practice: Nutrition. © 2000 National Center for Education in Maternal and Child Health.

plans for a variety of reasons, including cultural or religious beliefs, moral or environmental concerns, and health beliefs; as a means to restrict calories and/or fat intake; and as a means of exerting independence by adopting eating behaviors that differ from those of the teen's family. Regardless of the reason for consuming a vegetarian diet, the adolescent's diet should be thoroughly assessed for nutritional adequacy. As a rule, the more foods that are restricted in the diet, the more likely it is that nutritional deficiencies will result.

Vegetarian adolescents have been found to be shorter and leaner than omnivores during childhood and to enter puberty at a later age. On average, menarche occurs 6 months later in vegetarians than among omnivores.[20] After puberty, vegetarian adolescents are as tall or taller than omnivores and are generally leaner, although final adult height may be reached at a later age.[20,21]

Well-planned vegetarian diets can offer many health advantages to adolescents, such as a high intake of complex carbohydrates and relatively high intake of vitamins and minerals found in plant-based foods. Data suggest that vegetarian adolescents are twice as likely to consume fruits and vegetables, one-third as likely to consume sweets, and one-fourth as likely to consume salty snack foods once per day as omnivores.[19] When well planned, vegetarian diets can provide adequate protein to promote growth and development among pubescent adolescents, particularly if small amounts of animal-derived foods, such as milk or cheese, are consumed at least two times per week. If vegetarian diets restrict intake of all animal-derived food products, however, careful attention must be paid to assure adequate intakes of protein, calcium, zinc, iron, and vitamins D, B_6, and B_{12}.[21] Supplements of vitamin B_{12} are often required among vegans. A suggested dietary food

guide for adolescent vegetarians is given in Table 14.4. Also see Case Study 14.1.

Adolescents who consume vegan diets must also be assessed for adequacy of total fat and essential fatty acid intakes. Docosahexaenoic acid (DHA) is derived from alpha-linolenic acid. Although it is found in soy products, flaxseed, nuts, eggs, and canola oil, DHA intake is very low in the diets of vegans. Diets that are low in fat may not supply an adequate ratio of linoleic acid to alpha-linolenic acid in order to facilitate the metabolism of alpha-linolenic acid to DHA.[21] Therefore, particular attention should be paid to sources of fat in the diets of vegans and other vegetarians with low fat intakes. Some plant sources of alpha-linolenic acid are shown in Table 14.5.

Table 14.5 Plant sources of alpha-linolenic acid

Food Source	Alpha-Linolenic Acid, g
Flaxseed, 2 tb	4.3
Walnuts, 1 oz	1.9
Walnut oil, 1 tb	1.5
Canola oil, 1 tb	1.6
Soybean oil, 1 tb	0.9
Soybeans, ½ c cooked	0.5
Tofu, ½ c	0.4

SOURCE: Reprinted with permission. E. Haddad and P. Johnston, "Vegetarian Diets and Pregnant Teens." In M. Story and J. Stang (eds.), Nutrition and the Pregnant Adolescent: A Practical Reference Guide. Minneapolis, MN: Center for Leadership, Education, and Training in Maternal and Child Nutrition, University of Minnesota, 2000.

Case Study 14.1

Nutritional Issues and the Vegan Adolescent

Nikki is a 13-year-old female who is being seen by her family physician for her annual physical examination. Nikki's sexual maturation rating is estimated at between 2 and 3. Her body mass index (BMI) is plotted at just under the 20th percentile. During the visit, Nikki reports that she is following a strict vegan diet. She explains that she thinks it is a healthier way to eat and states that she avoids meat, poultry, fish, seafood, eggs, and dairy products.

Rubberball Productions

Questions

1. How might a vegetarian diet affect an adolescent's growth and maturation?
2. Why is protein a concern in Nikki's diet?
3. Which vitamins and minerals may be deficient in Nikki's diet? For each vitamin or mineral, list possible food sources that are appropriate for a vegan diet.
4. Is it necessary to assess Nikki for the presence of an eating disorder? Explain your answer.

(Answers are in Appendix D.)

Adolescents who consume a vegetarian diet, particularly if they report doing so for health- or weight-related reasons, should be carefully assessed for the presence of eating disorders, chronic dieting, and body image disturbances. Neumark-Sztainer and colleagues have found that vegetarian adolescents are somewhat more likely to report binge eating, almost twice as likely to report frequent or chronic dieting, four times more likely to report purging, and eight times more likely to report laxative use than nonvegetarian peers.[19] Rather than indicating the vegetarian diet as the cause of these behaviors, these results seem to stem from the fact that many individuals who are chronic dieters or who have eating disorders adopt a vegetarian diet as a means of self-denial or self-control.

Dietary Intake and Adequacy among Adolescents

Data on food intakes of U.S. adolescents suggest that many adolescents consume diets that do not match the Dietary Guidelines for Americans or the Food Guide Pyramid recommendations.[22,23] National data suggest that only 1% of teens consume diets that meet recommendations for all food groups in the Food Guide Pyramid, and 5% meet the recommendations for at least four food groups.[24] More than 45% of teens meet recommendations for one of the food groups, while 7% of male and 18% of female adolescents do not meet any of the Food Guide Pyramid recommendations. Inadequate consumption of dairy products, grain products, fruits, and vegetables is commonplace among adolescents (Table 14.6).

Few adolescents meet the 5 a Day recommendation for fruit and vegetable consumption. Data from the CSFII suggests that 32% of males and 18% of females met this recommendation over a 3-day period.[25] Krebs-Smith and colleagues found that 56.7% of male and 63.7% of female teens consumed less than 1 serving of fruit per day.[25] When all vegetables were included in analyses, 9% of males and 13% of females consumed less than 1 serving of vegetables per day. However, when french fries were excluded from analyses, 20% of males and 26% of females consumed less than 1 serving of vegetables per day. French fries alone account for 23% of all vegetables consumed by adolescents. Data from the Minnesota Adolescent Health Survey suggest a slightly different pattern of fruit and vegetable intake, where 28% of adolescents

Table 14.6 Percentage of adolescents meeting the recommended number of Food Guide Pyramid servings for select food groups

	Male, %	Female, %
Dairy products	49	22
Fruits	17	19
Vegetables	50	46
Grains	43	21

SOURCE: Data from K. A. Munoz, S. M. Krebs-Smith, R. Ballard-Barbash, and L. E. Cleveland, "Food Intakes of U.S. Children and Adolescents Compared with Recommendations," *Pediatrics* 101(5), 1998: 952–3, and *Pediatrics* 100 (1997): 323–9.

Table 14.7 Dietary reference intakes of selected nutrients for preadolescents and adolescents

Life-Stage Group	Calcium, mg/d	Phosphorus, mg/d	Magnesium, mg/d	Vitamin D, μg/d[a,b]	Iron, mg/d	Thiamin, mg/d	Riboflavin, mg/d	Niacin, mg/d[c]
Males								
9–13 years	1300*	1250	240	5*	8	0.9	0.9	12
14–18 years	1300*	1250	410	5*	11	1.2	1.3	16
19–30 years	1000*	700	400	5*	8	1.2	1.3	16
Females								
9–13 years	1300*	1250	240	5*	8	0.9	0.9	12
14–18 years	1300*	1250	360	5*	15	1.0	1.0	14
19–30 years	1000*	700	310	5*	15	1.1	1.1	14
Pregnancy								
≤18 years	1300*	1250	400	5*	27	1.4	1.4	18
19–30 years	1000*	700	350	5*	27	1.4	1.4	18
Lactation								
≤18 years	1300*	1250	360	5*	10	1.4	1.6	17
19–30 years	1000*	700	310	5*	9	1.4	1.6	17

NOTE: This table presents Recommended Dietary Allowances (RDAs) in **bold type** and Adequate Intakes (AIs) in ordinary type followed by an asterisk (*). RDAs and AIs may both be used as goals for individual intake. RDAs are set to meet the needs of almost all (97 to 98%) individuals in a group. For healthy and breastfed infants, the AI is the mean intake. The AI for other life-stage and gender groups is believed to cover needs of all individuals in the group, but lack of data or uncertainty in the data prevent being able to specify with confidence the percentage of individuals covered by this intake. Source: Reprinted with permission from Dietary Reference Intakes: Recommended Intakes for Individuals, © by the National Academy of Sciences. Courtesy of the National Academy Press, Washington, D.C.

[a] As cholecalciferol. 1 mg cholecalciferol = 40 IU vitamin D.
[b] In the absence of adequate exposure to sunlight.
[c] As niacin equivalents (NE). 1 mg of niacin = 60 mg tryptophan; 0–6 months = preformed niacin (not NE).

consumed fruit and 36% consumed vegetables on a less-than-daily basis.[26] Both surveys indicate that adolescents' fruit and vegetable intake is not adequate to promote optimal health and reduce risk of chronic diseases. Clearly, adolescents do not consume diets that comply with the national nutrition recommendations or provide the recommended level of intakes for all food groups.

Socioeconomic status (SES) appears to be related to food-intake patterns. Consumption of fruit and fruit juices, grain products, low-fat/skim milk and milk products, soft drinks, and sugars and sweets tends to increase as SES increases.[13,25] Consumption of vegetables is also positively related to SES, with the exception of starchy vegetables (white potatoes, dried beans, green peas, corn) and fried potatoes, which are negatively related to SES. Consumption of whole milk and total meat/poultry/fish intakes decreases among adolescents as SES increases.

Energy and Nutrient Requirements of Adolescents

Increases in lean body mass, skeletal mass, and body fat that occur during puberty result in energy and nutrient needs that exceed those of any other point in life. Energy and nutrient requirements of adolescence correspond with the degree of physical maturation that has taken place. Unfortunately, little available data define optimal nutrient and energy intakes during adolescence. Most existing data are extrapolated from adult or child nutritional requirements. Recommended intakes of energy, protein, and some other nutrients are based upon adequate growth as opposed to optimal physiological functioning. The Dietary Reference Intakes (DRIs) provide the best estimate of nutrient requirements for adolescents (Table 14.7). It should be noted, however, that these nutrient recommendations are classified according to chronological age, as opposed to individual levels of biological development. Thus, health care professionals must use prudent professional judgment based on SMR status, and not solely on chronological age, when determining the nutrient needs of an adolescent.

Nutrient intakes of U.S. adolescents suggest that many adolescents consume inadequate amounts of vitamins and minerals; this trend is more pronounced in females than males. It is not surprising, given that most adolescents do not consume diets that comply with the Food Guide Pyramid or the Dietary Guidelines for Americans. On average, adolescents consume diets inadequate in several vitamins and minerals, including folate; vitamins A, B_6, and E; iron; zinc; and calcium.[27] Dietary fiber intake among adolescents is also low. Diets consumed by many teens exceed current recommendations for total and

Table 14.7 Dietary reference intakes of selected nutrients for preadolescents and adolescents (continued)

Life-Stage Group	Vitamin B$_6$, mg/d	Folate, μg/d	Vitamin B$_{12}$, μg/d	Pantothenic Acid, mg/d	Biotin, μg/d	Vitamin A, μg/RAE/d	Vitamin C, mg/d	Vitamin E, mg/d	Selenium, μg/d
Males									
9–13 years	1.0	300	1.8	4*	20*	600	45	11	40
14–18 years	1.3	400	2.4	5*	25*	900	75	15	55
19–30 years	1.3	400	2.4	5*	30*	900	90	15	55
Females									
9–13 years	1.0	300	1.8	4*	20*	600	45	11	40
14–18 years	1.2	400^d	2.4	5*	25*	700	65	15	55
19–30 years	1.3	400^d	2.4	5*	30*	700	75	15	55
Pregnancy									
≤18 years	1.9	600	2.6	6*	30*	750	80	15	60
19–30 years	1.9	600	2.6	6*	30*	770	85	15	60
≤18 years	2.0	500	2.8	7*	35*	1200	115	19	70
19–30 years	2.0	500	2.8	7*	35*	1300	120	19	70

SOURCE: Food and Nutrition Board, Institute of Medicine, National Academies. *Dietary Reference Intakes: Recommended Intakes for Individuals.* Washington, DC: National Academy Press, 2000.

^{d} In view of evidence linking folate intake with neural tube defects in the fetus, it is recommended that all women capable of becoming pregnant consume 400 mg from supplements or fortified food until their pregnancy is confirmed and they enter prenatal care, which ordinarily occurs after the end of the periconceptional period—the critical time for formation of the neural tube.

saturated fats, cholesterol, sodium, and sugar. Data on nutrient intakes of adolescents taken from the 1994–1995 CSFII suggest that 47% to 62% of teens consume less than 75% of the DRI for calcium, 49% to 52% consume less than 75% of the DRI for zinc, and 36% to 55% consume less than 75% of the DRI for vitamins A and E. Similar findings have been reported from the School Nutrition Dietary Assessment Study (SNDAS).[28] More than one-third of females consume inadequate amounts of all of these nutrients.[27]

Based on growth and development of adolescents, as well as national findings on dietary intakes of foods and nutrients, adolescent diets should be assessed for adequacy of intake of vitamins and minerals, energy, protein, carbohydrate, and fiber. Nutrients of particular concern for teens are discussed in greater detail in the following sections.

Energy

Energy needs of adolescents are influenced by activity level, basal metabolic rate, and increased requirements to support pubertal growth and development. Basal metabolic rate (BMR) is closely associated with the amount of lean body mass of individuals. Because adolescent males experience greater increases in height, weight, and lean body mass, they have significantly higher caloric requirements than females do. The RDAs for total calories and

calories per centimeter of height by age group are listed in Table 14.8. Due to the great variability in the timing of growth and maturation among adolescents, the calculation of energy needs based on height provides a better estimate than that based on total caloric recommendation. Adolescents, especially females, may not meet the DRI for total energy intakes. Approximately 99% of teen males and 86% of teen females meet or exceed the DRI for energy.[13]

The DRI for energy is based upon the assumption of a light to moderate activity level. Therefore, adolescents who participate in sports, those who are in training to increase muscle mass, and those who are more active than average may require additional energy to meet their individual needs. Conversely, adolescents who are not physically active or those who have chronic conditions or physical disabilities that limit their mobility will require less energy to meet their needs. Physical activity has been found to decline throughout adolescence, with approximately one in four adolescents involved in no physical activity.[29] Therefore, caloric needs of older adolescents who have completed puberty and are less active may be significantly lower than those of younger, active, still-growing adolescents.

Physical growth and development during puberty is sensitive to energy and nutrient intakes. When energy intakes fail to meet requirements, linear growth may be retarded and sexual maturation may be delayed. The standard way to gauge adequacy of energy intake is to assess height, weight, and body composition. If, over time, height as well as weight-for-height continuously fall within the same percentiles when plotted on gender-appropriate National Center for Health Statistics growth charts, it can be assumed that energy needs are being met. If percentile of weight-for-height measurements begins to fall or rise, energy intake should be thoroughly assessed and adjustments in energy intake should be made accordingly. The use of body fat measurements, such as triceps and subscapular skinfold measurements, can provide useful information when weight-for-height does not remain consistent. Remember, however, that transient increases and decreases in body fat are commonly noted among adolescents during puberty due to the variation in timing of increases in height, weight, and accumulation of body fat and lean body mass. Repeated measurements of weight, height, and body composition over a several-month period are needed to accurately assess adequacy of growth and development.

Protein

Protein needs of adolescents are influenced by the amount of protein required to maintain existing lean body mass, plus allowances for the amount required to accrue additional lean body mass during the adolescent growth spurt. Because protein needs vary with the degree of growth and development, requirements based upon developmental age will be more accurate than absolute recommendations based upon chronological age. Recommended protein intakes based upon age, gender, and height are shown in Table 14.8. Protein requirements per unit of height are highest for females at 11 to 14 years, and for males at 15

to 18 years. These peak periods of protein need correspond to the usual timing of peak height velocity. The 2002 DRI report sets the RDA for protein intake for females and males aged 9–13 years at 0.95 g/kg/d, and at 0.85 g/kg/d for 14–18-year-olds. As with energy, growth is affected by protein intakes. When protein intakes are consistently inadequate, reductions in linear growth, delays in sexual maturation, and reduced accumulation of lean body mass may be seen.

Traditionally, U.S. adolescents consume more than adequate amounts of protein. National data suggest that on average teens consume about two times the recommended level of protein intake.[28] Subgroups of adolescents may be at risk for marginal or low protein intakes, however, including adolescents from food-insecure households, those who severely restrict calories, and those who consume semivegetarian or vegetarian diets, most notably vegans.

Carbohydrates

Carbohydrates provide the body's primary source of dietary energy. Carbohydrate-rich foods such as fruit, vegetables, whole grains, and legumes are also the main source of dietary fiber. Absolute requirements for carbohydrate intake among adolescents have not been established. Instead, dietary recommendations suggest that 50% or more of total daily calories should come from carbohydrate, with no more than 10% of calories derived from sweeteners, such as sucrose and high-fructose corn syrup. Data from a major study suggest that adolescents consume approximately 53% of their calories as carbohydrate.[28] Foods that contribute the most carbohydrate to the diets of adolescents include (in descending order) yeast bread, soft drinks, milk, ready-to-eat cereal, and foods such as cakes, cookies, quick breads, donuts, sugars, syrups, and jams.[14]

Sweeteners and added sugars provide approximately 16% of total calories to the diets of adolescents.[24] Soft drinks are a major source of added sweeteners in the adolescent diet, accounting for more than 12% of all carbohydrate consumed.[13] (See Table 14.9.) An estimated 86% of the soft drinks consumed by teens are sweetened, nondiet soft drinks.[30]

Dietary Fiber

Dietary fiber is important for normal bowel function and may play a role in the prevention of chronic diseases such as certain cancers, coronary artery disease, and type 2 diabetes mellitus. Adequate fiber intake is also thought to reduce serum cholesterol levels, moderate blood sugar levels, and reduce the risk of obesity. The American Academy of Pediatrics (AAP) Committee on Nutrition has recommended that dietary fiber intakes among children and adolescents should be 0.5 grams per kilogram of body weight.[31] This corresponds to average fiber intakes of

Table 14.8 Recommended caloric (kcal) and protein intakes for adolescents

LIFE-STAGE GROUP	CALORIES, kcal		PROTEIN, grams	
Age, years	Kcal/day	Kcal/cm*	Grams/day	Grams/cm
Females				
11–14	2200	14.0	46	0.29
15–18	2200	13.5	44	0.27
19–24	2200	13.4	46	0.28
Males				
11–14	2500	15.9	45	0.29
15–18	3000	17.0	59	0.34
19–24	2900	16.4	58	0.33

*2.54 cm = 1 in

SOURCE: Data taken from E. J. Gong and F. P. Heard, "Diet, Nutrition, and Adolescence." In M. E. Shils, J. A. Olson, and M. Shike (eds.), *Modern Nutrition in Health and Disease*, 8th ed. Philadelphia: Lea & Febiger, 1994; and *1989 Recommended Daily Allowances*, 10th ed. of the RDAs, Food and Nutrition Board, Commission on Life Sciences. Washington, DC: National Academy Press, 1989.

Table 14.9 Soft-drink consumption among adolescents

Ounces per Day	Adolescents, %
≥26 oz per day	22
13–25 oz per day	28
0.1–12 oz per day	32
0 oz per day	18

SOURCE: Data from L. Harnack, J. Stang, and M. Story, "Soft-Drink Consumption among U.S. Children and Adolescents: Nutritional Consequences," *J Amer Diet Assoc* 99 (1999): 436–41.

15.5 to 34.5 grams per day among 10- to 18-year-old males, and 16.0 to 28.5 grams per day among 10- to 18-year-old females. The AAP has further recommended that fiber intake not exceed 35 grams per day, because levels above this amount may reduce the bioavailability of some minerals. More recent recommendations by the American Health Foundation suggest that average fiber requirements for children and adolescents be based on an "age plus five" rule, where the individual's age is added to the number 5 to determine minimum fiber requirements.[32] A factor of 10 is added to age to determine the recommended upper limit of fiber intake. Thus, a 14-year-old would require 19–24 grams of fiber each day, in contrast to the AAP recommendation of 25 grams per day.

National data indicate that adolescent males consume 11.5 to 15.4 grams of fiber, while female adolescents consume 10.0 to 14.0 grams of fiber each day.[33,34] During adolescence, fiber intake among males increases slightly with age; among females, it decreases with age. Significant sources of fiber in the adolescent diet include breads, ready-to-eat cereal, potatoes, popcorn and related snack foods, tomatoes, and corn.[16] The low intake of fruit and vegetables among adolescents is the greatest contributing factor affecting fiber intake among adolescents. Adolescents who skip breakfast or do not routinely consume whole grain breads or ready-to-eat cereals are at the highest risk for having an inadequate consumption of fiber.

Fat

The human body requires dietary fat and essential fatty acids for normal growth and development. Current recommendations by the National Cholesterol Education Program (NCEP) suggest that children over the age of 2 years consume no more than 30% of calories from fat, with no more than 10% of calories derived from saturated fat.[35] The 2002 DRIs for fat intake for children and adolescents, which were released after the NCEP report, indicate that 4–18-year-olds should consume 25–35% of total calories from fat. The DRIs recommend that children and adolescents consume as little saturated and trans fats as possible while consuming an adequate diet.

Data on energy and macronutrient intakes among adolescents suggest that approximately 33% of total calories consumed are derived from fat, with 12% originating from saturated fat.[13] Approximately two-thirds of teens meet NCEP recommendations for total fat and saturated fat intake. Major sources of total and saturated fat intakes among adolescents include milk, beef, cheese, margarine, and foods such as cakes, cookies, donuts, and ice cream.[14] NCEP guidelines also suggest that adolescents consume no more than 300 milligrams of dietary cholesterol per day. This recommendation is exceeded by 48% of male and 32% of female teens.[36] Significant sources of cholesterol in adolescent diets include eggs, milk, beef, poultry, and cheese.

Calcium

Achieving an adequate intake of calcium during adolescence is crucial to physical growth and development. Calcium is the main constituent of bone mass. Because about half of peak bone mass is accrued during adolescence, calcium intake is of great importance in developing dense bone mass and reducing the lifetime risk of fractures and osteoporosis. Female adolescents appear to have the greatest capability to absorb calcium at about the time of menarche, with calcium absorption rates decreasing from then on.[37] Calcium absorption rates in males also peak during early adolescence, a few years later than in females. Young adolescents have been found to retain up to four times as much calcium as young adults do. By age 24 in females and 26 in males, calcium accretion in bone mass is almost nonexistent.[38] Clearly, an adequate intake of calcium is of paramount importance during adolescence.

The DRI for calcium for 9- to 18-year-olds is 1300 milligrams per day (Table 14.7). National data suggest that many adolescents, most notably females, do not consume the DRI for calcium. Adolescent females consume 536–815 mg of calcium per day, while adolescent males have been found to consume about 681–1146 mg of calcium each day.[6,39–41] These levels of dietary intake are not adequate to support the development of optimal bone mass. Supplements may be warranted for adolescents who do not consume adequate calcium from dietary sources.

Milk provides the greatest amount of calcium in the diets of adolescents, followed by cheese, ice cream, and frozen yogurt.[14] Adolescents increasingly consume their calcium in the form of fortified foods. One study indicated that more than half of teens reported drinking calcium-fortified juices, and 31% ate cereals fortified with calcium.[41] Other foods that may be supplemented with calcium include margarine and bread. These foods may become important sources of calcium for adolescents.

The consumption of soft drinks by adolescents may displace the consumption of more nutrient-dense beverages, including milk and fortified juices. A study by Harnack and colleagues showed an inverse relationship between the intake of carbonated beverages and the intake of milk and juice.[30] Because milk and fortified juices are significant sources of calcium in the diets of adolescents, interventions aimed at reducing consumption of soft drinks may be warranted. Such interventions are especially important in light of the growing body of evidence suggesting that carbonated beverage consumption increases the risk of bone fractures among children and adolescents.[15]

Calcium consumption drops as age increases among both male and female adolescents; however, males consume greater amounts of calcium at all ages than do females.[6,39–40] Calcium intakes among adolescents are highly correlated with energy intakes. When dietary calcium intake is adjusted for energy intake, no differences in calcium density of diets are found between males and females.[41] This fact suggests that females who restrict calories in an effort to control their weight are at particularly high risk for inadequate calcium intakes. The relationship between socioeconomic status (SES) and calcium intake is not clear. Some studies show a weak positive relationship, while others find no relationship.[6,36,42] Some variation in calcium intake follows ethnic/race categories among females: Cuban, Asian, and African American females consume less calcium on average than do Mexican American, Puerto Rican, and Caucasian females.

A recent study of knowledge regarding calcium and bone health revealed some interesting findings. When adolescents were questioned about their knowledge of the health benefits of calcium, 92% knew that it was needed to strengthen bones, 60% knew that it was required for "good" teeth, and 60% realized that adolescence was a critical time for the development of peak bone mass. Only 19% of teens knew that the recommended intake of milk, milk products, or fortified soymilk for their age group is 4 servings per day.[41] Nutrition education and interventions that target calcium consumption by older children and young teens are needed.

Iron

The rapid rate of linear growth, the increase in blood volume, and the onset of menarche during adolescence increase a teen's need for iron. The DRIs for iron for male and female adolescents are shown in Table 14.7. These recommendations are based on the amount of dietary iron intake needed to maintain a suitable level of iron storage, with additional amounts of iron added to cover the rapid linear growth and onset of menstruation that occur in male and female adolescents, respectively. Note that even

Serum Iron, Plasma Ferritin, and **Transferring Saturation** Measures of iron status obtained from blood plasma or serum samples.

Heme Iron Iron contained within a protein portion of hemoglobin that is in the ferrous state.

Nonheme Iron Iron contained within a protein of hemoglobin that is in the ferric state.

though DRIs are based on chronological age, the actual iron requirements of adolescents are based on sexual maturation level. Iron needs of an adolescent are highest during the adolescent growth spurt in males, and after menarche in females.

Estimates of iron deficiency among adolescents are 2.8% to 3.5% of 11- to 14-year-old females, 4.1% of 11- to 14-year-old males, 6.0% to 7.2% of 15- to 19-year-old females, and 0.6% of 15- to 19-year-old males.[33] The age-specific hemoglobin and hematocrit values used to determine iron-deficiency anemia are listed in Table 14.10. Hemoglobin and hematocrit levels, although commonly used to screen for the presence of iron-deficiency anemia, are actually the last serum indicators of depleted iron stores to drop. More sensitive indicators of iron stores include *serum iron, plasma ferritin,* and *transferring saturation.* Recent data suggest that 17% to 25% of female adolescents had at least two abnormal indexes of iron status. Thus, although the prevalence of iron-deficiency anemia may be relatively low among adolescents, a larger proportion may have inadequate iron stores. This finding is particularly relevant among adolescents from low-SES homes, because rates of iron deficiency tend to be higher in adolescents from low-income families.

The availability of dietary iron for absorption and utilization by the body varies by its form. The two types of dietary iron are *heme iron,* which is found in animal products, and *nonheme iron,* which is found in both animal and plant-based foods. Heme iron is highly bioavailable while nonheme iron is much less so. More than 80% of the iron consumed is in the form of nonheme iron. Bioavailability of nonheme iron can be enhanced by con-

Table 14.10 Hemoglobin and hematocrit cut-point values for iron-deficiency anemia in adolescents

Gender/Age[a]	Hemoglobin, <g/dL Less Than:	Hematocrit, <% Less Than:
Males and Females		
8–12 years	11.9	35.4
Males		
12–15 years	12.5	37.3
15–18 years	13.3	39.7
18+ years	13.5	39.9
Females[b]		
12–15 years	11.8	35.7
15–18 years	12	35.9
18+ years	12	35.7

[a]Age- and gender-specific cutoff values for anemia are based on the 5th percentile from the third National Health and Nutrition Examination Survey (NHANES III).
[b]Nonpregnant and lactating adolescents.
SOURCE: Abridged from Centers for Disease Control and Prevention. "Recommendations to Prevent and Control Iron Deficiency in the United States," *MMWR* 47 (no. RR-3): 1998.

suming it with heme sources of iron or vitamin C. This point is particular salient for adolescents who avoid animal foods as a means of restricting calories, and for those who consume few animal-based foods (semivegetarian) or vegetarian diets for moral or cultural reasons.

Dietary intakes of iron range from 10.0 to 12.5 mg per day in females.[33,40] Data suggest that 32% of male and 83% of female teens consume less than the DRI for iron.[36] The most common dietary sources of iron in diets of adolescents include ready-to-eat cereal, bread, and beef.[14]

Zinc

Zinc is particularly important during adolescence because of its role in the synthesis of RNA and protein, and its role as a cofactor in over 200 enzymes. The body's need for zinc, along with its ability to retain zinc, dramatically increases during the adolescent growth spurt. Zinc is required for sexual maturation to occur. Males who are zinc deficient experience growth failure and delayed sexual development. Zinc supplementation of both male and female zinc-deficient adolescents from developing countries often initiates accelerated growth and sexual development. Data on zinc nutrition of adolescents are limited, but evidence shows that serum zinc levels decline in response to the rapid growth and hormonal changes during adolescence. Serum zinc levels indicative of mild zinc deficiency (<10.71 µmol/L) have been found in 18% to 33% of female adolescents.[33]

The bioavailability of zinc from dietary sources is highly dependent upon the source of zinc. Zinc from animal sources is more bioavailable than that from plant-based sources. Undigestible fibers found in many plant-based sources of zinc can inhibit its absorption by the body. Zinc and iron compete for absorption, so elevated intakes of one can reduce the absorption of the other. Adolescents who take iron supplements may be at increased risk of developing mild zinc deficiency if iron intake is more than twice as high as zinc intake.

Dietary intakes of zinc among adolescent females range from 6.6 to 7.9 mg per day.[33,40] National surveys suggest that 75% of males and 81% of females consume less than the DRI for zinc, with 46% of males and 59% of females consuming less than 77% of recommended intakes.[36] The top five sources of dietary zinc consumed by adolescents include beef, milk, ready-to-eat cereal, cheese, and poultry.[14] Vegetarians, particularly vegans, and teens who do not consume many animal-derived products are at highest risk for low intakes of zinc.

Folate

Folate is an integral part of DNA, RNA, and protein synthesis. Thus, adolescents have increased requirements for folate during puberty. The DRI for folate is listed in Table 14.7. Folate in the form of folic acid is twice as bioavail-able as other forms of folate. For this reason, dietary folate equivalents (DFEs) are used in the DRIs. One microgram (mcg or µg) of folic acid is equivalent to approximately 2 DFEs, while 1 mcg of other forms of folate is equivalent to 1 DFE. Folic acid is the form of folate added to fortified cereals, breads, and other refined grain products.

Severe folate deficiency results in the development of megaloblastic anemia, which is rare among adolescents. Evidence, however, indicates that a significant proportion of adolescents have inadequate folate status. Twelve percent of adolescent females are mildly folate deficient, based on low serum folate levels, while 8% to 48% of female teens have been shown to have low red cell folate levels indicative of subclinical folate deficiency.[43,44] Serum folate levels drop during adolescence among females as sexual maturation proceeds, suggesting that increased folate needs during growth and development are not being met. For this reason, sexual maturation level should be used to identify folate needs as opposed to chronological age.

Poor folate status among adolescent females also presents an issue related to reproduction. Studies show that adequate intakes of folate prior to pregnancy can reduce the incidence of spina bifida and selected other congenital anomalies and may reduce the risk of Down syndrome among offspring.[45] The protective effects of folate occur early in pregnancy, often before a woman may know she is pregnant. Thus, it is imperative that all women of reproductive age (15–44 years old) consume adequate folic acid, preferably through dietary sources, or if needed, through supplements.

National data suggest that many adolescents do not consume adequate amounts of folate. Twenty-five percent of male and 43% of female adolescents consume less than the DRI for folate.[36] The top five sources of dietary folate consumed by adolescents include ready-to-eat cereal, orange juice, bread, milk, and dried beans or lentils.[14] Teens who skip breakfast or do not commonly consume orange juice and ready-to-eat cereals are at increased risk for having a low consumption of folate.

Vitamin A

Vitamin A deficiency is rare among adolescents in the United States; however, national studies have consistently shown low dietary intakes of this vitamin. It has been reported that 52% of teen males and 62% of teen females consume less than the DRI for vitamin A, while 34% of males and 44% of females consume less than 77% of the recommended amounts.[36] The DRI for vitamin A is shown in Table 14.7. The top five dietary sources of vitamin A in the diets of adolescents are ready-to-eat cereal, milk, carrots, margarine, and cheese. Beta-carotene, a precursor of vitamin A, is most commonly consumed by teens in carrots, tomatoes, spinach and other greens, sweet potatoes, and milk.[14] The low intake of fruits, vegetables, and milk and dairy products by adolescents contributes to their less-than-optimal intake of vitamin A.

Vitamin E

Vitamin E is well known for its antioxidant properties, a role that becomes increasingly important as body mass expands during adolescence. The DRIs for vitamin E for adolescents are shown in Table 14.7. Few data are available on the vitamin E status of adolescents. National nutrition surveys suggest that dietary intakes of vitamin E are well below recommended levels, which may be indicative of poor vitamin E status. Seventy-six percent of adolescent males and females consume less than the DRI for vitamin E, with 55% consuming less than 77% of recommended amounts.[36] Among adolescents the five most commonly consumed sources of vitamin E are (1) margarine; (2) cakes, cookies, quick breads, and donuts; (3) salad dressings and mayonnaise; (4) nuts and seeds; and (5) tomatoes.[14] Increasing adolescent intakes of vitamin E through dietary sources is a challenge, given that many of the sources of vitamin E are high-fat foods.

Vitamin C

Vitamin C is involved in the synthesis of collagen and other connective tissues. For this reason, vitamin C plays an important role during adolescent growth and development. Vitamin C intakes are marginally adequate within the adolescent population; however, 17% to 35% of teens consume less than 75% of the recommended amount.[27] The five most common sources of vitamin C among adolescents are orange and grapefruit juice, fruit drinks, ready-to-eat cereals, tomatoes, and white potatoes.[14] The DRIs for 9- to 13-year-old and 14- to 18-year-old adolescents are shown in Table 14.7.

Vitamin C acts as an antioxidant. Smoking increases the need for this antioxidant within the body because it consumes vitamin C in antioxidation reactions. Consequently, smoking results in reduced serum levels of vitamin C. Recommended levels of vitamin C intake are higher among smokers. Because more than one-third of adolescents consume less than 75% of the DRI for vitamin C, an even greater proportion of adolescents who smoke will likely not consume enough vitamin C to promote optimal health. On average, adolescents who use tobacco and other substances have poorer-quality diets and consume fewer fruits and vegetables, which are primary sources of vitamin C.[26]

Nutrition Screening, Assessment, and Interventions

The American Medical Association's Guidelines for Adolescent Preventive Services (GAPS) recommend that all adolescents receive annual health guidance related to healthy dietary habits and methods to achieve a healthy weight.[46] This health guidance begins by annually screening all adolescents for indicators of nutritional risk. Common concerns that should be investigated during nutrition screening include overweight, underweight, eating disorders, hyperlipidemia, hypertension, iron-deficiency anemia, food insecurity, and excessive intake of high-fat or high-sugar foods and beverages. Pregnant adolescents should also be assessed for adequacy of weight gain and compliance with prenatal vitamin-mineral supplement recommendations.

Nutrition screening should include an accurate measurement of height and weight, and calculation of body mass index (BMI). These data, plotted on age- and gender-appropriate National Center for Health Statistics 2000 growth charts, indicate the presence of any weight or other growth problems. Indicators of height and weight status are listed in Table 14.11.

Teens below the 5th percentile of weight-for-height or BMI-for-age are considered to be underweight and should be referred for evaluation of metabolic disorders, chronic health conditions, or eating disorders. Adolescents with a BMI above the 85th percentile but below the 95th percentile are considered to be at risk for overweight. They should be referred for a full medical evaluation to determine the presence or absence of obesity-related complications. Teenagers with a BMI greater than 95th percentile are considered to be overweight and should also be referred for a medical evaluation. Referral to a weight-management program specially designed to meet the needs of adolescents may also be warranted for overweight adolescents who have completed physical growth.[47]

Nutrition screening should also include a brief dietary assessment. Food frequency questionnaires, 24-hour recalls, and food diaries or food records are all appropriate for use with adolescents. Table 14.12 lists the advan-

Table 14.11	Indicators of height and weight status for adolescents	
Indicator	**Body Size Measure**	**Cutoff Values**
Stunting (low height-for-age)	Height-for-age	<3rd percentile
Thinness (low BMI-for-age)	BMI-for-age	<5th percentile
At risk for overweight	BMI-for-age	≥85th percentile, but <95th percentile
Overweight	BMI-for-age	≥95th percentile

SOURCE: From E. Haddad. © American Journal of Clinical Nutrition, 1943;59:12485–12545; and M. Story, et al., (Eds.) Bright Futures in Practice: Nutrition. © 2000. Reprinted by permission of the American Journal of Clinical Nutrition.

Table 14.12 Strengths and limitations of various dietary assessment methods used in clinical settings

	Strengths	Limitations	Applications
24-Hour Recall	• Does not require literacy • Relatively low respondent burden • Data may be directly entered into a dietary analysis program • May be conducted in person or over the telephone	• Dependent on respondent's memory • Relies on self-reported information • Requires skilled staff • Time consuming • Single recall does not represent usual intake	• Appropriate for most people as it does not require literacy • Useful for the assessment of intake of a variety of nutrients and assessment of meal patterning and food group intake • Useful counseling tool
Food Frequency	• Quick, easy, and affordable • May assess current as well as past diet • In a clinical setting, may be useful as a screening tool	• Does not provide valid estimates of absolute intake of individuals • Can't assess meal patterning • May not be appropriate for some population groups	• Does not provide valid estimates of absolute intake for individuals, thus of limited usefulness in clinical settings • May be useful as a screening tool; however, further development research is needed
Food Record	• Does not rely on memory • Food portions may be measured at the time of consumption • Multiple days of records provide valid measure of intake for most nutrients	• Recording foods eaten may influence what is eaten • Requires literacy • Relies on self-reported information • Requires skilled staff • Time consuming	• Appropriate for literate and motivated population groups • Useful for the assessment of intake of a variety of nutrients and assessment of meal patterning and food group intake • Useful counseling tool
Diet History	• Able to assess usual intake in a single interview • Appropriate for most people	• Relies on memory • Time consuming (60 to 90 minutes) • Requires skilled interviewer	• Appropriate for most people as it does not require literacy • Useful for assessing intake of nutrients, meal patterning, and food group intake • Useful counseling tool

SOURCE: M. Story and J. Stang (Eds.), Nutrition and the Pregnant Adolescent: A Practical Reference Guide. Minneapolis, MN: Center for Leadership, Education, and Training in Maternal and Child Nutrition, University of Minnesota, 2000. Used with permission.

tages and disadvantages of each dietary assessment method. Less formal dietary assessment questionnaires that target specific behaviors, such as consumption of savory snacks and high-sugar beverages, can also be used for initial nutrition screening. These "rate your plate" types of questionnaires can be completed quickly and may be used to determine those adolescents in need of additional dietary assessment and nutrition counseling.

Nutrition risk indicators that may warrant further nutrition assessment and counseling are listed in Table 14.13 beginning on the following page. Adolescents who have a poor-quality diet characterized by an excessive intake of high-fat or high-sugar foods and beverages or meal skipping should be provided with nutrition counseling that offers concrete examples of ways to improve dietary intake. Adolescents who have been found to have a nutrition-related health risk such as hyperlipidemia, hypertension, iron-deficiency anemia, overweight, or eating disorders should be referred for in-depth medical assessment and nutrition counseling. Pregnant adolescents may also benefit from in-depth nutrition assessment and counseling.

In-depth nutrition assessment should include a review of the full medical history, a review of psychosocial development, and evaluation of all laboratory data available. A complete and thorough dietary assessment should be performed, preferably using two dietary assessment methods. Most commonly, a food frequency questionnaire or a 3- to 7-day food record is combined with a 24-hour recall to provide accurate dietary intake data. Specific areas of nutrition concern can be identified during a complete nutrition assessment, and recommendations for nutrition education and counseling can be made accordingly.

Nutrition Education and Counseling

Providing nutrition education and counseling to teenagers requires a great deal of skill and a good understanding of normal adolescent physical and psychosocial development. When working with teens, it is important to treat them as individuals with unique needs and concerns. The initial component of the counseling session should involve getting to know the adolescent, including personal health

Table 14.13 Key indicators of nutrition risk for adolescents

Indicators of Nutrition Risk	Relevance	Criteria for Further Screening and Assessment
FOOD CHOICES		
Consumes fewer than 2 servings of fruit or fruit juice per day	Fruits and vegetables provide dietary fiber and several vitamins (such as A and C) and minerals. Low intake of fruits and vegetables is associated with an increased risk of many types of cancer. In females of childbearing age, low intake of folic acid is associated with an increased risk of giving birth to an infant with neural tube defects.	Assess the adolescent who is consuming less than 1 serving of fruit or fruit juice per day.
Consumes fewer than 3 servings of vegetables per day		Assess the adolescent who is consuming fewer than 2 servings of vegetables per day.
Consumes fewer than 6 servings of bread, cereal, pasta, rice, or other grains per day	Grain products provide complex carbohydrates, dietary fiber, vitamins, and minerals. Low intake of dietary fiber is associated with constipation and an increased risk of colon cancer.	Assess the adolescent who is consuming fewer than 3 servings of bread, cereal, pasta, rice, or other grains per day.
Consumes fewer than 3 servings of dairy products per day	Dairy products are a good source of protein, vitamins, and calcium and other minerals. Low intake of dairy products may reduce peak bone mass and contribute to later risk of osteoporosis.	Assess the adolescent who is consuming fewer than 2 servings of dairy products per day. Assess the adolescent who has a milk allergy or is lactose intolerant. Assess the adolescent who is consuming more than 20 ounces of soft drinks per day.
Consumes fewer than 2 servings of meat or meat alternatives (e.g., beans, eggs, nuts, seeds) per day	Protein-rich foods (e.g., meats, beans, dairy products) are good sources of B vitamins, iron, and zinc. Low intake of protein-rich foods may impair growth and increase the risk of iron-deficiency anemia and of delayed growth and sexual maturation. Low intake of meat or meat alternatives may indicate inadequate availability of these foods at home. Special attention should be paid to children and adolescents who follow a vegetarian diet.	Assess the adolescent who is consuming fewer than 2 servings of meat or meat alternatives per day or who consumes a vegan diet.
Has excessive intake of dietary fat	Excessive intake of total fat contributes to the risk of cardiovascular diseases and obesity and is associated with some cancers.	Assess the adolescent who has a family history of premature cardiovascular disease. Assess the adolescent who has a body mass index (BMI) greater than or equal to the 85th percentile.
EATING BEHAVIORS		
Exhibits poor appetite	A poor appetite may indicate depression, emotional stress, chronic disease, or an eating disorder.	Assess the adolescent if BMI is less than the 15th percentile or if weight loss has occurred. Assess if irregular menses or amenorrhea have occurred for 3 months or more. Assess for organic and psychiatric disease.
Consumes food from fast-food restaurants three or more times per week	Excessive consumption of convenience foods and foods from fast-food restaurants is associated with high fat, calorie, and sodium intakes, as well as low intake of certain vitamins and minerals.	Assess the adolescent who is at risk for overweight/obesity, or who has diabetes mellitus, hyperlipidemia, or other conditions requiring reduction in dietary fat.

Table 14.13 Key indicators of nutrition risk for adolescents (continued)

Indicators of Nutrition Risk	Relevance	Criteria for Further Screening and Assessment
Skips breakfast, lunch, or dinner/supper three or more times per week	Meal skipping is associated with a low intake of energy and essential nutrients, and, if it is a regular practice, could compromise growth and sexual development. Repeatedly skipping meals decreases the nutritional adequacy of the diet.	Assess the adolescent to ensure that meal skipping is not due to inadequate food resources or unhealthy weight-loss practices.
Consumes a vegetarian diet	Vegetarian diets can provide adequate nutrients and energy to support growth and development if well planned. Vegan diets may lack calcium, iron, and vitamins D and B₁₂. Low-fat vegetarian diets may be adopted by adolescents who have eating disorders.	Assess the adolescent who consumes fewer than 2 servings of meat alternatives per day. Assess the adolescent who consumes fewer than 3 servings of dairy products per day. Assess the adolescent who follows a low-fat vegetarian diet and experiences weight loss for eating disorder and adequacy of energy intake.
FOOD RESOURCES		
Has inadequate financial resources to buy food, insufficient access to food, or lack of access to cooking facilities	Poverty can result in hunger and compromised food quality and nutrition status. Inadequate dietary intake interferes with learning.	Assess the adolescent who is from a family with low income, is homeless, or is a runaway.
WEIGHT AND BODY IMAGE		
Practices unhealthy eating behaviors (e.g., chronic dieting, vomiting, and using laxatives, diuretics, or diet pills to lose weight)	Chronic dieting is associated with many health concerns (fatigue, impaired growth and sexual maturation, irritability, poor concentration, impulse to binge) and can lead to eating disorders. Frequent dieting in combination with purging is often associated with other health-compromising behaviors (substance use, suicidal behaviors). Purging is associated with serious medical complications.	Assess the adolescent for eating disorders. Assess for organic and psychiatric disease. Screen for distortion in body image and dysfunctional eating behavior, especially if adolescent desires weight loss, but BMI <85th percentile.
Is excessively concerned about body size or shape	Eating disorders are associated with significant health and psychological morbidity. Eighty-five percent of all cases of eating disorders begin during adolescence. The earlier adolescents are treated, the better their long-term prognosis.	Assess the adolescent for distorted body image and dysfunctional eating behaviors, especially if adolescent wants to lose weight, but BMI is less than the 85th percentile.
Has exhibited significant weight change in past 6 months	Significant weight change during the past 6 months may indicate stress, depression, organic disease, or an eating disorder.	Assess the adolescent to determine the cause of weight loss or weight gain (limited to too much access to food, poor appetite, meal skipping, eating disorder).
GROWTH		
Has BMI less than the 5th percentile	Thinness may indicate an eating disorder or poor nutrition.	Assess the adolescent for eating disorders. Assess for organic and psychiatric disease. Assess for inadequate food resources.
Has BMI greater than the 95th percentile	Obesity is associated with elevated cholesterol levels and elevated blood pressure. Obesity is an independent risk factor for cardiovascular disease and type 2 diabetes mellitus in adults. Overweight adolescents are more likely to be overweight adults and are at increased risk for health problems as adults.	Assess the adolescent who is overweight or at risk for becoming overweight (on the basis of present weight, weight-gain patterns, family weight history).

continued

Table 14.13 Key indicators of nutrition risk for adolescents (continued)

Indicators of Nutrition Risk	Relevance	Criteria for Further Screening and Assessment
PHYSICAL ACTIVITY		
Is physically inactive: engages in physical activity fewer than 5 days per week	Lack of regular physical activity is associated with overweight, fatigue, and poor muscle tone in the short term and a greater risk of heart disease in the long term. Regular physical activity reduces the risk of cardiovascular disease, hypertension, colon cancer, and type 2 diabetes mellitus. Weight-bearing physical activity is essential for normal skeletal development during adolescence. Regular physical activity is necessary for maintaining normal muscle strength, joint structure, and joint function; contributes to psychological health and well-being; and facilitates weight reduction and weight maintenance throughout life.	Assess how much time the adolescent spends watching television/videotapes and playing computer games. Assess the adolescent's definition of physical activity.
Engages in excessive physical activity	Excessive physical activity (nearly every day or more than once a day) can be unhealthy and associated with menstrual irregularity, excessive weight loss, and malnutrition.	Assess the adolescent for eating disorders.
MEDICAL CONDITIONS		
Has chronic diseases or conditions	Medical conditions (diabetes mellitus, spina bifida, renal disease, hypertension, pregnancy, HIV infection/AIDS) have significant nutritional implications.	Assess adolescent's compliance with therapeutic dietary recommendations. Refer to dietitian if appropriate.
Has hyperlipidemia	Hyperlipidemia is a major cause of atherosclerosis and cardiovascular disease in adults.	Refer adolescent to a dietitian for cardiovascular nutrition assessment.
Has iron-deficiency anemia	Iron deficiency causes developmental delays and behavioral disturbances. Another consequence is increased lead absorption.	Screen adolescents if they have low iron intake, a history of iron-deficiency anemia, limited access to food because of poverty or neglect, special health care needs, or extensive menstrual or other blood losses. Screen annually.
Has dental caries	Eating habits have a direct impact on oral health. Calcium and vitamin D are vital for strong bones and teeth, and vitamin C is necessary for healthy gums. Frequent consumption of carbohydrate-rich foods (e.g., lollipops, soda) that stay in the mouth a long time may cause dental caries. Fluoride in water used for drinking and cooking as well as in toothpaste reduces the prevalence of dental caries.	Assess the adolescent's consumption of snacks and beverages that contain sugar, and assess snacking patterns. Assess the adolescent's access to fluoride (e.g., fluoridated water, fluoride tablets).
Is pregnant	Pregnancy increases the need for most nutrients.	Refer the adolescent to a dietitian for further assessment, education, and counseling as appropriate.
Is taking prescribed medication	Many medications interact with nutrients and can compromise nutrition status.	Assess potential interactions of prescription drugs (e.g., asthma medications, antibiotics) with nutrients.

Table 14.13 Key indicators of nutrition risk for adolescents (continued)

Indicators of Nutrition Risk	Relevance	Criteria for Further Screening and Assessment
LIFESTYLE		
Engages in heavy alcohol, tobacco, and other drug use	Alcohol, tobacco, and other drug use can adversely affect nutrient intake and nutrition status.	Assess the adolescent further for inadequate dietary intake of energy and nutrients.
Uses dietary supplements	Dietary supplements (e.g., vitamin and mineral preparations) can be healthy additions to a diet, especially for pregnant and lactating women and for people with a history of iron-deficiency anemia; however, frequent use or high doses can have serious side effects. Adolescents who use supplements to "bulk up" may be tempted to experiment with anabolic steroids. Herbal supplements for weight loss can cause tachycardia and other side effects. They may also interact with	Assess the adolescent for the type of supplements used and dosages. Assess the adolescent for use of anabolic steroids and mega-doses of other supplements.

SOURCE: Adapted with permission from M. Story, Bright Futures in Practice: Nutrition. Copyright © 2000 National Center for Education in Maternal and Child Health.

or nutrition-related concerns. After establishing a rapport with the teen, the counselor should provide an overview of the events of the counseling session, including what specific nutrition topics will be discussed. Once again, the adolescent should be encouraged to add his or her own nutrition concerns to the list of topics to be discussed during the education session.

After the counselor and teen agree to a list of topic areas to be covered during the nutrition education session, a complete nutrition assessment should be performed. Upon completion of the assessment, the counselor and teen should work together to establish goals for improving dietary intake and reducing nutrition risk.

It is important to involve the adolescent in decision-making processes during nutrition counseling. Allowing teens to provide input as to what aspects of their eating habits they think need to be changed and what changes they are willing to make accomplishes several important tasks during the counseling session. First, the importance of the adolescent in the decision-making process is stressed, and she or he is encouraged to become involved in personal decisions about health. Second, a good rapport established between the health professional and the adolescent may lead to greater interaction between both parties. Finally, behavior change is more likely when the adolescent has suggested ways to accomplish this, thus expressing a willingness to change.

One or two goals during a counseling session is a reasonable number to work toward. Setting too many goals reduces the probability that the adolescent can meet all of the goals and may seem overwhelming. For each goal set, several behavior-change strategies should be mutually agreed upon for meeting that goal. These strategies should be concrete in nature and instigated by the teen. The adolescent and the counselor should also work together to decide how to determine when a goal is met. Frequent follow-up sessions also help to provide feedback and monitor progress toward individual goals.

Physical Activity and Sports

Regular physical activity leads to many health benefits. Physical activity improves aerobic endurance and muscular strength, may reduce the risk of developing obesity, and builds bone mass density.[29,48] Physical activity among adolescents is consistently related to higher levels of self-esteem and self-concept and lower levels of anxiety and stress. Thus, physical activity is associated with both physiological and psychological benefits, especially during adolescence, which offers opportunities to positively influence the adoption of lifelong activity patterns. Increasing physical activity among adolescents as a goal is important because regular physical activity declines during adolescence, and many American teens are inactive.

Physical activity is defined as any bodily movement produced by skeletal muscles that results in energy expenditure. This definition is distinguished from *exercise,* which is a subset of physical activity that is planned, structured, and repetitive and is done to improve or maintain physical fitness.[29] *Physical fitness* is a set of attributes that are either health or skill related. The International Consensus Conference on Physical Activity Guidelines for Adolescents recommends that all adolescents be physically active daily, or nearly every day, as part of play, games, sports, work, transportation, recreation, physical education, or planned exercise. Further, it is recommended that adolescents engage in three or more sessions per week of activities that last 20 minutes or more at a time and that require moderate to vigorous levels of exertion.[49] These recommendations are consistent with the surgeon general's report on physical activity and health, which advises people of all ages to include a minimum of 30 minutes of physical activity of moderate intensity (such as brisk walking) on most, if not all, days of the week. The report also acknowledges that greater health benefits can be obtained by engaging in physical activity of more vigorous intensity or of longer duration.

Despite common knowledge about the importance and benefits of physical activity, only about one-half of young people (aged 12–21 years) in the United States regularly participate in vigorous physical activity, and one-fourth report no vigorous physical activity.[29] Moreover, physical activity declines steadily throughout adolescence. For example, 81% of ninth-grade males report they exercised vigorously on at least three of the previous seven days, while only 67% of 12th-grade males did so. Among females, 61% of ninth-graders exercised vigorously on at least three of the preceding seven days, which declined to only 41% among 12th-grade girls. One explanation offered for the decline in adolescents' physical activity that occurs after the growth spurt may be related to the social demands of adolescence, changing interests, and the transition from school to work or college.

Factors Affecting Physical Activity

Individual, social, and environmental factors are associated with physical activity among adolescents. Females are less active than males, and among adolescent females, blacks are less active than whites.[48] Individual factors positively associated with physical activity among young people include confidence in one's ability to engage in exercise (i.e., self-efficacy), perceptions of physical or sports competence, having positive attitudes toward physical activity, enjoying physical activity, and perceiving positive benefits associated with physical activity (i.e., excitement, fun, adventure, staying in shape, improved appearance, weight control, improving skills). Social factors associated with engaging in physical activity are peer and family support. Environmental factors associated with physical activity are having safe and convenient places to play, sports equipment, and transportation to sports or fitness programs.

Schools offer an ideal setting for promoting physical activity through physical education classes. However, in recent years, overall daily attendance in physical education classes in grades 9–12 decreased significantly from 42% to 25%. Also, the percentage of high school students enrolled in physical education and who report being physically active for at least 20 minutes in physical education classes declined from about 81% to 70% in the 1990s.[29] Only 20% of all high school students report being physically active for 20 minutes or more in daily physical education classes.[48] Community programs are essential because most physical activity among adolescents occurs outside of schools.

The Centers for Disease Control and Prevention recently published Guidelines for School and Community Programs to Promote Lifelong Physical Activity among Young People, which provides a developmental framework for comprehensive school and community physical activity programs to be used by school districts, educators, health professionals, and policy makers.[48] The guidelines include recommendations (Table 14.14) to promote lifelong physical activity including school policies, physical and social environments that encourage and enable physical activity, developmentally appropriate physical education curricula and instruction, personnel training, family and community involvement, and program evaluation.

High levels of physical activity, combined with growth and development, increase adolescents' needs for energy, protein, and several vitamins and minerals. Participation in competitive sports often means an adolescent will participate in intense training and competition during an athletic season. If the athlete competes in several sports, energy and nutrient needs will remain relatively stable throughout the year. If an athlete participates in only one sport, however, energy and nutrient needs may fluctuate based on the timing of the sports season. Therefore, adolescents must be assessed for seasonal and yearly physical activity when energy and nutrient needs are determined.

The energy and nutrient needs of adolescent athletes are largely unknown. Many of the recommendations available are based on needs of young adult athletes or are extrapolated from usual nutrient needs of adolescents. The best method of assessing the nutrient needs of athletes is to begin with general dietary needs based on sexual maturation rating (SMR), with additional allowances based upon the unique needs of the individual and the intensity of physical activity in which she engages. To assess individual nutrient needs, health care professionals must gather information such as:

- What sport(s) does the adolescent engage in, and what is the duration of the competition season?
- What is the level of competition of the adolescent? Is participation recreational, competitive, or highly competitive?
- What kind of training does the adolescent engage in? The method(s), intensity, and duration of training activities should be noted.

Table 14.14 Recommendations for school health programs promoting healthy eating and physical activity

Healthy Eating	Physical Activity
• Adopt a coordinated school nutrition policy that promotes healthy eating through classroom lessons and a supportive school environment. • Implement nutrition education from preschool through secondary school as part of a sequential, comprehensive school health education curriculum designed to help students adopt healthy eating behaviors. • Provide nutrition education through developmentally appropriate, culturally relevant, fun participatory activities that involve social learning strategies. • Coordinate school food service with other components of the comprehensive school health program to reinforce messages on healthy eating. • Involve family members and the community in supporting and reinforcing nutrition education. • Evaluate regularly the effectiveness of the school health program in promoting healthy eating, and change the program as appropriate to increase the effectiveness.	• Establish policies that promote enjoyable lifelong physical activity among young people. • Provide physical and social environments that encourage and enable safe and enjoyable physical activity. • Implement physical education curricula and instruction that emphasize enjoyable participation in physical activity and that help students develop the knowledge, attitudes, motor skills, behavioral skills, and confidence needed to adopt and maintain physically active lifestyles. • Provide extracurricular physical activity programs. • Include parents and guardians in physical activity instruction and in extracurricular and community physical activity programs, and encourage them to support their children's participation in enjoyable physical activities. • Provide training for education, coaching, recreation, health care, and other school and community personnel that imparts the knowledge and skills needed to effectively promote enjoyable, lifelong physical activity among young people. • Assess physical activity patterns among young people, counsel them about physical activity, refer them to appropriate programs, and advocate for physical activity instruction and programs for young people. • Provide a range of developmentally appropriate community sports and recreation programs that are attractive to all young people. • Regularly evaluate school and community physical activity instruction, programs, and facilities.

SOURCE: Centers for Disease Control and Prevention. "Guidelines for School and Community Programs to Promote Lifelong Physical Activity among Young People, MMWR 45:1996; and "Guidelines for School Health Programs to Promote Lifelong Healthy Eating," MMWR 45:1996.

• Does the athlete typically sweat profusely or lose body weight during competition?

• Does the athlete follow a special diet or take supplements to improve athletic performance? Be sure to note the type, amount, and frequency of supplement use.

General energy and protein needs are shown in Table 14.8. These guidelines should provide the foundation for calculating protein and energy needs for athletes. Competitive athletes may require 500–1500 additional calories per day to meet their energy needs. Athletes and their parents should be encouraged to monitor weight stability. Any weight loss that is not transient (transient losses are often due to dehydration) signifies that the caloric intake is inadequate to support growth and development. A thorough assessment of energy and protein intakes, accompanied by measurements of body composition, should be taken when unexpected weight loss occurs. Protein should supply no more than 15% to 20% of calories in the diet.

Adolescents typically consume 1.5 to 2.5 times the recommended intake of protein, so additional protein is generally not required. Exceptions would include athletes who follow vegetarian diets or restrict caloric intake. When the main sources of protein are plant based, additional protein intake may be needed because plant-based sources of protein may be less bioavailable.

Dietary intakes of athletes should follow the Food Guide Pyramid recommendations, with the realization that the increased energy needs of athletes may require them to consume the upper limit of food group recommendations. Athletes should be encouraged to eat a pre-event meal at least 2 to 3 hours prior to exercise; eating too close to exercise may lead to indigestion and physical discomfort. Foods that are high in fat, high in protein, and high in dietary fiber should be avoided for at least 4 hours prior to exercise, because they take longer to digest and may cause physical discomfort during exercise. Protein and fat also displace complex carbohydrates, which are the most readily available source of energy during athletic events. Postevent meals should contain approximately 400–600 calories, and they should be comprised of high-carbohydrate foods and adequate amounts of noncaffeinated fluids.

Other nutrients of concern among adolescent athletes, particularly female athletes, are iron and calcium. Teens who are athletes are at a higher risk than nonathletes for developing iron-deficiency anemia, due in part to iron losses that occur during exercise.[50] Iron status should be closely monitored in athletes, especially among vegetarians and females who are postmenarcheal, in an effort to prevent iron-deficiency anemia. Calcium intakes have been shown to be below the DRIs in a significant proportion of adolescents, especially females. Athletes' increased risk for bone fractures makes adequate calcium intake extremely important. Recent data suggest that consumption of carbonated soft drinks by athletic females elevates their risk of developing bone fractures when compared to less active females who consume the same beverages.[16] Although the mechanism responsible for this tendency has not been identified, female adolescent athletes with low calcium consumption appear to be the highest-risk group of all adolescents for bone fractures; therefore, they should make every effort to consume adequate calcium in their diets. Teen athletes who cannot or will not consume calcium from dietary sources should be counseled to take a daily calcium supplement that meets their daily needs.

Promoting Healthy Eating and Physical Activity Behaviors

Meeting the challenge of improving the nutritional health of teenagers requires the integrated efforts of teenagers, parents, educators, health care providers, schools, communities, the food industry, and policy makers all working together to create more opportunities for healthful eating.

Effective Nutrition Messages for Youth

We need to rethink how we frame our messages to youth. Years ago, Leverton pointed out that too often, teenagers have been given the message that good nutrition means "eating what you don't like because it's good for you."[51] Rather, they should be told "eat well because it will help you in what you want to do and become." Teenagers are present oriented and tend not to be concerned about how their eating will affect them in later years. However, they are concerned about their physical appearance, achieving and maintaining a healthy weight, and having lots of energy.[52,53] Many are also interested in optimizing sports performance. Even though adolescents need to be aware of the long-term risks of an unhealthy diet and benefits of a more healthful one, focusing on the short-term benefits will have more appeal to them.

A change in the perception that healthful foods do not taste good is also needed. In a national Gallup survey, almost two-thirds (64%) of adolescents agreed either a lot or somewhat with the statement that "foods that are good for you do not taste good."[54] Almost three-fourths (71%) agreed with the statement that "your favorite foods are not good for you." Getting students involved in food preparation and tasting new foods may help counter this belief. Nutrition education and intervention programs for youth need to focus not only on attitudes and acquisition of knowledge but also on behavioral changes.[55,56] Adolescents need opportunities to learn and practice behavioral change skills. In the Gallup poll, 71% of adolescents who received nutrition information from doctors and nurses rated the advice as very useful.

Parent Involvement

Parents as well as teenagers should be targets for nutrition education because they fill the role of gatekeepers of foods and serve as role models for eating behavior. Even though parents may have little control over what their teenagers are eating outside the home, they have more control in the home environment. Teenagers tend to eat what is available and convenient. Parents can capitalize on this by stocking the kitchen with a variety of nutritious ready-to-eat foods and limiting the availability of high-sugar, high-fat foods within the home. The use of different creative settings and outlets to deliver innovative nutrition education programs to parents, including worksites, churches, community centers, libraries, supermarkets, restaurants, and parent education programs, should be explored.

School Programs

School-based programs can play important roles in promoting lifelong healthy eating and physical activity. Efforts to promote physical activity and healthful eating should be part of a comprehensive, coordinated school health program and include school health instruction (curriculum), school physical education, school food service, health services (screening and preventive counseling), school-site health promotion programs for faculty and staff, and integrated community efforts.[48] The Centers for Disease Control and Prevention recently published two complementary reports, "Guidelines for School Health Programs to Promote Lifelong Healthy Eating"[57] and "Guidelines for School and Community Programs to Promote Lifelong Physical Activity among Young People,"[48] which provide a developmental framework for comprehensive school nutrition and physical activity programs to be used by school districts, educators, health professionals, and policy makers. The guidelines (Table 14.14) include recommendations to promote healthy eating and lifelong physical activity including school policies, physical and social environments that encourage and enable physical activity and healthy eating, developmentally appropriate nutrition and physical education curricula and instruction, personnel training, family and community involvement, and program evaluation.

CLASSROOM NUTRITION EDUCATION

Contento and colleagues reviewed 43 school-based nutrition education programs published between 1980 and 1995.[58] Less than half of the studies (*n* = 20) were conducted with adolescents. Surprisingly little has been published regarding nutrition education for junior and senior high school students. Some form of nutrition education is taught in grades 7 through 9 in 92% of all public schools and in 86% of grades 10 through 12.[59] The percentage of schools with nutrition education requirements is substantially lower at each grade level than the percentage of schools that teach nutrition education. Although 50% or more of all K–8 schools have district or state requirements for students to receive nutrition education, only 40% of schools have requirements for grades 9–10, and about 20% have requirements for grades 11–12.

By the time they reach junior high, students are in the process of cognitive and social development changes that permit more advanced nutrition education concepts and activities. The ability for more abstract thinking coupled with the changing psychosocial terrain of young adolescents provides both a challenge and a unique opportunity for educators to offer new learning and teaching strategies to encourage them to make healthful food choices. Young adolescence is an ideal time to teach students how to assess their own behavior and set goals for change. As adolescents begin the social process of individuation, they become ready and eager to make their own decisions and show their individuality. Nutrition education often fails to take advantage of the social and cognitive transitions of adolescence to promote the adoption of more healthful behaviors. In addition, obstacles to implementing nutrition education programs persist, ranging from insufficient funding, to teacher ambivalence, to competition with other high-priority health concerns, such as HIV and prevention of substance use.[57]

Because knowledge alone is inadequate when students must decide which foods to eat and how to deal with peer and social influences—as well as with a widely available supply of high-fat, high-sugar foods—the focus of nutrition education and teaching methods should be on behavior-change strategies and skill acquisitions to make healthful food decisions. Characteristics of teaching methods found to be most effective in school health education curricula include use of discovery learning; use of student learning stations, small work groups, and cooperative learning techniques; cross-age and peer teaching; positive approaches that emphasize the intrinsic value of good health; use of personal commitment to change and goal setting; and provision of opportunities to increase self-efficacy in modifying health behaviors.[60] Most important,

adolescents need to be given repeated opportunities to develop, demonstrate, practice, and master the skills needed to make informed decisions and cope with social influences. Also, to be effective, programs must take into account cultural factors as well as the developmental processes of adolescents.

Teacher training in basic nutrition and instructional, motivational, and behavioral-change strategies increases the success of nutrition education curriculum. Training may be most effective if teachers have the opportunity to examine their own body images and assess their eating behaviors. Teacher training typically increases the time spent on teaching nutrition in the classroom.[61]

SCHOOL FOOD SERVICES The National School Lunch Program (NSLP) and School Breakfast Program (SBP) are federally sponsored nutrition programs administered by the U.S. Department of Agriculture (USDA) in conjunction with state and local education agencies. Youth from households with incomes at 130% to 185% of the poverty level receive meals at reduced rates; youth from households with incomes at 130% of poverty and below receive meals free. Almost 99% of public schools participate in the NSLP, and about 50% in the SBP. The National School Nutrition Dietary Assessment Study (SNDAS), funded by the USDA, found that youth who participate in the NSLP have greater nutrient intakes compared with those who do not participate.[62] The NSLP lunches provided one-third or more of the RDA for key nutrients.

The SNDAS, coupled with accumulating research on the relationship between diet and chronic disease risk, spurred drastic changes in the national school meal programs. In 1994, Congress passed legislation that required meals served through the NLSP and SBP to comply with the Dietary Guidelines for Americans. The legislation for school meals is only for USDA-reimbursable breakfast and lunch meals. Junior and senior high school students, however, have a variety of options for lunch in which high-fat foods are easily accessible, such as in à la carte food in the cafeteria, school stores, and vending machines.

The school lunch and breakfast programs can complement and reinforce what is learned in the classroom and serve as a learning laboratory for nutrition education. The synergism between the school lunch program and classroom learning should enhance the likelihood that adolescents will adopt healthful eating practices.

NUTRITION ENVIRONMENT OF THE SCHOOL

The school environment provides multiple food and nutrition activities and influences not only classroom nutrition education and school meals but also food sold in vending machines, school stores, and snack bars; food sold at fund-raising events; food given as rewards by teachers; corporate-sponsored nutrition education materials; and in-school advertising of food products.[63] Wolfe and

Campbell found that the nutrition experiences provided in schools tend to be fragmented and multicomponent, especially when multiple people are involved in decisions about them.[64] The result can be inconsistent nutrition messages. The growing stream of commercial messages, food advertisements, and easy access to high-fat and high-sugar food products in school are at cross-purposes and in direct conflict with the goals of nutrition education and may negate the efforts in the classroom and lunchroom to foster healthful eating practices.

Most schools offer students the opportunity to purchase foods that are not part of the NSLP or SBP, through the option of à la carte foods, canteens or school stores, or through vending machines. Data collected as part of the School Health Policies and Programs Study (SHPPS) demonstrated that 46% of elementary schools, 64% of middle/junior high schools, and 76% of high schools offered pizza, hamburgers, or sandwiches as à la carte food items.[65] In addition, 30% of elementary schools, 46% of middle/junior high schools, and 62% of high schools offered french fried potatoes.[65] This same study found that 20% of schools offered brand-name fast foods for sale.

Vending machines or school stores can be found in 43% of elementary, 74% of middle/junior high, and 98% of high school corridors;[65] 69% of schools allow students to purchase foods and beverages from these sources during school lunch periods. Most of the foods in school vending machines and school stores are high-fat or high-sugar items such as snack chips, candy, and soft drinks.[66] Healthy food choices or lower-fat alternatives are generally unavailable. In the promotion of a healthy nutrition environment, vending machines and school stores need to offer healthier choices and lower-fat alternatives.

Half of all school districts surveyed in the SHPPS reported having signed a contract with a soft-drink company allowing them exclusive rights to sell their product within the school.[65] Almost 80% of these school districts received a percentage of the profits generated by beverage sales, and almost one-third received other incentives such as the donation of sports equipment, bulletin boards, trophy cases, or cash incentives.

A growing trend of commercialism and marketing in schools uses in-school advertising and corporate-sponsored education materials. A study by Consumers Union Education Services found that direct advertising in schools has mushroomed.[63] Almost half of school districts that contract with soft-drink companies allow the placement of advertisments for these products on school grounds (including playgrounds), and 35% allow advertising within the school building.[65] Examples include school bus advertising for soft drinks and fast-food restaurants (2% of all school districts in the SHPPS); "free" textbook covers advertising candy, chips, and soft drinks; ads for high-sugar/high-fat products on wallboards and in hallways (7% of school districts in the SHPPS), in student publications such as newspapers and yearbooks, and on sports scoreboards; and product giveaways in coupons (6% of SHPPS schools). In addition, *Channel One,* the daily news program broadcast to millions of students in grades 6 through 12 in thousands of schools, devotes 2 minutes out of each daily 12-minute program to paid commercials for products that include candy bars, snack chips, and soft drinks.

Written nutrition policies provide needed guidelines for all food and nutrition activities and promotions in schools. Local and district policy initiatives can be instrumental in creating a supportive and integrated school nutrition environment with consistent health-promoting messages. Serious consideration should be given to restricting the sale of foods high in fat and sugar in schools. Corporate-sponsored education materials and programs should be carefully evaluated for nutrition accuracy, objectivity, completeness, and noncommercialism. Schools should also consider the Consumer's Union recommendation for making schools ad-free zones, where young people can pursue learning without commercial influences and messages.[67] Schools should be an environment where healthful eating behavior is normative, modeled, and reinforced. To improve the health of adolescents in the United States, schools should strive for an integrated, nutritionally supportive environment.

Community Involvement in Nutritionally Supportive Environments

Promoting lifelong healthy eating and physical activity behaviors among adolescents requires attention to the multiple behavioral and environmental influences in a community. Adolescents most likely to adopt healthy behaviors receive consistent messages through multiple channels (e.g., community, home, school, and the media) and from multiple sources (e.g., parents, peers, teachers, health professionals, and the media).[29]

Most physical activity occurs outside the school setting, making community sports and recreation programs essential for promoting physical activity among young people. Healthy eating can be integrated into these efforts by providing nutritious snacks. Community coalitions or task forces can be established to assess community needs and to develop, implement, and evaluate physical activity and nutrition programs for young people. Few studies have reported on nutrition education or physical activity programs outside of the school setting.[58]

Model Nutrition Program

One example of a model community-based nutrition intervention program is the California Adolescent Nutrition and Fitness (CANfit) Program.[68] Through competitive grants, the CANfit Program supports and empowers adolescent-serving, community-based organizations to develop and implement nutrition education and physical activity programs for ethnic adolescents from low-income communities. Using a capacity-building model, the CANfit Program attempts to change the community context by improving access to healthier food choices and safe, affordable physical activity opportunities, enabling adolescents to have the decision-making skills and social support necessary for making healthy nutrition and fitness choices. Examples of CANfit grantees' projects include (1) developing a 10-week curriculum for an after-school program for African American girls that focuses on self-esteem, body image, healthy eating, cooking, and physical activity (e.g., hip-hop dance); (2) developing a nutrition and physical activity program for adolescents and their parents attending Saturday Korean language schools in Los Angeles; and (3) Latino adolescents in a soccer league that worked with a local health department to train team coaches and parents in sports nutrition. Innovative programs using capacity-building models, such as the CANfit Program, can provide numerous benefits to other communities.

Resources

American Medical Association Adolescent Health Online

Links to health and nutrition-related Web sites specifically addressing adolescent health issues; information on key health issues that affect teens.
Web site: www.ama-assn.org/adolhlth

Bright Futures

View and download Bright Futures publications, including *Bright Futures in Practice: Nutrition* and *Bright Futures in Practice: Physical Activity.*
Web site: www.brightfutures.org

Centers for Disease Control and Prevention Nutrition and Physical Activity Programs

Facts on physical activity and nutrition among U.S. adults and youth, model programs, program guidelines; links to pediatric growth charts.
Web site: www.cdc.gov/nccdphp/dnpa

National Center for Education in Maternal and Child Health

Search databases on adolescent health; view and download publications and reports.
Web site: www.ncemch.org

The Vegetarian Resource Group

Information on choosing a healthy vegetarian diet; links to other resources and Web sites.
Web site: www.vrg.org/nutrition/teennutrition.htm

United States Department of Agriculture Food and Nutrition Information Center

Information on Dietary Guidelines for Americans, the Food Guide Pyramid, dietary supplements, food safety, the Healthy School Meals Resource System, and other sources of nutrition information.
Web site: www.nal.usda.gov/fnic

References

1. Gong EJ, Heard FP. Diet, nutrition, and adolescence. In: Shils ME, Olson JA, Shike M, eds. Modern nutrition in health and disease, 8th ed. Philadelphia: Lea & Febiger; 1994: 759–69.

2. Herman-Giddens ME, Slora EJ et al. Secondary sexual characteristics and menses in young girls seen in office practice: a study from the pediatric research in office settings network. Pediatrics 1997;99:505–11.

3. Barnes HV. Physical growth and development during puberty. Med Clin North Am 1975;59:1305–17.

4. Frisch RE. Fatness, puberty, and fertility: the effects of nutrition and physical training on menarche and ovulation. In: Brooks-Gunn J, Peterson AC, eds. Girls at puberty: biological and psychosocial perspectives. New York: Plenum Press; 1983: 29–49.

5. Frisch RE, McArthur JW. Menstrual cycles: fatness as a determinant of minimum weight for height necessary for their maintenance or onset. Science 1974;185:949–51.

6. Barr SI. Associations of social and demographic variables with calcium intakes of high school students. J Amer Diet Assoc 1994;94:260–6, 269; quiz 267–8.

7. Attie I, Brooks-Gunn J. Development of eating problems in adolescent girls: a longitudinal study. Dev Psychol 1989;25:70–9.

8. Killen JD, Hayward C et al. Is puberty a risk factor for eating disorders? Am J Dis Child 1992;146:323–5.

9. Wilson DM, Killen JD et al. Timing and rate of sexual maturation and the onset of cigarette and alcohol use among teenage girls. Arch Pediatr Adolesc Med 1994;148: 789–95.

10. Cusatis DC, Chinchilli VM et al. Longitudinal nutrient intake patterns of U.S. adolescent women: the Penn State Young Women's Health Study. J Adolesc Health 2000;26:194–204.

11. American School Health Association, Association for the Advancement of Health Education, Society for Public Health Education. The National Adolescent Student Health Survey, a report on the health of America's youth. Oakland, CA: Third Party Publishing; 1988: 1–178.

12. Bigler-Doughten S, Jenkins RM. Adolescent snacks: nutrient density and nutritional contribution to total intake. J Amer Diet Assoc 1987;87:1678–9.

13. Kennedy E, Powell R. Changing eating patterns of American children: a view from 1996. J Am Coll Nutr 1997;16:524–9.

14. Subar AF, Krebs-Smith SM, Cook A, Kahle LL. Dietary sources of nutrients among US children, 1989–1991. Pediatrics 1998;102:913–23.

15. Wyshak G. Teenaged girls, carbonated beverage consumption, and bone fractures. Arch Pediatr Adolesc Med 2000;154:610–3.

16. Lin BH, Guthrie J, Blaylock J. The diets of America's children: influences of dining out, household characteristics, and nutrition knowledge: U.S. Department of Agriculture Economic Report Number 746 (AER-746); 1996.

17. Siega-Riz AM, Carson T, Popkin B. Three squares or mostly snacks—what do teens really eat? A sociodemographic study of meal patterns. J Adolesc Health 1998;22: 29–36.

18. Nicklas TA, Myers L et al. Impact of breakfast consumption on nutritional adequacy of the diets of young adults in Bogalusa, Louisiana: ethnic and gender contrasts. J Amer Diet Assoc 1998;98:1432–8.

19. Neumark-Sztainer D, Story M, Resnick MD, Blum RW. Adolescent vegetarians: a behavioral profile of a school-based population in Minnesota. Arch Pediatr Adolesc Med 1997;151:833–8.

20. Sabate J, Llorca MC, Sanchez A. Lower height of lacto-ovo vegetarian girls at preadolescence: an indicator of physical maturation delay? J Amer Diet Assoc 1992; 92:1263–4.

21. Johnston P, Haddad E. Vegetarian diets and pregnant teens. In: Story M, Stang J,

eds. Nutrition and the pregnant adolescent: a practical reference guide. Minneapolis, MN: Leadership, Education and Training Program in Maternal and Child Nutrition, 2000; 135–45.

22. Dietary Guidelines Advisory Committee. Excerpt of the report of the Dietary Guidelines Advisory Committee on Dietary Guidelines for Americans, 2000.

23. USDA Center for Nutrition Policy and Promotion. Food Guide Pyramid 2000.

24. Munoz KA, Krebs-Smith SM, Ballard-Barbash R, Cleveland LE. Food intakes of US children and adolescents compared with recommendations [published erratum appears in Pediatrics May 1998;101(5):952–3]. Pediatrics 1997;100:323–9.

25. Krebs-Smith SM, Cook A et al. Fruit and vegetable intakes of children and adolescents in the United States. Arch Pediatr Adolesc Med 1996;150:81–6.

26. Neumark-Sztainer D, Story M, Resnick MD, Blum RW. Correlates of inadequate fruit and vegetable consumption among adolescents. Prev Med 1996;25:497–505.

27. Stang J, Story MT, Harnack L, Neumark-Sztainer D. Relationships between vitamin and mineral supplement use, dietary intake, and dietary adequacy among adolescents. J Amer Diet Assoc 2000;100:905–10.

28. Devaney BL, Gordon AR, Burghardt JA. Dietary intakes of students. Am J Clin Nutr 1995;61:205S–212S.

29. U.S. Department of Health and Human Services. Physical activity and health: a report of the surgeon general: U.S. Department of Health and Human Services, Centers for Disease Control and Prevention, National Center for Chronic Disease Prevention and Health Promotion; 1996.

30. Harnack L, Stang J, Story M. Soft-drink consumption among U.S. children and adolescents: nutritional consequences. J Amer Diet Assoc 1999;99:436–41.

31. American Academy of Pediatrics Committee on Nutrition. Carbohydrate and dietary fiber. In: Barness L, ed. Pediatric nutrition handbook, 3rd ed. Elk Grove Village, IL: American Academy of Pediatrics; 1993: 100–6.

32. Williams CL, Bollella M, Wynder EL. A new recommendation for dietary fiber in childhood. Pediatrics 1995;96:985–8.

33. Donovan UM, Gibson RS. Iron and zinc status of young women aged 14 to 19 years consuming vegetarian and omnivorous diets. J Am Coll Nutr 1995;14:463–72.

34. U.S. Department of Agriculture. Food consumption survey. Continuing survey of food intakes by individuals. Hyattsville, MD: USDA; 1985.

35. National Heart, Lung, and Blood Institute, National Cholesterol Education Program. Report of the Expert Panel on Blood Cholesterol Levels in Children and Adolescents. Bethesda, MD: National Institutes of Health; 1991.

36. Johnson R, Johnson D et al. Characterizing nutrient intakes of adolescents by sociodemographic factors. J Adolesc Health 1994;15:149–54.

37. Weaver CM. Better bones: girls who get plenty of calcium fare well later in life. The Island Packet, June 20, 1995.

38. Hui SL, Johnston CC Jr., Mazess RB. Bone mass in normal children and young adults. Growth 1985;49:34–43.

39. Albertson AM, Tobelmann RC, Marquart L. Estimated dietary calcium intake and food sources for adolescent females: 1980–92. J Adolesc Health 1997;20:20–6.

40. Briefel RR, McDowell MA, Alaimo K et al. Total energy intake of the U.S. population: the third National Health and Nutrition Examination Survey, 1988–1991. Am J Clin Nutr 1995;62:1072S–80S.

41. Harel Z, Riggs S, Vaz R et al. Adolescents and calcium: what they do and do not know and how much they consume. J Adolesc Health 1998;22:225–8.

42. Neumark-Sztainer D, Story M, Dixon LB et al. Correlates of inadequate consumption of dairy products among adolescents. J Nutr Educ 1997;29:12–20.

43. Clark A, Mossholder S, Gates R. Folacin status in adolescent females. Am J Clin Nutr 1987;46:302–6.

44. Hine R. Folic acid: contemporary clinical perspective. Perspect Appl Nutr 1993; 1:3–14.

45. Institute of Medicine. Nutrition during pregnancy: part I: weight gain; part II: nutrient supplements. Washington, DC: National Academy Press; 1990.

46. American Medical Association. Guidelines for adolescent preventive services. Chicago: American Medical Association, Department of Adolescent Health; 1992.

47. Barlow SE, Dietz WH. Obesity evaluation and treatment: Expert Committee recommendations. The Maternal and Child Health Bureau, Health Resources and Services Administration and the Department of Health and Human Services. Pediatrics 1998;102:E29.

48. Centers for Disease Control and Prevention. Guidelines for school and community programs to promote lifelong physical activity among young people. Morb Mortal Wkly Rep 1997;46:1–36.

49. Sallis J, Patrick K. Physical activity guidelines for adolescents: consensus statement. Pediatric Exercise Science 1994;6:302–14.

50. Clarkson PM. Minerals: exercise performance and supplementation in athletes. J Sports Sci 1991;9(suppl):91–116.

51. Leverton RM. The paradox of teenage nutrition. J Amer Diet Assoc 1968;53:13–6.

52. Olsen L. Food fight: a report on teenaged eating habits and nutritional status. Oakland, CA: Citizen's Policy Center; 1984.

53. Story M, Resnick MD. Adolescents' views on food and nutrition. J Nutr Educ 1986;18:188.

54. American Dietetic Association IFIC, the President's Council on Physical Fitness and Sports. Food, physical activity, and fun: what kids think. Washington, DC: The Gallup Organization; 1995.

55. Lytle L, Achterberg C. Changing the diet of America's children: what works and why? J Nutr Educ 1995;27:250–60.

56. Society for Nutrition Education, American Dietetic Association, ASFS. Association Joint Position of Society for Nutrition Education (SNE), the American Dietetic Association (ADA), American School Food Services Association: school-based nutrition programs and services. J Nutr Educ 1995;27:58–61.

57. Centers for Disease Control and Prevention. Guidelines for school health programs to promote lifelong healthy eating. Morb Mortal Wkly Rep 1996;45:1–37.

58. Contento L, Balch G, Bronner Y et al. The effectiveness of nutrition education and implications for nutrition education policy, programs, and research: a review of research. J Nutr Educ 1995;27:284–418.

59. National Center for Education Statistics. Nutrition education in public elementary and secondary schools: U.S. Department of Education, Office of Educational Research and Improvement; 1996.

60. Seffrin J. Why school health education? In: Wallace HM, Patrick K, Parcel G, Igbe JB, eds. Principles and practices of student health. Oakland, CA: Third Party Publishing; 1992.

61. Contento I, Manning AD, Shannon B. Research perspective on school-aged nutrition education. J Nutr Educ 1992;24: 247–60.

62. Burghardt J, Gordon A, Chapman N et al. The School Nutrition Dietary Assessment Study: school food service, meals offered, and dietary intakes. Princeton, NJ: Mathematica Policy Research, Inc.; 1993.

63. Consumers Union Education Services. Captive kids: commercial pressures on kids at school. Yonkers, NY: Consumers Union of United States; 1995.

64. Wolfe WS, Campbell CC. Nutritional health of school-aged children in upstate New York: what are the problems and what can schools do? New York: Cornell University; 1991.

65. Wechsler H, Brener ND, Kuester S, Miller C. Food service and foods and beverages available at schools: results from the School Health Policies and Programs Study 2000. J Sch Health 2001;71:313–24.

66. Story M, Hayes M, Kalina B. Availability of foods in high schools: is there cause for concern? J Amer Diet Assoc 1996;96: 123–6.

67. Consumers Union Education Services. Selling America's kids: commercial pressures on kids of the '90s. Yonkers, NY: Consumers Union of the United States; 1990.

68. Hinkle A. Community-based nutrition interventions: reaching adolescents from low-income communities. Ann N Y Acad Sci 1997;187:83–93.

> "Dramatic changes in body shape and size, heightened influence of peer perceptions, and the desire to adopt adultlike behaviors place adolescents at high risk for initiating health-compromising behaviors."
>
> Jamie Stang

Chapter 15

Adolescent Nutrition:
Conditions and Interventions

Chapter Outline

- Introduction
- Overweight and Obesity
- Special Concerns among Adolescent Athletes
- Adolescent Pregnancy
- Substance Use
- Dietary Supplements
- Iron-Deficiency Anemia
- Hypertension
- Hyperlipidemia
- Eating Disorders
- Children and Adolescents with Chronic Health Conditions

Prepared by **Jamie Stang, Dianne Neumark-Sztainer,** and **Mary Story**

Key Nutrition Concepts

1 Competitive adolescent athletes require an additional 500–1500 kcal per day to meet their energy needs. Additional protein may be required among adolescent athletes who are still growing.

2 Nutrient needs of young pregnant adolescents less than 2 years past menarche are greater than those of older adolescents because of continued growth and physical development.

3 Overweight adolescents are at increased risk for medical and psychosocial complications such as hypertension, hyperlipidemia, insulin resistance, type 2 diabetes mellitus, hypoventilation, orthopedic disorders, depression, and low self-esteem.

4 Approximately half of adolescent females and 15% of adolescent males are attempting weight loss.

Introduction

Multiple factors influence the nutritional needs and behaviors of adolescents. This chapter presents specific behaviors and nutrition concerns that affect significant numbers of adolescents, including low physical activity, participation in competitive sports, adolescent pregnancy, substance abuse, vegetarian diets, eating disorders, overweight, hypertension, and hyperlipidemia. Because overweight, sports participation, and eating disorders affect a larger group of adolescents than other conditions do, they are presented in greater detail.

Overweight and Obesity

The increase in the prevalence of overweight among adolescents mirrors that of adults over the past two decades. Exact reasons for this increase have not been identified. Although genetics is known to contribute to the occurrence of overweight, and having one or more overweight parent(s) increases a teen's risk of developing obesity, it alone clearly cannot account for the dramatic increase in overweight during the past two decades. Environmental factors, or interactions between genetic and environmental factors, are the most likely causes of the dramatic rise in overweight. Risk factors for the development of overweight among children and adolescents include having at least one overweight parent; coming from a low-income family; being the descendant of African American, Hispanic, or American Indian/Native Alaskan parents; and being diagnosed with a chronic or disabling condition that limits mobility. Table 15.1 provides prevalence estimates of overweight among adolescents in the United States by gender, race, and age. Inadequate levels of physical activity and consuming diets high in total calories and fat are additional risk factors among a significant proportion of

adolescents. These environmental factors increase the risk of developing overweight if an adolescent is genetically predisposed to obesity.

Weight status among adolescents should be assessed by calculating body mass index (BMI). BMI is calculated by dividing a person's weight (kg) by their height^2 (m^2). BMI values are compared to age- and gender-appropriate percentiles to determine the appropriateness of the individual's weight for height. Youth with BMI values greater than the 85th but lower than the 95th percentile are considered at risk for overweight; those with BMI values above the 95th percentile are considered overweight.[1] Growth curves based on BMIs for children and adolescents are available from the National Center for Health Statistics. An example of a BMI growth curve is shown in Illustration 15.1.

The persistence of overweight from childhood throughout adulthood has not been well quantified. Research suggests that the persistence of obesity from infancy to adulthood increases with age. More than 70% of overweight adolescents can be expected to remain overweight into adulthood.[2–4] The risk of persistence of obesity from childhood into adulthood increases if at least one parent is overweight. Risk of persistence of overweight is also

Table 15.1 Prevalence of overweight by age, race, and gender

Category	95th percentile %
6–11 years	
Both Genders	15.3
Boys	
White	12.0
Black	17.1
Mexican American	27.3
Total	16.0
Girls	
White	9.2*
Black	22.2
Mexican American	19.6
Total	14.5
12–17 years	
Both Genders	15.5
Boys	
White	12.8
Black	20.7
Mexican American	27.5
Total	15.5
Girls	
White	12.4
Black	26.6
Mexican American	19.4
Total	15.5

SOURCE: All data taken from NHANES 1999–2000 except datum marked with *, which is from NHANES III.

Illustration 15.1 CDC growth chart: United States.

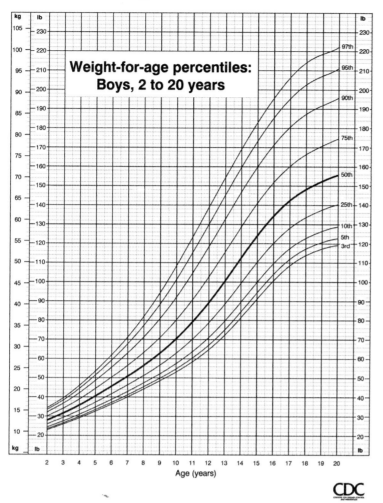

SOURCE: Developed by the National Center for Health Statistics in collaboration with the National Center for Chronic Disease Prevention and Health Promotion; 2000.

higher among the most overweight individuals, especially those whose weight is more than 180% of ideal weight.

Health Implications of Adolescent Overweight

A range of medical and psychosocial complications accompanies overweight among adolescents, including hypertension, dyslipidemia, insulin resistance, type 2 diabetes mellitus, sleep apnea and other hypoventilation disorders, orthopedic problems, body image disturbances, and lowered self-esteem.[5-8] Rates of type 2 diabetes mellitus are more common among overweight youth. In one study, a third of newly diagnosed diabetic patients under the age of 20 had type 2 diabetes mellitus; almost all of these patients had BMI values above the 90th percentile.[9]

All adolescents should be screened for appropriateness of weight-for-height on a yearly basis. Teens determined to be at risk for overweight require an in-depth medical assessment to diagnose any obesity-related complications.

Illustration 15.2 on the next page provides recommended screening and referral procedures for weight-for-height among adolescents.

Illustration 15.3 on the following page presents recommended weight goals based on BMI and age. Weight maintenance is recommended for adolescents who are at risk for overweight and have not yet completed puberty (unless medical complications are noted). Weight loss is recommended when medical complications are present in at-risk overweight youth, when the adolescent is determined to be overweight, and among older adolescents who have completed physical growth and development. Guidelines for treatment of overweight are given in Table 15.2 on page 357. Weight loss should be attempted only after the adolescent and his or her family show current weight can be maintained. For severely obese adolescents, as well as those with significant medical complications, rapid weight loss may be required. Several obesity treatment centers in the United States have health professionals experienced in the management of severe obesity with complications. Staff at these centers can guide other health professionals in the treatment of such youth when necessary. The Weight-Control Information Network (WIN) can help health professionals locate specialized programs when they are required.

No specific physical activities or dietary regimens are recommended for adolescent weight-management programs. Reducing or eliminating the intake of "problem" foods such as savory snacks or high-sugar beverages is often the first dietary change recommended. Health care providers generally should not recommend calorie or fat-gram counting, to avoid encouraging "good and bad food" thinking and to try to minimize the risk of developing disordered eating behaviors. Adolescents more readily accept replacing high-fat or high-sugar foods with healthier substitutes and monitoring portion control, resulting in a better chance of long-term behavior changes. Fast foods are a way of life for many adolescents. Fast-food establishments are prime employers of teens as well as a preferred gathering place for social activities. Therefore, asking teens to avoid fast-food restaurants is an unrealistic goal. Dietary counseling that teaches youth how to balance daily intake on days when fast food is consumed is useful, as are concrete examples of specific food choices that are lower in calories and added fats and sugars. Ideas on how to decrease sedentary pursuits and how to increase physical activity are beneficial. Referrals to community centers, athletic clubs, local parks and recreation programs, and community education programs that offer fun, noncompetitive physical activities—such as yoga, tai chi, swimming, weight lifting, and bicycling or walking clubs—point adolescents in a beneficial direction.

Illustration 15.2 Recommended screening procedures for overweight.

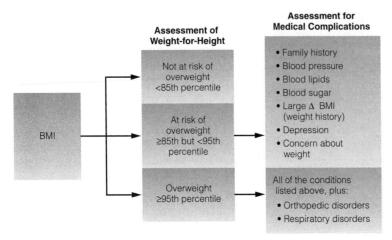

SOURCE: From E. Haddad, © American Journal of Clinical Nutrition, 1943;59:12485–12545; and M. Story, K. Holt, D. Sofka D. (Eds.) 2000. Bright Futures in Practice: Nutrition. Reprinted by permission of the American Journal of Clinical Nutrition.

Examples of how to incorporate physical activity into daily living, such as walking or biking to school, may be beneficial for students who do not see themselves as athletically inclined. (See Case Study 15.1 on page 358).

Special Concerns among Adolescent Athletes

Fluids and Hydration

Fluid intake is an important issue in sports nutrition for adolescents. Young adolescents and those who are pre-pubertal present a particular vulnerability to heat ill-nesses because their bodies do not regulate body temperature as well as those of older adolescents. Adolescents can become so mentally and physically involved in physical activities that they do not pay attention to physiological signals of fluid loss, such as excessive sweating and thirst. Some athletes commonly assume they do not need additional fluids if they are not actively moving all of the time during exercise. Other factors, such as ambient temperature and humidity levels and weight of equipment (helmets, padding, etc.) worn or utilized during exercise, also play a role. For instance, hockey goalies do not skate for great distances during a match, yet they may lose 5 or more pounds of body weight due to the weight of the padding and equipment they wear. Therefore, all athletes should be counseled to regularly consume fluids, even if they do not feel thirsty.

Athletes should consume 6–8 ounces of fluid prior to exercise, 4–6 ounces every 15–20 minutes during physical activity, and at least 8 ounces of fluid following exercise. Recommendations encourage athletes to weigh themselves periodically before and after exercise or competition to determine whether they have lost body weight. Each pound of body weight lost during an activity requires ingestion of 16 ounces of fluid following the activity to maintain proper hydration. Athletes should drink no more than 16 ounces of fluid every 30 minutes, however, to avoid potential side effects such as nausea.

The type of fluid an athlete drinks is affected by peer pressure and mass media more than by actual physiologi-

Illustration 15.3 Recommendations for weight goals.

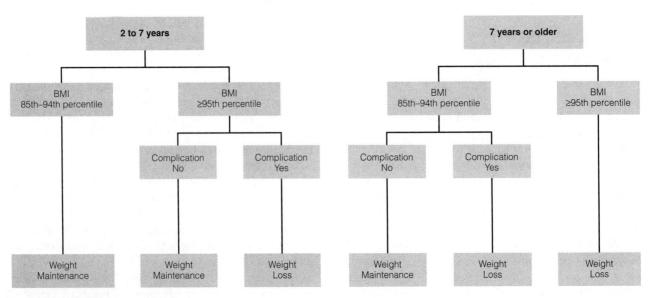

SOURCE: Reproduced with permission from Pediatrics, Vol. 102, Page 29, Copyright 1998 American Academy of Pediatrics.

Table 15.2 General guidelines for weight-management therapy

1. Early intervention is recommended, preferably before the child reaches the 95th percentile for BMI.

2. Parents should be informed of medical complications associated with childhood overweight, and all youth should be assessed for medical complications.

3. All family members and caregivers should be involved in the weight-management program.

4. All family members should be assessed to determine their readiness to make behavior changes, and treatment should not begin until all family members are ready to adopt behavior changes.

5. Weight-management programs should emphasize goals of improving eating and physical activity patterns as opposed to specific weight goals.

6. Programs should be skill based, with families taught to identify problem behaviors, monitor such behaviors, utilize behavior modification principles to address problem behaviors, and implement problem-solving skills when dealing with obstacles to behavior change.

7. Families should be involved in assessing current eating and activity patterns, deciding which behaviors need to be modified, setting goals for behavior changes, and determining how these goals will be achieved.

8. New behavior changes should not be instituted until previous changes have been accomplished and maintained.

9. Routine follow-up visits should be scheduled to monitor progress and prevent relapse to former eating and activity patterns.

SOURCE: J. Stang and L. Holladay, "Treatment of Childhood Obesity Involves the Family," American Dietetic Association Public Health Community Nutrition Practice Group newsletter. *The Digest* (Spring 2000). Adapted from S. E. Barlow and W. H. Dietz, "Obesity Evaluation and Treatment: Expert Committee Recommendations," *Pediatrics* 102 (1998):E29.

cal need. Sports drinks are very popular among teens, even those who do not participate in sports. Data on children suggest that even though water is an economical, easily available fluid, it may not provide optimal benefits for athletes who participate in physically intense events or those of great duration.[10] In such events, juice diluted at a ratio of 1:2 with water, or sports drinks containing no more than 6% carbohydrate, may allow for better hydration and physical performance.[11] Undiluted juices, fruit drinks, carbonated beverages, and drinks containing more than 6% carbohydrate are not recommended during exercise, because they may cause gastric discomfort. Their high carbohydrate content may also delay gastric emptying. Some carbonated soft drinks contain caffeine, which promotes diuresis and reduces their rehydration effectiveness.

Special Dietary Practices

Adolescent athletes may follow special diets or consume nutritional and nonnutritional supplements in an effort to improve physical performance and increase lean body mass. Even though data on the prevalence of supplements or special diets among adolescent athletes remain limited, data from surveys of collegiate athletes suggest that more than 10% of athletes use supplements.[12] Special diets that are noted among adolescent athletes include carbohydrate-loading regimens, high-protein diets, and "Zone" diets. Distance runners and other endurance athletes traditionally have used carbohydrate loading to improve the glycogen content of muscle. It involves the manipulation of training intensity and duration along with the carbohydrate content of meals to improve glycogen formation in muscle tissue. Carbohydrate loading is traditionally a weeklong process that begins with intense training 1 week prior to competition.[13] For the first 3 days of a carbohydrate-loading week, athletes choose low-carbohydrate foods, but continue to exercise in an attempt to deplete muscle glycogen stores. During the 3 days prior to competition, athletes rest, or exercise minimally, while consuming a high-carbohydrate diet to promote glycogen formation and storage.

High-protein diets may take many forms for teen athletes. In general, athletes who follow high-protein diets may consume three to four times the recommended protein intake, accompanied by a relatively low intake of carbohydrate. High-protein diets should be discouraged among athletes, for several reasons. First, many dietary protein sources are also sources of saturated fats, which may increase lifetime risk of coronary artery disease. Second, high protein and fat intakes delay digestion and absorption, limiting the amount of energy available for use during physical activity. Finally, more water is required for the breakdown of protein than either fat or carbohydrate due to the increased water loss that accompanies the excretion of nitrogen. This factor places an athlete at increased risk for dehydration, often accompanied by a decrease in physical performance. Adolescents should be reminded that the consumption of individual amino acids or high intakes of complete proteins do not increase physical performance, and may in fact decrease physical performance.

The popularity among adults of The Zone diet, which is a type of high-protein diet, has led to the adoption of this diet by adolescents. The Zone theory is based upon an intake of 40% carbohydrate, 30% fat, and 30% protein.[14] Proponents of The Zone diet suggest that a 40/30/30 ratio of macronutrient intakes allows the body to more efficiently burn calories, leading to loss of weight, reduction in body fat, and increases in lean body mass. Even though weight loss can occur with the intake of high-protein diets due to the increased loss of body water, reductions in body fatness and increases in lean body mass have not been scientifically proven with use of this diet. Another argument supporting the 40/30/30 diet plan claims that it reduces insulin levels and insulin resistance, which proponents

Case Study 15.1

Photo Disc

Adolescent Nutrition, Growth and Development, and Obesity

Anna is a 12-year-old female being seen in the pediatric and adolescent medicine clinic for a routine physical examination for summer camp. She reports no medical concerns. Her mother reports concern over recent weight gain. Anna's last visit was approximately 5 months earlier, when she was seen for an upper respiratory infection.

During the current visit, Anna's body mass index (BMI) value is calculated. Her hemoglobin and hematocrit levels are checked, and a blood pressure reading is taken. The primary health care provider also takes a medical history and determines Anna's sexual maturation rating (SMR). The following data are recorded:

Age: 12 years, 5.5 months   BMI: 23.3
SMR: early stage 3   Hematocrit: 35.1%
Menarche: not yet occurred   Hemoglobin: 11.9 g/dL
Height: 58 inches (1.47 meters)   Blood Pressure: 121/77
Weight: 111 pounds (50.35 kilograms)

Anna's BMI value is charted on the National Center for Health Statistics BMI reference growth chart for girls age 2–20 years, and is compared to previous data. On her previous medical visit, Anna's BMI was calculated to be 21.4. Her previous height was 56.0 inches (1.42 meters), and her previous weight was 95 pounds (43.09 kilograms).

Questions

1. What is the current BMI percentile for Anna, based on her age and gender?

2. What was her previous BMI percentile based on the BMI of 21.4 and age of 12.0 years?

3. What is her current classification for height-for-weight (BMI)? Normal, at risk for overweight, or overweight?

4. Given Anna's current weight classification based on BMI, what are the recommendations for weight management based on current recommendations?

   a. Maintain current weight and monitor food intake, physical activity levels, weight, height, growth, and development at 3- to 4-month intervals.

   b. Weight loss of 1–2 pounds per week.

   c. Weight gain of 1–2 pounds per week.

   d. No concerns or actions are identified at this time.

5. What do the SMR and the absence of menarche tell you about Anna's potential for linear growth and future weight gain?

6. Using the Centers for Disease Control and Prevention (CDC) guidelines, how would you classify Anna's blood pressure readings?

7. Using the CDC guidelines, how would you classify her hemoglobin and hematocrit levels?

(Answers are in Appendix D.)

Case Study 15.1 (continued)

Part Two

The primary health care provider refers Anna to a nutritionist at the clinic for nutrition counseling and weight management. The nutritionist conducts a 24-hour recall of food intake and asks about physical activity and sedentary activity. The results follow.

Food Intake in Past 24 Hours

Today:

7 a.m.	8 oz apple juice at home
10 a.m.	candy bar from school vending machine
12:15 p.m.	hamburger from school cafeteria and 12 oz of sweetened (nondiet) cola and 1-oz bag of potato chips from school vending machine
3:00 p.m.	tortilla chips with cheese sauce (approximately 3 oz chips and ½ c cheese sauce) and 12-oz nondiet cola at home

Previous evening:

6:30 p.m.	chicken rice casserole (approximately 2 c), 8 oz apple juice, 2 crescent rolls with 1 tb butter
8:30 p.m.	1 bag of butter flavored microwave popcorn (approximately 8 c)

Physical Activity in Past 24 Hours

Today:

7:45 to 2:45	walked between classes for 1–2 minutes, six times that day
1:00 p.m.	35 minutes of physical education (Anna played basketball for only 15 minutes due to need to rotate students on teams); physical education is offered 2 days per week in winter semester only

Previous evening:

5:00 p.m.	walked from friend's house to home (approximately 3 blocks)

Sedentary Activity in Past 24 Hours

Today:

7:45 to 2:45	sat in classroom for 35–40 minutes, five times that day
3:00 p.m.	on the computer in Internet chat rooms and e-mail for 40 minutes

Previous evening:

8:00 p.m.	watched television for 2 hours
10:45 p.m.	went to bed

Questions

1. List two nutrients that are likely to be consumed in inadequate amounts in Anna's diet.
2. In which of the Food Guide Pyramid food group(s) is Anna's diet inadequate?
3. List two recommendations that you would suggest to Anna to improve her food intake. (Try to use concrete examples because she is a young adolescent.)
4. List two recommendations that you would suggest to Anna to improve her physical activity or reduce her sedentary activity. (Try to use concrete examples because she is a young adolescent.)

(Answers are in Appendix D.)

claim result from consuming a high-carbohydrate diet. Teens should be told that insulin levels do increase temporarily after eating a high-carbohydrate meal, in order for lean body mass and other tissues to take up carbohydrate and store it as glycogen. Carbohydrate still provides the main source of energy available for physical activity.

Dietary Supplements

Supplements reportedly used by collegiate athletes, and perhaps also by adolescents, include creatine, individual amino acids or protein powders, carnitine, dehydroepiandrosterone (DHEA), beta-hydroxy-beta-methylbutyrate, anabolic steroids, and ephedra.[12] One survey found anabolic steroid use to be at 2% among high school athletes.[15] Adolescents should be reminded that few supplements are tested on adolescents, so dosing and safety information is not available. This point is particularly important because many supplements can have detrimental health effects when used inappropriately. Experience suggests that creatine, carnitine, DHEA, and chromium picolinate are among the most widely used supplements by adolescent athletes.

CREATINE Creatine is one of the most popular supplements used by athletes, particularly among athletes who compete in strength-related sports, such as football. Creatine, formed in the liver and kidney of the human body, can be obtained in more than adequate amounts from the consumption of meat. Creatine is thought to improve anaerobic metabolism and to promote weight and lean tissue gain. Studies of creatine in adults show mixed results.[16] It appears to be of no benefit to endurance athletes, and marginally beneficial during strength-related sports. Side effects of creatine use, which seem to be dose related, include abdominal pain and cramping, nausea, diarrhea, headache, increased tendency toward muscle strains, and muscle soreness.[17] No available data document the long-term health effects related to creatine use; however, chronic use may be associated with renal damage.[18] It is not recommended for use by adolescents at this time.

CARNITINE The human body forms carnitine from the amino acids lysine and methionine. Almost all carnitine in humans resides within muscle tissue. As its main function, it metabolizes fatty acid. Animal studies suggest that it may enhance fatty acid oxidation by arterial walls.[19] Studies with athletes during athletic performance have been inconclusive. The use of carnitine seems to result in few side effects, most notably diarrhea at high doses.[18] The use of carnitine is not recommended among adolescents, however, because commercially available forms often contain impurities.

DHEA Dehydroepiandrosterone (DHEA) is a precursor of testosterone and estrogen. Naturally produced in the human body by the adrenal glands, DHEA levels fall in humans as age increases. Its reputed effects include melting fat, decreasing insulin resistance, increasing immune system function, increasing lean body mass, and decreasing risk of osteoporosis; however, no scientific evidence backs such claims.[18] As a precursor hormone to sex hormones, DHEA carries significant side effects such as irreversible gynecomastia (breast enlargement) and prostate enlargement among males and hirsutism (facial hair) among females.[20] Adolescents primarily use this hormone to increase lean body mass, but no evidence supports this effect. Given the possibility of significant side effects of DHEA in pubescent adolescents undergoing hormonal changes, DHEA should not be used by adolescents.

CHROMIUM Studies of chromium picolinate supplementation among men, women, and athletes do not support claims that it burns fat, decreases fat deposition, increases lean body mass, and promotes weight loss. In fact, in a study of women, chromium supplementation increased body weight.[21,22] Chromium use among adolescent athletes is of concern because high doses can lead to renal failure,[23] cognitive impairment,[24] and increased risk of iron-deficiency anemia.

In the highly competitive world of sports, the adolescent athlete, the coach, the trainer, and the parents easily become vulnerable to nutrition misinformation. A widespread need exists for integration of sound nutrition into team sports and training regimens for athletes. Coaches especially should be targeted for nutrition education programs because high school athletes frequently rely on coaches for nutrition information.

Adolescent Pregnancy

The birthrate among adolescent females is approximately 51 births per 1000 adolescents.[25] Most teens who give birth are 18–19 years old; however, pregnancies can occur as early as 11–12 years of age. The birthrate for adolescents less than 15 years old is approximately 1.0 births/1000, while birthrates for 15- to 17-year-old and 18- to 19-year-old teens are 30.0 births/1000 and 82 births/1000, respectively. Adolescents at risk for pregnancy include teens from low-income families, teens from unstable family situations (divorce/remarriage or unstable income), daughters of adolescent mothers, and those who demonstrate poor scholastic performance or drop out of school.[26,27] Repeat childbearing also occurs among adolescents, with 17% of adolescent mothers bearing two or more children during adolescence.[28] Among births to adolescents less than 15 years of age, 3% are repeat pregnancies.

Energy and Nutrient Needs

Many factors bear consideration when determining the energy and nutrient needs of pregnant teens, including potential for continued growth and development for the adolescent, preconceptional nutritional status, food security, and the potential for substance use. *Gynecological age* (GA) signifies the degree of physiological maturity and the potential for continued growth and development of a pregnant teen. GA is calculated by subtracting the age of menarche (first menstrual cycle) from the chronological age of a female. Adolescents with a GA of <2 years are likely to experience continued growth and development during pregnancy. One line of thought is that the growing adolescent competes directly with her fetus for energy and nutrients.[29,30] Therefore, the energy and nutrient needs required for adolescent growth must be added to those required for optimal fetal development for pubescent adolescents.

Adolescents may have inadequate nutrient stores to support pregnancy at the time of conception. Females who are particularly high risk for poor preconceptional nutrition status include overweight or underweight teens, those who use substances, competitive athletes, females with eating disorders, chronic dieters, and vegetarians. Meal skipping is common among adolescent females and may lead to nutrient intakes that are inadequate to support a healthy pregnancy.

Warning signs of nutrition problems in pregnant adolescents are listed in Table 15.3. Nutrients most likely to be deficient in the diets of teen females include calcium, iron, zinc, magnesium, folate, and vitamins A, E, and B_6. Vegetarians may also enter pregnancy with limited intakes of vitamins D and B_{12}.

Alarmingly, 11% to 55% of pregnant adolescents report substance use.[31] Data suggest that approximately one-fourth of pregnant teens used tobacco, alcohol, and marijuana during the 6 months prior to conception, and almost half used at least one substance during pregnancy. Because substance use is associated with frequent meal skipping and overall poor-quality dietary intake, females who use substances are at increased risk for poor nutrition status at the time of conception, as well as throughout the course of pregnancy. (For specific information about nutritional effects of substance use, see the "Substance Use" section later in this chapter.)

> **Gynecological Age** Defined as chronological age minus age at menarch. For example, a female with the chronological age of 14 years minus age at first menstrual cycle of 12 equals a gynecological age of 2.

The DRIs for pregnant and lactating females are included in the table inside the front cover of this text. Energy needs of pregnant teens will vary according to the growth and development status of the teen, level of physical activity, body weight, and stage of pregnancy. Current recommendations for pregnant adult women suggest that an additional 300 calories per day are needed during the

---

Table 15.3 Warning signs of nutrition problems in pregnant adolescents

MEDICAL/OBSTETRIC FACTORS	NUTRITION/DIETARY FACTORS	PSYCHOSOCIAL FACTORS
• Adolescent <16 years or gynecological age <2* • A previous pregnancy • Closely spaced pregnancy • History of poor obstetric or fetal performance • Chronic systemic disease • Past or present eating disorder: anorexia nervosa or bulimia nervosa • Underweight or overweight prior to pregnancy • Inadequate weight gain during pregnancy • Excessive weight gain during pregnancy • Persistent nausea or vomiting during pregnancy • Iron-deficiency anemia or other nutritional deficiencies • Severe infections • Heavy smoker • Alcohol or drug use	• Inadequate refrigeration or cooling facilities • Lack of transportation or accessibility to grocery store • Cultural or religious dietary restrictions • Frequent eating away from home • Frequent snacking on low-nutrient-density foods • Poor appetite • Limited, monotonous, or highly processed diet • Irregular meal patterns (skipping meals) • History of frequent dieting • Exclusion of a major food group(s) • Binge-eating episodes • Eating of nonfood substances (pica) • Nontraditional dietary pattern (i.e., strict vegetarianism) • Overuse of nutritional supplements • Heavy caffeine intake	• Inadequate income • Living alone or in an unstable family or home environment • Little family, partner, or peer support • Denial or failure to accept the pregnancy • Significant emotional stress or depression

*Gynecological age = Current age − Age at menarche.
SOURCE: From Mary Story and Jamie Stang, Nutrition and the Pregnant Adolescent, 2000. Reprinted with permission from Mary Story.

second and third trimesters of pregnancy to support fetal growth and development.[32] Younger adolescents, who are likely to experience continued growth during pregnancy, and adolescents who are underweight at conception will have significantly higher energy needs than older females will. Adolescents who have completed puberty and enter pregnancy overweight may require fewer additional calories. The best determination of adequate energy intake is the sufficiency of maternal weight gain.

Protein requirements increase during pregnancy in order to support the growth of the fetus and the adolescent. The additional intake of 10 grams of protein recommended throughout pregnancy brings minimum protein requirements to 55 grams per day for most teens.[32] Carbohydrate and fat should be consumed at levels that support fetal growth and optimal weight gain while also adhering to the Food Guide Pyramid.

Fluid intake is especially important during pregnancy. As blood volume increases, so does the need for water and other fluids. Adolescents should be encouraged to drink 8 to 10 cups of fluids such as water, milk, and juice each day. Table 15.4 lists criteria for a healthy prenatal diet for teens.

Vitamin and Mineral Supplements

Vitamins and minerals that should be consumed at higher levels than those of nonpregnant adolescents include thiamin, riboflavin, niacin, folate, biotin, choline, pantothenic acid, iron, zinc, magnesium, iodine, selenium, and vitamins B_6, B_{12}, E, K, and C. A low-dose or prenatal vitamin-mineral supplement may be appropriate for pregnant teens who frequently skip meals, follow vegetarian diets, use substances, enter pregnancy with low nutrient stores, experience repeat pregnancy within 2 years, or have multiple gestations.[32]

Individual supplementation of 30 milligrams of elemental iron is recommended for adolescents to prevent iron-deficiency anemia. Calcium supplements are recommended for pregnant teens who do not consume at least 1300 milligrams of calcium per day. Chewable vitamins and minerals may be more to the liking of adolescents than nonchewable tablets.

Table 15.4 Criteria for a healthy prenatal diet

- Provides enough calories for adequate weight gain.
- Is well balanced and follows the Food Guide Pyramid.
- Tastes good and is enjoyable to eat.
- Spaces eating in intervals throughout the day.
- Provides adequate amounts of high-fiber foods.
- Includes 8 cups of fluid daily.
- Limits beverages that contain caffeine (2–3 servings or fewer daily).
- Is moderate in fat, saturated fat, cholesterol, sugar, and sodium.
- Excludes alcohol.

SOURCE: From Mary Story and Jamie Stang, Nutrition and the Pregnant Adolescent, 2000. Reprinted with permission from Mary Story.

Pregnancy Weight Gain

Recommended weight gains for pregnant teens can be found in Table 15.5. Inadequate weight gain increases the likelihood of delivery of a premature, low-birthweight, and small-for-gestational-age infant, especially among adolescents who enter pregnancy underweight.[32] Poor weight gain in the first 24 weeks of pregnancy is associated with an increased risk of delivery of a small-for-gestational-age infant, while inadequate gain after 24 weeks increases the risk of preterm delivery.[33]

Adolescents who are still growing gain 1 to 2 kilograms more weight during pregnancy than do those who have completed growth, yet they are more likely to deliver smaller infants. They should be counseled to achieve the upper limit of weight gain recommendations for optimal pregnancy outcomes. Table 15.6 lists evaluation and management criteria for inadequate gestational weight gain. Excessive gestational weight gain among adolescents may signal depression and alcohol consumption.[34] Criteria for evaluation and management of excessive gestational weight gain are listed in Table 15.7.

Table 15.5 Weight-gain recommendations for pregnant adolescents

PREPREGNANT BMI	TOTAL (lbs)	TRIMESTER I (lbs)	TRIMESTERS 2 & 3 (lbs/week)
Underweight	28–40	5	1.0+
Normal weight	25–35	3	1.0+
Overweight	15–25	2	0.66+
Obese	>15	1.5	0.5+

SOURCE: Reprinted with permission from Nutrition During Pregnancy, © National Academy of Sciences. Courtesy of the National Academy Press, Washington, D.C.

Table 15.6 Inadequate gestational weight gain

DEFINITION: <2 LBS PER MONTH AFTER THE FIRST TRIMESTER

Evaluation	
Measurement error	Inadequate food access
Excessive gain at previous visit (e.g., edema)	Homelessness
	Pica
Disordered eating	Substance abuse
Restrictive eating/dieting, meal skipping	Nausea, vomiting, or heartburn
Psychosocial stress	Inadequate sleep and rest
Social isolation	
Lack of partner or family support	High level of physical activity
Depression	Physically demanding job
Denial/rejection of pregnancy	Gestational diabetes
	Urinary ketones

Management	
5–6 nutrient-dense meals and snacks	Manage physical discomforts
Adequate rest and sleep	Refer to food-assistance programs
Decreased physical activity	
Stress management/ relaxation techniques	Psychosocial counseling
	Food journal

SOURCE: Reprinted with permission from Nutrition During Pregnancy, © National Academy of Sciences. Courtesy of the National Academy Press, Washington, D.C.

Table 15.7 Criteria for evaluation and management of excessive weight gain in adolescent pregnancy

DEFINITION: >6 LBS PER MONTH

Evaluation	
Measurement error	Multiple gestation
Weight loss at previous visit	Depression
Edema	Binge eating
Smoking cessation	Psychosocial stress
Alcohol use	Social isolation
Infrequent, large meals	Emotionally based eating
High fat and/or sugar intake	Pica
Physical inactivity	

Management	
Sensitive, supportive, and nonshaming manner	Alternatives to emotionally based eating
Moderate physical activity	Increased water and dietary fiber sources
Positive reinforcement for smoking cessation	Psychosocial counseling
Small, frequent meals (avoiding excess hunger and fullness)	Stress management/ relaxation techniques
Decreased fat and sugar intake	Continue to gain weight at expected rate; avoid weight loss
Healthy snack and fast-food choices	Food journal
Refer to food-assistance programs	Limit portions to average serving size

SOURCE: Reprinted with permission from Nutrition During Pregnancy, © National Academy of Sciences. Courtesy of the National Academy Press, Washington, D.C.

Substance Use

The use of substances such as tobacco, alcohol, and recreational drugs directly affects the nutritional status of adolescents. Data on smoking rates of teens suggest that 36% of white, 31% of Hispanic, and 16% of black adolescents smoke.[35] Studies document heavy smoking in 16% of white, 7% of Hispanic, and 2% of black teens. Traditionally, male adolescents were more likely to smoke than females; however, that tendency is no longer evident. Current data suggest that smoking rates are 1% to 2% higher among 8th- and 10th-grade females than among same-aged males.[36] Initiation of smoking among adolescents can begin as young as 9 years old, with 12–18 years of age being the highest-risk age group. Use of smokeless tobacco, also known as chewing tobacco, is also seen among adolescents. The prevalence of smokeless tobacco use is thought to be at least 6.5 times higher among males than females.[35]

Alcohol intake and substance use among adolescents increases with age. A national survey suggested that by the 12th grade, more than 80% of adolescents had tried alcohol[37] with more than one-third reporting binge drinking (drinking 5 or more alcoholic drinks during one occasion) at least one day during the past month. This survey also showed that almost one-quarter of adolescents reported using marijuana at least one time during the past month, 8% had tried cocaine, 10% had used inhalants, and 16% had used another form of recreational drug (such as LSD, PCP, speed, heroin, mushrooms, ecstasy, or methamphetamine).

Much of the data on differences in dietary intake and nutrient needs between substance users and nonusers comes from studies of adults. Data collected on adolescent substance users as part of the Minnesota Adolescent Health Survey and from high school students living in Israel provide some of the best data on dietary intakes among adolescent users. Adolescent substance users in these studies reported consuming diets that were less adequate than those of nonusers.[38,39] As a group they were more likely to skip breakfast, not eat a meal or snack at school, not eat three meals per day, report chronic dieting, and report purging than were their nonsubstance-using peers. All of these behaviors suggest that teens who use substances may be at high risk of poor nutritional status.

Substance use can result in depleted stores of vitamins and minerals, including thiamin, vitamin C, and iron.

Chronic ingestion of alcohol and drug use can result in a reduced appetite, leading to low dietary intakes of protein, energy, vitamins A and C, thiamin, calcium, iron, and fiber.[40,41] Other adverse nutritional affects of substance use are listed in Table 15.8.

Dietary Supplements

Vitamin-Mineral Supplements

Supplements may be used by adolescents for a variety of reasons, including improving health, treating iron-deficiency anemia, increasing energy, building muscle, and losing weight. National data suggest that the prevalence of vitamin-mineral supplement use among teens in the United States ranges from 16% to 33%.[42,43] More than half of adolescents who report using vitamin-mineral supplements take them occasionally, with slightly less than half using them daily. Data on racial differences on adolescent supplement use are equivocal; one study suggests that white female adolescents are most likely to report using vitamin-mineral supplements, followed by white males, Mexican American females, black females, Mexican American males, and finally black males; another study found no differences by race. Overall, approximately 28% of female adolescents and 24% of male adolescents use vitamin-mineral supplements.

Approximately half of vitamin-mineral supplements consumed by adolescents are multivitamins without minerals, 34% are individual vitamins or minerals, 18% are multivitamins with minerals, and 17% are iron with vitamin C tablets. Among the individual nutrient supplements used, vitamin C is the most common, followed by calcium, iron, vitamin E, and B-vitamin complex.[43] Among nutrients consumed at less than 75% of the RDA/DRI by the most adolescents are calcium and zinc, yet only one-third

of supplements consumed by adolescents contain minerals. Calcium is the second most commonly used individual nutrient supplement; however, it is the nutrient most likely to be consumed at levels less than 75% of the DRI.

Adolescents who take vitamin-mineral supplements tend to consume a more nutritionally adequate diet than those who do not.[43] Supplement users obtain a smaller proportion of calories from total and saturated fats, and more from carbohydrate compared to nonsupplement users. They also consume diets that tend to be more nutrient dense and higher in folate, calcium, iron, and vitamins E, C, and A than the diets of nonusers.

Herbal Remedies

Few data shed much light on the use of nonnutritional supplements such as herbs (including herbal weight-loss products) among adolescents. Experience suggests that adolescents may be likely to take herbal supplements for several reasons, including to lose weight, for treatment of attention deficit disorder, and to increase energy and stamina. Studies are needed to determine exactly what types of herbal products adolescents use, because many herbs are known to have potentially dangerous side effects, and few recommendations are available to guide the use of herbs by children or adolescents.

Iron-Deficiency Anemia

Iron-deficiency anemia is the most common nutritional deficiency noted among children and adolescents. Several risk factors are associated with its development among adolescents, including rapid growth, inadequate dietary intake of iron-rich foods or foods high in vitamin C, highly restrictive vegetarian diets, calorie-restricted diets, meal skipping, participation in strenuous or endurance sports, and heavy menstrual bleeding.[44] The effects of iron-deficiency anemia include delayed or impaired growth and development, fatigue, increased susceptibility to infection secondary to depressed immune system function, reductions in physical performance and endurance, and increased susceptibility to lead poisoning. Pregnant teens who are iron deficient in the early stages of gestation are at increased risk of preterm delivery and delivery of a low-birthweight infant.

Assessment of iron-deficiency anemia compares individual hemoglobin and hematocrit levels to standard reference values. Table 15.9 lists the 1998 Centers for Disease Control and Prevention criteria for determining anemia, based on age and gender. Adjustments to these values must be made for individuals who live at altitudes greater than 3000 feet and for smokers. An adjustment of +0.3 g/dL is required for adolescents who smoke.[45] Because adolescent males are not at high risk for iron-deficiency anemia, they do not need to be screened unless they exhibit one or more of the risk criteria listed. All adolescent females should be screened every 5 years for ane-

---

Table 15.8 Potential effects of substance use on nutrition status

- Appetite suppression
- Reduced nutrient intake
- Decreased nutrient bioavailability
- Increased nutrient losses/malabsorption
- Altered nutrient synthesis, activation, and utilization
- Impaired nutrient metabolism and absorption
- Increased nutrient destruction
- Higher metabolic requirements of nutrients
- Inadequate weight gain/weight loss
- Iron-deficiency anemia
- Decreased financial resources for food

SOURCE: From Mary Story and Jamie Stang, Nutrition and the Pregnant Adolescent, 2000. Reprinted with permission from Mary Story.

Table 15.9 Maximum hemoglobin concentration and hematocrit values for iron-deficiency anemia

GENDER/AGE[a]	HEMOGLOBIN, <g/dL	HEMATOCRIT, <%
Males and Females		
8 to <12 years	11.9	35.4
Males		
12 to <15 years	12.5	37.3
15 to <18 years	13.3	39.7
≥18 years	13.5	39.9
Females[b]		
12 to <15 years	11.8	35.7
15 to <18 years	12	35.9
≥18 years	12	35.7

[a]Age and gender-specific cutoff values for anemia are based on the 5th percentile from the third National Health and Nutrition Examination Survey (NHANES III).
[b]Nonpregnant and lactating adolescents.
SOURCE: Data taken from Centers for Disease Control and Prevention. "Recommendations to Prevent and Control Iron Deficiency in the United States," *MMWR* 47 (RR-3), 1998.

mia; those with one or more risk factors for anemia should be screened annually.

Treatment that follows a diagnosis of iron-deficiency anemia needs to include increased dietary intake of foods rich in iron and vitamin C as well as iron supplementa-tion. Adolescents under the age of 12 should be supplemented with 60 milligrams of elemental iron per day, and teenagers over the age of 12 should receive 60 to 120 milligrams of elemental iron per day.[45] These recommendations spark some controversy, however, given their high doses of elemental iron. Adolescents often report gastrointestinal side effects from iron supplementation, such as constipation, nausea, and cramping. These side effects can be lessened by giving smaller doses of iron more frequently throughout the day and counseling the adolescent to take the iron supplement at mealtimes or with food sources of vitamin C. Calcium supplements, dairy products, coffee, tea, and high-fiber foods may decrease absorption of iron supplements; these foods should be avoided within 1 hour of taking an iron supplement.

Hypertension

Adolescents are considered hypertensive if the average of three systolic and/or diastolic blood pressure readings exceeds the 95th percentile, based on age, gender, and height.[46] Blood pressure levels for the 95th percentiles for males and females are shown in Table 15.10. Classifications of blood pressure based on the average of three readings are

● Normal blood pressure: <90th percentile
● High-normal blood pressure: >90th and <95th percentiles
● Hypertension: >95th percentile

Table 15.10 Blood pressure levels for the 90th and 95th percentiles of blood pressure for boys and girls, aged 10 to 17 years

		SYSTOLIC BP (mm Hg), BY HEIGHT PERCENTILE FROM STANDARD GROWTH CURVES														DIASTOLIC BP (mm Hg), BY HEIGHT PERCENTILE FROM STANDARD GROWTH CURVES													
		Boys							Girls							Boys							Girls						
Age	BP Percentile*	5%	10%	25%	50%	75%	90%	95%	5%	10%	25%	50%	75%	90%	95%	5%	10%	25%	50%	75%	90%	95%	5%	10%	25%	50%	75%	90%	95%
10	90th	110	112	113	115	117	118	119	112	112	114	115	116	117	118	73	74	74	75	76	77	78	73	73	73	74	75	76	76
	95th	114	115	117	119	121	122	123	116	116	117	119	120	121	122	77	78	79	80	80	81	82	74	74	75	75	76	77	78
11	90th	112	113	115	117	119	120	121	114	114	116	117	118	119	120	74	74	75	76	77	78	78	74	74	75	75	76	77	77
	95th	116	117	119	121	123	124	125	118	118	119	121	122	123	124	78	79	79	80	81	82	83	78	78	79	79	80	81	81
12	90th	115	116	117	119	121	123	123	116	116	118	119	120	121	122	75	75	76	77	78	78	79	75	75	76	76	77	78	78
	95th	119	120	121	123	125	126	127	120	120	121	123	124	125	126	79	79	80	81	82	83	83	79	79	80	80	81	82	82
13	90th	117	118	120	122	124	125	126	118	118	119	121	122	123	124	75	76	76	77	78	79	80	76	76	77	78	78	79	80
	95th	121	122	124	126	128	129	130	121	122	123	125	126	127	128	79	80	81	82	83	83	84	80	80	81	82	82	83	84
14	90th	120	121	123	125	126	128	128	119	120	121	122	124	125	126	76	76	77	78	79	80	80	77	77	78	79	79	80	81
	95th	124	125	127	128	130	132	132	123	124	125	126	128	129	130	80	81	81	82	83	84	85	81	81	82	83	83	84	85
15	90th	123	124	125	127	129	131	131	121	121	122	124	125	126	127	77	77	78	79	80	81	81	78	78	79	79	80	81	82
	95th	127	128	129	131	133	134	135	124	125	126	128	129	130	131	81	82	83	83	84	85	86	82	82	83	83	84	85	86
16	90th	125	126	128	130	132	133	134	122	122	123	125	126	127	128	79	79	80	81	82	82	83	79	79	79	80	81	82	82
	95th	129	130	132	134	136	137	138	125	126	127	128	130	131	132	83	83	84	85	86	87	87	83	83	83	84	85	86	86
17	90th	128	129	131	133	134	136	136	122	123	124	125	126	128	128	81	81	82	83	84	85	85	79	79	79	80	81	82	82
	95th	132	133	135	136	138	140	140	126	126	127	129	130	131	132	85	85	86	87	88	89	89	83	83	83	84	85	86	86

*Blood pressure percentile determined by a single movement.
SOURCE: Adapted from the National High Blood Pressure Education Program Working Group on Hypertension Control in Children and Adolescents. *Update on the Task Force Report (1987) on High Blood Pressure in Children and Adolescents: A Working Group Report from the National High Blood Pressure Education Program.* NIH Publications, No. 97-3790. Bethesda, MD: National Heart, Lung, and Blood Institute, 1997.

Risk factors for hypertension among adolescents include a family history of hypertension, high dietary intake of sodium, overweight, hyperlipidemia, inactive lifestyle, and tobacco use.[46] Adolescents who display one or more of these risk factors should be routinely screened for hypertension. Nutrition counseling to decrease sodium intake, to limit fat intake to 30% or less of calories, and to consume adequate amounts of fruit, vegetables, whole grains, and low-fat dairy products should be provided when hypertension is diagnosed. Weight loss is recommended for adolescents who are hypertensive in the presence of overweight. If medications are prescribed, teens still must adhere to general dietary recommendations and should still be encouraged to reach and maintain a healthy weight for their height.

Hyperlipidemia

Approximately one in four adolescents in the United States has an elevated cholesterol level.[47] Table 15.11 provides the classification criteria for elevated cholesterol levels in children and adolescents. Risk factors for hypercholesterolemia include a family history of cardiovascular disease or high blood cholesterol levels, cigarette smoking, overweight, hypertension, diabetes mellitus, and low level of physical activity. Adolescents who exhibit these risk factors should be screened to determine causes of hyperlipidemia and should be referred for treatment as required.[48] Early intervention for adolescents who have high cholesterol levels may reduce their risk of coronary artery diseases later in life.

Dietary recommendations indicate that youth over the age of 5 years should obtain 25–35% of their calories from fat, with no more than 10% of calories derived from saturated fat.[48,49] Dietary cholesterol intakes of 300 milligrams per day or less have also been recommended. Counseling adolescents with hyperlipidemia to follow these guidelines can be challenging, given their frequent consumption of fast foods and their preferred food choices. Suggested dietary changes should take into account the eating habits of adolescents and emphasize healthier food choices as opposed to restriction of the teen's favorite foods. Health professionals can work with adolescents to make healthier choices at fast-food restaurants, to limit their portion sizes of high-fat food items such as high-fat snacks, and to consume adequate amounts of fruits, vegetables, grains, and low-fat dairy products.

Anorexia Nervosa An eating disorder characterized by extreme weight loss, poor body image, and irrational fears of weight gain and obesity.

Eating Disorders

The Continuum of Eating Concerns and Disorders

Eating concerns and disorders lie on a continuum ranging from mild dissatisfaction with one's body shape to serious eating disorders such as *anorexia nervosa, bulimia ner-*

Table 15.11	Classification of cholesterol levels in high-risk children and adolescents*	
	TOTAL CHOLESTEROL mg/dL	LDL CHOLESTEROL mg/dL
Acceptable	<170	<110
Borderline	170–199	110–129
High	≥200	≥130

*Children and adolescents from families with hypercholesterolemia or premature cardiovascular disease.
SOURCE: Data taken from the National Institutes of Health, National Heart, Lung, and Blood Institute, National Cholesterol Education Program. *Report of the Expert Panel on Blood Cholesterol Levels in Children and Adolescents.* Bethesda, MD: National Institutes of Health, 1991.

vosa, and *binge-eating disorder.* Along the continuum between these endpoints lie normative dieting behaviors and more severe disordered eating behaviors such as self-induced vomiting and binge eating (Illustration 15.4). Although engagement in anorexic behaviors and unhealthy dieting may not be frequent or intense enough to meet the formal criteria for being defined as an eating disorder, these behaviors may negatively affect health and may lead to the development of more severe eating disorders. All eating disorders present a serious public health concern in light of their prevalence and their potentially adverse effects on growth, psychosocial development, and physical health outcomes.[50]

Prevalence of Eating Disorders

An awareness of the prevalence of eating disorders is critical in effective planning for interventions aimed at their treatment and prevention. Conditions prevalent among youth warrant interventions that have the potential to reach large numbers of youth, such as community-based and school-based programs. The small percentage of the adolescent population affected by these types of conditions having severe health implications requires more intensive individual or small-group interventions. Estimates as to the prevalence of each of the eating disorders on the continuum are presented in Table 15.12.

Anorexia Nervosa

Anorexia nervosa and its impact on morbidity and mortality make it the most severe condition on the continuum of eating disorders. Among adolescent girls and young women, prevalence estimates of anorexia nervosa range from 0.2% to 1.0%.[51,52] Anorexia nervosa presents more frequently among females than among males; about 9 out of 10 individuals with anorexia nervosa are female. Only in recent years has attention been directed toward males with this condition, who may not be suspected of having anorexia nervosa and therefore may be diagnosed at later stages of the disease when treatment is more difficult.

Illustration 15.4 The continuum of weight-related concerns and disorders.

Characteristics of anorexia nervosa include self-starvation and strong fears of being fat. An adolescent may begin with simple dieting behaviors due to social pressures to be thin or comments by others about an adolescent's weight. However, as these behaviors lead to weight loss, feelings of control, and comments from others about weight loss, anorexia nervosa develops and the condition takes on a life of its own. Diagnostic criteria for anorexia nervosa are shown in Table 15.13. Key features of anorexia nervosa are refusal to maintain body weight over a minimal normal weight for age and height; intense fear of gaining weight or becoming fat, even though underweight; a distorted body image; and amenorrhea (in females).

The two types of anorexia nervosa are restricting and nonrestricting. In the restricting type, the individual does not regularly engage in binge-eating or purging behaviors. The nonrestricting type exhibits regular episodes of binge-eating and purging behaviors. However, both types present with a refusal to maintain a minimally normal body weight.

An estimated 10% to 15% of patients with anorexia nervosa die from their disease, although difficulties arise in assessing mortality rates from anorexia nervosa.[53] Reasons for fatality from anorexia include a weakened immune system due to undernutrition, gastric ruptures, cardiac arrhythmias, heart failure, and suicide.[54] The adolescent or the family commonly denies the condition, which delays the diagnosis and treatment, resulting in a poorer prognosis for recovery. Early recognition of possible signs of anorexia nervosa and seeking out of professional help significantly affect the time and intensity of treatment and improve chances for a successful recovery. Recovery rates are estimated at 40% to 50% for individuals with anorexia nervosa.[55]

Table 15.12 Estimated prevalence and brief description of weight-related concerns/disorders among adolescents

DISORDER	ESTIMATED PREVALENCE
Anorexia nervosa	Approximately 0.2% to 1.0% of adolescent females and young women
Bulimia nervosa	Approximately 1% to 3% of adolescent females and young women
Binge-eating disorder	Estimated 30% of weight-control population; 2% of general population
Anorexic/bulimic behaviors	Estimated 10% to 20% of adolescents although estimates vary
Dieting behaviors	Estimates vary and range from 44% of adolescent females and 15% adolescent males, to 50% to 60% of all adolescent females are attempting to lose weight
Body dissatisfaction	Estimates vary in accordance with type of measurement used and age, gender, and ethnicity of population: approximately 60% of girls and 35% of boys are not satisfied with their weight

Table 15.13 Diagnostic criteria for anorexia nervosa

- Refusal to maintain body weight at or above a minimally normal weight for age and height (e.g., weight loss leading to maintenance of body weight less than 85% of that expected; or failure to make expected weight gain during period of growth, leading to body weight less than 85% of that expected)
- Intense fear of gaining weight or becoming fat, even though underweight
- Disturbance in the way in which one's body weight or shape is experienced, undue influence of body weight or shape on self-evaluation, or denial of the seriousness of the current low body weight
- Amenorrhea in postmenarchal women; that is, the absence of at least three consecutive menstrual cycles (a woman is considered to have amenorrhea if her menstrual periods occur only following hormone-estrogen-administration)

Restricting Type: During the episode of anorexia nervosa, the person has not regularly engaged in binge eating or purging behavior (i.e., self-induced vomiting or the misuse of laxatives, diuretics, or enemas).

Binge-Eating/Purging Type: During the episode of anorexia nervosa, the person has regularly engaged in binge-eating or purging behavior (i.e., self-induced vomiting or the misuse of laxatives, diuretics, or enemas).

SOURCE: Reprinted with permission from the Diagnostic and Statistical Manual of Mental Disorders, Fourth Edition, Text Revision, Copyright 2000. American Psychiatric Association.

Bulimia Nervosa

Bulimia nervosa is an eating disorder characterized by the consumption of large amounts of food with subsequent purging by self-induced vomiting, laxative or diuretic abuse, enemas, and/or obsessive exercising. Whereas anorexia nervosa is characterized by severe weight loss, bulimia nervosa may show weight maintenance or extreme weight fluctuations due to alternating binges and fasts. In some individuals anorexia and bulimia nervosa overlap. Reliable estimates of bulimia nervosa range from 1.0% to 3.0%.[56] As with anorexia nervosa, about 90% of individuals with bulimia nervosa are female, probably due to the greater social pressures on women to be thin.

Diagnostic criteria for bulimia nervosa are shown in Table 15.14. Key features of bulimia nervosa include recurrent episodes of binge eating (rapid consumption of a large amount of food in a discrete period of time); a feeling of lack of control over eating behavior during the binge; self-induced vomiting; use of laxatives or diuretics; strict dieting or fasting; vigorous exercise in order to prevent weight gain; and persistent overconcern with body shape and weight. People with bulimia nervosa can be overweight, underweight, or of average weight for their height and body frame. Bulimia nervosa may be proceeded by a history of dieting or restrictive eating, which are thought to contribute to the binge-purge cycle. Mortality for bulimia nervosa appears to be lower than for anorexia nervosa. Based on a review of the existing literature in this area, Woodside has estimated that approximately 5% of patients die of their disease, usually due to heart failure resulting from electrolyte abnormality or suicide. Recovery rates for bulimia nervosa appear to be higher than for anorexia nervosa, with estimates of 50% to 60% for recovery.[55]

> **Bulimia Nervosa**   An eating disorder characterized by recurrent episodes of rapid, uncontrolled eating of large amounts of food in a short period of time. Episodes of binge eating are often followed by purging.
>
> **Binge-Eating Disorder (BED)**   An eating disorder characterized by periodic binge eating, which normally is not followed by vomiting or the use of laxatives. People must experience eating binges twice a week on average for over 6 months to qualify for this diagnosis.

Binge-Eating Disorder

Binge-eating disorder (BED) is a condition in which one engages in eating large amounts of food and feels that these eating episodes are not within one's control.[57] BED is defined by recurrent episodes of binge eating—at least 2 days a week for at least 6 months (Table 15.15). In addition, the person feels a subjective sense of a loss of control over binge eating, which is indicated by the presence of three of the following five criteria: eating rapidly, eating when not physically hungry, eating when alone, eating until uncomfortably full, and feeling self-disgust about bingeing. BED differs from bulimia nervosa in that binge eating is not followed by compensatory behaviors such as

Table 15.14 Diagnostic criteria for bulimia nervosa

A. *Recurrent episodes of binge eating.* An episode of binge eating is characterized by both of the following:
  - eating, in a discrete period of time (e.g., within any two-hour period), an amount of food that is definitely larger than most people would eat during a similar period of time and under similar circumstances.
  - a sense of lack of control over eating during the episode (e.g., a feeling that one cannot stop eating or control what or how much one is eating).

B. Recurrent inappropriate compensatory behavior in order to prevent weight gain, such as self-induced vomiting; misuse of laxatives, diuretics, enemas, or other medications; fasting; or excessive exercise.

C. The binge eating and inappropriate compensatory behaviors both occur, on average, at least twice a week for three months.

D. Self-evaluation is unduly influenced by body shape and weight.

E. The disturbance does not occur exclusively during episodes of anorexia nervosa.

Purging Type: During the current episode of bulimia nervosa, the person regularly engages in self-induced vomiting or the misuse of laxatives, diuretics, or enemas.

Nonpurging Type: During the current episode of bulimia nervosa, the person has used other inappropriate compensatory behaviors, such as fasting or excessive exercise, but has not regularly engaged in self-induced vomiting or the misuse of laxatives, diuretics, or enemas.

SOURCE: Reprinted with permission from the Diagnostic and Statistical Manual of Mental Disorders, Fourth Edition, Text Revision, Copyright 2000. American Psychiatric Association.

self-induced vomiting, as occurs in bulimia nervosa. BED is more prevalent among overweight clinical populations (30%) than among community samples (5% of females and 3% of males).[58] Studies on adolescents that assess the prevalence of binge eating include few that document prevalence rates of BED. In a college student sample, the rate of BED was 2.6%. In contrast to other weight-related conditions, significant differences were not found between male and female students. Further study of the prevalence and etiology of BED among adolescents seems critical in light of the increasing rates of obesity among youth.

Other Disordered Eating Behaviors

Some adolescents engage in anorexic or bulimic behaviors, but with less frequency or intensity than required for a formal diagnosis of an eating disorder. Behaviors typically

Table 15.15 Diagnostic criteria for binge-eating disorder

A. *Recurrent episodes of binge eating.* An episode of binge eating is characterized by both of the following:

- eating, in a discrete period of time (e.g., within any two-hour period), an amount of food that is definitely larger than most people would eat in a similar period of time and under similar circumstances.
- a sense of lack of control over eating during the episode (e.g., a feeling that one cannot stop eating or control what or how much one is eating).

B. The binge-eating episodes are associated with three (or more) of the following:

- eating much more rapidly than normal
- eating until feeling uncomfortably full
- eating large amounts of food when not feeling physically hungry
- eating alone because of being embarrassed by how much one is eating
- feeling disgusted with oneself, depressed, or guilty after overeating
- experiencing marked distress regarding binge eating
- occurring, on average, at least two days a week for six months

C. The method of determining frequency differs from that used for bulimia nervosa; future research should address whether the preferred method of setting a frequency threshold is counting the number of days on which binges occur or counting the number of episodes of binge eating.

D. The binge eating is not associated with the regular use of inappropriate compensatory behaviors (e.g., purging, fasting, excessive exercise) and does not occur exclusively during the course of anorexia nervosa or bulimia nervosa.

SOURCE: Reprinted with permission from the Diagnostic and Statistical Manual of Mental Disorders, Fourth Edition, Text Revision, Copyright 2000. American Psychiatric Association.

considered in this category include self-induced vomiting, laxative use, use of diet pills, fasting or extreme dieting, binge eating, and excessive physical activity. The heterogeneity of these behaviors makes it more difficult to estimate the prevalence of anorexic and bulimic behaviors.

In addition, the types of questions used to assess disordered eating behaviors may influence prevalence estimates. In a large national study of 6728 adolescents, disordered eating was reported by 13% of the girls and 7% of the boys.[59] In this study, disordered eating was assessed with the question: "Have you ever binged and purged (which is when you eat a lot of food and then make yourself throw up, vomit, or take something that makes you have diarrhea) or not?" In a large population-based study of Minnesota adolescents in grades 7 through 12, 12% of the girls reported that they had made themselves vomit for weight-control purposes at least once in their lives, and 2% reported having used laxatives or diuretics for weight-control purposes. Rates were lower among boys: 6% reported vomiting, and less than 1% reported laxative or diuretic use.[60] Binge eating (eating large amounts during a short period of time and feeling out of control while eating) was reported by 30% of the girls and 13% of the boys.[60]

Based on our own research and a review of the more reliable research findings, it seems reasonable to estimate that between 10% and 20% of adolescents have engaged in anorexic or bulimic behaviors. Disordered eating behaviors such as self-induced vomiting and binge eating have serious implications for health and may be precursors to full-blown eating disorders. Therefore, interventions aimed at their prevention are essential.

Dieting Behaviors

Dieting behaviors among adolescents, and in particular among adolescent girls, tend to be alarmingly high. In a study of weight-control behaviors among 459 adolescents (12–17 years of age) from four regions of the United States, current weight-control behaviors were reported by 44% of the adolescent girls and 37% of the adolescent boys.[61] Based on their review of the literature, Fisher and colleagues concluded that 50% to 60% of adolescent girls consider themselves overweight and have attempted to diet.[62] Of particular concern are the increasing rates of dieting behaviors among children and young adolescents. A review of studies focusing on youth between the ages of 9 and 12 found high rates of dieting among girls (16% to 50%) and high percentages of girls who expressed a desire to be thinner (33% to 58%).[63] Estimates of dieting behaviors tend to vary across studies, in accordance with how dieting is assessed.

Dieting behaviors among youth are of concern in that they are often used by youth who are not overweight. Furthermore, unhealthful dieting behaviors in which meals are skipped, energy intake is severely restricted, or food groups are lacking are common. Dieting behaviors have been found to be associated with inadequate intakes of essential nutrients such as calcium.[64] Dieting behaviors leading adolescents to experience hunger or cravings for specific foods may place them at risk for binge-eating episodes. Finally, dieting behaviors may be indicative of increased risk for the later development of eating disorders; Patton and colleagues found that the relative risk for dieters to develop an eating disorder was eight times higher than that for nondieters after a 1-year period.[65] Therefore, dieting should not be viewed as a normative and acceptable behavior, in particular among children and adolescents.

Body Dissatisfaction

During adolescence, body image and self-esteem tend to be closely intertwined; therefore, body image concerns should not be viewed as acceptable and normative components of adolescence. Furthermore, body dissatisfaction is probably the main contributing factor to dieting behaviors, disordered eating behaviors, and clinical eating disorders.[66,67] Body dissatisfaction is reported by a high percentage of adolescents. In a study of 36,000 adolescents in Minnesota, Story and colleagues found that less than 40% of adolescent girls and about 65% of adolescent boys were satisfied with their weight.[68] A study by Tienboon and colleagues looked at families with adolescents aged 14–15 years and reported that 41% of the girls and 14% of the boys considered themselves to be overweight.[69] Although actual weight status was directly associated with perceived weight status, a considerable number of girls who were not overweight perceived themselves as overweight.

Factors associated with increased body dissatisfaction include female gender, Caucasian ethnicity, and being overweight.[69] In working with overweight youth who express body dissatisfaction, health counselors are challenged to help them improve their body images while simultaneously working toward weight control. All adolescents, including overweight youth, should be encouraged to appreciate the positive aspects of their bodies. Overweight adolescents may need help in accepting that they may never achieve the thin ideal portrayed in the media, but can strive toward a leaner and healthier body that is realistic for them. Tips for fostering a positive body image among adolescents, regardless of their weight, are shown in Table 15.16.

Etiology of Eating Disorders

The etiology of eating disorders is multifactorial; that is, many factors contribute to their onset. Some of the major contributory factors include social norms emphasizing thinness, being teased about one's weight, familial relations (e.g., chaotic lifestyles, boundaries between family members, patterns of communication), physical and sexual abuse experiences, personal body shape and size, body

Table 15.16 Tips for fostering a positive body image among children and adolescents

CHILD OR ADOLESCENT	PARENTS	HEALTH PROFESSIONAL
• Look in the mirror and focus on your positive features, not the negative ones.	• Demonstrate healthy eating behaviors, and avoid extreme eating behaviors.	• Discuss changes that occur during adolescence.
• Say something nice to your friends about how they look.	• Focus on non-appearance-related traits when discussing yourself and others.	• Assess weight concerns and body image.
• Think about your positive traits that are not related to appearance.	• Praise your child or adolescent for academic and other successes.	• If a child or adolescent has a distorted body image, explore causes and discuss potential consequences.
• Read magazines with a critical eye, and find out what photographers and computer graphic designers do to make models look the way they do.	• Analyze media messages with your child or adolescent.	• Discuss how the media negatively affects a child's or adolescent's body image.
• If you are overweight and want to lose weight, be realistic in your expectations and aim for gradual change.	• Demonstrate that you love your child or adolescent regardless of what he weighs.	• Discuss the normal variation in body sizes and shapes among children and adolescents.
• Realize that everyone has a unique size and shape.	• If your child or adolescent is overweight, don't criticize her or his appearance—offer support instead.	• Educate parents, physical education instructors, and coaches about realistic and healthy body weight.
• If you have questions about your size or weight, ask a health professional.	• Share with a health professional any concerns you have about your child's or adolescent's eating behaviors or body image.	• Emphasize the positive characteristics (appearance- and non-appearance related) of children and adolescents you see.
		• Take extra time with an overweight child or adolescent to discuss psychosocial concerns and weight-control options.
		• Refer children, adolescents, and parents with weight-control issues to a dietitian.

SOURCE: M. Story, K. Holt, and D. Sofka, eds., *Bright Futures in Practice: Nutrition.* Arlington, VA: National Center for Education in Maternal and Child Health, 2000.

image, and self-esteem. These factors do not operate in isolation, but interact with each other to increase the adolescent's risk for engaging in potentially harmful eating and dieting behaviors. In considering etiological issues, it is essential to realize that different etiological pathways may lead to weight-related disorders in different adolescents. For some adolescents, family issues may be major factors, while for others social norms may be the key factors leading to the onset of a condition. Furthermore, different conditions tend to be influenced by different factors. That said, a number of factors play a major role across conditions and across individuals. Rosen and Neumark-Sztainer have grouped these potential contributory factors into socioenvironmental, personal, and behavioral domains.[70]

Socioenvironmental factors:

- Sociocultural norms (e.g., regarding thinness, eating, food preparation, roles of women)
- Food availability (e.g., type of food, amount of food)
- Familial factors (e.g., communication patterns, parental expectations, boundaries, weight concerns and dieting behaviors of parents and siblings, family meals)
- Peer norms and behaviors (e.g., dieting behaviors, eating patterns, weight concerns)
- Abuse experiences (e.g., by family members, other adults, rape experiences)
- Media influences (e.g., images portrayed in the media, roles assigned to thin actors as compared to fat actors)

Personal factors:

- Biological factors (e.g., genetic disposition, body mass index, age, gender, stage of development)
- Psychological factors (e.g., self-esteem, body image, drive for thinness, depression)
- Cognitive/affective factors (e.g., nutritional knowledge and attitudes, media internalization)

Behavioral factors:

- Eating behaviors (e.g., meal patterns, fast-food consumption, nutritional variety, bingeing)
- Dieting and other weight-management behaviors (e.g., dieting frequency, types of methods used, purging behaviors)
- Physical activity behaviors (e.g., TV viewing, sports involvement, daily activities)
- Coping behaviors (e.g., with dieting failures, with life frustrations)
- Skills (e.g., self-efficacy in dealing with harmful social norms, skills in food preparation, media advocacy skills)

An understanding of the etiology of eating disorders is essential to the development of effective interventions aimed at their treatment and prevention. In an individual clinical setting, counselors need to allow time to assess the factors leading to the onset of the condition for that particular adolescent. In developing prevention programs to reach larger groups of adolescents, it is more feasible to identify and address factors that may be contributing to the onset of weight-related behaviors and conditions for a broad sector of the targeted population. Although not all factors may be addressed within one intervention, it is important to be aware of the broad range of factors coming into play and the interactions among them.

Treating Eating Disorders

The complex etiology of eating disorders and their numerous potential psychosocial, physical, and behavioral consequences highlight the need for a multidisciplinary treatment approach. The health care team caring for an adolescent with an eating disorder will often include a physician, nutritionist, nurse, psychologist, and/or psychiatrist. The role of the nutritionist is paramount to the treatment of eating disorders at the stages of assessment, treatment, and maintenance. Initially an adolescent may be more willing to discuss his or her concerns with a nutritionist than with a psychologist. During treatment a major role of the nutritionist is to help the adolescent normalize eating patterns and to feel comfortable with these changes. For example, in the treatment of bulimia nervosa, some key goals of nutritional care include the following:

- Establishing a regular pattern of nutritionally balanced meals and snacks
- Ensuring adequate but not excessive levels of energy intake, with the goal of weight maintenance
- Ensuring adequate dietary fat and fiber intake to promote satiety
- Avoiding dieting behaviors and excessive exercise; gradually including formerly forbidden foods in the diet
- Dietary record keeping and review; stimulus control strategies to control high-risk situations; and weighing at scheduled intervals only.[71]

For some adolescents, denial of the condition or a lack of motivation for change make the work quite challenging. It is important for the nutritionist to work closely with other members of the health care team to ensure that the roles of different team members are clearly defined.

Preventing Eating Disorders

The high prevalence of eating disorders and their potentially harmful consequences point to a need for interventions aimed at their prevention. A pressing current public health issue that needs to be addressed concerns the potential for the occurrence of eating disorders (described in previous sections) as a result of efforts to prevent obesity. Even the prevention of a small percentage of these conditions, at a population level, returns huge benefits in relation to reducing physical, emotional, and financial burdens.

In developing interventions aimed at the prevention of eating disorders, it is essential to address factors that (1) contribute to the onset of these conditions for a large proportion of the targeted population, (2) are potentially modifiable, and (3) are suitable for addressing within the designated setting. For example, the use of media awareness and advocacy has been suggested as a suitable approach to preventing eating concerns and disorders.[72] Participants may learn about how the media influences one's body image and about techniques the media uses to improve the appearance of models, and then take action toward making changes in the media. This approach is suitable in that media influences, and the internalization of media messages, may contribute to weight concerns among a large sector of the adolescent population. These adolescent perceptions are potentially modifiable and suitable for addressing within clinical, community, and school-based settings where interventions may be implemented.

Any efforts toward prevention first must consider the target audience. An important question is whether to direct interventions to all adolescents or to adolescents at increased risk for eating disorders. Reasons for providing interventions for all adolescents include the high prevalence of eating concerns among adolescents, difficulties inherent in identifying and targeting high-risk individuals, and the advantages of developing positive social norms regarding eating issues within the peer group. Taking a more targeted approach offers the advantages of better use of limited resources, more intensive interventions, and interventions developed for specific high-risk groups (e.g., ballet dancers, adolescents with diabetes, or overweight girls). To be most effective in preventing eating disorders, both types of interventions seem necessary; more general approaches address the issues of the general adolescent population, while more refined approaches can better meet the needs of specific high-risk groups.

Prevention interventions may be implemented within clinical, community, and school-based settings. Clinical settings provide opportunities for identifying early signs of eating disorders and working toward their prevention. Questions about body concerns and eating patterns should be a part of routine health care visits, and appropriate channels should be made available for discussing concerns. Key factors to be assessed in screening for eating disorders are shown in Table 15.17.

Most prevention programs have been implemented within school settings.[73–76] Schools provide an excellent setting for implementing prevention interventions in that they reach all adolescents; they provide a captive audience within a learning atmosphere and numerous opportunities for social interactions.[77] Ideally, a school-based intervention includes various components aimed at reaching the general population of adolescents, adolescents at increased

Table 15.17 Screening elements and warning signs for individuals with eating disorders

SCREENING	WARNING SIGNS
Body image and weight history	• Distorted body image • Extreme dissatisfaction with body shape or size • Profound fear of gaining weight or becoming fat • Unexplained weight change or fluctuations greater than 10 lbs
Eating and related behaviors	• Very low caloric intake; avoidance of fatty foods • Poor appetite; frequent bloating • Difficulty eating in front of others • Chronic dieting despite not being overweight • Binge-eating episodes • Self-induced vomiting; laxative or diuretic use
Meal patterns	• Fasting or frequent meal skipping to lose weight • Erratic meal pattern with wide variations in caloric intake
Physical activity	• Participation in physical activity with weight or size requirement (e.g., gymnastics, wrestling, ballet) • Overtraining or "compulsive" attitude about physical activity
Psychosocial assessment	• Depression • Constant thoughts about food or weight • Pressure from others to be a certain shape or size • History of physical or sexual abuse or other traumatizing life event
Health history	• Secondary amenorrhea or irregular menses • Fainting episodes or frequent light-headedness • Constipation or diarrhea unexplained by other causes
Physical examination	• BMI <5th percentile • Varying heart rate, decreased blood pressure after arising suddenly • Hypothermia; cold intolerance • Loss of muscle mass • Tooth enamel demineralization

SOURCE: Used with permission. The Society for Nutrition Education. L. B. Adams and M. B. Shafer, "Early Manifestations of Eating Disorders in Adolescents: Defining Those at Risk," *J Nutr Educ* 20 (1988); American Medical Association. K. Perkins, N. Ferrari, and A. Rosas et al., "You Won't Know Unless You Ask: The Biopsychosocial Interview for Adolescents," *Clin Pediatr* 36(2), 1997; and *Guidelines for Adolescent Preventive Services (GAPS): Recommendation Monogragh*, 2nd ed. Chicago: American Medical Association, 1995.

Table 15.18 Suggested components of a comprehensive school-based program for preventing weight-related disorders

Staff training	1. Examination of their own body image and eating behaviors 2. Knowledge about weight-related disorders 3. Skills for working with youth and program implementation
Classroom interventions for the general student body	Modula specifically aimed at preventing weight-related concerns and disorders. Options for approaches include (1) feminist approach, (2) weight control and nutrition, and (3) promotion of life skills such as self-esteem and assertiveness
Integration of relevant material into existing curricula	Relevant information integrated into existing classes such as science, art, history, health, and physical education
Smaller and more intensive activities for high-risk students	Small-group work or individual counseling for overweight students or students with excessive weight preoccupation and unhealthy dieting behaviors
Referral system within school and between school and community	Training of school staff to be alert to warning signs; referral mechanism to ensure that students don't get overlooked and those in need get help
Opportunities for healthy eating at school	Options for attractive nutrient-dense and low-fat foods at affordable prices in the cafeteria, in vending machines, and at school events
Modifications in physical education and sport activities	Increased time being active in physical education classes; involvement of more students in after-school sports; increased sensitivity to needs of overweight youth in physical education classes
Outreach activities to the community	By students, staff, and parents (e.g., contact with local media)

SOURCE: Adapted from D. Neumark-Sztainer, "School-Based Programs for Preventing Eating Disturbances," *J School Health* (1996):66.

risk for eating disorders, and school staff. Interventions should aim not only for changes in levels of personal knowledge and attitudes but also for changes in social norms (e.g., peer norms promoting thinness and dieting), policy changes (e.g., regarding tolerance levels for weight-teasing), and environmental changes (e.g., food availability within the cafeteria). Suggested components for a comprehensive school-based program for preventing eating disorders are shown in Table 15.18.

Interventions aimed at preventing eating disorders may also be implemented within other community-based settings. Neumark-Sztainer and her colleagues utilized the Girl Scout framework to reach fifth- and sixth-grade girls with an intervention aimed at promoting a positive body image and preventing unhealthy dieting.[78] The program, Free to Be Me, focused on promoting media awareness and advocacy among participants. An outline of the program is provided in Table 15.19 on the next page.

Eating Disorders among Adolescents: Summing Things Up

Eating disorders may be viewed on a continuum ranging from body dissatisfaction or dieting behaviors to clinically significant eating disorders including anorexia nervosa, bulimia nervosa, and binge eating disorder. The high prevalence of eating disorders and their potential harmful physical and psychosocial consequences indicate the need for interventions aimed at their treatment and prevention. The numerous socioenvironmental, personal, and behavioral factors leading to the onset of eating disorders need to be addressed in interventions. Parents, peers, educators, and health care providers play an important role in decreasing the prevalence of eating disorders.

Children and Adolescents with Chronic Health Conditions

Approximately 18% of children and adolescents have a chronic condition or disability.[44] These children and adolescents are at increased risk for nutrition-related health problems because of (1) physical disorders or disabilities that may affect their ability to consume, digest, or absorb nutrients; (2) biochemical imbalances caused by long-term medications or internal metabolic disturbances; (3) psychological stress from a chronic condition or physical disorder that may affect a child's appetite and food intake; and/or (4) environmental factors, often controlled by parents who may influence the child's access to and acceptance of food.

Nutrition reports of children and adolescents with special health care needs estimate that as many as 40% have

Table 15.19 Outline of sessions in *Free to Be Me*

SESSION	ACTIVITIES
Body Truths I: Body Development	• *Let me Introduce You:* An introduction to the program and an ice-breaker activity based on the game Bingo. • *Stepping Stones:* Question-answer game in which girls learn about stages of body development. Each team advances to the next "stone" if question is answered correctly. • *Feelin' Good:* Take-home activity: Interviews with family members and friends about perceived positive traits.
Body Truths II: Working with the Ideal Image	• *Pin the Tail on the Timeline:* Girls attach ideal images of women throughout history to their appropriate spot on the timeline (e.g., Marilyn Monroe in the 1950s, Twiggy in the 1960s, and Tyra Banks in the 1990s). • *Sarah's Story:* Discussion of a story about a girl with poor body image who diets excessively. Discuss reasons for dieting, negative effects of skipping meals, and alternative approaches for Sarah. • *People Watching:* Take-home activity: Girls are asked to look for different body types in their community and then compare them to the body types they see in magazines.
Body Truths III: What Else Is Out There?	• *Looking for the Alternative (Stop-n-Go Posters):* Girls look through different teen magazines and make collages of the pictures that promote positive traits versus those that promote negative traits. Pictures are glued to either green poster board ("go") or red poster board ("stop"). • *Take-home activity:* Girls read "girl-friendly" magazine provided such as *New Moon* and discuss magazine with a parent.
Body Myths: Media Madness	• *Commercial Crazy:* Girls look at a variety of TV commercials and look for positive and negative media messages (body types shown, messages about healthy eating versus dieting). • *What do you see on TV?:* Take-home activity: Girls and their parent(s) watch 15 to 30 minutes of TV and look for positive and negative messages in the commercials.
Take Action I: What Can You Do to Impact the Media?	• *Write a Letter:* Each troop writes a letter to a company that has a positive or negative impact on dieting and body image. Letters are posted on Web site for other advocates to sign. • *Take-home activity:* Girls develop and practice their own "girl-friendly" skit or commercial promoting positive body image and healthy eating.
Take Action II: Spreading the Good Word	• *Lights, Camera, Action!:* Girls perform their "girl-friendly" commercials or skits in front of their parents and troop. • *Class overview.*

SOURCE: Reprinted with permission. D. Neumark-Sztainer, N. Sherwood, T. Coller, and P. J. Hannan, "Primary Prevention of Disordered Eating among Pre-Adolescent Girls: Feasibility and Short-Term Impact of a Community-Based Intervention," *J Am Dietetic Assoc.* In Press.

nutrition risk factors that warrant a referral to a dietitian.[44] Common nutrition problems in children and adolescents with special health care needs include the following:

• Altered energy and nutrient needs (e.g., inborn errors of metabolism, spasticity of movement, enzyme deficiencies)
• Delayed growth
• Oral-motor dysfunction (e.g., neurological disorders, swallowing disorders)
• Elimination problems
• Drug/nutrient interactions
• Appetite disturbances
• Unusual food habits (e.g., rumination)
• Dental caries, gum disease

Malnutrition has been implicated as a major factor contributing to poor growth and short stature in adolescents with a variety of diseases (e.g., chronic inflammatory bowel disease, cystic fibrosis). Factors such as inadequate nutrient and energy intakes, excessive nutrient losses, malabsorption, and increased nutrient requirements all lead to the chronic malnourished state. Studies have shown that the energy requirements for adolescents with cystic fibrosis or inflammatory bowel disease may be 30% to 50% higher than the RDA for adequate growth. In addition to the increased energy needs caused by malabsorption (or in the case of adolescents with cystic fibrosis, the increased work of breathing), fever, infection, and inflammation also increase energy requirements. Whereas undernourishment is frequently seen in adolescents with chronic illnesses, obesity is common among youth with gross motor limitations or immobility. Because of limited activity, caloric requirements are lower, and the balance between intake and expenditure is often difficult, resulting in obesity.

Consideration of nutrition needs of children with chronic disabling conditions or illnesses is complex and requires specialized, individualized care by an interdisci-

plinary team. Assessment of nutrition status followed by nutrition intervention, when necessary, and monitoring will help ensure the health and well-being of adolescents with chronic and disabling conditions. Also, during adolescence issues of personal responsibility and independent living skills related to food purchasing and preparation may need to be addressed.

Resources

American College of Sport Medicine
Web site: www.acsm.org

American Diabetes Association
Clinical practice, recommendations, nutrition, exercise, recipes, virtual grocery.
Web site: www.diabetes.org

American Heart Association
An office-based practical approach to the child with hypercholesterolemia.
Web site: www.americanheart.org/Scientific/pubs/hyperchol/toc.html
Nutrition tips and virtual cookbook for heart health.
Web site: www.deliciousdecisions.org

Diabetes
Betschart J, Thom S. *In Control: A Guide for Teens with Diabetes.* Minneapolis, MN: Chronimed Publishing, 1995.
Boland, E. *Teens Pumping It Up!* 2nd ed. Sylmar, CA: Minimed Inc., 1998.
Nutrition in the Fast Lane: A Guide to Nutrition and Dietary Exchange Values for Fast-Food and Casual Dining. Indianapolis, IN: Franklin Publishing, 2001.
Web site: www.FastFoodFacts.com

Family Voices
Health information for children and adolescents with special health needs.
Web site: www.familyvoices.org

Food Allergy Network
Educational materials and resources for youth with food allergies.
Web site: www.foodallergy.org

Juvenile Diabetes Association, International
Phone: 1-800-533-CURE

National Campaign to Prevent Teen Pregnancy
Web site: www.teenpregnancy.org

National Heart Lung and Blood Institute
Health information for public: cholesterol, obesity.
Health information for professionals: NCEP ATP III, overweight and obesity.
Web site: www.nhlbi.nih.gov

Weight Information Network
Publications and information on child and adolescent obesity treatment and prevention.
Phone: 1-800-946-8098
Web site: www.niddk.nih.gov//NutritionDocs.html

References

1. Barlow SE, Dietz WH. Obesity evaluation and treatment: Expert Committee recommendations. The Maternal and Child Health Bureau, Health Resources and Services Administration and the Department of Health and Human Services. Pediatrics 1998;102:E29.

2. Gortmaker SL, Must A, Perrin JM et al. Social and economic consequences of overweight in adolescence and young adulthood. N Eng J Med 1993;329:1008–12.

3. Guo SS, Roche AF, Chumlea WC et al. The predictive value of childhood body mass index values for overweight at age 35 y. Am J Clin Nutr 1994;59:810–9.

4. Whitaker RC, Wright JA, Pepe MS et al. Predicting obesity in young adulthood from childhood and parental obesity. N Eng J Med 1997;337:869–73.

5. Johnson AL, Cornoni JC, Cassel JC et al. Influence of race, sex, and weight on blood pressure behavior in young adults. Am J Cardiol 1975;35:523–30.

6. Loder RT, Aronson DD, Greenfield ML. The epidemiology of bilateral slipped capital femoral epiphysis. A study of children in Michigan. J Bone Joint Surg Am 1993;75:1141–7.

7. Mallory GB Jr., Fiser DH, Jackson R. Sleep-associated breathing disorders in morbidly obese children and adolescents. J Pediatr 1989;115:892–7.

8. Smoak CG, Burke GL, Webber LS et al. Relation of obesity to clustering of cardiovascular disease risk factors in children and young adults. The Bogalusa Heart Study. Am J Epidemiol 1987;125:364–72.

9. Pinhas-Hamiel O, Dolan LM, Daniels SR et al. Increased incidence of non-insulin-dependent diabetes mellitus among adolescents. J Pediatr 1996;128:608–15.

10. Boguslaw W, Bar-Or O. Effect of drink flow and NaCl on voluntary drinking and hydration in boys exercising in the heat. J Appl Physiol 1996;80:1112–7.

11. Steen SN. Nutrition for the school-age

child athlete. In: Berning JR, Steen SN, eds. Nutrition for sport and exercise, 2nd ed. Gaithersburg, MD: Aspen Publishers; 1998: 217–246.

12. National Collegiate Athletic Association. NCAA study of substance abuse habits of college student athletes. NCAA Committee on Medical Safeguards and Medical Aspects of Sports; 1997.

13. Coleman EJ. Carbohydrate: the master fuel. In: Berning JR, Steen SN, eds. Nutrition for sport and exercise. Gaithersburg, MD: Aspen Publishers; 1998: 21–44.

14. American College of Sports Medicine. Questioning 40/30/30: a guide to understanding nutrition advice. The American Dietetic Association, Women's Sports Foundation, and the Cooper Institute for Aerobics Research; 1998.

15. Schmalz K. Nutritional beliefs and practices of adolescent athletes. Journal of School Nursing 1993;9:18–22.

16. Clark JF. Creatine and phosphocreatine:

a review of their use in exercise and sport. Journal of Athletic Training 1997;32:45–51.

17. Hultman E, Soderlund K, Timmons JA et al. Muscle creatine loading in men. J Appl Physiol 1996;81:232–7.

18. Johnson WA, Landry GL. Nutritional supplements: fact vs. fiction. Adolesc Med State Art Rev 1998;9:501 13.

19. Dubelaar ML, Lucas CM, Hulsmann WC. Acute effect of L-carnitine on skeletal muscle force tests in dogs. Am J Physiol 1991;260:E189–93.

20. Burke ER. Nutritional ergogenic acids. In: Berning JR, Steen SN, eds. Nutrition for sport and exercise. Gaithersburg, MD: Aspen Publishers; 1998: 127–8.

21. Hasten DL, Rome EP, Franks BD, Hegsted M. Effects of chromium picolinate on beginning weight training students. Int J Sport Nutr 1992;2:343–50.

22. Lefavi RG. Response (letter). Int J Sport Nutr 1993;3:120–2.

23. Wasser WG, Feldman NS, D'Agati VD. Chronic renal failure after ingestion of over-the-counter chromium picolinate. Ann Intern Med 1997;126:410.

24. Huszonek J. Over-the-counter chromium picolinate. Am J Psychiatry 1993;150: 1560–1.

25. Ventura SJ, Matthews TJ, Curtin SC. Declines in teenage birthrates, 1991–1998: update of national and state trends. National vital statistics reports. Hyattsville, MD: National Center for Health Statistics; 1999.

26. Manlove J. The influence of high school dropout and school disengagement on the risk of school-age pregnancy. J Res Adolesc 1998;8:187–220.

27. Wu LL. Effects of family instability, income, and income instability on the risk of a premarital birth. Am Soc Rev 1996;61: 386–406.

28. Ventura SJ, Matthews TJ, Curtin SC. Declines in teenage birthrates, 1991–1997: national and state patterns. National vital statistics reports. Hyattsville, MD: National Center for Health Statistics; 1998.

29. Frisancho AR, Matos J, Bollettino LA. Influence of growth status and placental function on birthweight of infants born to young still-growing teenagers. Am J Clin Nutr 1984;40:801–7.

30. Naeye RL. Teenaged and pre-teenaged pregnancies: consequences of the fetal-maternal competition for nutrients. Pediatrics 1981;67:146–50.

31. Teagle SE, Brindis CD. Substance use among pregnant adolescents: a comparison of self-reported use and provider perception. J Adolesc Health 1998;22:229–38.

32. Institute of Medicine. Nutrition during pregnancy: part I: weight gain; part II: nutrient supplements. Washington, DC: National Academy Press; 1990.

33. Scholl TO, Hediger ML. A review of the epidemiology of nutrition and adolescent pregnancy: maternal growth during preg-

nancy and its effect on the fetus. J Am Coll Nutr 1993;12:101–7.

34. Suiter CW. Maternal weight gain: a report of an expert work group. Arlington, VA: National Center for Education in Maternal and Child Health; 1997.

35. Centers for Disease Control and Prevention. Tobacco use among high school students—United States, 1990. Morb Mortal Wkly Rep 1991;40:617–9.

36. Johnston LD, O'Malley PM, Backman JG. National survey results on drug use from Monitoring the Future study, 1975–1992, vol. 1. Rockville, MD: U.S. Department of Health and Human Services, Public Health Service, National Institute of Drug Abuse; 1993.

37. Kann L, Kinchen SA, Williams BI et al. Youth risk behavior surveillance—United States, 1997. State and Local YRBSS Coordinators. J Sch Health 1998;68:355–69.

38. Isralowitz RE, Trostler N. Substance use: toward an understanding of its relation to nutrition-related attitudes and behavior among Israeli high school youth. J Adolesc Health 1996;19:184–9.

39. Neumark-Sztainer D, Story M, Toporoff E et al. Covariations of eating behaviors with other health-related behaviors among adolescents. J Adolesc Health 1997;20:450–8.

40. Alton I. Substance use during pregnancy. In: Story M, Stang J, eds. Nutrition and the pregnant adolescent: a practical reference guide. Minneapolis, MN: Leadership, Education, and Training Program in Maternal and Child Nutrition; 2000: 125–134.

41. Marion IJ, Mitchell JL. Pregnant, substance-using women. Rockville, MD: U.S. Department of Health and Human Services, Public Health Service, Substance Abuse and Mental Health Services Administration, Center for Substance Abuse Treatment; 1993.

42. Balluz LS, Kieszak SM, Philen RM, Mulinare J. Vitamin and mineral supplement use in the United States. Results from the third National Health and Nutrition Examination Survey. Arch Fam Med 2000;9: 258–62.

43. Stang J, Story MT, Harnack L, Neumark-Sztainer D. Relationships between vitamin and mineral supplement use, dietary intake, and dietary adequacy among adolescents. J Am Diet Assoc 2000;100:905–10.

44. Story M, Holt K, Sofka D, eds. Bright futures in practice: nutrition. Arlington, VA: National Center for Education in Maternal and Child Health; 2000.

45. Centers for Disease Control and Prevention. Recommendations to prevent and control iron deficiency in the U.S. MMWR 1998;47.

46. National Heart, Lung, and Blood Institute, National High Blood Pressure Education Working Group on Hypertension Control in Children and Adolescents. Update on the task force report (1987) on high blood pressure in children and adolescents: a working group report from the National High Blood Pressure Education Program. Bethesda, MD: National Institutes of Health; 1997.

47. Williams CL, Bollella M. Guidelines for screening, evaluating, and treating children with hypercholesterolemia. J of Ped 1995; 9:153–62.

48. American Academy of Pediatrics CoN. Cholesterol in childhood. Pediatrics 1998; 101:141–7.

49. Reference Dietary Intakes for Energy, Carbohydrates, Fiber, Fat, Fatty Acids, Cholesterol, Protein, and Amino Acids. Washington DC: National Academies Press, 2002.

50. Neumark-Sztainer D. Excessive weight preoccupation: normative but not harmless. Nutr Today 1995;30:68 74.

51. Lucas AR, Beard CM, O'Fallon WM, Kurland LT. 50-year trends in the incidence of anorexia nervosa in Rochester, Minn.: a population-based study. Am J Psychiatry 1991;148:917–22.

52. Nylander I. The feeling of being fat and dieting in a school population: an epidemiological investigation. Acta Soc-Med Scand 1971;1:17–26.

53. Mott A, Lumsden D. Understanding eating disorders. Washington, DC: Tayler & Francis; 1994.

54. Kaplan A, Garfinkel P. Medical issues and eating disorders. New York: Brunner/ Mazel; 1993.

55. Woodside DB. A review of anorexia nervosa and bulimia nervosa. Curr Probl Pediatr 1995;25:67–89.

56. Stein DM. The prevalence of bulimia: a review of the empirical research. J Nutr Educ 1991;23:205–13.

57. American Psychiatric Association. American Psychiatric Association diagnostic and statistical manual of mental disorders, 4th ed. Washington, DC: American Psychiatric Association; 1994.

58. Spitzer RL, Yanovski S, Wadden T et al. Binge-eating disorder: its further validation in a multisite study. Int J Eat Disord 1993;13: 137–53.

59. Neumark-Sztainer D, Hannan PJ. Weight-related behaviors among adolescent girls and boys: results from a national survey. Arch Pediatr Adolesc Med 2000;154:569–77.

60. Neumark-Sztainer D, Story M, French SA et al. Psychosocial concerns and health-compromising behaviors among overweight and nonoverweight adolescents. Obes Res 1997;5:237–49.

61. Neumark-Sztainer D, Rock CL, Thornquist MD et al. Weight-control behaviors among adults and adolescents: associations with dietary intake. Prev Med 2000;30: 381–91.

62. Fisher M, Golden NH, Katzman DK et al. Eating disorders in adolescents: a background paper. J Adolesc Health 1995;16: 420–37.

63. Lacher Malvey L, Neumark-Sztainer D, Story M. Prevalence of dieting and weight concerns among adolescents: a review of the literature. Under review.

64. Neumark-Sztainer D, Story M, Dixon LB et al. Correlates of inadequate consumption of dairy products among adolescents. J Nutr Educ 1997;29:12–20.

65. Patton GC, Johnson-Sabine E, Wood K et al. Abnormal eating attitudes in London schoolgirls: a prospective epidemiological study: outcome at twelve-month follow-up. Psychol Med 1990;20:383–94.

66. French SA, Story M, Downes B et al. Frequent dieting among adolescents: psychosocial and health behavior correlates. Am J Public Health 1995;85:695–701.

67. Neumark-Sztainer D, Butler R, Palti H. Personal and socioenvironmental predictors of disordered eating among adolescent females. J Nutr Educ 1996;28:195–201.

68. Story M, French SA, Resnick MD, Blum RW. Ethnic/racial and socioeconomic differences in dieting behaviors and body image perceptions in adolescents. Int J Eat Disord 1995;18:173–9.

69. Tienboon P, Rutishauser IH, Wahlqvist ML. Adolescents' perception of body weight and parents' weight-for-height status. J Adolesc Health 1994;15:263–8.

70. Rosen DS, Neumark-Sztainer D. Review of options for primary prevention of eating disturbances among adolescents. J Adolesc Health 1998;23:354–63.

71. Rock CL, Curran-Celentano J. Nutritional management of eating disorders. Psychiatr Clin North Am 1996;19:701–13.

72. Levine MP, Smolak L. Media as a context for the development of disordered eating. In: Smolak L, Levine MP, Streigel-Moore R, eds. The developmental psychopathology of eating disorders. Mahwah, NJ: Erlbaum; 1996.

73. Killen J, Barr Taylor C, Hammer L et al. An attempt to modify unhealthful eating attitudes and weight regulation practices of young adolescent girls. Int J Eat Disord 1993;13:369–84.

74. Neumark-Sztainer D, Butler R, Palti H. Eating disturbances among adolescent girls. Evaluation of a school-based primary prevention program. J Nutr Educ 1995;27:24–31.

75. Piran M, Levine N. Steiner-Adair C. On the move from tertiary to secondary and primary prevention: working with an elite dance school. In: Piran N, Levine MP, Steiner-Adair C, eds. Preventing eating disorders: A handbook of interventions and special challenges. Philadelphia: Brunner/Mazel (1999).

76. Smolak L, Levine MP, Schermer F. A controlled evaluation of an elementary school primary prevention program for eating problems. J Psychosom Res 1998;44: 339–53.

77. Neumark-Sztainer D. School-based programs for preventing eating disturbances. J Sch Health 1996;66:64–71.

78. Neumark-Sztainer D, Sherwood N, Coller T, Hannan PJ. Primary prevention of disordered eating among pre-adolescent girls: feasibility and short-term impact of a community-based intervention. J Amer Diet Assoc; 2000. 1100(12):1466–73.

Chapter 16

"There is no dress
rehearsal for life."
Anonymous

Adult Nutrition

Chapter Outline

- Introduction
- Year 2010 Health Objectives
- Physiological Changes of Adulthood
- Maintaining a Healthy Body
- Dietary Recommendations
- Nutrient Recommendations
- Physical Activity Recommendations
- Nutrient Intervention for Risk Reduction

Prepared by **U. Beate Krinke**

Key Nutrition Concepts

1 Enjoyment of good food contributes to quality of life.

2 Food security means consistent access to safe, wholesome, and culturally acceptable food.

3 Good nutritional habits maximize peak physical and mental performance.

4 Hormones regulating reproduction also affect nutritional status.

5 Good nutrition now is an investment in healthy old age.

Introduction

Definition of Adulthood in the Life Cycle

With good luck, good genes, and good habits, adulthood covers a life span of roughly 60 years. What can be expected to affect nutritional health during this time? Dividing those 60 years into segments can help to address that question.

Early Adulthood: On reaching their twenties, adults have generally stopped growing. Some males grow slightly after age 20, men and women continue to develop bone density until roughly age 30, and muscle mass continues to grow as long as muscles are used. The primary tasks of adulthood include personal and career development and potentially reproduction. Challenges to good nutrition include juggling many competing demands, including work and personal schedules, travel, eating out, and exploring new relationships and community ties. Nutritional habits developed now are investments in future health.

Midlife: During their forties and fifties, most adults are reaching the peak of their career achievements. Physiologically, body composition slowly shifts somewhat; it is in tandem with hormonal shifts, but more probably due to decreased activity. On average, individuals start to gain weight after age 40. It is a good time to reassess earlier nutritional habits.

Old Age: In their sixties and beyond, adults harvest the fruits of earlier health habits. Good food and exercise habits practiced over a lifetime support continued enjoyment of sports and daily activities.

In this text, nutritional aspects of adulthood from age 20 to 64 will be covered in Chapters 16 and 17. Nutrition specifically for older adults will be covered in Chapters 18 and 19. This arbitrary division segments a natural continuum uniquely experienced by individuals. The focus of Chapter 16 is on how to develop the most powerful nutrition habits possible.

Importance of Nutrition

Nutrition is about the many roles of food. Food is fuel, enjoyment, comfort, and a symbol of traditions, rituals, and celebrations. Food serves as a connection for socializing. When things go well, we take food for granted. We expect to have sufficient quantities of safe, appetizing food whenever we are hungry. During adulthood, we are often too busy to pay much attention to food.

How does food and nutrition enhance life? Lifestyle factors have a greater influence on long life than do genetics, health care systems, and the environment; lifestyle factors account for approximately half of premature deaths.[1] After smoking, nutrition and exercise top the list of lifestyle factors; therefore, individuals have some control over whether food and nutritional habits will contribute to a long and healthy life. Even though the leading causes of death at age 21 (unintentional injuries, homicide, and suicide) are not nutrition related, good nutrition throughout adulthood will reduce the risk of the leading causes of death in later adulthood—namely, heart disease, cancer, stroke, and diabetes mellitus. Nutrition during adulthood supports an active lifestyle, contributes to maintenance of healthy weight, and promotes physical and mental health and well-being.

Year 2010 Health Objectives

Since the first Healthy People 2000 report[2] was generated under the guidance of the federal government in 1979, several goals have been met. Rates of heart disease, cancer, and stroke deaths have declined; fat intake (as percentage of total calories) has declined. However, rates of obesity and diabetes have increased, sugar intake has risen, and health care disparities still exist among ethnic and racial groups. Updated national health goals address these issues. Table 16.1 reflects dietary goals for disease prevention and health promotion for adults. The main focus of these goals is healthy weight maintenance.

Physiological Changes of Adulthood

Growth and maturation are complete by early adulthood, so that nutritional emphasis turns to maintaining physical status, continuing to build strength, and avoiding excess weight gain. Peak capacity for physical performance for most activities is reached during adulthood. Sports examples of athletes at the top of their form at age 20 or older include cycling, running, cross-country skiing, tennis (males), and soccer. The peak for endurance sports tends to be later in life than for sports based on speed (for example, marathon runners are likely to be older than sprinters). However, for the average individual who participates in sport as a leisure activity, the physiological changes of early and middle adulthood neither enhance nor limit participation in, or enjoyment of, sports.

Table 16.1 Healthy People 2010: Health promotion goals for adults compared to current levels

	Target	Population Percentage Baseline, Age 20+ Females	Males
Increase the number who are at a healthy weight.	60	45	38
Decrease the number who are obese (BMI 30+).	15	25	20

The following baseline information is for age 20–39
(Baselines for ages 40–59 are in parentheses)
Increase the proportion of persons who:

	Target	Females	Males
• Eat at least two daily servings of fruit	75	20 (26)	23 (28)
• Eat at least three daily servings of vegetables (at least one-third dark green or deep yellow)	50	4 (4)	3 (4)
• Eat at least six daily servings of grain products (at least three of those as whole grains)	50	4 (4)	10 (10)
• Eat less than 10% of calories from saturated fat[a]	75	41 (42)	32 (33)
• Eat no more than 30% of calories from fat[a]	75	38 (33)	29 (28)

The following baseline information is for age 20–49
Meet dietary recommendations (1000 mg) for calcium
(level increases to 1200 mg at age 50)

	Target	Females	Males
	75	40	64

The following baseline is for all adults, age 20+
Eat 2400 mg or less of sodium daily
(males consume more calories, so they consume more sodium)

	Target	Females	Males
	65	30	5
Increase the proportion of physician office visits made by patients with a diagnosis of cardiovascular disease, diabetes, or hyperlipidemia that includes counseling or education related to diet and nutrition	75	39	44

Increase food security and in so doing decrease hunger

	Target	Females	Males
• All households at or <130% of poverty	94	69[b]	69[b]
• All households >130% poverty	94	94[b]	94[b]

[a] Updated to recommend 20–35% of calories from fat, minimizing saturated and transfatty acid intake. [2002; Institute of Medicine]
[b] Adequate data to distinguish household food security levels for males and females is not available.
SOURCE: Food and Drug Administration and National Institutes of Health. *Healthy People 2010: National Health Promotion and Disease Prevention Objectives.* Washington, DC: U.S. Department of Health and Human Services, 2000.

Physiological changes of adulthood differ for women and men. Women asked about their health and nutrition concerns are likely to be concerned about "not gaining weight" and about "looking good." Men are likely to say that they do not think about nutrition and health at all; but upon probing, they will relate health concerns about strength, energy, and weight management. However, hormonal shifts related to reproductive capacity have the biggest impact on physiology and therefore on nutritional status and needs. Pregnancy and lactation were discussed in preceding chapters. Menopause in women and the climacteric change in men are discussed in the following sections.

Physiological Changes in Males: Climacteric

For men, a gradual decline in testosterone levels begins about age 40 to 50, although sperm is capable of fertilizing human eggs until much later. Decreased sperm production is linked to underweight; malnutrition is linked to declining libido. Alcohol use can result in defective sperm. For the general population, alcohol intake declines with age. Alcohol accounts for 5% of calorie intake between ages 20 to 34, but only 3% between ages 51 and 64.[3]

Based on cross-sectional data (see Table 16.4 on p. 385), males eat fewer calories as they get older. However, body weight rises slowly beginning around age 40, more likely due to declining activity levels than to changing hormones.

Physiological Changes in Females: Menopause

For 40 years, females have roughly 13 menstrual cycles per year (minus the ones missed during pregnancy or for other reasons such as extreme loss of body fat), totaling more than 500 menstrual cycles. After cessation of menstrual blood loss, the need for iron decreases from 18 to 8 milligrams daily.[4] Loss of estrogen leads to atrophy of tissues in the urinary tract and vagina, increased abdominal fat, and greater risk for chronic conditions such as osteoporosis and heart disease. Menopause is associated with, but not the cause of, weight gain and decreases in muscle mass.[5,6] Hormones regulate blood chemicals; that is, hormonal shifts affect nutritional status, as shown in Table 16.2.

Oral contraceptives lead to an 8% to 11% increase of total serum cholesterol, with increases in low-density lipoprotein (LDL) and triglycerides and concurrent decreases in high-density lipoprotein (HDL). Even though menopause negates the need for contraceptives, lack of estrogen and progesterone leads to a rise in total cholesterol. Hormone replacement therapy (HRT)—traditionally, estrogen combined with progesterone, given as tablets, patches, or creams—is no longer prescribed to maintain bone density, prevent height loss and vertebral fractures, maintain urogenital tissue structure, and to limit the risk of heart disease. HRT has also been linked to increases in blood clots and to breast cancer risk.[7] Women at low risk for heart disease do not benefit significantly from HRT.[8]

Nutritional Remedies for Symptoms of Menopause

Menopause does not lead to declining life satisfaction for women, nor do most women suffer from menopause.[9] Women who do experience menopausal effects such as hot flashes, fatigue, anxiety, sleep disturbances, and memory loss are bombarded with treatment options including hormone replacement therapy, herbs, teas, foods and dietary supplements, exercise regimens, creams, and ointments. A review of randomized, controlled trials of complementary treatments of menopausal symptoms included several nutritional approaches, herbs, dietary *phytoestrogens,* and vitamin E.[10] Although most of the trials so far have been small and of short duration, some promising approaches were identified. Black cohosh (*Cimicifuga racemosa*) was found to decrease menopausal symptoms, including hot flashes, but not consistently. Potential risks of long-term black cohosh use are unknown because none of the clinical trials lasted longer than 6 months. The active com-

Phytoestrogen A hormone-like substance found in plants, about 1/1000 to 1/2000 as potent as the human hormone, but strong enough to bind with estrogen receptors and mimic estrogen and antiestrogen effects.

pound in black cohosh was thought to be formononetin, an estrogenic isoflavone, which turned out to be absent in black cohosh extracts and also absent in Remifemin, a popular, standardized product taken for menopause symptoms. None of the other commonly used herbs (chaste tree berry, dong quai, ginseng, evening primrose oil, motherwort, red clover, and licorice), nor vitamin E (up to 800 IU daily), were found to decrease menopause symptoms. Potential problems in using nutritional remedies include increased bleeding after taking ginseng (*Panax ginseng* and other *Panax* species) and dong quai (*Angelica sinesis*), and photosensitization from the furocoumarins in dong quai. Evidence for phytoestrogens was mixed.

Benefits of Dietary Phytoestrogens

Phytoestrogen effects depend on gut flora and hormonal milieu. Lignins, found in whole grains, in beans and peas, and especially in flaxseed, are dietary precursors of phytoestrogens; gut flora convert them to usable forms. Isoflavones are especially concentrated in leguminous plants (plants that fix nitrogen in the soil, such as beans and peas). Soy contains isoflavones and has estrogenic activity. In the presence of high estrogen levels, such as before menopause, phytoestrogens link to estrogen receptors to prevent more potent human estrogens from binding with receptors. When estrogen levels are low, as in postmenopausal women, phytoestrogens bind with estrogen receptors to supplement endogenous estrogen.

This is one reason that soy, especially soy protein and soy isoflavones, is so popular during menopause. An epidemiological view suggests that in Asian cultures, where soy is a dietary staple, women do not report hot flashes. Lack of hot flashes and night sweats may be a result of early exposure to soy phytoestrogens.[11] Experiments using soybeans during menopause are still relatively new, and the evidence about soy effectiveness is promising but inconsistent.[10,12]

The North American Menopausal Society Recommendations state:

Although the observed health effects in humans cannot be clearly attributed to isoflavones alone, it is clear that foods or supplements that contain isoflavones have some physiological effects.[13]

Table 16.2 Hormonal shifts affect blood chemistries

Hormone	Triglyceride	LDL Cholesterol	HDL Cholesterol	Other Effects
Estrogen	Raises	Lowers	Raises	Slows calcium loss Increases insulin sensitivity Decreases homocysteine
Progesterone	Lowers	Raises	Lowers	. . . . . .

SOURCE: Adapted from I. Contreras and D. Parra, "Estrogen Replacement Therapy and the Prevention of Coronary Heart Disease in Postmenopausal Women." Available at www.medscape.com/ASHP/AJHP/2000/v57.n21/ajhp5721.01.cont-01.html, accessed 1/01.

Table 16.3 Nutrient content in 2 tablespoons (14 grams) of ground flaxseed

Kcal	70
Protein	2.8 g
Carbohydrate	4.9 g
Fiber	4.0 g
Fat	4.8 g
Omega-3 fatty acid	2.6 g

Also contains significant amounts of iron, zinc, magnesium, potassium, and some calcium. For fewer calories than in a slice of bread, 2 tablespoons of ground flaxseed provide the protein of half an egg and more fiber than an apple. (A small apple, with skin, contains 2.9 grams of dietary fiber.)

We do know that soy isoflavones function as antioxidants and may be especially useful to women at increased risk for heart disease, which occurs at menopause.[14] Soy protein, with or without isoflavones, increases HDL cholesterol and decreases LDL and total cholesterol as well as decreasing triglycerides.[15] In fact, the Food and Drug Administration (FDA) allows a health claim to be made for soy: "Diets low in saturated fat and cholesterol that include 25 grams of soy protein a day may reduce the risk of heart disease." Because one consequence of menopause is an increased risk of heart disease in women, soy may be a healthful dietary addition.

Here are a few ways to eat soy:

- Edamame, young green soybeans (sweet beans) are a popular bar food in Japan.
- Roasted mature soybeans look like roasted peanuts and can easily be flavored to make sweet or savory snacks (not yet an American bar food).
- Soy protein meat analogs can substitute for meats (burgers, bacon, sausage, turkey, and chicken) while providing soy nutrients.
- Soymilks do not taste like dairy milk, but are used as a beverage and in cooking. Soymilk production formulas vary; try several to find your favorite.
- Processed cereals, bars, candies, drinks, and other snack foods are convenient sources of isoflavones and soy protein.

Flaxseed has been recommended to reduce menopausal symptoms, but the evidence is inadequate for public health recommendations.[16] The proposed mechanism for flaxseed's effectiveness is thought to result from the intestinal bacteria's conversion of lignin precursors to weak estrogens, enterodiol and enterolactone. Flaxseed (see Table 16.3 for basic nutrition information) is a rich source of omega-3 or alpha-linolenic acid and of antioxidants. The recommendations[17] for alpha-linolenic acid are 1.6 grams for men and 1.1 grams for women daily, roughly the amount in 1 tablespoon of ground flax.

Flaxseed must be ground to make the nutrients available for metabolism. Flaxseeds can be heated to 300°F without losing nutritional value, but are most often used in chilled foods such as smoothies or sprinkled over cooked items such as cereal. Toasted flax has a nutty flavor; roast seeds until they start to pop. (Keep the pan cover handy; flaxseeds jump out of the pan like popcorn.)

Disadvantages of Adding Phytoestrogens to the Diet

Soy has an estrogenic effect. Estrogen increases the risk for breast cancer. It would be unwise to suddenly add several servings of soy to the diet of someone vulnerable to breast cancer development (females with family history of breast cancer).

The taste of soy, as in soy flour and soy powder, is unappealing to many people. Soy oil, which is versatile and mild flavored, does not contain enough phytoestrogens to affect menopausal symptoms. Tamari, or soy sauce, even in its reduced-sodium form, is quite salty (compare 540 mg of sodium in 1 tablespoon light soy sauce, 1230 mg in 1 tablespoon of regular soy sauce, and approximately 2300 mg of sodium in 1 teaspoon of table salt). In the small amounts used, soy sauce does not add much soy nutrition to the daily North American diet. Fermented soy products are good sources of soy protein and isoflavones but may not be widely available. Not many people are familiar with tempeh or miso, both fermented soy products.

Flax in larger amounts (1.5 oz or more) acts as a laxative. The high fiber content may interfere with medications such as digoxin.

Other Alternatives

Menopausal treatment effects such as increased well-being have been attributed to ginseng, but experimental evidence supporting such activity is lacking.[18] The edible tubers of wild yams (*Dioscorea villosa*) contain steroidal saponins that are used in progesterone production, leading to speculation about hormonal activity in tubers. Wild yams are primarily used in topical applications and have been shown ineffective as hormone replacement therapy.[10] Kava (*Piper methysticum*) is a relaxant and sleep enhancer; alcohol increases kava's effect. In Pacific Island cultures (Oceania) where kava is native, the root is made into a bitter ceremonial beverage. The kava beverage first numbs the mouth, then leads to greater sociability, decreased anxiety, less fatigue, and eventually deep sleep. Therapeutic uses include muscle relaxation, sleep induction, and reduction of nervous anxiety. Kava extract contains kavalactones, also called pyrones, accounting for kava's sedative effects. The FDA has issued warnings about kava's safety, citing liver toxicity. England banned kava as a food ingredient. Individuals with liver disease and those with Parkinson's are especially vulnerable to kava. And last, alcohol is thought to potentiate estrogen effects. In reviewing cancer and diet, Willett concluded, "Moderate alcohol appears to increase endogenous estrogen levels."[19] Table 16.11 (later in this chapter) contains additional information on herbal and other dietary supplements.

Maintaining a Healthy Body

Reproduction, weight management including maintenance of muscle mass while avoiding excess fat, and wellness are primary nutrition concerns of adulthood. Reproductive issues are discussed elsewhere in the text; the following sections focus on energy and other nutrients needed to maintain a healthy body.

Energy for Weight Management

The focus of adulthood is maintenance of physiological status rather than growth, although some males continue to gain height after age 20. Average calorie expenditures peak during late adolescence and early adulthood, and then decline roughly 20% over the course of a lifetime, although the size of this decline has not yet been clearly estimated because of limited longitudinal data.[20] In the mid-1960s, researchers at the Baltimore Longitudinal Study on Aging (BLSA) reported that caloric intake in men decreased 22%, from 2700 to 2100 calories, between age 30 and 80.[21] They suggested that the decrease was due to lowered metabolic rates as well as decreased activity levels. Subsequent research suggested that activity-related caloric expenditures are lower for women than for men, partly due to lower fat-free body mass.[22] Women's resting metabolic rate (energy expended when the body is not active) declined 2% to 4% after age 50. However, energy expenditures stayed constant as long as fat-free mass stayed constant. The Fels Longitudinal Study[23] examined weight gains over 20 years, starting at age 40, and calculated average annual increases:

> **Basal Metabolic Rate (BMR)**
> Measuring energy expenditure in an individual who has been awake less than 30 minutes and is still at *absolute* rest, has fasted for 10 hours or more, and is in a quiet room with normal, comfortable temperatures.
>
> **Resting Metabolic Rate (RMR)**
> Measuring energy expenditure in an individual who has fasted, had no vigorous physical activity prior to the test, has been given time to relax (e.g., rest) for 30 minutes before starting measurement, and is in a quiet, private room with privacy and normal, comfortable temperatures.

- Men gained 0.66 pounds per year (0.3 kg).
- Women gained 1.21 pounds per year (0.55 kg).

Determining Energy Needs

Changes in body weight are due to a complex system of interactions including gender, body size, energy intake, activity levels, health status, hormonal shifts, and individual variation.

- *Gender/body size/muscle mass:* Males in general use 5% to 10% more calories than females do, because

men have proportionately more lean body mass. Metabolism in muscle burns more oxygen and needs more energy than fat stores do. However, persons who are overweight or obese don't necessarily burn fewer calories or have a lower metabolic rate.

- *Activity levels:* Activity requires energy; but more than that, exercise increases subsequent resting energy expenditure.

- *Health status:* Illness affects energy needs. For example, a fever's higher body temperatures burn more energy—7% more per degree Fahrenheit (13% more per degree Celsius). Starvation slows energy expenditures by 20% to 30%.

- *Hormones:* The thyroid hormone thyroxin accelerates metabolism, while lack of thyroid hormones leads to weight gain. Growth hormone is associated with weight gain, as is estrogen loss.

- *Individual variation:* Individuals vary by as much as 20% in their calorie expenditures during light to moderate activity.[24] In addition, the efficiency of food metabolism (the thermic effect of food) varies from person to person. Each of following approaches to estimating calorie needs tries to accommodate aspects of individual variation.

THE HARRIS-BENEDICT EQUATION The Harris-Benedict equation was derived from 239 subjects, aged 15–74 years old, with a mean age of 27 years and 31 years for men and women, respectively.[25] Although Harris and Benedict called their measure a *basal metabolic rate (BMR),* by today's interpretations, the reported measurements really reflect *resting metabolic rates (RMR)* because subjects had to travel to the testing site.[26] That is, the measurements were not taken at *absolute* rest.

> **Harris-Benedict Energy Estimation Formula** (developed 1919)
> (Weight is in kilograms; height is in centimeters; age is in years.)
> **Males:** RMR = 66.5 + (13.75 × wt) + (5 × ht) − (6.8 × age)
> **Females:** RMR = 655 + (9.6 × wt) + (1.8 × ht) − (4.7 × age)

THE MIFFLIN-ST. JEOR EQUATION Healthy adults today have different body compositions and lifestyles than adults in the early 1900s had. New validation studies have have yielded newer estimation formulas.[27–29] Energy estimation formulas that are within ± 10% of "true" RMR are considered accurate. The Mifflin-St. Jeor equation accurately reports RMR 82% of the time in non-obese adults (i.e., BMI of 18.5–29.9 kg/m^2). If the estimation is wrong, the worst underestimation error is 18%, and the worst overestimation error is 15%. In contrast, an energy estimate using the Harris-Benedict equation is accurate 69% of the time.[30]

For example, if you calculate an individual's resting metabolic rate as 1500 kcal using the Mifflin-St. Jeor equation, the "true" RMR will be between 1350 to 1650 calories per day, and you will be right 82% of the time. If you were wrong when calculating a 1500-calorie RMR, the worst *under*estimation error will be to prescribe only 1500 kcal for an individual needing 1770 calories (a maximum *under*estimation error of 18% by the formula). If you were overestimating with the Mifflin-St. Jeor equation, the worst overestimation error is to prescribe 1500 calories per day when an indiviudal needs 1275 calories (i.e., a maximum 15% *over*estimation errror by the formula).[30]

Once RMR has been determined, physiological stress factors related to illnesses are added to estimate daily energy expenditure in clinical use. In healthy adults, however, an activity factor is added to arrive at the estimated calorie expenditure per day. Metric measurements were used in the original formulas. The following box describes the steps for using the Mifflin-St. Jeor equation to calculate the energy expenditure in an individual who is 67 inches tall (170.2 centimeters), weighs 150 pounds (68.1 kg), and is 55 years old.

Mifflin-St. Jeor Energy Estimation Formula (developed 1990)
(Weight is in kilograms; height is in centimeters; age is in years.)
Males: RMR = $(10 \times wt) + (6.25 \times ht) - (5 \times age) + 5$
Females: RMR = $(10 \times wt) + (6.25 \times ht) - (5 \times age) - 161$

Example of how to use estimation formulas

Example: Using the Mifflin-St. Jeor equation, calculate the resting metabolic rate for a 55-year-old female who weighs 150 pounds and is 5 ft 7-in.

Step 1: Identify which estimation formula will be most reliable for individual.
Selected Mifflin-St. Jeor.

Step 2: Convert English measurements into metric.
67 inches multiplied by 2.54 centimeter/inch = 170.2 centimeters
150 pounds divided by 2.2 kilograms/pound = 68.2 kilograms

Step 3: Use formula* based on gender.
   Step 3a: FEMALE
RMR = $(10 \times 68.2) + (6.25 \times 170.2) - (5 \times 55) - 161$
RMR = $(682) + (1064) - (275) - 161 = {\sim}1310$ calories/day
   Step 3b: MALE
RMR = $(10 \times wt) + (6.25 \times ht) - (5 \times age) + 5$
RMR = $(10 \times 68.2) + (6.25 \times 170.2) - (5 \times 55) + 5$
RMR = $(682) + (1064) - (275) + 5 = {\sim}1476$ calories/day

Step 4. RMR is multiplied by activity or injury and stress factors.

Sedentary or weight maintenance	1.2
Light activity	1.3
Moderately active, infection, healing	1.5
Very active, extreme stress, burns	2.0

*Weight is in kilograms; height is in centimeters; age is in years.

To explore differences, complete the same steps when using the Harris-Benedict equation. How different are the results between males and females? Comparing these two equations demonstrates why calorie calculations are called calorie estimates. Trial and error exists in calculating caloric need determinations because energy expenditures are determined by complex interactions. However, the technology for measuring oxygen consumption and carbon dioxide using an indirect calorimeter is advancing, offering another method for identifying energy expenditure.

Alternately, the Miller method[28,31] most closely approximates the weights at the midpoints of weight ranges in the Metropolitan Life Tables of 1983.[25] The Miller method is not accurate for males taller than 74 inches.

- Female: 119 pounds for the first 5 feet, plus 3 pounds for each additional inch
- Male: 135 pounds for the first 5 feet 3 inches, plus 3 pounds for each additional inch

Other weight-for-height measures are discussed with obesity issues in Chapter 17.

Energy for Weight Change

It takes approximately 3500 calories to gain or lose 1 pound of body weight. To gain or lose a pound a week, add or subtract 500 kcal daily. These 500 calories can be a combination of intake and activity either added or subtracted. For instance, walking 1 mile uses roughly 100 calories, cycling 5 miles uses 200, and eliminating 16 ounces of cola allows subtraction of 200 calories, summed to total 500 calories. Seven days of burning 300 calories and eating 200 fewer calories leads to a weight loss of approximately 1 pound.

"We're a culture that eats everywhere and all the time. Food is everywhere. When was the last time you looked for a snack and couldn't find one?"

Actual Energy Intake

Older adults eat fewer calories than younger adults do. Table 16.4 shows findings from data collected by the U.S.

Table 16.4 Comparing caloric intakes of adults, by gender, from CSFII (1994–1995) and NHANES III (1988–1994)

AGE	DAILY CALORIE INTAKE, CSFII		DAILY CALORIC INTAKE, NHANES III	
	Males	Females	Males	Females
20–29	2844	1828	3025	1957
30–39	2702	1676	2872	1883
40–49	2411	1680	2545	1764
50–59	2259	1583	2341	1629
60–69	2100	1496	2110	1578

Department of Agriculture's Continuing Survey of Food Intake by Individuals (CSFII) and by the Department of Health and Human Services in the National Health and Nutrition Examination Survey (NHANES).

Achieving Wellness: Linking Food, Nutrition, and Disease

It is not enough to know how many calories will maintain healthy weight. More important questions are: What should those calories look like? Do certain foods contribute to better health? Can some foods cause illness? Dietary guidelines help to answer these questions. And to understand dietary guides or recommendations, it is useful to know about nutritional risk factors: How are our dietary habits linked to wellness and the avoidance of disease?

Much as we try, we cannot always achieve the degree of health we would like. But we can reduce the risk of contracting both short-term and chronic illnesses. Chronic conditions with modifiable risk factors include five of the ten leading causes of death for adults: heart disease, cancer, stroke, diabetes, and hypertension. Short-term illnesses include food poisoning and infections. Table 16.5 provides a list of risk factors that can be addressed.

Table 16.5 Illnesses and their modifiable risk factors

Short-Term Illnesses	**Nutritional Risk Factors**
Food poisoning	Unsafe food-handling practices
	Lack of hand washing, especially after toilet use
	Cross-contamination, mixing raw meat and vegetable preparation
	Lack of refrigeration, food at room temperature longer than 2 hours
	Inadequate temperatures in maintaining hot foods
	Contaminants such as salmonella, *E. coli,* mercury in some fish, etc.
Lowered immunity, infections	Malnutrition, especially protein and calorie malnutrition
	Low antioxidant intake
	Low vitamin/mineral levels
	Low levels of zinc and excessive levels of zinc
	Dehydration
Chronic Conditions	**Nutritional Risk Factors** (Inactivity is associated with most)
Heart disease	Diet high in saturated fat
	More than 10% of calories from saturated fat (Healthy People 2010)
	Daily dietary cholesterol above 300 mg, on average
Atherosclerosis	Any trans fatty acids
	Low levels of monounsaturated fatty acids
	Low antioxidants, low fruit and vegetable intake
	Low intake of whole grains
	Low folic acid intake and high levels of homocysteine
	No or excess alcohol*
	Obesity (BMI ≥ 30); waist > 40 in. for men, > 35 in. for women
	Elevated plasma Apo B levels
	High levels of LDL cholesterol in men
	Low levels of HDL cholesterol in women
Cancer	Low fruit and vegetable intake
	Low level of antioxidants (especially vitamins A, C)
	Low intake of whole grains, especially fiber-rich grains
	High dietary fat intake
	Nitrosamines, burnt and charred food
	High intakes of pickled and fermented food
	Obesity
Hypertension, Stroke	High blood pressure: obesity, waist > 40 in. for men, > 35 in. for women
	High sodium
	Low potassium
	Excess alcohol
	Low levels of antioxidants
Diabetes	Obesity

*In middle-aged adults, moderate alcohol reduces risk of heart disease (for men, after age 45; for women, after age 55).

Nutritional risk factors are translated into dietary recommendations, such as the Dietary Guidelines for Americans and health claims on food packages, as well as other public documents including Healthy People 2010 and *Diet and Health,* the consensus report of the Food and Nutrition Board, National Academy of Sciences, that followed the 1988 Surgeon General's Report describing nutrition and disease links.

Diet and Health

The Diet and Health summary guidelines[33] include both food and nutrition advice for disease risk reduction. They were developed for professional use and need interpretation to be applied in planning food programs, health promotions, and other nutrition services. Using this type of national consensus document in nutrition programs sets standards and facilitates consistent message development, which may decrease consumer confusion. The Healthy People 2010 goals (Table 16.1) integrate some of the recommendations from the Committee on Diet and Health:

1. Reduce total fat intake to 30% or less of total kilocalories (kcal), limit saturated fat to less than 10% of total kcal (see Table 16.6 for calculations) and cholesterol to no more than 300 mg daily; limit polyunsaturated fats to no more than 10% of total kcal. *Comment:* The 2002 DRIs suggest a total fat intake range of 20 to 35% of daily calories, with 5 to 10% as linoleic acid and 0.6 to 1.2% as alpha-linolenic acid. The American Heart Association suggests limiting dietary cholesterol to an average of no more than 300 mg daily. Less emphasis is placed on dietary cholesterol limits because it is saturated and trans fats that raise blood cholesterol levels the most. For example, evidence from 128 feeding studies over 30 years suggests modest increases in dietary cholesterol lead to small, if any, increases in blood cholesterol of populations.[34] McNamara calculated that consuming 6 eggs per week raises serum cholesterol by 0.9 mg/dL (1994, Nutrition Close-Up, funded by the Egg Nutrition Center). Much individual variability is seen when it comes to response to any dietary intervention, and it seems that the general population is better off limiting saturated fat intake than counting cholesterol.

2. Each day, eat 5 or more servings of vegetables and fruits, especially dark green and deep yellow vegetables and citrus fruits; eat 6 or more servings of breads, cereals, or legumes. Increase of added sugars is not recommended. *Comment:* Matches Healthy People 2010 goals! The 2002 DRIs suggest limiting added sugars to 25% of total calories, whereas the World Health Organization suggests getting no more than 10% of calories from sugars, including fruit juices.

3. Maintain moderate protein intake, below twice the RDA.

4. Balance food intake and physical activity to maintain appropriate weight; avoid diets that are either excessive or severely restricted in kilocalories, and avoid large fluctuations in body weight.

5. The committee does not recommend alcohol consumption. For those who do drink alcoholic beverages, limit consumption to the equivalent of less than 1 ounce of pure alcohol in a single day. (That's 2 cans of beer.) *Comment:* Women are more sensitive to alcohol than men are. The Dietary Guidelines for Americans 2000 suggest that if you drink, do so with meals to slow absorption and keep it to one drink per day for women and two for men. (One drink: 12 oz regular beer, 5 oz wine, 1.5 oz of 80-proof distilled spirits; see Table 16.7.)

6. Limit daily intake of salt (sodium chloride or table salt) to less than 6 grams. *Comment:* 6 grams of

---

Table 16.6 How to translate 10% of calories from fat to grams of fat

1. Multiply total calories in the diet by 0.1 to determine calories equaling 10%.
2. Divide these 10% of calories by 9 (calories per gram of fat) to calculate the number of grams of saturated fat that represent the "10% of calories from saturated fat" limit suggested for reducing LDL cholesterol in blood.

Amount of fat that represents 10% of calories for selected calorie levels:

Total Calorie Intake	Fat Grams = 10%
1200	13
1500	17
1800	20
2800	31

Table 16.7 Alcohol content in beverages

AMOUNT	BEVERAGE	ALCOHOL, g*
12 oz	Regular beer	13
12 oz	Light beer	12
1.5 oz	Distilled spirits (gin, whiskey, rum)	15
4 oz	Red wine	12
4 oz	Dry white wine	11
2 oz	Sherry	9

*Count one drink as 10–15 grams (or about ½ oz) of alcohol.

sodium chloride correspond to 2400 mg sodium. "Na" is the chemical abbreviation for "sodium" seen on labels; 1 teaspoon of salt contains 2325 mg of Na. In 2004, the Institute of Medicine set the tolerable upper intake level (UL) for salt at 5.8 g per day, and an adequate intake at 3.8 g daily.

7. Maintain adequate calcium intake. Potential benefits of excess intake are not documented. *Comment:* Potential for harm is now set at the tolerable upper intake level of 2000 mg of calcium per day.[35]

8. Avoid taking dietary supplements in excess of the RDA. *Comment:* Newer evidence about folic acid and vitamin B_{12} suggest that they are better absorbed in synthetic form during gastric illnesses; 1000 micrograms is the tolerable upper limit for folic acid, with no determined upper limit for vitamin B_{12}.[4,35]

9. Maintain an optimal intake of fluoride, available in fluoridated water.

Dietary Recommendations

How would you like your advice?

"Eat a juicy crisp apple with the peel."

or

"Eat more fiber, an apple is a good source of fiber."

"Monounsaturated fats contribute to a lower risk for heart disease."

or

"Eating a salad with olive oil dressing makes your blood flow and gives you energy."

Governmental and private groups make food and nutrition recommendations according to their missions and goals. Three main perspectives underlie such dietary advice.

1. *Advocating for reduction of specific disease risk.* For example, the American Heart Association suggests reducing total dietary fat to decrease risk of heart disease. The National Cancer Institute (as well as the fruit and vegetable produce industry) urges us all to eat more fruits and vegetables to reduce risk of certain cancers.

2. *Ensuring adequate population intake of specific nutrients.* For example, FDA regulations mandate flour and grain product fortification with B vitamins, folic acid, and iron. Folic acid was added in the late 1990s to prevent neural tube defects. Here, FDA recommendations are more than "advice"; they are enacted into law. Enrichment and fortification laws are intended to improve the health status of populations.

3. *Offering guidance on what, and how much, to eat.* For example, the U.S. Department of Agriculture and U.S. Department of Health and Human Services together publish the Food Guide Pyramid and "Dietary Guidelines for Americans." On the nongovernmental side, many dairy councils have developed dietary guidance tools built on the basic food groups, including the cheese and milk group. Dietary guidance tools such as these can help consumers make meal plans.

Dietary Recommendations to Combat Nutritional Concerns

Dietary recommendations for the public translate and integrate documents such as *Diet and Health;* nutrient guidelines from the Food and Nutrition Board and the U.S. Department of Health and Human Services; and position statements from the various health advocacy coalitions, such as American Heart Association, National Cancer Institute, and National Osteoporosis Foundation, into a consensus document. This document spells out the national health and nutrition goals and priorities, makes evident the links among eating, drinking, and health outcomes, and includes meaningful food advice.

"Dietary Guidelines for Americans, 2000" addresses current health issues by outlining health goals, describing diet and disease linkages, and including the Food Guide Pyramid to provide advice on what to eat for good health. Dietary Guidelines 2000 continues a series first issued in 1980 and updated every 5 years. The latest version of health-promoting food advice reflects current nutrition science and policy. Food safety is a new topic in the 2000 guide because the number of incidents has increased. The 1995 statement about sugar changed from "choose a diet moderate in sugars" to "choose beverages and foods to moderate your intake of sugar" in order to reflect the rising intake of nonnutrient sugars, one of the contributors to obesity. Go back to Chapter 1 to review the specifics in Table 1.19, which shows the basic components of the Dietary Guidelines 2000, including specific food-based advice.

Food Advice

Food-guidance documents have been around at least since 1916, having been spurred by war and economic depression.[36] The current recommendations "Nutrition and Your Health: Dietary Guidelines for Americans" (Table 1.19) have grown from collaborations by the U.S. Department of Agriculture and U.S. Department of Health and Human Services, which each had separate guides until relatively recent times. What makes food advice useful? Deciding what to eat is harder than knowing what to avoid. The message "skip ice cream" tells an audience exactly what to do: avoid eating ice cream. But the message "eat less saturated fat" needs further knowledge to be translated into breakfast, lunch, snacks, or supper. To help individuals focus on foods to eat, a food guide needs several attributes:

- *Classify foods into relevant groups.* For example, what are the food groups for peanut butter on toast and a glass of juice?
- *Describe what constitutes a serving.* For example, do the lettuce, tomato, pickle, and ketchup on a burger count as 1 or more servings of vegetables?
- *Identify the number of servings needed from each group to promote health.* For instance, how many grain servings will it take to meet carbohydrate and fiber recommendations?

The Pyramid

Recommendations for following the Food Guide Pyramid, Table 16.8 (or see Illustration 1.9 in Chapter 1), are designed for individual flexibility by listing food groups, serving sizes, and number of servings for various calorie levels. (The Pyramid is also available in graphic form at www.usda.gov/cnpp). Consuming the recommended number of servings from each group to meet caloric needs over a period of 3 days on average, and 5–10 days at most, will meet individual overall nutrient needs.[32] When all the chosen foods from each group are lean, low in fat, and contain no added sugars, calorie levels of the basic five food groups range from 1200 to 2000 calories, with approximately 16% of the calories provided from fat. Fats, oils, and sweets make up the "other" food group "to be used spar-

ingly." Future revisions of the Food Guide Pyramid are planned to allow for greater individualization of dietary advice, to include exercise guidance, and to highlight caloric intake and weight control. An emerging understanding of **nutrigenomics** and the population's genetic patterns will be reflected in more flexible dietary guidance tools. For example, a female with the Apo E14 allele may be more prone to bone loss after menopause. She may be advised to build maximum bone density as a young girl.

Nutrigenomics The science of gene-nutrient interactions.

Food Advice Reflects Cultural Food Patterns

Food-group guidance comes in many forms other than the pyramid. For example, Canada's Food Guide to Healthy Eating is a rainbow (Illustration 16.1 on the next page). Inclusion of canned and frozen food pictures makes it obvious that processed foods count. The rainbow groups fruits and vegetables, lists metric and English amounts for serving sizes, and does not picture "other" foods.

The United States uses a pyramid, Canada uses a rainbow, and China uses a pagoda. China's pagoda highlights how food guides reflect their target audience's unique dietary needs and goals (Illustration 16.2 on page 391). Following pagoda guidance ensures that individuals eat from 1800 to 2800 calories per day. Additional advice to Chinese residents includes distributing food evenly throughout the day, making full use of local food sources, and making good nutrition a lifetime commitment. The gram-weights listed with food groups show that the Chinese diet is cereal based, includes greater quantities of vegetables than fruit,

Table 16.8 Food Guide Pyramid recommendations for active men and women

| | NUMBER OF SERVINGS PER kcal LEVEL | | |
Food Groups	1500–1700 (most women)	2100–2300 (most men, active women)	2700–2900 (active men)
Basic Five Food Groups			
Bread, cereal, rice, pasta (1 oz, 0.5 c cooked)	6	9	11
Vegetables (1 c leafy, 0.5 c raw, cooked)	3	4	5
Fruit (6 oz juice, 0.5 c canned, 1 med)	2	3	4
Milk, yogurt (1 c), cheese (1.5–2 oz)	2–3*	2–3*	2–3*
Meat, poultry, fish, eggs (2–3 oz); dry beans, tofu (0.5 c cooked); nuts, seeds (0.33 c)	2 (total 5 oz)	2 (total 6 oz)	3 (total 7 oz)
Fluids (1 c) (amounts based on 1 mL/kcal)	6	8	10
The "Other" Group			
Fats, oils, sweets (1 tsp oil, sugar, jam)	5	7	9
Alcohol: If using, moderation is 1 drink for women and 2 drinks for men per day			

*2 servings between ages 19 and 50, 3 servings after age 50.
SOURCE: Modified from USDA/HHS Food Guide Pyramid. Available at www.usda.gov, or 202-720-2600, voice and TDD, for Braille, large print, or alternative means of communication.

Illustration 16.1 Canada's Food Guide to Healthy Eating—the rainbow.

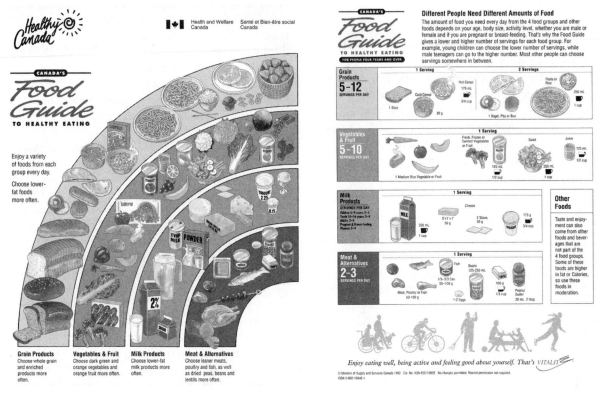

SOURCE: Health and Welfare Canada. Nutrition recommendations. Canada: Minister of Supply Services; 1990.

includes fish daily, and uses small portions of meat. The placement of milk/milk products and bean/bean products together in one group is also unique to the pagoda.

Dietary advice for health promotion, Japanese style, includes the following recommendations.

Make all activities pertaining to food and eating pleasurable ones:

1. use the mealtime as an occasion for family communications and
2. treasure family taste and home cooking.

Alcohol: Food, Drug, and Nutrient

Canadian nutrition recommendations are to limit alcohol calories to 5% or less of total calories or 2 drinks a day, whichever is less.[37] U.S. guidelines are "If you drink, do so in moderation," defined as no more than 2 drinks per day for males, and no more than 1 drink per day for females. Saving up drinks to have all 7 to 14 on the weekend is not "moderation." The Canadian guide states that having more than 4 drinks at a time is risky.

What is a drink? Equivalent alcohol contents for "a drink" are 12 ounces of regular beer, 5 ounces of wine, and 1.5 ounces of 80-proof distilled spirits (see Table

16.7). A drink contains roughly 13–15 grams of alcohol or 0.5 ounce of ethanol (ounces are listed on the blood alcohol charts describing legal limits for drivers). New Zealand gives guidelines in grams of alcohol: 20 or fewer grams of alcohol per day for women and 30 grams or less for men.

Alcohol is considered a drug for its central nervous system effects, but it is also a nutrient (although not a required one) because it yields energy; that is, 7 calories per gram. When developing a dietary assessment tool for adults, alcohol needs a separate line because it supplies a significant number of calories (3% to 4% of total calories) to the diet.[3,38] Males drink more alcohol than females do (12 grams versus 5 grams daily for the total population, all ages). Roughly one-third of adults abstain from alcohol.[39,40] Therefore, alcohol ingestion per person is really somewhat higher than the population averages might indicate. Current data reports average alcohol consumption for alcohol per se. Caloric contributions of alcoholic beverages are higher. Based on an analysis of the first (1977–1978) Nationwide Food Consumption Survey of just those individuals who drank alcoholic beverages, males consumed 17% and females 22% of total calories from mixed drinks, wine, and beer.[41]

Illustration 16.2
The Food Pagoda Guide for Chinese residents.

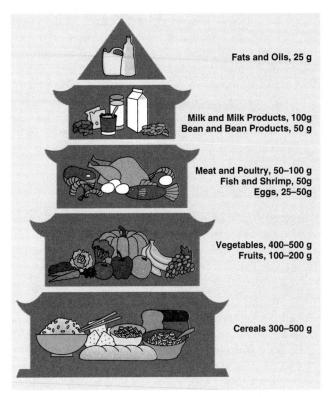

Fats and Oils, 25 g

Milk and Milk Products, 100g
Bean and Bean Products, 50 g

Meat and Poultry, 50–100 g
Fish and Shrimp, 50g
Eggs, 25–50g

Vegetables, 400–500 g
Fruits, 100–200 g

Cereals 300–500 g

SOURCE: Reprinted by permission of the Chinese Nutrition Society.

Fluids

Adults can survive roughly 60 days without food, but only 3 to 4 days without water.[6,42] Because fluid is such an integral part of metabolism, the body has many regulatory mechanisms to maintain hydration. Increased thirst, concentrating urine to reabsorb water, slowing down activities, and shutting down perspiration conserve water. Metabolism of foods and of body tissues also yields water. For instance, breaking down 1 kilogram of body tissue composed half and half of fat and non-fatty tissue yields nearly 1000 milliliters of water.[42] Someone who has stopped eating and drinking metabolizes stored tissues and will still produce urine.

Beyond thirst as a guide to maintain hydration, methods to determine fluid recommendations relate to caloric intake or body weight. The Food and Nutrition Board recommends 1 milliliter (mL) water per calorie of food ingested.[24] Therefore, 2000 mL of fluid would be recommended for an individual eating the 2000-calorie diet used as reference on food labels. Some calorie-based recommendations also add fluid intake, which should be at least 1500 mL daily. So getting a jump start on dieting with a 1200-calorie diet means drinking 1500 mL or more rather than drinking 1200 mL or more.

Estimates using body weight vary by activity and age, with the average adult needing 30–40 mL/kg.[42]

Young active adults,
 aged 16 to 30 years: 35–40 mL/kg
Adults aged 55–65 years: 30 mL/kg

Using the guideline of 30 mL of fluid per kg body weight, a 143-pound person weighing 65 kg would need 65 × 30 or 1950 mL of fluid daily. A 170-pound person weighing 77 kilograms would need 77 × 30 or 2310 mL of fluid per day.

A stepwise formula[43] that allows adjustment for increasing or decreasing body weight and ensures a 1500-mL minimum daily intake was developed for pediatric patients and is useful for smaller, lighter adults:

- 100 mL/kg body weight for the first 10 kg
- 50 mL/kg body weight for the next 10 kg
- 15 mL/kg body weight for the remaining kg

Applying this formula to a 143-pound (65-kg) person leads to a recommendation of just over 2 liters (8 cups), and roughly 10 cups for a 170-pound (77 kg) person.

Body weight of	143 pounds (65 kg)	170 pounds (77 kg)
10 kg × 100 =	1,000 mL	1000 mL
10 kg × 50 =	500 mL	500 mL
Remaining ___ kg × 15 =	45 × 15 = 675 mL	57 × 15 = 855 mL
Total =	2,175 mL	2355 mL

240 mL = 8 ounces = 1 cup; 4 cups = 1 quart, roughly equivalent to 1 liter

A simple, yet practical guide to adequate fluid intake that does not rely on numbers refers to the color of urine: when it is nearly colorless or pale yellow, the individual is drinking enough. (Note: This doesn't always work. Therapeutic amounts of riboflavin supplements as well as some medications and foods affect the color of urine; e.g., flavins are pigments.)

Diuretic Effects of Caffeine and Alcohol

What about fluids containing caffeine? Occasionally someone suggests that coffee, tea, or other caffeine-containing beverages not be counted as fluid intake because they have a diuretic effect on the body. Caffeine is a stimulant that relaxes the esophageal sphincter (leads to acid reflux), has a laxative effect, and increases urine production.[44] But no evidence connects drinking coffee or tea or other caffeine-containing-fluids with greater dehydration.[45] At this time, the issue of whether to count coffee, tea, or other caffeine-containing beverages in or out of the fluid allotment is a policy issue rather than a scientific one.

Several countries address caffeine in their dietary guidelines; none address whether caffeine-containing beverages should count as fluids.[36] South Africa suggests restricting caffeine (no amounts given) as part of dietary guidance. Canada is the only country to suggest an amount for caffeine consumption: "The Canadian diet should contain no more caffeine than the equivalent of four regular cups of coffee per day."[37]

iatrogenic Used in reference to disease, it is a condition induced by a medical treatment.

Fluids that contain alcohol increase urine output in the short term, followed by an antidiuretic phase. Most health professionals do not count alcoholic beverages as contributing fluids to the diet. With moderate alcohol use, the amounts of fluid in question are necessarily small.

International Guidance Related to Fluids

We tend to take water for granted and not emphasize it in our dietary guidelines. The Dietary Guidelines for Americans integrate fluids into the sugar and sodium advisories, suggesting "Choose sensibly to limit your intake of beverages and foods that are high in added sugars" and "Drink water often" (p. 33).[46] Part of the sodium advice is: "Drink water freely. It is usually very low in sodium. Check the label on bottled water for sodium content." (For example, 8 ounces of spring water typically contains 2 mL of sodium; 8 ounces of municipal tap water contains 7 mL; and 8 ounces of household softened water 9 mL of sodium. These levels vary dramatically by geographic region.) Here is how several countries address the need for adequate fluid in their dietary guidance documents:[36]

- New Zealand: Drink 1 to 2 liters per day (1000 to 2000 mL, about 8 cups) and choose from tap water and other beverages, including tea and coffee.
- South Africa: Drink at least 1 liter of clean, safe water or fluids.
- Switzerland: More fluids.
- Venezuela: Drink water.
- Hungary, Korea, Philippines, Singapore, and Switzerland suggest "milk every day" or "more milk" as part of their dietary recommendations. This suggestion may have more to do with increasing calcium and other nutrients than with fluid intake.

Nutrient Recommendations

Nutrient recommendations are established and regularly revised by the Institute of Medicine of the National Academies.[4,17,24] For instance, pre-2002 public health recommendations, such as published in Healthy People 2010 (see Table 16.1), were to eat fewer than 30% of total calories as fats; the newer recommendations are to consume from 20% to 35% of calories in the form of fat.[17] Dietary Reference Intake (DRI) values for all ages can be found on the inside covers of this text. DRIs for adults are included

in Table 16.9. Recommended nutrient levels used in food labeling as Daily Values (DV)[47] as well as the Tolerable Upper Intake Levels (UL), are included in Chapter 18. In many cases, excessive nutrient intake can be as detrimental to health as insufficient intake, so ranges of recommended intake are given where possible. The UL indicates a potentially toxic limit for vitamins and mineral intake. Diet alone is unlikely to get anyone to these UL amounts, although eating fortified foods and taking vitamin-mineral supplements might. Intake should stay below the UL to avoid *iatrogenic* illness (when what you do to be healthy makes you sick instead).

Nutrients of Public Health Concern

Saturated fat, fiber, folic acid (for women and all adults aged 60 and older), vitamin E, calcium, and magnesium, plus iron for women are of public health concern because intake levels do not meet recommendations (Table 16.9).

FAT In the late 1970s, 41% of calories in U.S. diets came from fat. That level has dropped to 33% of calories from fat (11% saturated fats) in 1999–2000.[48] Fat-intake levels seem to have stabilized. An analysis of 1988–1991 NHANES data shows that roughly 34% of the average population's caloric intake was supplied by fat, with 12% from saturated fat.[48] CSFII data from 1994–1995[49] reported 33% of total calories from fat and 11% from saturated fatty acids. Total fat intake is in now in the acceptable macronutrient distribution range,[17] and fat-intake guidelines no longer specify percentages of saturated fats for which to aim. The 11% level is close to the 10% recommended in Healthy People 2010 goals, and it is slightly more than the 7% recommended to lower LDL blood cholesterol levels by the National Cholesterol Education Program's Adult Treatment Panel III report.

FIBER Average daily dietary fiber recommendations are 38 grams for males and 25 grams for women, or 14 grams of dietary fiber per 1000 calories. Daily median intakes[17] are 16.5–19.5 grams for adult males and 12.1–13.8 grams for adult females. If fruit, vegetable, and whole grain intakes rise to meet food guide recommendations, adequate fiber intake will result. Chapter 18 includes examples for increasing fiber intake.

FOLIC ACID The evidence for folic acid in the prevention of neural tube defects is so strong that the FDA mandated folic-acid fortification of grains and grain products beginning in 1998. Dietary surveys show that fortification is working; mean intake has increased to 327 micrograms for 20- to 39-year-old women and 435 micrograms for 20- to 39-year-old men, compared to the 400-microgram DRI. Some evidence also supports folic acid as a way to lower risk of heart disease. (More on folic acid in Chapter 18.)

Table 16.9 Selected nutrient intakes of adults, CSFII 1994–1995 data collection, compared to recommendations

	INTAKE				
	ACTUAL				**RECOMMENDED**[a]
	20–29 years		40–49 years		31–50 years
Nutrient	Males	Females	Males	Females	Males/Females
Protein, g	106	68	96	64	63/50
Total fat, g	104	66	91	63	65 (DV)[b]
Saturated fat, g	36	23	31	21	20 (DV)[b]
Monounsaturated fat, g	40	25	35	24	—
Cholesterol, mg	349	226	339	225	300 or less
Fiber, g	18	12	18	14	25 (DV)
Sodium, mg	4627	2974	4061	2771	Up to 2400
Potassium, mg	3278	2240	3225	2411	3500
Vitamin A, mcg RE	1430	1149	1000	800	900/700
Carotenes, mcg RE	539	506	539	500	—
Vitamin E, mg TE	10.1	7.1	9.4	7.9	15
Vitamin C, mg	123	87	106	93	90/75
Vitamin B$_6$, mg	2.4	1.6	2.1	1.5	1.3/1.3
Vitamin B$_{12}$, mcg	6.2	3.8	6.2	4.4	2.4/2.4
Folate, mcg[c]	323	226	288	226	400/400
Calcium, mg	1005	890	717	643	1000/1000
Magnesium, mg	341	229	330	241	420/320[d]
Zinc, mg	15.2	9.7	13.2	9.4	11/8
Iron, mg	19.8	13.5	17.6	13.1	8/18
Copper, mcg	1600	1100	1500	1100	900

[a]Recommended intake according to DRI (see inside cover of text) or to DV nutrient reference amount used on food labels, relevant to a 2000-kcal diet.[47]

[b]Based on a 2000-kcal diet; for other caloric levels, recommendation is no more than 30% of calories from fat and 10% from saturated fat.

[c]Synthetic, prior to 1998 grain fortification with folic acid.

[d]Nonfood sources.

SOURCE: Consumption data from J. W. Wilson et al., *Data Tables: Combined Results from USDA's 1994 and 1995 Continuing Survey of Food Intakes by Individuals and 1994 and 1995 Diet and Health Knowledge Survey,* 1997.

VITAMIN E   This versatile antioxidant has been shown effective in prostate cancer, heart disease, and cataract risk reduction. Food sources include sunflower seeds and oil, walnuts, wheat germ, olive oil and mayonnaise, avocado, and kale. (More in Chapter 18.)

CALCIUM   Adults could meet calcium goals by adding a glass of milk, 8 ounces of yogurt, or 8 ounces of calcium-fortified orange juice to their diets daily. Eight ounces of milk contains roughly 300 milligrams of calcium, regardless of butterfat. Some milks and yogurt provide additional calcium (>30% of DV per serving) because milk solids have been added for extra body and flavor. See Table 16.10 on the next page for strategies that optimize calcium intake.

Calcium is a public health concern, for several reasons. Calcium plus vitamin D and magnesium help to develop and maintain bone density, which delays osteoporosis and reduces risk of bone fractures. Adequate calcium is a potential contributor to lowered risk of colon cancer. Milk as part of the DASH diet (see Chapter 19) has been successfully used as part of blood pressure reduc-

tion intervention.[50] Calcium's crucial role in metabolism is supported by physiological mechanisms to keep calcium in tight balance including calcium intake, amount of calcium absorbed, and calcium excreted through urine and feces. Of these factors, intake and fractional absorption of calcium account for 25% of the variance in calcium balance, while losses through urine account for another 50%.[51,52] Because protein and sodium promote urinary calcium excretion, eating a diet high in protein and sodium (like the typical adult diet) contributes to calcium losses through urine.[53] Other factors that decrease calcium absorption are ingestion of glucocorticoids (used to treat inflammatory diseases such as rheumatoid arthritis, asthma, and inflammatory bowel disease). As the body ingests higher levels of calcium, the percentage absorbed decreases. Conversely, as calcium intake drops, the percentage absorbed increases. A larger portion will be absorbed from a 500-milligram-per-day intake than from a 1500-milligram intake.

The tolerable upper limit for calcium is 2500 milligrams. It would be difficult to exceed this amount from

Table 16.10 Optimizing calcium intake for osteoporosis prevention

- **Vitamin D:** Adequate intake of Vitamin D (1,25-dihydroxyvitamin D, also called calcitriol) stimulates active transport of calcium in the small intestine and the colon.
- **Sodium:** Lower levels of sodium intake lead to lower levels of urinary calcium; that is, less calcium is excreted while consuming a low-sodium diet.
- **Lower protein intakes:** Excess protein leads to greater urinary excretion of calcium. In the Nurses Health Study, women (aged 35 to 59) who ate more than 95 grams protein per day suffered more osteoporotic forearm (but not hip) fractures over 12 years.[53]
- **Caffeine:** Two to three cups of coffee daily in postmenopausal women, when consumed with less than 800 mg/day of calcium, were associated with bone loss.[51]
- **Timing:** Drink milk and take supplements with a meal because food slows intestinal transit time and allows more calcium to be absorbed from the gut. Calcium carbonate is best taken with meals; calcium citrate may be taken at any time. In addition, stomach acidity enhances absorption; therefore, calcium is better absorbed when not consumed with antacids.
- **Moderation:** Consuming nutrients at recommended levels (enough but not too much) helps to synthesize and develop collagen matrix into which minerals are deposited during bone mineralization.[58]
- **Other vitamins:** Vitamin C plays a role in the development of the protein bone matrix (collagen), but has no defined role in treatment. Vitamin B_6 is a cofactor in stabilizing collagen cross-links and has been linked to preventing hip fractures. Vitamin K is required for the formation of proteins that stimulate osteoblasts to build new bone and attract osteoclasts to initiate bone reabsorption (e.g., glutamic acid residues are carboxylated to form gla-proteins; most studied is osteocalcin).

food sources alone. For example, the basic diet without dairy products equals 400 milligrams of calcium. Four 8-ounce glasses of milk = 1200 milligrams of calcium; an ounce of cheese = 250 milligrams. These sources add up to approximately 1850 milligrams calcium. However, supplement usage adds up more quickly.

Calcium carbonate is the cheapest form of calcium supplement. However, taking it can decrease or interfere with iron absorption. Citrate- or ascorbic acid–bound calcium supplements, on the other hand, lead to greater iron absorption. Citrate and ascorbic acid calcium supplements are more easily absorbed than calcium carbonate.

They usually cost more, too. Calcium and its binder are bulky, so multivitamin-mineral supplements tend to contain little calcium. Calcium is best absorbed when taken in divided doses. Take calcium carbonate with food, but take calcium citrate any time.

MAGNESIUM Poor diet, diabetes, or prolonged illness can cause magnesium deficiency, with adverse effects on bone strength and heart health. This issue tends to be problematic in older adults (see section in Chapter 18). Good magnesium sources are sunflower seeds, almonds, beans, milk, whole grains, spinach, and bananas.

IRON For women, iron needs drop from 18 to 8 milligrams per day at menopause. The benefits of sufficient iron are healthy blood cells that carry oxygen for metabolism and overall energy; the drawbacks of excess iron are greater need for antioxidants. High-iron foods include red meats, fortified cereals, strawberries, and dried beans as used in baked beans, bean soup, bean dip, and chili. Changing from a high-iron to a low-iron cereal during the menopausal transition can help to adjust for the dramatic shift in iron needs.

Actual Intake of Food

FOOD GROUPS AND PORTION SIZE How are Americans doing when it comes to meeting the pyramid guidelines? A pyramid top-heavy in sweets reflects the American plate. Fruit and vegetable servings totaled 3.5 rather than 5.0. But we are closing in on 5 daily servings of fruits and vegetables, although intake from the "other" group at the tip is still high.[3] CSFII tracks the percentage of individuals who consume foods from each food group. Among 20- to 29-year-old adults, fewer than half eat fruit, but four of five eat vegetables daily. Two-thirds (67.5% of males) to three-fourths (74.5% of females) consume some milk or milk products every day. On any given day, 7% to 8% eat nuts and seeds, 11% to 13% eat beans and other legumes, 40% to 50% eat sugars and sweets, and 62% to 68% consume carbonated soft drinks.[49] However, the movement to supersized restaurant and take-out foods (e.g., fries, drinks, rolls, and cookies) has led to "portion distortion." For example, a review of food intake over 19 years prior to 1997 found that portion sizes increased for salty snacks, desserts, soft drinks, and especially fast foods.[54] Within fast foods, the largest increase was 133 calories for Mexican foods. Try to estimate your next fast-food meal in pyramid portions as in Table 16.8. It is not easy to think "normal."

Using NHANES data, Kant calculated that 27% of the average individual's calories are eaten as energy-dense, nutrient-poor (EDNP) foods; that is, the "other" food group.[3] When we add to that the 3% to 4% of calories alcohol provides, it turns out that we rely on 69% to 70% of our calories for the bulk of vitamins and minerals.

EATING OCCASIONS More and more people are eating away from home. For example, 62% to 72% of 20- to 29-year-olds consume more than a third of their calories away from home. The National Restaurant Association[55] estimates that by 2010, 53% of food dollars will be spent away from home, and fast-food restaurants will be the fastest-growing type of operation. Greater numbers of fast-food meals are associated with higher total and fat calories.[56]

Supplements: Vitamins and Minerals

Adults can eat nutritiously without dietary supplements. Circumstances under which dietary supplements may enhance peak functioning are childbearing (vitamins and minerals for reproduction are covered elsewhere in this book), high-level athletic performance, illness, or when eating restricted diets—for example, a wheat-free diet due to allergies or a calorie-restricted diet for weight loss. Immune status enhancement from physiological doses (RDA, DRI levels) of supplements taken for a year or more has been found in older adults.[57,58] Younger adults might consider supplementing dietary gaps (potentially calcium, iron, magnesium, vitamin E, and folic acid, plus vitamins B_{12} and D for nonmeat eaters and vegetarians).

CSFII data reports that 42% of males and 56% of females aged 20 and older consume a vitamin or mineral supplement daily, nearly every day or "every so often."[49] Supplement use by males gradually rises with age (37% to 45% of individuals from age 20 to age 70+). Female usage rises from 51% of individuals at age 20 to a peak of 62% between ages 50 to 59 and then drops to 54% by age 70 and older. The greatest percentage of males who take supplements use a multivitamin. For females, the pattern shifts: a multivitamin is most popular until age 40, after which single vitamins and minerals are used more. At age 70 and older, 23.4% of women use a single vitamin or mineral and another 23.4% of women use a multivitamin. NHANES findings are slightly different from those of CSFII, showing that 40% of the U.S. population (2 months and older) took at least one vitamin or mineral supplement in the previous month (35% of males and 44% of females). The most commonly found nutrients in supplement products are ascorbic acid, vitamin B_{12}, vitamin B_6, niacin, thiamin, riboflavin, vitamin E, beta-carotene (vitamin A precursor), vitamin D, and folic acid.[59]

Supplements: Herbal Products

Herbal products are often grouped with other nutritional supplements. Table 16.11 on the next page shows examples of some of the most popular herbal remedies and suggests potential uses and side effects.

Herbs and foods with special nutritive and medicinal attributes (functional foods) comprise a rapidly growing and largely unregulated market. *When in doubt, throw it out* is a food safety education message that seems appropriate for herbal products as well.

The FDA has banned several herbs, although they occasionally appear on the market (sassafras, yohimbe/yohimbine, comfrey, chaparral, lobelia, ephedra). Others that are unsafe and/or ineffective include apricot pits (cyanide), belladonna, blue cohosh, coltsfoot, dong quai or angelica, eyebright, garcinia, life root, mistletoe, pokeroot (except the berries), wild yam, willowbark, and wormwood.[18] Additional information on the complex topic of herbal supplements can be found via the Web sources listed at the end of the chapter.

> "Mmm, mushrooms, a natural food. Morels, truffles, porcini, portobello, enoki, shiitake . . . they taste so good. Yet some will kill you."
>
> The author, daydreaming about edible natural things

Cross-Cultural Considerations

The health and nutrition goals outlined in Healthy People 2010 are designed to decrease health disparities among ethnic groups. Cultural groups that are part of the North American landscape can be role models for the way their traditional diets are designed. For instance, early Native Americans in the southwestern part of the United States thrived on corn, beans, and squash, the three "sister vegetables." They form a solid nutritional base.

Here are a few examples of dietary patterns from various cultures:

1. Asian diets tend to be high in carbohydrate and low in fat, especially in saturated fat. The diet includes many varieties of fish. Beans, including fermented soy products, supply protein, fiber, and antioxidants. Even Americanized Vietnamese restaurants serve many low-fat, vegetable-rich choices. Spring rolls with plum and chili sauce are a nonfried alternative to deep-fried egg rolls. Thai dishes are based on vegetables, noodles, and fish. Fish sauce, although salty, provides flavor and adds protein without fat. Peanut sauce and coconut milk are high in fat and flavor and are usually eaten in small amounts. Peanuts provide protein.

2. Indian menus offer many vegetarian choices and are low in saturated fats and high in fiber. Extensive use of spices provides flavor without added fat.

3. North African cuisine (for example, Amharic, Oromo, Ethiopian, Egyptian) relies heavily on beans and whole grains. Injera or buddeena, the flat breads that feel a bit like a pancake, hold dabs of lentils, split peas, lamb, beef, and yogurt. Sharing is customary. Each person pulls off a piece of the buddeena and uses it to pick up fillings such as peas and lentils (dal), beef, lamb, or vegetable mixtures. Picture the amount of filling you can pick up

Table 16.11 Dietary supplement and herbal remedies, proposed claims, potential side effects, and likely interactions

REMEDY	PROPOSED CLAIMS	POTENTIAL SIDE EFFECTS	POTENTIAL INTERACTION WITH PRESCRIPTION MEDICATIONS
Cholestin	Maintains desirable blood cholesterol levels	Safety of some ingredients unknown	
Creatine	Sport supplement (increased performance in short, high-intensity events); Inconclusive claims regarding use in congestive heart failure	Kidney disease; side effects possible and include vomiting and diarrhea	
DHEA	Improves physical well-being throughout aging	Increases risk of breast and endometrial cancer in women and prostate cancer in men	
Echinacea	Prevents and treats colds and sore throat; anti-inflammatory	Allergies to plant components in daisy family; affects oral cavity integrity; depresses immune system if taken longer than 6 weeks	Corticosteroids
Ephedra (ma huang)[a]	Promotes short-term weight loss	Insomnia, headaches, nervousness, stomach problems, heart palpitations, seizures, death	Corticosteroids, digoxin, monoamine oxidase inhibitors (MAOIs); oral hypoglycemics
Garlic	Lowers blood cholesterol; relieves colds and other infections, asthma	Heartburn, gas, blood thinner	Oral hypoglycemics, blood thinner[b]
Ginger	Calms stomach upset, fights nausea	Central nervous system depression, heart rhythm disturbances if using very large overdoses	Oral hypoglycemic medications; blood thinners
Ginkgo biloba	Increases mental skills, delays progression of Alzheimer's disease, increases blood flow, decreases depression	Nervousness, headache, stomachache; interacts with blood thinners	Thiazides; blood thinner (e.g., Warfarin); MAOIs
Ginseng	Increases energy, normalizes blood glucose, stimulates immune function, relieves impotence in males	Insomnia, hypertension, low blood glucose, menstrual dysfunction	Estrogens, insulin, oral hypoglycemics, blood thinner[b]; MAOIs
Glucosamine-chondroitin	Slows progression and symptoms of osteoarthritis and its pain	Gastrointestinal upset, fatigue, headache; allergic reactions in individuals with shellfish allergies	Blood thinner
Peppermint	Treats indigestion and flatulence, spasmolytic (relaxes muscles), antibacterial agent	Heartburn; allergic reactions	
Saint John's wort	Relieves depression	Dry mouth, dizziness; interacts with many drugs	Cyclosporin, digoxin, iron supplements, oral contraceptives, selective serotonin reuptake inhibitors; MAOIs
SAMe	Relieves mild depression, pain relief for arthritis	May trigger manic excitement, nausea	
Saw palmetto	Improves urine flow, reduces urgency of urination in men with prostate enlargement	Nausea, abdominal pain	Estrogens

[a]Shekelle et al., "Efficacy and Safety of Ephedra," 2003.
[b]Blood thinners include aspirin, Warfarin, coumarin. Blank boxes denote unknown.
SOURCES: Table compiled from P. G. Shekelle, M. L. Hardy, and S. C. Morton et al., "Efficacy and Safety of Ephedra and Ephedrine for Weight Loss and Athletic Performance: A Meta-Analysis," *J Am Med Assoc.* 289(12), 2003:1537–45; *AACE Guidelines.* Medical guidelines for the clinical use of dietary supplements and nutraceuticals. *Endocrine Prac.* 2003;9(5):418–70, Table 7; J. E. Brown, *Nutrition Now.* 3rd ed. Belmont, CA: West/Wadsworth Publishing Company, 2002, Table 24.4; S. Stupay and L. Sivertsen, "Herbal and Nutritional Supplement Use in the Elderly," *Nurs Pract.* 25(9), 2000:56–67.

with a piece of soft bread. You would be eating pyramid style—mainly grain complemented with a little animal protein.

4. Caribbean dishes are made with lots of potatoes and other vegetables such as spinach, with fruit, and with small portions of meat, often chicken.

5. Traditionally prepared Mexican food is based on corn, grains, beans, vegetables, and small amounts of meat and fat. Loads of cheese and sour cream on chips with some pickled jalapeño slices are an American restaurant invention.

When population groups migrate to another country, nutritional problems may arise because acculturation requires rebalancing traditional lifestyles. For instance, California and Minnesota saw a large influx of Hmong refugees, who had been displaced from their farms in Cambodia and Laos after the Vietnam War. Hmong (pronounced "mung") youths (but not their elders) quickly adopted American food habits; but families were no longer pursuing their old physically active lifestyles. Overweight and obesity have increased dramatically in Hmong youths since coming to America. In a study of eating habits of mostly first-generation Korean immigrants (average age 41 years; time in United States 15 years), Korean foods were consumed nearly twice as often as American foods. Consumption of Korean foods declined with adoption of American culture.[60]

Cross-Cultural Dietary Guidance

Several years ago, public health nutritionists Pat Splett and Karen Zeleznak worked with the Supplemental Nutrition Program for Women, Infants, and Children (WIC). Splett and Zeleznak thought about ways to teach health-promoting eating habits to pregnant and breastfeeding women, some of whom did not know English. So they asked potential clients about how they perceived food and what they might wish to learn about foods. People wanted to know, "What will this food do for me?" So these nutritionists developed a set of posters (see Illustration 16.3) that grouped foods according to function:

1. *Staple foods.* Grains, rice, cereals, and starchy roots provide energy and bulk to the diet. They are the filling basis of the diet and, barring bad luck, are readily available.

2. *Bodybuilding foods.* Beans, lentils, peas, nuts, seeds, eggs, milk, cheese, fish, poultry, and red meats are all high in protein and provide amino acids for tissue building and repair.

3. *Protective foods.* Vegetables and fruits contribute vitamins and minerals, antioxidants, and other phytochemicals to keep the individual healthy.

In these nutritionists' WIC program, grouping food according to function dealt with the milk/dairy dilemma faced by some Asian and African American audiences. Milk, cheese, calcium-fortified soymilk, or yogurt can all be placed in the protein or "bodybuilding" group. The "staple" group depicted grains and energy foods for several local cultures. In cross-cultural nutrition work, it is important to find out what the other person or group thinks and believes. Go ahead. Ask.

Vegetarian Diets

How many people consider themselves to be vegetarian? Among individuals aged 18 and older, 3% to 7% consider themselves "vegetarian."[61] Fewer than 1% of the population eats no animal products, roughly 6% eat no red meat,

Illustration 16.3 Foods grouped according to function for cross-cultural nutrition counseling.

Photos courtesy of General Mills, Inc.

Table 16.12 Classification of vegetarians by food groups consumed

Type of Vegetarian	Foods Not Eaten	Foods Eaten
Vegan	No animal foods of any kind, no honey	Grains, beans, nuts, seeds, nut butters, fruits, vegetables, sugar, molasses, oils, margarine, soda, alcohol, soy analogs (e.g., textured vegetable bacon, burgers, "meats")
Lacto-vegetarian	No meat, poultry, fish	Above plus milk and other dairy products, cheese, yogurt, butter
Lacto-ovo vegetarian	No meat, poultry, fish	Above plus eggs of any sort
Vegetarian	No beef, pork, lamb, venison, buffalo, other red meat; "No red meat"	Depends on individual interpretation: fish, both fin fish and shellfish; poultry and game birds may be allowed

4% never eat eggs, 4% never eat seafood, and 3% never eat poultry. One difficult thing about polling people about vegetarianism is defining just exactly what it means to be a vegetarian. Table 16.12 gives an overview.

Individuals can achieve a high-quality diet whether or not they eat meat. Can adopting a vegetarian diet make one healthier? Often people eating vegetarian diets are health conscious in other areas of their lives as well. So far, studies haven't consistently controlled for all lifestyle factors that potentially contribute to health. A well-chosen vegetarian diet is associated with decreased mortality and morbidity; diet is clearly a contributing factor.[62] In a relatively small study (80 women) that controlled for body composition, aerobic fitness, and psychological well-being, Nieman and colleagues found that except for zinc, eating or not eating meat makes no difference to immune status.[63] Higher intakes of zinc (3–25 grams) negatively affected immune status.

Potential health benefits of vegetarian diets outlined in the American Dietetic Association (ADA) Position Statement[62] include

● Disease-specific benefits

   Decreased risk of mortality and symptoms of heart disease

   Lower incidence of hypertension

   Decreased risk of some colorectal cancer; in some cultures, also less breast cancer

● General diet quality

   Higher intake of vegetables and fruits improves overall nutritional risk picture

   Diet tends to be lower in saturated fats and may also be lower in calories

● Environmental benefits

   Vegetable foods are lower on the food chain than meat, fish, and poultry

   Depending on packaging, transportation, food production uses fewer resources

Nutrients singled out for evaluation in vegetarian diets include protein, vitamins B_{12} and D, calcium, iron, and linolenic acid.

1. *Protein.* Eating a variety of amino acids throughout the day ensures amino acid balance. Eating complementary proteins at each meal is not necessary.[64] Protein intake of vegetarians appears to be adequate despite lower overall protein quality and intake.[63] See Table 16.13 for nutrient content of several high-protein foods from the Food Guide Pyramid.

2. *Vitamin B_{12}.* Plant foods are not reliable sources of vitamin B_{12}; most of the B_{12} in sea vegetables and fermented soy is an inactive analog. B_{12} is found in fortified ready-to-eat breakfast cereals, soymilks, nutritional yeast, and meat analogs. (Check the label, they vary!)

3. *Vitamin D.* One good source of D is sunshine (all year in the south; during spring, summer, and fall in northern United States and Canada). Vitamin D formation is blocked by sunscreen. Another source is Vitamin D–fortified milk, providing 2.5 micro-

Table 16.13 Examples of nonmeat protein foods and serving sizes used in the Food Guide Pyramid

Protein Food Serving	Protein Content grams	Total fat, grams	kcal
Almonds, 2 tb whole	3.5	9.3	105
Peanuts, 2 tb dry roasted	4.8	9.0	106
Peanut butter, 2 tb	8.1	16.5	191
Sesame seeds, 2 tb	4.2	8.8	94
Egg, 1 large	6.2	5.0	74
Kidney beans, 0.5 c cooked	7.7	0.4	112
Soymilk, 1 c	6.6	4.6	79
Lentils, 0.5 c cooked	8.9	0.4	115
Tofu, firm and soft, 0.25 c	5.7	3.4	54
Cheese, 1.5 oz	10.6	14.1	171

Case Study 16.1

Run, Kristen, Run

Kristen, who was active in competitive sports throughout high school, has decided to run a marathon with some of her college friends. She is 25 years old, 5 ft 8 in. tall, and weighs 135 pounds. She eats all sorts of foods, likes fruits and vegetables, but tries to avoid greasy foods. She doesn't like sweets, although she keeps ice cream in her freezer. A family history notes that her mother needed angioplasty to treat occluded arteries shortly after menopause and that her father is not at risk for any chronic conditions. Although she would eventually like to have children, Kristen is not pregnant now.

Photo Disc

An analysis of a 24-hour dietary recall shows the following:

2090 kcal	34 mg iron
377 gm carbohydrate (41 gm total fiber)	158 mg vitamin C
98 gm protein	213 IU of vitamin D
33 gm fat (7 gm saturated fat, 1 gm trans fat,	35 IU of vitamin E
1.5 gm omega-3 fatty acid, 99 mg cholesterol)	1548 mcg of folic acid
3343 mg sodium	6 mg of pantothenic acid
958 gm calcium	

Questions

1. How many calories does Kristen need to maintain her weight? Is she eating enough to support daily workouts?
2. Describe three health-promoting aspects of Kristen's diet.
3. Make three suggestions that could improve Kristen's diet.
4. What types of performance-enhancing products might complement this diet?

(Answers are in Appendix D.)

---

grams (100 IU) per cup of the 10 micrograms (400 IU) needed per day. Fortified cereals, soymilk, dairy milk, and vitamin tablets are reliable sources of vitamin D.

4. *Calcium.* Absorption varies among foods. The calcium content in vegetables such as spinach and fruits such as rhubarb looks impressive in a food table, but doesn't reach the bones to the extent that calcium from dairy sources such as yogurt or cheese does. Some dark leafy greens such as collards are low in oxalic acid so that the calcium can be absorbed. No general rule fits calcium absorption. Fortunately, vegetarians do not tend to have excessive protein intakes. If they also keep sodium intakes low, their body needs less calcium.

5. *Iron.* Enriched flour and breads, beans, and fortified cereals, especially when eaten in the presence of vitamin C and not washed down with coffee or tea, can provide all the iron a healthy adult needs. An example of how to reach the 10 milligrams of iron needed daily by all adults except premenopausal women is to

include two slices whole wheat bread (1.8 gm), 1 cup tomato juice (1.3 gm), 1 cup cooked chard (3.8 gm), 0.5 cup kidney beans (1.5 gm), and 2 tablespoons of pumpkin seeds (2.5 gm) in the day's intake.

6. *Fats.* For vegetarians who consume eggs and dairy, chosing low-fat dairy products will help to keep the saturated fat within dietary guidelines.[62] Most plant foods do not contain significant amounts of omega-3 (alpha-linolenic) fatty acids as fish and eggs do. However, good plant sources of omega-3 fatty acids are ground flaxseed and *linseed* oil, walnuts, walnut oil, and canola and soybean oils.[62]

The American Dietetic Association suggests that a health-promoting vegetarian diet include a variety of foods from each basic food group. Whole, unrefined foods are more nutrient dense than highly processed ones, containing phytochemicals that were not lost in processing and about

> **Linseed** From the flax plant, *linum*; *linseed* is another name for flaxseed. Linseed oil is used in paints, varnishes, and inks but is also produced in food form for its rich nutrient content.

which we know too little to replace; for example, red grapes contain resveratrol, a compound implicated in heart health and in reducing cancer risk. Real grapes contain resveratrol, but manufactured imitation grape beverages do not.

Physical Activity Recommendations

"I could have danced all night!"

Who can improve on the following paragraph from Healthy People 2010?

> Because the highest risk of death and disability is found among those who do no regular physical activity, engaging in any amount of physical activity is preferable to none. Physical activity should be encouraged as part of a daily routine. While moderate physical activity for at least 30 minutes is preferable, intermittent physical activity also increases caloric expenditure and may be important for those who cannot fit 30 minutes of sustained activity into their daily schedules.

In addition, resistance training (at least 30 minutes per day, two to three times per week) develops muscle mass, strength, and endurance.[65] See Table 16.14 for current activity goals for adults.

"Shall we dance?"

Nutrition Intervention for Risk Reduction

A Model Health-Promotion Program

Nutrition intervention occurs not only by providing counseling and education to help improve nutritional status; it occurs through the provision of food. Food pantries, community kitchens, and the 5 a Day for Better Health Campaign are all examples of nutrition interventions that mobilize individuals and communities to improve nutritional status.

An example of a health-promotion program for adults illustrating some of the concepts of nutrition intervention is the Health Partners Better Health Restaurant Challenge. Health Partners is a large midwestern health maintenance organization (HMO).

- *Health goal:* Improve nutritional health by enabling HMO members to choose tasty, health-promoting, low-fat foods when eating away from home.
- *Target audience:* Health Partners sees its mission as serving the whole community, so the entire metropolitan community was eligible to participate in the program.
- *Theoretical basis:* Health behavior change depends on social support as well as on individual goals and commitments. Health Partners wanted to build social networks that enabled healthy choices.
- *Intervention strategies:* Participating restaurants (sit-down and fast-food) were challenged to feature health-promoting items on the menu. Diners voted on favorite items, and the HMO compiled and publicized the restaurants and the winning items. The HMO provides additional publicity for restaurants in its publications and through local media advertising, offering discount coupons for members to visit participating restaurants.
- *Evaluation:* Member satisfaction increased program participation from year to year, and a growing list of restaurants wished to join the program. Member surveys addressed the nutritional goal of enabling access to tasty, health-promoting menu items by asking about restaurant ordering patterns and satisfaction with featured menu items. Diner surveys revealed that 90% of diners would order low-fat items again, and 50% thought low-fat items tasted better than expected.

Public Food and Nutrition Programs

One of the U.S. national health goals is to minimize health disparities.[66] Living in poverty is linked to poor diets and to adverse health outcomes such as increased rates of obesity.[66,67] The Census Bureau has determined that approximately 10% of adult U.S. residents aged 18 to 64 live in poverty. The poverty threshold is derived from calculating the cost of foods needed for basic dietary requirements (according to the Thrifty Food Plan) and multiplying that cost by three. This poverty index was developed in 1963 by Mollie Orshansky of the Social Security Administration; at that time, food costs made up about one-third of the average household budget. For 2003, the poverty threshold was $9,573 for an adult under age 65 and

Table 16.14 Healthy People 2010 physical activity goals and baseline activity levels

	BASELINE LEVELS		TARGET
	Age 18 and older	Age 25–44	All adults
Engage in no leisure time physical activity	40%	34%	20%
Engage in regular physical activity at least 20 minutes, three times per week	15%	15%	30%
Perform strengthening activities two or more days per week	19%	22%	30%
Perform stretching and flexibility activities	30%	34%	40%

$18,979 for a family of four. Percent of after-tax income spent on food by households ranges from 9–34% depending on household income, with an average of 14% in 1996. Households living in poverty spend the larger percentage of income on food.

Poverty guidelines are a simplified version of the poverty threshold, used for administrative purposes rather than statistical measures. The guidelines are published annually in the Federal Register by the Department of Health and Human Services. In 2003, a person with an annual income of $8,980 or less lived in poverty.[68] Poverty guidelines are determined for various household sizes and adjusted for higher costs in Alaska and Hawaii. For a four-person household, the poverty guideline is $18,400 in the contiguous states and District of Columbia, $21,160 in Alaska, and $23,000 in Hawaii (http://aspe.gov/poverty/03poverty.htm). Food-support programs, such as the Food Stamp Program, use poverty guidelines to establish eligibility. Benefit levels can be adjusted to meet varied household needs. For instance, The Child Nutrition Program (school lunch and breakfast), which can help whole families stretch their food budget, provides free lunches to children living at 130% or less of poverty. The program provides reduced-cost meals when family income is between 130% and 185% of poverty. Families must apply for these programs.

Many other programs work to help hungry individuals and families gain food security. For example:

- governmental extension programs teach budgeting, shopping, and food-handling skills;
- the Second Harvest food bank keeps food items out of the waste stream, coordinates charitable giving programs, and supplies food shelves and community kitchens;
- soup kitchens and shelters provide hot meals and snacks for hungry and homeless people; and
- Meals-on-Wheels programs serve homebound adults.

Together, governmental and private organizations help individuals gain consistent access to safe, wholesome foods that are culturally acceptable. Such access is the basis of food security, which is one of the national health goals (see Table 16.1).

Nutrition and Health Promotion

What are the most powerful nutritional habits that we can develop? The general message is to follow the principles of variety and moderation in choosing a diet that will achieve and maintain a healthy body weight when combined with activity. More specifically, nutrition educators are developing scoring systems to quantify that general message.[69] The Diet Quality Index Revised (DQIR) is an example of a scoring system that reflects key nutritional habits for promoting health and reducing overall disease risk. DQIR system designers recognized that no single magic bullet (e.g., "eat less fat") will make us all well. Instead, the system is designed to help track the many aspects that contribute to diet quality. The following 10 nutritional habits (each scored at 10 points) were used to measure diet quality before the Institute of Medicine published newer Acceptable Macronutrient Distribution Ranges:

1. Total fat = < 30% of calories (now 20 to 35%)
2. Saturated fat = < 10% of energy intake (now low in saturated fat, trans fat, and cholesterol)
3. Dietary cholesterol < 300 mg daily (on average; moving to greater emphasis on omega-6 and omega-3 fats ratio)
4. 2–4 servings fruit per day
5. 3–5 servings vegetables per day
6. 6–11 servings grain per day (especially whole grains)
7. Calcium intake to meet age and gender recommendations
8. Iron intake to meet age and gender recommendations
9. Dietary diversity score: one-half or more servings of each of 23 food component groupings including seven grain groups, seven vegetable groups, two fruit groups, and seven meat/dairy groups during a 2-day period
10. Dietary moderation score composed of four items; teaspoons of added sugar with 1 teaspoon = 4 grams carbohydrate (12 ounce cola = 38 grams carbohydrate, thus contributing 9.5 teaspoons of sugar); discretionary fat grams (25 grams or less per day); sodium intake below 2400 milligrams per day; alcohol in moderation.

Knowing does not automatically lead to doing. To make health-promoting behavior changes takes knowledge of desired behaviors, skills to practice them, belief in one's own ability to carry them out, and the intention or commitment to do so.

Resources

American Dietetic Association

Provides information, advice, and links.

Web site: **www.eatright.org**

American Heart Association

Site includes assessments, eating advice, and definitions.

Web site: **www.americanheart.org**

Calorie Control Council

Lists overall diet and fitness information. (Don't look for this site to sponsor the health-at-any-size movement.)

Web site: **www.caloriecontrol.org**

Centers for Disease Control and Prevention

The main site also includes morbidity and mortality data at www.cdc.gov/mmwr.

Web site: **www.cdc.gov**

Consumer Lab

Check out vitamin and mineral supplements. (Also use www.fda.gov.)

Web site: **www.consumerlab.com**

Environmental Protection Agency

Among other things, ratings for local water supplies are available.

Web site: **www.epa.gov/safewater**

Food and Drug Administration

Resource for labeling information, health claims on foods and supplements, fortification rules, and more.

Web site: **www.nutrition.gov**

U.S. government's gateway site for nutrition information.

Web site: **www.fda.gov**

U.S. Census Bureau

The site to visit to discover who we are and where we live.

Web site: **www.census.gov/population/estimats/ nation/intfil2-1.txt**

U.S. Department of Agriculture

Wonder about your diet? This site includes a diet self-assessment. Follow the HEI icon to USDA's interactive Healthy Eating Index. Type in what you've eaten and see how your intake stacks up against the Food Pyramid recommendations. For another view, go to www.cyberdiet.com.

Web site: **www.usda.gov/cnpp**

Vegetarian Resource Group

Provides information and recipes.

Web site: **www.vrg.org**

References

1. Mokdad AH, Marks JS, Stroup DF, Gerberding JL. Actual causes of death in the United States, 2000. JAMA 2004; 291: 1238–1245.

2. Healthy People 2000: National Health Promotion and Disease Prevention Objectives. DHHS Publication No. (PHS) 91-50212. Washington, DC: U.S. Department of Health and Human Services; 1991.

3. Kant AK. Consumption of energy-dense, nutrient-poor foods by adult Americans: nutritional and health implications. The Third National Health and Nutrition Examination Survey, 1988–1994. Am J Clin Nutr 2000;72:929–36.

4. Trumbo P, Yates AA, Schlicker S, Poos M. Dietary reference intakes: vitamin A, vitamin K, arsenic, boron, chromium, copper, iodine, iron, manganese, molybdenum, nickel, silicon, vanadium, and zinc. J Am Diet Assoc 2001;101:294–301.

5. Crawford SL, Casey VA, Avis NE, McKinlay SM. A longitudinal study of weight and the menopause transition: results from the Massachusetts Women's Health Study. Menopause 2000;7:96–104.

6. Forbes GB. Human body composition. Growth, aging, nutrition, and activity. New York: Springer-Verlag; 1987: 31.

7. Reid IR. Pharmacological management of osteoporosis in postmenopausal women: a comparative review. Drugs and Aging 1999;15:349–63.

8. Col NF, Pauker SG, Goldberg RJ et al. Individualizing therapy to prevent long-term consequences of estrogen deficiency in post-menopausal women. Arch Intern Med 1999;159:1458–66.

9. Dennerstein L, Dudley E, Guthrie J, Barrett-Connor E. Life satisfaction, symptoms, and the menopausal transition. Available at www.medscape.com/Medscape/ WomensHealth/journal/2000/v05.n04/ wh7254.denn-01.html, accessed 1/01.

10. Kronenberg F, Fugh-Berman A. Complementary and alternative medicine for menopausal symptoms: a review of randomized, controlled trials. Ann Intern Med 2002;137(10):805–13.

11. Barnes S. Phytoestrogens and breast cancer. Baillieres Clin Endocrinol Metab 1998; 12:559–79.

12. Messina MJ. Legumes and soybeans: overview of their nutritional profiles and health effects. Am J Clin Nutr 1999;70: 439S–50S.

13. The role of isoflavones in menopausal health: consensus opinion of the North American Menopause Society. Menopause 2000;7:215–29. Available at www .menopause.org, accessed 1/14/01.

14. Wiseman H, O'Reilly JD, Adlercreutz H et al. Isoflavone phytoestrogens consumed in soy decrease F(2)-isoprostane concentrations and increase resistance of low-density lipoprotein to oxidation in humans. Am J Clin Nutr 2000;72:395–400.

15. Schryver T. Soy update 2000. Today's Dietitian 2000;2:2–7.

16. Fragakis AS. The health professional's guide to popular dietary supplements. 2nd ed. Chicago: American Dietetic Association; 2003.

17. Institute of Medicine of the National Academies. Dietary Reference Intakes: energy, carbohydrate, fiber, fat, fatty acids, cholesterol, protein, and amino acids. Washington, DC: National Academies Press; 2002.

18. Foster S, Tyler VE. Tyler's honest herbal. New York: Haworth Herbal Press; 1999.

19. Willett WC. Diet, nutrition, and the prevention of cancer. In: Shils ME et al., eds. Modern nutrition in health and disease. Baltimore, MD: Williams and Wilkins; 1999: 1250.

20. Harper EJ. Changing perspectives on aging and energy requirements: aging and energy intakes in humans, dogs, and cats. J Nutr 1998;128:2623S–26S.

21. McGandy RB, Barrows CH, Spanias A et al. Nutrient intakes and energy expenditure in men of different ages. J Gerontol 1966;21:581–7.

22. Poehlman ET, Toth MJ, Bunyard LB et al. Physiological predictors of increasing total and central adiposity in aging men and women. Arch Intern Med 1995;155:2443–8.

23. Guo SS, Zeller C, Chumlea WC, Siervogel RM. Aging, body composition, and lifestyle: the Fels Longitudinal Study. Am J Clin Nutr 1999;70:405–11.

24. Food and Nutrition Board. National Academy of Sciences (Institute of Medicine)

NRC, Subcommittee on the 10th edition of the RDAs, Commission on Life Sciences. Recommended dietary allowances. Washington, DC: National Academy Press; 1989.

25. Harris JA, Benedict FG. Standard basal metabolism constants for physiologists and clinicians: a biometric study of basal metabolism in man. Washington, DC: Carnegie Institute of Washington; 1919.

26. Frankenfield D, Muth ER, Rowe WA. The Harris-Benedict studies of human basal metabolism: history and limitations. J Am Diet Assoc 1998;98:439–45.

27. Mifflin MC, St. Jeor ST, Hill LA, Scott BJ, Daugherty SA, Koh YO. A new predictive equation for resting energy expenditure in healthy individuals. Am J Clin Nutr 1990;51(2):241–47.

28. Owen OE, Kavle E, Owen RS, Polansky M et al. A reappraisal of the caloric requirements in healthy women. Am J Clin Nutr 1986;44:1–19.

29. Owen OE, Holup JL, D'Allesio DA et al. A reappraisal of the caloric requirements in healthy women. Am J Clin Nutr 1987; 46:875–85.

30. American Dietetic Association. Let the evidence speak: indirect calorimetry and weight-management guides. Presentation given at the Food & Nutrition Conference & Exhibition, San Antonio, TX. October 27, 2003.

31. Miller MA. A calculated method for determination of ideal body weight. Nutr Support Serv 1985;5:31–3.

32. Roth JH. What is optimum body weight? J Am Diet Assoc 1995;95:856–7.

33. Committee on Diet and Health (Food and Nutrition Board; National Research Council). Diet and health: implications for reducing chronic disease risk. Washington, DC: National Academy Press; 1989.

34. McNamara DJ. Dietary cholesterol and the optimal diet for reducing risk of atherosclerosis. Can J Cardiol 1995;11 Suppl G:123G–6G.

35. Yates AA, Schlicker SA, Suitor CW. Dietary Reference Intakes: the new basis for recommendations for calcium and related nutrients, B vitamins, and choline. J Am Diet Assoc 1998;98:699–706.

36. Welch S. Nutrient standards, dietary guidelines, and food guides. In: Ziegler EE, Filer LJ, eds. Present knowledge in nutrition. Washington, DC: ILSI Press; 1996: 630–46.

37. Health and Welfare Canada. Nutrition recommendations: the report of the Scientific Review Committee. Canada: Minister of Supply Services; 1990: 6.

38. McDowell MA, Briefel RR, Alaimo K et al. Energy and macronutrient intakes of persons ages two months and over in the United States: Third National Health and Nutrition Examination Survey, Phase 1, 1988–1991. Adv Data 1994:1–24.

39. National Center for Health Statistics. Health, United States, 2000. Washington, DC: U.S. Dept of Health and Human Services, 2000. Available at www.cdc.gov/nchs, accessed 12/11/00.

40. Parker DR, McPhillips JB, Derby CA et al. High-density-lipoprotein cholesterol and types of alcoholic beverages consumed among men and women. Am J Public Health 1996;86:1022–7.

41. Windham CT, Wyse BW, Hansen RG. Alcohol consumption and nutrient density of diets in the Nationwide Food Consumption Survey. J Am Diet Assoc 1983;82:364–70, 373.

42. Zeman, FJ. Clinical nutrition and dietetics. 2nd ed. New York: Macmillan; 1991.

43. Hodgkinson B, Evans D, Wood J. Maintaining oral hydration in older people. A systematic review. York, UK: NHS Centre for Review and Dissemination; 2001: 1–66.

44. Coffee break or caffeine fix? Environmental Nutrition 2000;23:1,6.

45. Grandjean AC, Reimers KJ, Bannick KE, Haven MC. The effect of caffeinated, noncaffeinated, caloric, and non-caloric beverages on hydration. J Am Coll Nutr 2000; 19:591–600.

46. Nutrition and your health: dietary guidelines for Americans, 5th ed. Washington, DC: U.S. Department of Agriculture, U.S. Department of Health and Human Services; 2000.

47. Pennington JA, Hubbard VS. Derivation of daily values used for nutrition labeling. J Am Diet Assoc 1997;97:1407–12.

48. Wright JD, Wang CY, Kennedy-Stephenson J, Ervin RB. Dietary intake of ten key nutrients for public health, United States: 1999–2000. Advance data from vital and health statistics (no. 334). Hyattsville, MD: National Center for Health Statistics; 2003.

49. Wilson JW, Enns CW, Goldman KS et al. Data tables: combined results from USDA's 1994 and 1995 Continuing Survey of Food Intakes by Individuals and 1994 and 1995 Diet and Health Knowledge Survey. ARS Food Surveys Research Group. June 2, 1997. Available at www.barc.USDA.gov/bhnrc/foodsurvey/home.htm.

50. Appel LJ, Moore TJ, Obarzanek E et al. A clinical trial of the effects of dietary patterns on blood pressure. DASH Collaborative Research Group. N Eng J Med 1997; 336:1117–24.

51. Krall EA, Dawson-Hughes B. Osteoporosis. In: Shils ME, Olson JA, Shike M, Ross AC, eds. Modern nutrition in health and disease. Philadelphia: Lippincott, Williams & Wilkins; 1999: 1353–64.

52. National Institutes of Health. Optimal calcium intake. Consensus Statement. 1994; 12:1–31.

53. Feskanich D, Willett WC, Stampfer MJ, Colditz GA. Protein consumption and bone fractures in women. Am J Epidemiol 1996; 143:472–9.

54. Nielsen SM, Popkin BM. Patterns and trends in food portion sizes, 1977–1998. J Am Med Assoc 2003;289(4):450–3.

55. National Restaurant Association. Restaurant industry pocket factbook; 2000.

56. Jeffery RW, French SA. Epidemic obesity in the United States: are fast foods and television viewing contributing? Am J Public Health 1998;88:277–80.

57. Bogden JD, Bendich A, Kemp FW et al. Daily micronutrient supplements enhance delayed-hypersensitivity skin test responses in older people. Am J Clin Nutr 1994;60: 437–47.

58. Chandra RK. Effect of vitamin and trace-element supplementation on immune responses and infection in elderly subjects. Lancet 1992;340:1124–7.

59. Balluz LS, Kieszak SM, Philen RM, Mulinare J. Vitamin and mineral supplement use in the United States. Results from the Third National Health and Nutrition Examination Survey. Arch Fam Med 2000; 9:258–62.

60. Lee S-K, Sobal J, Frongillo EA. Acculturation, food consumption, and diet-related factors among Korean Americans. J Nutr Ed 1999;31:321–30.

61. Vegetarian Resource Group. Available at www.vrg.org, accessed 11/22/00.

62. Messina VK, Burke KI. Position of the American Dietetic Association: Vegetarian diets. J Am Diet Assoc 1997;97:1317–21.

63. Nieman DC, Butterworth DE, Henson DA et al. Animal product intake and immune function. Vegetarian Nutr 1997; 1:5–11.

64. Young VR, Pellett PL. Plant proteins in relation to human protein and amino acid nutrition. Am J Clin Nutr 1994;59: 1203S–12S.

65. Pollock ML, Franklin BA, Balady GJ et al. AHA Science Advisory. Resistance exercise in individuals with and without cardiovascular disease: benefits, rationale, safety, and prescription: an advisory from the Committee on Exercise, Rehabilitation, and Prevention, Council on Clinical Cardiology, American Heart Association. Position paper endorsed by the American College of Sports Medicine. Circulation 2000;101:828–33.

66. Healthy People 2010: national health promotion and disease prevention objectives. Washington, DC: U.S. Department of Health and Human Services; 2000.

67. Kamimoto LA, Easton AN, Maurice E et al. Surveillance for five health risks among older adults—United States, 1993–1997. Mor Mortal Wkly Rep CDC Surveill Summ 1999;48:89–130.

68. U.S. Department of Health and Human Services. The 2001 HHS poverty guidelines. Federal Register February 16, 2001;66: 10695-7. Available at http://aspe.hhs.gov/poverty/01poverty.htm, accessed 11/06/03.

69. Haines PS, Siega-Riz AM, Popkin BM. The Diet Quality Index revised: a measurement instrument for populations. J Am Diet Assoc 1999;99:697–704.

"Don't tell me what I can't eat. Tell me about the health value of food."
Margaret Meter

Chapter 17

Adult Nutrition:
Conditions and Interventions

Chapter Outline

- Introduction
- Cancer
- Cardiovascular Diseases: Coronary Heart Disease
- Overweight and Obesity
- Diabetes Mellitus
- HIV/AIDS

Prepared by **U. Beate Krinke**

Key Nutrition Concepts

1 Dietary intake, body weight and composition, and physical activity influence changes in health status with age.

2 Optimal nutrition interventions for nutrition-related diseases and disorders rely on good information and a support system for follow-through.

3 Nutritional care practiced in the context of wellness is dynamic; it is evaluated by changing relationships as well as absolute values of health measures.

4 Foods often assume extra signifance when a person is ill.

5 Relevant diet counseling considers the many roles of foods in life.

Introduction

Currently, more people in the United States die of lifestyle-related diseases and disorders than from any other cause. Diseases that develop during the adult years may fully or partially result from the cumulative effects of diets high in saturated fats and low in vegetables, fruits, and fiber; smoking, physical inactivity, and other lifestyle factors insidiously influence health on a day-to-day basis. For example, four of the five leading causes of death (cancer, heart disease, stroke, and diabetes) of 45- to 64-year-old adults in the United States are related to diet and other lifestyle behaviors.[1,2] Because they are related to lifestyles, these diseases can be prevented, in part, by healthful changes in diet. One risk factor for lifestyle diseases is obesity, which some people in the health care field also consider to be a disease. This chapter addresses a number of nutrition-related diseases and disorders that affect adults.

Cancer

Adults are improving their lifestyles related to cancer: smoking rates, alcohol consumption, and dietary fat intake (as a percentage of total calories) have all fallen.[3] Because cancer prevalence increases with age, additional information on cancer is presented in Chapter 19.

Definition

Cancer is defined as a group of diseases in which abnormal cellular growth affects specific organ systems such as the skin (melanoma), lungs, reproductive organs, or the gastrointestinal system (stomach or colon cancer).

Prevalence/Etiology

The four most common cancers in adults are prostate, breast, colorectal, and lung cancers. For women aged 50–64 years, a slight increase in breast cancer has occurred in recent years.[3] Blacks have the highest rate of new cancers, and American Indians/Alaska Natives have the lowest. Social, cultural, behavioral, and environmental factors affect the initiation and progression of cancer.

Risk Factors

Although smoking is the most-recognized contributor to cancer occurrence (estimates are that smoking causes about 30% of all U.S. deaths from cancer), obesity, low intakes of fruits and vegetables, and excess alcohol consumption are the main nutrition-related risks.

Nutrition Interventions for Cancer

Cancer prevention depends, in part, on minimizing nutritional risks:[3,4]

- Maintaining or reaching a healthy weight
- Eating 5 or more servings of fruits and vegetables daily
- Eating a low-fat diet (less than 30% of kcalories from fat)
- Consuming alcohol in moderation, if at all

Research on how fats and types of fat alter cancer risk continues. What is needed is a reliable and precise test that accurately measures the influences of fat on cancer risk in the body.

After a diagnosis of cancer, good nutrition supports treatment strategies (surgery, chemotherapy, radiation) or palliative care in terminal patients. Health care professionals provide guidance on dealing with weight loss, taste aversions, types of foods likely to be tolerated for dry or sore mouth and throat, treating constipation and diarrhea, and nutritional therapy to maintain overall stamina. The American Cancer Society has synthesized evidence-based guidance on nutrition during and after cancer, such as suggesting nutrient-dense supplements to provide adequate calories and protein for recovery.[5] In some cases, complementary treatments can supplement or enhance traditional cancer therapies.

Alternative Medicine and Cancer Treatment

The National Cancer Institute maintains a Physicians Database Query (PDQ), with information available at http://cancer.gov/cancerinfo/pdq/cancerdatabase. This database provides up-to-date information regarding complementary and alternative medicine summaries for many purported anticancer remedies. Specific details about treatment options for adult cancers can be found at http://cancer-gov/cancerinfo/pdq/adulttreatment. An example of the types of information included in a PDQ summary is given for cartilage (bovine or shark) in Illustration 17.1. Good information helps cancer patients and their families to balance the benefits and risks of various therapies.

Illustration 17.1 Example of PDQ summary.

CARTILAGE (Bovine and Shark)

Date first reported to have an *angiogenesis inhibitor* tendency: 1980

How sold, and number of brands sold in the United States: As a dietary supplement; there are 40 different brand names. No FDA approval is needed for dietary supplements unless specific disease prevention or treatment claims are made.

Number of human studies using cartilage products: 9 clinical studies; three studies have been published in peer-reviewed, scientific journals.

Doses and cartilage treatment duration: Varies because of the many cartilage products available and the wide array of administration methods (i.e., topical, oral, enema, or as subcutaneous injection).

Adverse effects: taste changes, nausea, heartburn, fever, fatigue, weight loss

Overall level of evidence for cartilage: Reported data have been from controlled and uncontrolled research trials; they are inconclusive.

Recommendation: Use of cartilage (bovine or shark) as a treatment for cancer cannot be recommended outside the context of well-designed clinical trials.

SOURCE: National Cancer Institute Web site. Available at www.Cancer.gov. Link to Cancer Info and PDQ. Final Web site location is http://cancer.gov/cancerinfo/pdq/cam/cartilage, accessed 12/7/03.

Cardiovascular Diseases: Coronary Heart Disease

Cardiovascular diseases (CVD) consist of coronary heart disease (CHD, often called "heart disease"), *stroke,* and other diseases and disorders related to the heart, blood vessels, and circulation. CHD is the number one cause of death in men and women in the United States. Table 17.1 lists nutrition-related health objectives for the nation for decreasing rates of heart disease as well as stroke.

CHD and stroke are characterized by atherosclerosis, or hardening of the *arteries* due to plaque buildup in the walls of arteries. Atherosclerosis occuring in blood vessels of the heart is related to CHD, whereas atherosclerosis in the cerebral artery of the brain is related to stroke. Atherosclerosis narrows arteries, increasing the risk that a blood clot will form, shut off blood flow, and cause a heart attack (myocardial infarction) or stroke.

Mortality rates from CHD and stroke fell by 24% and 10%, respectively, in 2000.[1] Still, the American Heart Association (AHA) estimates that close to 1 million deaths occur annually due to heart disease.[6] Heart disease is the third leading cause of death in 25- to 44-year-olds, the second in 45- to 64-year-olds, and the first in adults aged 65 years and older. Table 17.2 on the next page shows the annual number of deaths from heart disease by ethnic group. The prevalence of CVD in American adults increases with age, from 30% for ages 45–54 to about 50% for ages 55–64.

Table 17.1 Healthy People 2010 nutrition-related health objectives to reduce heart disease and stroke among adults

	Percentage of All Adults	
	Target	Baseline
Increase the proportion of adults with high blood pressure whose blood pressure is under control.	50%	18%
Increase the proportion of adults with high blood pressure who are taking action (for example, losing weight or reducing sodium intake) to help control their blood pressure.	95%	72%
Reduce the mean total blood cholesterol levels among adults.	199 mg/dL	206 mg/dL
Reduce the proportion of adults with high total blood cholesterol levels (240 mg/dL or greater).	17%	21%
Eat 2400 mg or less of sodium daily.	65%	21%

SOURCE: *Healthy People 2010: National Health Promotion and Disease Prevention Objectives.* Washington, DC: U.S. Department of Health and Human Services, 2000: Sections 12 and 19.

Etiology of CHD

Atherosclerosis begins when nutrient substances such as cholesterol, fatty acids, and calcium become part of tissues that form over injured arterial wall cells. Blood lipids, such as LDL cholesterol and fatty acids, become incorporated into plaque over time, increasing the extent of atherosclerosis.

Atherosclerosis is a multifactorial disease that develops in the following way:

Angiogenesis Inhibitor Angiogenesis is the formation of new blood vessels. An angiogenesis inhibitor slows or stops vessel formation. Tumors cannot grow or expand without additional blood vessels to carry oxygen and other nutrients.

Stroke The event that occurs when a blood vessel in the brain becomes occluded due to a clot or ruptures cutting off blood supply to a portion of the brain. Also called a cerebral vascular accident.

Arteries Blood vessels that carry oxygenated blood to cells.

- Fatty streaks in the smooth muscle of the artery wall
- Vessel thickening as result of endothelial cell damage
- Formation of fibrous plaques (deposits of fats, cholesterol, collagen, muscle, and other cells and metabolites) inside the artery walls at the point of injury or lipid accumulation
- Calcification of fibrous plaques into lesions that eventually deteriorate, become infected, or in general weaken the artery wall

High blood levels of homocysteine, inflammatory diseases, abnormal blood clotting factors, and other conditions also

Table 17.2 Deaths from heart disease and stroke compared to national goals

NATIONAL GOALS TO REDUCE MORTALITY

	Per 100,000 Population	
	Target	Baseline, 1997
Reduce coronary heart		
disease deaths	166	216
American Indian or		
Alaska Native	166	134
Asian or Pacific Islander	166	125
Black or African American	166	257
Hispanic or Latino	166	151
White	166	214
Reduce stroke deaths	48	60
American Indian or		
Alaska Native	48	39
Asian or Pacific Islander	48	55
Black or African American	48	82
Hispanic or Latino	48	40
White	48	60

SOURCE: *Healthy People 2010: National Health Promotion and Disease Prevention Objectives*. Washington, DC: U.S. Department of Health and Human Services, 2000: Tables 12-15 and 12-18.

influence the development of atherosclerosis.[7–9] The progression of atherosclerosis can be slowed, neutralized, or partially reversed by dietary and lifestyle modifications.

Effects of CHD

The effects of CHD can range from shortness of breath after exertion to chest pain (angina) and death from a heart attack. Atherosclerosis decreases blood circulation to the heart, resulting in decreased energy, decline in organ function, and the inability to perform activities of daily living. Buildup of plaque and lesions inside the blood vessels leave less room for blood flow. Consequently, the heart has to work harder to pump blood through this narrower space to reach all parts of the body, leading to higher blood pressure levels. An analysis of data from more than 12,000 middle-aged men in the Seven Countries Study found that increases in blood pressure (diastolic and systolic) led to increased heart disease, just as higher levels of blood cholesterol can lead to atherosclerosis.[10]

Risk Factors for CHD

Risk factors for CHD include the following:[11–17]

- Excess weight for height; for example, BMI >30kg/m^2
- Excess abdominal fat: waist circumference of 35 inches or more for females; 40 inches or more for males
- Blood pressure above 140/90 mmHg

- High-density lipoprotein (HDL) cholesterol levels less than 40 mg/dL, especially in women
- Low-density lipoprotein (LDL) levels of 130–159 mg/dL (borderline high), 160–189 (high), 190+ (very high)
- Total cholesterol levels of 200–240 mg/dL (moderate risk) or 240 mg/dL or more (high risk)
- Elevated blood triglyceride (TG) of >150 mg/dL
- Elevated plasma apolipoprotein B (an atherogenic lipoprotein)
- Diabetes, especially if uncontrolled; elevated fasting plasma insulin levels
- High saturated fats and trans fatty acid intake
- Consumption of few vegetables and fruits
- Consumption of few whole grain products
- Inadequate folate intake; high blood homocysteine levels
- Infrequent intake of fish (low omega-3 fatty acid intake)
- Smoking cigarettes and cigars; chewing tobacco
- Lack of physical activity
- Unresolved emotional stress, hostility, angry personality

Other risk factors for CHD include

- Family history of, genetic disposition toward CHD
- Male gender; females after menopause
- Old age

Nutrition Interventions for CHD

Nutrition interventions for heart disease include risk reduction and management of symptoms related to atherosclerosis. These interventions should begin early in adulthood to enhance life expectancy. Abnormal blood lipid levels are the strongest risk factors for heart disease, and the strongest evidence for atherosclerotic plaque reduction supports limiting saturated and trans fatty acids in the diet. However, no single dietary change or other action guarantees freedom from heart disease.

The National Heart, Lung, and Blood Institute initiated the National Cholesterol Education Program (NCEP) in 1985 in order to decrease the population's average blood cholesterol, which was considered to be the most modifiable nutrition-related risk factor for heart disease. The NCEP Expert Panel on Detection, Evaluation, and Treatment of High Blood Cholesterol developed population-based strategies to reduce heart disease[12,13] through identification of cut-points for diagnosis of risk-associated blood lipid levels, development of corresponding dietary and pharmaceutical treatment suggestions, and implementation of health promotion campaigns. Population levels of cholesterol have decreased. (See Case Study 17.1.)

Case Study 17.1

Joleen Celebrates 50

As part of Joleen's fiftieth birthday celebration, she and some friends have made reservations for a spa weekend. Joleen is really looking forward to time away because she has been so tired lately. Work has been ultra-busy, her husband has seemed crabby, and she hasn't had the energy to keep up her routines at the gym. So when the card comes from her doctor to suggest that it is time for a physical, she thinks, "Why not? Another sign that I'm getting old." Here are some of the results from Joleen's clinic visit: weight of 145 pounds, height of 5 ft 3 in., waist measurement of 32 inches, blood pressure of 120/77, HDL cholesterol of 35, LDL cholesterol of 137. Dr. Williams suggests prescribing a cholesterol-lowering drug and schedules Joleen for an osteoporosis evaluation.

Questions

1. From what you've seen of Joleen's health indicators, would you agree that she is at risk of cardiovascular disease (CVD)?
2. What are her risk factors?
3. List three or more nutritional interventions that Joleen could consider to decrease her risk of CVD.
4. Do you agree with Chandra, Joleen's friend, who tells her, "Oh, don't worry. You're just getting older. I remember my sister; when she turned 50, she was depressed for weeks! You just need more spa time."

(Answers are in Appendix D.)

Originally, the NCEP's dietary guidelines were made in the form of the Step I and Step II diets.[12] Since then, both NCEP[13] and AHA[11] have released new guidelines. The Nutrition Committee of AHA based its report on three underlying philosophies: (1) Diet and lifestyle practices can be safely followed throughout life; (2) individual intake is to be evaluated over an extended time rather than a single meal; and (3) guidelines are a population framework into which individual needs are to be integrated. The AHA dietary guidelines underscore the potential benefits of eating a diet that emphasizes vegetables, fruits, whole grain products and other high-fiber foods, nonfat dairy products, nuts, fish, and lean meats. (The AHA diet is similar to the DASH diet described in Chapter 19.) The following NCEP guidelines result in reduction of LDL-cholesterol levels and reduced risk of CHD. They are called Therapeutic Lifestyle Changes (TLC):[13]

- Saturated fat intake should be less than 7% of total calories, monounsaturated fatty acids should contribute up to 20% of calories, and polyunsaturated fatty acids should not total more than 10% of calories.
- Total fat intake should range between 25 and 35% of calories.

- Dietary cholesterol intake should be less than 200 mg per day.
- Carbohydrates should consist of 50–60% of total calories.
- Dietary fiber intake should be 20–30 grams per day, with 10–25 grams coming from soluble fibers.
- Plant stanols or sterols (2 grams per day) should be consumed (e.g., from spreads).
- Weight reduction should be attempted if person is overweight or obese.
- At least 200 calories per day should be expended in physical activity.

Further guidance about fat intake comes in the form of Dietary Reference Intakes (DRIs; see the macronutrient distribution guidelines in Chapter 1). The DRI guidelines recommend that saturated fatty acid intake be kept as low as possible while consuming a nutritionally adequate diet and that trans fatty acid intake be as low as possible. The DRIs further recommend that 5 to 10% of calories come from linoleic acid (an omega-6 fatty acid) and that 0.6 to 1.2% of calories come from alpha-linolenic acid (an omega-3 fatty acid). These fat intake levels are meant to promote health in general and heart health in particular.[17]

Fatty fish are a good source of alpha-linolenic acid (lake whitefish, chinook salmon, Atlantic herring, rainbow trout), as are ground flaxseed, walnuts, and to a lesser extent, pecans. Purslane, treated as a garden weed by some, is a good plant source; brussels sprouts, broccoli, and kale also contain some omega-3 fatty acids. But while flax has long been an oil source (linseed oil), the cabbage family is low in fat. You would need to eat about 5 cups of brussels sprouts to ingest 1 gram of alpha-linolenic acid.

Eating as few trans fats as possible means limiting processed foods, especially those made with hydrogenated oils. French fries, pot pies, breaded and fried fish or chicken, doughnuts, cakes, cookies, biscuits, and crackers are examples of foods high in trans fatty acids. Foods that contain very few trans fatty acids are vegetables; cereal grains such as rice, corn and oats; oils; and fruits. A bowl of salad greens tossed with herbs, onion, olive oil, and a splash of vinegar and topped with grilled chicken, poached salmon, or hard-cooked eggs is an example of a meal that is free of trans fatty acids.

Dietary Supplements and CHD

Herbal products are being used in an attempt to reduce the risk of heart disease.[18,19] Because they are considered dietary supplements, the efficacy of these products does not have to be demonstrated. Examples of herbs used for heart disease include

- *Hawthorn:* The leaves, flowers, and fruits contain antioxidants (flavonoids and procyanidins), which are used in the early stages of congestive heart failure and arrhythmias.
- *Garlic:* Cloves of garlic contain allicin and antioxidants. Garlic has antibiotic properties, decreases blood cholesterol, and inhibits platelet aggregation or blood clotting. Recommended doses range from 1 clove (4 grams) to 5–20 cloves of fresh garlic per day; chopping and drying release the active compound allicin, although special processing can result in active dried garlic for supplement use. The German Commission E approves garlic to treat hyperlipoproteinemia and to slow *arteriosclerosis.*

> **Arteriosclerosis** Age-related thickening and hardening of the artery walls, much like an old rubber hose that becomes brittle or hard.

- *Green tea:* Green tea and its extract contain antioxidants (polyphenols and proanthocyanidins). Regular consumption of green tea is associated with decreased LDL cholesterol and triglycerides, and increased HDL-cholesterol levels.
- *Red yeast (Monascus):* Chinese records from 800 AD report that red yeast is useful for treating diarrhea, indigestion, and poor stomach health and blood circulation. Red yeast is a fungus used in rice wine fermentation and is the food coloring in Peking duck. Red yeast rice is used to treat hyper- lipoproteinemia due to its content of monacolin K, also known as the cholesterol-lowering drug lova-statin (with the same side effects). The monacolin K in red yeast blocks cholesterol formation. Although red yeast has a long tradition of use, it also has the potential to interfere with statin drugs.
- *Coenzyme* Q_{10} *(ubiquinone):* Present in most cells (it is ubiquitous), Q_{10} facilitates electron transport in mitochondria. It is not an essential nutrient. There is no conclusive evidence that it reduces heart disease symptoms, but it is recommended in congestive heart failure and ischemia (inadequate blood supply to tissues, including muscle and brain) when a patient does not respond to conventional therapy.
- *Carnitine:* Derived from amino acids (primarily meat and dairy as in "carniverous"), carnitine functions in long-chain fatty acid metabolism. Some evidence supports its use for cardiac performance and reduction of hyperhomocysteinemia.[18]

Overweight and Obesity

"The second day of a diet is always easier than the first. By the second day you're off it."

Jackie Gleason

Casual conversations about food and weight tend to reveal that men and women are aware of their own weight and whether it is currently up or down from the usual. Most people are likely to have dieted at some time or other. Despite the high level of awareness of body weight and knowledge about weight maintenance, the number of overweight and obese individuals is steadily increasing. Obesity is now considered an epidemic.[20]

Overweight and obesity are often examined together because obesity is a degree of overweight. In the 1980s, health professionals used to use the term *ideal body weight.*[21] Ideal weights were based on survivorship of healthy adults applying for life insurance, published as the 1983 Metropolitan Life Insurance Company actuarial tables. These weights were also called "normal," "suggested" and "acceptable" weight. In the mid-1980s, the term *ideal body weight* was replaced with the term *desirable weight,* that was subsequently replaced with the term *healthy body weight* in the 1995 and 2000 *Dietary Guidelines.* In general, a healthy weight is one that can be maintained through a health-promoting lifestyle. Weight becomes "overweight" when it reaches a level associated with higher risk for disease, disability, and death. The word *obesity* was not introduced into the Dietary Guidelines until 2000.[21,22]

Definition of Overweight and Obesity

The Obesity Education Initiative of the National Institute of Health classifies overweight and obesity by body mass index (BMI, kg body weight divided by height in meters, squared).[23] BMI can be calculated by following these steps:

1. Multiply body weight in pounds by 704.5.
2. Divide that number by height in inches.
3. Divide that number by height once more.

For example, here is the calculation of the BMI for a 5 ft 7 in. individual weighing 150 pounds:

1. 150 pounds × 704.5 = 105,675
2. 105,675 divided by 67 inches = 1577.239
3. 1577.239 divided by 67 inches = 23.54 BMI

Based on this single number expressing the relationship between weight and height, overweight in adults is defined by BMIs of 25.0–29.9 and obesity by BMIs of 30 or greater.[23] A BMI of 30 is roughly equivalent to being 30 pounds overweight.

Body fat content is a more important indicator of health than is weight-for-height. Although BMI approximates body fat for most healthy individuals, a BMI value is not the same as a measurement of body fat. For instance, a heavily muscled football player who is 6 feet tall and weighs 200 pounds has a BMI of 27.2 but may have a low body fat content. For the following individuals, BMI measures don't accurately represent healthy weights:

- Athletes or others with greater-than-average percentages of muscle mass
- Individuals with little muscle mass
- Individuals with dense, large bones
- Dehydrated individuals

Prevalence of Overweight and Obesity

Obesity rates have increased rapidly in the United States in the last two decades. Nearly one-third of U.S. adults are obese (31%), with black women having the highest rate (50%) and white men the lowest (27.8%).[2,24] No adult population group meets the Healthy People 2010 target of 15% obesity prevalence. Using the criterion of BMI between 18.5–24.9, only 34% of U.S. adults are at a healthy weight. For gender differences in overweight, obesity, and healthy weight, see Table 17.3.[2]

Table 17.3 Percentage of the adult population that is at healthy weight, overweight, or obese, by gender			
	Healthy Weight	Overweight	Obese
Males	32%	67%	28%
Females	35%	62%	34%

SOURCE: Data from National Center for Health Statistics. Hyattsville, Maryland: 2003.[2] Percentages do not total 100% due to rounding and lack of data for "thin."

Etiology of Overweight and Obesity

Between 1970 and 1990, average caloric intake of adults in the United States increased by as much as 300 calories daily.[24] Mean caloric intake ranges from 2028 to 2825 for 20- to 39-year-old females and males, respectively.[25] Small changes can add up: eating or not eating 100 calories per day adds up to approximately 10 pounds per year gained or lost. Two out of every five U.S. adults do not participate in leisure-time physical activity, and only 15% participate in regular physical exercise (at least 20 minutes, three times per week). An abundant food supply (3600 calories daily, available for every man, woman, and child in the United States) coupled with sedentary lifestyles contribute to the population's rising weight.

Overweight and obesity may appear to be simply a matter of intake exceeding output. But if it were so simple, the weight-loss industry would hardly be so huge. Overweight and obesity are complex and chronic conditions, stemming from multiple causes. Environmental, genetic, physiological, psychological, socioeconomic, and cultural factors all play a role in the development of obesity.

Effects of Overweight and Obesity

The effects of excess body weight are a threat to the public's health, and one that is serious enough that the surgeon general convened a conference on obesity to plan a nationwide antiobesity campaign.[26] Overweight and obesity are associated with an increased risk of hypertension, dyslipidemia, coronary heart disease, type 2 diabetes, stroke, gallbladder disease, osteoarthritis, sleep apnea, and endometrial, breast, prostate, and colon cancers.[23] As anyone who has ever been obese can tell you, social stigma also accompanies obesity. Weight loss in overweight and obese persons decreases the presence and severity of obesity-related health problems.

Nutrition Interventions for Overweight and Obesity

A successful weight-loss program includes physical activity and a diet that allows for lifelong personal adherence, an acceptable diet that is safe and that results in long-term maintenance of weight lost.[27] Compared to normal-weight individuals or those who regain lost weight, individuals who successfully maintain weight loss use more behavioral strategies to support weight loss and maintenance. These behaviors include consistently controlling caloric intake, exercising more often and more strenuously, tracking weight, and eating breakfast (see Table 17.4 on the next page for details).[27–29]

Keeping excess weight off over time is often harder than losing it in the first place. This is illustrated by the widely varying results of a meta-analysis of weight-loss treatments.[30] After 1 year, average weight change with diet therapy and/or physical activity ranged from a 1.9-kg weight

Table 17.4 Dietary weight-loss strategies of individuals in the National Weight Control Registry, by gender*

	Women	Men
Number sampled	629	155
Age (years)	44.4 ± 11.5	49.1 ± 11.9
Weight change (kg)	28.7 ± 13.6	35.4 ± 20.7
Weight loss duration (years)	5.5 ± 6.8	5.8 ± 6.9
Maintenance Diet		
Calories/day	1296 ± 454	1725 ± 647
Fat (%)	24 ± 9	23 ± 8
Protein (%)	19 ± 4	18 ± 4
Carbohydrate (%)	55 ± 10	56 ± 10
Eating frequency/day	4.9 ± 3.2	4.5 ± 1.6
Weight-Loss Dietary Strategies		
Restricted intake of certain type or classes of food	88%	87%
Ate all foods but limited quantity	47%	32%
Counted calories	45%	3%
Limited percentage of daily energy from fat	31%	37%

*To join the National Weight Loss Registry (www.lifespan.org), an individual must be 18 or older, have lost at least 30 pounds, and have maintained a weight loss of ≥30 pounds for 1 year or more.
SOURCE: Adapted from M. L. Klem, R. R. Wing, M. T. McGuire, H. M. Seagle, and J. Hill, "A Descriptive Study of Individuals Successful at Long-Term Maintenance of Substantial Weight Loss," *Am J Clin Nutr* 66 (1997): 239–46.

Type I Diabetes A disease characterized by high blood glucose levels resulting from destruction of the insulin-producing cells of the pancreas. In the past, this type of diabetes was called juvenile-onset diabetes and insulin-dependent diabetes.

Type 2 Diabetes A disease characterized by high blood glucose levels due to the body's inability to use insulin normally, or to produce enough insulin. In the past, this type of diabetes was called adult-onset diabetes and non-insulin-dependent diabetes.

gain to an 8.8-kg weight loss.[30] Counseling for low-calorie diets (1000–1200 kcal per day) reduced body weight by an average of 8% over 3 to 12 months and decreased abdominal fat. Counseling for physical activity led to weight losses of 2% to 3% and reduced abdominal fat.

Components of a successful weight-management program include the following:

1. *Realistic goals.* Identifies a healthy weight and a feasible rate of loss (0.5–1 pound/week), with the ability to self-monitor progress.
2. *Caloric deficit.* Develops an individualized meal plan with sufficient calories to lose weight gradually.
3. *The meal plan.* Builds meal plans around a variety of foods that can be readily obtained and enjoyed by the entire household.
4. *Long-term effectiveness.* The weight-management plan is built around learning and practicing behaviors that can be maintained for a lifetime.
5. *Practice problem-solving techniques.* Develops strategies to anticipate and solve potential weight-management problems.
6. *Stress management.* Teaches strategies other than eating to deal with stressful situations.
7. *Maintenance.* Makes available support for weight loss and maintence of the loss.
8. *Regular exercise.* Includes strength training and aerobic exercise.
9. *Cultivates self-image.* Builds self-confidence while guarding against disordered eating.

Although weight loss can benefit health, the way in which it is lost can undo that benefit. Inappropriate eating habits, potentially hazardous herbal remedies (such as ephedra), and diet drugs are harmful to health. Using the preceding checklist can provide some assurance that the proposed weight-loss approach will be helpful rather than harmful. If weight loss were easy, Americans would not be spending more than $33 billion per year on the weight-loss industry. See Case Study 17.2.

Diabetes Mellitus

"What AIDS was in the last 20 years of the 20th century, diabetes is to be in the first 20 years of this century."

Paul Zimmet, International Diabetes Institute[31]

There are three major forms of diabetes: *type 1, type 2,* and gestational *diabetes.* Type 2 diabetes is the most common by far and is fueling the diabetes epidemic. Rates of type 2 diabetes are escalating worldwide, due in part to rising rates of obesity. Although most often diagnosed in people over the age of 40, type 2 diabetes is becoming increasingly common in children and adolescents.[32]

Diabetes exists when blood glucose levels are above normal. Type 1 and 2 diabetes are diagnosed when fasting levels of blood glucose are 126 mg/dl and higher.[33]

Prevalence of Diabetes Mellitus

Diabetes affects approximately 200 million individuals worldwide, including 17 million people in the United States.[31] Less than 1% of the U.S. population was diagnosed with diabetes in 1960. The figure has grown to 6.5% of adults (aged 18 years and over) in 2002.[34] Prevalence increases with age, from 2% of 18- to 44-year-olds to nearly 8% of 45- to 54-year-olds.[34] Ethnic background affects prevalence. The age-adjusted prevalence of diagnosed diabetes was higher among non-Hispanic black persons (10%) and Hispanic persons (9.2%) than among non-Hispanic white persons (5.7%).[34] The rise in diabetes correlates with

Case Study 17.2

Maintaining a Healthy Weight

Adam is 5 ft 11 in. tall and weighs 190 pounds. He lives alone. The commute to his software development job takes about 90 minutes each day. He likes his coworkers and the employment environment and is happy that his workplace provides a cafeteria so he doesn't have to bring a lunch. But he is trying to work from home occasionally to cut car expenses. His main hobby is golf; the course he plays encourages carts. He's an avid football fan. In his spare time, he is restoring an old car.

Photo Disc

Questions

1. Calculate Adam's current BMI. How would you classify his weight status based on the NIH classifications?
2. What would it take for Adam to achieve a BMI of 24? Calculate an energy level and estimate the number of weeks it would take at that level.
3. What would you consider a "healthy weight" for Adam?
4. What are some suggestions you would discuss with Adam in order to decrease his BMI?

(Answers are in Appendix D.)

---

a rise in overweight adults. In a review of data on overweight and obese adults from the Third National Health and Nutrition Examination Survey (NHANES III; 1988–1994, 45- to 74-year-olds), 10.8% had undiagnosed diabetes, 22.6% had prediabetes, and only 54% had normal glucose metabolism.[35] It is anticipated that due to rising rates of obesity, 800,000 new cases of type 2 diabetes may develop each year in the United States.[36]

Etiology of Diabetes Mellitus

The central defect in diabetes is an elevated blood glucose level, caused by an inadequate supply or ineffective utilization of insulin. Insulin is a hormone produced by the beta cells of the pancreas. Insulin performs many functions, one of which is to lower increases in blood glucose level after meals by facilitating the passage of glucose into cells. If insulin is produced in insufficient amounts, or if insulin receptors on cell membranes are too few or not sensitive to the action of insulin, cells become starved for glucose. Functional levels of multiple tissues and organs in the body degrade as a result. There are adverse side effects of high levels of blood glucose on the body, too, such as elevated blood levels of triglycerides, increased blood pressure, and hardening of the arteries.

Effects of Diabetes Mellitus

Health effects of diabetes vary depending on how well blood glucose levels are controlled and on the presence of other health problems such as hypertension or heart dis-

ease. In the short run, untreated or poorly controlled diabetes produces blurred vision, frequent urination, weight loss, increased susceptibility to infection, delayed wound healing, and extreme hunger and thirst. In the long run, diabetes may contribute to heart disease, hypertension, blindness, kidney failure, stroke, and the loss of limbs due to poor circulation. The number one cause of death among people with diabetes is heart disease. Many of the side effects of diabetes can be prevented or delayed if blood glucose levels are maintained within the normal range.[36]

ETIOLOGY OF TYPE 2 DIABETES People who develop type 2 diabetes usually have a condition known as *prediabetes* years before type 2 diabetes is diagnosed. Elevated fasting blood glucose levels that are somewhat below the cut-point used to diagnose type 2 diabetes characterize prediabetes. Approximately 6% of U.S. adults, and 314 million people worldwide are at risk of type 2 diabetes due to this condition.[32,36] Most people diagnosed with prediabetes have a condition known as *insulin resistance.*

> **Prediabetes** A condition in which blood glucose levels are higher than normal but not high enough for the diagnosis of diabetes. It is characterized by impaired glucose tolerance, or fasting blood glucose levels between 110 and 126 mg/dl.
>
> **Insulin Resistance** A condition is which cells "resist" the action of insulin in facilitating the passage of glucose into cells.

Insulin Resistance Obesity, low levels of physical activity, and a genetic predisposition are common risk factors for insulin resistance. Abdominal obesity (an apple-shaped body or a waist circumference >40 inches for males, >35 inches for females) is a particularly potent risk factor for insulin resistance. Insulin resistance reduces glucose passage into cells. This in turn prompts the beta cells of the pancreas to produce and secrete additional insulin. Higher than normal blood levels of insulin are generally sufficient to keep blood glucose transfer into cells and blood levels of glucose under control for a number of years. Cells in the pancreas may become exhausted, however, from the years of overwork. In such cases, production of insulin slows down or stops and, as a result, glucose accumulates in blood.[36]

> **Metabolic Syndrome** A constellation of metabolic abnormalities that increase the risk of heart disease. Metabolic syndrome is characterized by insulin resistance, abdominal obesity, high blood pressure and triglyceride levels, low levels of HDL cholesterol, and impaired glucose tolerance. It is also called Syndrome X and insulin resistance syndrome.

Insulin resistance is related to the development of a spectrum of metabolic abnormalities that have far-reaching effects. Collectively, the adverse effects of insulin resistance are included in a disorder called *metabolic syndrome.*

Metabolic Syndrome Metabolic syndrome is related to a cluster of metabolic abnormalities that increase the risk of type 2 diabetes and heart disease and include

- High levels of central body fat
- High blood insulin levels
- High blood pressure (130/85 mm/Hg or higher)
- Elevated blood triglyceride levels (150 mg/dl or higher)
- Low levels of protective HDL cholesterol (less than 50 mg/dl in women and 40 mg/dl in men)
- High blood glucose levels (110 mg/dl or higher)

The diagnosis of metabolic syndrome is made when three or more abnormalities are identified.[37] Individuals with four or five metabolic abnormalities have a 3.7 times greater risk of heart disease, and a 25 times higher risk for type 2 diabetes than do people with no abnormalities.[38] It is estimated that 25% of adults in the United States and 15% in Canada have metabolic syndrome, with the bulk of cases being made up of overweight and obese inactive adults.[37] Weight loss and aerobic exercise are key components of prevention and management of metabolic syndrome.

Risk Factors for Metabolic Syndrome Type 2 diabetes, insulin resistance, and metabolic syndrome share a set of risk factors that include obesity, high levels of central body fat, physical inactivity, and eating patterns providing few whole grains and little fiber. There are genetic components to these disorders, as evidenced by the fact that they track in families and are more likely to occur in certain groups (Hispanic American, African Americans, Asian and Pacific Islanders, and Native Americans) than others.[32]

Effects of weight loss and exercise on the prevention of type 2 diabetes can be quite dramatic.[39] In one large study

that took place over a 3-year period, people with prediabetes reduced their risk of developing type 2 diabetes by over 50% through losses in body weight of around 7% and 150 minutes a week of exercise.[40] Diets rich in whole grain and high-fiber foods are protective against the development of type 2 diabetes and appear to aid weight loss.[41] High-fiber, whole grain foods raise blood glucose levels to a lower extent than do refined grain products, and they appear to provide nutrients and other biologically active substances that lessen the risk of this disease.

Nutrition Interventions for Type 2 Diabetes

Diet and exercise are the cornerstones of the treatment of type 2 diabetes.[42,43] Modest weight loss alone (5–10% of body weight) has repeatedly been shown to significantly improve blood glucose control in overweight and obese people with type 2 diabetes.[44]

In general, diets developed for diabetes emphasize:

- Whole grain breads and cereals and other high-fiber foods, vegetables, fruits, nonfat and low-fat milk, lean meats, and fish
- Unsaturated fats
- Regular meals and snacks[43,45]

Chromium supplements (500–1000 micrograms per day) are sometimes recommended. It appears that the essential mineral chromium improves blood glucose and lipid levels in many people with type 2 diabetes.[46]

Dietary management of type 2 diabetes should focus on heart disease risk reduction as well as blood glucose control. Food sources of monounsaturated fats—such as vegetable oils, nuts, seeds, lean meats, and seafoods—are recommended over foods high in saturated or trans fats. Monounsaturated fats tend to reduce LDL-cholesterol levels while not also reducing blood levels of HDL cholesterol; they improve insulin resistance and lower blood glucose levels somewhat.[43] Diet and weight-loss interventions may be supplemented by oral medications that decrease insulin resistance and blood lipids, and by insulin if needed.[47] Dietary recommendations for type 2 diabetes are currently not consistent across developed countries, indicating that scientific consensus is yet to be reached on a number of important issues related to diet and diabetes.[47,48]

HIV/AIDS

Advances in drug therapy have shifted HIV/AIDS in the United States away from a terminal disease characterized by malnutrition and severe weight loss toward being a chronic condition that can be managed.

Definition

AIDS, or acquired immunodeficiency syndrome, is caused by the human immunodeficiency virus (HIV). This virus invades and kills specific immune cells, causing increased susceptibility to infection and cancer. Individuals infected with the virus have HIV, whereas people with symptoms of disease have AIDS.

Prevalence

The United Nations, which tracks the worldwide epidemic of HIV/AIDS, reports that 40 million people live with HIV/AIDS. Of these, 37 million are adults and 2.5 million are children under 15. Approximately half a million (511,000) people with HIV/AIDS live in the United States. Worldwide, AIDS has caused the deaths of an estimated 3 million people.[49]

In the United States, the estimated number of diagnoses of AIDS through 2002 is 886,575.[50] Rates of new cases of AIDS are declining. According to the U.S. Centers for Disease Control and Prevention, AIDS affects nearly seven times more African Americans and three times more Hispanics than whites.[50]

Etiology

HIV is transmitted through blood and body fluid exchange. Unprotected sex and sharing of contaminated needles are routes of viral transmission among adults. Women can transmit HIV to their babies during pregnancy, at birth, or by breastfeeding. Approximately one-quarter to one-third of all untreated pregnant women infected with HIV will pass the infection to their babies. HIV also can be spread to babies through the breast milk of mothers infected with the virus. If the mother takes the drug AZT during pregnancy, she can significantly reduce the chances that her baby will get infected with HIV.[51]

Effects

HIV is able to make its own DNA and replicate itself by using genetic material from the host's cells. It penetrates the body's immune cells and eventually destroys the cells. Decreased functional level of the immune system increases the risk that people with HIV will develop infections and cancer, although not all individuals with HIV go on to develop AIDS. The CDC is responsible for tracking the spread of AIDS in the United States.

Nutrition-related symptoms common in people with AIDS affect dietary intake and include

- Coughing and shortness of breath
- Seizures and lack of coordination
- Difficult or painful swallowing
- Mental disturbances such as confusion and forgetfulness
- Severe and persistent diarrhea
- Fever
- Nausea, abdominal cramps, and vomiting
- Weight loss and extreme fatigue

AIDS progression often begins when HIV mutates and drugs used to treat the virus become ineffective. AIDS progression is marked by decreased appetite, increased nutrient needs, nutrient deficiencies, weight loss, and tissue wasting.[52] HIV/AIDS raises nutrient requirements as long as the infection is present, and may cause other nutrition-related conditions, such as nutrient malabsorption due to infection in the gastrointestinal tract.[52]

Drug therapy for HIV/AIDS causes changes in body shape due to the redistribution of fat stores. Fat stores shift from the arms and legs to the central part of the body. Increased central body fat stores elevate blood triglyceride and cholesterol levels, and may increase the risk of heart disease and insulin resistance in the long term.[53]

Nutrition Interventions for HIV/AIDS

Nutrition interventions for people with HIV/AIDS center on maintaining weight and nutritional adequacy as well as on reducing blood lipid levels and insulin resistance related to drug treatment. During the early phase of the disease, adequate nutrient intakes improve immune function and decrease susceptibility to infection. Even the best nutritional advice and self-care cannot prevent the eventual progression of AIDS and restore immune function, however. Weight maintenance and nutritionally adequate diets can help people with the disease increase their level of control and sense of well-being.[52]

"Eat smart. Play hard."

Nutrition Education Promotion Campaign of USDA and American Dietetic Association

Resources

Diabetes and Metabolic Syndrome Resources
Go to the "Conditions Center" on the WebMD gateway site; select "diabetes" and hit "go." Get the latest information on treatment and a risk assessment for diabetes. Site leads to reliable information on insulin resistance and metabolic syndrome as well.
Web site: **www.webmd.com**

Search the keywords *insulin resistance, diabetes, metabolic syndrome,* and *hypoglycemia* to find reliable reports.
Web site: **www.healthfinder.gov**

AIDS/HIV Resources
Pan American Health Organization (PAHO) For current statistics on the number of reported AIDS cases in North, Central,

and South America, contact the regional World Health Organization office for the Americas at 525 23rd Street, N.W., Washington, D.C. 20037.

Telephone: 202-861-4346

AIDSinfo provides referrals and current information on federally and privately funded clinical trials for AIDS patients and others infected with HIV. AIDSinfo is primarily Web based and can be found at http://aidsinfo.nih.gov. AIDSinfo also operates a telephone service from 12:00 p.m. to 5:00 p.m.

Eastern Time, Monday through Friday. English- and Spanish-speaking health information specialists are available to answer questions about HIV/AIDS, treatment options, and navigating the Web site.

Telephone: 800-HIV-0440 (1-800-448-0440)

International: 301-519-0459

TTY/TDD: 888-480 3739

E-mail: **ContactUs@aidsinfo.nih.gov**

References

1. Healthy People 2010: national health promotion and disease prevention objectives. Washington, DC: U.S. Department of Health and Human Services; 2000.

2. National Center for Health Statistics. Hyattsville, MD: 2003.

3. U.S. Department of Health and Human Services, Public Health Service, National Institutes of Health, National Cancer Institute. 2001 Cancer Progress Report. NIH Pub No. 02-5045. Bethesda, MD. Available at http://progressreport.cancer.gov, accessed 12/6/03.

4. Byers T, Nestle MM, McTiernan A et al. American Cancer Society guidelines on nutrition and physical activity for cancer prevention: reducing risk of cancer with healthy food choices and physical activity. CA Cancer J Clin 2002; 52:92–119.

5. Brown JK, Byers T, Doyle C et al. Nutrition and physical activity during and after cancer treatment: an American Cancer Society guide for informed choices. CA Cancer J Clin 2003;53:268–91.

6. The American Heart Association. American Heart Association heart and stroke statistics. 2003 update. Dallas, TX: 2003. Available at www.americanheart.org, accessed 11/03.

7. Gauthier GM, Keevil JG, McBride PE. The association of homocysteine and coronary artery disease. Clin Cardiol. 2003; 16(12):563–8.

8. Stanger O, Herrmann W, Pietrzik P et al., on behalf of the DACH-LIGA Homocystein e.V. DACH-LIGA Homocystein (German, Austrian, and Swiss Homocysteine Society): Consensus paper on the rational clinical use of homocysteine, folic acid, and B-vitamins in cardiovascular and thrombotic diseases: guidelines and recommendations. Clin Chem Lab Med. 2003;41(11): 1392–1403.

9. Ridker PM, Hennekens CH, Buring JE, Rifai N. C-reactive protein and other markers of inflammation in the prediction of cardiovascular disease in women. N Eng J Med 2000;342:836–43.

10. Van den Hoogen PC, Feskens EJ, Nagelkerke NJ et al. The relation between blood pressure and mortality due to coronary heart disease among men in different parts of the world. Seven Countries Study

Research Group. N Eng J Med 2000; 342:1–8.

11. Krauss RM, Eckel RH, Howard B et al. AHA Dietary Guidelines, revision 2000: a statement for healthcare professionals from the Nutrition Committee of the American Heart Association. Circulation 2000;102: 2284–99.

12. Report of the National Cholesterol Education Program Expert Panel on Detection, Evaluation, and Treatment of High Blood Cholesterol in Adults. The Expert Panel. Arch Intern Med 1988;148:36–69.

13. Third report of the National Cholesterol Education Program (NCEP) Expert Panel on Detection, Evaluation, and Treatment of High Blood Cholesterol in Adults (Adult Treatment Panel III). J Am Med Assoc 2001;285:2486–97; Available at www.nhlbi.nih.gov/guidelines/cholesterol/ atp_iii.htm, accessed 11/21/03.

14. Westerveld HT, van Lennep JE, van Lennep HW et al. Apolipoprotein B and coronary artery disease in women: a cross-sectional study in women undergoing their first coronary angiography. Arterioscler Thromb Vasc Biol 1998;18:1101–7.

15. Lamarche B, Tchernof A, Mauriege P et al. Fasting insulin and apolipoprotein B levels and low-density lipoprotein particle size as risk factors for ischemic heart disease. J Am Med Assoc 1998;279:1955–61.

16. Giles WH, Croft JB, Greenlund KJ et al. Association between total homocyst(e)ine and the likelihood for a history of acute myocardial infarction by race and ethnicity: results from the Third National Health and Nutrition Examination Survey. Am Heart J 2000;139:446–53.

17. National Academy of Sciences. Food and Nutrition Board. Dietary Reference Intakes for energy, carbohydrates, fiber, fat, fatty acids, cholesterol, protein, and amino acids. Institute of Medicine, National Academy of Sciences, Washington, DC: National Academies Press; 2002. Available at http://books .nap.edu/books/0306085373/html, accessed 5/5/03.

18. American Association of Clinical Endocrinologists. Medical guidelines for the clinical use of dietary supplements and nutraceuticals. Endocrine Prac 2003;9(5): 418–70. Available at http://aace.com/clin/ guidelines/nutraceuticals, accessed 11/03.

19. Robbers JE, Tyler VE. Tyler's herbs of choice. New York: Haworth Herbal Press; 1999.

20. Mokdad AH, Serdula MK, Dietz WH et al. The spread of the obesity epidemic in the United States, 1991–1998. J Am Med Assoc 1999;282:1519–22.

21. Flegal KM, Troiano RP, Ballard-Barbash R. Aim for a healthy weight: what is the target? J Nutr 2001;131:440S–450S.

22. U.S. Department of Agriculture, U.S. Department of Health and Human Services. Nutrition and your health: dietary guidelines for Americans, 5th ed.; 2000.

23. NHLBI Obesity Education Initiative Expert Panel on the Identification Evaluation, and Treatment of Overweight and Obesity in Adults. Clinical guidelines on the identification, evaluation, and treatment of overweight and obesity in adults: the evidence report. Bethesda, MD: National Institutes of Health; 1998. Available at www.nhlbi.nih.gov/guidelines/obesity/ ob_home.htm, accessed 11/18/03.

24. Flegal KM et al. Prevalence and trends in obesity among U.S. adults, 1900–2000. J Am Med Assoc 2002;288(14):1723–7.

25. Wright JD, Wang CY, Kennedy-Stephenson J, Ervin RB. Dietary intake of ten key nutrients for public health, United States: 1999–2000. Advance data from vital and health statistics; no. 334. Hyattsville, MD: National Center for Health Statistics; 2003.

26. Satcher D. Surgeon General's Conference on Obesity, a year-long process to develop a weight reduction plan, co-sponsored by National Institutes of Health, Centers for Disease Control and Prevention, and the Office of Public Health and Science; January 8, 2001.

27. Position of the American Dietetic Association: weight management. J Am Diet Assoc 2002;102(8):1145–55.

28. McGuire MT, Wing RR, Klem ML, Hill JO. Behavioral strategies of individuals who have maintained long-term weight losses. Obes Res 1999;7:334–41.

29. Wyatt HR, Grunwald GK, Mosca CL et al. Long-term weight loss and breakfast in subjects in the National Weight Control Registry. Obes Res 2002;10:78–82.

30. McTigue KM, Harris R, Hemphill B et

al. Screening and interventions for obesity in adults: summary of the evidence for the U.S. Preventive Services Task Force. Ann Intern Med 2003;139:933–49.

31. World seen facing diabetes catastrophe: impact may outpace AIDS. International Diabetes Federation Conference, Paris 2003. Available at www.medscape.com, accessed 9/6/03.

32. Cowie CC. MMWR, CDC Surveillance System 2003:52:833–7; and Diabetes rose slightly in the 1990s. Centers for Disease Control Press Release, 9/12/03.

33. Gale EAM et al. Can we change the course of beta-cell destruction in type 1 diabetes? N Engl J Med 2002;346:1740–1.

34. Centers for Disease Control (CDC). Early release of selected estimates based on data from the 2002 National Health Interview Survey. Release date June 18, 2003.

35. Benjamin SM, Valdez R, Geiss LS et al. Estimated number of adults with prediabetes in the U.S. in 2000. Diab Care 2003; 26(3):645–49.

36. Diabetes overview. Available at www .intelihealth.com/diabetes, accessed 6/03.

37. Park Y-W et al. The metabolic syndrome: prevalence and associated risk factors. Arch Int Med 2003;163:427–36.

38. Men with 3 of 5 metabolic abnormalities risk diabetes, heart disease. Available at www.nlm.nih.gov/medlineplus/heart disease .html, accessed 7/03.

39. Sheard NF. Moderate changes in weight and physical activity can prevent or delay the development of type 2 diabetes mellitus in susceptible individuals. Nutr Rev 2003;61:76–9.

40. Diabetes Prevention Program Research Group. Reduction in the incidence of type 2 diabetes with lifestyle intervention or metformin. N Engl J Med 2002;346;393–403.

41. McKeown NM, Meigs JB, Liu S et al. Whole grain is favorably associated with metabolic risk factors for type 2 diabetes and cardiovascular disease in the Framingham Offspring Study. M J Clin Nutr. 2002; 76:390–8.

42. Brand-Miller J et al. Low glycemic index foods and glycemic control in diabetics. Diabetes Care 2003;26:2261–7.

43. Solomon CG. Reducing cardiovascular risk in type 2 diabetes. N Engl J Med 248:2003;457–60.

44. Heymsfield S et al. Effects of weight loss with Orlistat on glucose tolerance and progression to type 2 diabetes in obese adults. Arch Inter Med 2000;160:1321–6.

45. Franz MJ et al. Evidence-based nutrition principles and recommendations for the treatment and prevention of diabetes and related complications. Diabetes Care 2002;25:148–66.

46. Wong Z et al. Chromium supplements and glucose sensitivity in type 2 diabetes. 18th IDF Congress, abstract 154, 756, 762; August 28, 2003.

47. Nathan DM. Initial management of glycemia in type 2 diabetes mellitus. N Engl J Med 2002;347:1242–50.

48. Foster-Powell K, Holt SH, Brand-Miller JC. International table of glycemic index and glycemic load values: 2002. Am J Clin Nutr 2002;76:5–56.

49. Joint United Nations Programme on HIV/AIDS (UNAIDS)/World Health Organization (WHO) Annual Report. AIDS Epidemic Update 2003. December 2003. Geneva, Switzerland. Available at www .unaids.org, accessed 11/24/03.

50. National Center for Health Statistics. Centers for Disease Control. Division of HIV/AIDS Prevention-Surveillance and Epidemiology, National Center for HIV, STD, and TB Prevention. Atlanta, GA. Available at www.cdc.gov/hiv/stats, accessed 11/03.

51. Gayle HP, Janssen RS. Successful implementation of perinatal HIV prevention guidelines: a multi-state surveillance evaluation. Morbid & Mortal Week Rep 2001; 50(RR-6):15–28.

52. Kline DA. AIDS in the 21st century. Today's Dietitian 2003;July:12–16.

53. Kuritzkes DR, Currier J. Cardiovascular risk factors and antiretroviral therapy. N Engl J Med 2003;348:679–80.

"Scientific evidence increasingly supports that good nutrition is essential to the health, self-sufficiency, and quality of life of older adults."

American Dietetic Association[1]

Chapter 18

Nutrition and the Elderly

Chapter Outline

Prepared by U. Beate Krinke

Key Nutrition Concepts

1 Eating and enjoying a varied diet contributes to mental and physical well-being.

2 Generalizations relative to health status changes with aging are unwise because "older adults" are a heterogeneous population.

3 Diseases and disabilities are *not* inevitable consequences of aging.

4 Functional status is more indicative of health in older adults than chronological age.

5 Body composition changes that occur with aging can alter lifestyle; these have the greatest impact on nutritional needs.

Introduction

"An ounce of prevention is worth a pound of cure."

Traditional

In "normal" aging, inevitable and irreversible physical changes occur over time, and some diseases are more prevalent. The leading causes of death are heart disease, cancer, and stroke; these are linked with nutrition in complex ways. Most older people consider themselves to be healthy. More than anything, they want to remain healthy and independent and certainly do not want to be a burden to others.[2] Older adults feel that good nutrition and exercise are the most important health habits they can maintain in order to avoid losing autonomy and independence.[2]

Just exactly what is good nutrition for older adults? They can meet their decreasing energy requirements by choosing more nutrient-dense foods. They can drink more water to stay hydrated, even when not thirsty. By eating adequate vegetables, fruits, and whole grains, keeping fats in balance, and drinking alcohol only in moderation, older adults can reduce their risks of getting chronic diseases. Diet quality is linked to the longevity of older men and women.[3,4] Older adults can live longer and better (i.e., postponing disability and shortening the period of decreased functional capabilities at the end of life) with good health habits.[5] This chapter defines aging and provides information about the nutrient requirements, dietary recommendations, and food and nutrition programs designed to support healthy aging.

Longevity Length of life; it is a measure of life's duration in years.

What Counts as Old?

Many chronological ages have been used as cut-points to mark the beginning of "old age." While no biological benchmark signals a person's becoming old, there are soci-etal and governmental definitions for *old*. Societal definitions label someone as "old" at a fairly young age. Turning 50 makes you eligible to join the American Association for Retired Persons (AARP). Views on "normal" retirement ages are changing, however. At age 60, many cafeterias, movie theaters, and shops give senior discounts. What counts as "old" depends on who is counting.

Governmental definitions of old age include the following categories suggested by the U.S. Census Bureau:

- 65 to 74 years is "young old"
- 75 to 84 years is "aged"
- 85 and older is "oldest old"

The Elderly Nutrition Program, first funded under the Older Americans Act in 1972, calls "older adults" those aged 60 years and older. The Social Security program identifies people 65 years old as being eligible for retirement benefits (62 years is "early retirement"). These various age cut-points are important because services and programs that support nutrition for the elderly do not use the same age limits. The arbitrarily set retirement age of 65 years is most commonly used in textbooks, and we will use that label here. Readers will also learn that many factors besides chronological age affect nutrition in the elderly.

Chronological age is a marker of where we are on life's path. Our functional status, defined as how well we accomplish the desired tasks of daily living, is more indicative of health than chronological age. Rather than ask, "How old are you?" we should ask, "What can you do?" Nonetheless, chronological age is an easy measure, so we continue to ask that question. Our perception is that age is a proxy for predicting health status and functional abilities.

Food Matters: Nutrition Contributes to a Long and Healthy Life

"We found that tomatoes were the primary food associated with higher functional status in centenarians ... even when controlled for other factors such as illness, depression, and gastrointestinal problems."

Dr. Mary Ann Johnson, commenting on findings that elderly nuns with higher functional status also had higher blood lycopene levels[6]

In our search for magic bullets, tomatoes will probably not turn out to be nature's perfect food. The cumulative effects of lifelong dietary habits determine nutritional status in old age. Good nutrition throughout life contributes to optimal growth, to appropriate weight, and to nutrient levels in blood and other tissues that provide disease resistance. In trying to assess the contribution good nutrition can make to longer life, the Centers for Disease Control and Prevention (CDC) suggest that *longevity* depends 19% on genetics, 10% on access to high-quality health care, 20% on environmental factors such as pollution, and 51% on lifestyle factors.

Besides not smoking, diet and exercise are estimated to be the lifestyle factors contributing most to decreased mortality, or longer life.[7] In a longitudinal study including diet monitoring, older women who ate the best diets were 30% less likely to die than the women who ate few whole grains, fruits, vegetables, low-fat dairy products, and lean meats.[4]

The role of food and nutrition often changes during aging. Besides reducing risk of disease and delaying death, diet plays a role in health and longevity by contributing to wellness. Wellness means having the energy and ability to do the things one wants to do and to feel in control of one's life. Being able to choose, purchase or prepare, and eat a satisfying diet every day; enjoying traditional foods at holidays, birthdays, and other special occasions; and having the resources to purchase desired foods on a regular basis all contribute to independence and a higher quality of life. Good nutrition, as defined by dietary guidelines covered later in this chapter, can help to "add life to years" as well as add years to life.

A Picture of the Aging Population: Vital Statistics

More and more of us are growing old and older. During Roman times, fewer than 1% of the population reached age 65; but today, roughly 12% of the North American population reaches age 65. (The figure is slightly higher at 15% in Italy, the UK, Germany, and France.) The number of people aged 65 and over will be around 40 million in 2010, expand to 70 million in 2030, and may reach 82 million in 2050![8]

Persons aged 85 years and older, having increased from 122,000 in 1900 to 4.4 million in 2001, are the fastest-growing segment of the population, accounting for 2% of the population in 2000.[8] A White House conference on aging called this age wave "a demographic revolution" and predicted that it will change the twenty-first century much as information technologies revolutionized the twentieth century and the industrial revolution affected the nineteenth century.[9]

Global Population Trends: Life Expectancy and Life Span

Today, life expectancy of older adults is quite different from 1900, when average life expectancy was 47 years. The National Center for Health Statistics (NCHS) reports that women born in 1995 in the United States can expect to live 79 years, whereas men can expect to live 72 years (see Table 18.1). For babies born in 2050, life expectancy is expected to increase to 84 years for women and 80 years for men.[10] Life expectancy in the United States lags behind that of many other countries, primarily because infants in the United States have higher mortality rates than those in other nations. Other contributors to life

Table 18.1 Examples of life expectancy for people born 1990–1995 in selected countries

COUNTRY	FEMALE LIFE EXPECTANCY AT BIRTH, YEARS	COUNTRY	MALE LIFE EXPECTANCY AT BIRTH, YEARS
Japan	82.9	Japan	76.4
France	82.6	Sweden	76.2
Switzerland	81.9	Israel	75.3
Sweden	81.6	Canada	75.2
Spain	81.5	Switzerland	75.1
Canada	81.2	Greece	75.1
Australia	80.9	Australia	75.0
Italy	80.8	Norway	74.9
Norway	80.7	Netherlands	74.6
Netherlands	80.4	Italy	74.4
United States	78.9	Costa Rica	73.0
Chile	77.8	Finland	72.8
Costa Rica	77.8	United States	72.5
Kenya	57.2	Kenya	54.2
Nigeria	52.0	Nigeria	48.8

SOURCES: *Healthy People 2010: National Health Promotion and Disease Prevention Objectives,* Washington, D.C.: U.S. Department of Health and Human Services, 2000 (cohort born in 1995); National Center for Health Statistics, 2000, at http://cdc.gov/nchs, Table K, accessed 10/1/00 (life expectancy cohorts from 1990–1995, including estimates from United Nations data); *Health, United States 2003 with Chartbook in Trends in the Health of Americans,* Washington, D.C.: U.S. Government Printing Office, 2003. Numbers do not include recalculated expectancies; see S. Tuljapurkar et al.[12]

expectancy estimates include rates of childhood mortality, infectious and chronic diseases, and death from violence and accidents.

Since the early 1900s, immunizations and other risk-reduction measures, treatment of disease, decreased infant and childhood mortality rates, and clean water and safe food have increased average *life expectancies,* which are getting closer to the potential human *life span.* For instance, from 1970 to 1990 life expectancy increased by more than 4 years, largely due to fewer deaths from heart disease, stroke, and accidents among older adults.[10] Over-all, the United States has the second highest number of people 80 years and over (China ranks first).

Although life expectancy is rising and the population is aging, human life span remains stable at around 110 to 120 years.[11] Few people currently survive to age 120. However, Jeanne Calment, the oldest known person who lived to age 122, is an exception. Nonetheless, the number of centenarians (persons living to age 100 and beyond) in the United States has increased markedly and may add to our understanding of current limits on the human life span. Demographers are predicting that by 2030, there will be 381,000 centenarians in the United States—up from 50,364 in the year 2002.

Life Expectancy Average number of years of life remaining for persons in a population cohort or group; most commonly reported as life expectancy from birth.

Life Span Maximum number of years someone might live; human life span is projected to range from 110 to 120 years.

Nutrition: A Component of Health Objectives for the Older Adult Population

Behaviors that can enhance the health of an aging population are reflected in national health objectives. Table 18.2 identifies dietary goals related to disease prevention and to health promotion for older adults. The current set of health objectives emphasizes public education about health consequences of overweight and obesity. The greatest dietary improvement in the older adult population would be to eat more vegetables and grains, especially whole grain products.

Theories of Aging

What triggers aging? Theories explaining aging grow from human desire to understand the biological processes that determine how long and how well we live. Aging theory tries to explain the mechanisms behind loss of physical resilience, decreased resistance to disease, and other physical and mental changes that accompany aging.

Biological systems are much too complex to have one theory that is robust enough to explain the mechanisms of aging.[13] Basic biological processes involved in aging are largely determined by genetics. However, environmental factors influence expression of the genetic code by exacerbating certain traits. For example, height and weight are

Table 18.2 Healthy People 2010 goals for older adults

PERCENTAGE OF POPULATION

	Target	Current, Age 60+ Females	Males
Increase the number who are at a healthy weight	60	37	33
Decrease the number who are obese (BMI 30+)	15	26	21
Increase the proportion of persons who:			
• Eat at least 2 daily servings of fruit	75	35	40
• Eat at least 3 daily servings of vegetables (at least one-third dark green or deep yellow)	50	6	5
• Eat at least 6 daily servings of grain products (at least 3 of those as whole grains)	50	4	11
• Eat less than 10% of calories from saturated fat	75	47	42
• Eat no more than 30% of calories from fat	75	40	34
Baseline information is for age 50 and older			
Meet dietary recommendations (1200 mg) for calcium	75	27	35
Baseline information is for age 65+			
Increase the proportion of physician office visits made by patients with a diagnosis of cardiovascular disease, diabetes, or hyperlipidemia that include counseling or education related to diet and nutrition	75	33[a]	33[a]
Increase food security and in so doing, decrease hunger			
Households <130% of poverty, with elderly persons	94	85[a]	85[a]
Households >130% poverty, with elderly persons	94	98[a]	98[a]

[a]Adequate data to distinguish levels for males and females is not available.

SOURCE: *Healthy People 2010: National Health Promotion and Disease Prevention Objectives,* Washington, D.C.: U.S. Department of Health and Human Services, 2000.

genetically programmed, but diet and other environmental exposures can moderate these outcomes.[14] Persons with a family history of high levels of LDL cholesterol (i.e., bad cholesterol) and early death from cardiovascular disease can moderate the outcomes suggested by their genetic programming through weight loss, a diet low in saturated fat, adequate exercise, and no smoking. Someone with a family history of type 2 diabetes can reduce his or her risk of developing the disease with a healthy diet, exercise, and maintenance of normal weight.

Theories of aging fall into three major categories: (1) programmed aging theories, (2) "wear-and-tear" theories, and (3) caloric restriction. Categories of aging theories overlap to some extent because it is clear that neither genetic programming nor environmental exposures fully account for changes related to aging.

Programmed Aging

HAYFLICK'S THEORY OF LIMITED CELL REPLICATION   Hayflick proposed that all cells contain a genetic code that directs them to divide a certain number of times during their life span.[11] After cells divide according to their programmed limit, and barring disease or accident, cells begin to die. For example, if individual cells of a fly have a 3-day life span and replicate 15 times, a fly can live 45 days. Using this theory, Hayflick calculated the potential human life span to be in the range of 110 to 120 years, estimating that human cells replicate from 40 to 60 times. Although most human cells can regenerate (e.g., blood, liver, kidney, and skin cells reproduce themselves), not all cells have that capacity (e.g., spinal cord, nerves, and brain cells). Hayflick's theory is difficult to prove in humans because we die from age-associated chronic disease more often than from old age itself.

MOLECULAR CLOCK THEORY   Another theory of programmed aging is that of the molecular clock. *Telomeres* that cap the ends of chromosomes act as clocks, becoming a bit shorter with each cell division. Eventually loss of telomeres stops the ability of chromosomes to replicate, and they become *senescent*. Loss of chromosomal replication may produce signs of aging because new cells cannot be formed and the function of existing cells declines with time. A major thrust of current research is to identify ways to limit loss of telomeres and thus prolong cell replication.[15,16]

Wear-and-Tear Theories of Aging

Wear-and-tear theories are built on the concept that things wear out with use. Mistakes in the replication of cells or buildup of damaging by-products from biological processes eventually destroy the organism. Cytotoxicity (poisoning of the cell) results when damaged cell components accumulate and become toxic to healthy cells. According to this theory, the accumulation of damaged cells and waste by-products leads to aging.

OXIDATIVE STRESS THEORY   Oxygen is an integral and versatile part of metabolic processes; it can both accept and donate electrons during chemical reactions. One cause of aging is thought to be oxidative stress due to the buildup of reactive (unstable) oxygen compounds. Unstable oxygen, formed normally during metabolism (e.g., hydroxyl radi-

cals) can also damage cells by initiating reactions that break down cell membranes and modify the normal metabolic processes that protect people from disease. Exposure to oxidizing agents is increased by smoking, ozone, solar radiation, and environmental pollutants. Unstable oxygen compounds are neutralized, however, when they combine with an antioxidant. This prevents them from interfering with normal cell functions. The body produces antioxidant enzymes (such as catalases, glutathione, peroxidase reductases, and superoxide dismutase), but part of our need for antioxidants is met from the diet. Dietary antioxidants include selenium, vitamins E and C, and other phytochemicals. Remember, these are plant substances such as beta-carotene, lycopene, flavonoids, lutein, zeaxanthin, resveratrol, and isoflavones that contribute to normal metabolism. For example, flavonoids found in grapes, apples, broccoli, and onions act as antioxidants.

RATE OF LIVING THEORY   The rate of living theory is similar to the oxidative stress theory in that it suggests that "faster" living results in faster aging.[17] For example, higher metabolic rate and energy expenditure leads to greater turnover of all body tissues. Theoretically, fast-paced living shortens life span, whereas living more slowly leads to a longer life. Scientists have not adequately examined old people, including centenarians, to fully understand how this theory works.

> **Telomere**   A cap-like structure that protects the end of chromosomes; it erodes during replication.
>
> **Senescent**   Old to the point of nonfunctional.

Calorie Restriction and Longevity

Animal studies (e.g., of fruit flies, water fleas, spiders, guppies, mice, rats, and other rodents) show that an energy-restricted diet that meets micronutrient needs can prolong healthy life.[18,19] For example, laboratory mice and rats fed calorie-restricted diets live longer and have fewer age-associated diseases than their counterparts whose diets are unrestricted. In the 1930s, McCay and colleagues suggested that delays in aging result after food restriction, due to slowed growth and development.[20] But since then, rodent studies have shown that instituting caloric restrictions in mid-life, after growth and development were completed, results in longer life spans.[19,21]

Caloric restriction is also being examined in primates.[22,23] Because nonhuman primate life expectancy is roughly 40 years, it will be years until we learn whether life extension through caloric restriction might be applied to non-human primates. Stay tuned.

Could calorie-restricted diets also extend human life? Experimental findings in small animals have led some individuals, such as Dr. Roy Walford of Biosphere 2, to personally adopt very low calorie diets. Walford coordinated the calorie-restricted diets of eight normal-weight people living in Biosphere 2. However, a study lasting only 6 months is too brief to determine human life span extension results from caloric restriction. Proposed anti-aging effects of caloric restriction in humans are thought to result from an evolutionary response to famine and feasting cycles.[24] The weakness in this theory is that humans' need for reproductive capacity is greater than our need for longevity.[11]

From an ecological view, people in France and Japan have lower caloric intake than do people in the United States, and people in both of those countries have longer life expectancies.[25] Physiologically, we know that nutrition affects human longevity by moderating risks of developing chronic diseases, ameliorating certain chronic conditions, and contributing to healing in a variety of acute conditions.[10,26] Severe caloric restriction during famine leads to malnutrition and starvation, with poor outcomes in human reproduction, growth, development, immune status, and healing. Obesity leads to early death, and so does starvation. An example of reducing calories might be adding vegetables to a diet in order to decrease chronic disease risk. Eating nutrient-dense foods and avoiding obesity enhances prospects for longevity. Scientists search for the perfect range of energy and nutrient intake; one that will maintain a weight range associated with optimal health, longevity, and quality of life.

> **Resilience** Ability to bounce back, to deal with stress and recover from injury or illness.
>
> **Lean Body Mass** Sum of fat-free body tissues: muscle, mineral as in bone, and water.

"Everything in moderation. Including moderation."

Joyce Hendley, food writer

Physiological Changes

Normal aging is associated with shifts in body composition and subsequent loss of physical *resilience*. Physiological system changes commonly associated with healthy aging are described in Table 18.3. As scientists come to understand the human aging process, they will learn to sort through age-associated physiological changes to be able to distinguish exactly which changes are due to genetic factors and which are due to poor diets, inactivity, or other lifestyle-related factors. Yet aging is not all loss or decline. Rather, healthy aging is associated with continuing psychosocial, personal, moral, cognitive, and spiritual development. Diseases that affect older persons' health are discussed further in Chapter 19.

Body Composition Changes

LEAN BODY MASS (LBM) AND FAT Individual shifts in body composition are common but neither inevitable nor irreversible.[27] Of all physiologic changes that do occur during aging, the biggest effect on nutritional status is due to the shifts in the musculoskeletal system (Table 18.3). Many individuals experience a decline in

Table 18.3 Age-associated physiological system changes that affect nutritional health*

Cardiovascular System
- Reduced blood vessel elasticity, stroke volume output
- Increased blood pressure

Endocrine System
- Reduced levels of estrogen, testosterone
- Decreased secretion of growth hormone
- Reduced glucose tolerance
- Decreased ability to convert provitamin D to previtamin D in skin

Gastrointestinal System
- Reduced secretion of saliva and of mucus
- Missing or poorly fitting teeth
- Dysphagia or difficulty in swallowing
- Reduced secretion of hydrochloric acid and digestive enzymes
- Slower peristalsis
- Reduced vitamin B_{12} absorption

Musculoskeletal System
- Reduced lean body mass (bone mass, muscle, water)
- Increased fat mass
- Decreased resting metabolic rate
- Reduced work capacity (strength)

Nervous System
- Blunted appetite regulation
- Blunted thirst regulation
- Reduced nerve conduction velocity affecting sense of smell, taste, touch, cognition
- Changed sleep as the wake cycle becomes shorter

Renal System
- Reduced number of nephrons
- Less blood flow
- Slowed glomerular filtration rate

Respiratory System
- Reduced breathing capacity
- Reduced work capacity (endurance)

*Some of these age-associated changes, such as the increase in blood pressure, are usual but not normal.

lean body mass with aging (see Table 18.4). On average, fat-free mass decreases by 15% in the 50 years from the mid-twenties to the mid-seventies.[28] and is known as sarcopenia. These composition changes are associated with lower levels of physical activity, food intake, and hormonal changes in women.[29,30] Loss of mineral and muscle mass is also accompanied by loss of body water. Overall, older people have lower mineral, muscle, and water reserves to call upon when needed.

Table 18.4 Comparison of body composition of a young and an old adult

	20 to 25 Years	70 to 75 Years
Protein/cell solids	19%	12%
Water	61%	53%
Mineral mass	6%	5%
Fat	14%	30%

SOURCE: Data from R. Chernoff, ed., Geriatric Nutrition: The Health Professional's Handbook. Gaithersburg, MD: Aspen, 1999, p. 391; based on N. W. Shock, *Biological Aspects of Aging,* New York, NY, Columbia University Press, 1962.

At the same time, many older adults gain body fat. Compared to males in their 20s, males in their 70s have roughly 24 pounds less muscle (a decrease from 53 to 29 lb on average) and 22 pounds more fat (an increase from 33 to 55 lb on average).[31] Think about losing 24 pounds of muscle: It is roughly equivalent to the muscle mass that girls or boys gain throughout the puberty growth spurt. Over 50 years, 24 pounds of muscle slowly goes away to be replaced by 22 pounds of fat. Of course, the extra fat does provide a reserve of energy for periods of low food intake, recovery from illness or surgery, as an insulator in cold weather, and to cushion falls. These shifts in body composition tend to occur even when weight is stable.[29]

MUSCLES: USE IT OR LOSE IT Many older people expect decreases in physiologic function with increasing age. However, physical activity contributes to staying strong, no matter the age.[28] For example, the HEalth, RIsk Factors, Training, And GEnetics (HERITAGE) Family Study compared the effects of a 20-week strength-training program on older and younger men and women, finding that the training response differed by gender and by race, but *not* by age.[32] Training exercises led to increases in fat-free mass; decreases in total, subcutaneous, and visceral fat mass; and to weight loss. Weight-bearing and resistance exercise increase lean muscle mass and bone density.[28,33] Because muscle tissue contains more water than fat tissue, building muscle also results in more stored water. Regular physical activity, including strengthening and flexibility exercise, contributes to maintenance of *functional status.*

> "I would not wish to imagine a world in which there were no games to play and no chance to satisfy the natural human impulse to run, to jump, to throw, to swim, to dance."
>
> Sir Roger Bannister, 1989

WEIGHT GAIN Weight gain, although not inevitable, tends to accompany aging. Obesity is a problem for 24% of adults aged 60 years and older.[10] Reasons for age-associated gains are uncertain, but longitudinal studies are showing that lack of exercise could be a factor.[27] For example, men in the Baltimore Longitudinal Study on Aging decreased their energy expenditure by 17 to 24 calories per day after age 55, but gained weight.[34] In the Fels Longitudinal Study, men gained 0.7 pounds per year and women gained an average of 1.2 pounds as they aged.[29] Subjects were 40 to 66 years old when entering the study and were followed for up to 20 years. Weight gains were concurrent with decreases in lean body mass and increases in body fat. This overall weight and body composition shift was moderated by physical activity. For example, the two groups with moderate or high physical activity levels (as opposed to the least active group) increased lean body mass and decreased total and percentage of body fat as age increased. Physical activity effects differed by gender. In women, higher levels of physical activity were associated with higher levels of lean body mass. However, lack of estrogen seems to promote fat accumulation, and total weight increased regardless of the group's activity level. (The menopausal transition was covered in Chapter 16.) Men in the highest physical activity groups in the Fels Longitudinal Study slowed their total body weight and body fat gains.

Changing Sensual Awareness: Taste and Smell, Chewing and Swallowing, Appetite and Thirst

TASTE AND SMELL Although there is some argument about the extent to which aging affects the sense of taste, there is general agreement that taste and smell senses eventually decline.[36–39] Eating is a sensuous activity, involving taste buds and olfactory nerves. In healthy adults, aging is associated with a decline in ability to identify smells beginning sometime around age 55 for men and age 60 or later for women.[40,41] Some smells are perceived more easily than others. A blunted sense of smell can lead to a blunted sense of enjoyment of food as well as decreased ability to detect spoiled or overcooked foods. Women's abilities to identify smells remain higher than men's throughout the life span.

The controversy around taste is that the number and structure of taste buds are not significantly altered during aging. In addition, taste perception for sucrose does not decline with age. Bartoshuk[42] argues that taste is so important to biological survival that the body has developed "redundancy in the mediation of taste," meaning that several pathways (nerves and receptors) control taste mechanisms. All of the pathways would have to be damaged before the ability to identify sweet, bitter, salty, sour, or savory tastes is lost.

> **Functional Status** Ability to carry out the activities of daily living, including telephoning, grocery shopping, food handling and preparation, and eating.

Disease and medications do affect taste and smell more than does age itself. Roughly three of four adults have had a temporary smell loss at some time, most often due to colds, flu, or allergies.[41] Yet age makes a difference. For example, during illness or with the use of medications, younger individuals maintained greater ability to detect salty, bitter, sour, sweet, and savory tastes than older ones.[43]

ORAL HEALTH: CHEWING AND SWALLOWING

Being able to chew and swallow contributes to eating enjoyment. Oral health depends on several organ systems working together: gastrointestinal secretions (saliva), the skeletal system (teeth and jaw), mucus membranes, muscles (tongue, jaw), taste buds, and olfactory nerves for smelling and tasting. Disturbances in oral health are associated with, but not necessarily caused by aging. Poor dietary habits are a modifiable risk factor that may contribute to caries and potential tooth loss. The Healthy People 2010 oral health objective for adults[10] aged 65–74 years, is to reduce the percentage of individuals who have lost all their teeth from 26% to 20%. The prevalence of missing teeth varies by population and has been found to be as high as 42% in Native American elders.[44] Reduction of periodontal disease and early cancer detection are other public health goals relating to older adults.[10]

Missing teeth and ill-fitting dentures can make chewing difficult and have a negative impact on eating.[45] An oral health assessment asks questions about soft tissues, teeth, and other factors that may affect dietary intake.[46] One such factor is functional status, which may affect the ability to brush and floss, potentially resulting in periodontal disease.

What and when we eat affects oral health. Oral health, in turn, affects what and how we eat. Healthy teeth are protected by an enamel layer, but bacterial action on the breakdown products of food slowly erodes tooth enamel. For about 15 minutes after we ingest food or drink, oral bacteria feast on the food breakdown products, especially those of sucrose. Foods such as caramels or raisins stick around longer, especially when they lodge between teeth. Frequent eating and drinking of sugary beverages provides a continuous substrate for bacteria. The acid in carbonated beverages adds to the corrosive potential of food.

Saliva, which lubricates the mouth and begins the digestive process (amylase in "spit" begins starch breakdown), also helps to keep tooth enamel clean. However, saliva seems to become thicker and more viscous with age. Lack of saliva slows nutrient absorption. Lack of sufficient and effective saliva, especially in the presence of gingivitis and periodontal disease, also makes the oral cavity more sensitive to temperature extremes and coarse textures, resulting in pain while eating.

Pain and discomfort with chewing foods can result in eating fewer fruits, vegetables, and whole grains. A loss of self-esteem associated with missing teeth and worry regarding how to pay for dental care can affect quality of life. Edentulous older people are less likely to visit the dentist for oral care (denture adjustment, periodontal disease management) than are those individuals having their natural teeth.[47]

APPETITE AND THIRST Hunger and satiety cues are weaker in older than in younger adults. Roberts and colleagues examined the ability of 17 young men (mean age 24) and 18 older men (mean age 70) to adjust caloric intake after periods of overeating and of undereating.[48] All men were healthy and not taking medications. Food intake and weight were monitored for 10 days, and then men were overfed by roughly 1000 calories, or underfed by about 800 calories, for 21 days. Periods of over- or underfeeding were followed by 46 days of "ad-lib" intake during which all men were free to eat as much or as little as desired. After the periods of over- and underfeeding, young men adjusted their caloric intake to get back to their initial calorie intake level and weight. Older men kept overeating if they had been in the overfed group, and undereating if they had been in the underfed group. The authors suggest that older adults may need to be more conscious of food-intake levels because their appetite-regulating mechanism may be blunted. Whereas healthy young people adjust to cycles of more and less food intake, healthy older people's inability to adapt to these changes may lead to overweight or anorexia.

Elderly people don't seem to notice thirst as clearly as younger people do. A small study demonstrated that the thirst-regulating mechanism of older adults was less effective than that of younger individuals.[49–51] Researchers compared thirst response to fluid deprivation in a group of seven 20- to 31-year-old men, and seven 67- to 75-year-old men. Subjects lost 1.8% to 1.9% of body weight during 24 hours without fluids. Both groups were asked about feeling thirsty, mouth dryness, and how pleasant it would be to drink something. After fluid deprivation, the younger group reported being thirsty and having dry mouth. The older group, however, reported no change in thirst or mouth dryness. Both the older and younger groups thought that it would be pleasant to drink something after fluid deprivation. Blood measures showed that older men lost more blood volume than younger men did, indicated by their plasma concentrations of sodium. Researchers also measured how much water the men drank in the hour after their 24-hour period of fluid deprivation. Older men drank less water than their younger counterparts did. Younger people made up for fluid loss in 24 hours; older people did not drink enough to achieve their prior state of hydration. It appears that dehydration occurs more quickly after fluid deprivation and that rehydration is less effective in older men.

Nutritional Risk Factors

Identifying nutritional risk factors before chronic illness occurs is basic to health promotion. Decreasing risk factors forms the basis of dietary guidance. For instance, the leading causes of death for older adults are heart disease, cancer, and stroke. In all adults, dietary risk factors that increase the likelihood of developing these diseases are consuming a high-fat diet that is also high in saturated fat; low intake of vegetables, fruits, and whole grain products; and excessive calorie intake leading to obesity. Healthy People 2010, the Food Guide Pyramid, and the Dietary Guidelines for Americans emphasize eating patterns that reduce the risk of the leading killer diseases.

Another approach to nutritional risk factor identification is to compare adequacy of current dietary intake to dietary intake recommendations such as the Recommended Dietary Allowances and the Dietary Reference Intakes. According to national intake data[52] older adults are not consuming enough protein, vitamin E, folic acid, magnesium, and calcium to meet recommended nutrient levels.

A third approach is to examine a population and determine how environmental factors combine with dietary factors to predict nutritional health. This approach was used by a consortium of care providers, policy makers, and researchers to develop the Nutrition Screening Initiative (NSI).[53] The American Academy of Family Physicians, the American Dietetic Association (ADA), and the National Council on the Aging, Inc., sponsored the development of this health-promotion campaign. The NSI consortium used literature review, expert discussion, and a consensus process to generate a list of warning signs of poor nutritional health in older adults. See Table 18.5 for a condensed version of the acronym (DETERMINE) it promotes.[54,55]

These warning signs were integrated into a screening tool, the NSI DETERMINE checklist (Illustration 18.1 on the next page), and pilot-tested for use in community settings. Ten risk factors remained after testing.[56] The revised tool is used by community agencies, educators, and service providers to screen program participants and the aged public for risk of malnutrition.

The nutritional risk factors identified during the NSI process[54] are reflected in the list of risk factors identified in the ADA position on nutrition and aging.[1] Presence of any of the following conditions places older adults at greater nutritional risk:

- Hunger
- Poverty
- Inadequate food and nutrient intake
- Functional disability
- Social isolation
- Living alone
- Urban and rural demographic areas
- Depression
- Dementia
- Dependency
- Poor dentition and oral health; chewing and swallowing problems
- Presence of diet-related acute or chronic diseases or conditions

Table 18.5 DETERMINE: Warning signs of poor nutritional health
Disease. Any disease, illness, or chronic condition (i.e., confusion, feeling sad or depressed, acute infections) that causes changes in the way you eat, or makes it hard for you to eat, puts your nutritional health at risk.
Eating poorly. Eating too little, too much, or the same foods day after day, or not eating fruits, vegetables, and milk products daily will cause poor nutritional health.
Tooth loss/mouth pain. It is hard to eat well with missing, loose, or rotten teeth, or dentures that do not fit well or cause mouth sores.
Economic hardship. Having less (or choosing to spend less) than $34.30 (female) to $38.60 (male) weekly for groceries makes it hard to get the foods needed to stay healthy (www.usda.gov/cnpp).
Reduced social contact. Being with people has a positive effect on morale, well-being, and eating.
Multiple medicines. The more medicines you take, the greater the chance for side effects such as change in taste, increased or decreased appetite and thirst, constipation, weakness, drowsiness, diarrhea, nausea, and others. Vitamins or minerals taken in large doses can act like drugs and can cause harm.
Involuntary weight loss or gain. Losing or gaining a lot of weight when you are not trying to do so is a warning sign to discuss with your health care provider.
Needs assistance in self-care. Older people who have trouble walking, shopping, and buying and cooking food are at risk for malnutrition.
Elder years above 80. As age increases, risk of frailty and health problems also rises.

SOURCE: Warning signs adapted from Nutrition Screening Initiative, a project of the American Academy of Family Physicians, the American Dietetic Association, and the National Council on the Aging, Inc., and funded by a grant from Ross Products Division, Abbott Laboratories, Inc.; dollar amounts under economic hardship inserted by author using September 2003, USDA low-cost food plan.

- Polypharmacy (use of multiple medications)
- Minority status
- Advanced age

Why are factors such as poverty and minority status included in a list of nutritional risk factors? Economic security contributes to food security, one of the health goals for the nation. Lack of food security leads to food intakes below 67% of the RDA, especially in older persons, who eat fewer calories and less magnesium, calcium, zinc, and vitamins E, C, and B_6 when food insecure.[57] Although older people on average are less likely to live in poverty than are children, they are a heterogeneous population, and several groups are at high risk of poverty.

The Warning Signs of poor nutritional health are often overlooked. Use this checklist to find out if you or someone you know is at nutritional risk.

Read the statements below. Circle the number in the yes column for those that apply to you or someone you know. For each yes answer, score the number in the box. Total your nutritional score.

DETERMINE YOUR NUTRITIONAL HEALTH

	Yes
I have an illness or condition that made me change the kind and/or amount of food I eat.	2
I eat fewer than 2 meals per day.	3
I eat few fruits or vegetables, or milk products.	2
I have 3 or more drinks of beer, liquor, or wine almost every day.	2
I have tooth or mouth problems that make it hard for me to eat.	2
I don't always have enough money to buy the food I need.	4
I eat alone most of the time.	1
I take 3 or more different prescribed or over-the-counter drugs a day.	1
Without wanting to, I have lost or gained 10 pounds in the last 6 months.	2
I am not always physically able to shop, cook, and/or feed myself.	2
TOTAL	

Total your nutrition score. It's . . .

0–2   Good! Recheck your nutritional score in 6 months.

3–5   You are at moderate nutritional risk.
See what you can do to improve your eating habits and lifestyle. Your office on aging, senior nutrition program, senior citizens center, or health department can help. Recheck your nutritional score in 3 months.

6 +   You are at high nutritional risk.
Bring this checklist the next time you see your doctor, dietitian, or other qualified health care or social service professional. Talk with them about any problems you may have. Ask for help to improve your nutritional health.

Remember: warning signs suggest risk, but do not represent diagnosis of any condition.

These materials developed and distributed by the Nutrition Screening Initiative, a project of: AMERICAN ACADEMY OF FAMILY PHYSICIANS, THE AMERICAN DIETETIC ASSOCIATION, and NATIONAL COUNCIL ON AGING, INC.

Illustration 18.1 Determine Your Nutritional Health checklist.

SOURCE: Reprinted by permission by the Nutrition Screening Initiative, a project of the American Academy of Family Physicians, The American Dietetic Association and the National Council on the Aging, Inc., and funded by a grant from Ross Products Division, Abbott Laboratories, Inc.

Minority status is related to economic status. In 2001 nearly 60% of the black and Hispanic population aged 65 and over were either poor or near poor (199% of the poverty threshold; see Chapter 16).[8] Furthermore, minority populations are more likely to be in fair or poor health while non-Hispanic whites are most likely to be in excellent or good health. High health care costs may contribute to food insecurity. All of us can think of a story about seniors having to choose between drugs and food.

Poverty is one risk factor for malnutrition; polypharmacy is another. Reported medication use increases with age. In a study of prescription and over-the-counter (OTC) drugs in community-living older adults, 75% took at least one prescription drug. The average number of medications (prescription, OTCs, vitamin/mineral or herbal supplements) was 12–15 per day.[58] The top five most commonly used prescription and OTC drugs by men and women are identified in Table 18.6.[59]

Taken individually, the risk factors identified in the DETERMINE acronym and in the ADA position statement on aging are not unique to older adults. But each is more likely to lead to nutritional problems in a frail, vulnerable population. For instance, functional disability can affect dietary intake at any age, but very old people are more likely to live alone and to have fewer resources to compensate for any type of lost function. Consequently, diet quality may decline. Fewer 65-year-olds live alone than 75- and 85-year-olds, and more women than men live alone. Race and ethnicity affect living situations; Hispanic men and women are least likely to live alone, whereas white women over age 85 are most likely to live alone; 65% of older white women live alone, compared to 47% of African American and 35% of Hispanic women.

A common perception is that older adults living alone eat poorly. Although living alone is a nutritional risk factor, it's not clear whether the issue is eating alone or living alone. On average, meals eaten with other people last longer and supply more calories than do meals eaten alone.[60] The effect of living alone starts at a younger age for men and affects nutrition more extensively than for women.[61] Women aged 75 and older eat less protein and also less sodium than those living with others; men aged 65 and older eat less protein, beta-carotene, vitamin E, phosphorus, calcium, zinc, and fiber. Men and women living alone did not consume greater amounts of any of the 23 nutrients (alcohol was not included) examined than did men and women living with someone.

Table 18.6 Top 5 most commonly used medications in men and women aged 65 and older

Men	Women
Aspirin (39%)	Acetaminophen (27%)
Acetaminophen (16%)	Aspirin (23%)
Furosemide (12%)	Conjugated estrogens (17%)
Digoxin (9%)	Ethinyl estradiol (13%)
Warfarin sodium (8%)	Hydrochlorothiazide (12%)

SOURCE: D. W. Kaufman, J. P. Kelly, L. Rosenberg et al. "Recent Patterns of Medication Use in the Ambulatory Adult Population of the United States: The Slone Survey," *J Am Med Assoc* 287(3), 2002:337–44.

The purpose of a screening tool is to identify aging individuals at risk for diseases and provide interventions that reduce or eliminate risk. The intensive push for elderly persons to receive a flu shot is an example of this type of tool. In secondary prevention, such as cholesterol screening or obtaining a bone scan to evaluate bone health, early symptoms of the disease are identified and treated to prevent symptoms or worsening of the condition. Thus, the question remains, "Does the nutrition screening initiative tool prevent and/or identify those older persons at nutritional risk?" The results are mixed. The cumulative score resulting from using the NSI tool weakly predicted mortality in an aging population of mostly white, educated adults.[62] In another study, the NSI checklist did not identify all those individuals who had poor health or low nutrient consumption.[56] Another screening tool suggested is use of the Mini-Nutritional Assessment (MNA), which is an 18-question survey that includes mid-arm circumference and calf circumference measurements and questions related to lifestyle, medication and mobility, dietary intake, and self-perception of health.[63] A large study compared the NSI and the MNA using dietary intake, anthropometrics, and blood biochemistries and found that each tool was of limited value.[64] Use of the NSI checklist to screen frail individuals who had been newly admitted into a nursing home identified problem areas, linked problems to care protocols, and resulted in expedited treatment.[65]

> Phil: "There's so much advice out there, how do I know what's right for me?"
> Harriet: "I know, it's all so confusing. How are you supposed to know what to eat?"

Dietary Recommendations

Sometimes, in all the discussions about nutritional effects on health and disease status of older adults, we forget that old age is not a disease. However, recommendations for specific nutrients do change with age, so the food on one's plate should also change over time.

Food-Based Guidance: The Pyramid

The Food Guide Pyramid is easily recognized as dietary advice. Pyramid adaptations for older adults have been developed by several groups, including the American Dietetic Association,[66] the Senior Nutrition Awareness Project (SNAP) of Connecticut and Rhode Island, and Tufts University researchers (see Illustration 18.2). The American Dietetic Association pyramid for persons aged 50 and over depicts fluids, but does not include vitamin-mineral supplements. In the Tufts "elderly" pyramid, authors portray the need for fewer calories with a narrower pyramid base.[67] They also emphasize adequate fluid intake by depicting 8 glasses of water at the pyramid base, but this depiction is confusing for some older audiences who interpret the water glasses to mean that water is the only fluid that counts. Potentially higher need for supplemental calcium and vitamins D, and B_{12} is flagged at the top of the pyramid. Symbols depicting fat, sugar, and fiber in foods are intended to help older adults recognize sources of those food constituents and subsequently choose a more nutrient-dense diet.

The pyramid shows food groups and corresponding serving sizes to be used for meal plan development. Table 18.7 on the next page lists food-group suggestions for

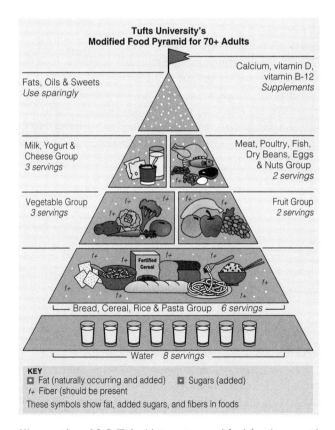

Illustration 18.2 Tufts University modified food pyramid for 70+ adults.

SOURCE: Tuft's University Modified Food Pyramid for 70+ Adults. Reprinted by permission from Dr. Robert M. Russell, Tufts University. ©1999 Tufts University.

Table 18.7 Meeting Food Guide Pyramid recommendations at selected calorie levels

	NUMBER OF SERVINGS		
FOOD GROUPS/SERVING SIZE	1500–1700 cal	1800–2000 cal	2100–2300 cal
Water and other nonalcoholic fluid (1 cup)	6–8	7–8	8
Bread, cereal, rice, pasta (1 oz, 0.5 cup cooked)	6	7–8	8–9
Vegetables (1 cup leafy, 0.5 cup raw, cooked)	3	3	4
Fruit (6 oz juice, 0.5 cup canned, 1 med)	2	2	3
Milk or yogurt (1 cup), cheese (1.5–2 oz)	3	3	3
Meat, poultry, fish, eggs (ounces); dry beans, tofu (0.5 cup cooked); nuts, seeds (0.33 cup)	5*	5*	6*
The "Other" Group			
Fats, oils, sweets (1 tsp oil, sugar, jam)	5	6	7
Alcohol: if using, moderation is 1 drink for women and 2 drinks for men per day			

*Ounces or meat equivalent servings
SOURCE: Adapted from Senior Nutrition Awareness Program (SNAP) sponsored by the Connecticut and Rhode Island Family Nutrition Program. The Senior Food Guide Pyramid can be downloaded from www.canr.uconn.edu/nusci/outrch/GNCsrfdpyr.pdf. University of CT and University of RI or by calling SNAP at 1-800-595-0929. Used by permission.

common calorie levels, which can be modified by choosing the more or less nutrient-dense foods from each group. For instance, citrus and berries are among the most nutrient-dense fruits. A baked potato has more vitamins and minerals per bite than french fries. Among grain-based products, whole grain breads and cereals provide fiber as well as the usual nutrients found in grain. Consistently choosing nutrient-dense foods decreases caloric intake. Fortification and enrichment, while enhancing nutrient intake, complicate nutrient density calculations. For instance, unless whole grain products are also fortified with folic acid, they may actually be considered to be less nutrient-dense than their plain white, but enriched, counterparts. That is one reason variety remains a part of dietary advice. Regularly choosing a variety of foods from each group enhances overall diet quality.

Actual Food Group Intake

A picture of the eating habits of older Americans can be developed from records of consumption according to food group, by analyzing occasions for eating (e.g., snacks) and by nutrients consumed (see Tables 18.8 and 18.9).

Eating Occasions

EATING OUT   Older adults eat away from home at a lower rate than younger persons. Whereas 62–72% of individuals in their twenties consume more than one-third of their calories away from home, 39–50% of individuals in their 60s consume one-fifth of their calories away from home, and approximately 28% of people aged 70 and older consume one-eighth of their calories away from home. Food eaten away from home supplies relatively more protein and fat than carbohydrate. For example, the population aged 70 years and older obtains the following

percentages of the day's calories with foods eaten away from home:[52]

- 13–14% of protein
- 14–15% of fat
- 11–12% of carbohydrate

Table 18.8 Daily consumption of named food groups by adults ages 60–69 years and ≥70 years using Continuing Survey of Food Intake by Individuals (CSFII), 1996 data collection

	NUMBER OF SERVINGS EATEN PER DAY			
	60–69 Years		70+ Years	
Food Group	Males	Females	Males	Females
Grain products	6.9	5.3	6.0	4.8
Whole-grain	1.1	0.8	1.2	1.0
Vegetables	3.8	2.9	2.9	2.8
Potatoes	1.0	0.7	0.7	0.7
Tomatoes	0.5	0.4	0.4	0.4
Fruits	2.0	1.6	2.1	1.7
Citrus fruits, melons, berries	0.9	0.8	0.9	0.8
Milk and milk products	1.3	1.0	1.2	1.1
Meat, poultry, fish, (Ounces or lean meat equivalents)	5.4	3.9	4.1	3.4
Eggs	0.6	0.4	0.4	0.3
Legumes (dried peas and beans)	0.3	0.1	0.2	0.1
Nuts and seeds	0.1	0.1	0.2	0.1

SOURCE: Actual intake amounts are taken from 1996 Continuing Survey of Food Intakes by Individuals, 2-day average. USDA, ARS, Beltsville Human Nutrition Research Center, Food Surveys Research Group.

Table 18.9 Caloric intake comparison of younger and older adults, by gender, from CSFII (Continuing Survey of Food Intake by Individuals, 1996 data collection)

Age	Actual Daily Calorie Intake Males	Actual Daily Calorie Intake Females	Recommended Calorie Intake Males	Recommended Calorie Intake Females
20–29	2620	1774	2900	2200
50–59	2216	1559	2300	1900
60–69	2019	1438	2300	1900
70 and older	1779	1393	2300	1900

SOURCE: Data from *1996 Continuing Survey of Food Intakes by Individuals, 2-day Average.* USDA, ARS, Beltsville Human Nutrition Research Center, Food Surveys Research Group.

SNACKING   For older adults, snacking is less common than in other age groups. Snack foods do not appear to be nutrient dense. Two-thirds of people aged 70 and older (65% of women, 67% of men) eat snacks that supply 12% of their food energy but only 6–10% of vitamins, minerals, and potassium consumed. Here are the energy-nutrient contributions of snacks for this group:

- 7% of protein consumed
- 10–11% of fat consumed
- 13–14% of carbohydrate consumed

Information about nutrient intake of older adults (see Tables 18.10 and 18.11, on the following page) helps to answer questions such as, "How does Uncle Al's diet compare to that of other men his age?" and "Are there unusual patterns in Al's diet that might reflect some potential problem?" Although change is good when it helps people to stay flexible, unexplained changes in diet and nutritional status need to be evaluated.

Nutrient Recommendations

Nutrient recommendations change as scientists learn more about the effect of foods on human function. Specific nutrient levels for population groups above age 51 were first established in 1997 as the Dietary Reference Intakes (DRIs).[70] The complete tables, including Tolerable Upper Intake (UL), are found on the inside covers of this text. This section first addresses the macronutrient recommendations (carbohydrates, protein, and fat) and then the micronutrient recommendations relevant to older adults.

Estimating Energy Needs

The main goal for energy calculations is to maintain a healthy body weight. Decrease of physical activity and basal metabolic rate from early to late adulthood results in 20% fewer calories needed for weight maintenance.[71] Total energy expenditure is primarily determined from basal metabolic rate (see Chapter 16), diet-induced thermogenesis, and physical activity energy expenditure. Genetics, gender, and body composition also influence metabolic rates, resulting in broad ranges of energy needs for older populations. Basal metabolic rates (BMRs), using doubly-labeled water measurements, have ranged from 1004 kcal/d to 2060 kcal/d in older individuals with normal body mass index (BMI).[72] A small international study of the oldest-old (i.e., >90 years) who were living at home demonstrated variability in physical activity and energy expenditures. Mean resting metabolic rates were 1523 kcal for women and 1944 kcal for men with physical activity ratios (i.e., the measured total energy expenditure/resting metabolic rate) of 1.2, ranging up to 1.7.[73]

Traditionally, the Harris-Benedict equation has been used to estimate individual calorie needs (see Chapter 16). However, validation was done with mostly young adults.

Table 18.10 Selected macronutrient intakes of adults ages 70 years and over, CSFII, 1994–1996, 1999–2000, data collection, with relevant recommendations

	ACTUAL INTAKE Caucasian Males	ACTUAL INTAKE Caucasian Females	ACTUAL INTAKE Black Males	ACTUAL INTAKE Black Females	RECOMMENDED INTAKE All ethnicities Males	RECOMMENDED INTAKE All ethnicities Females
Protein, g	80	59	72	54	56	46
Total fat, g	75	52	66	50	65	65 (DV)
Saturated fat, g	25	17	21	16	20	20 (DV)
Monounsaturated fat, g	29	20	26	19	Up to 20% calories	
Polyunsaturated fat, g	16	11	13	11	Up to 10% calories	
Cholesterol, mg	285	198	335	228	300 or less	

SOURCES: Actual intake amounts are taken from *1994–1996 Continuing Survey of Food Intakes by Individuals, 2-day Average.* USDA, ARS, Beltsville Human Nutrition Research Center, Food Surveys Research Group, with the exception of cholesterol, which was taken from 1999–2000 data in J. D. Wright, C. Y. Wang, J. Kennedy-Stephenson, and R. B. Erving, "Dietary Intake of Ten Key Nutrients for Public Health, United States: 1999–2000," *Advance Data from Vital and Health Statistics;* no. 334. Hyattsville, MD: National Center for Health Statistics, 2003. Recommended intake data are based on Dietary Reference Intake (DRI) Institute of Medicine of the National Academies. *Dietary Reference Intakes: Energy, Carbohydrate, Fiber, Fat, Fatty Acids, Cholesterol, Protein, and Amino Acids* (Washington, DC: National Academies Press, 2002) and the nutrient reference amounts used on the Nutrition Facts Panel food labels for sodium; Daily Values (DV) used on the Nutrition Facts Panel food labels for fat and saturated fat.

Table 18.11 Selected micronutrient intakes of older adults, CSFII, 1994–1996, 1999–2000 with relevant recommendations for adults aged 60 and older

	ACTUAL INTAKE		RECOMMENDED INTAKE	
	Males	Females	Males	Females
Sodium, mg	3447	2532	2400 (DV)	2400 (DV)
Vitamin A, mcg RE	1282	939	900	700
Vitamin E, mg	9.4	7.2	15	15
Vitamin B_{12}, mcg	6.0	4.5	2.4	2.4
Folate, mcg	387	312	400	400
Calcium, mg	754	587	1200	1200
Magnesium, mg	286	224	420	320
Zinc, mg	12.2	9.3	11	8
Iron, mg	17.3	12.8	8	8

SOURCES: Actual intake amounts of A, E, B_{12} taken from 1994–1996 Continuing Survey of Food Intakes by Individuals (CFSII) conducted by the Agricultural Research Service of the U.S. Department of Agriculture and available online; the remaining micronutrient actual intakes are based on 1999–2000 data in J. D. Wright, C. Y. Wang, J. Kennedy-Stephenson, and R. B. Erving, "Dietary Intake of Ten Key Nutrients for Public Health, United States: 1999–2000," *Advance Data from Vital and Health Statistics*; no. 334. Hyattsville, MD: National Center for Health Statistics, 2003. Recommended intake data are based on Dietary Reference Intake (DRI) Institute of Medicine of the National Academies. *Dietary Reference Intakes: Energy, Carbohydrate, Fiber, Fat, Fatty Acids, Cholesterol, Protein, and Amino Acids* (Washington, DC: National Academies Press, 2002) and the nutrient reference amounts used on the Nutrition Facts Panel food labels for sodium.

Newer energy estimation formulas are being developed to account for greater individual variation. The Mifflin-St. Jeor formula most closely predicts energy requirements of adults in general (see Chapter 16), and validation did include individuals up to age 80; but, it was not specifically developed for older adults.

Arcerio and colleagues developed formulas for use with older adults.[74–76] The equation used with older adult females includes a factor for hormonal status (e.g., menopause). Males completed the Minnesota Leisure Time Activity survey as well as a set of three chest skinfold measurements. Using these formulas requires training; they are not in common use.

Table 18.9 shows that on average, adults do eat fewer calories as they age. Women aged 60 and older eat fewer than 1500 calories and are especially vulnerable to malnutrition, even though approximately one in four is obese (BMI = 30 or more). Calorie intake and weight status is an area that illustrates the dangers of generalizing from populations to individuals, or the reverse. For older adults, it is difficult to meet vitamin and mineral needs at caloric levels below 1200–1500; but adding activity can maintain caloric needs. This is often easier to say than to do.

Nutrient Recommendations: Macro- and Micronutrients of Concern

Potentially problematic nutrients for older adults are presented in Tables 18.10 and 18.11, listing both actual and recommended intake. Each listed nutrient plays a role in normal, healthy aging and is eaten in amounts different from the recommended level (except for dietary cholesterol and zinc). The reasons that each nutrient is of concern follow next in the text.

CARBOHYDRATE AND FIBER Adequate carbohydrate intake of 45% to 65% of calories is generally not a problem.[72] Of individuals aged 70 and over, males eat 51% of their calories from carbohydrate and women eat 54%.[52]

Following the Food Guide Pyramid recommendations can ensure that the carbohydrate quantity and quality is adequate and that fiber guidelines are met. For example, men who need 2300 calories/day would need eat between 288 to 345 grams of carbohydrate per day to meet the recommended 50% to 60% of calories from carbohydrate sources; women would need 188 to 285 grams. A listing of foods that provides at least 50% carbohydrates of a 1500-calorie diet is presented in Table 18.12 (see Table 18.7 for serving sizes).

Depending on caloric intake, a range of 21–30 grams dietary fiber (14 grams/1000 calories) daily is recommended for adults aged 51 and older.[72] Males eat 18 grams and females eat 14 grams of fiber daily, which is less than the recommended levels (see Chapter 1).

Table 18.13 shows how common portions of foods eaten by older adults can supply adequate fiber. Reasons to eat adequate fiber include the strong associations between fiber intake and incidence of diverticular disease in men, non-insulin-dependent diabetes in women, coronary heart disease in men, and hypertension in women and men.[77] The role of fiber in gastrointestinal problems is discussed in Chapter 19.

When increasing fiber intake, additional fluids are needed to process the fiber. Slowly adding both fiber and fluids allows the intestinal system to adapt to the additional bacterial substrate.

PROTEIN Although most adults in North America eat sufficient or even excess amounts of protein, older adults (especially those over 75 years old, living alone, inactive) may be eating too little.[61] Inadequate protein intake contributes to muscle wasting (sarcopenia), a weakened immune status, and delayed wound healing. Protein

Table 18.12 Using the pyramid groups to estimate adequate carbohydrate and fiber for daily intake

BASIC FOOD GROUPS

Number of Servings/ Grams of Carbohydrate per Serving	Carbohydrate grams	Approximate Total Fiber Content, grams
6 servings of grain at 15 g each	90	12
2 servings of fruit at 15 g each	30	6
3 servings of milk at 12 g each	36	0
3 servings vegetable at 5 g each	15	9
Total from basic groups	**171**	**27**

OTHER CARBOHYDRATE-CONTAINING FOODS

1 tb sugar for coffee or tea	12	0
2 Fig Newton cookies	20	1
Total, including "Other" group	**203**	**28**

Mixed dishes such as soups, sandwiches, and salads count as partial servings from their contributing food groups.

guidelines for individuals aged 18 years and older are currently set at 56 grams per day for males and 46 grams for females.[72] Some researchers found that older persons need higher levels of protein intake to maintain nitrogen balance.[28,78] Nitrogen-balance studies used to determine current recommendations of 0.8 grams per kilogram (or 0.36 gram per pound) were done with young males who have proportionately more muscle mass than older adults, male or female, and are more efficient in maintaining nitrogen balance.

Nitrogen balance is also easier to achieve when the protein eaten is of high quality (such as meat, milk, and eggs; see Table 18.14), is ingested with adequate calories, and when an individual is doing resistance exercises.[31]

Table 18.13 An example of fiber-containing foods that might comprise one day's intake

Food Item	Grams of Fiber
Oatmeal (½ cup) with wheat germ (¼ cup)	8
Banana	2
Peanut butter (2 tb)/whole wheat sandwich (2 sl)	6
Orange	3
Baked potato with skin	4
Green beans (½ cup)	3
Bran muffin (1 med)	2
Pear	4
Total dietary fiber	**32**

Note: Meat, poultry, fish, eggs, milk, sugar, and oils do not contain dietary fiber.

During resistance training, less nitrogen is excreted and thus nitrogen balance improves. Consuming a low-calorie diet, as many older adults do, leads to a greater need for protein in order to maintain nitrogen balance. Evans and Cyr-Campbell recalculated nitrogen-balance study data and did new laboratory tests with older adults.[28] They found that older adults on average maintain nitrogen balance eating 0.91 grams of high-quality protein per kilogram body weight per day (see Table 18.14 for quality scores of common proteins). They also calculated that 1.25 g protein/kg body weight/day would be a safe intake for 97.5% of the older adult population, recommending that older adults eat 1 to 1.25 g of protein/kg body weight daily. This translates to 0.45 to 0.57 g protein/lb body weight. This protein range echoes an earlier recommendation by Campbell:[78] "A safe protein intake for older people would be 1.0 to 1.25 gm protein per kg per day of high quality protein." Proportionately less lean body mass in an older person does not lead to lower protein requirements. In fact, someone who is losing muscle due to inactivity requires extra protein.

Leading protein sources for all adults, based on NHANES II surveys,[79] are (1) beef steak, roast; (2) hamburgers; (3) white bread, rolls; (4) whole milk; and (5) pork. The chemical protein scores of four of the five leading protein sources are 100 (see Table 18.14), but the digestibility scores of the five leading protein sources are all in the 90s (see the same table).

Table 18.14 Protein sources and protein quality measures: examples of protein scores used by the Institute of Nutrition of Central America and Panama (INCAP)

Protein Source	Chemical Quality Score*	True Digestibility in Humans
Meat, eggs, milk	100	95
Beans	80	78
Soy protein isolate	97	94
Rice	73	88
Oats, oatmeal	63	86
Lentils	60	Not available
Corn	50	85
Wheat	44	Refined, 96; whole wheat, 86

*Amino acid score relative to amino acid score of egg, the reference protein.
Source: Data from B. Torun, M. T. Menchu, and L. G. Elias, *Recomendaciones dieteticas diarias del INCAP.* Guatemala: INCAP, 1994 [INCAP Publicacion ME/0571]; and M. E. Shils, J. A. Olson, and M. Shike et al., *Modern Nutrition in Health and Disease.* Philadelphia: Lippincott, Williams & Wilkin, 1999.

People eat mixed diets, so individual protein scores tell only part of the story. For example, using the human protein digestibility scale, the average American and Chinese mixed diets score 96, the rural Mexican diet scores 80, and an Indian rice-and-bean diet scores 78. However, for older adults who are following modified diets, who are cutting back on meat because they can't afford it or lack energy to prepare it, or who may eat too few calories, protein quality can make the difference between a good and a poor diet.

On average, older adults eat slightly more protein as percentage of total calories than adults in general (16.1% and 16.0 % of calories from protein for males and females, compared to 14.9% and 14.6% percent in younger persons).[69] But when the population is heterogeneous, average data are not enough. The following questions address protein adequacy for older individuals:

- Based on height and weight, how much protein will meet the individual's need?
- Are enough calories eaten so that protein does not have to be used for energy?
- If marginal amounts of protein are eaten, is the protein of high quality?
- Are there additional needs: wound healing, tissue repair, surgery, fracture, infection?
- Is the individual exercising? It is harder to achieve nitrogen balance while sedentary.

Because many older adults are frail, it seems wise to keep within the range of 1–1.25 g/kg daily recommended by Evans & Cyr-Campbell.[28] The average older woman weighs 143 pounds, and the average older man weighs 170 pounds. Eating 1 gram of protein per kilogram of body weight would mean eating 65 grams/day for women and 77 grams/day for men, significantly more than the 50 grams recommended as the basis for the Daily Value food-product nutrition labels.[80] Fifty grams of protein would be adequate for an older adult weighing 50 kilograms or 110 pounds.

FATS AND CHOLESTEROL   The need for fat does not seem to change with age; high saturated fat and trans fatty acid intake continues to be a risk factor for chronic disease. Minimizing the amount of saturated fat in the diet and keeping total fat between 20 and 35% of calories is a reasonable goal for older adults to maintain a beneficial blood cholesterol ratio. Cholesterol intake for older adults is already below recommendations for the nation (see Table 18.10). Eggs, which have high cholesterol content, are a nutrient-dense, convenient, and safe food for most people (those without lipid disorders; e.g., high triglyceride and high serum cholesterol). McNamara, a professor of nutrition sciences and director of the Lipid Metabolism Laboratory at the University of Arizona, calculated that

eating 7 eggs per week would raise total serum cholesterol by 1.3 mg/dL.[81] For most people, normal dietary cholesterol does not measurably increase blood cholesterol, and[82] they can enjoy high-cholesterol foods such as shrimp, eggs, and liver for their high-quality protein and other nutrients, without added risk of CVD. Even individuals who are hyperresponders to dietary cholesterol have been shown, albeit in a small study, to maintain their LDL:HDL ratio after high dietary cholesterol intake.[82]

Recommendations for Fluid

Water as percentage of total body weight decreases with age, resulting in a smaller water reservoir and leaving a smaller safety margin for staying hydrated. Drinking 6 or more glasses of fluid per day prevents dehydration (and subsequent confusion, weakness, and altered drug metabolism) in individuals whose thirst mechanism may no longer be very sensitive.

To individualize fluid recommendations, provide 1 milliliter (mL) of fluid per calorie eaten, with a minimum of 1500 mL. For a 2000-calorie diet, that would be 2000 mL or 2 liters of fluid, roughly 8 cups. Foods such as stews, puddings, fruits, and vegetables contribute significant amounts of fluid to the diet but are not counted as part of the fluid allowance in healthy individuals. The Tufts (see Illustration 18.2) and Connecticut adaptations of the Food Guide Pyramid for healthy older adults show 8 glasses of water, which is adequate for a 2000-calorie diet. Individuals who need additional calories can use milk, juice, shakes, and soups as nutrient-dense fluids.

Age-Associated Changes in Metabolism: Nutrients of Concern

The nutritional health of older adults depends on modifying dietary habits that address age-associated changes in absorption and metabolism. The nutrients discussed in the following sections are of special concern for older adults because of age-associated metabolic changes or low dietary intake.

VITAMIN D, CALCIFEROL   Age-related metabolic changes affect vitamin D status, independent of dietary intake, primarily due to decreased ability of the skin to synthesize previtamin D_3 from its precursor, 7-dehydrocholesterol.[83–85] Declining photochemical production may be compounded by limited exposure to sunlight due to use of sunscreen or pollution, institutionalization, or being homebound. Furthermore, in northern regions (above 42° north, the latitude of Boston and Chicago) between November and February, ultraviolet (UV) light is not powerful enough to synthesize vitamin D in exposed skin.[86] The sun's rays are even weaker in Edmonton, Canada (at 52° N latitude), so that previtamin D_3 is not synthesized between mid-October and mid-April. Older individuals who live in these northern latitudes are at greater risk for vitamin D deficiency because they need more vitamin D than younger people do. How far south does one have to go for the winter sun to convert vitamin D precursors? Tests in Los Angeles, at 34° N, showed vitamin D production in skin even in January. The sun's rays in Puerto Rico (latitude of 18° N) were even more effective.[86] Fortunately for people living in northern regions, the body stores vitamin D, so that getting summer sun builds a winter reserve.

Another reason that vitamin D is of concern for older adults (who use more drugs than younger persons) is that commonly used medications interfere with vitamin D metabolism. Examples include barbiturates, cholestyramine, phenytoin (Dilantin), and laxatives.

Intake recommendations for vitamin D intake are 10 mcg (400 international units—IU) for persons aged 51 to 70 and 15 mcg (600 IU) for individuals older than 70 years. Potentially toxic doses are those above 50 mcg (2000 IU), the Tolerable Upper Intake Level (UL) set by the National Academy of Sciences. Symptoms of toxicity are hypercalcemia (high blood calcium levels), anorexia, nausea, vomiting, general disorientation, muscular weakness, joint pains, bone demineralization, and calcification (calcium deposits) of soft tissues. Toxicity is rare from food sources. The relatively few good food sources of vitamin D are fortified cereal, milk (but not cheese), eggs, liver, salmon, tuna, catfish, and herring. Mushrooms also contain a small amount of vitamin D. Cod liver and other fish oils contain medicinal levels of Vitamin D (about 21 mcg, or 840 IU per teaspoon). Although vitamin D intake has not been tracked in national food-monitoring surveys (because technically, humans can derive vitamin D from nonfood sources; i.e., UV light), researchers have assessed individual intake. For instance, Dawson-Hughes et al.[87] found that individuals 65 years old and older consumed 4.5 to 5.0 mcg (180–200 IU) on average.

For healthy aging, the critical function of vitamin D is maintenance of blood calcium levels through intestinal absorption or through bone resorption when the diet is insufficient in calcium.[77] Osteoporosis and osteomalacia (adult rickets) prevention are covered in Chapter 19.

VITAMIN B$_{12}$ Although population intake of vitamin B$_{12}$ or serum cobalamine is higher than the DRI level of 2.4 mcg (see Table 18.11), vitamin B$_{12}$ status suggests that many older adults are unable to use B$_{12}$ efficiently. Vitamin B$_{12}$ blood levels decrease with age even in healthy adults. Compared to 18% of healthy 22- to 63-year-olds in the Framingham Heart Study, 40% of 65- to 99-year-olds had low serum B$_{12}$ levels. Despite adequate food intake, an estimated 30% of older adults suffer from atrophic gastritis. In atrophic gastritis, a bacterial overgrowth in the stomach leads to inflammation, and decreased secretion of hydrochloric acid and pepsin with or without decreased secretion of intrinsic factor and subsequent inability to split Vitamin B$_{12}$ from its food protein carrier. It takes years to develop a B$_{12}$ deficiency; but once developed, the neurological symptoms are irreversible. Symptoms include deterioration of mental function, change in personality, and loss of physical coordination.

"Food first" is usually sound advice regarding nutritional needs, but B$_{12}$ is one of the two vitamins better absorbed in synthetic or purified form. Folic acid is the other one. Synthetic, not protein-bound, vitamin B$_{12}$ is found in fortified foods such as cereals and soy products. Protein-bound B$_{12}$ is found in all animal products, although poultry is a surprisingly poor source (2.4% of B$_{12}$ in adult diets is derived from poultry, compared to 27.3% from beef). The leading food sources of vitamin B$_{12}$ among U.S. adults are beef, milk, fish (excluding canned tuna), and shellfish.[79]

VITAMIN A   Older adults are more likely to overdose with vitamin A than to be deficient in the nutrient. They eat more than the DRI of 900 mcg for males and 700 mcg RE for females (see Table 18.11). Plasma levels and liver stores of vitamin A increase with age. This may be due to increased absorption but is more likely due to decreased clearance of vitamin A metabolites (retinyl esters) from the blood.[88] Kidney disease further elevates serum vitamin A levels because retinol-binding protein, another vitamin A metabolite, can no longer be cleared from blood. Thus older adults are more vulnerable to vitamin A toxicity and possible liver damage than younger individuals are. The UL for vitamin A is 3000 mcg (3 mg).[89] Older adults have a smaller margin of safety for vitamin A than do younger adults.

The plant precursor of vitamin A, beta-carotene, will not damage the liver, although supplements used as antioxidants to prevent cardiovascular disease have been linked to higher all-cause mortality.[90] Excess dietary beta-carotene, because it is water soluble, may give old skin a yellow-orange tint; but it will not lead to hair loss, dry skin, nausea, irritability, blurred vision, or weakness as excess vitamin A would. Leading food sources of vitamin A and beta-carotene among U.S. adults are carrots, ready-to-eat cereals, and milk.[79] Rich vitamin A sources are liver, milk, cheese, eggs, and fortified cereals.

IRON   The need for iron decreases with aging for women after menopause. Like vitamin A, iron is stored more readily in the old than in the young; high intakes of vitamin C enhance absorption. On average, older adults eat more iron than the DRI of 8 mg (males and females; see Table 18.11). Excess iron contributes to oxidative stress, increasing the need for antioxidants to deal with oxidant overload. The UL for iron is 45 mg per day.[89]

Older adults are a heterogeneous population, and not every older person has adequate iron stores. Reasons for inadequate iron status include blood loss from disease or medication (e.g., aspirin), poor absorption due to antacid interference or decreased stomach acid secretion, and low caloric intake. Leading sources of iron in the American diet are ready-to-eat cereals, yeast bread, and beef.[79]

Low Dietary Intake: Nutrients of Concern

VITAMIN E   Also known as tocopherol, vitamin E is a potent antioxidant. It is a problematic nutrient because dietary intake is well below the recommended 15 mg or 15 IU alpha-tocopherol equivalents (TE; see Table 18.11). Vitamin E plays a special role in the health of older adults due to its antioxidant functions, such as hindering development of cataracts[91] and heart disease.[90] Vitamin E is associated with enhanced immune function[91,92] and cognitive status, although not with reduced cardiovascular disease risk.[90,94] The UL is 1000 mg (or IU, seen on supplement labels) alpha TE. Even though vitamin E is fat soluble and can be stored in the body, it seems safe in doses up to 800 IU. At higher doses, vitamin E is linked to longer blood-clotting times and increased bleeding. Aspirin, anticoagulants, and fish oil supplements also increase blood-clotting time and are incompatible with high doses of vitamin E intake. Vitamin E is thought to be more effective in its "d" (rather than "dl") form.

Leading food sources of vitamin E are salad dressings/mayonnaise, margarine, and ready-to-eat cereals. Other good sources of vitamin E are oils (especially sunflower and safflower oils), fats, whole grains, wheat germ, leafy green vegetables, tomatoes, nuts, seeds, and eggs. Food sources cannot supply the optimal dose of 200 IU vitamin E found most effective in modulating heart disease and immune status.

FOLATE, FOLIC ACID   For persons with low serum folate levels, dietary increases of folic acid (100 to 400 mcg) can lower serum homocysteine levels and subsequent risk of heart disease.[95] The DRI for males and females is 400 mcg folate (the food source) per day, with a UL of 1000 mcg of folic acid (the synthetic form). On average, older adults do not meet recommended levels (see Table 18.11). Leading food sources for U.S. adults are ready-to-eat cereals, yeast bread, and orange/grapefruit juice.[79] The folate fortification of grain products begun in the late 1990s has increased older adults' average daily intake; grain products are likely to remain the leading dietary sources of folic acid.

Despite intakes below recommended levels, folate deficiency as measured by serum levels is uncommon in healthy older adults. Depletion is measured as serum folate <3 ng/mL and red blood cell folate <160 ng/mL[87] and is estimated to exist in 11 to 28% of all elderly persons.[96] Absorption of folate, like vitamin B_{12}, may be impaired by atrophic gastritis. Moreover, alcoholism is associated with folate deficiency and subsequent pernicious anemia. Folic-acid deficiency can also be secondary to vitamin B_{12} deficiency, which is more common in the elderly than folic-acid deficiency. Folic-acid deficiency should be treated only with folic acid, or treatment may mask potential B_{12} deficiency, which will continue to cause irreversible neurological damage. Medications commonly used by older adults—such as antacids, diuretics, phenytoin (Dilantin), sulfonamides, and anti-inflammatory drugs—affect folate metabolism.

CALCIUM   A 1994 consensus conference sponsored by the National Institutes of Health[97] recommended calcium intake for women depending on estrogen status; namely, 1500 mg for women aged 50 to 64 not taking supplementary estrogen and 1000 mg calcium per day for postmenopausal women taking estrogen. After age 65, it was recommended that all men and women consume 1500 mg calcium per day. This recommendation of 1500 mg was lowered in 1997, when the National Academy of Sciences set the DRI for adults aged 51 and older at 1200 mg daily, independent of gender or hormone status. The Canadian 2002 Clinical Practice Guidelines outline the highest level of evidence to support these recommendations in addition to stating that calcium and vitamin D should not be used as the sole treatment of osteoporosis; rather, they are adjuncts.[98] Recommended levels are met by a small portion of the population; on average, calcium intake of older men and women ranges between 600 and 800 mg per day, with the lowest intake among older black women.[52,68]

Low calcium intake has been linked to colon cancer, overweight, and hypertension. Study results are strongest for the protective effects of calcium intake in the development of hypertension.[99] For example, blood pressure decreases in a subgroup of hypertensive individuals with higher calcium intake. On average, participants who adhered to the Dietary Approaches to Stop Hypertension (DASH) diet reduced their blood pressure. The diet includes 2 or more servings of low-fat dairy products (1265 mg calcium in the experimental group); 10 servings of fruits and vegetables; and reduced fat, saturated fat, and cholesterol.

The National Academy of Sciences has set the UL for calcium at 2500 mg per day. Individuals should not consume supplements in excess of this amount. Adverse effects of excess calcium (reported at 4 grams per day) include high blood calcium levels, kidney damage, and calcium deposits in soft tissues and outside the bone matrix, such as bone spurs on the spine.[89] High calcium intake, resulting in elevated urinary excretion of calcium, could also lead to new kidney stones in individuals with a history of kidney stones.

MAGNESIUM   Of adults aged 70 and older, approximately one in four meet 100% or more of their magnesium DRI. Adequate magnesium intake (see Table 18.11) is needed for bone and tooth formation, nerve activity, glucose utilization, and synthesis of fat and proteins. Magnesium deficiency can result not only from low intake but also from malabsorption due to gastrointestinal disorders, chronic alcoholism, and diabetes. Signs of deficiency include personality changes (irritability, aggressiveness), vertigo, muscle spasms, weakness, and seizures. An indicator of the wide-ranging functions of magnesium is that it plays a part in over 300 enzyme systems.[100]

Age does not seem to affect magnesium metabolism. However, drugs commonly used by older adults, such as magnesium-based antacids and cathartics, may lead to magnesium overdose. The DRI is 420 mg for males and 320 mg daily for females, with a UL of 350 mg from nonfood (supplementary) sources. Signs of magnesium toxicity are diarrhea, dehydration, and impaired nerve activity. Magnesium from food sources does not result in toxicity. Leading food sources for adults are milk, yeast breads, coffee, ready-to-eat cereals, beef, and potatoes.[79]

ZINC The role of zinc is changing from problematic nutrient because of insufficient intake to potentially problematic because of too much zinc in the diet. Few diets supplied adequate zinc before the requirement was adjusted downward to 8 mg daily for females and 11 mg for males by the 2001 DRI. Fewer than 20% of adults aged 70 and older ingested 100% of the zinc recommended in 1989 (12–15 mg for females and males). However, most people meet the current DRI (see Table 18.11). The UL is 40 mg; excess dietary zinc results in decreased immune response, decreases in HDL cholesterol, and copper deficiency. Leading food sources of zinc among U.S. adults are beef, ready-to-eat cereals, milk, and poultry.[79]

Healthy older individuals maintain zinc balance as well as younger ones do, even though they absorb less zinc.[92] Zinc absorption is affected by conditions of the intestinal tract such as infections, surgery, muscle-wasting diseases, pancreatic insufficiency, and alcoholism, which all increase the need for zinc. Medications including diuretics, antacids, laxatives, and iron supplements also increase the need for zinc.

Of particular interest to older adults are the relationships among zinc deficiency and delayed wound healing, decreased taste acuity and immune response, and increased susceptibility to dermatitis. Zinc supplementation seems to be beneficial only when zinc status is depleted; supplementation in the absence of zinc deficiency does not improve wound healing, taste acuity, or dermatitis. Studies dealing with improved immune status following zinc supplementation are inconsistent.[101]

Nutrient Supplements: Why, When, Who, What, and How Much?

WHEN TO CONSIDER SUPPLEMENTS Can older adults benefit from taking dietary supplements? Yes. Dietary supplements can decrease infection-related illness.[102,103] Population surveys show that diets of many older persons fall short of meeting recommended nutrient levels (see Table 18.15). A boost from appropriately used supplements completes the nutritional balance of an inadequate regular diet.[104] Writing in the Clinician's Corner of the *Journal of the American Medical Association*, Fletcher and Fairfield recommend that all adults take one multivitamin daily.[105]

Table 18.15 Dietary supplements potential used by older adults for health conditions

Condition or Health Status	Dietary Supplement/Herbals
Poor appetite or dieting, leading to intake below 1200–1600 calories	Multivitamin/mineral
Weight loss, chronic underweight	Add high-calorie/protein foods/fats as oils
Vegetarian or vegan	Vitamins B_{12}, D, calcium, zinc, iron
Arthritis*	Antioxidants potentially useful (vitamins C, E), lactobacillus, fiber
Cataracts	Vitamin C
Constipation	Fiber (cellulose, bran, psyllium) Fluid to accompany fiber
Diarrhea	Fluid, multivitamin/mineral
Energy boosters	Evaluate total nutrient intake, adequate calories, iron if blood levels are low
Immune status enhancement	Multivitamin/mineral (DRI dosage)
Memory aids	None proven at this time
Osteoporosis	Vitamins D, K (avoid excess K if on blood thinners), calcium, fluoride, magnesium
Sleep aids	Milk and a sweet at bedtime
Stress reduction	Eat well and play in the sunshine (not available in a pill or tonic)

*K. Hanninen, A. L. Rauma, and M. Nenonen et al., "Antioxidants in Vegan Diet and Rheumatic Disorders," *Toxicology* 155 (2000):45–53.

Any discussion about supplements is based on the assumption that whole foods are the ideal source of nutrients, and supplements are meant to boost a marginal diet. Sometimes it turns out that, as is the case with beta-carotene, the pill form of a nutrient is harmful while the food form promotes health. Furthermore, the interactions among nutrients and the composition of plants and animals making up our food supply are much too complex to replicate in supplements. However, vitamin B_{12} and folic acid are two nutrients better absorbed in a synthetic form than in their protein-bound food form; but this becomes important only when normal metabolic processes fail. It is possible for healthy older adults to live well without dietary supplements.

> **USP (United States Pharmacopeia)** A nongovernmental, nonprofit organization (since 1820); establishes and maintains standards of identity, strength, quality, purity, processing, and labeling for health care products.
>
> **NF (National Formulary)** A uniformity standard for herbs and botanicals.

Age-associated nutritional risk factors such as poverty, isolation, chronic illness, or lack of appetite highlight circumstances when dietary intake might be inadequate and further assessment and intervention, potentially including a supplement, may be useful. Such factors include:

- Lack of appetite resulting from illness, loss of taste or smell, or depression
- Diseases or bacterial overgrowths in the gastro-intestinal tract that prevent absorption
- Poor diet due to food insecurity, loss of function, dieting, or disinterest in food
- Avoidance of specific food groups such as meats, milk, or vegetables
- Contact with substances that affect absorption or metabolism: smoke, alcohol, drugs

WHO TAKES SUPPLEMENTS? The most likely individuals to take supplements are non-Hispanic white females, as are individuals with more education and higher incomes.[106] Women are more likely than men to use supplements.

WHAT TO TAKE? The question: "What should I take?" may not have a simple answer. A multivitamin was the most commonly used vitamin/mineral preparation, and around 30% of elderly men and women use multivitamins.[106] Vitamin E was the second highest vitamin/mineral supplement taken by men (i.e., 14%), whereas calcium was next for women (23%). Finally, the third most often used supplement was vitamin C, with 12% and 19% in men and women, respectively. Most of the nearly 2500 products reported in NHANES III were vitamin-mineral combinations or vitamin–single nutrient combinations.[106] The most common supplement ingredients in this survey were vitamin C, vitamin B_{12}, vitamin B_6, niacin, thiamin, riboflavin, vitamin E, beta-carotene, cholecalciferol or vitamin D, and folic acid. In cases where older adults feel that a supplement will prevent cancer or cardiovascular disease, the U.S. Pre-

ventive Services Task Force (USPSTF) concluded that "the evidence is insufficient to recommend for or against the use of supplements of vitamins A, C, or E; multivitamins with folic acid; or antioxidant combinations for the prevention of cancer or cardiovascular disease."[107]

Considerations that guide supplement choices are the same for older as for younger adults, and address the following five questions:

1. Does the supplement contain a balance of vitamins and minerals?
2. When all supplements and fortified foods are combined, is the dose still safe?
3. Does the supplement contain the missing nutrients?
4. Does the supplement carry a *USP* (U.S. Pharmacopeia, a mark that indicates that the manufacturer followed recognized standards when making the product) or *NF* (National Formulary) code to assure potency and purity?
5. Is the supplement safe? ("Natural" does not mean safe.)

HOW MUCH TO TAKE? In general, multivitamin/mineral supplements should be used in physiologic rather than high-potency doses to maintain a balance of vitamins and minerals. Physiologic dose formulas unique for older adults are available with little or no iron and additional vitamins B_{12} and E.

Generously fortified breakfast cereals, "power" bars, and drinks count as vitamin or mineral supplements. As a group, older adults need the vitamins B_{12}, E, D, and folic acid. The others are superfluous and possibly unsafe (e.g., beta-carotene supplements). Balluz et al.[106] report that adults use more than 300 nonvitamin and nonmineral products, including some that have toxic effects at normal doses (e.g., blue cohosh, chaparral). Because older adults eat fewer calories and are less resilient physically, they also have less tolerance for mistakes in dietary supplement use.

Why? Older people are motivated by wellness and want to take responsibility for their own health, wishing to increase their energy, improve memory, and reduce susceptibility to illness and stress.[106,108,109] Table 18.15 on the facing page is a summary of some of the vitamins, minerals, and other dietary supplements of special interest to older adults.

Dietary Supplements, Functional Foods, Nutraceuticals: Special Interest for Older Adults

> "Eat leeks in March and wild garlic in May,
> And all the year after, physicians may play."
>
> Welsh rhyme

The growing availability of functional foods can be a boon for older adults, who can benefit from the convenience of

Table 18.16 Vitamin and mineral recommendations for nutrition labeling (Daily Reference Values or DV) compared with Dietary Reference Intakes (DRI) and Tolerable Upper Intake Levels (UL) for older adults

MANDATORY VITAMIN AND MINERAL COMPONENTS OF THE NUTRITION LABEL

Nutrient	Daily Values (1993, 1995) All Adults	Dietary Reference Intake (2001) Adults over 70 Males	Females	Upper Limit (UL)
Vitamin A, IU or RE (5 IU = 1 RE = 1 RAE, Retinol Activity Equivalent)	5000 IU	900 RE	700 RE	3000 RE
Vitamin C, mg	60	90	75	2000
Calcium, mg	1000	1200	1200	2500
Iron, mg	18	8	8	45

VOLUNTARY VITAMIN AND MINERAL COMPONENTS OF THE NUTRITION LABEL

Nutrient	Daily Values (1993, 1995) All Adults	Males	Females	Upper Limit (UL)
Vitamin D, mcg (1 mcg = 40 IU)	400 IU	15 mcg	15 mcg	50 mcg
Vitamin E, mg (1 mg = 1 TE = 1 IU)	30 IU	15 mg	15 mg	1000 mg
Vitamin K, mcg	80	120	90	—
Thiamin, mg	1.5	1.2	1.1	—
Riboflavin, mg	1.7	1.3	1.1	—
Niacin, mg	20	16	14	35
Vitamin B_6, mg	2.0	1.7	1.5	100
Folic acid, mcg	400	400	400	1000[a]
Vitamin B_{12}, mcg	6.0	2.4	2.4	—
Biotin, mcg	300	30	30	—
Pantothenic acid, mg	10	5.0	5.0	—
Phosphorus, mg	1000	700	700	3000
Iodine, mcg	150	150	150	1100
Magnesium, mg	400	420	320	350[b]
Zinc, mg	15	11	8	40
Selenium, mg	70	55	55	400
Copper, mcg	2000	900	900	10,000
Manganese, mg	2.0	2.3	1.8	11
Chromium, mcg	120	30	20	—
Molybdenum, mcg	75	45	45	2000
Chloride, mg	3400	2300	2300	—
Potassium, mg	3500	4700	4700	—
Choline, mg	—	550	425	3500
Fluoride, mg	—	4.0	3.0	10

[a]synthetic
[b]nonfood sources
SOURCE: J. A. Pennington and V. S. Hubbard, "Derivation of Daily Values Used for Nutrition Labeling," *J Am Diet Assoc* 97 (1999):1407–12; P. Trumbo et al., "Dietary Reference Intakes: Vitamin A, Vitamin K, Arsenic, Boron, Chromium, Copper, Iodine, Iron, Manganese, Molybdenum, Nickel, Silicon, Vanadium, and Zinc," *J Am Diet Assoc* 101 (2001):294–301; 2004 DRIs for Macronutrients.

nutrient-dense foods such as calcium-fortified orange juice, yogurt-juice beverages, fortified high-fiber cereals, yogurt with live cultures, soymilks, breakfast powders, and various fortified chews, bars, and drinks. However, regular users of functional foods, no matter what age, need to add up all intake in order to avoid potentially toxic nutrient levels (see Table 18.16 for recommended and potentially toxic intake levels). Advice that "you can't overdose on nutrients from foods" does not apply to fortified foods. High nutrient doses can act like drugs, and should be treated as such. Just like drugs, nutrient supplements interact with each other (and with medications); side effects may outweigh the desired benefits.

Several other "ingestibles" besides vitamins and minerals affect older adults' health status. Examples include quasi-functional foods, herbs, stimulants, and other nonvitamin or mineral food components used by older adults to promote health and ward off chronic disease. *Herbal therapy* is defined as the use of plants for medicinal purposes rather than for food consumption, and it comes in many forms: capsules, pills, infusions, tea; tinctures or extracts; and oils and salves. Different parts of the plant are used to produce supplements.[110] Herbal therapies used alongside traditional medical prescriptions are called complementary, and those used alone are called alternative.

Here is a guide for common botanicals and nutraceuticals:

1. *Herbs and spices.* Rosemary, sage, thyme, cinnamon, clove, and ginger enhance the taste of food and protect against oxidative stress.

2. *Caffeine.* A stimulant to which older people may become increasingly sensitive.

3. *Black and green teas (Camellia sinensis).* Contribute fluid *and* phytochemicals, especially the antioxidant catechins and flavonols.

4. *Garlic.* Contains allicin, which has antibacterial activity, and ajoene, which enhances blood thinning. Eating garlic has been linked to colon cancer protection and better heart health. Individuals taking aspirin or blood thinners shouldn't take garlic pills or eat large amounts of garlic (one or two garlic cloves are considered a small amount).

5. *Herbal treatments.* Widely used and often effective, but herbs used as medicinal treatment can also be dangerous. Herbs should help, not hinder health! Rule #1 in using herbs is also the first rule of medicine: "Do no harm." A list of unsafe herbs is found in Chapter 16.

6. *Phytochemicals.* These plant-based compounds (such as reservatrol, flavonoids, carotenoids, indoles, isoflavones, lignans, and salicylates) are of special interest in aging because they have been linked to reduction of chronic disease. Eating fruits, vegetables, and whole grains automatically increases phytochemical intake.

7. *Pre- and probiotics.* Prebiotics are nondigestible food ingredients that feed health-promoting colon bacteria, and probiotics are live health-promoting bacterial cultures such as lactobacillus acidophilus and bifidobacterium in yogurt. After a course of antibiotic treatment, biotics can reestablish intestinal bacterial life.

8. *Plant stanols.* Plant products that compete with cholesterol in the small intestine because they are similar in structure to cholesterol. Found in corn, wheat, oats, rye, and some other foods; also processed from wood. Eating margarine spreads containing stanols results in decreased levels of LDL but stable HDL.

9. *Hormones.* DHEA (dehydroepiandrosterone) is taken to increase muscle mass and immune function, pregnenolone to enhance memory, and melatonin to enhance sleep. There is no evidence that preg-

nenolone enhances human memory or improves concentration, but there is some evidence that the other two can work. Despite equivocal evidence for melatonin, it is popularly used to induce sleep and reduce jet lag.[111] Secretion of this biorhythm regulator normally decreases with age. Studies are now examining whether melatonin supplements help people with Alzheimer's syndrome sleep better.

Nutrient Recommendations: Using the Food Label

The Nutrition Facts Panel on food packages is structured to provide nutrient content information in relation to nutrient needs. The Food and Nutrition Board of the National Academies of Sciences, National Research Council, worked with scientists from Canada to establish an integrated set of reference values for specific nutrients.

The recommended nutrient amounts for older adults (aged 70 years and older) are slightly different from the amounts used as references for the nutrition label and that are recommended for "all" adults (see Table 18.16). For example, older adults need more vitamin C and calcium than the reference amounts listed in the mandatory components of the label, and they need less iron and zinc than the label value. Values also differ for vitamins D, E, and B_{12}. Table 18.16 lists the levels used for food labels and the current DRI values for older adults. The UL is listed because with today's food supply, it is easy to overdose vitamins and minerals.

Older adults can still use the percentages on food labels for dietary guidance, as long as they adjust them to get more than 100% calcium and vitamins D and C and less than 100% for iron and zinc. The vitamin B_{12} label recommendation is higher than the DRI, but poor absorption in older adults makes unsafe intake from foods unlikely. In nutrition labeling and dietary guidance, "one size does not fit all."

Cross-Cultural Considerations in Making Dietary Recommendations

Food habits develop in cultural contexts, and we can learn about them through various ways. Travel through North America would allow us to observe cultures that make up our society. Visiting ethnic restaurants, stores, and farmer's markets can be another way to get a glimpse of cultural food diversity. Cookbooks, films, talking with individuals about their food history, and participating in ethnic celebrations are other sources of insight into food patterns of various cultures. Each new immigrant wave adds unique food traditions to the country's mix. Older adults may be stronger advocates for upholding traditional food patterns than young people are. In working with the elderly on food issues, it is useful (and interesting) to determine whether food and lifestyle habits are patterned on specific cultural considerations.

National food-monitoring programs survey the population with proportionately larger samples drawn from minority groups in order to develop a balanced picture of the whole population. North America is home to rapidly growing Hispanic, Asian, Russian, and African immigrant groups. The U.S. census tracks minority groups, but it is completed only once every 10 years. The local Area Agencies on Aging and Senior Nutrition Programs also track population trends, and are likely to have greater insight about some of the smaller ethnic population groups in their unique communities and regions.

Cultural differences are reflected in approaches to dietary guidance. For example, Chile has separate guidelines for older and younger adults. In New Zealand, older adults are encouraged to socialize at mealtimes to improve appetite. In France, South Korea, and Japan, people of all ages are encouraged to enjoy mealtimes and take pleasure in eating. Other unique guidelines are those of China, which suggest people eat 20 to 25 grams (nearly an ounce) of fish daily. Guatemala uses a bean pot as a nutritional icon. India has developed one set of guidelines for the rich (i.e., overall energy intake should be restricted to levels commensurate with the sedentary occupations of the affluent, so that obesity is avoided; total fat intake is not to exceed 20% of total energy; and use of clarified butter, a prized Indian culinary ingredient, should be restricted to special occasions) and one for the poor (addressing the fact that at least one-third of the households in India are not able to afford even the minimum nutritional requirements, even though they are spending 80% of their income on food, and so recommendations identify food combinations that are most likely to meet recommended dietary intakes).

Communicating effectively and avoiding misinterpretation in intercultural settings is probably the most important thing a nutritionist can learn to do when working with older adults from various cultures; it requires skill in transferring information, developing and maintaining relationships, and gaining compliance.[112] Developing individual skills takes time, commitment, and practice. On the other hand, nutrition education and guidance tools have been developed in many languages and for diverse cultures, although not typically for elders of ethnic groups. Culturally appropriate resources can be found through local extension services, cross-cultural education centers, diabetes education programs, public health agencies, and some commodity groups such as the Dairy Council. Multilingual versions of the Food Guide Pyramid can be downloaded from the Nutrition Education for New Americans Project in the Department of Anthropology and Geography at Georgia State University (http://multiculturalhealth.org). Food pyramids in nearly 40 languages are available, including Amharic, Russian, Somali, and Vietnamese.

Food Safety Recommendations

Foodborne illness risk defines a biological, chemical or physical property of food (a hazard) that may cause illness or injury under a given set of conditions.[113] Older adults are vulnerable to foodborne illness because they often have a compromised immune status. How widespread this problem is is basically unknown, because many foodborne illnesses are not reported when individuals think it is "the flu."

Poor storage, thawing, and food-handling practices leading to microorganism growth are generally to blame for these illnesses. Bacteria and viruses, especially campylobacter, salmonella, and Norwalk-like viruses, are among the most prevalent causes of foodborne illnesses. Signs and symptoms may appear within half an hour of eating a contaminated food or may not develop for up to 3 weeks. They include gastrointestinal distress, diarrhea, vomiting, and fever.

Leading practices that put an older person at risk are as follows:

- Improper holding temperatures of foods
- Poor personal hygiene
- Contaminated food preparation equipment (cutting boards, knives)
- Inadequate cooking time

The new Dietary Guidelines for Americans provide the following suggestions to keep food safe:

- Wash hands and surfaces often.
- Separate raw, cooked, and ready-to-eat foods while shopping, preparing, or storing.
- Cook foods—especially raw meat, poultry, fish, and eggs—to a safe temperature.
- Refrigerate or freeze perishable or prepared foods within 2 hours.
- Follow the label for food-safety preparation and storage instructions.
- Serve hot foods hot (140°F or above) and cold foods cold (40°F or below).
- When it doubt, throw it out!

Physical Activity Recommendations

> "There is no segment of the population that can benefit more from exercise than the elderly."
>
> William J. Evans[114]

Exercise is a true fountain of youth. Physical activity builds lean body mass, helps to maintain balance and flexibility, contributes to aerobic capacity and to overall fitness,[28,115] and boosts immune status.[116] Only resistance

Table 18.17 Healthy People 2010 physical activity goals and baseline activity levels comparing older and younger adults

TYPE OF ACTIVITY	BASELINE LEVELS		TARGET
	Age 75+	Age 18–24	All Adults
Engage in no leisure-time physical activity	65%	31%	20%
Engage in regular physical activity at least 20 min, 3 times/wk	23%	36%	30%
Perform strengthening activities 2 or more days/wk	8%	30%	30%
Perform stretching and flexibility activities	21%	39%	40%

SOURCE: *Healthy People 2010: National Health Promotion and Disease Prevention Objectives,* Washington, D.C.: U.S. Department of Health and Human Services, 2000.

exercises; for example, weight lifting, build muscle and bone. All physical activities count toward energy expenditure and health maintenance.

National health goals encourage increased physical activity levels for older adults (see Table 18.17). On average, older adults are less active than younger ones. Lower activity levels as well as deteriorating strength, endurance, and sense of balance are associated with, but not caused by, increasing age.

Older people benefit from exercise even more than younger people do because strength training is the only way to maintain and build muscle mass. When evaluating strength in men and women, muscle mass (not function) is the major determinant of age- and gender-related differences in strength, independent of where muscle is located.[28] Besides making a person stronger, increased muscle mass also leads to higher caloric turnover and less likelihood of weight gain.

Age does not hinder training effects. For example, in a comparison of changes in body composition between young and old individuals during 20 weeks of an endurance training program, all individuals gained total body density and fat-free mass while losing fat mass and abdominal fat. Responses to training differed by gender and by race, but *not* by age.[32]

Exercise Guidelines

How can one predict whether physical activity will exacerbate existing medical conditions? Physician screening or assessment by completing a questionnaire like the one in Table 18.18 can identify potential problem areas.[28] Kligman and colleagues identify a more detailed assessment and recommend ways to evaluate cardiovascular fitness, strength, function, balance, flexibility, body composition, bone den-

Table 18.18 Keep moving—fitness after 50 chart

A. Do I get chest pains while at rest and/or during exercise?

B. If the answer to question A is "yes": Is it true that I have not had a physician diagnose these pains yet?

C. Have I ever had a heart attack?

D. If the answer to question C is "yes": Was my heart attack within the last year?

E. Do I have high blood pressure?

F. If you do not know the answer to question E, answer this: Was my last blood pressure reading more than 150/100?

G. Am I short of breath after extremely mild exertion and sometimes even at rest or at night in bed?

H. Do I have any ulcerated wounds or cuts on my feet that do not seem to heal?

I. Have I lost 10 lb or more in the past 6 months without trying and to my surprise?

J. Do I get pain in my buttocks or the back of my legs—my thighs and calves—when I walk? (This question is an attempt to identify persons who suffer from intermittent claudication. Exercise training may be extremely painful; however, it may also provide relief from pain experienced when performing lower-intensity exercise.)

K. When at rest, do I frequently experience fast irregular heartbeats or, at the other extreme, very slow beats? (Although a low heart rate can be a sign of an efficient and well-conditioned heart, a very low rate can also indicate a nearly complete heart block.)

L. Am I currently being treated for any heart or circulatory condition, such as vascular disease, stroke, angina, hypertension, congestive heart failure, poor circulation to the legs, valvular heart disease, blood clots, or pulmonary disease?

M. As an adult, have I ever had a fracture of the hip, spine, or wrist?

N. Did I have a fall more than twice in the past year (no matter what the reason)? (Many older persons have balance problems and at the initiation of a walking program will have a high chance of falling. These persons may benefit from balance training and resistance exercise before beginning a walking program.)

O. Do I have diabetes?

SOURCE: From W. J. Evans, D. Cyr-Campbell, "Nutrition, Exercise, and Healthy Aging," JADA 1 997; 97:632–8. Copyright the American Dietetic Association. Reprinted by permission from the author.

Case Study 18.1

JT—Spiraling Out of Control?

JT, a retired computer company executive, eats out four times a week since his wife died last year. Meals at home consist of microwave dinners or supreme pizza. He belongs to a health club, which he visits three times a week. After working out and socializing, JT and his friends go for beer. He developed type 2 diabetes five years ago. Last week, he visited the clinic for his annual check-up. He was measured at 5 feet 9 inches and weighed 235 pounds.

Photo Disc

Questions

1. If you were his nutritionist, what nutrition remedies would you prescribe for JT?
2. What sort of advice would you give JT about weight management?
3. What would you ask JT about his food and fitness routine, and how would you convince him that he needs an aggressive nutrition and fitness program?
4. What sort of fluid recommendation would you make?

(Answers are in Appendix D.)

---

sity, and lipid levels.[117] Overall, an evaluation by an individual's physician is needed to identify contraindications to exercise. Individuals cleared for participation find that even small increases in exercise add up. Even patients with heart disease can be compliant with exercise recommendations.[118]

Many types of physical activity are good. Muscle mass is built through resistance or weight-bearing activities such as walking, running, and jumping, which move the body's own weight against gravity. Playing tennis and jumping rope are examples of games or training activities that incorporate weight-bearing exercise. Lifting weights is another example. Water exercises can be done either as resistance training, by pushing webbed gloves or empty plastic containers against the water, or as aerobic activity.

Aerobic exercise builds endurance and contributes to cardiac fitness, even when mobility is limited. Nonimpact aerobics and chair aerobics classes can help people with arthritis or hip replacements, or those who are wheelchair bound, to get and stay strong. An explicit integrative review on types of interventions to promote physical activity in older adults found that some older adults are able to increase activity levels. However, there is no single intervention component that dramatically or consistently produces successful outcomes.[119] Elaine Souza, RD, MPH, has developed the following guide for planning effective exercise sessions:

- Decide on frequency: 2–3 times per week is effective for strength training, using 8–10 different exercises with 8–12 repetitions each, the whole routine to be done in 20–30 minutes.

- For general health, exercise for 30 minutes on most days of the week.
- Drink water when exercising (before, during, and after exercise).
- Do warm-up and cooldown activities of 5 to 10 minutes each.

Individuals can take charge of their own healthy aging by developing appropriate and effective exercise habits. Simple and, occasionally challenging, fitness advice: Eat good food, drink lots of water, and play hard! See Case Study 18.1

Nutrition Policy and Intervention for Risk Reduction

Nutrition policy promotes health by combining nutrition education for individuals and population interventions. The ultimate goal of nutrition intervention is to produce a better health outcome.

Nutrition Education

"The human mind, once stretched to a new idea, never goes back to its original dimension."

Oliver Wendell Holmes

Contrary to some beliefs, older people do learn and change. Someone born in 1920 has seen the invention of fast-food restaurants, microwaves, television and TV dinners, and a whole host of computer-controlled kitchen appliances. To age is to grow and adapt. Learning new nutrition habits is part of aging. Nutrition education is different from education in general because its goal is changed dietary behaviors. Nutrition education consists of a set of learning experiences to facilitate voluntary adoption of nutrition-related behaviors that are conducive to health and well-being. Several requirements must be met for nutrition education—that is, behavior change—to occur (see Illustration 18.3). Think of it as the 4 C's of nutrition education.[120]

1. *Commitment.* Commitment means being motivated to adopt health-promoting behaviors and intending to adopt and maintain a new food behavior.
2. *Cognitive processing.* Understanding how a food behavior contributes to health and planning how it will fit into your life constitutes cognitive processing.
3. *Capability.* Acquiring the skills to practice new food behaviors is part of nutrition education. An example is learning to identify whole grain breads or to prepare vegetables when intending to adopt a high-fiber diet.
4. *Confidence.* "Nothing breeds success like success!" The best predictor that someone will practice new dietary habits is their personal confidence in being able to do so.

Educational sessions for older learners are best designed around their potential limitations, such as declines in visual acuity and hearing loss.[120] Adaptation for written material includes:

- Larger type size
- Serif lettering (Helvetica is a typeface with no serifs; Times Roman is a typeface with serifs.)
- Bold type
- High contrast (black on white)
- Non-glossy paper to decrease glare

Illustration 18.3 Four essential elements to achieve and maintain individual dietary behavior change.

SOURCE: Adapted from U. B. Krinke, "Effective Nutrition Education Strategies to Reach Older Adults," in R. R. Watson (ed.), *Handbook of Nutrition in the Aged.* Boca Raton, FL: CRC Press, 2001.

- Avoid blue, green, and violet (paper and print color) due to decreased ability to discriminate among these colors
- Reading level of 5th to 8th grade

Educational strategies can result in better diets for individuals, but these alone may not be sufficient to enhance the dietary patterns of populations. Cultural environments support, ignore, or punish desired behavior change. An example is a peer group that values health and fitness; group members will support each other's health-promoting behaviors.

Food and nutrition policies arise from public values, beliefs, and opinions that define the cultural context in which dietary behaviors exist. Policies can be overt or unspoken. Public policies supporting the health of older adults are evident in the Social Security program, which provides financial support for postretirement living, and the Food Stamp program, which makes grocery money available when living resources are inadequate.

Model Programs Exemplify Intervention Goals

NUTRITION PARTNERS Nutrition Partners is a model nutrition intervention program for free-living older adults in a large midwestern metropolitan community. Nadine Reiser, RD, project director of the Senior Dining Program, applied for and received a grant to pilot a home-visiting program to assist elders with their nutritional problems. Dietetic staff of the nutrition program make home visits in response to referrals from community agencies, physicians, and program site staff; staff members do dietary assessments, develop and prioritize treatment plans together with the client, and make return visits periodically to monitor the intervention. Clients make changes appropriate to their specific lifestyles. For instance, to avoid dehydration and side effects from medications, one client decided to keep a large bottle of water in the refrigerator, refilling it each morning and drinking until it was empty at the end of the day. A daily mark on the calendar tracking refills helps to ensure that lack of thirst does not lead to dehydration.

STORE-TO-DOOR After "retiring" in the 1980s, Dr. Dave Berger surveyed his community to see where he might do some good. Many older adults told him that getting to the grocery store was impossible, and that even when they could get to the store, bringing the bags home was difficult. Winter ice and snow made things worse, for people feared falling. There had been a grocery delivery service, but it closed because profits were not meeting shareholder expectations. So Dr. Berger joined forces with his wife Fran and with friends, colleagues, and a lot of volunteers to start up the nonprofit Store-to-Door, a home-delivered grocery program for older people and those who are disabled. For a small fee, volunteers will shop for you. They buy items that discount grocers offer: food, of course, but also greeting cards, medicines, paper goods, and cleaning supplies, although no alcohol or cigarettes. Customers can get credit for coupons. Volunteers deliver the groceries throughout the year. After starting Store-to-Door in the Upper Midwest, Dr. Berger started another in Portland, Oregon. When last heard from, he was in Ventura, California, running Shop Ahoy, his latest home-delivered grocery program.

Community Food and Nutrition Programs

Elderly Nutrition Programs

Governmental programs for older adults include the USDA's Food Stamp and extension programs, Adult Day Services food programs, Nutrition Assistance Programs for Seniors (NAPS), Meals-on-Wheels and other home-delivered meal programs, and the Senior Nutrition Program of the Older Americans Act.

Nongovernmental home health programs provide food and nutrition services as part of a broader range of screening and assessment, nursing, and other support services. For instance, home health aides will shop for and prepare food, and clean up the kitchen afterward. Home care services allow individuals to receive the necessary support to stay in their homes for as long as they wish. Remaining in one's home indefinitely is sometimes referred to as "aging in place." A broader definition of this concept is found in the Position of the American Dietetic Association.[1]

Aging in place has many definitions and does not necessarily mean living in one setting or one home for a lifetime. Ideally, aging in place offers choices from a spectrum of living options and medical and supportive services customized to accommodate those who are fully active and have no impairments, those who require limited assistance, and those with more severe impairments who require care in long-term care facilities.

Other food and nutrition programs that contribute to the continuum of nutrition services include commodity foods, food pantries and soup kitchens, cooperative buying groups such as Fare for All, and screening and referral services.

Senior Nutrition Program

Congress first appropriated funds under Title VII of the Older Americans Act of 1965 to begin the Senior Nutrition Program, also called the Elderly Nutrition Program (ENP). The Senior Nutrition Program was created to alleviate poor nutritional intake and reduce social isolation among older adults. It was based on evidence that older adults do not eat adequately because of the following:

1. Lack of income limits ability to purchase food.
2. Lack of skills limits ability to select and prepare nourishing meals.
3. Limited mobility affects shopping and meal preparation.
4. Feelings of isolation and loneliness decrease the incentive to eat well.

Senior Nutrition, now Title IIIC of the Older Americans Act, is a community-based nutrition program that provides meals (congregate and home-delivered), increased social contact, nutrition screening and education, and information and linkages to other support programs and services, as well as volunteer opportunities. Anyone who is 60 or more years of age (and spouse regardless of age) is eligible to participate in the congregate dining program; home-delivered meal clients must be homebound and unable to prepare their own meals. Typically, $1.00 of Title III funds spent on congregate services is supplemented by an additional $1.70 from other sources; the average cost of an ENP meal is $5.17, and a home-delivered one is $5.31.[121] Title VI grant programs are similar to Title III funds and were established to help deliver social and nutrition services to older American Indians, Alaskan Natives, and Native Hawaiians. About 25% of participants are minorities, almost twice the national percentage of minority adults over age 60.[121]

Today there is less poverty among younger seniors; but there are greater nutritional and social needs among frail elderly and individuals with low incomes, chronic health conditions, limited mobility, and limited English-speaking ability, and among minority and isolated elders. About one-third of Title III congregate meal participants and more than one-half of Title VI meal participants have incomes at or below the Department of Health and Human Services (DHHS) poverty threshold.[122]

Senior dining sites are targeted to neighborhoods where older, frail, impoverished seniors live. Dining sites are located in community centers, senior centers, civic buildings, subsidized housing units, schools, and other accessible locations. Meals are delivered to the homes of individuals who are 60 years of age or older, homebound by reason of illness or disability, and unable to prepare meals. Services are adapted to meet each unique community's setting. For instance, meal vouchers for use at local cafes or diners are available in some small communities.

Other services to meet the needs of frail older seniors include multiple meals, weekend meals, take-home snacks, liquid supplements, nutrition screening and education, and one-to-one nutrition counseling. Also available are special diets for medical reasons and special meals for Jewish and ethnic elders.

Senior or elderly nutrition programs have successfully brought together millions of people to socialize and enjoy nutritious meals. In 2002, about 250 million congregate and home-delivered meals were served to approximately

2.6 million older adults.[121] The Older Americans Act 2000 Amendment, Section 339 (Nutrition) (H.R. 782) states that nutrition projects shall use a dietitian (or person with compatible expertise) to provide meals that comply with the Dietary Guidelines for Americans. Nutrition program meals were found to exceed the one-third RDA standard. Compared to nonparticipants, participants had up to 31% higher intake of recommended nutrients. In other words, the program is working. Surveys show that targeting those who need services most is successful, and a national evaluation of the senior nutrition program found that the program is targeted to those most in need.[122] Increasing socialization was one of the original program goals and continues to be one of the outcomes. Participants have 16–18% more social contacts per month than nonparticipants do. Relative to the general older population, participants are older and more likely to be female, to belong to an ethnic minority, to live alone, and to have incomes well below poverty level.

Grocery shopping assistance is an important service for frail seniors. A variety of models are used, including volunteer escorts to the supermarket, bus rides, and grocery delivery to the door. However, Title IIIC funds may not currently pay for grocery shopping assistance.

The Promise of Prevention: Health Promotion

"Grow old along with me, the best is yet to be!"

Robert Browning

Although good nutritional habits make a greater impact when started early in life,[25] sometimes individuals are not motivated to pursue these risk-reduction strategies until later in life or after experiencing a health problem. Successful strategies to reach an older audience reflect their specific needs and interests. It is time for the belief that an 80-year-old is too old to learn and practice health-promotion strategies to become an outdated myth.

Resources

AARP

Information related to growing old in America: finances, politics, health issues, travel, and population statistics.
Web site: www.aarp.org/visit/brpromo.htm

Center for Disease Control and Prevention

Mortality and morbidity data.
Web site: www.cdc.gov/mmwr

Florida National Policy and Resource Center on Nutrition and Aging

Aging Policy and education center.
Web site: www.fiu.edu/~nutreldr

National Heart, Lung, and Blood Institute

Gateway site for consumers and health professionals; links and information materials about heart health and more.
Web site: www.nhlbi.nih.gov

Oral Health in America

A report of the Surgeon General.
Web site: www.nidr.nih.gov/sgr/sgrohweb/home.htm

Tufts University Center on Aging

Popular gateway to information on aging.
Web site: www.navigator.tufts.edu

U.S Administration on Aging

Fact sheet, news, links to other resources.
Web site: www.aoa.dhhs.gov

References

1. Position of the American Dietetic Association: nutrition, aging, and the continuum of care. J Am Diet Assoc 2000;100:580–95.

2. Maloney SK, Fallon B, Wittenberg CK. Executive summary. Aging and health promotion: market research for public education. Washington, DC: Office of Disease Prevention and Health Promotion, U.S. Department of Health and Human Services, PHS; 1984.

3. Nube M, Kok FJ, Vandenbroucke JP et al. Scoring of prudent dietary habits and its relation to 25-year survival. J Am Diet Assoc 1987;87:171–5.

4. Kant AK, Schatzkin A, Graubard BI et al. A prospective study of diet quality and mortality in women. J Am Med Assoc 2000;283:2109–15.

5. Vita AJ, Terry RB, Hubert HB et al. Aging, health risks, and cumulative disability. N Engl J Med 1998;338:1035–41.

6. MA Johnson. Achieving 100 candles: the Georgia Centenarian Study lights the way. USDA 2000 Millenium Lecture Symposium. September 28, 2000.

7. Mokdad AH, Marks JS, Stroup DF, Gerberding JL, Actual causes of death in the United States 2000. 2004; 291:1238–1245.

8. Health, United States 2003, with chartbook in trends in the health of Americans. Washington, DC: U.S. Government Printing Office; 017-022-01546-3, December 2003. Available at www.cdc. gov/nchs, accessed 10/03.

9. 1995 White House Conference on Aging. The road to aging policy for the 21st century. Washington, DC: U.S. National Commission on Libraries; 1996.

10. Healthy People 2010: national health promotion and disease prevention objectives. Washington, DC: U.S. Department of Health and Human Services; 2000.

11. Hayflick L. How and why we age. Exp Gerontol 1998;33:639–53.

12. Tuljapurkar S, Li N, Boe C. A universal pattern of mortality decline in the G7 countries. Nature 2000;405:789–92.

13. Bernarducci MP, Owens NJ. Is there a fountain of youth? A review of current life extension strategies. Pharmacotherapy 1996;16:183–200.

14. Forbes GB. Human body composition: growth, aging, nutrition, and activity. New York: Springer-Verlag; 1987:31.

15. Buys CH. Telomeres, telomerase, and cancer. N Engl J Med 2000;342:1282–3.

16. Bodnar AG, Ouellette M, Frolkis M et al. Extension of life-span by introduction of telomerase into normal human cells. Science 1998;279:349–52.

17. Parsons PA. The limit to human longevity: an approach through a stress theory of ageing. Mech Ageing Dev 1996; 87:211–8.

18. Masoro EJ. Hormesis and the antiaging action of dietary restriction. Exp Gerontol 1998;33:61–6.

19. Weindruch R, Sohal RS. Seminars in medicine of the Beth Israel Deaconess Medical Center. Caloric intake and aging. N Engl J Med 1997;337:986–94.

20. McCay C, Crowell M, Maynard L. The effect of retarded growth upon the length of life and upon ultimate size. J Nutr 1935;10:63–79.

21. Masoro EJ. Nutrition and aging in animal models. In: Munro HN, Danford DE, eds. Nutrition, aging, and the elderly. New York: Plenum Press; 1989: 25–41.

22. Weindruch R. The retardation of aging by caloric restriction: studies in rodents and primates. Toxicol Pathol 1996;24:742–5.

23. Roth GS, Ingram DK, Lane MA. Calorie restriction in primates: will it work and how will we know? J Am Geriatr Soc 1993;47:896–903.

24. Holliday R. Food, reproduction, and longevity. Bioessays 1989;10:125–7.

25. Weisburger JH. Approaches for chronic disease prevention based on current understanding of underlying mechanisms. Am J Clin Nutr 2000;71:1710S–4S; discussion 1715S–9S.

26. Walker AR, Walker BF. Nutritional and non-nutritional factors for "healthy" longevity. J R Soc Health 1993;113:75–80.

27. Forbes GB. Longitudinal changes in adult fat-free mass: influence of body weight. Am J Clin Nutr 1999;70:1025–31.

28. Evans WJ, Cyr-Campbell D. Nutrition, exercise, and healthy aging. J Am Diet Assoc 1997;97:632–8.

29. Guo SS, Zeller C, Chumlea WC et al. Aging, body composition, and lifestyle: the Fels Longitudinal Study. Am J Clin Nutr 1999;70:405–11.

30. Poehlman ET, Toth MJ, Bunyard LB et al. Physiological predictors of increasing total and central adiposity in aging men and women. Arch Intern Med 1995;155:2443–8.

31. Carter WJ. Macronutrient requirements for elderly persons. In: Chernoff R, ed. Geriatric nutrition: the health profes-sional's handbook. Gaithersburg, MD: Aspen; 1999.

32. Wilmore JH, Despres JP, Stanforth PR et al. Alterations in body weight and composition consequent to 20 weeks of endurance training: the HERITAGE Family Study. Am J Clin Nutr 1999;70: 346–52.

33. Fiatarone MA, Marks EC, Ryan ND et al. High-intensity strength training in nonagenarians: effects on skeletal muscle. J Am Med Assoc 1990;263:3029–34.

34. Elahi VK, Elahi D, Andres R et al. A longitudinal study of nutritional intake in men. J Gerontol 1983;38:162–80.

35. Chernoff R, ed. Geriatric nutrition: the health professional's handbook. Gaithersburg, MD: Aspen; 1999.

36. Kaneda H, Maeshima K, Goto N et al. Decline in taste and odor discrimination abilities with age, and relationship between gustation and olfaction. Chem Senses 2000;25:331–7.

37. Schiffman SS. Taste and smell losses in normal aging and disease. J Am Med Assoc 1997;278:1357–62.

38. McDonald RB. Influence of dietary sucrose on biological aging. Am J Clin Nutr 1995;62:284S–292S; discussion 292S–293S.

39. Cain WS, Stevens JC. Uniformity of olfactory loss in aging. Ann N Y Acad Sci 1989;561:29–38.

40. Ship JA, Pearson JD, Cruise LJ et al. Longitudinal changes in smell identification. J Gerontol A Biol Sci Med Sci 1996; 51:M86–91.

41. Wysocki CJ, Gilbert AN. National Geographic Smell Survey. Effects of age are heterogenous. Ann NY Acad Sci 1989; 561:12–28.

42. Bartoshuk LM. Taste. Robust across the age span? Ann NY Acad Sci 1989; 561:65–75.

43. Schiffman S. Changes in taste and smell: drug interactions and food preferences. Nutr Rev 1994;52(II):S11–4.

44. Jones DB, Niendorff WJ, Broderick EB. A review of the oral health of American Indian and Alaska Native elders. J Public Health Dent. 2000;60(suppl 1):256–60.

45. Appollonio I, Carabellese C, Frattola A et al. Influence of dental status on dietary intake and survival in community-dwelling elderly subjects. Age Ageing 1997;26:445–56.

46. Mobley C, Saunders MJ. Oral health screening guidelines for nondental health care providers. J Am Diet Assoc 1997; 97(suppl 2):S123–S126.

47. Vargas CM, Kramarow EA, Yellowitz JA. The oral health of older Americans. Aging Trends; no. 3. Hyattsville, MD: National Center for Health Statistics; 2001.

48. Roberts SB, Fuss P, Heyman MB et al. Control of food intake in older men. J Am Med Assoc 1994;272:1601–6.

49. Rolls BJ. Regulation of food and fluid intake in the elderly. Ann NY Acad Sci 1989;561:217–25.

50. Phillips PA, Johnston CI, Gray L. Disturbed fluid and electrolyte homoeostasis following dehydration in elderly people. Age Ageing 1993;22:S26–33.

51. Phillips PA, Rolls BJ, Ledingham JG et al. Reduced thirst after water deprivation in healthy elderly men. N Engl J Med 1984; 311:753–9.

52. Wilson JW, Enns CW, Goldman KS et al. 1997, June 2. Data tables: combined results from USDA's 1994 and 1995 Continuing Survey of Food Intakes by Individuals and 1994 and 1995 Diet and Health Knowledge Survey, 1997. Available at www.barc.usda.gov/bhurc/foodsurvey/home.htm, accessed 1/00.

53. White JV, Dwyer JT, Posner BM et al. Nutrition screening initiative: development and implementation of the public awareness checklist and screening tools. J Am Diet Assoc 1992;92:163–7.

54. A Consensus Conference sponsored by the Nutrition Screening Initiative. Report of nutrition screening I: toward a common view, Washington, DC, April 8–10, 1991. The Nutrition Screening Initiative.

55. White JV, Ham RJ, Lipschitz DA et al. Consensus of the Nutrition Screening Initiative: risk factors and indicators of poor nutritional status in older Americans. J Am Diet Assoc 1991;91:783–7.

56. Posner BM, Jette AM, Smith KW et al. Nutrition and health risks in the elderly: the nutrition screening initiative. Am J Public Health 1993;83:972–8.

57. Rose D, Oliveira V. Nutrient intakes of individuals from food-insufficient households in the United States. Am J Public Health 1997;87:1956–61.

58. French D. Avoiding adverse drug reactions in the elderly patient: issues and strategies. Nurs Pract 1996;21(9):90–105.

59. Kaufman DW, Kelly JP, Rosenberg L, Anderson TE, Mitchell AA. Recent patterns of medication use in the ambulatory adult population of the United States. The Slone Survey. J Am Med Assoc 2002;287(3): 337–44.

60. De Castro JM. Social facilitation of food intake: people eat more with other people. Food Nutr News 1994; 66:29–30.

61. Gerrior SA, Guthrie JF, Fox JJ et al. How does living alone affect dietary quality? U.S. Department of Agriculture, ARS, Home Economics Research Report No. 51 in Fam Econ Nutr Rev 1995;8:44–6.

62. Sahyoun NR, Jacques PF, Dallal GE et al. Nutrition Screening Initiative checklist may be a better awareness/educational tool than a screening one. J Am Diet Assoc. 1997;97:760–64.

63. Garry PJ, Owen GM, Eldride TO. The New Mexico Aging in Process Study. 1980–1997. Albuquerque: University of New Mexico Health Sciences Center; 1997.

64. DeGroot LC, Beck AM, Schroll M, van Staveren WA. Evaluating the DETERMINE Your Nutritional Health Checklist and the Mini-Nutritional Assessment as tools to identify nutritional problems in elderly Europeans. Eur J Clin Nutr 1998;52(12):877–83.

65. Crogan NL, Corbett CF, Short RA. The minimum data set: predicting malnutrition in newly admitted nursing home residents. Clin Nurs REs. 2002;11(3):341–353.

66. American Dietetic Association Foundation. Nutrition and health for older Americans toolkit, 1998. Available at www .eatright.org/olderamericans/, accessed 10/14/03.

67. Russell RM, Rasmussen H, Lichtenstein AH. Modified Food Guide Pyramid for people over seventy years of age. J Nutr 1999;129:751–3.

68. The National Academies Press. Dietary reference intakes for energy, carbohydrate, fiber, fat, fatty acids, cholesterol, protein, and amino acids. 1994–1996, 1998. Appendix E. USDA, ARS, Beltsville Human Nutrition Research Center, Food Surveys Research Group, Washington, DC: National Academies Press, 2002.

69. Wright JD, Wang CY, Kennedy-Stephenson J, Erving RB. Dietary intake of ten key nutrients for public health, United States: 1999–2000. Advance Data from Vital and Health Statistics; no. 334. Hyattsville, MD: National Center for Health Statistics; 2003.

70. Yates AA, Schlicker SA, Suitor CW. Dietary Reference Intakes: the new basis for recommendations for calcium and related nutrients, B vitamins, and choline. J Am Diet Assoc 1998;98:699–706.

71. Harper EJ. Changing perspectives on aging and energy requirements: aging and energy intakes in humans, dogs, and cats. J Nutr 1998;128:2623S–26S.

72. Institute of Medicine of the National Academies. Dietary Reference Intakes: energy, carbohydrate, fiber, fat, fatty acids, cholesterol, protein, and amino acids. Washington, DC: National Academies Press pre-publication; 2002. Available at www.nap .edu/openbook/0309085373/html/93, accessed 11/15/03.

73. Rothenberg EM, Bosaeus IG, Westerterp KR, Steen BC. Resting energy expenditure, activity energy expenditure, and total energy expenditure at age 91–96 years. Br J Nutr. 2000;319–24.

74. Arcerio PJ, Goran MI, Gardner AW et al. A practical equation to predict resting metabolic rate in older men. Metabolism 1993;42(8):950–7.

75. Arcerio PJ, Goran MI, Gardner AW et al. A practical equation to predict resting metabolic rate in older females. J Am Geriatric Soc 1994;41:389–95.

76. Taylor HL, Jacobs DR Jr, Schucker B et al. A questionnaire for the assessment of leisure time physical activities. J Chronic Dis 1978;31:741–55.

77. Fuchs CS, Giovannucci EL, Colditz GA et al. Dietary fiber and the risk of colorectal cancer and adenoma in women. N Engl J Med 1999;340:169–76.

78. Campbell WW. Dietary protein requirements of older people: is the RDA adequate? Nutr Today 1996;31:192–7.

79. Subar AF, Krebs-Smith SM, Cook A et al. Dietary sources of nutrients among US adults, 1989 to 1991. J Am Diet Assoc 1998;98:537–47.

80. Pennington JA, Hubbard VS. Derivation of daily values used for nutrition labeling. J Am Diet Assoc 1997;97:1407–12.

81. McNamara DJ. Increase in egg intake minimally affects blood cholesterol levels (editorial). Washington, DC: Nutrition Close-Up Egg Nutrition Center; 1994.

82. Heron KL, Vega-Lopez S, Conde K et al. Pre-menopausal women, classified as hypo- or hyper-responders, do not alter their LDL/HDL ratio following a high dietary cholesterol challenge. J Am Coll Nutr 2002;21:250–8.

83. Dawson-Hughes B, Harris SS, Dallal GE. Plasma calcidiol, season, and serum parathyroid hormone concentrations in healthy elderly men and women. Am J Clin Nutr 1997;65:67–71.

84. Holick MF. McCollum Award Lecture, 1994: vitamin D—new horizons for the 21st century. Am J Clin Nutr 1994;60:619–30.

85. MacLaughlin J, Holick MF. Aging decreases the capacity of human skin to produce vitamin D_3. J Clin Invest 1985;76:1536–8.

86. Webb AR, Kline L, Holick MF. Influence of season and latitude on the cutaneous synthesis of vitamin D_3: exposure to winter sunlight in Boston and Edmonton will not promote vitamin D_3 synthesis in human skin. J Clin Endocrinol Metab 1988;67:373–8.

87. Dawson-Hughes B, Harris SS, Krall E, et al. Effect of calcium and vitamin D supplementation on bone density in men and women 65 years of age or older. N Engl J Med 1997;337:670–6.

88. Krasinski SD, Russell RM, Otradovec CL et al. Relationship of vitamin A and vitamin E intake to fasting plasma retinol, retinol-binding protein, retinyl esters, carotene, alpha-tocopherol, and cholesterol among eldery people and young adults: increased plasma retinyl esters among vitamin A–supplement users. Am J Clin Nutr 1989;49:112–20.

89. Trumbo P, Yates AA, Schlicker S et al. Dietary reference intakes: vitamin A, vitamin K, arsenic, boron, chromium, copper, iodine, iron, manganese, molybdenum, nickel, silicon, vanadium, and zinc. J Am Diet Assoc 2001;101:294–301.

90. Vivekananthan DP, Penn MS, Sapp SK, Hsu A, Topol EJ. Use of antioxidant vitamins for the prevention of cardiovascular disease: meta-analysis of randomised trials. Lancet 2003;361:2017–23.

91. Meydani SN, Meydani M, Blumberg JB, et al. Vitamin E supplementation and in vivo immune response in healthy elderly subjects. A randomized controlled trial. J Am Med Assoc 1997;277:1380–6.

92. Meydani M. Effect of functional food ingredients: vitamin E modulation of cardiovascular diseases and immune status in the elderly. Am J Clin Nutr 2000;71:1665S–8S; discussion 1674S–5S.

93. Jacques PF. The potential preventive effects of vitamins for cataract and age-related macular degeneration. Int J Vitam Nutr Res 1999;69:198–205.

94. Sano M, Ernesto C, Thomas RG et al. A controlled trial of selegiline, alpha-tocopherol, or both as treatment for Alzheimer's disease. The Alzheimer's Disease Cooperative Study. N Engl J Med 1997;336:1216–22.

95. Malinow MR, Duell PB, Hess DL et al. Reduction of plasma homocyst(e)ine levels by breakfast cereal fortified with folic acid in patients with coronary heart disease. N Engl J Med 1998;338:1009–15.

96. Keane EM, O'Broin S, Kelleher B et al. Use of folic acid–fortified milk in the elderly population. Gerontology 1998;44:336–9.

97. Optimal calcium intake. National Institutes of Health Consensus Statement 1994; 12:1–31.

98. Brown JP, Josse RG, for the Scientific Advisory Council of the Osteoporosis Society of Canada. 2002 Clinical practice guidelines for the diagnosis and management of osteoporosis in Canada. Can Med Assoc J. 2002;167(10 suppl):S1–S33.

99. Appel LJ, Moore TJ, Obarzanek E et al. A clinical trial of the effects of dietary patterns on blood pressure. DASH Collaborative Research Group. N Engl J Med 1997;336:1117–24.

100. Shils ME. Magnesium. In: Ziegler EE, Filer LJ, eds. Present knowledge in nutrition. Washington, DC: ILSI Press; 1996: 256–64.

101. Fosmire GJ. Trace metal requirements. In: Chernoff R, ed. Geriatric nutrition: the health professional's handbook. Gaithersburg, MD: Aspen; 1999: pp 84–106.

102. Girodon F, Lombard M, Galan P et al. Effect of micronutrient supplementation on infection in institutionalized elderly subjects: a controlled trial. Ann Nutr Metab 1997; 41:98–107.

103. Chandra RK. Effect of vitamin and trace-element supplementation on immune responses and infection in elderly subjects. Lancet 1992;340:1124–7.

104. Fairfield KM, Fletcher RH. Vitamins for chronic disease prevention in adults: scientific review. J Am Med Assn. 2002;287:3116–26.

105. Fletcher RH, Fairfield KM. Vitamins for chronic disease prevention in adults: clinical applications. J Am Med Assn. 2002; 287:3127–9.

106. Balluz LS, Kieszak SM, Philen RM et al. Vitamin and mineral supplement use in the United States. Results from the Third National Health and Nutrition Examination Survey. Arch Fam Med 2000;9:258–62.

107. U.S. Preventive Services Task Force. Routine vitamin supplementation to prevent cancer and cardiovascular disease: recommendations and rationale. Ann Intern Med 2003;139:51–5. Available at www.preventiveservices.ahrq.gov, accessed 11/18/03.

108. National Council for Reliable Health Information. Prevention magazine assesses dietary supplement use. Newsletter 2000; 23: cites BA Johnson in Herbalgram 2000; 48:65 and Natural Health Line, 3/10/00.

109. Eliason BC, Huebner J, Marchand L. What physicians can learn from consumers of dietary supplements. J Fam Pract 1999; 48:459–63.

110. Stupay S, Sivertsen L. Herbal and nutritional supplement use in the elderly. Nurs Pract 2000;25(9):56–66.

111. Fragakis AS. The health professional's guide to popular dietary supplements. 2nd edition. Chicago: American Dietetic Association; 2003.

112. Kittler PG, Sucher K. Food and culture in America. Belmont, CA: Wadsworth; 1998.

113. Position of the American Dietetic Association: food and water safety. J Am Diet Assoc 1997;97:184–9.

114. Evans WJ. Exercise training guidelines for the elderly. Med Sci Sports Exerc 1999; 31:12–7.

115. Pollock ML, Franklin BA, Balady GJ et al. AHA Science Advisory. Resistance exercise in individuals with and without cardiovascular disease: benefits, rationale, safety, and prescription. An advisory from the Committee on Exercise, Rehabilitation, and Prevention, Council on Clinical Cardiology, American Heart Association; Position paper endorsed by the American College of Sports Medicine. Circulation 2000;101:828–33.

116. Ventrakaman JT, Fernandes G. Exercise, immunity and aging. Aging Clin Exp Res 1997;9:42–56.

117. Kligman EW, Hewitt MJ, Crowell DL. Recommending exercise to healthy older adults. The Physician Sportsmed 1999;27(11):1–11.

118. Papadopoulou SK, Papadopoulou SD, Zerva A et al. Health status and socioeconomic factors as determinants of physical activity level in the elderly. Med Sci Monit 2003;9(2):79–83.

119. Conn VS, Minor MA, Burks KJ et al. Integrative review of physical activity intervention research with aging adults. J Am Geriatr Soc 2003;51:1159–68.

120. Krinke UB. Effective nutrition education strategies to reach older adults. In: Watson RR, ed. Handbook of nutrition in the aged. Boca Raton, FL: CRC Press; 2001: 319–31.

121. Wellman NS, Rosenzweigh LY, Lloyd JL. Thirty years of the Older Americans Nutrition Program. J Am Diet Assoc. 2002:102(3):348–50.

122. Ponza M, Ohls J, Millen B. Serving elders at risk. The Older Americans Act Nutrition Programs—National Evaluation of the Elderly Nutrition Program, 1993–1995. Washington, DC: Mathematica Policy Research, Inc., and Administration on Aging; 1996.

"Knowing is not enough;
we must apply.
Willing is not enough;
we must do."
Goethe

Chapter 19

Nutrition and the Elderly:
Conditions and Interventions

Chapter Outline

- Introduction
- Nutrition and Health
- Nutrition and Oral Health
- Heart Disease/Cardiovascular Disease
- Stroke
- Hypertension
- Cancer
- Diabetes Mellitus
- Metabolic Syndrom
- Obesity
- Osteoporosis
- Gastrointestinal Diseases
- Inflammatory Diseases
- Mental Health and Cognitive Disorders
- Low Body Weight/Underweight
- Dehydration
- Bereavement

Prepared by **U. Beate Krinke**
with **Lori Roth-Yousey**

Key Nutrition Concepts

1 Multiple health problems put older adults at higher nutritional risk; interventions with the greatest potential for benefit are targeted to treat specific conditions.

2 There are nutrient thresholds: dietary nutrients that are added to treat a deficiency have little effect when the diet is adequate, and are dangerous when consumed in excess of recommendations.

3 Successful nutrition interventions complement other lifestyle choices to enhance physical and mental resilience and to stabilize physiological changes.

Introduction

Aging adults want to stay healthy until death. Yet those who have chronic illnesses or age-related changes might not meet their *health* goals. Nonetheless, good food choices can limit illnesses, reduce risks, and contribute to *quality of life.* Medical nutrition therapy can be a part of treatment that encourages better food choices once diseases have occurred. On January 1, 2002, a federal bill was enacted, specifying that registered dietitians could receive reimbursement for *medical nutrition therapy (MNT)* provided to Medicare Part B beneficiaries with diabetes mellitus and kidney disease.[1] Registered dietitians are encouraged to follow nationally recognized nutrition protocols and evidence-based practice guidelines.[2]

Foundations for the bill's passage were numerous studies that supported the link between nutritional status and health care utilization by older adults.[3–7] The Lewin Group researchers estimated that covering MNT for older managed care clients who had cardiovascular disease, diabetes, or renal disease would recover Medicare costs after 3 years and would begin to save system dollars by the fourth year.[3] Malnourished older patients have higher postoperative complication rates and longer hospital stays, therefore incurring greater health care costs.[8] In free-living older adults, nutritional risk status was found to be the most important predictor of total number of physi-

Health More than the absence of disease, health is a sense of well-being. Even individuals with a chronic condition may properly consider themselves to be healthy. For instance, a person with diabetes mellitus whose blood sugar is under control can be considered healthy.

Quality of Life A measure of life satisfaction that is difficult to define, especially in heterogeneous aging population. Quality of life measures include factors such as social contacts, economic security, and functional status.

Medical Nutrition Therapy (MNT) Comprehensive nutrition services by registered dietitians to treat the nutritional aspects of acute and chronic diseases.

Xerostomia Dry mouth, or xerostomia, can be a side effect of medications (especially antidepressants), of head and neck cancer treatments, and of diabetes, and is also a symptom of Sjogren's syndrome, which is an autoimmune disorder for which no cure is known.

cian visits, visits to physicians in the emergency room, and hospitalization rates.[9] Aging adults use proportionately more health care services and products than younger persons do; therefore, nutrition interventions are particularly important.

Nutrition and Health

When asked, older people generally say they feel good about their health. Only 10% say their health is fair or poor, while 90% consider their health to be good, very good, or excellent.[10] This happens even when many are troubled with a chronic health problem. In contrast to older adults' perceptions, public health professionals objectively monitor health by measuring leading causes of death (mortality) and leading diagnoses of health conditions (morbidity).

Heart disease and cancer are the leading causes of death for all persons aged 65 and older,[11] followed by three other diseases:

Heart disease	32%
Cancer	22%
Cerebrovascular disease	8%
Chronic lower respiratory disease	6%
Influenza and pneumonia	3%

Most (87%) older Americans have one or more chronic conditions, such as high blood cholesterol, hypertension, osteoporosis, obesity, and overweight.[12] Eating patterns of older adults contribute to the incidence and course of maladies such as hypertension, heart disease, and cancer; this in turn affects functional ability. For example, an overweight individual with heart disease may continue to overeat and become obese, further complicating arthritis management. In turn, arthritis limits functional abilities and may limit access to food. Overall, nutritional status is a major factor in disease prevention, treatment, and recovery of health.

Nutrition and Oral Health

"We can't have a healthy mouth, a great smile, or a good conversation without it. Saliva or 'spit' lubricates living."

Dr. Nelson Rhodus, Director of Oral Pathology, University of Minnesota

Certain changes in oral health, such as cavities and missing teeth, are common in older adults. However, some disturbances can be associated with diseases, medical treatment, and medications. For example, head and neck cancer treatment can cause dry mouth (known as *xerostomia*); diuretic treatment for hypertension leads to less salivary secretion (i.e., makes the mouth dry). Lack of saliva for any reason gives bacteria a better environment for building plaque. There is also a tendency to have dysgeusia (loss of taste) and pain in the tongue (glossodynia). In another example, high blood sugars associated with diabetes can accelerate periodontal disease and make the mouth more susceptible to yeast infection (candidiasis).

Nutrition interventions to improve blood sugars and provide adequate nutrients can promote increased immune status and, with proper dental care, decreased oral infections. In addition, recommendations that benefit oral health can also benefit nutritional status. For example, fighting dry mouth with sips of water can improve oral health and hydration status at the same time. Successful nutrition interventions complement other lifestyle choices to enhance physical and mental resilience and to stabilize physiological changes.

Heart Disease/ Cardiovascular Disease

Heart disease (cardiovascular disease or CVD) is the leading cause of death in older adults and is potentially reversible by adopting a healthy lifestyle. The adult risk factors and course of heart diseases have been discussed in Chapter 17. Specifics for older adults are highlighted in this section, including stroke and hypertension.

Heart disease prevalence varies by race and gender (see Table 19.1).[13] Approximately one-third of adults aged 65 and older have some form of heart disease.[11,13] Of all deaths from heart disease each year, over three-fourths (78%) occur in elderly persons. In adults over age 85 who are hospitalized, 53% are admitted with heart failure as the principal diagnosis.[10] In addition, 23% of residents admitted into nursing homes in 1999 had a primary diagnosis of CVD.[14]

Risk Factors

Risk factors for cardiovascular disease in old age remain the same as in younger adults, except that the factors have less predictive value in old age.[15] Of adults aged 65 years and older examined in the NHANES III survey, 86% had one or more modifiable cardiovascular risk factors. These include hypertension (140 mmHg/90 mmHg), elevated LDL cholesterol (at least 130 mg/dL), and/or diabetes mel-

litus (physician diagnosis or fasting plasma glucose greater than 126 mg/dL).[15,16] Race is associated with risk, and older African Americans are especially vulnerable. They are nearly three times more likely than the average population to have one of the three cardiovascular risk factors.

Obesity increases the risk of CVD in aging adults. For example, 93% of those with a body mass index (BMI) over 30 (i.e., obesity) had one or more nutrition-related CVD risk factors (see Chapter 17). In the overweight category (BMI of 25 to 29.9), 87% had one or more risk factors.

Nutrition Interventions for Cardiovascular Diseases

Assertive treatment can modify the course of heart disease at any age. Assessment of personal motivation regarding diet and health, functional status, and lifestyle habits is among the special considerations for older adults contemplating dietary changes. Is the individual interested in making needed diet changes? If willing to change, does the individual have the knowledge and skill to adopt new dietary behaviors? What motivated this individual to seek nutrition advice? What is the individual's physical and mental functional level? Is assistance needed to shop for groceries or to prepare meals? Does the individual eat alone, with family and friends, at home, or in restaurants? Factors such as these influence nutritional intake and help to predict whether lifestyle changes will be adopted and maintained.

Dietary guidelines to fight cardiovascular disease in adults of all ages include those from the National Cholesterol Education Program (NCEP)[17] and the American Heart Association (AHA).[18] Cardiac rehabilitation programs adapting these guidelines were described in Chapter 17 and age-related considerations are described in Table 19.2 on the following page.

Stroke

The American Heart Association defines *stroke* as a cardiovascular disease that affects the blood vessels supplying blood to the brain. A stroke can occur in several forms, including cerebral thrombosis (a blood clot, or thrombus, blocks blood flow to the brain), cerebral embolism (an embolus, or wandering blood clot, lodges in an artery and blocks blood flow in or to the brain), or hemorrhage (a blood vessel bursts, and part of the brain goes without oxygen as a result).

Factors that can lead to a stroke include blocked arteries, easily clotting blood cells, and weak heartbeats that are unable to keep blood circulating through the body, allowing pools of blood to form and clot. Hypertension is another stroke contributor because the force of blood may break weak vessels.

Of adults aged 70 and older, 8% of females and 10% of males have had a stroke.[13] Stroke is one of the leading causes of death for older adults.

Table 19.1 Prevalence of heart diseases in adults aged 65 and older

	Total (%)	GENDER (%)		RACE (%)	
		Male	Female	White	Black
Heart disease	30.8	36.2	26.9	31.5	26.1
Hypertension	40.3	34.9	44.2	39.5	53.3
Cerebrovascular diseases	7.1	7.9	6.5	7.1	*
Atherosclerosis	4.1	4.5	3.9	4.5	*

*Information not available.

Table 19.2 Treatment factors for older adults with heart disease

Target Area	Adults (≥65 years)
Decrease amount and type of fat.	Focus on 1–2 items to decrease fat intake in individual's regular diet rather than change all things.
• Use lean meats.	Ensure adequate protein intake.
• Substitute saturated fatty acids with polyunsaturated fatty acids and monounsaturated fatty acids.	Focus on oils currently using and suggest one to change, if appropriate. Decrease trans fatty acids.
• Decrease trans fatty acids.	Consider giving a brief description of trans fatty acids and sources—margarine, vegetable shortening, cookies, pastries, and other processed fats—and base on mental awareness and readiness to change.
Reduce cholesterol intake.	Focus on 1–2 food items; research is conflicting on the role of cholesterol in older adults; liver makes less.
Increase fiber, fruits, and vegetables.	Work with fruits and vegetables that the indiviudal can chew (e.g., if dentures, do they fit?).
Cook healthy meals.	May not be controllable if receiving Meals-on-Wheels; overall goal is adequate nutrient intake.
Limit salt.	Focus on "no added salt," and no salt shaker on the table.
Read food labels.	May be difficult if eyesight is poor; consider financial limits; know bargain strategies.
Exercise regularly.	Obtain doctor's approval prior to starting; emphasize health benefits that include mobility, agility, and strength; emphasize that walking is exercise.
Maintain healthy weight.	Strongly influenced by functional status of individual; emphasize adequate nutrient intake.
Reduce stress.	Encourage exercise, relaxation, and socialization with friends.
Quit smoking.	Refer to smoking cessation program; continue the no smoking followed while in hospital; discuss potential for weight gain.

SOURCE: Adapted from Anne F. Gerlach, "Principles in a Cardiac Rehabilitation Program." Guest lecturer for Nutrition for Adults and the Elderly, University of Minnesota, Minneapolis, MN, 2002.

Effects of Stroke

Strokes deprive the brain of needed oxygen and other nutrients. Brain and nerve cells deprived of oxygen for only a few minutes can die. Brain cells do not regenerate. As a result, stroke leads to loss of function (speaking, walking, talking, and eating) for parts of the body controlled by the oxygen-deprived nerves. Quick recognition of stroke results in faster treatment and better recovery. Although brain cells do not regenerate, new nerve pathways can develop in the gray-matter reservoirs of the brain. This provides hope for successful rehabilitation therapies. Nonetheless, rehabilitation from a stroke is a slow, arduous process. Learning how to feed oneself, chew, and swallow are among functions that may have to be relearned during rehabilitation.

Carotid Artery Disease Condition in which the arteries that supply blood to the brain and neck become damaged.

Atrial Fibrillation Degeneration of the heart muscle, causing irregular contractions.

Transient Ischemic Attack (TIA) Temporary and insufficient blood supply to the brain.

Risk Factors for Stroke

Gender is not a risk factor for stroke, although more women die from a stroke than men do. The following factors place an individual at higher risk for stroke:

- Family history
- African American, Asian, and Hispanic ethnic groups
- Having had a prior stroke
- Long-term high blood pressure (either systolic or diastolic)
- Cigarette smoking
- Diabetes mellitus
- *Carotid artery disease, atrial fibrillation, transient ischemic attacks (TIAs)*
- High red blood cell count
- Sickle-cell anemia
- Living in poverty
- Excessive use of alcohol, use of cocaine and intravenous drugs

Nutrition Interventions for Stroke

The focus of dietary advice in stroke prevention is to normalize blood pressure.[19] Other dietary goals are to reduce overweight and obesity (particularly abdominal fat) and to moderate alcohol intake.

The Dietary Approaches to Stop Hypertension (DASH) diet has been a promising strategy to decrease blood pressure[19] and risk of stroke in adults under age 65 (see Table 19.3). It has also been shown to enhance perceptions of quality of life.[20] Theoretically, the DASH diet should also work for older adults. Other nondrug interventions have successfully lowered the blood pressure of older adults (e.g., using weight reduction and/or sodium restriction of 1800 mg per day over 30 months).[21,22] See Case Study 19.1 on the following page.

Hypertension

In Western societies, blood pressure increases with age and stabilizes around age 50 to 60 for men. A blood pressure greater than or equal to 140/90 mmHg is defined as stage 1 hypertension. Systolic blood pressure of 120–139 mmH or diastolic pressure of 80–89 is defined as prehypertension. Although it has been suggested that older adults can tolerate higher blood pressure and may even benefit from increased blood flow to the brain, old age does not change the diagnosis criteria for high blood pressure. Higher blood pressure puts more force on potential vessel blockages and increases chances of blood vessel breakage. An individual who controls high blood pressure with medication is still considered to have hypertension.

Analysis of the NHANES III data shows that 40% of older adults have hypertension.[13] This is almost two times greater than the prevalence among all adults (40% vs. 24%). Prevalence increases with age and differs by gender. In adults aged 75 years old and over, 64% of males and 77% of females have hypertension. Over half of individuals with hypertension (53%) control it with prescribed medications. Uncontrolled hypertension is a public health challenge; only 34% of individuals with hypertension have it under control.

Table 19.3 The DASH eating plan for blood pressure control

	Servings per Day	Serving Sizes of Foods within the Food Group
Grains and Grain Products Especially whole grain[a]	7–8	Breads: 1 slice or 1 oz Cereal: ½ cup cooked or dry Rice, pasta: ½ cup cooked
Vegetables Fresh, frozen, no-salt-added canned	4–5	Raw, 1 cup; cooked, ½ cup
Fruits Fresh, frozen, or canned in juice	4–5	Juice: 6 oz Fresh: 1 med piece Mixed or cut: ½ cup Juice: 6 oz Dried: ¼ cup
Dairy Foods Skim or 1% milk, fat-free dairy products	2–3	Milk: 8 oz Yogurt: 1 cup Cheese: 1½ oz
Meats, Poultry, and Fish	Up to 2	3 oz, cooked
Nuts, Seeds, Dry Beans	4–5 per week	⅓ cup or 1½ oz nuts 2 tb or ½ oz seeds ½ cup cooked beans (legumes)
Fats and Oils[b] Select olive, canola, corn, and safflower oils	2–3	1 tsp soft margarine, oils, mayonnaise 1 tb low-fat mayonnaise 2 tb light salad dressing
Sweets	Up to 5 per week	1 tb jam, jelly, syrup, or sugar ½ oz fat-free candy, jelly beans, or 12-oz sweetened beverage

[a]Whole grain is the entire edible part of wheat, corn, rice, oats, barley, and other grains. Whole grain bread has the words "whole grain" before the type of flour is listed; whole grain breakfast cereals include the word "whole" or "whole grain" before the grain name (e.g., whole grain wheat).
[b]One serving is equivalent to 5 grams of fat.
SOURCE: Adapted from a 2000-calorie eating plan, the National High Blood Pressure Education Program's HeartFile, Winter 1999, National Heart, Lung, and Blood Institute and the National Dairy Council's "DASH TO THE DIET," 2000.

Case Study 19.1

Photo Disc

Bridget Doyle Remembers Laura

Just because she lives at Lenoir Manor, a continuing care retirement facility, Laura, a petite (4 ft. 8 in., 97 pounds) widow of the local college dean, does not consider herself as old. She is 87. She has had no major nutritional or health problems and her appetite is good. She had been a good cook and entertained graciously, but in the residential care facility, meals are prepared for her. Because she has had slight fluid retention over the past year, she no longer adds salt to her meals. She tells Bridget Doyle, her nutritionist, that yes, occasionally she does not like her meals and misses cooking for herself.

One Monday morning, Laura is found in bed with her left side paralyzed. The diagnosis is a right-sided stroke, resulting in 3 weeks of hospitalization. Back at the skilled care wing of Lenoir, Laura needs a nasalgastric tube for feeding. She is alert and knows people, but is limited in speech. Overnight, Laura's care has changed from an individual needing routine nutritional monitoring to someone with many interrelated problems:

- Inability to communicate her overall medical and nutritional concerns clearly
- Weight loss of 9 pounds during the 3-week hospital stay
- Intense dislike of the nasal tube, as demonstrated by repeated attempts to pull it out, leading to restraints of her hands

Questions

1. What nutritional parameters should be assessed and monitored now that Laura is back at Lenoir Manor?
2. What disciplines should be involved in Laura's care plan, and why?
3. The interdisciplinary care team wants to meet Laura's needs in a dignified and respectful manner. How can the care team address both clinical and ethical concerns?
4. How could Bridget ensure that Laura's nutritional needs are met?
5. What are strategies young adults can adopt to reduce their risk of stroke?

(Answers are in Appendix D.)

Effects of Hypertension

Prolonged high blood pressure puts extra tension on blood vessels and organs in the body, wearing them out faster than the natural aging process would. Damaged kidneys are a common sign of uncontrolled hypertension.

RISK FACTORS FOR HYPERTENSION Nutritional risk factors are drinking alcohol to excess, high-saturated-fat diets leading to dyslipidemia and atherosclerosis, lifestyles resulting in overweight and obesity, and a diet low in calcium.[23] Family history and ethnic background increase the risk of hypertension; African Americans are most likely to have hypertension. Salt intake can also contribute to hypertension. Researchers with the Intersalt Study calculated that over time, 20% of hypertension in Western societies is attributable to salt intake.

Nutrition Interventions for Hypertension

The DASH diet described in the earlier section on stroke is an important nutrition intervention for persons with a history of hypertension. Two strategies that achieved the greatest reduction in systolic blood pressure were achieving and maintaining a healthy weight and reducing sodium intake to no more than 2400 mg/day.[23] The Dietary Guidelines for Americans also suggest limiting dietary sodium to 2400 mg per day. Estimated intake of 70-year-olds (not counting salt added at the table) was 3122 mg daily for males and 2376 mg for females.[24] In the DASH sodium study, the greatest overall blood pressure reduction occurred in the subjects with the strictest sodium intake limit (1500 mg a day). Blood pressure reduction occurred whether individuals were normo- or hypertensive. Because most of dietary sodium is contributed by processed foods, choosing a baked potato rather than eating more highly processed potato chips limits sodium intake. Adding salt at the table increases sodium intake by roughly one-fourth.[25]

Other recommendations for reducing blood pressure were to use moderation in drinking alcohol and to maintain adequate potassium, magnesium, and calcium intakes.

Cancer

Cancer is a name for a group of conditions resulting from uncontrolled growth of abnormal cells. It is a complex progression through several stages: activation, initiation (injury or insult by a carcinogen), promotion (damaged DNA divides during a lag period, potentially over 10 to 30 years), progression (growth and spread), and possible remission (successful treatment or reversal). See Table 19.4.

Half of all new cancer cases are diagnosed in individuals aged 65 and older, with the most common sites being breast, colon/rectum, lung and bronchial system, and prostate.[11] In the case of prostate cancer, 80% of all cases occur in men aged 65 and older.[26] Cancers of the breast, skin, intestine, genitals, lungs, and other respiratory sites are present in 7.4% of individuals aged 65 and older (roughly 1 of 14). Diet-related cancers include cancer of the stomach, breast, prostate, and colon.

Table 19.4 The phases leading to transformation of a normal cell into a cancerous cell

1. *Activation.* Certain chemicals and/or radiation can trigger a cellular change. In normal processes, an individual's body removes harmful substances; in some cases, the substance remains and attaches to the DNA within a cell.
2. *Initiation.* The changed or mutated DNA within the cell is copied. If it occurs within a specific DNA region, this makes the cell more sensitive to the harmful substances and/or radiation.
3. *Promoters.* When cells become sensitive, promoters encourage the cells to divide rapidly. If the normal sequences of DNA are damaged, a clump of abnormal cells bind together to form a mass, or tumor.
4. *Progression.* The cells continue to multiply and spread into nearby tissues. If they enter the lymph system, the abnormal cells are transported to other body organs.
5. *Reversal.* The goal of reversal is to prevent the cancer's progression, or to block any of the first four stages.

SOURCE: Excerpted from "Cancer—The Intimate Enemy." In E. N. Marieb, *Human Anatomy and Physiology.* Redwood City, CA: Benjamin Cummings Publishing, 1996.

Genetics contribute to the development of cancer; some cancers are hormone sensitive (e.g., uterine, breast, prostate, and ovarian cancer). As humans live longer, it becomes more likely that some mistake or insult will adversely affect the DNA replication process and ultimately cause cancer. Environmental factors can largely determine whether an individual develops cancer. Diet alone is estimated to account for 35% of all cancers, while smoking and dietary habits together have been estimated to cause 50% or more of cancers.[25,27]

Although most older adults do not have cancer, it is the second leading cause of death after heart disease. Compared to the death rate from heart disease (1808 per 100,000), the death rate from cancers (listed as malignant neoplasms in some charts) is 1131 per 100,000.[11] After age 85, cancer is responsible for 10% of deaths among women and 16% among men.[27]

Effects of Cancer

Cancer development is age associated but not age dependent; development is neither consistent nor linear. In a healthy, resilient individual, initiation may be repaired and subsequent cancer avoided or delayed. In someone with impaired immunity or suffering major physiological stress, initiation may proceed through promotion or progression. For older adults, cancer treatment presents a greater nutritional challenge than cancer prevention because of the lag times for development.

Cancer treatment affects all aspects of nutrition. Side effects include mouth sores and altered sense of taste and smell, both of which affect appetite and eating. Chemotherapy, surgery, and radiation may lead to changes in intake, digestion, and nutrient absorption. For example, treatment may leave the cancer patient tired and lacking immunity to fight common diseases. In addition, ability to socialize over meals or to shop for groceries may be curtailed. Medications also interact with nutrient absorption. Overall, cancer treatment effects cause significant amounts of stress and nutritional insult in older adults.

Risk Factors for Cancer

Diet can refocus the role of genetics in cancer development. Although dietary patterns make the biggest impact when begun early in life, older adults show greater interest in risk reduction. Vegetable and fruit consumption is the primary diet strategy for cancer risk reduction. Vegetable, fruit, and antioxidant intake are related to reductions in cancer initiation and progression. Extensive reviews of diet and cancer studies[26,28–30] conclude that high-level vegetable and fruit consumption is associated with decreased risk of cancer, including oral, esophageal, stomach, colorectal, and lung cancers. The American Cancer Society Expert Committee has judged that the scientific evidence suggests a possible benefit from increasing vegetable and fruit intake to prevent breast, colorectal, and prostate cancer and for overall survival in breast, colorectal, and lung cancer.[31] This expert committee also concluded that increasing fiber intake may benefit overall survival in colorectal cancer and omega-3 fatty acids in lung cancer. However, insufficient evidence exists for increasing fiber, omega-3 fatty acids, and soy intake to improve overall survival in breast, lung, and prostate cancer. Research is currently under way to examine how the antioxidant components in green tea hinder the body's abilities to develop cancer-promoting chemicals.

Table 19.5 offers an overview of specific types of cancers and their link to risk factors. These studies do not deal specifically with older adults, but with adults of all ages. Scientists continue to study the role of lycopene (a flavonoid found in tomatoes) and flaxseed (containing a concentrated source of phytoestrogens) as possible dietary strategies to decrease prostate cancer risk and lengthen survival.

Nutrition Interventions for Cancer

The time to worry about a low-fat diet is not during cancer treatment. The focus of cancer treatment is to maximize nutritional status and keep up intake to avoid weight loss and dehydration. This can be a challenging task. Anticancer medications and radiation treatments are associated with nausea, vomiting, diarrhea or constipation, fatigue, and weight loss. Finding food items that are tolerated, especially if they are calorie dense, is one way family and friends can help medical care providers support a cancer patient's treatment.

Treating cancer is a process that tends to generate quests for alternative methods. The hope for remission and cure is a powerful motivator to try herbs, new nutrient combinations, and complementary healing practices. Herbal products have potentially useful roles in cancer treatment to ameliorate nausea and common symptoms; for example, chamomile tea for gastrointestinal discomfort and peppermint tea as a digestive aid. These other herbs may be considered:

- *Ginger.* Has possible antinausea properties, but it may not be as effective during chemotherapy treatment because it acts on the stomach and chemotherapy nausea is triggered in the central nervous system, yet the taste and smell of small amounts of ginger may be calming.
- *Garlic.* Possible antibacterial or antifungal effects of garlic with possible cancer prevention in moderate amounts; excess consumption via garlic supplements can cause stomach pain and gas.

The first rule of treatment, "Do no harm," is especially relevant during the vulnerable period of cancer treatment. A product that can be enjoyed or perhaps tolerated by a healthy person may exacerbate illness in someone with health problems. Discussing all contemplated treatments with a physician, registered dietitian, and pharmacist will help to prevent dangerous interactions. A registered dietitian can also assist with providing individualized recommendations. Aging, food preferences, dietary intakes, the effects of the tumor, anticancer treatments, and side effects (that include pain) are all considered when planning therapy for elderly cancer patients.[32]

Diabetes Mellitus

Two diabetes mellitus classifications relate to older adults: type 1 and type 2. Type 1 diabetes was formerly labeled insulin-dependent diabetes mellitus (IDDM) or juvenile-onset diabetes because this type often begins in youth. Type 2 was formerly referred to as non-insulin-dependent

Table 19.5 Nutritional risk factors for the most common cancers

Type of Cancer	Nutritional Risk Factors with Convincing or Probable Evidence
Lung	Low vegetable and fruit intake
Stomach	Low vegetable and fruit intake Lack of refrigeration for food High salt intake High use of smoked, cured, and pickled foods (salt and nitrites)
Breast	Low vegetable and fruit intake Increased obesity (especially in postmenopausal women) Increased alcohol
Colon/rectum	Low vegetable and fruit intake High meat consumption (especially fat from red meats) Excess alcohol (folate may help offset detrimental effects)
Mouth/pharynx/ nasopharynx	Low vegetable and fruit intake Excess alcohol, together with tobacco (smoke or chew) High salted fish intake (nasopharynx)
Liver	Excess alcohol Eating contaminated (especially aflatoxins[a]) foods
Cervix	Low vegetable and fruit intake
Esophagus	Low vegetable and fruit intake Diets generally deficient in nutrients High alcohol intake
Prostate	High meat or meat fat or dairy fat intake

[a]Aflatoxins are toxins produced by molds in some foods such as peanuts.
SOURCE: Adapted from World Cancer Research Fund and American Institute for Cancer Research, Food, Nutrition, and Prevention of Cancer, p. 542.

diabetes mellitus (NIDDM) or adult-onset diabetes because it was generally diagnosed after age 40. Older women who had impaired glucose tolerance or gestational diabetes (GDM) during pregnancy are at increased risk for developing type 2 diabetes.

Diabetes diagnosis criteria for older adults are the same as for younger adults—a fasting blood glucose of 126 mg/dL or higher. A random nonfasting blood sugar of >200 mg/dL would be confirmed by either (1) fasting (no caloric intake for at least 8 hours) blood sugar of 126 mg/dL or greater, (2) an abnormal blood glucose (equal to or greater than 200 mg/dL following a drink of 75 grams glucose), or (3) another casual blood sugar of ≥200 mg/dL. Fasting blood sugars ≧100 to 125 mg/dL (i.e., impaired fasting glucose or IFG) or a 2-hour plasma glucose following an oral glucose tolerance test (OGTT) of ≥140 to 199 mg/dL suggests impaired glucose tolerance (IGT).[33] Nondiabetic adults maintain fasting blood sugars below 100 mg/dL. A third classification, called prediabetes, is defined as overweight people (BMI ≥25 kg/m²) aged ≥45 years with IFG or IGT.[34]

Approximately 45% (nearly half!) of the U.S. diabetic population is aged 65 years and older.[35] The Centers for Disease Control and Prevention estimate that 10.3 million individuals in the United States have been diagnosed with diabetes. Most (90–95%) of new diagnoses are type 2 diabetes, making it the most common type for older adults.

In the National Health and Nutrition Examination Survey (NHANES) Epidemiologic Follow-Up Study (1971–1992), it was found that 18.7% of adults aged 65 or older have diabetes, and this rate decreases to 11% in the population aged 70 years and older.[36] Prevalence is higher in some ethnic groups. For example, 29.4% of Mexican American and 27.3% of African American older adults have diabetes. Using a different database, the prevalence of diabetes in Native Americans and Alaska Natives increased from 18.2% in 1990 to 22.8% in 1997 for adults aged 65 years and older.[37] Non-Hispanic black and Hispanic individuals are twice as likely to have diabetes as whites are.[35] For every two people diagnosed with diabetes, it is estimated that another one to two individuals have diabetes without knowing it. A reanalysis of NHANES III (1988–1994) data estimated that 10.8% of adults 45–74 years had undiagnosed diabetes, and 22.6% had prediabetes.[34]

Effects of Diabetes

Diabetes affects all organs. Older adults who are experiencing declines in organ function are likely to be more affected by diabetes. For instance, blood glucose levels above a threshold of roughly 120 mg/dL exceed the kidney's capacity for reabsorption, and subsequently glucose starts spilling into the urine. With aging, there are fewer nephrons, less blood flow, and a slowed glomerular filtration rate, further slowing glucose reabsorption.

In four of five older people, diabetes occurs as one of several coexisting conditions that include heart disease, hypertension, elevated blood lipids, and obesity. Individuals with diabetes are at greater risk for heart disease and its complications; diabetes itself is an independent risk factor for atherosclerosis. Furthermore, individuals with diabetes are more likely to die after suffering a myocardial infarction (MI) than those without diabetes.[38,39]

Diabetes leads to a tenfold greater risk of amputations, macular degeneration, visual loss, cataracts, glaucoma, and neuropathies (nerve damage, pain, or tingling) of the hands and feet. Other effects of hyperglycemia may include sodium depletion and dehydration, trace mineral depletion (zinc, chromium, magnesium), insomnia, nocturia, blurred vision, increased platelet adhesiveness related to atherosclerosis, increased infection and decreased wound healing, and aggravated peripheral vascular disease.

RISK FACTORS FOR DIABETES   Risk factors for type 1 diabetes are primarily genetic and mediated by viral and other diseases. Obesity is rarely a risk factor. Type 1 diabetes is less common in old age. Risk factors for type 2 diabetes are more lifestyle related (e.g., obesity) and also include

- *Body weight and abdominal fat.* Higher body fat levels (BMI over 25, but even more so over 30) are associated with higher incidence of type 2 diabetes, especially when fat is stored in the abdominal area (waist circumference greater than 35 inches or more in females, 40 inches or more in males).

- *Race/ethnicity.* Blacks have a higher risk of developing type 2 diabetes than whites do at normal or low body weight. Risks of developing diabetes are similar for blacks and whites at higher body weights.[36] NHANES data (30- to 74-year-olds) collected from 1971 to 1992 yielded risk calculations for developing type 2 diabetes relative to BMI (weight for height). Cumulative risk increases with BMI in all race and sex groups. At a BMI of 22 (i.e., desirable weight), the probability of developing diabetes was 1.87 for blacks compared to the probability of 1.76 (p <.01) for whites.[36] At BMIs of 32 (obesity), the probability of developing diabetes was not significantly different for blacks and whites. The difference in risk may be associated with visceral adiposity (fat storage in midsection). Greater visceral adiposity affects blacks more than whites at normal weight (BMIs of 22), but not at levels defined as overweight or obese (BMIs of 25 or greater).

- Hyperhomocysteinemia, which is linked to increased risk of heart disease, is a stronger risk factor for overall mortality in individuals with type 2 diabetes, independent of CVD and other risk factors, than in individuals without diabetes.[40]

Nutrition Interventions for Diabetes

The goal of nutritional interventions is to achieve the best possible glycemic (blood sugar) control and reduction of cardiovascular disease risk factors without jeopardizing the quality of life, health, or safety of the patient[41] while also providing for psychosocial needs. Diabetes care providers tend to agree that the following components comprise the nutritional management of both types of diabetes.[38,41–44]

1. Maintain blood glucose levels as close to normal as possible by balancing food intake, exercise, and insulin/medications.
2. Choose health goals similar to the USDA/DHHS "Nutrition and Your Health: Dietary Guidelines for Americans 2000";[45] these include recommendations for physical activity.
3. Individualize dietary recommendations according to the patient's ethnic group.
4. Many practitioners suggest that a stable weight (even if overweight) is better than weight cycling, because weight cycling is associated with greater risk for complications of diabetes (such as cardiovascular disease). Weight loss in overweight or obese individuals improves blood glucose control.
5. Vitamin and mineral supplementation benefits most older adults.
6. Interactions among carbohydrates, fat, fiber, and fluids in foods make it impractical to assign useful glycemic scores to individual foods for population-wide use. However, individuals can adjust their blood sugar response according to the glycemic loads of their diets. The *glycemic index* (GI) of food indicates that some readily digested starches cause greater rises in blood sugar than do certain simple sugars. The glycemic index and glycemic load (GL) of foods are tools to help individuals find dietary patterns to moderate insulin responses and thus control blood sugar.

Controversy over the usefulness and accuracy of the GI is an example of applying guidelines into practice. An area of controversy in meal planning for aged individuals with Type 2 diabetes revolves around the role of simple sugars.

Glycemic Index (GI) A measure of the extent to which blood glucose levels are raised by a specific amount of carbohydrate-containing food compared to the same amount of glucose or white bread.

For example, if older adults are receiving fixed daily insulin doses or meds, variable simple sugar intakes can lead the body to secrete endogenous insulin. If the insulin is defective and unable to assist sugars to move into the cells to be metabolized for energy, the elevated sugar and insulin (hyperinsulinemia) circulating in the blood can damage organs and nerves. Hyperinsulinemia also promotes weight gain.[46] Yet what if an older adult looks forward to coffee and cookies? Are we ever old enough to stop watching what we eat? Knowing the effects of foods on health helps older adults make informed decisions regarding their quality of life.

Constant attention to meal planning in diabetes management has led many individuals to search for foods that will not affect blood sugar. One example is a vegetable called Jerusalem artichoke, which is not related to the globe artichoke or artichoke hearts seen in recipes and restaurants. Jerusalem artichokes contain inulin, a fructose polymer, which does *not* turn to insulin. Inulin may be absorbed more slowly than other starches. Although Jerusalem artichokes are a perfectly fine vegetable, eating them does not lower blood sugar.

A high-carbohydrate diet is also likely to elevate fiber intake. Eating for blood sugar control and decreased risk of cardiovascular disease means eating high-fiber foods; oats, rye, and other whole grains; and fruits and vegetables. Fibers such as pectins and gums in citrus, apples, oats, barley, and dried beans have been shown to slow glucose absorption and lower serum cholesterol.[47] Evidence is inconclusive about whether fiber is most effective as soluble or insoluble fiber and whether it can be supplemental or must be eaten as part of a mixed diet. Older individuals have difficulty getting enough fiber from foods (i.e., 21–30 grams daily for women and men aged 70 and older).[10]

Moderate increases in carbohydrate (raising percent of energy from 45 to 55%) have been well tolerated in adults (42–79 years old) with diabetes when given as breakfast cereal over a 6-month period.[48] If an individual's blood triglyceride level increases during a 3-month trial of a higher-carbohydrate meal plan, this demonstrates the individual has sensitivity to a high-carbohydrate diabetic diet.[38] Carbohydrate energy can then be reduced to 45% of total calories while maintaining other heart-healthy features of the meal plan.

Overall, the American Diabetes Association recommends that carbohydrate and monosaturated fat together should provide 60–70% of energy intake, making adaptations according to the individual's metabolic profile.[44] Because older adults are more likely to be undernourished than overnourished, consuming needed nutrients from a mixed diet is generally solid nutrition advice for older adults with type 1 or type 2 diabetes.

The American Diabetes Association suggests that a daily multivitamin supplement may be appropriate for older adults in order to complement inadequate intake or unique metabolic needs.[44] Chromium picolinate supplements have shown some promise for reducing insulin needs and oral hyperglycemic medications when chromium deficiency exists.[49] Chromium-rich foods include brewer's yeast, oysters, meats (especially liver), whole grains, skin-on potatoes, and apples.

Magnesium deficiency has been linked with insulin resistance, carbohydrate intolerance, and abnormal blood lipid metabolism. Magnesium may be lost through the urine when excess glucose is being excreted, or intake may be insufficient, or intestinal absorption may be limited. Magnesium supplementation should be monitored closely to avoid magnesium toxicity, especially if kidney function is impaired.

Occasionally, complementary medicine is used to treat diabetes. This is especially true in Mexican American and Native American populations. Evening primrose oil, milk thistle, fenugreek seeds, and prickly pear cactus (Spanish *nopales*) are some examples of botanical treatments. Safe, complementary nutritional remedies can enhance the standard nutritional therapies.

Individualized medical nutrition therapy (MNT) for older adults takes into account food availability, food preferences, and the client's treatment goals. Diabetes educators provide ongoing self-management education and care that supports quality of life. Cost-benefit analyses have shown that tight control of blood sugars (maintaining *glycosylated hemoglobin* below 7%) can lead to better quality of life for the older patient with diabetes and can result in fewer complications from diabetes.[50–52]

Metabolic Syndrome

Metabolic syndrome, also known as Syndrome X,[53] describes a combination of atherogenic risk factors.[53,54] The American Association of Clinical Endocrinologists has created a new ICD-9 code for "Dysmetabolic Syndrome" and developed a position statement on identification and treatment of this syndrome, which is discussed in Chapter 17.[54] Management strategies for older adults take into account common comorbidities. For example, substituting whole grains (whole wheat breads, oat cereals, barley, and brown rice) for refined carbohydrates can benefit older adults who also have gastrointestinal problems. Overweight or obese adults are encouraged to lose weight using a diet with moderate levels of carbohydrate (see Table 19.6) and low in saturated fats and trans fatty acids.[55–57] When an individual has elevated triglycerides, carbohydrate calories can be decreased by including a greater percentage of protein and fat (up to 40%) calories.[46,53]

Research continues to explore how total amounts and types of dietary (mono- and polyunsaturated) fat affect insulin sensitivity.[57] No matter what the diet, muscle-building exercise enhances insulin sensitivity at any age.

Obesity

The National Heart, Lung, and Blood Institute[58] and the World Health Organization[59] define *obesity* as a BMI of 30 or higher and extreme obesity as a BMI of 40 or higher. In the United States, approximately one-third of older adults are not just overweight, but obese. Obesity rates have risen rapidly since the 1980s. At last count, 74% of men and 68% of women aged 60 years and older were classified as overweight with a BMI of 25 or greater.[60] Within this group, 32% of men and 35% of women were obese. The highest prevalence of obesity by gender among adults ≥60 years of age occurs in non-Hispanic black women (50%) and non-Hispanic white men (34%). The lowest prevalence of obesity in this age group occurs in non-Hispanic black men (26%) and non-Hispanic white women (33%).[60]

Glycosylated Hemoglobin A laboratory test that measures how well the blood sugar level has been maintained over a prolonged period of time; also called Hemoglobin A_1C.

Current evidence supports the notion that the body mass index (BMI) associated with the lowest mortality falls within the range of 18.5–24.9 in men and women aged 20–74. However, whether age affects the association between BMI and deaths is controversial.[61] Based on Ernsberger's analysis,[62] morbidity and mortality is not any higher, and sometimes is lower, in older people who are at the high end of the BMI continuum. A study of older, community-dwelling Canadians found that an increased BMI was associated with lower mortality.[63] Conversely, weight cycling is associated with higher mortality, and evidence suggests that the main emphasis should be in primary prevention of obesity through lifelong weight maintenance.[64] In cases of extreme obesity, gradual weight loss is warranted.

Table 19.6 Comparing dietary approaches to insulin resistance with the American Heart Association Dietary Guidelines

Macronutrient	American Heart Association, Krauss 2000[18]	Syndrome X, Reaven 2000[53]	Insulin Resistance, Coulston 1997[56]	Low Calorie, Higher-Protein Diet, Holtmeier 2000[46]
Carbohydrate	50–60	45	50	45
Protein	10–25	15	15–20	25
Fat	25–30	40	30–35	30

For older adults, extra weight during illness episodes, especially hospitalizations, seems to be protective. Materials developed in 1999 by the American Dietetic Association Long-Term Care Task Force in conjunction with Health Care Finance Administration (HCFA, now the CMS or Centers for Medicare and Medicaid Services) suggest a BMI range of 19–27 as an acceptable and health-promoting weight range for older adults.[65] This range is similar, yet broader than that used by the Nutrition Screening Initiative, where BMI cut-points are 22–27. The point is that health care providers are more flexible when it comes to BMI recommendations for older than for younger adults.

NUTRITION INTERVENTION FOR OBESITY

Older obese individuals should undertake a healthy eating program based on enough nutrient-dense calories to support a slow weight loss. Age is not a factor in formulas to calculate calorie levels supporting desired activity and weight loss, although comorbidities are. Ensuring adequate nutrient intake by eating and drinking a balance of servings from basic food groups—for instance, as outlined in the Food Guide Pyramid or the DASH eating pattern—promotes health in older as well as younger adults.

All adults benefit from physical activities designed to ensure functional independence. The Healthy People 2010 goal is for 40% of the population to perform physical activities that enhance and maintain flexibility. Approximately 22% of adults aged 65 years or older achieved this goal. Health care providers need to spend more time promoting exercise and physical activity for older people.[66] Approximately 5 out of 10 Americans ages 65–74 are sedentary, and this figure increases to 6 out of 10 persons for individuals aged 75 years and older. Due to functional limitations of older adults, health professionals will need to acquire special skills and coordinate activities between health care and community or senior center programs. The U.S. Assistant Surgeon General summarized the issue: "Public health interventions to decrease obesity prevalence must apply the same kind of multifaceted and coordinated approach that reduced tobacco use in order to change individual behavior patterns and effectively address the environmental barriers to physical activity and healthful food choices."[67]

Osteoporosis

> "Watch it when you hug Grandma."
>
> Mary Nelsestuen, afraid that a strong hug might
> break her frail mother's osteoporotic ribs

Osteoporosis means "porous bone," which results from reduction in bone mass and disruption of bone architecture. Osteoporosis can develop rapidly or slowly, depending on the homeostatic mechanism involved. An accelerated phase of bone loss (also called Type I) occurs due to estrogen or testosterone loss. Bone mass loss is sharper for women, who are most vulnerable to bone loss in the 3 to 5 years past menopause due to estrogen decline (Type I osteoporosis).[68] Men develop osteoporosis later than women do. Their testosterone levels decline roughly 40% between ages 40 and 70, leading to declining bone mass.[69] Men's bone mass losses double after surgical or hormonal castration (a form of treatment for prostate cancer).

A slow phase of bone loss, also called Type II or senile osteoporosis, occurs due to decline in calcium absorption. Osteoporosis is an age-associated disease because effects of slow bone loss are usually not seen until late adulthood.

Diagnosis of osteoporosis can be aided by assessment of clinical risk and biochemical markers and by bone densitometry measurement; for example, dual-energy X-ray absorptiometry (DEXA or DXA).[70]

Just as bones grow strong over time, they become brittle slowly. In addition, not all bone tissue is the same. Up to 50% of trabecular or spongy bone (wrist, vertebrae, and ends of long bones) may be lost, and up to 35% of cortical or compact bone (shafts of long bones) will be lost during a lifetime.[71] Average bone loss is 0.5–1% per year after approximately age 50; when bone density is below the "young normal" mean, osteoporosis is diagnosed.[72]

Osteoporosis is four times more common in women than men (80% compared to 20%). Men have larger bodies and denser bones than women, and thus have greater peak and total bone mass.[68] Blacks have denser bones and less osteoporosis than whites do.[73]

Osteoporosis is dissimilar in men and women. Women tend to break hips and backbones, while wrist fractures are less common. Men with osteoporosis are also likely to have hip fractures, but suffer fractures of or near the wrist more often than women do. One out of every two women and one out of every eight men over age 50 will have an osteoporosis-related fracture in their lifetimes.[74]

Etiology of Osteoporosis

Bone mass is gained primarily during growth periods, with peak bone density reached between ages 18 and 30.[73] Subsequently, bone mass remains stable until about age 40 to 50 for women and about age 60 for men. Inadequate building of peak bone mass coupled with significant bone loss leads to a low bone density and increased risk for fractures.

INADEQUATE BONE MASS   Although osteoporosis is seen most often in the elderly, the risk for developing osteoporosis in later years begins during childhood and adolescence. Development of osteoporosis is delayed when an individual develops bigger, denser bones during youth.[75] For example, an epidemiological study in Yugoslavia showed that higher calcium intake in youth led to higher peak bone mass, independent of exercise and other factors. Higher bone mass has also been associated with slower decline in later life.[76] More recent studies in the United States have also shown that getting enough calcium during growth spurts (between ages 11 and 17) reduces the risk of osteoporosis. In an intervention study, girls aged 11 to 12 years who received calcium supplements (500 mg calcium citrate-malate) gained an additional 1.3% bone mineral density per year compared to controls.[77] In a cross-sectional study of 18- to 31-year-old women, positive correlations were found between adolescent milk intake in childhood and bone mineral density (total skeleton, spine, femoral neck, radius).[78] While there are many actions that build bones in older adults, preventing brittle bones is best done early in life.

Lack of exercise or inactivity leads to osteoporosis. Bed rest, hospitalization, and sedentary lifestyles make bones weak. Weight-bearing or resistance exercises are needed to develop bone mass because bone grows in response to pressure on the bone tissue. The more often and the harder you push on the bone (not enough to break it, of course), the more the body will respond by depositing minerals (calcium, magnesium, phosphorous, fluoride, and boron) into the bone matrix. For instance, tennis players have significantly higher bone mass in their playing arms than do nonathletic controls.[79] However, even the controls had more bone mass in their dominant arm than in the less-used one. Exercise also stimulates growth hormone, which in turn stimulates bone development. "Use it or lose it" applies here. Lack of exercise leads to loss of bone and eventually a fragile skeleton. Fear of falling and breaking a bone keeps some older adults from getting much-needed exercise, contributing to a vicious circle.

INCREASED BONE LOSS   The skeleton acts as structural support and as a calcium reservoir for the body. Bone tissue includes jawbones and teeth. Bones and teeth contain about 99% of the calcium in an adult, roughly 2.2 to 3.3 pounds.[79] The remaining 1% of calcium is found linked with protein in blood, soft tissues, and extracellular fluids. This reservoir is needed for nerve transmission, muscle contraction, and enzyme systems such as those controlling blood clotting. Maintaining nerve transmissions takes physiologic priority over maintaining bone structure. In order to be consistently available to perform many functions, calcium is tightly regulated by hormone systems. When calcium levels in the blood fall, the body responds by secreting more parathyroid hormone (PTH). PTH acts on bone to release calcium and thus raise the blood calcium levels. Too much calcium in the blood stimulates calcitonin secretion. The hormone calcitonin slows release of stored calcium.[79] Bone mineral reserves are dissolved (resorption) and rebuilt constantly to maintain adequate calcium levels for these messenger functions of calcium.

A consistent dietary supply of bone-building minerals (i.e., calcium, magnesium, phosphorus, fluoride, boron) and vitamins (primarily D and K) coupled with regular weight-bearing exercise helps maintain the skeleton reserves and supply calcium when needed. When some portion of this build-dissolve-rebuild cycle is malfunctioning, the body's first priority is to maintain blood calcium levels for nerve, muscle, and enzyme functions. Bone loss results from inadequate nutrient levels or calcium absorption, or from excess calcium excretion.

Osteoporosis can also develop from a shortage of the mineral phosphate during bone mineralization. A balanced calcium-phosphorus ratio as provided in a varied diet allows both nutrients to be used by the bone-building cells. Inadequate phosphorus promotes release of calcium from the skeleton.[80] Although phosphorus is abundant in the food supply, some antacids bind with phosphorus, making it unusable by the body. In the absence of phosphorus, bone mineral formation is delayed until more phosphate is available. Shortage of vitamin D also delays bone mineral formation. Finally, the skin's ability to make vitamin D from the sun is less efficient with aging.

EFFECTS OF OSTEOPOROSIS   Avoiding fractures is an important goal for older adults because fractures contribute to earlier death. Twelve percent of older women who broke a hip were dead within 6 months.[81] The rate for men who did not survive 6 months after breaking a hip was almost double (25%).[81] Death is not due to the fracture itself, but to complications resulting from the break. One of these complications is impaired mobility, complicating all the activities of daily living (including eating and exercising). If an older adult has also had a stroke, impaired mobility becomes the leading cause of institutionalization in the United States. Furthermore, 50% of individuals who fall and break a bone are likely to need assistance walking, and 25% are permanently disabled.[73]

Shrinking Height, Kyphosis   In contrast to hip fractures, most vertebral fractures (67%) are asymptomatic. Postmenopausal women with compression and/or a bone fracture in the spinal column have a condition known as "shrinking height," leading to dowager's hump (also known as *kyphosis,* meaning a bent upper spine). Shrinking in height is slow and usually not painful. For example, a woman who was 5 feet 6 inches tall at age 30 may measure only 5 feet at age 83. She may not be aware of the gradual height loss until someone else comments, or until she notices that clothes no longer fit.

Risk Factors for Osteoporosis

A typical osteoporosis patient is a petite elderly white female. Brittle bones develop from a complex array of physiological factors, including nutrition and exercise. Major risk indicators are male and female glucocorticoid treatment and past history of fractures. Table 19.7 on the next page lists risk factors related to osteoporosis.

Table 19.7 Risk factors associated with osteoporosis in older adults
Not Modifiable
Female, number of pregnancies, length of time between pregnancies
Age, age at which pregnancy(ies) occurred, breastfeeding
Caucasian, Asian
Thin, small-boned rather than large-boned
Family history of osteoporosis
Inadequate bone mass achieved during youth
Low body weight
Premature menopause
History of amenorrhea
Hypogonadism
Glucocorticoid use
Maternal history of hip fracture
Potentially Modifiable
Lack of weight-bearing exercise
Cigarette smoking
Long-term dietary phosphorus deficiency (e.g., use of phosphorus-binding antacids)
Heavy alcohol consumption
Underweight
Malnourished
Inadequate dietary calcium (<1200 mg) and vitamin D (<400/600 IU) intake
Still Controversial or Not Yet Clear
Diet high in phosphorus while low in calcium (mixed evidence)
Inadequate fluoride, boron, and magnesium in diet
Consistently high protein and/or sodium intake
Eating soy products (soymilk, textured protein, etc.) for their estrogen-like activity

Nutrition Interventions for Osteoporosis

The first remedy for osteoporosis is optimal diet, including intake of the recommended daily allowance of calcium for older adults—1200 mg per day, not to exceed the tolerable upper limit of 2500 mg (or 1500 mg calcium per day in Canada).[82] The dietary intake goal is to provide enough available calcium and vitamin D through diet or supplements so that despite declining absorption rates, bone loss is minimized. Calcium retention increases with increasing calcium intake up to each person's threshold. Beyond that threshold, greater intake of calcium does not lead to higher levels of calcium retention. A sample meal plan that provides the Dietary Reference Intake (DRI) for calcium in older adults is shown in Table 19.8.

In older persons, taking calcium supplements without other bone-building activity has not been shown to increase bone density. In fact, an individual at bed rest or otherwise immobilized loses bone mass rapidly. Thus, Canadian practice guidelines for nonpharmacologic interventions in osteoporosis recommend physical activity ≥30 minutes per day three times per week.[82]

Several dietary components (protein, sodium, caffeine, vitamins) are closely linked to calcium metabolism, and their intake can interfere with appropriate supplementation. High protein intakes result in less calcium being available, because protein leads to greater excretion of calcium into the urine. In the Nurses Health Study, women (aged 35 to 59, followed for 12 years) who ate more than 95 grams of protein per day suffered more osteoporotic forearm (but not hip) fractures.[83] High sodium intake leads to higher levels of urinary calcium; that is, more calcium is excreted when higher levels of sodium are eaten. Caffeine to equal 2 to 3 cups of coffee daily (in postmenopausal women), consumed with less than 744 mg/day of calcium, has been associated with bone loss.[80] Along with supplementation, a number of nutritional remedies can improve calcium intake and absorption:

- Drink milk and take supplements with a meal, because food slows intestinal transit time and allows more calcium to be absorbed from the gut.

Table 19.8 A day's intake for an older adult: consuming at least 1200 mg calcium (1520 mg)	
	Amount of Calcium
Food	mg
Oatmeal made with milk, 1 cup total	266
Banana, one medium	6
Coffee, 10 ounces with 1 ounce (2 tb) evaporated milk	87
Turkey sandwich on whole wheat bread, lettuce, mayonnaise	54
Cheese added to sandwich, 1 ounce cheddar	148
Canned fruit cocktail, ½ cup	9
Iced tea, plain	0
Orange juice, calcium fortified, 8 ounces	289
Roasted almonds, 2 tb	33
Pasta with chicken, 1½ cups	54
Tomato slices, 2	2
Milk, 1%, 8 ounces	279
Sugar cookie, 1 medium	5
Chocolate milk, 8 ounces	287

- Divide calcium supplements to be taken at a different time than antacids, because stomach acidity enhances absorption; for older adults who have decreased acidity, calcium *citrate* is considered more soluble.[84]

- Consume foods that are rich in bone-building vitamins (C, D, B$_6$, K) at recommended levels. This will help to synthesize bone and develop the collagen matrix, into which minerals are deposited during bone mineralization.[85]

- *Vitamin C* plays a role in the development of the protein bone matrix (collagen), but has no defined role in treatment.

- *Vitamin D* (1,25-dihydroxyvitamin D, also called calcitriol) stimulates active transport of calcium in the small intestine and the colon; in those states where the elderly are unlikely to get adequate sun exposure, ingested forms of vitamin D are encouraged to meet adequate intake of 10–15 mcg of vitamin D$_3$ (see inside front cover).[82]

- *Vitamin K* is required for the formation of proteins that stimulate *osteoblasts* to build new bone and attract *osteoclasts* to initiate bone resorption (see Table 19.9). This process has the potential to slow bone loss; further research is needed to understand its association with osteoporosis.

Although Vitamin K has been linked to building a stronger bone matrix, it also plays an important role in the blood clotting process. Older adults with a history of strokes are placed on anticlotting medication. Nutrition counselors advise people who are taking anticlotting medication to maintain a stable Vitamin K intake.

Two forms of vitamin K are found in foods. Phylloquinone, also known as vitamin K$_1$, is naturally occurring in plants; and menatetrenone, also known as vitamin K$_2$, is found in meat, cheese, and fermented products. Consuming big portions of broccoli and greens for a few weeks in summer could increase chances of blood clotting, which can increase the chances of causing another stroke. In addition, when a vitamin K–containing supplement is added to the diet (e.g., taking 2–3 Viactiv calcium chews as per recommendation), 150% of vitamin K is instantly provided. One can easily get too much of a good thing.

How much is too much? As yet, no tolerable upper levels are set for vitamin K. But for fat-soluble vitamins, a general rule (although one that does not apply to vitamin E) is to stay under 500% of the RDA or DRI.

HORMONES AND OSTEOPOROSIS Estrogen, testosterone, growth hormone, insulin-like growth factor-I, and parathyroid hormone increase calcium absorption rate. Hormone replacement, with or without additional calcium, has been shown to increase bone mineral density in early postmenopausal women (4.5% and 1.5% respectively).[86] Pines and colleagues also found that the women

Table 19.9 Amounts of Vitamin K in selected foods (DRI for ages 51 and older is 120 mcg/day for males, 90 mcg/day for females)

Food	Amount, Size	Vitamin K mcg
Kale, chopped	1 cup	547
Broccoli, cooked	1 small stalk, 5 in.	378
Swiss chard, chopped	1 cup	299
Pumpkin, canned	½ cup	196
Broccoli, raw	1 cup	180
Spinach, raw	1 cup	120
Cabbage, raw	1 cup shredded	102
Butterhead lettuce	1 cup chopped	67
Lettuce	½ cup shredded	59
Parsley, raw	1 tablespoon	22
Canola oil	1 tablespoon	20
Asparagus	4 small spears, <5 in.	19
Soybeans, dry roasted	¼ cup	16
Mayonnaise	1 tablespoon	11
Plums, raw	1 med, 2-⅛ in. diameter	8
Peanut butter, smooth	2 tablespoons	3
Cucumber without skin	½ cup slices	1
Apple without skin	1 med, 2-¾ in. diameter	0.5

SOURCE: USDA provisional table, Vitamin K; available at www.nal.usda.goiv/fnic/foodcomp, accessed August 2000.

not receiving hormone therapy, either with or without calcium supplementation, lost bone mineral density (1.4% and 3.7% respectively). However, in 2002, the Women's Health Initiative reported increased rates of breast cancer, coronary heart disease, stroke, and venous thromboembolism in addition to decreased rates of hip fracture and colorectal disease.[87] This information was quickly disseminated to health professionals and women. The risks and benefits for use of hormone replacement therapy must be carefully considered.

Individuals choosing not to take hormone replacements may try soy (e.g., milk, textured protein, bars, soy nuts) for its estrogen-like activity (e.g., phytoestrogens). Naturally occuring phytoestrogens are weak estrogen-like chemicals. More information is available on synthetic phytoestogen, known as ipriflavone. Evidence is not definitive that soy will mimic the effects of estrogen in bone development, though; to obtain a definitive answer, studies will need to ensure that there were adequate intakes of calcium and vitamin D.

Osteoblasts Bone cells involved with bone formation; bone-building cells.

Osteoclasts Bone cells that absorb and remove unwanted tissue.

EXERCISE AND OSTEOPOROSIS Although there is a relationship among calcium intake, exercise, and bone mass, it is not known whether exercise increases absorption or whether it enhances bone mineralization.

In summary, the best osteoporosis prevention strategy is exercise and adequate diet in young people when bones are first growing. For older individuals who have brittle bones, exercise supplemented with calcium and vitamin D strengthens bone mass. As much as possible, stay active, enjoy sunny days, and drink milk!

Gastrointestinal Diseases

The gastrointestinal system, which is roughly the length of a football field, serves many functions. It should thus not be surprising to learn that by late adulthood, it occasionally malfunctions. It seems miraculous that we so consistently eat what we like, without much thought, and our body converts that food to energy for daily living. Parts of the gastrointestinal system most likely to malfunction in old age are

1. *The esophageal-stomach juncture.* Weakened muscle results in gastroesophageal reflux disease (GERD).
2. *The stomach.* Decreased acidity leads to changes in nutrient absorption or increased acidity causes ulcers.
3. *The intestines.* Decreased motility or hyperactivity results in constipation, diarrhea, and some food intolerance.

Often these problems are secondary to other diseases. No matter what the cause, older adults are at higher risk for some of the gastrointestinal conditions discussed next, any of which may impair older adults' activity.

Gastroesophageal Reflux Disease (GERD)

Gastroesophageal reflux disease occurs when stomach contents flow back into the esophagus. Approximately 19 million Americans, or one of five older adults, have GERD.[88,89]

EFFECTS OF GERD It is not clear whether acid in the esophagus leads to a weakened *lower esophageal sphincter (LES),* or a weakened sphincter leads to GERD.[90] The main symptoms of GERD are heartburn and acid regurgitation. Stomach contents, which are highly acidic, spill back into the esophagus, resulting in irritation and pain. Other effects are chest pain, trouble swallowing, nausea and vomiting, and belching.

> **Lower esophageal sphincter (LES)**
> The muscle enabling closure of the junction between the esophagus and stomach.

RISK FACTORS FOR GERD Alcohol in excess of seven drinks per week, obesity, and smoking are consistently linked to GERD episodes.[88,89] In addition, regular and decaffeinated coffee are associated with heartburn.[89]

NUTRITIONAL INTERVENTIONS FOR GERD The primary dietary remedy is to omit foods that are chemically or mechanically irritating. There is little consistency about which foods these are. However, general guidelines are to choose a low-fat diet and nonspicy foods. Chew thoroughly and eat slowly. Finally, to take advantage of gravity and make it difficult for the stomach acids to reflux upward, don't lie down after eating.

VITAMIN B$_{12}$ DEFICIENCY The normal process by which the body absorbs vitamin B$_{12}$ from foods requires the following conditions: The stomach environment needs to be acidic and producing enzymes (especially pepsin), and the stomach cell lining needs to be secreting an intrinsic factor (IF). Stomach acids and enzymes split off and transfer vitamin B$_{12}$ from foods to carrier proteins (mostly secreted by the salivary gland in the mouth). Then, the vitamin B$_{12}$–carrier protein complex moves into the small intestine, where the vitamin B$_{12}$ will again be broken off and bound to the IF (that was produced in the stomach and migrated to the small intestine). The vitamin B$_{12}$–IF linked complex then binds to surface receptors in the lining of the small intestine. From there, vitamin B$_{12}$ is absorbed into epithelial cells and then transported through blood to tissue cells (see Illustration 19.1).

The two primary types of vitamin B$_{12}$ deficiency in older adults are (1) pernicious anemia due to lack of IF being released from the stomach cell wall; and (2) inadequate dietary intake (i.e., lack of extrinsic factor), but more often poor absorption of vitamin B$_{12}$.

Pernicious Anemia Pernicious anemia results when IF is lacking. IF is released from the stomach wall and needed for the dietary vitamin B$_{12}$ to be absorbed in the small intestine. Physical signs of pernicious anemia (macrocytic megaloblastic anemia) include large, undeveloped red blood cells, increased redness and swelling in the mouth (glossitis), and tongue fissures. Individuals with pernicious anemia also tend to have shrunken stomach glands and mucosa, leading to decreased enzyme and hydrochloric acid secretion. Lack of IF leads to vitamin B$_{12}$ deficiency.

- **Effects** Effects of pernicious anemia are glossitis and irreversible neurological damage. Other symptoms may also include fatigue, anorexia, weight loss, nausea and vomiting, diarrhea or constipation, and tachycardia.[89] Pernicious anemia is often accompanied by inadequate stomach acidity (hypochlorhydria). Less stomach acid leads to impaired iron absorption.

- **Prevalence** Pernicious anemia takes 5 to 6 years to develop and is rarely seen before age 35. It is found in approximately 2% of the population, most commonly in women over age 65.[89]

- **Risk Factors** Risk factors are a history of *Helicobacter pylori (H. pylori)* infections, decreased stomach acid production, and a family history of pernicious anemia.

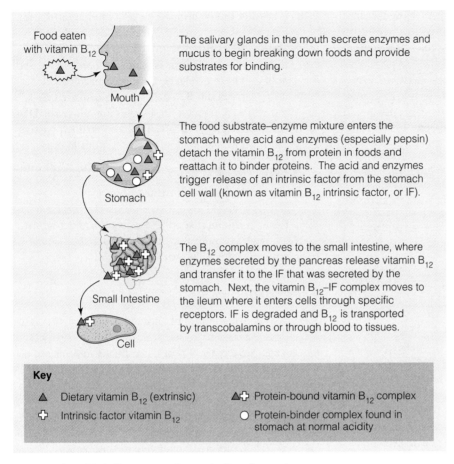

The salivary glands in the mouth secrete enzymes and mucus to begin breaking down foods and provide substrates for binding.

The food substrate–enzyme mixture enters the stomach where acid and enzymes (especially pepsin) detach the vitamin B_{12} from protein in foods and reattach it to binder proteins. The acid and enzymes trigger release of an intrinsic factor from the stomach cell wall (known as vitamin B_{12} intrinsic factor, or IF).

The B_{12} complex moves to the small intestine, where enzymes secreted by the pancreas release vitamin B_{12} and transfer it to the IF that was secreted by the stomach. Next, the vitamin B_{12}–IF complex moves to the ileum where it enters cells through specific receptors. IF is degraded and B_{12} is transported by transcobalamins or through blood to tissues.

Key

▲ Dietary vitamin B_{12} (extrinsic)	▲✚ Protein-bound vitamin B_{12} complex
✚ Intrinsic factor vitamin B_{12}	◯ Protein-binder complex found in stomach at normal acidity

Illustration 19.1 Overview of vitamin B_{12} absorption.

tritis. Older adults with atrophic gastritis still absorb some dietary vitamin B_{12}, but they will become deficient over time. Bacterial overgrowth (usually related to infection with *H. pylori*) will also use vitamin B_{12} and diminish the amount available for absorption. Finally, antacid treatment in individuals neutralizes stomach acid (i.e., pH increases), and this rise in stomach pH does not allow all of the available vitamin B_{12} to be split from the food carrier, even though the intrinsic vitamin B_{12} factor is present. Symptoms of vitamin B_{12} deficiency begin to appear after 3–6 years of poor absorption.[89,91]

- **Prevalence** The prevalence of vitamin B_{12} deficiency caused by poor absorption is greater in older adults than that caused by pernicious anemia. Protein-bound vitamin B_{12} is estimated to occur in 20% of people over 69 years old (known or undiagnosed).[92] In reviewing studies, Nilsson-Ehle found prevalence rates of up to 80%, but 20–30% was typical.

- **Nutritional Interventions** Vitamin B_{12} may be given orally or by injection. Injections are administered daily until lab values stabilize, then six to eight times per year throughout life.[70] In Scandinavian countries, injections are prescribed until pernicious anemia is in remission. Then, maintenance doses of oral vitamin B_{12} are used. Roughly 80% of Scandinavian therapy costs are for oral vitamin B_{12} administration rather than injections. One reason that oral maintenance therapy works is that approximately 1–2% of synthetic vitamin B_{12} (not food bound) is absorbed passively, removing the need for either IF or carrier protein.

Foods high in vitamin B_{12} are unlikely to maintain adequate blood levels in pernicious anemia. However, foods fortified with crystalline or synthetic vitamin B_{12} increase the likelihood of passive absorption. Other dietary recommendations include eating proteins of high biological value and, if there is an iron deficiency, iron.

Inadequate Dietary Intake or Absorption (Protein-Bound Vitamin B_{12} Deficiency) The most common cause of vitamin B_{12} deficiency in older adults is related to abnormal stomach function. For instance, after prolonged inflammation, the stomach mucosa shrinks and secretes less acid and enzymes, a condition known as atrophic gas-

Serum *cobalamin* levels are used to detect deficiency, although the cut-off limit has been set too low for the elderly.[93] It is more effective to measure the blood levels of intermediate breakdown products in pathways where vitamin B_{12} is needed. For example, vitamin B_{12} is needed to change *methylmalonic acid (MMA)* into a component used in the *Krebs cycle*. When adequate amounts of vitamin B_{12} are unavailable, blood levels of MMA rise. Because B_{12} is the only coenzyme to catalyze this reaction, a test for levels of MMA is specific to vitamin B_{12} deficiency.

Another test is to measure blood levels of the coenzyme *homocysteine.* If vitamin B_{12} isn't available to complete the pathway that

Cobalamin Another name for vitamin B_{12}. Important roles of cobalamin are fatty acid metabolism, synthesis of nucleic acid (i.e., DNA, a complex protein that controls the formation of healthy new cells), and formation of the myelin sheath that protects nerve cells.

Methylmalonic Acid (MMA) An intermediate product that needs vitamin B_{12} as a coenzyme to complete the metabolic pathway for fatty acid metabolism. Vitamin B_{12} is the only coenzyme in this reaction; when it is absent, the blood concentration of MMA rises.

Krebs Cycle A series of metabolic reactions that produce energy from the proteins, fats, and carbohydrates that constitute food.

Homocysteine Another intermediate product that depends on vitamin B_{12} for complete metabolism. However, both vitamin B_{12} and folate (another B vitamin) are coenzymes in the breakdown of certain protein components in this pathway. Thus, elevated homocysteine levels can result from vitamin B_{12}, folate, or pyridoxine deficiencies.

forms DNA, then homocysteine (intermediate metabolite) levels become elevated. Because folate is also needed in this metabolic pathway, lack of either folate or vitamin B_{12} can lead to high levels of homocysteine.

Therefore, elevated MMA and homocysteine levels confirm a vitamin B_{12} deficiency. However, if MMA levels are normal but homocysteine levels are high, folic acid deficiency or another cause of increased homocysteine levels should be investigated.

- **Effects** Vitamin B_{12} deficiency leads to irreversible neurological damage, walking and balance disturbances, and cognitive impairment (including confusion and mood changes). High levels of homocysteine are known to be a risk factor for heart and peripheral vascular disease.[94]

- **Risk Factors** Risk factors include advanced age, gastrointestinal disorders, genetic family patterns, medications, and (to some extent) inadequate food intake. An example of aging as a risk factor comes from Nilsson-Ehle, who found that 1.5% of 50-year-olds have achlorhydria and B_{12} deficiency as compared to 18% of 80-year-olds.[92] Gastrointestinal disorders that affect vitamin B_{12} absorption include atrophic gastritis (with higher risks in diabetes mellitus and autoimmune thyroid disorders), partial stomach removal, and *H. pylori* infection. Medications that suppress stomach acid secretion or impair absorption are associated with the risk of deficiency. Examples of these types of medicines are oral biguanides (e.g., metformin used to treat type 2 diabetes mellitus), modified-release potassium preparations, anesthesia, hydrogen-receptor antagonists (e.g., Cimetidine or Tagamet) and proton pump inhibitors (e.g., omeprazole/Prilosac given for gastroesophogeal reflux disease). In genetically linked cases, individuals often have antibodies for the intrinsic factor.

Vegetarians not taking a B_{12} supplement are at risk of vitamin B_{12} deficiency. For example, a study of 245 Australian Seventh Day Adventist ministers found most (70%) had a vitamin B_{12} deficiency due to a low dietary intake.[95] Even with intrinsic vitamin B_{12} factor present, 53% of the group had serum cobalamin levels below the reference level for this study (171–850 picomoles per liter). Some ethnic and gender differences may exist for developing a vitamin B_{12} deficiency. Elderly white men had the highest prevalence rates, whereas black and Asian women had the lowest rates among community-dwelling seniors (aged 60 or more) in Los Angeles.[96] In contrast, Zeitlin and colleagues[97] concluded that there were no age- or gender-related differences in the mean levels of serum B_{12} after studying approximately 440 subjects in the Bronx Aging Study (mean age = 79 years).

Diverticulitis Infected "pockets" within the large intestine.

- **Nutrition Interventions** The DRI of vitamin B_{12} for men and women over 50 years is 2.4 mcg/day,[98] and the UL is not determined.[99] Good food sources include meats, shellfish (shrimp, crab, mussels), and milk and milk products. Oral pharmacological doses for vitamin B_{12}–deficient patients are 0.2–1 mg (200–1000 mcg); these doses are above the DRI because roughly 1–2% is absorbed through passive diffusion in the small intestine.[92] Foods high in folate and iron are also necessary in cases where folate deficiency and/or iron deficiency has been diagnosed. In 1999, fortified flour and breakfast cereals supplied almost two-thirds of the total folate in the food supply.[100] Thus, for older adults, a bowl of vitamin B_{12}– and folate-fortified whole grain cereal is a "power food."

Constipation

There are many definitions of "normal bowel pattern." Individuals can consistently experience from two or three bowel movements per day to two or three bowel movements per week. When two or fewer bowel movements per week is used as the definition of constipation, the prevalence of constipation was 3.8% in older adults aged 60–69 years and 6.3% in older adults aged 80 years and older.[101] In a midwestern community of people ages 65 years and older, as many as 40.1% reported some type of constipation.[102]

The physiological causes of constipation in older adults are emerging.[103] Animal studies indicate that aging intestinal muscles respond less to triggers and that the brain-muscle transmitters are either inadequate or less responsive. Older adults' thirst mechanisms decline. Lower levels of stomach secretions and potentially less muscle strength affect peristalsis. Due to chewing problems, there is a potential for eating fewer fiber-rich foods. Low-fiber diets can also exacerbate *diverticulitis*.

Persons with laxation difficulties are prone to be more anxious and focused on bowel movements as an aspect of health.

RISK FACTORS FOR CONSTIPATION Risk factors for constipation include:

- Poor water intake
- Eating small amounts of food
- Medications, such as nonsteroidal anti-inflammatory drugs
- High-iron mineral supplement

NUTRITION INTERVENTIONS FOR CONSTIPATION Remedies are to encourage increased dietary fiber in tandem with an increased fluid intake. Bran in breads and cereals is one way to add to fecal bulk that stimulates persistalsis to move wastes through the colon. It is, again, important to add water when increasing fiber. One caution: Find out if there is potential for fecal impaction due to disease; that is, colon cancer. Increased exercise may relieve individuals who worry about bloating and flatulence.

Inflammatory Diseases

Examples of inflammatory disease are osteoarthritis, rheumatoid arthritis, atrophic gastritis (typically due to *H. pylori* infection), celiac disease (**gluten** intolerance), irritable bowel syndrome (IBS), diverticulitis (an infection in the large intestine), and asthma. Of these conditions, arthritis affects the greatest number of older individuals.

Osteoarthritis is the most common form of arthritis, affecting roughly 20 million people in the United States.[104] Prevalence of osteoarthritis at the knee is twice as common as at the hip (6% compared to 3%); the knee and hip are the two most common sites for persons aged 30 and older. Prevalence increases with age and peaks between 70 and 79 years. More men than women have osteoarthritis before age 50; after age 50, women are more often affected. (Study results are inconsistent about the potential benefits of estrogen treatment in osteoarthritis.)

Etiology of Inflammatory Diseases

Cartilage loss, bone changes such as bone outgrowth, hardening of soft tissues, and inflammation all lead to tissue damage. Some wear and tear of aging also contributes to the complex disease of osteoarthritis. Variability in prevalence and progression of knee, hip, and hand osteoarthritis suggests that the disease may be a group of different and unique conditions. Rheumatoid arthritis is a collagen disease characterized by inflammation and increased protein turnover.

Effects of Inflammatory Diseases

Pain occurring with joint movement is common, but the cause of pain is still unexplained.[104] Osteoarthritis often leads to disability; in the words of Felson and colleagues:[104]

> The risk for disability (defined as needing help walking or climbing stairs) attributable to osteoarthritis of the knee is as great as that attributable to cardiovascular disease and greater than that due to any other medical condition in elderly persons.

Individuals with osteoarthritis tend to have pain, depressive symptoms, muscle weakness, and poor aerobic capacity.

Risk Factors for Inflammatory Diseases

Obesity, continuous overexposure to oxidants, and possibly low vitamin D levels are risk factors for developing osteoarthritis. Low intake of vitamins C and D is a risk factor for its progression. Risk factors may turn out to be different for knee or hip and hand osteoarthritis.

Nutrition Interventions for Inflammatory Diseases

Weight loss is the first remedy advised. Felson and colleagues' summary report[104] found that women in the Framingham study who lost 11 pounds cut their risk for knee osteoarthritis in half. However, the connection between weight loss and relief of symptoms of osteoarthritis has not yet been well studied. Evidence to date suggests weight loss helps to reduce symptoms.

Gluten A protein found in wheat, oats, barley, rye and triticale (all in the genus *Triticum*); gliadin is the toxic fraction of gluten.

ANTIOXIDANTS Individuals with the highest levels of vitamin C intake had significantly slower (threefold) disease progression and less knee pain than did individuals with the lowest intakes. Results of increased vitamin E and beta-carotene intake were inconsistent.

VITAMIN D Progression at higher intakes and serum levels is roughly one-third slower than at the lowest levels. The DRI for older adults is 400 to 600 IU, with a UL of 2000 IU.

FLAVONOIDS This is a large group of phytochemicals (found in vegetables, fruits, tea, and whole grains) with antioxidant and anti-inflammatory properties that act to maintain cell membranes. Higher levels of antioxidants are required to scavenge oxidized metabolites and free radicals in inflammatory diseases.[105]

CHONDROITIN AND GLUCOSAMINE These two substances naturally occur in the body as substrate for cartilage repair. Although the mechanisms by which they act are not known, most clinical trials of their use demonstrate favorable effects, with a slightly larger effect for chondroitin sulfate.[106]

SAM-E S-adenosylmethionine may reduce pain and stiffness and possible depression; in large doses it may also cause nausea.

CAPSAICIN The compound that provides the heat in peppers is used in a topical cream for pain relief.

OTHER TREATMENTS Fatty acids and oils are other anti-inflammatory therapies with potential to lessen signs and symptoms in a variety of conditions including osteo- as well as rheumatic arthritis.[107] Borage seed and evening primrose oils contain gamma-linolenic acid (GLA), which plays a role in prostaglandin synthesis (GLA competes with other fatty acids, limiting the production of inflammatory omega-6 fatty acids) and can decrease pain of arthritis after several months of use.[105] Reductions in pain, morning stiffness, and swelling of rheumatoid arthritis follow intake of omega-3 fatty acids (fish oil, marine fatty acids) for at least 12 weeks.[78] Flaxseed and purslane also contain a type of omega-3 fatty acid. Traditional Native American medical therapy used echinacea (three varieties) for pain relief, rheumatism, and arthritis; ginseng for asthma and rheumatism; garlic for asthma; and evening primrose oil for obesity. Today echinacea and ginseng are marketed to boost immune function. Evening primrose oil is marketed as an antioxidant and for its role in decreasing premenstrual pain.[108]

> **Memory Impairment** Moderate or severe impairment is present when four or fewer words can be recalled from a list of 20.

VEGETARIAN DIETS Plants are rich in antioxidants that may play a role in managing inflammatory diseases. A review of the roles of vegetarian diets[109] found that it is difficult to design good experiments to answer the question: "Do vegetarians have less inflammatory disease?"

FOOD ALLERGIES Avoiding gluten has a clear role in celiac disease, which is an inflammatory disease. Food allergies have also been suggested as a cause for inflammatory diseases such as irritable bowel syndrome, Crohn's disease, and rheumatoid arthritis; but evidence from well-done studies that can point to allergens or mechanisms is not yet available.[80]

Traditional and Alternative Therapies

Prescription medications are medicines that physicians or other health care providers, acting within the scope of their license, can order. Prescription drugs, according to health care plans, do not include vitamins, herbal medicines, or over-the-counter (OTC) medicines. OTC medications are any pills, liquids, salves, creams, and supplements that are purchased at a pharmacy, discount, or food store without a prescription. Most complementary medicines such as botanicals and herbs are sold as OTCs.

A high prevalence of ongoing health problems leaves older adults using more medications, both OTC and prescription types, than younger adults. Pain and suffering lead to greater vulnerability to use of unproven, potentially dangerous therapies. Disease progression, drug effects, and functional limitations, such as poor eyesight or memory, all increase risks for using a drug incorrectly and for noncompliance.[110] The larger the number of drugs used by an individual, the greater the chance of error.

Effects of Medications

Medications may interfere with appetite, food digestion, and absorption and general alertness. Inability to eat or forgetting to eat lead to unintentional weight loss, while some medications lead to undesired weight gain. Table 19.10 identifies the nutritional implications associated with medications commonly taken with chronic conditions in older

adults.[111] The expense of needed medications may reduce the amount of money available for health-promoting foods.

Mental Health and Cognitive Disorders

A group of disorders called dementia or cognitive disorders includes depression, Alzheimer's disease, and *memory impairment*. Often considered an expected (and dreaded) aspect of "getting old," this "dementia of aging" is being more carefully defined as knowledge grows about the mechanisms of normal aging. It is important to note that Alzheimer's disease and dementia symptoms are not a part of normal aging. Definitions of these conditions are as follows:

- Depression is a brain disorder characterized by "being down in the dumps" and not wanting to socialize with others over a long period of time.
- In Alzheimer's disease, the brain atrophies, leading to memory loss, behavior and personality changes, and a decline in ability to think. There is a slow and steady loss of sense of self and a meaningful relationship with the world.
- Dementia is a clinical state in which intellectual level declines; it usually involves deterioration in memory and one or more of the other intellectual functions such as language, spatial thinking and orientation, judgment, and abstract thought.

Because Alzheimer's is difficult to diagnose, prevalence statistics are approximate estimates. In 1997, the estimated prevalence of Alzheimer's disease in the United States was 2.32 million cases (estimates range from 1.09–4.58 million).[89] The risk of developing Alzheimer's disease doubles every 5 years after age 65. Among people 65 years and older, as many as 3 in 100 suffer from clinical depression.[89] Table 19.11 lists prevalence data for other cognitive disorders.

Etiology of Cognitive Disorders

Cognitive disorders stem from brain damage including tumors, broken blood vessels, and neurological diseases such as Alzheimer's. What causes Alzheimer's disease? No one knows. We do know that aluminum cookware does not cause Alzheimer's disease. Aluminum, copper, carnitine, and choline have been examined as potential causes because of the role they play in neurological functioning. Vitamin B_{12} deficiency can lead to neurological damage and cognitive decline, but it is not an underlying factor in Alzheimer's disease.

Effects of Cognitive Disorders

Weight loss is a nutritional feature associated with the diagnosis of Alzheimer's disease. Short-term memory loss, difficulty in decision making, depression, and changes in

Table 19.10 Medications associated with chronic conditions in older adults: nutritional implications

Coronary Heart Disease

- Cardiac glycosides (e.g., digitalis) may result in anorexia and/or nausea.
- Statins may result in elevated enzymes.
- High doses of niacin may be associated with flushing, hyperglycemia, hypotension, hypoalbuminemia, upper GI distress, and liver enzyme elevation.

Hypertension

- Diuretics may result in depletion of sodium, calcium, magnesium, and/or potassium.
- Centrally acting antihypertensives may result in a decline in food intake due to sedation, confusion, and depression.
- Medications such as beta blockers may cause constipation and delayed gastric emptying.
- Potential related side effects are dizziness, dry mouth, or mouth pain.

Diabetes Mellitus

- Medications to treat diabetes may cause hypoglycemia, especially if nutritional intake is erratic and/or if there is increased or decreased appetite or diarrhea.
- Oral hypoglycemics (e.g., sulfonylureas such as glipizide) may cause heartburn, nausea, hypoglycemia, decreased appetite; biguanides (such as Glucophage) may cause decreased appetite, diarrhea, and vomiting.

Dementia

- Cholinesterase inhibitors (e.g., rivastigmine or galantamine) may cause nausea, diarrhea, or weight loss.
- Antipsychotic/antidepressants with anti-cholinergic side effects may lead to dry mouth, delayed stomach emptying, and constipation.
- Antidepressants may enhance appetite in depressed patients, but selective serotonin reuptake inhibitors (SSRIs) may cause a decrease in appetite.

Congestive Heart Failure

Diuretics (e.g., Diuril) may lead to electrolyte abnormalities, especially sodium and potassium and/or thiamine deficiency (e.g., Furosemide).

Cancer

Radiation, chemotherapy, and/or surgery can negatively affect nutritional status and metabolism.

Chronic Obstructive Pulmonary Disease (COPD)

Xanthine derivatives (e.g., theophylline) may result in anorexia and nausea.

Lower Respiratory Infections

- Antibiotics (e.g., amoxicillin) may result in nausea, anorexia, diarrhea, and stomatitis.
- Radiation, chemotherapy, and/or surgery can negatively affect nutritional status and metabolism.

SOURCE: Reprinted with permission by the Nutrition Screening Initiative, a project of the American Academy of Family Physicians and The American Dietetic Association, funded in part by a grant from Ross Products Division, Abbott Laboratories, Inc.

smell and taste are a few aspects of this disease that make good nutritional habits difficult to maintain. Although individuals may be able to maintain an adequate diet early in the disease, they will need nutritional assistance when the disease progresses to later stages.

Nutrition Interventions for Cognitive Disorders

Additional calories may be needed for increased energy expended by individuals who pace and wander. Vitamin E and selenium may be useful[89] for their antioxidant properties; choline (in soybeans and eggs) is a building block for acetylcholine but has not been proven useful in controlled research. Developing routines and eliminating distractions during meals, allowing plenty of time to eat, serving foods one course at a time, offering finger foods, and tracking fluid intake to ensure adequacy

Table 19.11 Percentage of persons aged 65 years and older with cognitive disorders

Age	PERCENTAGE WITH MODERATE TO SEVERE MEMORY IMPAIRMENT		PERCENTAGE WITH SEVERE DEPRESSIVE SYMPTOMS	
	Males	Females	Males	Females
65–74 years	15.4	9.7	22.4	35.2
75–84 years	39.0	30.6	27.5	39.8
85+ years	37.3	35.0	22.5	23.0

are some of the nutritional strategies in treating the disease. Safe use of kitchen tools and equipment and food safety are two food-related considerations in evaluating dementia patients.

Low Body Weight/Underweight

There is no consensus or universal definition for underweight in the frail elderly. The most common methods that measure changes in nutrition status are body mass index (BMI) and unplanned weight loss within the past 3 to 6 months. A BMI or weight found in the lowest percentiles of a reference standard in a comparable population is a starting point. Because of the National Health and Nutrition Examination Survey III (NHANES), we now have weight percentiles for older adults by decade (beginning at 50 years), gender, and ethnic-racial groups and are able to recognize underweight in older adults (see Table 19.12).[112] Individuals with weights falling at the 5th or lower percentile of this population can be defined as "underweight."

Another approach is to compare an individual's current weight to their "usual" body weight. Terms such as *sarcopenia* (muscle loss), *anorexia* (no appetite), and *cachexia* (another way to say "no appetite") are related to undernutrition in older adults and are associated with becoming frail.

The National Heart, Lung, and Blood Institute (NHLBI) defines underweight as a BMI <18.5 kg/meter squared for all adults.[113] The World Health Organization (WHO) further defines levels of underweight as grades of "thinness."[114]

BMI 17.0–18.49   indicates grade 1 thinness
BMI 16.0–16.99   indicates grade 2 thinness
BMI <16.00   indicates grade 3 thinness

Approximately one-third of older adults are underweight.

Table 19.12 Potential for malnutrition: mean weight at the 5th percentile, by age and gender, of older adults residing in the United States

Age	WEIGHT IN POUNDS Male	Female
60–69 years	134.5	109.0
70–79 years	128.7	100.5
≥80 years	114.3	92.0

SOURCE: Third National Health and Nutrition Examination Survey (1988–1994). Available at www.cdc.gov/nchs/about/major/nhanes/Anthropometric%20Measures.htm, accessed July 2003.

Etiology of Underweight

Underweight is not considered problematic when the individual has had a lifelong low weight. However, weight cycling is problematic. In addition, unintentional weight loss is likely due to disease. A loss of 10% or more total body weight lost in a 6-month period is associated with increased mortality. Intended weight loss should be consistent with reasonable weight-for-height standards.

For older adults, underweight is much more serious than overweight. Being thin has been related to increased incidence of diseases, but it is impossible to tell from the data whether thin precedes or follows incidence of disease. Overall, malnutrition affects immune response, muscle and respiratory function, and wound healing; malnutrition is associated with, but not caused by, aging.

Protein-calorie malnutrition leads to underweight. Underlying causes may be illness, poverty, or functional decline. See Case Study 19.2.

Nutrition Interventions for Underweight

Avoiding weight loss is desirable, but not always possible. Weight loss in the elderly is intertwined with disease-related biochemical and physiologic mechanisms, which in turn, affect functional status and appetite. In a systematic review of studies to evaluate the effect of nutrition treatment in protein-energy malnutrition in connection with multiple disorders in the elderly, 20 studies noted an improvement in anthropometric or biochemical measures and 10 studies reported an improvement in function. In contrast, there was insufficient evidence to determine how nutrition treatment should be formulated, because inconsistent or uncertain treatment adherence and inability to separate treatment effects further complicates nutrition monitoring.[115] In a systematic review of 22 studies reports by the Cochrane Library, protein and energy supplementation appears to produce a small but consistent weight gain (2–4%), reduced mortality, and shorter length of hospital stays.[116] Supplementation was associated with nausea, diarrhea, and other gastrointestinal disturbances. Overall, medical nutrition therapy (MNT) for a frail, elderly, malnourished person should occur in consultation with an experienced registered dietitian. Refeeding and rehydration are done gradually:

- *Calories:* Eat and exercise to build muscle mass, strength.
- *Protein:* From 1 to 1.5 grams of protein per kilogram body weight is adequate. 1.5–2 g/kg/day is recommended for severe depletion by the American Dietetic Association's Consultant Dietitians in Health Care Facilities Practice Group.[117] Exceptions are patients with renal or liver failure, who may need a protein restriction.
- *Water:* Drink 1 mL per calorie; rehydrate slowly (see section on dehydration).

Photo Disc

Case Study 19.2

Ms. Wetter—Surviving a Bad Stretch

Ms. Elizabeth Wetter, about to turn 81, is 5 feet 6 inches tall and weighs 106 pounds. She has had Parkinson's disease for 5 years, but that is not what concerns her. Her problem is pain from arthritis and lack of energy. She saw an ad on television for a vitamin-mineral supplement with ginseng that promises "more energy." Her son also told her to take a liquid dietary supplement to "feel better." Eighteen months ago, Ms. Wetter had successful surgery for colon cancer, which was followed by chemotherapy treatments. She is now free of cancer. After the cancer treatment, she fell and broke a hip. This healed well, but serious leg pains started shortly afterward. There seems to be no cause for the pain, and a cure is unavailable. She is no longer able to take her walks through the neighborhood or tend her prize-winning garden. She would like to weigh 118 pounds again (her "usual") and is seeking nutritional counseling to try to regain some of her energy.

Questions

1. What are some of the nutritional issues Ms. Wetter is facing? (Hint: Calculate her current weight to usual body weight as a percentage.)
2. How would you prioritize these issues in a nutritional care plan?
3. Calculate Ms. Wetter's energy needs, and suggest strategies she might use to regain some energy.
4. What other information would you want to know in order to counsel Ms. Wetter?

(Answers are in Appendix D.)

Dehydration

Dehydration is the physiological state in which cells lose water to the point of interfering with the metabolic processes. Normal urination does not cause dehydration. Phillips and colleagues[118] defined dehydration as losing nearly 2% of initial body weight; this can occur after avoiding all fluid and eating only dry foods for 24 hours.

There are three types of dehyration categories (isotonic, hypotonic, and hypertonic), and they are related to the balance of proportional sodium and water losses. Abnormally high serum sodium levels (>150 milliequivalents per liter) or a high ratio of blood urea nitrogen to creatinine (>25) can also be used to diagnose "significant dehydration."[119] Hypertonic dehydration (i.e., serum sodium is >145 mmol/L) is the type seen in iatrogenic cases (i.e., fluid deprivation with possible neglect). It can also be seen in individuals with fever.

Dehydration can be measured as percentage of body weight lost when normal body weight is known. In a continuum of dehydration shown by Briggs and Calloway[120] (Table 19.13 on the next page), originally designed for the National Aeronautics and Space Administration (NASA), the indicators are somewhat different from the indicators

that Gross and colleagues found. The NASA continuum shows how the human body responds to water losses.

Weight loss of 4% normal weight would be hard to ignore unless the individual had a compromised mental or cognitive status. Flushed skin, nausea, and apathy or lack of energy occur when 4% body weight is lost due to dehydration.

Although the human body is quite resilient in that it can lose water up to 20% of its weight before death, dehydration is common for older people. Prevalence data are hard to gather because dehydration is usually temporary, and there is no simple or commonly used standard definition for diagnosing dehydration in the elderly. For example, patients coming to a nursing home from a hospital may have intravenous (IV) tubes, and nursing notes may simply say, "The patient looks well hydrated."

Etiology of Dehydration

Aging itself does not cause dehydration, even though the percentage of total body water shrinks from infancy to old age. But dehydration occurs more often in the elderly as a result of illness or other problems. Older people are less sensitive in detecting thirst than younger people are, and

Table 19.13 Effects of dehydration, based on percentage of initial weight lost

Percent Lost	Physiological Signs
1%	Thirst (true for young people, not necessarily in older men or women)
4–6%	Economy of movement; flushed skin; sleepiness; apathy; nausea; tingling in arms, hands, feet; headache; heat exhaustion in fit men; increases in body temperature, pulse rate, respiratory rate
8%	Dizziness, slurred speech, weakness, confusion
12%	Cognitive signs: wakefulness, delirium
20%	Bare survival limit

So, for someone who normally weighs 160 pounds:

1% loss means weight down to 158.4 lb

4% loss means weight down to 153.6 lb

6% loss means weight down to 150.4 lb

20% loss means weight down to 128.0 lb

SOURCE: Adapted from G. M. Briggs and D. H. Calloway, *Bogert's Nutrition and Physical Fitness.* New York: Holt, Rinehart and Winston, 1984; originally designed for NASA in 1967.[120]

therefore they may forget to drink.[118] Once fluids are consumed, aging kidneys may lose the ability to concentrate urine and the *antidiuretic hormone* may become less effective. Swallowing problems, depression, or dementia may cause individuals to forget to eat or drink. Decreased mobility impairs older adults' access to water, and subsequent mobility to the bathroom. Fear of incontinence, in general, is another reason leading to decreased fluid intake and subsequent dehydration.

Antidiuretic Hormone Hormone that causes the kidneys to dilute urine by absorbing more water.

Effects of Dehydration

In older adults, thirst and skin turgor (pinch a skin fold on the forearm, forehead, or over the breastbone and observe it fall back) are not good indicators of dehydration.[119] In the same review of 38 potential indicators of dehydration in older adults (61 to 98 years old, median age 82; $n = 55$; half were free living, half were admitted to the hospital emergency department from extended care), seven signs and symptoms were strongly related to dehydration (p < .01 or p < .001) and *not* to patient age:

1. Upper body muscle weakness
2. Speech difficulty
3. Confusion

4. Dry mucous membranes in nose and mouth
5. Longitudinal tongue furrows
6. Dry tongue
7. Sunken appearance of eyes in their sockets

These signs were also confirmed in a systematic review of maintaining oral hydration in older people.[121] Although regulation of body temperature is one of the functions of body water, fever or elevated temperature did not identify persons with dehydration status in Gross's study of individuals admitted to emergency rooms. However, fever or elevated temperature may be an indicator of impending dehydration.

Dehydration increases susceptibility of developing urinary tract infection, pneumonia, and pressure ulcers; it also leads to confusion, disorientation, and dementia.[122] Because confusion and delirium are signs of—as well as risk factors for—dehydration, not getting enough fluids can become a vicious circle for someone at risk for cognitive decline.

Nutrition Interventions for Dehydration

The Food Guide Pyramid for older adults suggests drinking 8 glasses of fluid daily. Some health professionals suggest 1500 to 2500 mL per day, or approximately 6 to 10 cups.[123] The ultimate beverage is *water*—tap or flavored, *not* sugared. Water is generally accessible, adds no calories to the diet, does provide traces of minerals needed for metabolism, and is very low in sodium, even when softened. When a water beverage provides calories, sugar or some other carbohydrate was added. Pure water doesn't provide energy, but lacking water to the point of dehydration can dramatically reduce one's energy.

Many beverages contribute nutrients in addition to providing fluid:

1. Tea, especially green tea, has been linked to cancer risk reduction; the flavonoids in tea act as antioxidants.[124]
2. Low-fat milk provides calcium, protein, riboflavin, and vitamin D and plays a role in treating hypertension and weight maintenance.
3. Regular use of cranberry juice reduces urinary tract infection in older women.[125]
4. Fruit and vegetable juices can be counted as part of the 5-a-day recommended fruit and vegetable servings.

REHYDRATE SLOWLY To treat dehydration in older adults, replace fluids slowly. Guidelines are to provide roughly one-fourth to one-third of the overall fluid deficit each day in the form of water or a 5% glucose solution (when the individual's blood values are stable).[126] For individuals with swallowing problems, or dysphagia, thickened liquids count as fluid. When bedridden older adults are offered fluids hourly and also with medication, they achieve higher levels of of hydration.[121]

DEHYDRATION AT END OF LIFE   Lack of hydration can be an issue for people with a terminal disease or who are near death. Some individuals stop eating and/or drinking hours, days, or even weeks before death. Laboratory values for blood and urine are likely to become abnormal. Treatment for dehydration at the end of life may differ from treatment of dehydration during an acute disease episode. Suggestions for treating dehydration in a dying person are to integrate four C's:[127]

1. *Common sense.* Use approaches that benefit the whole patient; ask, "What does the patient want?"
2. *Communication.* Respect and acknowledge the sadness felt by friends and family.
3. *Collaboration.* With the patient's permission, bring in experienced people such as hospice workers to guide treatment.
4. *Caring.* Listen, respond with love and compassion.

Dehydration at the end of life contributes to an overall slowing down of body systems, including production of body fluids, resulting in less congestion, less edema and ascites or water retention, and less gastrointestinal action. A person experiencing dehydration at the end of life may experience slight thirst, although many dying patients are not thirsty. They may experience dry mouth that can be alleviated by sucking on ice chips or using artificial saliva. Decreased urine output can be a benefit because there is less need to go to the toilet. The most commonly reported symptoms in the last week of life of an individual with advanced progressive disease are loss of appetite, asthenia (loss of strength), dry mouth, confusion, and constipation.[128] Dehydration may also lead to increasing levels of confusion and drowsiness, which can reduce fear and anxiety related to dying.

Bereavement

Bereavement is the loss felt when someone who is personally significant dies. Losses of friends and family members happen more often in the lives of older persons. Grief, a very powerful emotion, is a natural response to bereavement. The grieving process, with its stages of shock and denial, disorganization, volatile reactions, guilt, loss and loneliness, relief, and reestablishment,[129] diverts attention from normal activities. Shopping and food preparation, eating, and drinking may get lost in the grieving process. Any loss of long-shared relationships through death, dementia, or moving brings about lack of interest in activities surrounding meal planning, preparation, shopping, and eating. People who are in mourning are vulnerable to malnutrition.

Widowhood has been shown to trigger disorganization and changes in daily routine, especially related to food preparation and eating.[130] Widowed persons who are able to enjoy mealtimes, have good appetites, have high-quality diets, and receive social support are able to work through the grieving process with fewer health consequences. They demonstrate healthy aging, which is, in the words of Dr. Tamara Harris, Chief of Geriatric Epidemiology at the National Institutes of Health, "the ability of the individual to be resilient, to be adaptive, to be flexible, and to mobilize compensatory areas as they face adversities in all areas associated with health, disease, and decline in old age."

Resources

Alzheimer's Disease, Education and Referral Center
Site provides information about Alzheimer's disease and related disorders. It is a service of the National Institute of Aging (NIA).
Web site: www.alzheimers.org

American Diabetes Association
Site with areas for professionals and for the public; includes links.
Web site: www.diabetes.org

American Dietetic Association
Site offers food and nutrition tips, fact sheets, and search tips to find a registered dietitian.
Web site: www.eatright.org

Choices in Dying
Information about hydration and artifical nutrition during end-of-life decision making.
Address: 1035 30th Street NE, Washington, DC 20007
Telephone: 1-800-989-9455
Web site: www.choices.org.

Food and Drug Administration and AARP
Provides information for seniors about safe handling of foods.
Web site: http://vm.cfsan.fda.gov/~dms/seniorsd.html

Fowkes WC. Prolonging Death—An American Tragedy.
Long Beach, CA: The Archstone Foundation. 562-590-8655
Debates the use of life support.

National Center for Complementary and Alternative Medicine
Site provides fact sheets for complementary medicine associated with dietary supplements, cancer prevention and treatment, and other dietary components. The Web site is supported by the National Institutes of Health (NIH).
Web site: www.nccam.nih.gov

National Institute of Diabetes and Digestive and Kidney Diseases
Site that offers health information for diabetes, metabolic illnesses, and kidney disease. Research and clinical trial information is also available at this site. The site provides online interactive diet planning interviews to develop client-

personalized food recommendations. A written handout worksheet can be printed.
Web site: www.niddk.nih.gov

National Institute of Health, Office of Dietary Supplements
Web site: http://ods.od.nih.gov/showp

National Institute of Mental Health
Major depression is most prevalent in developed nations. This site provides a link for the public with a specific topic discussing depression. A further link, titled "older adults," pro-

vides comprehensive information on definitions of depression and the treatment options available.
Web site: www.nimh.nih.gov

National Osteoporosis Foundation
Web site: www.nof.org

Partnership for Caring
Web site: www.partnershipforcaring.org/homepage/index.html

References

1. Federal Register, 42 CFR, Part 405 et al., vol. 66, no. 212, November 1, 2001; medicare program; revisions to payment policies and five-year review of and adjustments to the relative values units under the physician fee schedule for calendar year 2002: Final Rule, pages 55275–81.

2. Lacey K, Pritchett E. Nutrition care process and model: ADA adopts road map to quality care and outcomes management. J Am Diet Assoc 2003;103(8):1061–72.

3. Sheils JF, Rubin R, Stapleton DC. The estimated costs and savings of medical nutrition therapy: the Medicare population. J Am Diet Assoc 1999;99:428–35.

4. Chima CS, Barco K, Dewitt ML et al. Relationship of nutritional status to length of stay, hospital costs, and discharge status of patients hospitalized in the medicine service. J Am Diet Assoc 1997;97:975–8; quiz 979–80.

5. Barents Group. The clinical and cost-effectiveness of medical nutrition therapy: evidence and estimates of potential Medicare savings from the use of selected nutrition interventions. Washington, DC: KMPG Peat Marwick LLP; 1996.

6. American Dietetic Association Position Paper. Cost effectiveness of medical nutrition therapy. J Am Diet Assoc 1995;95:88–91.

7. Warnold I, Lundholm K. Clinical significance of preoperative nutritional status in 215 noncancer patients. Ann Surg 1984; 199:299–305.

8. Sullivan DH, Sun S, Walls RC. Protein-energy undernutrition among elderly hospitalized patients: a prospective study. J Am Med Assoc 1999;281:2013–9.

9. Wolinsky FD, Coe RM, Miller DK et al. Health services utilization among the noninstitutionalized elderly. J Health Soc Behav 1983;24:325–37.

10. U.S. Department of Health and Human Services. (Conference edition, in two volumes.) Healthy People 2010. Washington, DC: January 2000. Available at www.health.gov/healthypeople/document/html.

11. National Center for Health Statistics. Health, United States, 2003 with chartbook in trends in the health of Americans. Washington, DC: U.S. Department of Health and Human Services; 2003.

12. Institute of Medicine. Role of nutrition in maintaining health in the nation's elderly: Evaluating coverage of nutrition services for the Medicare population. Washington, DC: National Academy Press; 2000. Available at http://nap.edu.books, accessed 11/2/2003.

13. Desai MM, Zhang P, Hennessy CH. Surveillance for morbidity and mortality among older adults—United States, 1995–1996. Mor Mortal Wkly Rep CDC Surveill Summ 1999;48:7–25.

14. American Heart Association. Heart disease and stroke statistics—2003 update. Dallas, TX: American Heart Association; 2002.

15. Erlinger TP, Pollack H, Appel LJ. Nutrition-related cardiovascular risk factors in older people: results from the Third National Health and Nutrition Examination Survey. J Am Geriatr Soc 2000;48:1486–9.

16. American Diabetes Association. Report of the Expert Committee on the Diagnosis and Classification of Diabetes Mellitus. Diabetes Care 2001;24:S5–S20.

17. Expert Panel on Detection, Evaluation, and Treatment of High Blood Cholesterol in Adults. Executive summary of the third report of the National Cholesterol Education Program (NCEP) Expert Panel in detection, evaluation, and treatment of high blood cholesterol in adults (adult treatment panel III). J Am Med Assoc 2001;16:2486–97.

18. Krauss RM, Eckel RH, Howard B et al. American Heart Association dietary guidelines: Revision 2000: A statement for healthcare professionals from the Nutrition Committee of the American Heart Association. Circulation 2000;102:2284–99.

19. Appel LJ, Moore TJ, Obarzanek E et al. A clinical trial of the effects of dietary patterns on blood pressure. DASH Collaborative Research Group. N Eng J Med 1997;336:1117–24.

20. Plaisted CS, Lin PH, Ard JD et al. The effects of dietary patterns on quality of life: a substudy of the Dietary Approaches to Stop Hypertension trial. J Am Diet Assoc 1999;99:S84–9.

21. Whelton PK, Appel LJ, Espeland MA et al. Sodium reduction and weight loss in the treatment of hypertension in older persons: a randomized controlled trial of nonpharmacologic interventions in the elderly (TONE).

TONE Collaborative Research Group. J Am Med Assoc 1998;279:839–46.

22. Sacks FM, Svetkey LP, Vollmer WM et al. Effects on blood pressure of reduced dietary sodium and the Dietary Approaches to Stop Hypertension (DASH) diet. DASH-Sodium Collaborative Research Group. N Engl J Med 2001;344:3–10.

23. The seventh report of the Joint National Committee on Prevention, Detection, Evaluation, and Treatment of High Blood Pressure. (JNC 7). NIH Pub No. 03-5233. May 2003. Available at www.nhlbi.nih.gov/guidelines/, accessed 11/2/03.

24. Wilson JW, Enns CW, Goldman KS et al. Data tables: combined results from USDA's 1994 and 1995 Continuing Survey of Food Intakes by Individuals and 1994 and 1995 Diet and Health Knowledge Survey; 1997. Available at www.barc.usda.gov/bhnrc/foodsurvey/home.htm.

25. Food and Nutrition Board. National Academy of Sciences (Institute of Medicine) NRC, Subcommittee on the 10th Edition of the RDAs, Commission on Life Sciences. Recommended dietary allowances. Washington, DC: National Academy Press; 1989.

26. World Cancer Research Fund & American Institute for Cancer Research. Food, nutrition, and the prevention of cancer: A global perspective. Washington, DC; 1997.

27. Kramarow E, Lentzner H, Rooks R et al. Health United States, 1999. With health and aging chartbook. Hyattsville, MD: National Center for Health Statistics; 1999.

28. Steinmetz KA, Potter JD. Vegetables, fruit, and cancer prevention: a review. J Am Diet Assoc 1996;96:1027–39.

29. Steinmetz KA, Potter JD. Vegetables, fruit, and cancer. I. Epidemiology. Cancer Causes Control 1991;2:325–57.

30. Steinmetz KA, Potter JD. Vegetables, fruit, and cancer. II. Mechanisms. Cancer Causes Control 1991;2:427–42.

31. Brown JK, Byers T, Doyle C, Courneya KS et al. Nutrition and physical activity during and after cancer treatment: an American cancer society guide for informed choices. CA Cancer J Clin 2003;53:28–291.

32. deCicco M, Bortolussi R, Fantin D et al. Supportive therapy of elderly cancer

patients. Crit Rev Oncol Hematol 2002; 42(2):189–211.

33. Kahn R. The Expert Committee on the Diagnosis and Classification of Diabetes Mellitus. Follow-up report on the diagnosis of diabetes mellitus. Diabetes Care 2003; 26(11):3160–7.

34. Benjamin SM, Valdez R, Geiss LS et al. Estimated number of adults with prediabetes in the U.S. in 2000. Diabetes Care 2003; 26(3):645–9.

35. Centers for Disease Control and Prevention. Unrealized prevention opportunities: reducing the health and economic burden of chronic disease. Atlanta, GA: Centers for Disease Control and Prevention, National Center for Chronic Disease Prevention and Health Promotion; 1997.

36. Resnick HE, Valsania P, Halter JB et al. Differential effects of BMI on diabetes risk among black and white Americans. Diabetes Care 1998;21:1828–35.

37. Burrows NR, Geiss LS, Engelgau MM et al. Prevalence of diabetes among Native Americans and Alaska Natives, 1990–1997: an increasing burden. Diabetes Care 2000;23:1786–90.

38. Powers MA. Handbook of diabetes medical nutrition therapy. Gaithersburg, MD: Aspen; 1996.

39. Haffner SM, Lehto S, Ronnemaa T et al. Mortality from coronary heart disease in subjects with type 2 diabetes and in nondiabetic subjects with and without prior myocardial infarction. N Eng J Med 1998; 339:229–34.

40. Hoogeveen EK, Kostense PJ, Jakobs C et al. Hyperhomocysteinemia increases risk of death, especially in type 2 diabetes: 5-year follow-up of the Hoorn Study. Circulation 2000;101:1506–11.

41. Nuttall FQ, Chasuk RM. Nutrition and the management of type 2 diabetes. J Fam Pract 1998;47:S45–53.

42. Stanley K. Assessing the nutritional needs of the geriatric patient with diabetes. Diabetes Educ 1998;24:29–30,35–7.

43. Terpstra TL. The elderly type II diabetic: a treatment challenge. Geriatr Nurs 1998;19:253–9; quiz 259–60.

44. Franz MJ, Bantle JP, Beebe CA et al. Evidence-based nutrition principles and recommendations for the treatment and prevention of diabetes and related complications. Diabetes Care 2003;26(supp 1):S51–S61.

45. U.S. Department of Agriculture, U.S. Department of Health and Human Services. Nutrition and your health: dietary guidelines for Americans, 5th ed. 2000.

46. Holtmeier KB, Seim HC. The diet prescription for obesity. What works? Minn Med 2000;83:28–32.

47. Chandalia M, Garg A, Lutjohann D et al. Beneficial effects of high dietary fiber intake in patients with type 2 diabetes mellitus. N Eng J Med 2000;342:1392–8.

48. Tsihlias EB, Gibbs AL, McBurney MI et al. Comparison of high- and low-glycemic-index breakfast cereals with monounsaturated fat in the long-term dietary management of type 2 diabetes. Am J Clin Nutr 2000;72: 439–49.

49. Preuss HG, Anderson RA. Chromium update: examining recent literature 1997–1998. Curr Opin Clin Nutr Metab Care 1998;1:509–12.

50. The effect of intensive treatment of diabetes on the development and progression of long-term complications in insulin-dependent diabetes mellitus. The Diabetes Control and Complications Trial Research Group. N Eng J Med 1993;329:977–86.

51. Effect of intensive blood-glucose control with metformin on complications in overweight patients with type 2 diabetes (UKPDS 34). UK Prospective Diabetes Study (UKPDS) Group. Lancet 1998;352:854–65.

52. Intensive blood-glucose control with sulphonylureas or insulin compared with conventional treatment and risk of complications in patients with type 2 diabetes (UKPDS 33). UK Prospective Diabetes Study (UKPDS) Group. Lancet 1998;352:837–53.

53. Reaven GM, Strom TK, Fox B. Syndrome X: overcoming the silent killer that can give you a heart attack. New York: Simon & Schuster; 2000.

54. Baumgartner J. "Syndrome X" (hypertension, diabetes, coronary heart disease). National Maternal Nutrition Intensive Course, Minneapolis, MN; July 10–13, 1996.

55. The American Association of Clinical Endocrinologists position statement on the insulin resistance syndrome. Endrocr Pract 2003;9(no. 3):239–52.

56. Coulston AM. Insulin resistance: its role in health and disease and implications for nutrition management. Topics in Nutrition no. 6: Hershey Foods Corporation; 1997.

57. Vessby B. Dietary fat and insulin action in humans. Br J Nutr 2000;83(suppl 1): S91–96.

58. National Heart, Lung, and Blood Institute Expert Panel on the Identification, Evaluation, and Treatment of Overweight and Obesity in Adults. Clinical guidelines on the idenification, evaluation, and treatment of overweight and obesity in adults: the evidence report. Obes Res 1998;6(suppl 2): 51S–209S.

59. World Health Organization Consultation on Obesity. Obesity: preventing and managing the global epidemic. Geneva, Switzerland: World Health Organization; 2000; WHO Technical Report Series 894.

60. Flegal KM, Carroll MD, Ogden CL, Johnson CL. Prevalence and trends in obesity among U.S. Adults, 1999–2000. J Am Med Assoc 2002;288(14):1723–7.

61. Stevens J. Impact of age on associations between weight and mortality. Nutr Rev 2000;58:129–37.

62. Ernsberger P, Koletsky RJ. Biomedical rationale for a wellness approach to obesity: an alternative to a focus on weight loss. J Soc Issues 1999;55:221–59.

63. Ostbye T, Steenhuis R, Wolfson C et al. Predictors of five-year mortality in older Canadians: the Canadian Study of Health and Aging. J Am Geriatr Soc 1999;47: 1249–54.

64. Hegman K. Weight loss and mortality in adults. Abstract and commentary for Pamuk ER, Williamson DF, Serdula MK et al. Weight loss and subsequent death in a cohort of U.S. adults. Ann Intern Med 1993;119(7-pt2):744–8. In: ACP Journal Club, May–June, 1994: 120–81.

65. The American Dietetic Association Long-Term Care Task Force. Nutrition risk assessment form, guides, strategies and interventions. Chicago: American Dietetic Association; 1999.

66. Centers for Disease Control. Physical activity trends—United States, 1990–1998. Morbidity & Mortality Weekly Report 2001;50:166–9.

67. Blumenthal SJ, Hendi JM, Marsillo L. A public health approach to decreasing obesity. J Am Med Assoc 288(17):2178.

68. Looker AC, Orwoll ES, Johnston CC Jr. et al. Prevalence of low femoral bone density in older U.S. adults from NHANES III. J Bone Miner Res 1997;12:1761–8.

69. Niewohner K. Conference on Osteoporosis. Minnesota Department of Health, Minnesota Board on Aging. Midwest Dairy Council, St. Paul, MN; September, 1998.

70. Madhok R, Allison T. Bone mineral density measurement in the management of osteoporosis: a public health perspective. In: Fordham JN, ed. Manual of bone densitometry measurements. An aid to the interpretation of bone densitometry measurements in a clinical setting. New York: Springer; 2000: 1–16.

71. Faine MP. Dietary factors related to preservation of oral and skeletal bone mass in women. J Prosthet Dent 1995;73:65–72.

72. Nordin BE, Need AG, Steurer T et al. Nutrition, osteoporosis, and aging. Ann NY Acad Sci 1998;854:336–51.

73. Chumlea WC, Guo SS. Body mass and bone mineral quality. Curr Opin Rheumatol 1999;11:307–11.

74. McGowan J. Osteoporosis research, prevention and treatment. Senate Special Committee on Aging; 1999.

75. Rubin LA, Hawker GA, Peltekova VD et al. Determinants of peak bone mass: clinical and genetic analyses in a young female Canadian cohort. J Bone Miner Res 1999;14:633–43.

76. Matkovic V, Kostial K, Simonovic I et al. Bone status and fracture rates in two regions of Yugoslavia. Am J Clin Nutr 1979;32:540–9.

77. Lloyd T, Andon MB, Rollings N et al. Calcium supplementation and bone mineral density in adolescent girls. J Am Med Assoc 1993;270:841–4.

78. Teegarden D, Lyle RM, McCabe GP et al. Dietary calcium, protein, and phosphorus are related to bone mineral density and content in young women. Am J Clin Nutr 1998; 68:749–54.

79. Heaney RP. Bone biology in health and disease. In: Shils ME, Olson JA, Shike M et al., eds. Modern nutrition in health and disease. Philadelphia: Lippincott, Williams & Wilkins; 1999: 1327–38.

80. Shils ME, Olson JA, Shike M et al. Modern nutrition in health and disease. Philadelphia: Lippincott, Williams & Wilkins; 1999.

81. Economos C. Osteoporosis facts. Conference on Osteoporosis. Minnesota Department of Health, Minnesota Board on Aging, Midwest Dairy Council, St. Paul, MN; September 1998.

82. Brown JP, Josse RG for the Scientific Advisory Council of the Osteoporosis Society of Canada. 2002 clinical practice guidelines for the diagnosis and management of osteoporosis in Canada. CMAJ 2002;167(10 suppl):S1–S34.

83. Feskanich D, Willett WC, Stampfer MJ et al. Protein consumption and bone fractures in women. Am J Epidemiol 1996; 143:472–9.

84. Levenson DI, Bockman RS. A review of calcium preparations. Nutr Rev 1994;52(7): 221–32.

85. Weber P. The role of vitamins in the prevention of osteoporosis—a brief status report. Int J Vitam Nutr Res 1999;69:194–7.

86. Pines A, Katchman H, Villa Y et al. The effect of various hormonal preparations and calcium supplementation on bone mass in early menopause. Is there a predictive value for the initial bone density and body weight? J Intern Med 1999;246:357–61.

87. Writing Group for the Women's Health Initiative Investigators. Risks and benefits of estrogen plus progestin in healthy postmenopausal women. Principal results from the Women's Health Initiative randomized controlled trial. J Am Med Assoc 2002;288:321–33.

88. Locke GR, III, Talley NJ, Fett SL et al. Risk factors associated with symptoms of gastroesophageal reflux. Am J Med 1999; 106:642–9.

89. Brookmeyer R, Gray S, Kawas C. Projections of Alzheimer's disease in the United States and the public health impact of delaying disease onset. Am J Pub Health 1998; 88:1337–42.

90. Castell DO. Gastroesophageal reflux and abnormal esophageal pressures: cause or effect? J Clin Gastroenterol 2000;30:3.

91. Kasper H. Vitamin absorption in the elderly. Int J Vitam Nutr Res 1999;69: 169–72.

92. Norberg B. Turn of tide for oral vitamin B_{12} treatment. J Intern Med 1999; 246:237–8.

93. Nilsson-Ehle H. Age-related changes in cobalamin (vitamin B_{12}) handling. Implications for therapy. Drugs Aging 1998; 12:277–92.

94. Lindenbaum J, Rosenberg IH, Wilson PW et al. Prevalence of cobalamin deficiency in the Framingham elderly population. Am J Clin Nutr 1994;60:2–11.

95. Hokin BD, Butler T. Cyanocobalamin (vitamin B-12) status in Seventh-Day Adventist ministers in Australia. Am J Clin Nutr 1999;70:576S–578S.

96. Carmel R, Green R, Jacobsen DW et al. Serum cobalamin, homocysteine, and methylmalonic acid concentrations in a multiethnic elderly population: ethnic and sex differences in cobalamin and metabolite abnormalities. Am J Clin Nutr 1999;70:904–10.

97. Zeitlin A, Frishman WH, Chang CJ. The association of vitamin B_{12} and folate blood levels with mortality and cardiovascular morbidity incidence in the old old: the Bronx aging study. Am J Ther 1997;4: 275–81.

98. Yates AA, Schlicker SA, Suitor CW. Dietary Reference Intakes: the new basis for recommendations for calcium and related nutrients, B vitamins, and choline. J Am Diet Assoc 1998;98:699–706.

99. Trumbo P, Yates AA, Schlicker S et al. Dietary Reference Intakes: vitamin A, vitamin K, arsenic, boron, chromium, copper, iodine, iron, manganese, molybdenum, nickel, silicon, vanadium, and zinc. J Am Diet Assoc 2001;101:294–301.

100. Bente LB, Gerrior SA. Selected food and nutrient highlights of the 20th century: U.S. food supply series. Family Econ Nutr Rev 2002;14(1):43–51.

101. Harari D, Gurwitz JH, Avorn J et al. Bowel habits in relation to age and gender. Findings from the National Health Interview Survey and clinical implications. Arch Intern Med 1996;156:315–20.

102. Talley NJ, Fleming KC, Evans JM et al. Constipation in an elderly community: a study of prevalence and potential risk factors. Am J Gastroenterol 1996;91:19–25.

103. Camilleri M, Lee JS, Viramontes B et al. Insights into the pathophysiology and mechanisms of constipation, irritable bowel syndrome, and diverticulosis in older people. J Am Geriatr Soc 2000;48:1142–50.

104. Felson DT, Lawrence RC, Dieppe PA et al. Osteoarthritis: new insights. Part 1: the disease and its risk factors. Ann Intern Med 2000;133:635–46.

105. Galperin C, German BJ, Gershwin ME. Nutrition and diet in rheumatic diseases. In: Shils ME, Olson JA, Shike M et al., eds. Modern nutrition in health and disease. Philadelphia: Lippincott, Williams & Wilkins; 1999: 1339–51.

106. Felson DT, Lawrence RC, Hochberg MC et al. Osteoarthritis: new insights. Part 2: treatment approaches. Ann Intern Med 2000;133:726–37.

107. Kremer JM. n-3 fatty acid supplements in rheumatoid arthritis. Am J Clin Nutr 2000;71:349S–51S.

108. Borchers AT, Keen CL, Stern JS et al. Inflammation and Native American medicine: the role of botanicals. Am J Clin Nutr 2000;72:339–47.

109. Johnston PK. Nutritional implications of vegetarian diets. In: Shils ME, Olson JA, Shike M et al., eds. Modern nutrition in health and disease. Philadelphia: Lippincott, Williams & Wilkins; 1999: 1755–67.

110. Lamy PP. Compliance and the elderly. Amer Pharmacy 1982;NS22(5):4382.

111. American Academy of Family Physicians, Nutrition Screening Initiative, American Dietetic Association. A physicians guide to nutrition in chronic disease management for older adults. Waldorf, MD; 2002. Can be ordered by e-mail at nsi@gmmb.com.

112. Kuczmarski MF, Kuczmarski RJ, Najjar M. Descriptive anthropometric reference data for older Americans. J Am Diet Assoc 2000;100:59–66.

113. Clinical guidelines on the identification, evaluation, and treatment of overweight and obesity in adults—the evidence report. National Institutes of Health. Obes Res 1998;6(suppl 2):51S–209S.

114. The World Health Organization/Food and Agricultural Organization/United Nations University. Physical status: the use and interpretation of anthropometry. Report of a WHO Expert Committee. Geneva, Switzerland: World Health Organization; 1995.

115. Akner G, Cederholm. Treatment of protein-energy malnutrition in chronic nonmalignant disorders. Am J Clin Nutr 2001;74:6–24.

116. Milne AC, Potter J, Avenell A. Protein and energy supplementation in elderly people at risk from malnutrition. The Cochrane Database of Systematic Reviews. 2002;(no. 4):1–71. Available online via Ovid, accessed 11/14/02.

117. Consultant Dietitians in Health Care Facilities, Niedert K, Dorner B. Nutrition care of the older adult. 2nd edition. Chicago: American Dietetic Association; 2004.

118. Phillips PA, Rolls BJ, Ledingham JG et al. Reduced thirst after water deprivation in healthy elderly men. N Eng J Med 1984;311:753–9.

119. Gross CR, Lindquist RD, Woolley AC et al. Clinical indicators of dehydration severity in elderly patients. J Emerg Med 1992;10:267–74.

120. Briggs GM, Calloway DH. Bogert's nutrition and physical fitness. New York: Holt, Rinehart and Winston; 1984.

121. Hodgkinson B, Evans D, Wood J. Maintaining oral hydration in older people. The Joanna Briggs Institute for Evidence Based Nursing and Midwifery; Systematic Review 2001(no. 12):1–66.

122. Chidester JC, Spangler AA. Fluid intake in the institutionalized elderly. J Am Diet Assoc 1997;97:23–8; quiz 29–30.

123. Chernoff R, ed. Geriatric nutrition: the

health professional's handbook. Gaithersburg, MD: Aspen; 1999.

124. Bushman JL. Green tea and cancer in humans: a review of the literature. Nutr Cancer 1998;31:151–9.

125. Fleet JC. New support for a folk remedy: cranberry juice reduces bacteriuria and pyuria in elderly women. Nutr Rev 1994; 52:168–78.

126. Carter WJ. Macronutrient requirements for elderly persons. In: Chernoff R, ed. Geri-

atric nutrition: the health professional's handbook. Gaithersburg, MD: Aspen; 1999.

127. Fordyce M. Dehydration near the end of life. Ann Long-Term Care 2000;8. Available at www.mmhc.com/nhm/articles/NHM0005/fordyce.html, accessed 5/31/2000.

128. Conill C, Verger E, Henriquez I et al. Symptom prevalence in the last week of life. J Pain Symptom Manage 1997;14:328–31.

129. Leming MR, Dickinson GE. Under-

standing dying, death, and bereavement. Fort Worth, TX: Holt, Rinehart and Winston; 1990.

130. Rosenbloom CA, Whittington FJ. The effects of bereavement on eating behaviors and nutrient intakes in elderly widowed persons. J Gerontol 1993;48:S223–9.

CDC Growth Charts

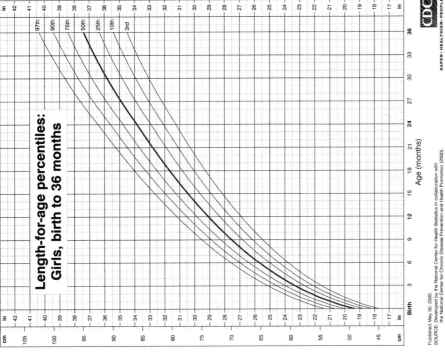

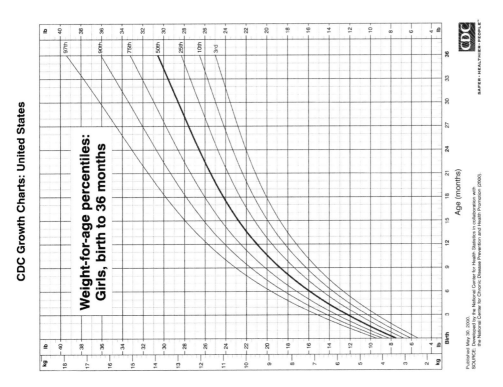

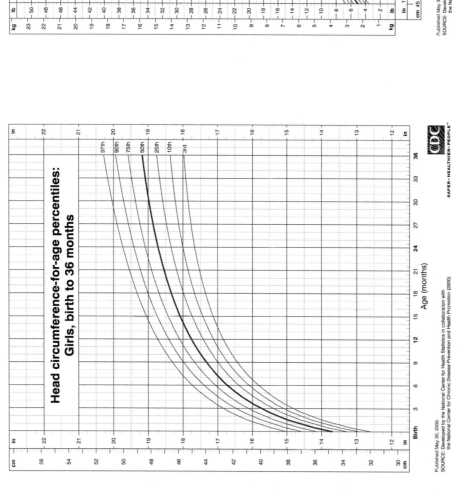

CDC Growth Charts: United States

Weight-for-length percentiles: Girls, birth to 36 months

CDC Growth Charts: United States

Head circumference-for-age percentiles: Girls, birth to 36 months

CDC Growth Charts: United States

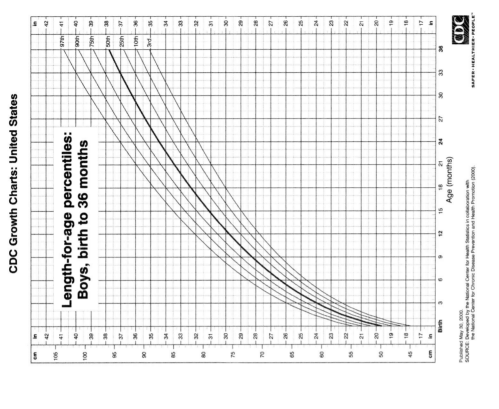

CDC Growth Charts: United States

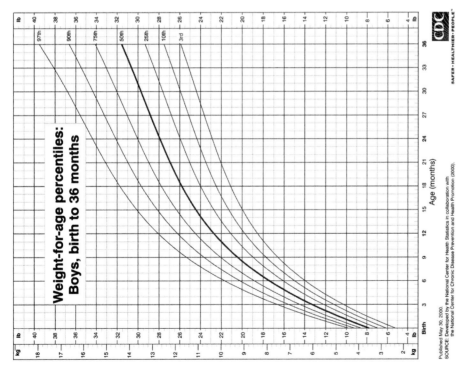

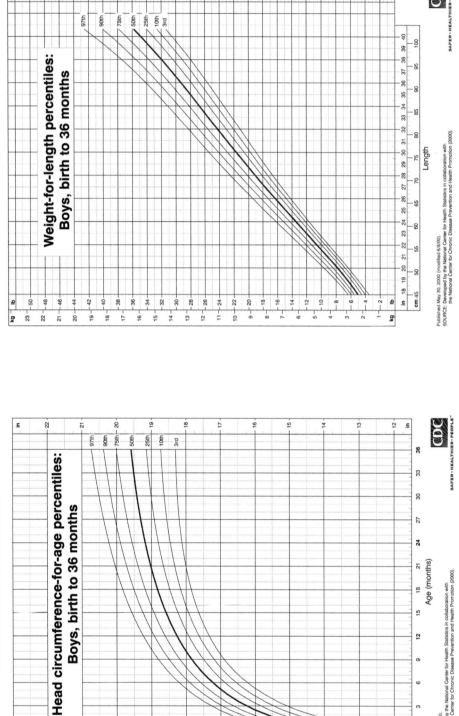

CDC Growth Charts: United States

Weight-for-length percentiles: Boys, birth to 36 months

CDC Growth Charts: United States

Head circumference-for-age percentiles: Boys, birth to 36 months

CDC Growth Charts: United States

Stature-for-age percentiles:
Girls, 2 to 20 years

Published May 30, 2000.
SOURCE: Developed by the National Center for Health Statistics in collaboration with
the National Center for Chronic Disease Prevention and Health Promotion (2000).

CDC Growth Charts: United States

Weight-for-age percentiles:
Girls, 2 to 20 years

Published May 30, 2000.
SOURCE: Developed by the National Center for Health Statistics in collaboration with
the National Center for Chronic Disease Prevention and Health Promotion (2000).

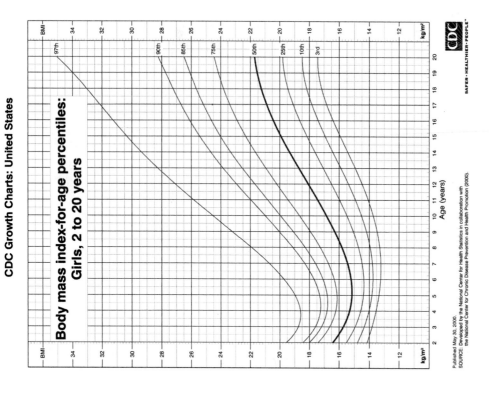

CDC Growth Charts: United States

Body mass index-for-age percentiles:
Girls, 2 to 20 years

Published May 30, 2000.
SOURCE: Developed by the National Center for Health Statistics in collaboration with
the National Center for Chronic Disease Prevention and Health Promotion (2000).

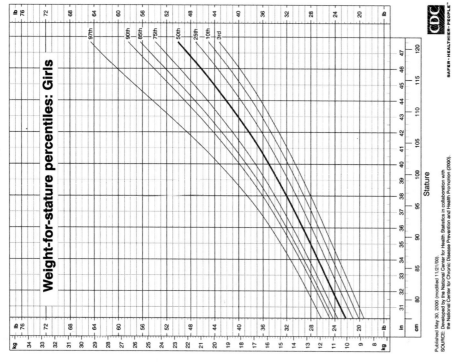

CDC Growth Charts: United States

Weight-for-stature percentiles: Girls

Published May 30, 2000 (modified 11/21/00).
SOURCE: Developed by the National Center for Health Statistics in collaboration with
the National Center for Chronic Disease Prevention and Health Promotion (2000).

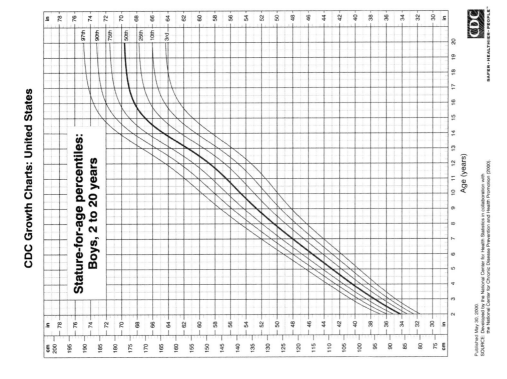

CDC Growth Charts: United States

Stature-for-age percentiles: Boys, 2 to 20 years

Age (years)

Published May 30, 2000.
SOURCE: Developed by the National Center for Health Statistics in collaboration with
the National Center for Chronic Disease Prevention and Health Promotion (2000).

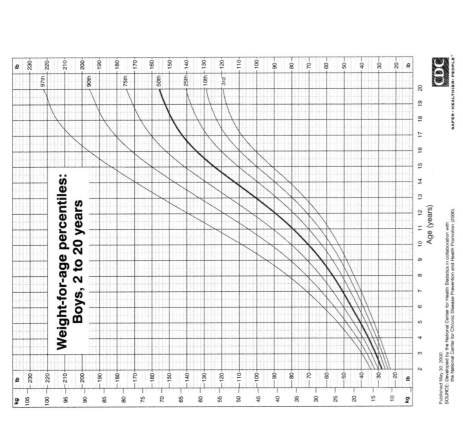

CDC Growth Charts: United States

Weight-for-age percentiles: Boys, 2 to 20 years

Age (years)

Published May 30, 2000.
SOURCE: Developed by the National Center for Health Statistics in collaboration with
the National Center for Chronic Disease Prevention and Health Promotion (2000).

CDC Growth Charts: United States

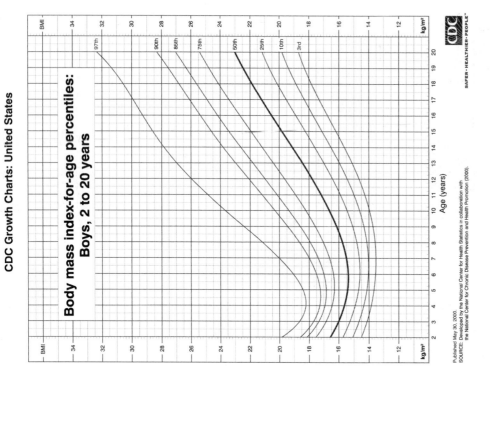

Body mass index-for-age percentiles:
Boys, 2 to 20 years

Published May 30, 2000.
SOURCE: Developed by the National Center for Health Statistics in collaboration with
the National Center for Chronic Disease Prevention and Health Promotion (2000).

CDC Growth Charts: United States

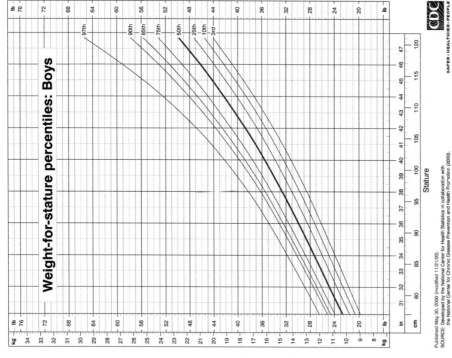

Weight-for-stature percentiles: Boys

Published May 30, 2000 (modified 11/21/00).
SOURCE: Developed by the National Center for Health Statistics in collaboration with
the National Center for Chronic Disease Prevention and Health Promotion (2000).

Nutrient Intakes of Adults Aged 70 and Older

From CSFII[48]

NUTRIENT	MALES	FEMALES
ENERGY, kcal	1854	1377
PROTEIN, gm	74	57
TOTAL FAT, gm	69	48
Saturated fatty acids, gm	23	16
Monounsaturated fatty acids, gm	27	18
Polyunsaturated fatty acids, gm	14	10
CHOLESTEROL, mg	274	185
TOTAL CARBOHYDRATE, gm	233	184
DIETARY FIBER, gm	18	14
Vitamin A, mcg RE	1430.0	1149.0
Carotenes, mcg RE	648.0	602.0
Vitamin E, mg alpha TE	9.0	6.0
Vitamin C, mg	99.0	96.0
Thiamin, mg	1.6	1.2
Riboflavin, mg	2.0	1.5
Niacin, mg	22.2	17.5
Vitamin B_6, mg	2.0	1.5
Folate, mcg	283.0	234.0
Calcium, mg	754.0	587.0
Phosphorus, mg	1211.0	912.0
Magnesium, mg	286.0	224.0
Iron, mg	16.7	12.3
Zinc, mg	11.8	8.4
Copper, mg	1.3	1.0
Sodium, mg	3122.0	2376.0
Potassium, mg	2825.0	2824.0

Measurement Abbreviations and Equivalents

Metric or SI Unit

Unit	Abbreviation
kilogram	kg
gram	g
milligram	mg
microgram	μg, mcg
nanogram	ng
meter	m
centimeter	cm
millimeter	mm
liter	L
deciliters	dL
milliliter	mL
millimole	mmol
micromole	μmol
picomole	pmol

Nonmetric

Unit	Abbreviation
ounce	oz
pound	lb
tablespoon	Tabl, tb
teaspoon	tsp
cup	c
pint	pt
quart	qt
gallon	gal
inch	in.
foot	ft
yard	yd

Equivalents

Weight: Metric

1 kilogram	=	2.2 pounds, 1000 grams
1 gram	=	0.035 ounce; 1000 milligrams
1 milligram	=	1000 micrograms
1 microgram	=	1000 nanograms

Weight: Nonmetric

1 ounce	=	28.35 grams
1 pound	=	0.45 kilograms; 454 grams

Linear

1 millimeter	=	0.039 inch
1 centimeter	=	0.01 meter; 0.39 inch
1 meter	=	100 centimeters; 39.4 inches; 3.28 feet
1 inch	=	2.54 centimeters; 0.025 meter
1 foot	=	30.5 centimeters; 0.31 meter
1 yard	=	3 feet; 0.91 meters

Fluid Volume

1 teaspoon	=	5 mL
1 Tablespoon	=	15 mL; 0.5 ounce; 3 teaspoons
1 ounce	=	30 mL; 29.57 grams; 6 teaspoons; 2 tablespoons
1 cup	=	240 mL; 8 ounces; 48 teaspoons; 16 tablespoons
1 pint	=	480 mL; 2 cups; 16 ounces; 1 pound ("A pint is a pound the whole world round.")
1 quart	=	0.95 liter; 2 pints; 4 cups; 32 ounces; 2 pounds
1 liter	=	1.06 quarts; 1000 mL
1 gallon	=	3.79 liters; 4 quarts; 8 pints; 16 cups; 8 pounds ("Two cups in a pint, two pints in a quart, four quarts in a gallon.")

Convention Units and SI Units

a. To convert conventional units to SI Units, multiply by the conversion factor.

b. To convert from SI units to conventional units, divide by the conversion factor.

	CONVENTIONAL UNITS	CONVERSION FACTOR	SI UNIT
Calcium	mg/dL	0.25	mmol/L
Cholesterol	mg/dL	0.0259	mmol/L
HDL cholesterol	mg/dL	0.0259	mmol/L
Folate	mg/mL	2.266	nmol/L
Glucose	mg/dL	0.0555	mmol/L
Hematocrit	%	0.01	Proportion of 1.0
Hemoglobin	g/dL	10.0	g/L
Homocysteine	mg/dL	7.397	μmol/L
Insulin	μIU/mL	6.945	pmol/L
Iron	μg/dL	0.179	μmol/L
LDL cholesterol	mg/dL	0.0259	mmol/L
Lipoprotein (a)	mg/dL	0.0357	μmol/L
Triglycerides	mg/dL	0.0113	mmol/L

Case Study Questions and Answers

Case Study 2.1
Cyclic Infertility with Weight Loss and Gain

Questions

1. Was Tonya underweight or normal weight based on BMI when she weighed 107 pounds? (Use the BMI chart on the inside, back cover of the text to answer this question.)
2. Can you determine Tonya's body fat content based on her BMI?
3. What likely happened to Tonya's average estrogen level when her weight decreased from 121 to 107 pounds?
4. What are two likely reasons Tonya was advised to gain weight to improve her changes of conception rather than being given Clomid or another ovulation-inducing drug?

Answers

1. Underweight.
2. No. (BMI correlates with body fat content in groups of people, but does not indicate an individualís level of body fat.)
3. It likely decreased.
4. Fertility-enhancing drugs may not induce ovulation in underweight women; becoming pregnant while underweight increased the likelihood of adverse pregnancy outcomes; and the initial treatment approach recommended for weight-related amenorrhea is weight gain.

There could be several different reasons why Tonya stopped menstruating. One reason could be her loss of body fat and alterations in reproductive hormone levels that are sensitive to body fat content. The case for this being the fact is strengthened by the return of menstruation and ovulation after Tonya gained weight.

This case is not the only clinical picture observed in women experiencing amenorrhea after weight loss. In some cases, FSH is low and LH release and levels normal; other cases are characterized by elevated estrogen levels; and so on. Each case must be considered individually.

Case Study 3.1
Anna Marie's Tale

Questions

1. What is likely the reason for Anna Marie's infertility when she was inactive?
2. Give an example of a hormonal change that may have occurred after Anna Marie began exercising regularly.
3. Name a possible health consequence related to Anna Marie's high weight and lack of menstrual cycles.

Answers

1. Excess body fat can alter estrogen, LH, and FSH levels.
2. Since Anna Marie's weight did not change when she swam, she likely gained lean body mass by exercising and lost some body fat. The return of menstrual cycles could be related to lower levels of body fat and their effect on sex hormone levels.

Case Study 3.2
Celiac Disease

Questions

1. What should have provided the first clue that Cloe might have celiac disease?
2. What facts provide other clues to the possibility of celiac disease?
3. How long will Cloe have to stay on a gluten-free diet?

Answers

1. The lack of a diagnosed cause for her stomach cramps, diarrhea, constipation, and iron-deficiency anemia during childhood.
2. Lack of menstrual periods, normal gynecological exam results, and the lack of response to fertility-enhancing drugs.
3. All of her life.

Case Study 4.1
Vegan Diet During Pregnancy

Questions

1. Is Ms. L consuming enough protein?
2. Based on the information presented, which nutrients are consumed in amounts that are below the DRI standard for pregnancy?
3. Suggest three types of food Ms. L could consume to bring up her intake of nutrients identified for question number 2.

Answers

1. No. The RDA for protein for pregnant adult women is 71 grams/day, but Ms. L needs about 30% more protein than that because all of her protein sources are from plants.
2. Protein, n-3 fatty acids, Vitamin B_{12}, and Vitamin D intakes may be low.
3. Types of food that would bring up her intake of nutrients listed in the answer to question 2 include vitamin D and B_{12} fortified soy products and meat substitute products, fortified cereal products, algae-derived n-3 fatty acids, and DHA/EPA fortified oils and meat substitutes. Increasing Ms. L's intake of n-3 fatty acids from flaxseed oil, walnuts, and green vegetables would likely not provide sufficient DHA, because conversion of linolenic acid in the body to DHA is very inefficient. She may also be encouraged to expose her face and arms to 15 minutes of direct sunlight every other day to increase vitamin D production in the skin.

Case Study 5.1
A Case of Preeclampsia

Questions

1. What clinical finding might be involved in Susan's previous miscarriage of a malformed fetus, her high homocysteine level, and the severity of her preeclampsia?
2. Name three disorders Susan is at risk of developing due to her history of preeclampsia.
3. If you were to develop a nutrition intervention plan for Susan and could only focus on one health problem, what would you focus on?

Answers

1. Inadequate folate status.
2. Susan is at risk of developing hypertension, type 2 diabetes, stroke, atherosclerosis, and heart attack.
3. It would likely be most productive to focus on increasing Susan's sensitivity to insulin (or to decreasing her level of insulin resistance).

Improvements in insulin sensitivity would improve blood lipid profile, decrease blood levels of insulin, and probably decrease her blood pressure. Because weight loss and exercise would be part of the intervention that focused on increasing insulin sensitivity, Susan's health and her sense of well-being may be benefited in additional ways.

Case Study 5.2
Elizabeth's Story: Gestational Diabetes

Questions

1. Does Elizabeth have insulin resistance?
2. What's the primary reason why Elizabeth lost weight during her first week on her new diet plan?
3. If Elizabeth continues to eat a healthful diet, exercise, and not gain weight, will she be able to prevent type 2 diabetes from developing?

Answers

1. Yes, she has insulin resistance. It is a feature of gestational diabetes.
2. Weight loss lowers blood glucose levels in women with gestational diabetes.
3. Chances are good that she will not develop type 2 diabetes. The risk for type 2 diabetes is substantially reduced by a healthful diet, exercise, and normal weight status.

Case Study 6.1
Breastfeeding and Adequate Nourishment

Questions

1. What factors make Molly high risk for early termination of breastfeeding?
2. What factors indicate Daniel is getting adequate nourishment?
3. What concerns do you have about Molly's diet? What advice would you give her about her weight loss plans and eagerness to return to exercise? Do Molly or Daniel need any vitamin-mineral supplements?
4. If Molly lived in your community, what resources are available for help and support for breastfeeding mothers?
5. What steps can Molly take to continue successful breastfeeding when she returns to work in two months?

Answers

1. Mother fatigue and insecurity; concerns of husband and mother-in-law; time constraints; worry

about returning to work; lack of information about weight loss and exercise while breastfeeding.

2. Six to 8 wet diapers and 2–3 very loose stools each day; infant alertness, pediatrician examination.

3. Is fluid intake adequate? What is Molly's usual and current diet? Perform a complete nutrition assessment; then, with Molly, develop a realistic diet and exercise plan to support breastfeeding and gradual weight loss of no more than 1 pound per week. Daniel will need 200 IU vitamin D supplement per day. Supplement recommendations for Molly should be based on an evaluation of her diet and medical history. Well-nourished breastfeeding women do not need routine vitamin or mineral supplements.

4. The answer to this question varies; investigate and evaluate resources in your community and on the Web.

5. Molly should not introduce a bottle until breast-feeding is well established, usually around 4 weeks, or approximately 10 days before she returns to work. Her child care arrangements should be supportive of breastfeeding. If it is possible to express milk at work, Molly should arrange for a place to express milk, a breast pump, and refrigerator to store expressed milk. If it is not possible to express milk at work or to leave work to nurse her infant, Molly will need to train her body to produce milk at times of day when she is not working.

Case Study 7.1
Chronic Mastitis

Questions

1. Name the causes of engorgement.

2. Name at least 2 recommendations the lactation consultant might make to decrease engorgement.

3. What measures (other than taking the medications prescribed by her physician) should Barbara Ann take to manage the mastitis?

4. How might Barbara Ann decrease the flow rate of her milk?

Answers

1. Early accumulation of milk and increased vascularity of the breast.

2. Recommendations might include (a) frequent breast feeding, (b) cold packs after a feeding to reduce the swelling, (c) breast massage in a warm shower before nursing the infant, (d) ibuprofen, and (e) cool cabbage leaves applied to the breast after feeding.

3. Mastitis can be managed through frequent feeding, moist heat, rest, local massage, and appro-

priate nipple care to prevent cracking. Also make sure the breast is emptied completely at feedings.

4. To decrease flow rate of milk, nurse while reclining, or pump for a few minutes before putting infant to breast.

Case 7.2
Breastfeeding Premature Infants

Questions

1. Describe the appropriate steps that a mother of premature infants should take to establish a good milk supply by pumping.

2. What are the causes of plugged ducts?

3. What are the treatments for plugged ducts?

4. Describe how Mom would go about feeding a baby only hindmilk.

Answers

1. Steps to establish a good milk supply: (a) Begin pumping as soon as the mother is medically able. (b) Begin slowly, and increase the time of pumping over the first week. (c) Pump frequently (at least 5 times per day) until the breast is empty or the flow is reduced to drops. (d) Get adequate rest, including 6 hours of uninterrupted sleep. (e) Encourage skin-to-skin care. (f) Use a double pump to reduce the time of pumping. (g) Have a goal of 750 to 100 ml by about 10 days postpartum.

2. Incomplete emptying of the breast; infrequent feeding; poor-fitting bra.

3. Manual massage before complete emptying of the breast; warm compresses.

4. Pump for a few minutes before putting the baby to breast. Adjust the mother's pumping time according to the goals of feeding and the volume that the infant takes.

Case Study 8.1
Baby Samantha Will Not Eat

Questions

1. What are the signs Samantha is giving that shows she needs comforting rather than hunger?

2. How might Kathy change her routine to give Samantha more attention and meet the needs of her older daughter?

3. At 8 months, is Samantha too young to overeat out of emotional needs?

4. Should Kathy stop or continue breastfeeding to improve Samantha's eating?

Answers

1. Signs include wanting to be held, not sitting in high chair, and not eating offered food.

2. Kathy could bring a snack for the older child with her from her work and briefly stop at a park on the way home, especially in nice weather. Then Kathy can hold Samantha quietly while her sister has a snack, away from home where Kathy feels she has so many other things to do. This gives mother and baby the quiet time that Samantha needs before she can show an interest in eating dinner. Her older sister may associate eating with returning home from day care, but it may be that she also wants attention from her mother more than she wants a snack.

3. Kathy is encouraging an upset baby to eat, which is not a good idea. She is inadvertently learning to overeat when her emotional needs are not met.

4. Samantha can easily accept breast milk while at day care, and it still benefits her, so there is no need to stop breastfeeding.

Case Study 9.1

Premature Birth in an At-Risk Family

Questions

1. Did Eric's early birth account for his slow growth later?

2. Did nutrition impact when his developmental delay probably started; it was diagnosed near age 3?

3. What are the signs that Eric can outgrow his problems?

Answers

1. No, Eric had catch-up growth, and his small head size and possible diagnosis of fetal alcohol syndrome better explain his growth pattern than his premature birth.

2. The developmental delay probably started early in pregnancy with the formation of the brain being affected by alcohol and drugs, so nutrition was not directly involved.

3. Eric is getting all the best interventions available in a loving family and enriched early education, but it is likely that he will outgrow his problems. He is still at risk for learning problems and mental retardation.

Case Study 9.2

Noah's Cardiac and Genetic Condition

Questions

1. How did breast-feeding benefit Noah?

2. How do you know what is good growth in such a case when the standard growth chart may not fit?

3. Why didn't the family want to have WIC or early intervention services since they were eligible?

Answers

1. The same way that breastfeeding benefits any newborn. It cannot correct the unusual growth or the genetic defect, but it was valuable in the family's viewpoint, and may have helped them accept their son's special needs.

2. Growth recommendations are customized for the special needs of the child. There were other indicators that Noah had sufficient nutrition in having good fat stores and gaining consistently, even if at a slower rate than that of other children.

3. The parents did not agree that their baby needed any services. This coping style of denial is very common and did not prevent them from enjoying being parents. Noah's parents will likely accept more services when they realize Noah needs them; but it may be several years, especially for the learning problems that come with this diagnosis.

Case Study 10.1

Developing Appropriate Feeding Behaviors

Questions

1. Identify some of the inappropriate eating habits that Lindsey's parents have allowed her to develop.

2. Considering her stage of development, what advice would you give her parents to try and increase the number of foods that she will eat?

3. What types of food preparation activities would be appropriate for Lindsey's parents to have her participate in? Why is this important?

4. What suggestions do you have for snack food items for Lindsey?

5. Would you advise Lindsey's parents to give her a daily multivitamin supplement?

6. What advice would you give the family in regards to physical activity?

Answers

1. (a) Lindsey's parents need to stop giving Lindsey free access to apple juice and snack foods between meals. (b) She should not be allowed to "graze" throughout the day. If she is allowed to build up an appetite for her meals and snacks, she will be more interested in eating and in the food being served. (c) Lindsey's parents need to eliminate the distraction of letting her watch TV and cartoons during mealtime. (d) Lindsey's mother needs to stop preparing special items for Lindsey.

2. Food jags are common at this age. Lindsey's parents can serve her favorite basic foods along with a small portion of a new food. Lindsey may need repeated exposure to a new food, particularly vegetables, before accepting the new food. Lindsey needs to see her parents eating the same foods that they are encouraging her to eat.

3. Simple food preparation activities are appropriate for Lindsey at this age. These include activities such as snapping beans, wiping the tabletop, and tearing lettuce. These types of activities teach young children about foods and food preparation. It may also make them more interested in eating these foods at mealtime.

4. The cheese and crackers that Lindsey's parents have available for her snacks are appropriate foods. It would be good for them to limit the amount of cookies their daughter eats. Other appropriate snack food items for Lindsey would be canned or fresh fruit, yogurt, and fresh vegetables with dip. Snack foods should be appropriate for Lindsey's age, and they should always be given under adult supervision. Lindsey's parents should avoid giving Lindsey foods that may cause choking. Some foods can be modified to make them safer, such as cutting grapes lengthwise.

5. According to the American Academy of Pediatrics guidelines, a multivitamin supplement is recommended only for children who are at risk of developing or have one or more nutrient deficiencies. It would be better for Lindsey's parents to help her establish good eating behaviors, including eating a variety of foods, rather than relying on a multivitamin supplement to meet her nutritional needs.

6. It is not too early to encourage her parents to engage Lindsey in active play every day and to limit the amount of time she is allowed to watch TV or videos.

Case Study 11.1
A Picky Eater

Questions

1. What are the signs that Greg's feeding problem may be related to his speech?

2. Since he is growing well and meeting his calorie needs, why not just wait for him to mature to accept other foods?

3. Was his pediatrician wrong to say that he will when he is hungry?

Answers

1. Greg likes soft, easy-to-eat foods. If he cannot control his tongue and jaw for speech, he may have similar difficulties with foods. Ability to speak and eat is controlled by the same oral muscles and nerves, but not the same parts of the brain.

2. Greg's unusual eating behaviors are consistent with his developmental problems. He will not outgrow the eating problems automatically with increasing age. He will not mature like others with age.

3. This standard advice works well for 3-year-old children with typical development, which is what the pediatrician thinks is the case, based on Greg's low medical risk factors. This advice is not appropriate if the child does have developmental delay. The results of testing would go back to the pediatrician.

Case Study 11.2
Early Intervention Services for a Boy at Risk for Nutrition Support

Questions

1. What are the signs that Robert needed gastrostomy feeding?

2. Could the gastrostomy placement been prevented if he had gained weight?

3. Can he enjoy life if he cannot eat?

Answers

1. Robert choked when being fed, refused meals, and had a known medical history from before birth that caused long-term neuromuscular problems. Finding aspiration with swallowing requires the gastrostomy.

2. Not if the parts of Robert's brain controlling his swallowing were damaged by the infection before birth. He may be stronger with weight gain, but that does not predict recovery.

3. Yes; Robert would feel better being fed by gastrostomy than he would in having the sensation of choking when trying to eat. He probably enjoys many things more, because he has more energy due to good nutritional status.

Case Study 12.1
Pediatric Overweight

Questions

1. What is your assessment of Timothy's body size based on his weight for age, height for age, and BMI for age percentiles?
2. What suggestions do you have for Timothy's parents about improving his eating habits?
3. What suggestions do you have for Timothy's parents for increasing his physical activity level?
4. Is it significant that Timothy's mother also has a weight problem?

Answers

1. With his weight-for-age at the 95th percentile and his height-for-age at between the 75th and 90th percentiles, Timothy is large for his age as compared to other boys the same age. The most significant finding is a BMI-for-age greater than the 95th percentile. This places him in the "overweight" category.
2. Timothy's parents should limit the amount of money they give him for extra food items at school. They should provide healthier choices of foods, such as fruit and vegetables, in the home for after-school snacks. It would be good for the whole family to adopt healthier eating behaviors, especially since his mother is obese.
3. (a) Timothy's parents could talk to his PE teacher at school and make sure that he is being encouraged to be active during PE class. (b) His parents could look into the possibility of having Timothy participate in an after-school program where physical activity and active play would be encouraged. They could investigate available community programs and resources that would facilitate physical activity. (c) Eating in front of the TV should be discouraged. (d) The whole family could be engaged in a more active lifestyle.
4. Having an obese parent does increase the likelihood that a child will have a weight problem. Having two obese parents further increases the risk of overweight. Researchers are realizing that pediatric overweight is a multifaceted problem, involving genetic as well as environmental factors. Lifestyle issues such as overeating and inactivity are also contributing factors.

Case Study 13.1
Adjusting Caloric Intake for a Child with Spina Bifida

Questions

1. Since Sam does not care about his size or being overweight, why is a diet plan necessary?
2. What are the risks from Sam's weight since he is only at the 75%ile for his age on the standard growth chart?
3. Will he grow taller when he goes though puberty and be able to eat more calories each day?

Answers

1. At his age, Sam cannot take the responsibility for managing his weight. The long-term consequences of overweight with spina bifida are the basis for weight control being part of his medical management.
2. Sam cannot expend calories by moving his lower body, and so he makes fat more easily. He may have scoliosis when he is older (this is common with spina bifida) or other complications that are more difficult to manage with overweight.
3. The level of the spina bifida is a factor in how Sam will grow in puberty. He will never need as many calories as will others his age.

Case Study 13.2
Peanut Allergy

Questions

1. Why is it the school's job to check for peanuts when other parent's are sending snacks?
2. The parents seem over-protective. Can the teacher transfer Judy to another classroom?
3. What are the chances that Judy will outgrow the peanut allergy by next year?

Answers

1. Judy's food allergies are a health condition, so federal laws entitle her to closer oversight from public schools.
2. No; 504 Accommodations require that a physician's order be submitted to the school, so all teachers have to follow the order. However, the teachers tend to negotiate with parents in finding a solution all can accept.
3. It is unlikely that Judy will outgrow the allergy in one year, but each person is different. She may have it all her life, or she may outgrow it by late childhood. Severe peanut allergies can be life-threatening at any age.

Case Study 14.1

Nutritional Issues and the Vegan Adolescent

Questions

1. How might a vegetarian diet effect an adolescent's growth and maturation?

2. Why is protein a concern in Nikki's diet?

3. Which vitamins and minerals may be deficient in Nikki's diet? For each vitamin/mineral, list possible food sources that are appropriate for a vegan diet.

4. Is it necessary to assess Nikki for the presence of an eating disorder? Explain your answer.

Answers

1. Vegetarian adolescents have been found to be shorter and leaner than omnivores during childhood and to enter puberty at a later age. On average, menarche occurs 6 months later in vegetarian adolescents than among omnivores. After puberty, vegetarian adolescents are as tall or taller than omnivores, although final adult height may be reached at a later age.

2. Plant proteins have lower biological values than proteins derived from animal products. Therefore, vegan diets must include a wide variety of protein sources in order to provide all of the amino acids needed for a high-quality protein diet. This is achieved by consuming complementary proteins, which means that the essential amino acids deficient from one food's protein are supplied by the protein of another food in the same meal or in the next. For example, many legumes do not provide enough methionine, and many cereals are limited in lysine. When a combination of these two foods is eaten, the cereals provide the methionine that is deficient in the legumes and the legumes provide the lysine that is deficient in the cereals, thereby supplying the body with adequate amounts of both amino acids.

3. (a) Calcium: Fortified soybean milk, fortified orange juice, calcium-rich tofu, and fortified cereals. Green leafy vegetables and nuts provide additional calcium, but the calcium from these foods often is either found in small amounts or not well absorbed. (b) Zinc: Whole grains, nuts, and legumes. Phytic acid in these foods limits zinc absorption. Grains provide the greatest amount of zinc when leavened (as in yeast breads) because this process reduces the influence of phytic acid. (c) Iron: Whole grains, dried fruits and nuts, and legumes. The iron in these foods is not absorbed as well as the iron found in animal foods, but a good source of vitamin C taken with sources of iron will enhance absorption. (d) Vitamin D: Regular exposure to the sun and adding fortified margarine to diet. (e) Vitamin B_6: Whole grains are a good source of vitamin B_6. However, vitamin B_6 is lost during the refining of grains and is not one of the vitamins added during enrichment. Other sources of vitamin B_6 appropriate for vegan diets include bananas, spinach, avocados, potatoes, and sunflower seeds. (f) Vitamin B_{12}: Vitamin B_{12} occurs naturally only in animal foods. Foods containing vitamin B_{12} that are appropriate for a vegan diet include fortified soybean milk and special yeast grown on media rich in vitamin B_{12}.

4. Studies have found that vegetarian adolescents are more likely to report binge eating, frequent or chronic dieting, purging, and laxative use than are their nonvegetarian peers. Therefore, adolescents who consume a vegetarian diet, particularly if they report doing so for health- or weight-related reasons, should be assessed for the presence of eating disorders, chronic dieting, and body image disturbances. Also, given Nikki's low BMI percentile and her late entrance into puberty, an assessment of eating behaviors, growth, and potential eating disorders is warranted.

Case Study 15.1

Adolescent Nutrition, Growth and Development, and Obesity
PART ONE

Questions:

1. What is the current BMI percentile for Anna, based on her age and gender?

2. What was her previous BMI percentile based on the BMI of 21.4 and age of 12.0 years?

3. What is her current classification for height-for-weight (BMI)? Normal, at-risk for overweight, or overweight?

4. Given her current weight classification based on BMI, what are the recommendations for weight management based on current recommendations?

   a. Maintain current weight and monitor food intake, physical activity levels, weight, height, growth, and development at three- to four-month intervals.

   b. Weight loss of 1–2 pounds per week.

   c. Weight gain of 1–2 pounds per week.

   d. No concerns or actions are identified at this time.

5. What do the SMR and the absence of menarche tell you about Anna's potential for linear growth and future weight gain?

6. Using the Centers for Disease Control and Prevention (CDC) guidelines, how would you classify Anna's blood pressure readings?

7. Using the CDC guidelines, how would you classify her hemoglobin and hematocrit levels?

Answers

1. Approximately 89th to 90th percentile.

2. Approximately 80th percentile.

3. At risk for overweight.

4. (a) Maintain weight and monitor every 3–4 months.

5. Because Anna has gained almost 17 pounds and grown 2 inches in the past 5 months and she has not yet had her first menstrual cycle, she has most likely not completed her peak growth spurt. She can be expected to gain several more inches in height over the next 6–12 months. She can be also be expected to continue gaining weight, but the rate of weight gain will slow down. Her weight gain may be excessive given her increases in height, however, so suggestions to improve her diet and increase physical activity are recommended at this time.

6. Anna has normal systolic and diastolic pressure for her age and gender.

7. Hemoglobin is normal, but hematocrit is slightly low.

PART TWO

Questions

1. List two nutrients that are likely to be consumed in inadequate amounts in Anna's diet.

2. In which of the Food Guide Pyramid food group(s) is/are Anna's diet inadequate?

3. List two recommendations that you would suggest to Anna to improve her food intake. (Try to use concrete examples because she is a young adolescent.)

4. List two recommendations that you would suggest to Anna to improve her physical activity or reduce her sedentary activity. (Try to use concrete examples because she is a young adolescent.)

Answers

1. Folic acid, calcium, iron, zinc, vitamins A, C, E, and D.

2. Fruits, vegetables, dairy products, breads/cereals/pasta.

3. Suggested responses may include: Eat breakfast so as not to need a mid-morning vending machine snack; choose school lunch options such as salad bars and main dishes instead of a la carte and vending machine foods; drink low-fat milk or water at meals instead of juice or cola; limit cola to 12 ounces or less, 1–2 times per week; reduce portion sizes of main dishes and supplement with vegetables and fruits if still hungry; increase fruit and vegetable consumption by choosing fresh fruit for breakfast, lunch, and dinner as well as snacks; choose baked tortilla chips and salsa instead of fried tortilla chips and cheese sauce; increase iron, calcium, zinc, and folic acid contents of diet by choosing fortified cereals at breakfast and snack times.

4. Suggested responses may include: Walk to or from school if possible and safe; when possible, walk to a friend's house instead of calling on the telephone or chatting by e-mail; walk the dog each afternoon; join a local community center or health club that offers fun physical activities such as classes in yoga, martial arts, dance, aerobics, and swimming, or walking/hiking paths; take walks with friends or family after school or in the evenings; limit television viewing to 1 hour or less per day; limit Internet access to 1 hour or less per day; sign up for additional physical education classes at school when offered; buy or borrow a pedometer and try to walk at least 10,000 steps each day.

Case Study 16.1
Run Kristen Run

Questions

1. How many calories does Kristen need to maintain her weight? Is she eating enough to support daily workouts?

2. Describe 3 health-promoting aspects of Kristen's diet.

3. Make 3 suggestions that could improve Kristen's diet.

4. What types of performance-enhancing products might complement this diet?

Answers

1. Number of calories Kristen needs: 135 lb × 15 = 2025; 135 lb × 18 = 2430; she is barely eating enough to support daily workouts. One way to judge adequate intake would be to monitor her weight.

2. This diet has many health-promoting aspects, including the following: (a) Low daily intake of saturated and trans fats, less than the 300 mg

cholesterol limit, and in the acceptable range for omega-3 fatty acids (1.55 g is 0.7% of 2090 kcal). But, at 14% of calories, total fat is less than the macronutrient recommendation of 20–35% of calories from fat. (b) Fiber intake of 41 g suggests adequate fruit and vegetable and whole grain intake; one would need to look at actual foods eaten to count the number of servings. The fiber recommendation for a 25-year-old female is 14 g per 1000 kcal; the daily value on food labels is 25 g. Either way, Kristen eats enough fiber. (c) Reported micronutrient intake meets or exceeds guidelines (without exceeding the tolerable upper limit, with the exception of pantothenic acid). Her folic acid intake is quite high; this is a good thing, because folic acid reduces neural tube defects, and she could become pregnant. Kristen's diet also looks to be heart healthy.

3. Using a 24-hour recall limits any dietary assessment, but results suggest where to look further. Kristen's caloric intake is at the lower end of adequate, so her diet may lack the calories to support a rigorous training program. The overall advice would be to ensure a stable weight (she appears to eat nutrient-dense foods, so increasing amounts will likely result in a balanced diet, except for the low fat intake). Here are other suggestions for improving Kristen's diet: (a) Ask Kristen about her fluid/water intake. Typically, a young woman can rely on thirst to ensure adequate fluid intake. Hydration is especially important during training. (b) Ask Kristin if her 33-g fat intake (14% of calories) is typical. To eat 20% or more of calories from fat, she would consume at least 46 g of fat. More meat in her diet may also help her to increase the pantothenic acid intake. (c) Her pantothenic acid intake is 60% of the Daily Value recommendation. Although this vitamin is widespread in the food supply, and deficiencies are rare, pantothenic acid's role in protein and energy metabolism is important to an athlete's diet. Intake of at least 67% of the DRI will optimize her diet. Table 1.8 in Chapter 1 discusses this vitamin and suggests food sources.

4. Potential performance-enhancing products include protein powders, shakes, Power Bars, and dietary supplements. Kristen ate power bars as a matter of convenience before the race, but her protein intake of 98 g exceeds the recommended 0.8 g/kg or 49 g for a 61-kilogram person. There is no evidence that amino acid supplementation increases muscle size in healthy subjects, and such supplements may even have adverse effects on the brain and central nervous

system (Christen and Smith, *Food Chemistry: Principles and Applications,* Science Technology System, 2000, p. 127).

Case Study 17.1
Joleen Celebrates 50
Questions

1. From what you've seen of Joleen's health indicators, would you agree that she is at risk of cardiovascular disease?
2. What are her risk factors?
3. List 3 or more nutritional interventions that Joleen could consider to decrease her risk of CVD.
4. Do you agree with Chandra, Joleen's friend, who told her: "Oh, don't worry. You're just getting older. I remember my sister; when she turned 50, she was depressed for weeks! You just need more spa time."

Answers

1. Yes, Joleen is at increased risk of heart disease. Even without a physical, postmenopausal status increases the risk of heart disease for women. Menopause typically occurs around age 50.
2. A BMI of 25.7 makes Joleen slightly overweight; her waist circumference is not a risk factor, because it would have to be 35 in. or greater. (Although she may have abdominal obesity. With a hip measurement of 40 in., she has a waist-to-hip ratio of 0.8; there is some research indicating that a waist-to-hip ratio above 0.8 for women places them at increased risk of chronic disease.) Her HDL of 35 (below 40) increases risk, and her LDL cholesterol is borderline high (130–159). Her blood pressure is good.
3. Nutritional interventions: (a) Add fatty fish to Joleen's diet to get adequate omega-3 fatty acids (if she eats fish); also try flaxseed. Omega-3 fats should comprise 0.6 to 1.2% of total calories. (b) Lose weight to get BMI below 25; building body mass would help so that she can keep up a higher caloric intake level. (c) Resistance exercise would build both muscle and bone, which is important because postmenopausal status increases bone loss. Exercise helps to increase HDL cholesterol. (d) Schedule a dietary consult! Properly balanced dietary fats (limited saturated and trans fats, adequate omega-3 fats and monounsaturated fats) can help to lower LDL cholesterol. Alcohol in moderation raises HDL levels, but it isn't known if Joleen drinks. Stanol esters, such as those found in some margarines and orange juice, lower blood cholesterol.

(e) Weight loss alone improves blood cholesterol profiles. If dietary strategies do not bring results in a few months, there is still time for medications.

4. Lack of energy or being tired can be signs of heart disease. Occluded arteries prevent oxygenated blood, which we need to give us energy, from reaching cells and organs.

Case Study 17.2
Maintaining a Healthy Weight

Questions

1. Calculate Adam's current BMI. How would you classify his weight status based on the NIH classifications?
2. What would it take for Adam to achieve a BMI of 24? Calculate an energy level and estimate the number of weeks it would take at that level.
3. What would you consider a "healthy weight" for Adam?
4. What are some suggestions you would discuss with Adam in order to decrease his weight, and therefore his BMI?

Answers

1. To easily calculate Adam's BMI, given that he is 5 ft 11 in. tall and weighs 190 pounds, use the non-metric formula:

    190 pounds $\times$ 704.5 = 133,855; divide by height in inches (71) = 1885.282; then divide by height again = 26.55. A BMI of 25 or greater but less than 30 classifies Adam as overweight.

2. At a BMI of 24, Adam would weigh 172 pounds (do the math, or check the text chart providing BMIs). To reach that weight, he would have to lose 18 pounds. To lose 1 pound of body weight a week, Adam would have to achieve a caloric deficit of 3500 kcal in one week, or 500 calories per day. Walking to play 18 holes of golf covers at least 4 miles but more likely 5–6, easily using 500 calories. By eating 300 fewer calories per day and burning 200 kcal through activity, he would achieve a daily caloric deficit of 500 kcal. Adam could lose 18 pounds in 18 weeks.

3. A healthy weight for Adam would fall into the BMI equivalent range of 18.5–24.9, see BMI charts in the appendix.

4. Potential suggestions for Adam to decrease his weight, and therefore his BMI: (a) Begin by asking him what he is willing and able to do. (b) Brainstorm ideas together, such as walking to play golf, working at home and using some of

the time saved to exercise more, and shopping for and preparing foods that are lower in calories. (c) Discuss snacking patterns: Does he eat while watching football? Does he snack at night? At the office? What does he choose at the cafeteria?

Case 18.1
JT: Spiraling Out of Control?

Questions

1. If you were his nutritionist, what nutrition remedies would you prescribe for JT? What other information would you ask for before working with JT?
2. What sort of advice would you give him about weight management?
3. What would you ask him about his food and fitness routine and how would you convinced him that he needs an aggressive nutrition and fitness program?
4. What sort of fluid recommendation would you make?

Answers

1. Information about JT's diabetes should include glycemic control measures (such as glycosylated hemoglobin), blood sugar values, blood lipids including LDL and HDL cholesterol and triglycerides, blood pressure, measures of kidney function and a vitamin B_{12} measure to ensure he is not at risk for irreversible cognitive losses.

2. Calculate current BMI (it is 34.8; a BMI >30, means JT is obese) and prescribe weight loss of 10% of body weight. Although a weight of 211 pounds will not put JT in the normal weight range (BMI would be 31), it will improve his health status. It is possible for JT to make lifestyle changes to achieve a loss of 24 pounds in 6 months or 27 weeks; if he is willing and able to do this, you can continue to work with him to get his weight down to 190 pounds, or a BMI of 28, which would be solidly "overweight." Examples of successful weight control strategies from the National Weight Control Registry (Table 17.4) could serve as a start. A moderately high-protein diet is likely to enhance glycemic control. JT needs heart-healthy nutritional strategies because diabetes increases cardiovascular disease (CVD) risk.

3. Ideally, motivational interviewing would help JT's counselor to generate a care plan. Strategies that will improve JT's health include losing weight; eating a diet with a low glycemic load;

consuming adequate levels of vegetables for their high nutrient density, antioxidant activity, and fiber contributions; choosing whole grains rather than refined grain; eating fish and other sources of omega-3 fatty acids, and low-fat dairy products; and using fatty acids in balance. He may benefit from a moderate-carbohydrate, high-protein diet, especially if his waist circumference is 40 inches or more, indicating abdominal obesity. Daily exercise is one of the behavioral changes reported by members of the Weight Control Registry; the fact that JT belongs to a health club indicates a desire for improved fitness. Can he schedule his exercise and activity so that drinks afterward are not an option (e.g., early in the day, before lunch)?

4. Fluid recommendations would be 1 mL per calorie consumed, so if JT were eating 2100 kcal, he'd be drinking approximately 2 liters. This could include coffee (count the cream or sugar if used). An occasional beer can be fit into a weight-loss diet; but calories add up, even if it's a light or low-carb beer. One hundred calories a day equals 10 pounds per year—gained or lost!

Case Study 19.1
Bridget Doyle Provides Nutrition Care for Laura

Questions

1. What nutritional parameters should be assessed and monitored now that Laura is back at Lenoir Manor?

2. What disciplines should be involved in Laura's care plan, and why?

3. How can the care team address both clinical and ethical concerns?

4. How could Bridget ensure that Laura's nutritional needs are met?

5. What are strategies young adults can adopt to reduce their risk of stroke?

Answers

1. Caloric and overall nutrient intake, body weight, and weight-for-height status

2. The interdisciplinary care team wants to meet Laura's needs in a dignified and respectful manner.

3. It is obvious that Laura does not want to be fed by a tube. But without a living will to the contrary, the staff cannot withhold food. Restraints are used to prevent starvation. The clinical care team can work with Laura to be sure she consumes adequate calories through small bites of pureed food and nectar-thickened liquids. Speech

pathology can work with Laura to improve her speech. A liquid multivitamin/mineral supplement can be added to her diet, and a plan should be in place to help her regain her pre-stroke weight and learn to walk with assistance.

4. Bridget should monitor Laura's nutritional intake, ensuring that calories are adequate for weight gain, that protein levels support healing, and that vitamin and mineral intake optimize metabolism and absorption. Laura obviously does not want to be tube-fed, so getting the texture and consistency of her diet right is crucial to adequate intake. Good food safety precautions and general care are needed to help Laura avoid infections or other diseases secondary to being institutionalized.

5. The DASH diet, which consists of basic wholesome foods (see Chapter 19), has been shown to reduce risk of high blood pressure and stroke. Limiting sodium further reduces risk.

Case Study 19.2
Ms. Wetter: A Senior Surviving Through a Bad Stretch

Questions

1. What are some of the nutritional issues faced by Ms. Wetter? (Hint: Calculate her current weight to usual body weight as a percentage.)

2. How would you prioritize these in a nutritional care plan?

3. Calculate her energy needs and suggest strategies she might use to regain some energy.

4. What other information would you want to know in order to counsel Ms. Wetter?

Answers

1. (a) Parkinsonism is a progressive neurological disease; limiting protein intake and adjusting carbohydrate to protein balance throughout the day can improve drug action. Leg pain could be associated with Parkinson's disease. (b) Colon cancer may return; a lack of energy should not be dismissed as "oh, it's probably nothing." (c) Arthritis pain can be managed so that Ms. Wetter is more likely to exercise; chondroitin (see Table 16.11 on Dietary Supplements) may be useful, but needs to be evaluated by her physician because of several comorbid conditions. (d) Osteoporosis can be treated, so Ms. Wetter's care provider should evaluate her diet for calcium, Vitamin D and K, and magnesium; current and potential bone-and muscle-building activities (to prevent further falls) should also be

evaluated. Increasing strength slowly will improve energy. (e) Weight loss: At 106 pounds, Ms. Wetter has a BMI of 17, classified as Grade I thinness (but still above the 5th percentile for her age; see Table 19.12). At 118, her BMI would be 19—in the "healthy weight" category. Ms. Wetter weighs 90% of her usual weight; she is at risk of malnutrition.

2. Gaining weight would be the first priority for Ms. Wetter. Motivational information to use: Slowly increasing calories while building strength will result in more energy. (Check if she is sleeping adequately. Good sleep patterns support weight and energy gain.) Snacks that would not interfere with the Parkinson's meds but are energy dense: dried fruits such as apricots, cherries, and bananas; nuts—especially walnuts, which are a good omega-3 fat source (but not too many nuts, because they contain significant protein; almonds are a good calcium source); whole grains like crackers, popcorn, or corn chips and salsa (for antioxidants); and chocolate (calories and antioxidants). Specialized nutritional supplements such as Polycose or Benecalorie can be added to foods.

3. Energy needs for Ms. Wetter would be "weight times 10–13 kcal per pound (sedentary)" plus "500 calories per day" for a weight gain of 1 pound per week. ($106 \times 10 = 1060$; $106 \times 13 = 1378$, plus 500 total of 1600–1900 kcal per day). (Alternately, do a dietary assessment and add 500 calories to current intake; a dietary assessment allows you to discuss intervention points with Ms. Wetter.) Strategies can include eating more frequently, adding oils or other fat sources to meals, using caloric beverages instead of plain water, and using dietary supplements. Fish oils may decrease inflammation. Involvement in activities to the highest level tolerated will increase her appetite and build muscle (or at least slow its loss), giving her more strength to pursue activities. Energy for activities can be a vicious or a beneficial cycle. Bed rest leads to calcium loss from bones, muscle wasting, and lack of energy. Engaging in activities and exercise retards loss and can build strength. It is hard to get the energy for action when you feel worn out.

4. Other information that would aid counseling includes exploring Ms. Wetter's perception of the conditions from which she suffers. What is she willing and able to do to enjoy an acceptable quality of life? What sort of support system can she call on? If she agrees that muscle building is something she would like to do, is she willing to pay for physical therapy out of pocket if her insurance does not cover it? Or, would she exercise by herself? Is she willing to overcome her personal "no snacking" rule to gain weight? Can you work with her Parkinson's care team to track the interaction between diet and absorption of the medication? Are other physiological or mental health concerns more important to her than the weight gain or lack of energy? For example, if she is worried about her leg pains, getting a good diagnosis will set her mind at ease and allow her to do what she can under the circumstances.

Body Mass Index (BMI)

Height	18	19	20	21	22	23	24	25	26	27	28	29	30	31	32	33	34	35	36	37	38	39	40
											Body Weight (pounds)												
4'10"	86	91	96	100	105	110	115	119	124	129	134	138	143	148	153	158	162	167	172	177	181	186	191
4'11"	89	94	99	104	109	114	119	124	128	133	138	143	148	153	158	163	168	173	178	183	188	193	198
5'0"	92	97	102	107	112	118	123	128	133	138	143	148	153	158	163	168	174	179	184	189	194	199	204
5'1"	95	100	106	111	116	122	127	132	137	143	148	153	158	164	169	174	180	185	190	195	201	206	211
5'2"	98	104	109	115	120	126	131	136	142	147	153	158	164	169	175	180	186	191	196	202	207	213	218
5'3"	102	107	113	118	124	130	135	141	146	152	158	163	169	175	180	186	191	197	203	208	214	220	225
5'4"	105	110	116	122	128	134	140	145	151	157	163	169	174	180	186	192	197	204	209	215	221	227	232
5'5"	108	114	120	126	132	138	144	150	156	162	168	174	180	186	192	198	204	210	216	222	228	234	240
5'6"	112	118	124	130	136	142	148	155	161	167	173	179	186	192	198	204	210	216	223	229	235	241	247
5'7"	115	121	127	134	140	146	153	159	166	172	178	185	191	198	204	211	217	223	230	236	242	249	255
5'8"	118	125	131	138	144	151	158	164	171	177	184	190	197	203	210	216	223	230	236	243	249	256	262
5'9"	122	128	135	142	149	155	162	169	176	182	189	196	203	209	216	223	230	236	243	250	257	263	270
5'10"	126	132	139	146	153	160	167	174	181	188	195	202	209	216	222	229	236	243	250	257	264	271	278
5'11"	129	136	143	150	157	165	172	179	186	193	200	208	215	222	229	236	243	250	257	265	272	279	286
6'0"	132	140	147	154	162	169	177	184	191	199	206	213	221	228	235	242	250	258	265	272	279	287	294
6'1"	136	144	151	159	166	174	182	189	197	204	212	219	227	235	242	250	257	265	272	280	288	295	302
6'2"	141	148	155	163	171	179	186	194	202	210	218	225	233	241	249	256	264	272	280	287	295	303	311
6'3"	144	152	160	168	176	184	192	200	208	216	224	232	240	248	256	264	272	279	287	295	303	311	319
6'4"	148	156	164	172	180	189	197	205	213	221	230	238	246	254	263	271	279	287	295	304	312	320	328
6'5"	151	160	168	176	185	193	202	210	218	227	235	244	252	261	269	277	286	294	303	311	319	328	336
6'6"	155	164	172	181	190	198	207	216	224	233	241	250	259	267	276	284	293	302	310	319	328	336	345

Under-weight (<18.5) | Healthy Weight (18.5–24.9) | Overweight (25–29.9) | Obese (≥30)

Find your height along the left-hand column and look across the row until you find the number that is closest to your weight. The number at the top of that column identifies your BMI. The area shaded in green represents healthy weight ranges. The figure below presents silhouettes of various BMI.

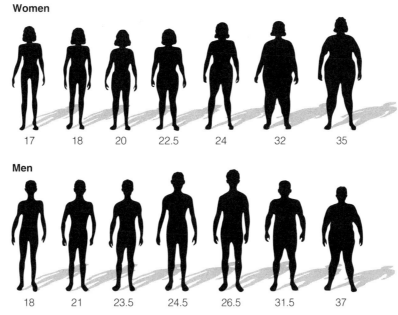

Women

17   18   20   22.5   24   32   35

Men

18   21   23.5   24.5   26.5   31.5   37

SOURCE: Reprinted from material of the Dietitians of Canada.

Glossary

Adiposity or BMI Rebound A normal increase in body mass index that occurs after BMI declines and reaches its lowest point at 4 to 6 years of age.

AIDS = Acquired Immunodeficiency Syndrome.

Allergy Hypersensitivity to a physical or chemical agent.

Alveoli Rounded or oblong cavities present in the breast (singular = alveolus).

Amenorrhea Absence of menstrual cycle.

Amino Acids The "building blocks" of protein. Unlike carbohydrates and fats, amino acids contain nitrogen.

Amniotic Fluid The fluid contained in the amniotic sac that surrounds the fetus in the uterus.

Amylophagia Compulsive consumption of laundry starch or cornstarch.

Anaphylaxis Sudden onset of a reaction with mild to severe symptoms, including a decrease in ability to breathe, which may be severe enough to cause a coma.

Androgens Types of steroid hormones produced in the testes, ovaries, and adrenal cortex from cholesterol. Some androgens (testosterone, dihydrotestosterone) stimulate development and functioning of male sex organs.

Anemia A reduction below normal in the number of red blood cells per cubic mm in the quantity of hemoglobin, or in the volume of packed red cells per 100 ml of blood (hematocrit). This reduction occurs when the balance between blood loss and blood production is disturbed.

Anencephaly Condition initiated early in gestation of the central nervous system in which the brain is not formed correctly, resulting in neonatal death.

Angiogenesis Inhibitor Angiogenesis is the formation of new blood vessels. An angiogenesis inhibitor slows or stops vessel formation. Tumors cannot grow or expand without additional blood vessels to carry oxygen and other nutrients.

Anorexia Nervosa A disorder characterized by extreme underweight, malnutrition, amenorrhea, low bone density, irrational fear of weight gain, restricted food intake, hyperactivity, and disturbances in body image.

Anovulatory Cycles Menstrual cycles in which ovulation does not occur.

Anthropometry The science of measuring the human body and its parts.

Antidiuretic Hormone Hormone that causes the kidneys to dilute urine by absorbing more water.

Antioxidants Chemical substances that prevent or repair damage to cells caused by exposure to oxidizing agents such as oxygen, ozone, and smoke and to other oxidizing agents normally produced in the body. Many different antioxidants are found in foods; some are made by the body.

Appropriate for Gestational Age (AGA) Weight, length, and head circumference are between the 10th and 90th percentiles for gestational age.

Arteries Blood vessels that carry oxygenated blood to cells.

Arteriosclerosis Age-related thickening and hardening of the artery walls, much like an old rubber hose that becomes brittle or hard.

Asthma Condition in which the lungs are unable to exchange air due to lack of expansion of air sacs. It can result in a chronic illness and sometimes unconsciousness and death if not treated.

Atherosclerosis A type of hardening of the arteries in which cholesterol is deposited in the arteries. These deposits narrow the coronary arteries and may reduce blood flow to the heart.

Athetosis Uncontrolled movements of the large muscle groups as a result of damage to the central nervous system.

Atrial Fibrillation Degeneration of the heart muscle, causing irregular contractions.

Attention Deficit Hyperactivity Disorder (ADHD) Condition characterized by low impulse control and short attention span, with and without a high level of overall activity.

Autism Condition of deficits in communication and social interaction with onset generally before age 3 years, in which mealtime behavior and eating problems occur along with other behavioral and sensory problems.

B-lymphocytes White blood cells that are responsible for producing immunoglobulins.

Baby-Bottle Tooth Decay Dental caries in young children caused by being put to bed with a bottle or allowed to suck from a bottle for extended periods of time. Also called baby- or nursing-bottle dental caries.

Basal Metabolic Rate (BMR) Measuring energy expenditure in an individual who has been awake less than 30 minutes and is still at absolute rest, has fasted for 10 hours or more, and is in a quiet room with normal, comfortable temperatures.

Binge-Eating Disorder (BED) An eating disorder characterized by periodic binge eating, which normally is not followed by vomiting or the use of laxatives. People must experience eating binges twice a week on average for over 6 months to qualify for this diagnosis.

Body Mass Index An index that correlates with total body fat content or percent body fat, and is an acceptable measure of adiposity or body fatness in children and adults. It is calculated by dividing weight in kilograms by the square of height in metters (kg/m^2).

Bone Age Bone maturation; correlates well with stage of pubertal development.

Bronchopulmonary Dysplasia (BPD) Condition in which the underdeveloped lungs in a preterm infant are damaged so that breathing requires extra effort.

Bulimia Nervosa A disorder characterized by repeated bouts of uncontrolled, rapid ingestion of large quantities of food (binge eating) followed by self-induced vomiting, laxatives or diuretic use, fasting, or vigorous exercise in order to prevent weight gain. Binge eating is often followed by feelings of disgust and guilt. Menstrual cycle abnormalities may accompany this disorder.

Calorie A unit of measure of the amount of energy supplied by food. Also known as the "kilocalorie," or the "large Calorie."

Carotenemia A condition, caused by ingestion of high amounts of carotenoids (or carotenes) from plant foods, in which the skin turns yellowish orange.

Carotid Artery Disease Condition in which the arteries that supply blood to the brain and neck become damaged.

Catch-Up Growth Period of time shortly after a slow growth period when the rate of weight and height gains is likely to be faster than expected for age and gender.

Celiac Disease Malabsorption with fatty stools (steatorrhea) due to an inherited sensitivity to the prolamin portion of gluten in wheat, rye, barley, and perhaps oats. It is often responsible for iron, folate, and zinc deficiencies. Also called celiac sprue and nontropical sprue.

Cerebral Palsy   A group of disorders characterized by impaired muscle activity and coordination present at birth or developed during early childhood.

Cerebral Spinal Atrophy   Condition in which muscle control declines over time as a result of nerve loss, causing death in childhood.

Children with Special Health Care Needs   A federal category of services for infants, children, and adolescents with, or at risk for physical or developmental disability, or with a chronic medical condition caused by or associated with genetic/metabolic disorders, birth defects, prematurity, trauma, infection, or perinatal exposure to drugs.

Cholesterol   A fat-soluble, colorless liquid found in animals but not plants.

Chronic Condition   Disorder of health or development that is the usual state for an individual and unlikely to change, although secondary conditions may result over time.

Cleft Lip and Palate   Condition in which the upper lip and roof of the mouth are not formed completely and are surgically corrected, resulting in feeding, speaking, and hearing difficulties in childhood.

Cobalamin   Another name for vitamin B_{12}. Important roles of cobalamin are fatty acid metabolism, synthesis of nucleic acid (i.e., DNA, a complex protein that controls the formation of healthy new cells), and formation of the myelin sheath that protects nerve cells.

Coenzymes   Chemical substances that activate enzymes.

Cognitive Function   The process of thinking.

Colic   A condition marked by a sudden onset of irritability, fussiness, or crying in a young infant between 2 weeks and 3 months of age who is otherwise growing and healthy.

Colostrum   The milk produced in the first 2–3 days after the baby is born. Colostrum is higher in protein and lower in lactose than milk produced after the milk supply is established.

Commodity Program   A USDA program in which food products are sent to schools for use in the child nutrition programs. Commodities are usually acquired for farm price support and surplus removal reasons.

Competitive Foods   Foods sold to children, in food service areas during mealtimes, that compete with the federal meal programs.

Congenital Abnormality   A structural, functional, or metabolic abnormality present at birth. Also called congenital anomalies. These may be caused by environmental or genetic factors, or by a combination of the

two. Structural abnormalities are generally referred to as congenital malformations, and metabolic abnormalities as inborn errors of metabolism.

Congenital Anomaly   Condition evident in a newborn that is diagnosed at or near birth, usually as a genetic or chronic condition, such as spina bifida or cleft lip and palate.

Corpus Luteum   (*corpus* = body, *luteum* = yellow) A tissue about 12 inches in diameter formed from the follicle that contained the ovum prior to its release. It produces estrogen and progesterone. The "yellow body" derivation comes from the accumulation of lipid precursors of these hormones in the corpus luteum.

Critical Periods   Preprogrammed time periods during embryonic and fetal development when specific cells, organs, and tissues are formed and integrated, or functional levels established. Also called sensitive periods.

Cystic Fibrosis   Condition in which a genetically changed chromosome 7 interferes with all the exocrine functions in the body, but particularly pulmonary complications, causing chronic illness.

Daily Values (DVs)   Scientifically agreed-upon standards for daily intakes of nutrients from the diet developed for use on nutrition labels.

Development   Progression of the physical and mental capabilities of an organism through growth and differentiation of organs and tissues, and integration of functions.

Developmental Delay   Conditions represented by at least a 25% delay by standard evaluation in one or more areas of development, such as gross or fine motor, cognitive, communication, social, or emotional development.

Developmental Disabilities   General term used to group specific diagnoses together that limit daily living and functioning and occur before age 21.

Diaphragmatic Hernia   Displacement of the intestines up into the lung area due to incomplete formation of the diaphragm in utero.

Dietary Fiber   Complex carbohydrates and lignins naturally occurring and found mainly in the plant cell wall. Dietary fiber cannot be broken down by human digestive enzymes.

Dietary Reference Intakes (DRIs)   Quantitative estimates of nutrient intakes, used as reference values for assessing the diets of healthy people. DRIs include Recommended Dietary Allowances (RDAs), Adequate Intakes (AI), Tolerable Upper Intake Level (UL), and Estimated Average Requirement (EAR).

Dietary Supplements   Any product intended to supplement the diet, including vitamin and mineral supplements, proteins, enzymes, amino acids, fish oils, fatty acids, hormones and hormone precursors, and herbs and other plant extracts. In the United States, such products must be labeled "Dietary Supplement."

Differentiation   Cellular acquisition of one or more characteristics or functions different from that of the original cells.

DiGeorge Syndrome   Condition in which chromosome 22 has a small deletion, resulting in a wide range of heart, speech, and learning difficulties.

Diplegia   Condition in which the part of the brain controlling movement of the legs is damaged, interfering with muscle control and ambulation.

Disproportionately Small for Gestational Age (dSGA)   Newborn weight is ≤10th percentile of weight for gestational age; length and head circumference are normal. Also called asymmetrical SGA.

Diverticulitis   Infected "pockets" within the large intestine.

Doula   An individual who gives psychological encouragement and physical assistance to a mother during pregnancy, birth, and lactation; the doula may be a relative, friend, or neighbor and is usually but not necessarily female.

Down Syndrome   Condition in which three copies of chromosome 21 occur, resulting in lower muscle strength, lower intelligence, and greater risk for overweight.

Dysmenorrhea   Painful menstruation due to abdominal cramps, back pain, headache, and/or other symptoms.

Early Intervention Services   Federally mandated evaluation and therapy services for children in the age range from birth to 3 years under the Individuals with Disabilities Education Act.

Edema   Swelling (usually of the legs and feet, but can also extend throughout the body) due to an accumulation of extracellular fluid.

Embryo   The developing organism from conception through 8 weeks.

Empty-Calorie Foods   Foods that provide an excess of calories relative to their nutrient content.

Endocrine   A system of ductless glands, such as the thyroid, adrenal glands, ovaries, and testes, that produces secretions that affect body functions.

Endometriosis   A disease characterized by the presence of endometrial tissue in abnormal locations, such as deep within the uterine wall, in the ovary, or in other sites within the body. The condition is quite painful and is associated with abnormal menstrual cycles and infertility in 30–40% of affected women.

Endothelium The layer of flat cells lining blood and lymph vessels. These cells produce a variety of proteins that play a role in blood pressure regulation and body fluid distribution.

Enrichment The replacement of thiamin, riboflavin, niacin, and iron that are lost when grains are refined.

Enteral Feeding Fluid or food being delivered into the gastrointestinal system. The delivery can be by mouth or through a tube that is placed into the stomach or intestines.

Epididymis Tissues on top of the testes that store sperm.

Epithelial Cells Cells that line the surface of the body.

EPSDT The Early Periodic Screening, Detection, and Treatment Program is a part of Medicaid and provides routine checkups for low-income families.

Essential Fatty Acids Components of fat that are a required part of the diet (i.e., linoleic and alpha-linolenic acids). Both contain unsaturated fatty acids.

Essential Nutrients Substances required for growth and health that cannot be produced, or produced in sufficient amounts, by the body. They must be obtained from the diet.

Exposure Index The average infant milk intake per kilogram body weight per day × (the milk to plasma ratio divided by the rate of drug clearance) × 100. It is indicative of the amount of the drug in the breast milk that the infant ingests and is expressed as a percentage of the therapeutic (or equivalent) dose for the infant.

Extremely Low Birthweight Infant (ELBW) An infant weighing ,1000 g or 2 lb 3 oz at birth.

Failure to Thrive (FTT) Condition of inadequate weight or height gain thought to result from a caloric deficit, whether or not the cause can be identified as a health problem.

Familial Hyperlipidemia A condition that runs in families and results in high levels of serum cholesterol and other lipids.

Fatty Acids The fat-soluble components of fats in foods.

Fecundity Biological ability to bear children.

Fertility Actual production of children. The word best applies to specific vital statistic rates, but is commonly taken to mean the ability to bear children.

Fetal-Origins Hypothesis The theory that exposures to adverse nutritional and

other conditions during critical or sensitive periods of growth and development can permanently affect body structures and functions. Such changes may predispose individuals to cardiovascular diseases, type 2 diabetes, hypertension, and other disorders later in life. Also called *metabolic programming* and the *Barker Hypothesis*.

Fetus The developing organism from 8 weeks after conception to the moment of birth.

Fine Motor Skills Development and use of smaller muscle groups demonstrated by stacking objects, scribbling, and copying a circle or square.

Fluorosis Permanent white or brownish staining of the enamel of teeth caused by excessive ingestion of fluoride before teeth have erupted.

Food Allergy (Hypersensitivity) Abnormal or exaggerated immunologic response, usually immunoglobulin E (IgE) mediated, to a specific food protein.

Food Insecurity Limited or uncertain availability of safe, nutritious foods.

Food Intolerance An adverse reaction involving digestion or metabolism but not the immune system.

Food Security Access at all times to a sufficient supply of safe, nutritious foods.

Fortification The addition of one or more vitamins or minerals to a food product.

Full-Term Infants Infants born between 37 and 42 weeks of gestation.

Functional Fiber Nondigestible carbohydrates including plant, animal, or commercially produced sources, that have beneficial effects in humans.

Functional Foods Generally taken to mean food, fortified foods, and enhanced food products that may have health benefits beyond the effects of essential nutrients they contain.

Functional Status Ability to carry out the activities of daily living, including telephoning, grocery shopping, food handling and preparation, and eating.

Galactosemia A rare genetic condition of carbohydrate metabolism in which a blocked or inactive enzyme does not allow breakdown of galactose. It can cause serious illness if not identified and treated soon after birth.

Gastroesophageal Reflux (GER) Movement of the stomach contents backward into the esophagus, due to stomach muscle contractions. The condition may require treatment depending on its duration and degree. Also known as *gastro-esophageal reflux disease (GERD)*.

Gastrostomy Feeding Form of enteral nutrition support for delivering nutrition by

tube placement directly into the stomach, bypassing the mouth through a surgical procedure that creates an opening through the abdominal wall and stomach.

Geophagia Compulsive consumption of clay or dirt.

Gestational Diabetes Carbohydrate intolerance with onset or first recognition in pregnancy.

Glucogenic Amino Acids Amino acids such as alanine and glutamate that can be converted to glucose.

Gluten A protein found in wheat, oats, barley, rye and triticale (all in the genus Triticum); gliadin is the toxic fraction of gluten.

Glycemic Index (GI) A measure of the extent to which blood glucose levels are raised by a specific amount of carbohydrate-containing food compared to the same amount of glucose or white bread.

Glycemic Load (GL) A measure of the extent to which blood glucose levels are raised by a specific amount of carbohydrate-containing food. It is calculated by multiplying the carbohydrate content of an amount of food consumed by the glycemic index of the

Glycerol A component of fats that is soluble in water. It is converted to glucose in the body.

Glycosylated Hemoglobin A laboratory test that measures how well the blood sugar level has been maintained over a prolonged period of time; also called Hemoglobin A_1C.

Gravida Number of pregnancies a woman has experienced.

Gross Motor Skills Development and use of large muscle groups as exhibited by walking alone, running, walking up stairs, riding a tricycle, hopping, and skipping.

Growth Increase in an organism's size through cell multiplication (hyperplasia) and enlargement of cell size (hypertrophy).

Growth Velocity The rate of growth over time.

Gynecological Age Defined as chronological age minus age at menarch. For example, a female with the chronological age of 14 years minus age at first menstrual cycle of 12 equals a gynecological age of 2.

Health More than the absence of disease, health is a sense of well-being. Even individuals with a chronic condition may properly consider themselves to be healthy. For instance, a person with diabetes mellitus whose blood sugar is under control can be considered healthy.

Heart Disease The leading cause of death and a common cause of illness and disability in the United States. Coronary heart disease, the principal form of heart disease, is caused by buildup of cholesterol deposits in the coronary arteries that feed the heart.

Hematocrit An indicator of the proportion of whole blood occupied by red blood cells. A decrease in hematocrit is a late indicator of iron deficiency.

Heme Iron Iron contained within a protein portion of hemoglobin that is in the ferrous state.

Hemoglobin A protein that is the oxygen-carrying component of red blood cells. A decrease in hemoglobin concentration in red blood cells is a late indicator of iron deficiency.

Hemolytic Anemia Anemia caused by shortened survival of mature red blood cells and inability of the bone marrow to compensate for the decreased life span.

Hemolytic Uremic Syndrome (HUS) A serious, sometimes fatal complication associated with illness caused by *E. coli* O157:H7, which occurs primarily in children under the age of 10 years. HUS is characterized by renal failure, *hemolytic anemia,* and a severe decrease in platelet count.

High-Poverty Neighborhoods Neighborhoods where 40% or more of the people are living in poverty.

HIV = human immunodeficiency virus.

Homeostasis Constancy of the internal environment. The balance of fluids, nutrients, gases, temperature, and other conditions needed to ensure ongoing, proper functioning of cells and, therefore, all parts of the body.

Homocysteine Another intermediate product that depends on vitamin B_{12} for complete metabolism. However, both vitamin B_{12} and folate (another B vitamin) are coenzymes in the breakdown of certain protein components in this pathway. Thus, elevated homocysteine levels can result from vitamin B_{12}, folate, or pyridoxine deficiencies.

Hydrolyzed Protein Formula Formula that contains enzymatically digested protein, or single amino acids, rather than protein as it naturally occurs in foods.

Hyperbilirubinemia Elevated blood levels of bilirubin, a yellow pigment that is a by-product of the breakdown of fetal hemoglobin.

Hypertonia Condition characterized by high muscle tone, stiffness, or spasticity.

Hypoallergenic Foods or products that have a low risk of promoting food or other allergies.

Hypocalcemia Condition in which body pools of calcium are unbalanced, and low

levels are measured in blood as a part of a generalized reaction to illnesses.

Hypogonadism Atrophy or reduced development of testes or ovaries. Results in immature development of secondary sexual characteristics.

Hypothyroidism A condition characterized by growth impairment and mental retardation and deafness when caused by inadequate maternal intake of iodine during pregnancy. Used to be called cretinism.

Hypotonia Condition characterized by low muscle tone, floppiness, or muscle weakness.

Iatrogenic Used in reference to disease, it is a condition induced by a medical treatment.

Immunoglobulin A specific protein that is produced by blood cells to fight infection.

Immunological Having to do with the immune system and its functions in protecting the body from bacterial, viral, fungal, or other infections and from foreign proteins (i.e., those proteins that differ from proteins normally found in the body).

Indirect Calorimetry Measurement of energy requirements based on oxygen consumption and carbon dioxide production.

Infant Health and Development Program (IHDP) Growth charts with percentiles for VLBW (,1500-g birthweight) and LBW (,2500-g birthweight).

Infant Mortality Death that occurs within the first year of life.

Infant Mortality Attributable to Birth Defects (IMBD) Category used in tracking infant deaths in which specific diagnoses have a high mortality.

Infecundity Biological inability to bear children after 1 year of unprotected intercourse.

Infertility Commonly used to mean a biological inability to bear children.

Innocenti Declaration The Innocenti Declaration on the Protection, Promotion, and Support of Breastfeeding was produced and adopted by participants at the WHO/UNICEF policymakers' meeting on "Breastfeeding in the 1990s: A Global Initiative," held at the Spedale degli Innocenti, in Florence, Italy, on August 1, 1990. The Declaration established exclusive breastfeeding from birth to 4–6 months of age as a global goal for optimal maternal and child health.

Insulin Hormone usually produced in the pancreas to regulate movement of glucose from the bloodstream into cells within organs and muscles.

Insulin Resistance A condition in which cells "resist" the action of insulin in facilitating the passage of glucose into cells.

Intrauterine Growth Retardation (IUGR) Fetal undergrowth from any

cause, resulting in a disproportionality in weight, length, or weight-for-length percentiles for gestational age. Sometimes called intrauterine growth restriction.

Iron Deficiency A condition marked by depleted iron stores. It is characterized by weakness, fatigue, short attention span, poor appetite, increased susceptibility to infection, and irritability.

Iron-Deficiency Anemia A condition often marked by low hemoglobin level. It is characterized by the signs of iron deficiency plus paleness, exhaustion, and a rapid heart rate.

Jejunostomy Feeding Form of enteral nutrition support for delivering nutrition by tube placement directly into the upper part of the small intestine.

Juvenile Rheumatoid Arthritis Condition in which joints become enlarged and painful as a result of the immune system; generally occurs in children or teens.

Kernicterus or Bilirubin Encephalopathy The end result of very high untreated bilirubin levels. Excessive bilirubin in the system is deposited in the brain, causing toxicity to the basal ganglia and various brain-stem nuclei.

Ketogenic Diet High-fat, low-carbohydrate meal plan in which ketones are made from metabolic pathways used in converting fat as a source of energy.

Ketones Metabolic by-products of the breakdown of fatty acids in energy formation. b-hydroxybutyric acid, acetoacetic acid, and acetone are the major ketones, or "ketone bodies."

Klinefelter's Syndrome A congenital abnormality in which testes are small and firm, legs abnormally long, and intelligence generally subnormal.

Krebs Cycle A series of metabolic reactions that produce energy from the proteins, fats, and carbohydrates that constitute food.

Kwashiorkor A disease syndrome in children, primarily caused by protein deficiency. It is generally characterized by edema (or swelling), loss of muscle mass, fatty liver, rough skin, discoloration of the hair, growth retardation, and apathy.

***L. Monocytogenes,* or Listeria** A foodborne bacterial infection that can lead to preterm delivery and stillbirth in pregnant women. Listeria infection is commonly associated with the ingestion of soft cheeses, unpasteurized milk, ready-to-eat deli meats, and hot dogs.

Lactation Consultant A health care professional who provides education and management to prevent and solve breastfeeding problems and to encourage a social environment that effectively supports the breastfeeding mother-infant dyad. Those who successfully complete the International Board of Lactation Consultant Examiners (IBLCE) certification process are entitled to use IBCLC (International Board Certified Lactation Consultant) after their names (www.iblce.org/).

Lactiferous Sinuses Larger ducts for storage of milk behind the nipple.

Lactogenesis Another term for human milk production.

Lactose A form of sugar or carbohydrate composed of galactose and glucose.

Large for Gestational Age (LGA) Weight for gestational age exceeds the 90th percentile for gestational age. Also defined as birthweight greater than 4500 g ($\geq$10 lb) and referred to as excessively sized for gestational age, or macrosomic.

LDL Cholesterol Low-density lipoprotein cholesterol, the lipid most associated with atherosclerotic disease. Diets high in saturated fat, trans fatty acids, and dietary cholesterol have been shown to increase LDL-cholesterol levels.

Le Leche League An international, nonprofit, nonsectarian organization dedicated to providing education, information, support, and encouragement to women who want to breastfeed. It was founded in 1956 by seven women who had learned about successful breastfeeding while nursing their own babies. (www.lelecheleague.org).

Lean Body Mass Sum of fat-free body tissues: muscle, mineral as in bone, and water.

Leptin A protein secreted by fat cells that, by binding to specific receptor sites in the hypothalamus, decreases appetite, increases energy expenditure, and stimulates gonadotropin secretion. Leptin levels are elevated by high, and reduced by low, levels of body fat.

Life Expectancy Average number of years of life remaining for persons in a population cohort or group; most commonly reported as life expectancy from birth.

Life Span Maximum number of years someone might live; human life span is projected to range from 110 to 120 years.

Lignin Noncarbohydrate polymer that contributes to dietary fiber.

Linseed From the flax plant, linum; linseed is another name for flaxseed. Linseed oil is used in paints, varnishes, and inks but is also produced in food form for its rich nutrient content.

Liveborn Infant The World Health Organization developed a standard definition of liveborn to be used by all countries when assessing an infant's status at birth. By this definition, a liveborn infant is the outcome of delivery when a completely expelled or extracted fetus breathes, or shows any sign of life such as beating of the heart, pulsation of the umbilical cord, or definite movement of voluntary muscles, whether or not the cord has been cut or the placenta is still attached.

Lobes Rounded structures of the mammary gland.

Long-Chain Fats Carbon molecules that provide fatty acids with 12 or more carbons, which are commonly found in foods.

Longevity Length of life; it is a measure of life's duration in years.

Low-Birthweight Infant (LBW) An infants weighing ,2500 g or ,5 lb 8 oz at birth.

Lower esophageal sphincter (LES) The muscle enabling closure of the junction between the esophagus and stomach.

Macrobiotic Diet This diet falls between semivegetarian and vegan diets and includes foods such as brown rice and other grains, vegetables, fish, dried beans, spices, and fruits.

Macrocephaly Large head size for age and gender as measured by centimeters (or inches) of head circumference.

Macrophages A white blood cell that acts mainly through phagocytosis.

Malnutrition Poor nutrition resulting from an excess or lack of calories or nutrients.

Mammary Gland The source of milk for offspring, also commonly called the breast. The presence of mammary glands is a characteristic of mammals.

Maple Syrup Urine Disease Rare genetic condition of protein metabolism in which breakdown by-products build up in blood and urine, causing coma and death if untreated.

MCT Oil A liquid form of dietary fat used to boost calories; composed of medium-chain triglycerides.

Meconium Dark green mucilaginous material in the intestine of the full-term fetus.

Medical Neglect Failure of parent or caretaker to seek, obtain, and follow through with a complete diagnostic study or medical, dental, or mental health treatment for a health problem, symptom, or condition that, if untreated, could become severe enough to present a danger to the child.

Medical Nutrition Therapy (MNT) Comprehensive nutrition services by registered dietitians to treat the nutritional aspects of acute and chronic diseases.

Medicinal Herbs Plants used to prevent or remedy illness.

Medium-Chain Fats Carbon molecules that provide fatty acids with 6–10 carbons, again not typically found in foods.

Memory Impairment Moderate or severe impairment is present when four or fewer words can be recalled from a list of 20.

Menarche The occurrence of the first menstrual cycle.

Meningitis Viral or bacterial infection in the central nervous system that is likely to cause a range of long-term consequences in infancy, such as mental retardation, blindness, and hearing loss.

Menopause Cessation of the menstrual cycle and reproductive capacity in females.

Menses The process of menstruation.

Menstrual Cycle An approximately 4-week interval in which hormones direct a buildup of blood and nutrient stores within the wall of the uterus and ovum maturation and release. If the ovum is fertilized by a sperm, the stored blood and nutrients are used to support the growth of the fertilized ovum. If fertilization does not occur, they are released from the uterine wall over a period of 3 to 7 days. The period of blood flow is called the menses, or the menstrual period.

Mental Retardation Substantially below average intelligence and problems in adapting to the environment, which emerge before age 18 years.

Metabolic Syndrome A constellation of metabolic abnormalities that increase the risk of type 2 diabetes and heart disease. It is characterized by insulin resistance, abdominal obesity, high blood pressure and triglyceride levels, low levels of HDL cholesterol, and impaired glucose tolerance. Also called Syndrome X and insulin-resistance syndrome.

Metabolism The chemical changes that take place in the body. The conversion of glucose to energy or body fat is an example of a metabolic process.

Methylmalonic Acid (MMA) An intermediate product that needs vitamin B_{12} as a coenzyme to complete the metabolic pathway for fatty acid metabolism. Vitamin B_{12} is the only coenzyme in this reaction; when it is absent, the blood concentration of MMA rises.

Microcephaly Small head size for age and gender as measured by centimeters (or inches) of head circumference.

Middle Childhood Children between the ages of 5 and 10 years; also referred to as school-age.

Milk/Plasma Drug Concentration Ratio (M/P Ratio) The ratio of the concentration of drug in milk to the concentration of drug in maternal plasma. Since the ratio varies over time, a time-averaged ratio provides more meaningful information than data obtained at a single time point. It is helpful in understanding the mechanisms of drug transfer and should not be viewed as a predictor of risk to the infant as it is the concentration of the drug in milk, and not the M/P ratio, is critical to the calculation of infant dose and assessment of risk.

Miscarriage Generally defined as the loss of a conceptus in the first 20 weeks of pregnancy. Also called spontaneous abortion.

Mitochondria Intracellular unit in which fatty acid breakdown takes place and many enzyme systems for energy production inside cells are regulated.

Monounsaturated Fats Fats in which only one pair of adjacent carbons in one or more of its fatty acids is linked by a double bond (e.g., –C–C=C–C–).

Monovalent Ion An atom with an electrical charge of +1 or –1.

Morbidity The rate of illnesses in a population.

Mortality Rate The rate of death.

Myoepithelial Cells Specialized cells that line the alveoli and can contract to cause milk to be secreted into the duct.

Necrotizing Enterocolitis (NEC) Condition with inflammation or damage to a section of the intestine, with a grading from mild to severe.

Neonatal Death Death that occurs in the period from the day of birth through the first 28 days of life.

Neural Tube Defects (NTDs) Spina bifida and other malformations of the neural tube. Defects result from incomplete formation of the neural tube during the first month after conception.

Neurobehavioral Pertains to control of behavior by the nervous system.

Neuromuscular Term pertaining to the central nervous system's control of muscle coordination and movement.

Neuromuscular Disorders Conditions of the nervous system characterized by difficulty with voluntary or involuntary control of muscle movement.

Neutrophils A class of white blood cells that are involved in protecting against infection.

NF (National Formulary) A uniformity standard for herbs and botanicals.

Nonessential Nutrients Nutrients required for growth and health that can be produced by the body from other components of the diet.

Nonheme Iron Iron contained within a protein of hemoglobin that is in the ferric state.

Nonorganic Failure to Thrive Inadequate weight or height gain without an identifiable biological cause, so that an environmental cause is suspected.

Nutrient-Dense Foods Foods that contain relatively high amounts of nutrients compared to their caloric value.

Nutrients Chemical substances in foods that are used by the body for growth and health.

Nutrigenomics The science of gene-nutrient interactions.

Nutrition Support Provision of nutrients by methods other than eating regular foods or drinking regular beverages, such as directly accessing the stomach by tube or placing nutrients into the bloodstream.

Obesity BMI-for-age greater than the 95th percentile with excess fat stores as evidenced by increased triceps skinfold measurements above the 85th percentile.

Oral-Gastric (OG) Feeding A form of enteral nutrition support for delivering nutrition by tube placement from the mouth to the stomach.

Organic Failure to Thrive Inadequate weight or height gain resulting from a health problem, such as iron-deficiency anemia or a cardiac or genetic disease.

Osmolarity Measure of the number of particles in a solution, which predicts the tendency of the particles to move from high to low concentration. Osmolarity is a factor in many systems, such as in fluid and electrolyte balance.

Osteoblasts Bone cells involved with bone formation; bone-building cells.

Osteoclasts Bone cells that absorb and remove unwanted tissue.

Osteoporosis Condition in which low bone density or weak bone structure leads to an increased risk of bone fracture.

Ova Eggs of the female produced and stored within the ovaries (*singular* = ovum).

Overweight Body mass index at or above the 95th percentile.

Oxytocin A hormone produced during letdown that causes milk to be ejected into the ducts.

Pagophagia Compulsive consumption of ice or freezer frost.

Parenteral Feeding Delivery of nutrients directly to the bloodstream.

Parity The number of previous deliveries experienced by a woman; *nulliparous* = no previous deliveries, *primiparous* = one previous delivery, *multiparous* = two or more previous deliveries. Women who have delivered infants are considered to be "parous."

Pediatric AIDS Acquired immunodeficiency syndrome in which infection-fighting abilities of the body are destroyed by a virus.

Pelvic Inflammatory Disease (PID) A general term applied to infections of the cervix, uterus, fallopian tubes, or ovaries. Occurs predominantly in young women and is generally caused by infection with a sexually transmitted disease, such as gonorrhea or chlamydia, or with intrauterine device (IUD) use.

Periconceptional Period Around the time of conception, generally defined as the month before and the month after conception.

Perinatal Death Death occurring at or after 20 weeks of gestation and through the first 28 days of life.

Phytochemicals (*Phyto* = plants) Chemical substances in plants, some of which affect body processes in humans that may benefit health.

Phytoestrogen A hormone-like substance found in plants, about 1/1000 to 1/2000 as potent as the human hormone, but strong enough to bind with estrogen receptors and mimic estrogen and anti-estrogen effects.

PICA An eating disorder characterized by the compulsion to eat substances that are not food.

PKU (Phenylketonuria) An inherited error in phenylalanine metabolism most commonly caused by a deficiency of phenylalanine hydroxylase, which converts the essential amino acid phenylalanine to the nonessential amino acid tyrosine.

Placenta A disk-shaped organ of nutrient and gas interchange between mother and fetus. At term, the placenta weighs about 15% of the weight of the fetus.

Platelets A component of the blood that plays an important role in blood coagulation.

Polycystic Ovary Syndrome (PCOS) (*polycysts* = many cysts; i.e., abnormal sacs with membranous linings) A condition in females characterized by insulin resistance, high blood insulin and testosterone levels, obesity, menstrual dysfunction, amenorrhea, infertility, hirsutism (excess body hair), and acne.

Polyunsaturated Fats Fats in which more than one pair of adjacent carbons in one or more of its fatty acids are linked by two or more double bonds (e.g., –C–C=C–C=C–).

Postictal State Time after a seizure of altered consciousness; appears like a deep sleep.

Pouring Rights Contracts between schools and soft-drink companies whereby the schools receive a percentage of the profits of soft-drink sales in exchange for the school offering only that soft-drink company's products on the school campus.

Prader-Willi Syndrome Condition in which partial deletion of chromosome 15 interferes with control of appetite, muscle development, and cognition.

Preadolescence The stage of development immediately preceding adolescence; 9 to 11 years of age for girls and 10 to 12 years of age for boys.

Prebiotics Certain fiberlike forms of indigestible carbohydrates that support the growth of beneficial bacteria in the lower intestine. Nicknamed "intestinal fertilizer."

Prediabetes A condition in which blood glucose levels are higher than normal but not high enough for the diagnosis of diabetes. It is characterized by impaired glucose tolerance, or fasting blood glucose levels between 110 and 126 mg/dl.

Preloads Beverages or foods such as yogurt in which the energy/macronutrient content has been varied by the use of various carbohydrate and fat sources. The preload is given before a meal or snack and subsequent intake is monitored. This study design has been employed by Birch and Fisher in their studies of appetite, satiety, and food preferences in young children.

Premenstrual Syndrome (*premenstrual* = the period of time preceding menstrual bleeding; *syndrome* = a constellation of symptoms) A condition occurring among women of reproductive age that includes a group of physical, psychological, and behavioral symptoms with onset in the luteal phase and subsiding with menstrual bleeding. Also called premenstrual dysphoric disorder (PMDD).

Preschool-Age Children Children between the ages of 3 and 5 years, who are not yet attending kindergarten.

Preterm Infants Infants born at or before 37 weeks of gestation.

Primary Malnutrition Malnutrition that results directly from inadequate or excessive dietary intake of energy or nutrients.

Probiotics Strains of Lactobacillus and bifidobacteria that have beneficial effects on the body. Also called "friendly bacteria."

Programming The process by which exposure to adverse nutritional or other conditions during sensitive periods of growth and development produces long-term effects on body structures, functions, and disease risk.

Prolactin A hormone that stimulates milk production.

Proportionately Small for Gestational Age (pSGA) Newborn weight, length, and head circumference are ≤10th percentile for gestational age. Also called symmetrical SGA.

Prostacyclins Biologically active substances produced by blood vessel walls that inhibit platelet aggregation (and therefore blood clotting), dilate blood vessels, and reduce blood pressure.

Prostaglandins A group of physiologically active substances derived from the essential fatty acids. They are present in many tissues and perform such functions as the constriction or dilation of blood vessels, and stimulation of smooth muscles and the uterus.

Psychostimulant Classification of medication that acts on the brain to improve mental or emotional behavior.

Puberty The time frame during which the body matures from that of a child to that of a young adult, and becomes biologically capable of reproduction.

Pulmonary Related to the lungs and their movement of air for exchange of carbon dioxide and oxygen.

Quality of Life A measure of life satisfaction that is difficult to define, especially in heterogeneous aging population. Quality of life measures include factors such as social contacts, economic security, and functional status.

Recommended Dietary Allowances (RDAs) The average daily dietary intake levels sufficient to meet the nutrient requirements of nearly all (97% to 98%) healthy individuals in a population group. RDAs serve as goals for individuals.

Recumbent Length Measurement of length while the child is lying down. Recumbent length is used to measure toddlers less than 24 months of age, and those between 24 and 36 months who are unable to stand unassisted.

Reflex An automatic (unlearned) response that is triggered by a specific stimulus.

Registered Dietitian An individual who has acquired knowledge and skills necessary to pass a national registration examination and who participates in continuing professional education.

Resilience Ability to bounce back, to deal with stress and recover from injury or illness.

Resting Energy Expenditure The amount of energy needed by the body in a state of rest.

Resting Metabolic Rate (RMR) Measuring energy expenditure in an individual who has fasted, had no vigorous physical activity prior to the test, has been given time to relax (e.g., rest) for 30 minutes before starting measurement, and is in a quiet, private room with privacy and normal, comfortable temperatures.

Rett Syndrome Condition in which a genetic change on the X chromosome results in severe neurological delays, causing children to be short, thin appearing, and unable to talk.

Rooting Reflex Action that occurs if one cheek is touched, resulting in the infant's head turning toward that cheek and the infant opening his mouth.

Saturated Fats Fats in which adjacent carbons in the fatty acid component are linked by single bonds only (e.g., –C–C–C–C–).

Scoliosis Condition in which the vertebral bones in the back show a side-to-side curve, resulting in a shorter stature than expected if the back were straight.

Scrotum A muscular sac containing the testes.

Secondary Condition Common consequence of a condition, which may or may not be preventable over time.

Secondary Malnutrition Malnutrition that results from a condition (e.g., disease, surgical procedure, medication use) rather than primarily from dietary intake.

Secondary Sexual Characteristics Physiological changes that signal puberty, including enlargement of the testes, penis, and breasts and the development of pubic and facial hair.

Secretory Cells Cells in the acinus (milk gland) that are responsible for secreting milk components into the ducts.

Secretory Immunoglobulin A One of the proteins found in secretions that protect the body's mucosal surfaces from infections. It may act by reducing the binding of a microorganism with cells lining the digestive tract. It is present in human colostrum but not transferred across the placenta.

Seizures Condition in which electrical nerve transmission in the brain is disrupted, resulting in periods of loss of function that vary in severity.

Self-Efficacy The ability to make effective decisions and to take responsible action based upon one's own needs and desires.

Semen The penile ejaculate containing a mixture of sperm and secretions from the testes, prostate, and other glands. It is rich in zinc, fructose, and other nutrients. Also called seminal fluid.

Senescent Old to the point of nonfunctional.

Sensorimotor An early learning system in which the infant's senses and motor skills provide input to the central nervous system.

Serum Iron, Plasma Ferritin, and Transferring Saturation Measures of iron status obtained from blood plasma or serum samples.

Sex Hormone Binding Globulin (SHBG) A protein that binds with the sex hormones testosterone and estrogen. Also called steroid hormone binding globulin, because testosterone and estrogen are produced from cholesterol and are thus considered to be steroid hormones. These hormones are inactive when bound to SHBG, but are available for use when needed. Low levels of SHBG are related to increased availability of testosterone and estrogen in the body.

Short-Chain Fats   Carbon molecules that provide fatty acids less than 6 carbons long, as products of energy generation from fat breakdown inside cells. Short-chain fatty acids are not usually found in foods.

Shoulder Dystocia   Blockage or difficulty of delivery due to obstruction of the birth canal by the infant's shoulders.

Small for Gestational Age (SGA)   Newborn weight is ≤10th percentile for gestational age. Also called small for date (SFD).

Social Marketing   A marketing effort that combines the principles of commercial marketing with health education to promote a socially beneficial idea, practice, or product.

Spastic Quadriplegia   A form of cerebral palsy in which brain damage interferes with voluntary muscle control in both arms and legs.

Stature   Standing height.

Steroid Hormones   Hormones such as progesterone, estrogen, and testosterone produced primarily from cholesterol.

Stroke   The event that occurs when a blood vessel in the brain becomes occluded due to a clot or ruptures cutting off blood supply to a portion of the brain. Also called a cerebral vascular accident.

Subfertility   Reduced level of fertility characterized by unusually long time to conception (over 12 months) or repeated, early pregnancy losses.

Subscapular Skinfold Thickness   A skinfold measurement that can be used with other skinfold measurements to estimate percent body fat; the measurement is taken with skinfold calipers just below the inner angle of the scapula, or shoulder blade.

Suckle   A reflexive movement of the tongue moving forward and backward; earliest feeding skill.

T-lymphocytes   White blood cells that are active in fighting infection may also be called T-cells; the T in T-cell stands for thymus). These cells coordinate the immune system by secreting hormones that act on other cells.

***T. Gondii*, or Loxoplasmosis**   A parasitic infection that can impair fetal brain devel-opment. The source of the infection is often hands contaminated with soil or the con-tents of a cat litter box; or raw or partially cooked pork, lamb, or venison.

Telomere   A cap-like structure that pro-tects the end of chromosomes; it erodes during replication.

Teratogenic   Exposures that produce mal-formations in embryos or fetuses.

Testes   Male reproductive glands located in the scrotum. Also called testicles.

Thromboxanes   Biologically active sub-stances produced in platelets that increase platelet aggregation (and therefore promote blood clotting), constrict blood vessels, and increase blood pressure.

Toddlers   Children between the ages of 1 and 3 years.

Tolerable Upper Intake Levels   Highest level of daily nutrient intake that is likely to pose no risk of adverse health effects to almost all individuals in the general popula-tion; gives levels of intake that may result in adverse effects if exceeded on a regular basis.

Total Fiber   Sum of dietary fiber and functional fiber.

Tracheoesophageal Atresia   Incomplete connection between the esophagus and the stomach in utero, resulting in a shortened esophagus.

Trans Fatty Acids   Fatty acids that have unusual shapes resulting from the hydro-genation of polyunsaturated fatty acids. Trans fatty acids also occur naturally in small amounts in foods such as dairy prod-ucts and beef.

Transient Ischemic Attack (TIA)   Tem-porary and insufficient blood supply to the brain.

Transpyloric Feeding (TP)   Form of enteral nutrition support for delivering nutrition by tube placement from the nose or mouth into the upper part of the small intestine.

Triceps Skinfold   A measurement of a double layer of skin and fat tissue on the back of the upper arm. It is an index of body fatness and measured by skinfold calipers. The measurement is taken on the back of the arm midway between the shoulder and the elbow.

Type I Diabetes   A disease characterized by high blood glucose levels resulting from destruction of the insulin-producing cells of the pancreas. In the past, this type of dia-betes was called juvenile-onset diabetes and insulin-dependent diabetes.

Type 2 Diabetes   A disease characterized by high blood glucose levels due to the body's inability to use insulin normally, or to produce enough insulin. In the past this type of diabetes was called adult-onset dia-betes and non-insulin-dependent diabetes.

Unsaturated Fats   Fats in which adjacent carbons in one or more fatty acids are linked by one or more double bonds (e.g., –C=C–C=C–).

USP (United States Pharmacopeia)   A nongovernmental, nonprofit organization (since 1820); establishes and maintains standards of identity, strength, quality, purity, processing, and labeling for health care products.

Vegan Diet   The most restrictive of vege-tarian diets, allowing only plant foods.

Venous Thromboembolism   A blood clot in a vein.

Very Low Birthweight Infant (VLBW)   An infant weighing <1500 g or <3 lb 5 oz at birth.

Waist-to-hip Ratio   The ratio of the waist circumference, measured at its narrowest, and the hip circumference, measured where it is widest. This ratio is an easy way to measure body fat distribution, with a higher ratio indicative of an abdominal fat pattern. A high waist-to-hip ratio is associ-ated with a high risk of chronic disease.

Weaning   Discontinuation of breastfeed-ing or bottle-feeding and substitution of food for breast milk or infant formula.

Work of Breathing (WOB)   A common term used to express extra respiratory effort in a variety of pulmonary conditions.

Working-Poor Families   Families where at least one parent worked 50 or more weeks a year and the family income was below the poverty level.

Xerostomia   Dry mouth, or xerostomia, can be a side effect of medications (espe-cially antidepressants), of head and neck cancer treatments, and of diabetes, and is also a symptom of Sjogren's syndrome, which is an autoimmune disorder for which no cure is known.

Index